BIRKHÄUSER

Progress in Probability
Volume 60

In and Out of Equilibrium 2

Vladas Sidoravicius
Maria Eulália Vares
Editors

Birkhäuser
Basel · Boston · Berlin

Editors:

Vladas Sidoravicius
Instituto de Matemática Pura e Aplicada
(IMPA)
Estrada Dona Castorina, 110
Jardim Botanico
CEP 22460-320 Rio de Janeiro, RJ
Brasil
e-mail: vladas@impa.br

Maria Eulália Vares
Centro Brasileiro de Pesquisas Físicas
Rua Dr. Xavier Sigaud, 150
Urca
22290-180 Rio de Janeiro, RJ
Brasil
e-mail: eulalia@cbpf.br

2000 Mathematics Subject Classification: 03D10, 14Q99, 20K01, 32G15,37A50, 82BXX, 82CXX, 86A10, 90B22, 96A17, 60FXX, 60GXX, 60H99, 60JXX, 60KXX, 62GXX, 62MXX, 62P12

Library of Congress Control Number: 2008927086

Bibliographic information published by Die Deutsche Bibliothek.
Die Deutsche Bibliothek lists this publication in the Deutsche Nationalbibliografie; detailed bibliographic data is available in the Internet at http://dnb.ddb.de

ISBN 978-3-7643-8785-3 Birkhäuser Verlag AG, Basel · Boston · Berlin

Basel · Boston · Berlin
P.O. Box 133, CH-4010 Basel, Switzerland
Part of Springer Science+Business Media
Cover picture: Courtesy Tati Howell
Printed on acid-free paper produced from chlorine-free pulp. TCF ∞
Printed in Germany

ISBN 978-3-7643-8785-3 e-ISBN 978-3-7643-8786-0

9 8 7 6 5 4 3 2 1 www.birkhauser.ch

Contents

Preface

This volume celebrates the tenth edition of the Brazilian School of Probability (EBP), held at IMPA, Rio de Janeiro, from July 30 to August 4, 2006, jointly with the 69th Annual Meeting of the Institute of Mathematical Statistics. It was indeed an exceptional occasion for the local community working in this field. The EBP, first envisioned and organized in 1997, has since developed into an annual meeting with two or three advanced mini-courses and a high level conference.

This volume grew up from invited or contributed articles by researchers that during the last ten years have been participating in the Brazilian School of Probability. As a consequence, its content partially reflects the topics that have predominated in the activities during the various editions of the School, with a strong appeal that comes from statistical mechanics and areas of concentration that include interacting particle systems, percolation, random media and disordered systems. All articles of this volume were peer-refereed.

The scientific committee of the 10th EBP comprised D. Brillinger, C. Newman, Y. Peres, C.-E. Pfister, I.M. Skovgaard and V. Yohai. The organizing committee consisted of F. Cribari, C.C. Dorea, L.R. Fontes, M. Fragoso, A.C. Frery, S.R. Lopes, and ourselves. We are indebted to all of them, for their help in shaping and organizing the school. We are also grateful to the colleagues responsible for the IMS meeting. Having such a large meeting jointly with the school required a significant effort and patience for structuring a reasonable schedule. We acknowledge the support constantly given by IMPA and also by "Instituto do Milenio" (IM-AGIMB) to the school, and the financial support given by the sponsors: Capes, CNPq, Faperj, Fapesp, Finep, ICTP, NSF, Prosul and RDSES. We are also grateful to the staff of IMPA, for their constant help before and during the joint event.

Rio de Janeiro, February 2008

Vladas Sidoravicius
*Insitituto de Matemática
Pura e Aplicada*

Maria Eulália Vares
*Centro Brasileiro
de Pesquisas Físicas*

Progress in Probability, Vol. 60, 1–19

Poisson Approximations via Chen-Stein for Non-Markov Processes

Miguel Abadi

Abstract. Consider a stationary stochastic process $X_1, \dots, X_t$. We study the distribution law of the number of occurrences of a string of symbols in this process. We consider three different ways to compute it. 1) When the string is the first block generated by the process, called N_t. 2) When the string is the first block generated by an independent copy of the process, called M_t. 3) When the string $\underset{\sim}{a}$ is previously fixed, called $N_t(\underset{\sim}{a})$. We show how this laws can be approximated by a Poisson law with an explicit error for each approximation. We then derive approximations for the distribution of the first occurrence time of the string, according also to the above three different ways to chose it. The base of the proofs is the Chen-Stein method which is commonly used to prove approximations in total variation distance in the Markov context. We show how to apply it in non-Markovian mixing processes.

Mathematics Subject Classification (2000). 60F05, 60G10, 60G55, 37A50.

Keywords. Number of occurrences, occurrence time, Poisson approximation, exponential approximation, Chen-Stein method, mixing processes.

1. Introduction

A very well-known result in probability theory is the convergence of the binomial distribution with parameters n, p to the Poisson distribution with parameter λ as $np \to \lambda$. That is, if X_n are independent identically distributed random variables (r.v.) with $\mathbb{P}(X_n = 1) = p = 1 - \mathbb{P}(X_n = 0)$ for $n \geq 1$ and if $N_n = \sum_{i=1}^{n} \mathbb{1}_{\{X_n=1\}}$ then

$$\lim_{np \to \lambda} \mathbb{P}\left(N_n = k\right) = \frac{e^{-\lambda}\, \lambda^k}{k!} \,.$$

In the lasts decades there was a lot of work generalizing this limit in several senses, motivated for applications:

MA is partially supported by CNPq, grant 308250/2006-0.

A. dependent processes,
B. occurrence of fixed observables other than singletons,
C. rate of convergence of the above limit.

Our motivation is to present results which extends previous ones in the above directions plus a new one. More specifically, with respect to B, we consider not a fixed observable but a random one, chosen according to the first outcomes of the process (Definition 5) or according to the first outcomes of an independent identically distributed process (Definition 4).

More specifically, we prove that the number of occurrences of a string of n symbols, properly re-scaled, can be well approximated by a Poisson law. Actually, we present three results which differ in the way the string is chosen. In the first two ways the string is randomly chosen as follows:

i. The string is the first block generated by the process.
ii. The string is the first block generated by an independent copy the process.

These results are based on the third case.

iii. The string $\underset{\sim}{a}$ is previously fixed.

Yet, with respect to A, our result holds for two kind of dependent processes. We consider firstly ψ-mixing processes (Definition 1). Secondly, we consider β-mixing processes (Definition 2). The result in the second case has a larger error but it is obtained under a much weaker hypothesis.

ψ-mixing process includes for instance, irreducible and aperiodic finite state Markov chains (with exponential function ψ.), also Gibbs states with a potential with exponential variations are exponentially ψ-mixing. See Bowen (1975) for definitions and properties. Also, chains of infinite order are exponentially ψ-mixing (see Fernández et al. (2001) for definition). We refer to Doukhan (1995) and Bradley (2005) for examples and properties of the many definitions of mixing processes, among them ψ-mixing and β-mixing (which are not ψ-mixing) decaying at any rate.

With respect to C, we use the Chen-Stein method to get an upper bound of the total variation distance between the true law of counting occurrences and a Poisson law.

This method is a powerful tool for proving convergence to Poisson distribution. It was firstly introduced by Stein (1972) for proving convergence of a sum of dependent random variables in a generalization of the Central Limit Theorem. Chen (1975) adapted the ideas of Stein to study the question of convergence in law to a Poisson law for dependent r.v. As far as we know, the Chen-Stein method is the first one who provided error terms for the rate of convergence to the Poisson law. Some advantages of using it include:

- it applies to a very general kind of process,
- not even stationarity is required,
- it is very easy to use,

- it provides sharp error terms,
- small modifications of the proof adapt to other cases.

The main difficulty in applying the Chen-Stein method is to compute two quantities. They are usually called b_2, which measure short correlations of the observable and b_3 which measures large ones. The goal is then, to find conditions (on the process and/or the observable) in which the bounds are small.

There are a number of works which apply the Chen-Stein method for Markov process where these quantities can be well controlled. These references can be found in Alvarez (2005). Arratia, Goldstein and Gordon (1989) present a clear introduction to the Chen-Stein method. In a further paper, the same authors (1990) apply this method to a wide range of examples in which b_3 is zero, that is the processes have finite memory.

In this paper we show how to apply it for non-Markovian processes with infinite memory. We apply the method firstly to ψ-mixing processes. Then, even when we get larger bounds, we also show how to apply it under a much weaker condition, the so-called weak Bernoulli or β-mixing condition.

The Chen-Stein method allows to present an important enhance with respect to previous works. It is the *velocity* of convergence (with respect to the total variation distance). In our results, this error is explicitly expressed in terms of the measure of the observable and the rate of mixing of the process.

We obtain as immediate corollaries that the first occurrence time of a string chosen also as in: i. denoted by R_n; ii. denoted by W_n and iii. denoted by $\tau_{\underset{\sim}{a}}$ (when properly re-scaled) can be well approximated by an exponential law. For $\tau_{\underset{\sim}{a}}$ related results in different mixing contexts can be found in Galves and Schmitt (1997), Hirata et al. (2001), Abadi (2004). A survey can be found in Abadi and Galves (2001). For W_n and R_n there is only one result we are aware (Wyner (1999)) proved actually for two-folded ψ-mixing processes with a larger error bound.

R_n and W_n where also studied in other context by Ornstein and Weiss (2002) and Shields (1996). They proved respectively that

$$\lim_{n\to\infty} -\frac{1}{n}\log R_n = h \qquad \text{and} \qquad \lim_{n\to\infty} -\frac{1}{n}\log W_n = h\ ,$$

where h is the entropy of the process. Both limits are in almost-sure sense. A significant difference between them is that the first case holds for any ergodic source while the second one holds for weak Bernoulli processes but fails for more general processes (for instance, the so-called *very* weak Bernoulli processes.) This results are now also consequence of our corollaries by a direct application of the Borel-Cantelli lemma.

The paper is organized as follows. In Section 2 we establish our framework. In Section 3 we briefly describe the Chen-Stein method following Arratia, Goldstein and Gordon (1989). In Section 4 we state our results. In Section 5 we give the proofs.

2. The framework

Let $\mathcal{C}$ be a finite set. Put $\Omega = \mathcal{C}^{\mathbb{N}}$. For each $\omega = (\omega_m)_{m\in\mathbb{N}} \in \Omega$ and $m \in \mathbb{N}$, let $X_m : \Omega \to \mathcal{C}$ be the mth coordinate projection, that is $X_m(\omega) = \omega_m$. We denote by $T : \Omega \to \Omega$ the one-step-left shift operator.

For a finite sequence $\underset{\sim}{x} = (x_0, \ldots, x_{n-1}) \in \mathcal{C}^n$ we say that the set

$$\{X_0 = x_0, \ldots, X_{n-1} = x_{n-1}\} ,$$

is a string.

We denote by $\mathcal{F}$ the σ-algebra over Ω generated by strings. Moreover we denote by $\mathcal{F}_I$ the σ-algebra generated by strings with coordinates in I, $I \subseteq \mathbb{N}$.

We consider a stationary probability measure $\mathbb{P}$ over $\mathcal{F}$. We shall assume that there are no singletons of probability 0.

Let j be an integer. Due to stationarity and in order to simplify notation we will write $\mathbb{P}(\underset{\sim}{x})$ instead of $\mathbb{P}(X_j^{j+n-1} = \underset{\sim}{x})$. We will only write $\mathbb{P}(X_j^{j+n-1} = \underset{\sim}{x})$ when it is needed to make the text clearer.

Definition 1. *Consider the sequence* $\psi = (\psi(l))_{l\geq 0}$ *defined by*

$$\psi(l) = \sup_{B,C} \left| \frac{\mathbb{P}(C \mid B)}{\mathbb{P}(C)} - 1 \right| .$$

The supremum is taken over the sets B *and* C, *such that* $B \in \mathcal{F}_{\{0,.,n\}}, C \in \mathcal{F}_{\{i|i\geq n+l\}}$ *and* $\mathbb{P}(B)\mathbb{P}(C) > 0, n \in \mathbb{N}$.

The process $(X_m)_{m\in\mathbb{N}}$ *is* ψ-mixing *if the sequence* $\psi = (\psi(l))_{l\geq 0}$ *is decreasing and converges to zero.*

A weaker notion of dependence is the following.

Definition 2. *Consider the sequence* $\beta = (\beta(l))_{l\geq 0}$ *defined by*

$$\beta(l) = \frac{1}{2} \sup_{\mathcal{B},\mathcal{C}} \sum_{B\in\mathcal{Q},C\in\mathcal{R}} |\mathbb{P}(B \cap C) - \mathbb{P}(B)\mathbb{P}(C)| .$$

The supremum is taken over all the finite partitions $\mathcal{Q}$ *of* $\mathcal{F}_{\{0,.,n\}}$ *and* $\mathcal{R}$ *of* $\mathcal{F}_{\{i|i\geq n+l\}}, n \in \mathbb{N}$.

The process $(X_m)_{m\in\mathbb{N}}$ *is* β-mixing *or* weak Bernoulli *if the sequence* $\beta = (\beta(l))_{l\geq 0}$ *is decreasing and converges to zero.*

We shall use the following notation. For any two measurables A and B, $\mathbb{P}(A \,;\, B) = \mathbb{P}(A \cap B)$. The conditional measure of B given A is denoted as usual $\mathbb{P}(B \mid A) = \mathbb{P}(B \,;\, A)/\mathbb{P}(A)$. We shall write

$$\{X_r^s = x_r^s\} = \{X_r = x_r, \ldots, X_s = x_s\} .$$

For $\underset{\sim}{x} = x_0^{n-1}$ we write $T^{-k}(\underset{\sim}{x}) = \{X_k^{k+n-1} = x_0^{n-1}\}$.

As usual, the mean of a random variable X will be denoted by $\mathbb{E}[X]$. Wherever it is not ambiguous we will write C and c for different positive constants even in the same sequence of equalities/inequalities.

3. The Chen-Stein method

Definition 3. *Let X and Y to r.v. with distribution concentrated in the non-negative integers. We define the* total variation distance *between the distribution laws of X and Y as the following quantity:*

$$||\mathcal{L}(X) - \mathcal{L}(Y)|| = \sup_{K \in \mathbb{N}} |\mathbb{P}(X \in K) - \mathbb{P}(Y \in K)| \ .$$

Let I be a finite or countable set (of indexes). For each $\alpha \in I$, let V_α be a Bernoulli r.v. with $0 < \mathbb{P}(V_\alpha = 1) = p_\alpha = 1 - \mathbb{P}(V_\alpha = 0) < 1$. Let $W = \sum_{\alpha \in I} V_\alpha$. Assume that $0 < \lambda = \mathbb{E}W = \sum_{\alpha \in I} p_\alpha < \infty$. For each $\alpha \in I$ take a set $B_\alpha \subseteq I$ with $\alpha \in B_\alpha$. Let Z be a Poisson r.v. with $\mathbb{E}Z = \lambda$. Define

$$b_1 = \sum_{\alpha \in I} \sum_{\beta \in B_\alpha} p_\alpha p_\beta \ ,$$

$$b_2 = \sum_{\alpha \in I} \sum_{\alpha \neq \beta \in B_\alpha} \mathbb{E}\left[V_\alpha V_\beta\right] \ ,$$

and

$$b_3 = \sum_{\alpha \in I} \mathbb{E}\left|\mathbb{E}\left[X_\alpha - p_\alpha \mid \mathcal{F}_{B_\alpha^c}\right]\right| \ .$$

Denote with $||.||$ the total variation distance. The following theorem holds.

Theorem 1 (Arratia, Goldstein and Gordon, 1989). *Let $W = \sum_{\alpha \in I} V_\alpha$ be the number of occurrences of dependent events. Let Z be a r.v. with Poisson distribution. Suppose that $\mathbb{E}Z = \mathbb{E}W = \lambda$. then*

$$||\mathcal{L}(W) - \mathcal{L}(Z)|| \leq 2\left(1 \wedge 1.4\lambda^{-1/2}\right)(b_1 + b_2 + b_3) \ .$$

Terms b_1 and b_2 refer to the size of the "neighborhood" B_α around α (for each $\alpha \in I$) in which the process has strong dependence. The term b_3 refers to the dependence of the process outside the neighborhoods. The best bound is obtained by a trade off between the size of the neighborhoods and the loss of memory of the process.

Yet, the term b_2 also refers to short correlations of the event, that is, inside the neighborhoods. This term can be controlled only if the event does not recur very fast.

4. Results

4.1. Poisson statistics

In this section we present the (approximated) Poisson statistics for the counting M_t, N_t and $N_t(\underset{\sim}{a})$ described above. The definitions and results follow. In what follows Z denotes a r.v. with Poisson distribution of mean 1.

Due to the bound of the Chen-Stein method we introduce the following notation used in our theorems.

$$f(t) = t\left(1 \wedge 1.4t^{-1/2}\right) = t \wedge 1.4\sqrt{t}\ .$$

Definition 4. *Let $(X_m)_{m\in\mathbb{N}}$ and $(Z_m)_{m\in\mathbb{N}}$ be two independent, identically distributed processes. For $t > 0$ integer define*

$$M_t = \sum_{i=1}^{t} \mathbb{1}_{\{X_i^{i+n-1} = Z_0^{n-1}\}}\ .$$

So, M_t counts the number of occurrences of the random sequence Z_0^{n-1}, in the independent process $(X_m)_{m\in\mathbb{N}}$, up to time t.

We have the following theorem.

Theorem 2 (First independent random string). *Let $(X_m)_{m\in\mathbb{N}}$ be a stationary process. Then*

$$\left\|\mathcal{L}(M_{t/\mathbb{P}(Z_0^{n-1})}) - \mathcal{L}(Z)\right\| \le Cf(t)\varepsilon(n)\ , \tag{1}$$

where

- *if the process is ψ-mixing*

$$\varepsilon(n) := \inf_{\Delta\ge n}\left\{\Delta \max_{\underset{\sim}{a}\in\mathcal{C}^n} \mathbb{P}(\underset{\sim}{a}) + \psi(\Delta)\right\}\ ,$$

- *if the process is β-mixing*

$$\varepsilon(n) := \inf_{\Delta\ge n}\left\{\Delta \inf_{0\le g\le n-1}\left\{\max_{a_g^{n-1}\in\mathcal{C}^{n-g}} \{\mathbb{P}(a_g^{n-1})\} + \beta(g)\right\} + \beta(\Delta)\right\}\ .$$

Definition 5. *Let $(X_m)_{m\in\mathbb{N}}$ be a process. For $t > 0$ integer define*

$$N_t = \sum_{i=1}^{t} \mathbb{1}_{\{X_i^{i+n-1} = X_0^{n-1}\}}\ .$$

So, N_t counts the number of occurrences of the initial random sequence X_0^{n-1}, up to time t.

Theorem 3 (First random string). *Let be a stationary ψ-mixing or β-mixing process. Then*

$$\left\|\mathcal{L}(N_{t/\mathbb{P}(X_0^{n-1})}) - \mathcal{L}(Z)\right\| \le Cf(t)\varepsilon(n)\ .$$

$\varepsilon(n)$ is defined in Theorem 2, according to the case.

The main results of this paper are Theorem 2 and Theorem 3. The basic tool for their proofs is the next theorem which refers to the Poisson statistics of a fixed n-string.

Definition 6. *Let $\underset{\sim}{a} \in \mathcal{C}^n$. For $t > 0$ integer let*

$$N_t(\underset{\sim}{a}) = \sum_{i=1}^{t} \mathbb{1}_{\{X_i^{i+n-1} = \underset{\sim}{a}\}} \,.$$

So, $N_t(\underset{\sim}{a})$ counts the number of occurrences of the fixed *string $\underset{\sim}{a}$ up to time t.*

To compute the error for the Poisson statistics of $N_t(\underset{\sim}{a})$ we need the following definition.

Definition 7. *Let $\underset{\sim}{a} \in \mathcal{C}^n$. We define the first overlapping position or* periodicity *of $\underset{\sim}{a}$ as the number $\mathcal{O}(\underset{\sim}{a})$ defined as follows:*

$$\mathcal{O}(\underset{\sim}{a}) = \min\left\{k \in \{1, \ldots, n\} \mid a_0^{n-1-k} = a_k^{n-1}\right\} \,.$$

(We consider $a_0^{-1} = a_n^{n-1}$ by definition.)

Remark 1. *For future purpose it is useful to notice the following. Suppose n can be written as $n = \mathcal{O}(\underset{\sim}{a})[n/\mathcal{O}(\underset{\sim}{a})] + r$, then we have two cases: If $\mathcal{O}(\underset{\sim}{a}) \leq n/2$ then*

$$a_0^{\mathcal{O}(\underset{\sim}{a})-1} = a_{\mathcal{O}(\underset{\sim}{a})}^{2\mathcal{O}(\underset{\sim}{a})-1} = \cdots = a_{([n/\mathcal{O}(\underset{\sim}{a})]-1)\mathcal{O}(\underset{\sim}{a})}^{[n/\mathcal{O}(\underset{\sim}{a})]\mathcal{O}(\underset{\sim}{a})-1} \quad \text{and} \quad a_{[n/\mathcal{O}(\underset{\sim}{a})]\mathcal{O}(\underset{\sim}{a})}^{n-1} = a_0^{r-1} \,.$$

If $\mathcal{O}(\underset{\sim}{a}) > n/2$ then

$$a_0^{n-\mathcal{O}(\underset{\sim}{a})-1} = a_{\mathcal{O}(\underset{\sim}{a})}^{n-1} \,.$$

We now can state the Poisson approximation theorem.

Theorem 4 (Fixed string). *Consider the stationary process $(X_m)_{m\in\mathbb{N}}$ with $\underset{\sim}{a} \in \mathcal{C}^n$. Then*

$$\left|\left|\mathcal{L}(N_{t/\mathbb{P}(\underset{\sim}{a})}) - \mathcal{L}(Z)\right|\right| \leq C f(t) \varepsilon(\underset{\sim}{a}) \,,$$

where

- *if the process is ψ-mixing*

$$\varepsilon(\underset{\sim}{a}) := \inf_{\Delta \geq n} \{\Delta \mathbb{P}(\underset{\sim}{a}) + \psi(\Delta)\} + n\mathbb{P}\left(a_{n-\mathcal{O}(\underset{\sim}{a})}^{n-1}\right) \,,$$

- *if the process is β-mixing*

$$\varepsilon(\underset{\sim}{a}) \quad := \quad n \inf_{0 \leq g \leq n-1} \left\{ \frac{\beta(g)}{\mathcal{O}(\underset{\sim}{a})\mathbb{P}(\underset{\sim}{a})} + \mathbb{P}(a_{n-\mathcal{O}(\underset{\sim}{a})+g}^{n-1}) \right\} \,.$$

Remark 2. *Consider $\varepsilon(\underset{\sim}{a})$ in the ψ-mixing case. The factor in brackets consists of two terms. A priori it is not possible to say which term is smaller. For instance, assume the system loses memory exponentially fast with a large positive constant, and assume $\mathcal{O}(\underset{\sim}{a})$ is small. Then the error is basically $n\mathbb{P}(a_{n-\mathcal{O}(\underset{\sim}{a})}^{n-1})$. On the other hand, assume the system loses memory polynomially fast. Assume also that $\mathcal{O}(\underset{\sim}{a}) = n$. Then the error is given basically by $\inf_{\Delta \geq n}\{\Delta \mathbb{P}(\underset{\sim}{a}) + \psi(\Delta)\}$.*

Remark 3. $\mathbb{P}(a_{n-\mathcal{O}(\underset{\sim}{a})}^{n-1})$ *is small when* $\mathcal{O}(\underset{\sim}{a})$ *is large. (Recall the meaning of* $\mathcal{O}(\underset{\sim}{a})$ *in Definition* 7.) *Collet et al.* (1999) *proved that for exponentially* ψ*-mixing processes there exist positive constants* C *and* c *such that*

$$\mathbb{P}(\underset{\sim}{a} \in \mathcal{C}^n \ ; \ \mathcal{O}(\underset{\sim}{a}) \leq n/3) \leq C \exp^{-cn} \ .$$

Abadi (2001) *extended the above inequality to* ϕ*-mixing processes when* $1/3$ *is replaced by some constant* $s \in (0,1)$. *Abadi and Vaienti* (2008) *proved the above inequality for* ψ*-mixing processes for* any *value of* s *(with* $c = c(s)$.) *This shows that Theorem* 4 *holds for typical (in the sense of* $\mathcal{O}(\underset{\sim}{a})$*) strings. Taking limit on the length of the strings along infinite sequences, we get that the Poisson limit law holds almost everywhere.*

4.2. Corollaries: occurrence times

In this section we derive a number of corollaries about the distribution of the first occurrence of a string. They are direct consequences of the Poisson approximation.

Consider the zero-level sets of M_t, N_t and $N_t(\underset{\sim}{a})$. It corresponds to the law of the first visit to the observable considered. Denote them W_n, R_n and $\tau_{\underset{\sim}{a}}$ respectively. The zero-level set of the Poisson law is the exponential law.

Definition 8. *Given* $n \in \mathbb{N}$, *we define the* waiting time $W_n : \Omega \to \mathbb{N} \cup \{\infty\}$ *as the following r.v. defined on the product space* $(\Omega, \mathcal{F}, \mathbb{P}) \times (\Omega, \mathcal{F}, \mathbb{P})$.

$$W_n = \inf\{i \geq 1 : X_i^{i+n-1} = Z_0^{n-1}\} \ .$$

So, W_n is the position of the first time the string z_0^{n-1} of the process $(Z_m)_{m\in\mathbb{N}}$ appears in $(X_m)_{m\in\mathbb{N}}$. We recall that $(X_m)_{m\in\mathbb{N}}$ and $(Z_m)_{m\in\mathbb{N}}$ are independent and identically distributed processes.

Corollary 1 (Waiting time). *Under the conditions of Theorem* 2, *the following inequality holds*

$$\left|\mathbb{P}\left(W_n \mathbb{P}(Z_0^{n-1}) > t\right) - e^{-t}\right| \leq C f(t) \varepsilon(n) \ .$$

Definition 9. *Given* $n \in \mathbb{N}$, *we define the* repetition time $R_n : \Omega \to \mathbb{N} \cup \{\infty\}$ *as*

$$R_n = \inf\{i \geq 1 : X_i^{i+n-1} = X_0^{n-1}\} \ .$$

So, R_n is the position of the first repetition of the initial string x_0^{n-1}.

Corollary 2 (Recurrence time). *Under the conditions of Theorem* 3, *the following inequality holds*

$$\left|\mathbb{P}\left(R_n \mathbb{P}(X_0^{n-1}) > t\right) - e^{-t}\right| \leq C f(t) \varepsilon(n) \ .$$

Definition 10. *Given* $\underset{\sim}{a} \in \mathcal{C}^n$, *we define the* hitting time $\tau_{\underset{\sim}{a}} : \Omega \to \mathbb{N} \cup \{\infty\}$ *as*

$$\tau_{\underset{\sim}{a}} = \inf\{k \geq 1 : X_i^{i+n-1} = \underset{\sim}{a}\} \ .$$

Corollary 3 (Hitting time). *Under the conditions of Theorem* 4, *the following inequality holds*

$$\left|\mathbb{P}\left(\tau_{\underset{\sim}{a}} \mathbb{P}(\underset{\sim}{a}) > t\right) - e^{-t}\right| \leq C f(t) \varepsilon(\underset{\sim}{a}) \ .$$

5. Proofs

5.1. Short correlations

In order to control error term b_2 in the Chen-Stein method, we need to compute certain short correlations which are controlled by Lemma 1 and Lemma 2 below. Before that we make the following remark.

Remark 4. *In the proofs we will use the following inequalities which are direct consequences of Definition 1 and Definition 2 respectively. Take any $A \in \mathcal{F}_{\{0,\dots,n\}}$ and $B \in \mathcal{F}_{\{i|i\geq m\}}$. For a ψ-mixing process*

$$\mathbb{P}(A;B) \leq C\mathbb{P}(A)\mathbb{P}(B) ,$$

with $C = 1+\psi([n-m]_+)$ (and $[b]_+ = max\{b,0\}$.) For a β-mixing process

$$\mathbb{P}(A;B) \leq \beta([n-m]_+) \ + \ \mathbb{P}(A)\mathbb{P}(B) \ .$$

Lemma 1 (Short correlation of a string). *Let $(X_m)_{m\in\mathbb{N}}$ be a stationary process. Let $\underset{\sim}{a} \in C^n$. Let Δ be any positive integer. Then the following inequality holds*

$$\sum_{j=1}^{\Delta+n-1} \mathbb{E}\left[\mathbb{1}_{\{X_0^{n-1}=\underset{\sim}{a}\}}\mathbb{1}_{\{X_j^{j+n-1}=\underset{\sim}{a}\}}\right] \leq C\delta(\underset{\sim}{a}) \ .$$

where

- *if the process is ψ-mixing*

$$\delta(\underset{\sim}{a}) := \mathbb{P}(\underset{\sim}{a})\left[\Delta\mathbb{P}(\underset{\sim}{a}) + n\mathbb{P}(a_{n-\mathcal{O}(\underset{\sim}{a})}^{n-1})\right] ,$$

- *if the process is β-mixing*

$$\delta(\underset{\sim}{a}) \ := \ n \inf_{0\leq g\leq \mathcal{O}(\underset{\sim}{a})-1}\left[\frac{\beta(g)}{\mathcal{O}(\underset{\sim}{a})} + \mathbb{P}(a_{n-\mathcal{O}(\underset{\sim}{a})+g}^{n-1})\mathbb{P}(\underset{\sim}{a})\right] .$$

Remark 5. *Both terms that defines $\delta(\underset{\sim}{a})$ in the ψ-mixing case control short correlations. The second one dominates the very short returns which are ruled by the overlap properties of $\underset{\sim}{a}$. The first one dominates the short, but not so much returns. That is short returns but without the influence of the overlap properties of $\underset{\sim}{a}$. For typical strings, $\mathcal{O}(\underset{\sim}{a}) = n$ (see Remark 3) and $\delta(\underset{\sim}{a})$ reduces to the first term. In the β-mixing case, the largest contribution comes from the very short returns.*

Proof of Lemma 1. The ψ-mixing case. Assume first $\Delta > n$. We decompose the above sum in terms with $j = 1,\dots,\mathcal{O}(\underset{\sim}{a})-1$; $j = \mathcal{O}(\underset{\sim}{a}),\dots,n-1$; and $j = n,\dots,\Delta+n-1$. One has for any j

$$\mathbb{E}\left[\mathbb{1}_{\{X_0^{n-1}=\underset{\sim}{a}\}}\ \mathbb{1}_{\{X_j^{j+n-1}=\underset{\sim}{a}\}}\right] = \mathbb{P}\left(X_0^{n-1} = \underset{\sim}{a} \ ; \ X_j^{j+n-1} = \underset{\sim}{a}\right) .$$

Thus, by the definition of $\mathcal{O}(\underset{\sim}{a})$ we have that

$$\sum_{j=1}^{\mathcal{O}(\underset{\sim}{a})-1} \mathbb{E}\left[\mathbb{1}_{\{X_0^{n-1}=\underset{\sim}{a}\}}\ \mathbb{1}_{\{X_j^{j+n-1}=\underset{\sim}{a}\}}\right] = 0 \ .$$

For $j = n, \ldots, \Delta + n - 1$, by Remark 4

$$\mathbb{P}\left(X_0^{n-1} = \underset{\sim}{a}\ ;\ X_j^{j+n-1} = \underset{\sim}{a}\right) \leq C\mathbb{P}(\underset{\sim}{a})^2\ .$$

Then

$$\sum_{j=n}^{\Delta+n-1} \mathbb{P}\left(X_0^{n-1} = \underset{\sim}{a}\ ;\ X_j^{j+n-1} = \underset{\sim}{a}\right) \leq C\Delta\mathbb{P}(\underset{\sim}{a})^2\ . \tag{2}$$

For $j = \mathcal{O}(\underset{\sim}{a}), \ldots, n-1$, and also by Remark 4

$$\begin{aligned}
\mathbb{P}\left(X_0^{n-1} = \underset{\sim}{a}\ ;\ X_j^{j+n-1} = \underset{\sim}{a}\right) &= \mathbb{P}\left(X_0^{n-1} = \underset{\sim}{a}\ ;\ X_n^{j+n-1} = a_{n-j}^{n-1}\right) \\
&\leq C\mathbb{P}(\underset{\sim}{a})\mathbb{P}\left(a_{n-j}^{n-1}\right) \\
&\leq C\mathbb{P}(\underset{\sim}{a})\mathbb{P}\left(a_{n-\mathcal{O}(\underset{\sim}{a})}^{n-1}\right)\ .
\end{aligned}$$

Thus

$$\sum_{j=\mathcal{O}(\underset{\sim}{a})}^{n} \mathbb{P}\left(X_0^{n-1} = X_j^{j+n-1} = \underset{\sim}{a}\right) \leq Cn\ \mathbb{P}(\underset{\sim}{a})\mathbb{P}\left(a_{n-\mathcal{O}(\underset{\sim}{a})}^{n-1}\right)\ . \tag{3}$$

If $\Delta \leq n$ the same bound holds. This ends the proof of this case.

The β-mixing case. Let us fix an index $j \in I$. By definition of $\mathcal{O}(\underset{\sim}{a})$ one has

$$\left\{X_j^{j+n-1} = \underset{\sim}{a}\right\} \bigcap \left\{X_{j+i}^{j+i+n-1} = \underset{\sim}{a}\right\} = \emptyset\ ,$$

for all $1 \leq i \leq \mathcal{O}(\underset{\sim}{a}) - 1$. So

$$\bigcup_{i=0}^{\mathcal{O}(\underset{\sim}{a})-1} \left\{X_{j+i}^{j+i+n-1} = \underset{\sim}{a}\right\}\ , \tag{4}$$

and its complement are a partition of $\mathcal{F}_{\{j,\ldots,j+\mathcal{O}(\underset{\sim}{a})-1+n-1\}}$. Thus, we divide the n indexes of $j = \mathcal{O}(\underset{\sim}{a}), \ldots, n-1+\mathcal{O}(\underset{\sim}{a})$, in blocks of length $\mathcal{O}(\underset{\sim}{a})$. Then we re-write the left-hand side of (2) as

$$\sum_{\substack{m=i\mathcal{O}(\underset{\sim}{a}) \\ i=1,\ldots,[n/\mathcal{O}(\underset{\sim}{a})]+1}} \sum_{j=m}^{m+\mathcal{O}(\underset{\sim}{a})-1} \mathbb{P}\left(X_0^{n-1} = \underset{\sim}{a}\ ;\ X_j^{j+n-1} = \underset{\sim}{a}\right)\ . \tag{5}$$

We consider only the first $\mathcal{O}(\underset{\sim}{a})$ symbols of the first occurrence of $\underset{\sim}{a}$. The above probability is bounded by

$$\mathbb{P}\left(X_0^{\mathcal{O}(\underset{\sim}{a})-1} = a_0^{\mathcal{O}(\underset{\sim}{a})-1}\ ;\ X_j^{j+n-1} = \underset{\sim}{a}\right)\ .$$

Now we open a gap of length g in the first occurrence of $\underset{\sim}{a}$. The inner sum in (5) is bounded by

$$\sum_{j=m}^{m+\mathcal{O}(\underset{\sim}{a})-1} \mathbb{P}\left(X_0^{\mathcal{O}(\underset{\sim}{a})-1-g} = a_0^{\mathcal{O}(\underset{\sim}{a})-1-g} \; ; \; X_j^{j+n-1} = \underset{\sim}{a} \right) .$$

The second set in the above probability belongs to the partition (4). Therefore, by Remark 4 and stationarity, the above sum is bounded by

$$\beta(g) + \mathbb{P}(X_0^{\mathcal{O}(\underset{\sim}{a})-1-g} = a_0^{\mathcal{O}(\underset{\sim}{a})-1-g})\mathcal{O}(\underset{\sim}{a})\mathbb{P}(\underset{\sim}{a}) .$$

The cardinal of the first sum in (5) is $[n/\mathcal{O}(\underset{\sim}{a})]+1$. Then the left-hand side of (2) is bounded by

$$2n \left[\frac{\beta(g)}{\mathcal{O}(\underset{\sim}{a})} + \mathbb{P}(a_0^{\mathcal{O}(\underset{\sim}{a})-1-g})\mathbb{P}(\underset{\sim}{a}) \right] .$$

Just to unify notation we note the following. Since for positive integers $n_1 \leq n_2$ one has by stationarity

$$\sum_{j=n_1}^{n_2} \mathbb{P}\left(X_0^{n-1} = \underset{\sim}{a}; X_j^{j+n-1} = \underset{\sim}{a} \right) = \sum_{j=-n_2}^{-n_1} \mathbb{P}\left(X_{-(j+n-1)}^{-j} = \underset{\sim}{a}; X_{-(n-1)}^{0} = \underset{\sim}{a} \right) ,$$

we can repeat the above argument exchanging the roles of the first and second occurrences of $\underset{\sim}{a}$. Then we can change $\mathbb{P}(a_0^{\mathcal{O}(\underset{\sim}{a})-1-g})$ by $\mathbb{P}(a_{n-\mathcal{O}(\underset{\sim}{a})+g}^{n-1})$ in the above bounds. This ends the proof of the lemma. □

Lemma 2 (Short correlation of the process). *Let $(X_m)_{m\in\mathbb{N}}$ be a stationary process. Let Δ be any positive integer. Then the following inequality holds*

$$\sum_{j=1}^{\Delta} \mathbb{E}\left[\mathbb{1}_{\{X_0^{n-1}=X_j^{j+n-1}\}} \right] \leq C\delta(n) ,$$

where

- *if the process is ψ-mixing*

$$\delta(n) = \Delta \max_{\underset{\sim}{a}\in\mathcal{C}^n} \mathbb{P}(\underset{\sim}{a}) ,$$

- *if the process is β-mixing*

$$\delta(n) = \Delta \inf_{1\leq g\leq n-1} \{\beta(g) + \max_{a_g^{n-1}\in\mathcal{C}^{n-g}} \mathbb{P}(a_g^{n-1})\} .$$

Proof of Lemma 2. The ψ-mixing case. Consider the equation

$$X_0^{n-1} = X_j^{j+n-1} . \tag{6}$$

We divide the proof in $j=1,\dots,n/2$, $j=n/2,\dots,n-1$ and $j=n,\dots,\Delta$. Let us start with $j=n,\dots,\Delta$. We condition on the first n-block. Thus

$$\mathbb{E}\left[\mathbb{1}_{\{X_0^{n-1}=X_j^{j+n-1}\}} \right] = \mathbb{E}\left[\mathbb{E}\left[\mathbb{1}_{\{X_0^{n-1}=X_j^{j+n-1}\}} | X_0^{n-1} \right] \right] .$$

The last expression is

$$\sum_{a_0^{n-1}\in\mathcal{C}^n} \mathbb{P}(X_j^{j+n-1} = a_0^{n-1} | X_0^{n-1} = a_0^{n-1}) \mathbb{P}(a_0^{n-1}) \ . \tag{7}$$

The leftmost probability is bounded by $\mathbb{P}(\underset{\sim}{a})$ using the ψ-mixing property. Thus, the previous sum is bounded by

$$\max_{a_0^{n-1}\in\mathcal{C}^n} \{\mathbb{P}(\underset{\sim}{a})\} \ .$$

Suppose now that $j = 1, \ldots, n/2$. Consider (6) and write $n+j = qj + r$, with q, j, r integers and $0 \le r < j$. Then X_0^{n+j-1} consists of repeating the first block of length j up to complete $n+j$ symbols, namely

$$X_0^{j-1} = X_j^{2j-1} = \cdots = X_{(q-1)j}^{qj-1} \qquad \text{and} \qquad X_{n+j-r}^{n+j-1} = X_0^{r-1} \ .$$

Thus, we condition on the first j-block

$$\mathbb{E}\left[\mathbb{1}_{\{X_0^{n-1}=X_j^{j+n-1}\}}\right] = \mathbb{E}\left[\mathbb{E}\left[\mathbb{1}_{\{X_0^{n-1}=X_j^{j+n-1}\}} | X_0^{j-1}\right]\right] \ . \tag{8}$$

The last expression is

$$\sum_{a_0^{j-1}\in\mathcal{C}^j} \mathbb{P}(X_j^{2j-1} = \cdots = X_{(q-1)j}^{qj-1} = a_0^{j-1}; X_{n+j-r}^{n+j-1} = a_0^{r-1} | X_0^{j-1} = a_0^{j-1}) \mathbb{P}(a_0^{j-1}).$$

Again, the leftmost probability is bounded by

$$C\mathbb{P}(X_j^{2j-1} = \cdots = X_{(q-1)j}^{qj-1} = a_0^{j-1} \ ; \ X_{n+j-r}^{n+j-1} = a_0^{r-1}) \ ,$$

which is the probability of an n-string.

Similarly to the previous case, when $j = n/2, \ldots, n-1$, equation (6) means that X_0^{n+j-1} has its first, last and central $n-2j$-blocks equals and the two remaining j-blocks equals too. Namely

$$X_0^{n-j-1} = X_j^{n-1} = X_{2n-j}^{n+j-1} \qquad \text{and} \qquad X_{n-j}^{j-1} = X_n^{2n-j-1} \ .$$

Thus, we condition on the first j-block and (8) still holds but is this case it is

$$\sum_{a_0^{j-1}\in\mathcal{C}^j} \mathbb{P}(X_0^{n-j-1} = X_j^{n-1} = X_{2n-j}^{n+j-1}; X_{n-j}^{j-1} = X_n^{2n-j-1} | X_0^{j-1} = a_0^{j-1}) \mathbb{P}(a_0^{j-1}).$$

Now, the leftmost probability is bounded using the ψ-mixing property by

$$C\mathbb{P}(X_j^{2j-1} = \cdots = X_{(q-1)j}^{qj-1} = a_0^{j-1} \ ; \ X_{n+j-r}^{n+j-1} = a_0^{r-1}) \ ,$$

which again is the probability of an n-string. This finishes this case.

The β-mixing case. We can re-write (7) as

$$\sum_{a_0^{n-1}\in\mathcal{C}^n} \mathbb{P}(X_0^{n-1} = a_0^{n-1} \ ; \ X_j^{j+n-1} = a_0^{n-1}) \ ,$$

which is bounded opening a gap at position j of length $1 \leq g \leq n-1$ by

$$\sum_{a_0^{n-1} \in \mathcal{C}^n} \mathbb{P}(X_0^{n-1} = a_0^{n-1} \; ; \; X_{j+g}^{j+n-1} = a_g^{n-1}) \; .$$

We bound the last sum using Remark 4 by

$$\begin{aligned} \beta(g) \quad &+ \sum_{a_0^{n-1} \in \mathcal{C}^n} \mathbb{P}(X_0^{n-1} = a_0^{n-1}) \mathbb{P}(X_{j+g}^{j+n-1} = a_g^{n-1}) \\ \leq \beta(g) \quad &+ \max_{a_g^{n-1} \in \mathcal{C}^{n-g}} \mathbb{P}(a_g^{n-1}) \; . \end{aligned}$$

Now consider $j = 1, \ldots, n-1$. This case is completely similar to the previous one. Equation (8) is equal to

$$\sum_{a_0^{j-1} \in \mathcal{C}^j} \mathbb{P}(X_0^{j-1} = a_0^{j-1} \; ; \; X_j^{j+n-1} = a_0^{n-1}) \; ,$$

which is bounded opening a gap at position j of length $g \leq n-1$ by

$$\sum_{a_0^{j-1} \in \mathcal{C}^j} \mathbb{P}(X_0^{j-1} = a_0^{j-1} \; ; \; X_{j+g}^{j+n-1} = a_g^{n-1}) \; .$$

We bound the last sum using again Remark 4 by

$$\begin{aligned} &\beta(g) + \sum_{a_0^{j-1} \in \mathcal{C}^j} \mathbb{P}(X_0^{j-1} = a_0^{j-1}) \mathbb{P}(X_{j+g}^{j+n-1} = a_g^{n-1}) \\ \leq \; &\beta(g) + \max_{a_g^{n-1} \in \mathcal{C}^{n-g}} \mathbb{P}(a_g^{n-1}) \; . \end{aligned}$$

This ends the proof of the lemma. □

5.2. Proofs for Poisson statistics

We first proceed with the proof of Theorem 4 since it is the base of the proofs of Theorem 2 and Theorem 3.

Proof of Theorem 4. *The ψ-mixing case.* It is enough to prove upper bounds for the quantities b_1, b_2 and b_3 in the Chen-Stein method. Therefore, we define the quantities involved in it. Let the set of indexes I be $\{1, \ldots, t/\mathbb{P}(\underset{\sim}{a})\}$. For any $j \in I$ put

$$V_j = \mathbb{1}_{\{X_j^{j+n-1} = \underset{\sim}{a}\}} \; .$$

Since the measure is stationary we have $p_j = \mathbb{E}[V_j] = \mathbb{P}(\underset{\sim}{a})$ for every $j \in I$. Moreover, fix an integer $\Delta \geq n$. Take $B_j = \{j - (\Delta + n), \ldots, j + \Delta + n\} \cap I$. Hence

$$b_1 \leq \frac{t}{\mathbb{P}(\underset{\sim}{a})} \, (2\Delta + 2n + 1) \, \mathbb{P}(\underset{\sim}{a})^2 \leq C \, t \, \Delta \, \mathbb{P}(\underset{\sim}{a}) \; . \tag{9}$$

Further, also by stationarity

$$\begin{aligned} b_2 &= \sum_{j=1}^{t/\mathbb{P}(\underset{\sim}{a})} \sum_{i\in B_j\setminus\{j\}} \mathbb{E}\left[\mathbb{1}_{\{X_i^{i+n-1}=\underset{\sim}{a}\}}\mathbb{1}_{\{X_j^{j+n-1}=\underset{\sim}{a}\}}\right] \\ &\le \frac{t}{\mathbb{P}(\underset{\sim}{a})}\max_{j\in I}\sum_{\substack{i\neq j\\ i=j-\Delta-n}}^{j+\Delta+n} \mathbb{P}\left(X_i^{i+n-1}=\underset{\sim}{a}\ ;\ X_j^{j+n-1}=\underset{\sim}{a}\right) \\ &= \frac{t}{\mathbb{P}(\underset{\sim}{a})}\, 2\sum_{i=1}^{\Delta+n}\mathbb{P}\left(\underset{\sim}{a}\ ;\ X_i^{i+n-1}=\underset{\sim}{a}\right) . \end{aligned} \tag{10}$$

The quantity above is bounded using Lemma 1 by

$$C\,t\,\left[\Delta\mathbb{P}(\underset{\sim}{a}) + n\mathbb{P}\left(a_{n-\mathcal{O}(\underset{\sim}{a})}^{n-1}\right)\right] .$$

Finally

$$b_3 \le \frac{t}{\mathbb{P}(\underset{\sim}{a})}\max_{j\in I}\mathbb{E}\left|\mathbb{E}\left[\mathbb{1}_{\{X_j^{j+n-1}=\underset{\sim}{a}\}} - \mathbb{P}(\underset{\sim}{a})\mid \mathcal{F}_{B_j^c}\right]\right| .$$

For the sake of simplicity denote $B = B_j$. Put $Y = \{X_i\}_{i\notin B}$. The above expectation is

$$\begin{aligned} &\sum_{Y=y}\mathbb{P}(Y=y)\left|\mathbb{E}\left[\mathbb{1}_{\{X_j^{j+n-1}=\underset{\sim}{a}\}} - \mathbb{P}(\underset{\sim}{a})|Y=y\right]\right| \\ =& \sum_y \mathbb{P}(y)\left|(1-\mathbb{P}(\underset{\sim}{a}))\mathbb{P}(X_j^{j+n-1}=\underset{\sim}{a}|y) - \mathbb{P}(\underset{\sim}{a})\mathbb{P}(X_j^{j+n-1}\neq\underset{\sim}{a}|y)\right| \\ =& \sum_y \left|\mathbb{P}(X_j^{j+n-1}=\underset{\sim}{a}\ ;\ y) - \mathbb{P}(\underset{\sim}{a})\mathbb{P}(y)\right| , \end{aligned}$$

We can write

$$\{Y=y\} = \{Y_l = y_l\}\ \cap\ \{Y_r = y_r\} ,$$

where $Y_l = \{X_t\}_{t<-|B|}$ and $Y_r = \{X_t\}_{t>|B|}$ (respectively y_l and y_r). We get the following triangular inequality

$$\begin{aligned} \left|\mathbb{P}(\underset{\sim}{a}\ ;\ y) - \mathbb{P}(\underset{\sim}{a})\mathbb{P}(y)\right| &\le \left|\mathbb{P}(\underset{\sim}{a}\ ;\ y) - \mathbb{P}(\underset{\sim}{a}\ ;\ y_l)\mathbb{P}(y_r)\right| \\ &+ \left|\mathbb{P}(\underset{\sim}{a}\ ;\ y_l) - \mathbb{P}(\underset{\sim}{a})\mathbb{P}(y_l)\right|\mathbb{P}(y_r) \\ &+ \mathbb{P}(\underset{\sim}{a})\left|\mathbb{P}(y_l)\mathbb{P}(y_r) - \mathbb{P}(y)\right| . \end{aligned} \tag{11}$$

The three terms on the right-hand side of the above inequality can be bounded using the ψ-mixing property. For the first one we have

$$\left|\mathbb{P}(\underset{\sim}{a}\ ;\ y) - \mathbb{P}(\underset{\sim}{a}\ ;\ y_l)\mathbb{P}(y_r)\right| \le \psi(\Delta)\mathbb{P}(\underset{\sim}{a}\ ;\ y_l)\mathbb{P}(y_r) .$$

In the same way we bound the modulus in the second one

$$\left|\mathbb{P}(\underset{\sim}{a}\ ;\ y_l) - \mathbb{P}(\underset{\sim}{a})\mathbb{P}(y_l)\right| \le \psi(\Delta)\mathbb{P}(\underset{\sim}{a})\mathbb{P}(y_l) .$$

And finally the modulus in the third one is bounded by

$$|\mathbb{P}(y_l)\mathbb{P}(y_r) - \mathbb{P}(y)| \le \psi(2\Delta)\mathbb{P}(y_l)\mathbb{P}(y_r) \ .$$

The above three inequalities give that the left-hand side of (11) is bounded by

$$\psi(\Delta)\mathbb{P}(\underset{\sim}{a}\ ;\ y_l)\mathbb{P}(y_r) + 2\psi(\Delta)\mathbb{P}(\underset{\sim}{a})\mathbb{P}(y_l)\mathbb{P}(y_r) \ .$$

Thus summing on $Y = y$

$$\sum_{y_l}\sum_{y_r} \left|\mathbb{P}(\underset{\sim}{a}\ ;\ y) - \mathbb{P}(\underset{\sim}{a})\mathbb{P}(y)\right| \le 3\psi(\Delta)\mathbb{P}(\underset{\sim}{a}) \ .$$

Summing over all $j \in I$ we have

$$b_3 \le 3\,t\,\psi(\Delta) \ .$$

This ends the proof of this case.

The β-mixing case. Take $B_j = \left\{j - (n + \mathcal{O}(\underset{\sim}{a})), \dots, j + n + \mathcal{O}(\underset{\sim}{a})\right\} \cap I$. b_1 is bounded identically to the ψ-mixing case by $Ctn\mathbb{P}(\underset{\sim}{a})$.

Due to Lemma 1 we have

$$b_2 \le \frac{t}{\mathbb{P}(\underset{\sim}{a})} n \inf_{0 \le g \le \mathcal{O}(\underset{\sim}{a}) - 1} \left[\frac{\beta(g)}{\mathcal{O}(\underset{\sim}{a})} + \mathbb{P}(a_{n-\mathcal{O}(\underset{\sim}{a})+g}^{n-1})\mathbb{P}(\underset{\sim}{a})\right] \ .$$

For b_3 consider the sum on y of the three terms on the right-hand side of (11). Each one is bounded using the β-mixing condition by $\beta(\mathcal{O}(\underset{\sim}{a}))$ and we get $b_3 \le t\beta(\mathcal{O}(\underset{\sim}{a}))/\mathbb{P}(\underset{\sim}{a})$. Thus, we get that $b_1 \le b_2$ and $b_3 \le b_2$. This ends the proof of the theorem. □

Proof of Theorem 2. *The ψ-mixing case.* Let $K \subseteq \mathbb{N}$. Conditioning M_t on the initial sequence Z_0^{n-1} of $(Z_m)_{m\in\mathbb{N}}$ and since it has the same distribution as $(X_m)_{m\in\mathbb{N}}$

$$\mathbb{P}(M_{t/\mathbb{P}(Z_0^{n-1})} \in K) = \sum_{\underset{\sim}{a}\in\mathcal{C}^n} \mathbb{P}(M_{t/\mathbb{P}(Z_0^{n-1})} \in K\ |Z_0^{n-1} = \underset{\sim}{a})\mathbb{P}(X_0^{n-1} = \underset{\sim}{a}) \ .$$

Since $(X_m)_{m\in\mathbb{N}}$ is independent of $(Z_m)_{m\in\mathbb{N}}$ one has

$$\mathbb{P}(M_{t/\mathbb{P}(Z_0^{n-1})} \in K\ |Z_0^{n-1} = \underset{\sim}{a}\ ) = \mathbb{P}(N_{t/\mathbb{P}(\underset{\sim}{a})}(\underset{\sim}{a}) \in K) \ .$$

By Theorem 1, the left-hand side of (1) is bounded by

$$\begin{aligned} &\sum_{\underset{\sim}{a}\in\mathcal{C}^n} \left|\mathbb{P}(N_{t/\mathbb{P}(\underset{\sim}{a})}(\underset{\sim}{a}) \in K) - \mathbb{P}(Z \in K)\right| \mathbb{P}(\underset{\sim}{a}) \\ \le\ &\sum_{\underset{\sim}{a}\in\mathcal{C}^n} (b_1(\underset{\sim}{a}) + b_2(\underset{\sim}{a}) + b_3(\underset{\sim}{a}))\mathbb{P}(\underset{\sim}{a}) \ . \end{aligned}$$

Here we denote with $b_i(\underset{\sim}{a}), i = 1, 2, 3$ the b_i's of Theorem 1 applied to each $\underset{\sim}{a} \in \mathcal{C}^n$. Using (10) we get that

$$\sum_{\underset{\sim}{a} \in \mathcal{C}^n} b_2(\underset{\sim}{a}) \mathbb{P}(\underset{\sim}{a}) \le \sum_{\underset{\sim}{a} \in \mathcal{C}^n} \frac{t}{\mathbb{P}(\underset{\sim}{a})} 2 \sum_{i=1}^{\Delta} \mathbb{P}\left(\underset{\sim}{a}\ ;\ X_i^{i+n-1} = \underset{\sim}{a}\right) \mathbb{P}(\underset{\sim}{a})$$

$$= 2t \sum_{i=1}^{\Delta} \sum_{\underset{\sim}{a} \in \mathcal{C}^n} \mathbb{P}\left(\underset{\sim}{a}\ ;\ X_i^{i+n-1} = \underset{\sim}{a}\right) = 2t \sum_{i=1}^{\Delta} \mathbb{P}\left(X_0^{n-1} = X_i^{i+n-1}\right) .$$

And we can bound the last term with Lemma 2.

Further, by (9) we have that

$$\sum_{\underset{\sim}{a} \in \mathcal{C}^n} b_1(\underset{\sim}{a}) \mathbb{P}(\underset{\sim}{a}) \le \Delta \max_{\underset{\sim}{a} \in \mathcal{C}^n} \mathbb{P}(\underset{\sim}{a}) .$$

Finally, $b_3(\underset{\sim}{a})$ is independent of $\underset{\sim}{a}$. This ends the proof of this case.

The β-mixing case. $\sum_{\underset{\sim}{a}} b_1(\underset{\sim}{a}) \mathbb{P}(\underset{\sim}{a})$ and $\sum_{\underset{\sim}{a}} b_2(\underset{\sim}{a}) \mathbb{P}(\underset{\sim}{a})$ are bounded identically to the previous case. To bound $\sum_{\underset{\sim}{a}} b_3(\underset{\sim}{a}) \mathbb{P}(\underset{\sim}{a})$ we use (11) and thus

$$\sum_{\underset{\sim}{a}} b_3(\underset{\sim}{a}) \mathbb{P}(\underset{\sim}{a}) \le t \sum_{y_l} \sum_{\underset{\sim}{a}} \sum_{y_r} \left| \mathbb{P}(y_l\ ;\ X_j^{j+n-1} = \underset{\sim}{a}\ ;\ y_r) - \mathbb{P}(\underset{\sim}{a}) \mathbb{P}(y_l\ ;\ y_r) \right| . \quad (12)$$

Now consider the general term in the above sum. The triangle inequality gives

$$\begin{aligned} & \left| \mathbb{P}(y_l\ ;\ X_j^{j+n-1} = \underset{\sim}{a}\ ;\ y_r) - \mathbb{P}(\underset{\sim}{a}) \mathbb{P}(y_l\ ;\ y_r) \right| \\ \le & \left| \mathbb{P}(y_l\ ;\ X_j^{j+n-1} = \underset{\sim}{a}\ ;\ y_r) - \mathbb{P}(y_l\ ;\ X_j^{j+n-1} = \underset{\sim}{a}) \mathbb{P}(y_r) \right| \\ + & \left| \mathbb{P}(y_l\ ;\ X_j^{j+n-1} = \underset{\sim}{a}) - \mathbb{P}(\underset{\sim}{a}) \mathbb{P}(y_l) \right| \mathbb{P}(y_r) \\ + & \mathbb{P}(\underset{\sim}{a}) \left| \mathbb{P}(y_l) \mathbb{P}(y_r) - \mathbb{P}(y_l\ ;\ y_r) \right| \end{aligned}$$

We apply the β-mixing property in the sum $\sum_{y_l} \sum_{\underset{\sim}{a}} \sum_{y_r}$ of each of the three terms in the right-hand side of the above inequality. The first one over the partitions $\{Y_l = y_l, X_j^{j+n-1} = \underset{\sim}{a}\}$ and $\{Y_r = y_r\}$. The second one over the partitions $\{Y_l = y_l\}$ and $\{X_j^{j+n-1} = \underset{\sim}{a}\}$. The third one over the partitions $\{Y_l = y_l\}$ and $\{Y_r = y_r\}$. Thus (12) is bounded by $3t\beta(\Delta)$. This ends the proof of the theorem. □

Proof of Theorem 3. The ψ-mixing case. We partition $\{N_{t/\mathbb{P}(X_0^{n-1})} \in K\}$ in eventually short and only long occurrences of X_0^{n-1}. We first need the following notation. For any $n \le s \le t$ denote

$$M_s^t = \sum_{i=s}^{t} \mathbb{1}_{\{X_i^{i+n-1}=Z_0^{n-1}\}} , \quad N_s^t = \sum_{i=s}^{t} \mathbb{1}_{\{X_i^{i+n-1}=X_0^{n-1}\}} ,$$

$$N_s^t(\underset{\sim}{a}) = \sum_{i=s}^{t} \mathbb{1}_{\{X_i^{i+n-1}=\underset{\sim}{a}\}} .$$

Then we have the following decomposition

$$\mathbb{P}(N_t \in K) = \mathbb{P}(N_\Delta^t = N_t \in K) + \mathbb{P}(N_\Delta^t < N_t \in K) .$$

Consider firstly the rightmost probability. Since $N_\Delta^t < N_t$ it means that there is some occurrence of $\underset{\sim}{a}$ in $\{1, \dots, \Delta - 1\}$. Thus

$$\mathbb{P}(N_\Delta^t < N_t \in K) \le \mathbb{P}(N_1^\Delta \ge 1) \le \sum_{i=1}^{\Delta} \mathbb{P}(X_0^{n-1} = X_i^{i+n-1}) .$$

The last term is bounded by Lemma 2. Thus, we only consider the most-left probability in the above decomposition.

$$\begin{aligned} \mathbb{P}(N_\Delta^t \in K) - \mathbb{P}(N_\Delta^t = N_t \in K) &= \mathbb{P}(N_\Delta^t \in K; N_1^\Delta \ge 1) \\ &\le \mathbb{P}(N_1^\Delta \ge 1) . \end{aligned}$$

The last probability is bounded again by Lemma 2. So, consider only $\mathbb{P}(N_\Delta^t \in K)$. Now decompose N_Δ^t and M_Δ^t on the initial sequences X_0^{n-1} and Z_0^{n-1} respectively. We have

$$\mathbb{P}(N_\Delta^t \in K) = \sum_{\underset{\sim}{a} \in \mathcal{C}^n} \mathbb{P}(N_\Delta^t(\underset{\sim}{a}) \in K \, ; \, X_0^{n-1} = \underset{\sim}{a}) ,$$

and

$$\mathbb{P}(M_\Delta^t \in K) = \sum_{\underset{\sim}{a} \in \mathcal{C}^n} \mathbb{P}(N_\Delta^t(\underset{\sim}{a}) \in K \, ; \, Z_0^{n-1} = \underset{\sim}{a}) = \sum_{\underset{\sim}{a} \in \mathcal{C}^n} \mathbb{P}(N_\Delta^t(\underset{\sim}{a}) \in K) \mathbb{P}(\underset{\sim}{a}) .$$

Thus, by the ψ-mixing property

$$\left| \mathbb{P}(N_\Delta^t \in K) - \mathbb{P}(M_\Delta^t \in K) \right| \le \psi(\Delta) \mathbb{P}(M_\Delta^t \in K) . \tag{13}$$

Further, the proof of Theorem 2 shows that,

$$\left| \mathbb{P}(M_\Delta^{t/\mathbb{P}(X_0^{n-1})} \in K) - \sum_{\underset{\sim}{a} \in \mathcal{C}^n} P(Z_{t-\Delta\mathbb{P}(\underset{\sim}{a})} \in K) \mathbb{P}(\underset{\sim}{a}) \right| \le \varepsilon(n) .$$

Here, we denote with Z_λ a Poisson r.v. with mean λ. The triangle inequality and the Mean Value Theorem give the inequalities

$$\begin{aligned} & \left| e^{-(t-\Delta\mathbb{P}(\underset{\sim}{a}))} (t - \Delta\mathbb{P}(\underset{\sim}{a}))^k - e^{-t} t^k \right| \\ \le & \left| e^{-(t-\Delta\mathbb{P}(\underset{\sim}{a}))} (t - \Delta\mathbb{P}(\underset{\sim}{a}))^k - e^{-(t-\Delta\mathbb{P}(\underset{\sim}{a}))} t^k \right| + \left| e^{-(t-\Delta\mathbb{P}(\underset{\sim}{a}))} t^k - e^{-t} t^k \right| \\ \le & \; e^{-(t-\Delta\mathbb{P}(\underset{\sim}{a}))} \Delta\mathbb{P}(\underset{\sim}{a}) (k t^{k-1} + t^k) . \end{aligned}$$

Summing on $k \in K$ and dividing by the $k!$ of the Poisson law we conclude that

$$\left|P(Z_{t-\Delta\mathbb{P}(\underset{\sim}{a})} \in K) - P(Z_t \in K)\right| \leq C\Delta\mathbb{P}(\underset{\sim}{a}) \, .$$

Therefore

$$\sum_{\underset{\sim}{a} \in \mathcal{C}^n} \left|P(Z_{t-\Delta\mathbb{P}(\underset{\sim}{a})} \in K) - P(Z_t \in K)\right| \mathbb{P}(\underset{\sim}{a}) \leq \Delta \max_{\underset{\sim}{a} \in \mathcal{C}^n} \mathbb{P}(\underset{\sim}{a}) \, .$$

This ends the proof of this case.

The β-mixing case. The unique difference between this proof and the previous one is that in this case the left-hand side of (13), by the β-mixing property, is bounded by $\beta(\Delta)$. This ends the proof of the theorem. □

5.3. Proofs for corollaries of occurrence times

Proof of Corollary 1. It is enough to note that

$$\left\{M_{t/\mathbb{P}(Z_0^{n-1})} = 0\right\} = \left\{W_n \mathbb{P}(Z_0^{n-1}) > t\right\} \, .$$

This ends the proof of the corollary. □

Proof of Corollary 2. It is enough to note that

$$\left\{N_{t/\mathbb{P}(X_0^{n-1})} = 0\right\} = \left\{R_n \mathbb{P}(X_0^{n-1}) > t\right\} \, .$$

This ends the proof of the corollary. □

Proof of Corollary 3. It is enough to note that

$$\left\{N_{t/\mathbb{P}(\underset{\sim}{a})}(\underset{\sim}{a}) = 0\right\} = \left\{\tau_{\underset{\sim}{a}} \mathbb{P}(\underset{\sim}{a}) > t\right\} \, .$$

This ends the proof of the corollary. □

Acknowledgment

The author kindly thanks the referee for a number of suggestions and comments to improve a previous version of the paper. The present work was done with the support of CAPES, institution of the Brazilian government for human resources.

References

[1] M. Abadi (2001). Exponential approximation for hitting times in mixing processes. *Math. Phys. Elec. J.* **7** No. 2.

[2] M. Abadi (2004). Sharp error terms and necessary conditions for exponential hitting times in mixing processes. *Ann. Probab.* **32**, 1A, 243–264.

[3] M. Abadi and S. Vaienti (2008). Large Deviations for Short Recurrence. *Discrete and Continuous Dynamical Systems Series A*, **21**, No. 3, July 2008.

[4] M. Abadi and A. Galves (2001). Inequalities for the occurrence times of rare events in mixing processes. The state of the art. *Markov Proc. Relat. Fields.* **7** 1, 97–112.

[5] D. Restrepo Álvarez (2005). Ocorrência de Eventos Sucessivos em Cadeias de Markov Estacionárias. *Ph. D. thesis, IME-USP.*

[6] D. Arratia, M. Goldstein and Gordon (1989). Two moments suffice for Poisson approximations: the Chen-Stein method *Ann. Prob.* **17**, 9–25.

[7] D. Arratia, M. Goldstein and Gordon (1990). Poisson approximation and the Chen-Stein method. With comments and a rejoinder by the authors. *Stat. sci.* **5**, 403–434.

[8] R. Bowen (1975). Equilibrium states and the ergodic theory of Anosov diffeomorphisms. *Lecture Notes in Math.* **470**. Springer-Verlag, New York.

[9] R.C. Bradley (2005). Basic Properties of Strong Mixing Conditions. A Survey and Some Open Questions *Probability Surveys Vol. 2* 107–144.

[10] L. Chen (1975). Poisson approximation for dependant trials. *Ann. Probability* **3**, 534–545.

[11] P. Collet, A. Galves and B. Schmitt (1999). Repetition times for gibbsian sources. *Nonlinearity* **vol. 12** , 1225–1237.

[12] P. Doukhan (1995). Mixing. Properties and examples. *Lecture Notes in Statistics* **85**. Springer-Verlag.

[13] R. Fernández, P. Ferrari and A. Galves (2001). Coupling, renewal and perfect simulation of chains of infinite order.
Can be downloaded from http://www.ime.usp.br/~ galves/papers/ebp5.ps

[14] A. Galves and B. Schmitt (1997). Inequalities for hitting times in mixing dynamical systems. *Random Comput. Dyn.* **5**, 337–348.

[15] M. Hirata, B. Saussol and S. Vaienti (1999). Statistics of return times: a general framework and new applications. *Comm. Math. Phys.* **206**, 33–55.

[16] D. Ornstein, B. Weiss (2002). Entropy and recurrence rates for stationary random fields. *Special issue on Shannon theory: perspective, trends, and applications. IEEE Trans. Inform. Theory* **48**, No. 6, 1694–1697.

[17] P.C. Shields (1996). The ergodic theory of discrete sample paths. *AMS, Providence RI.*

[18] Ch. Stein (1972). A bound for the error in the normal approximation to the distribution of a sum of dependent random variables. *Proceedings of the sixth Berkley Symposium on Mathematical Statistics and Probability* **2**, 583–602.

[19] A.J. Wyner (1999). More on recurrence and waiting times. *Ann. Appl. Probab.* **9**, No. 3, 780–796.

Miguel Abadi
IMECC B.P. 6065
University of Campinas
13083-859 Campinas, SP, Brasil
e-mail: `miguel@ime.unicamp.br`

Progress in Probability, Vol. 60, 21–30

An Inequality for Oriented 2-D Percolation

Enrique D. Andjel and Mariela Sued

Abstract. We consider a Bernoulli oriented percolation model on the subgraph of $\mathbb{Z}^2$ consisting of the points whose coordinates add up to an even number. In this model, each point (m, n) of this graph is the origin of two oriented bonds whose end points are $(m+1, n+1)$ and $(m-1, n+1)$. We show that the probability that there is an open path from $(0, 0)$ to $(2i, 2n)$ decreases strictly as i increases from 0 to n.

Mathematics Subject Classification (2000). 60K35.

Keywords. Oriented percolation, strict inequalities.

1. Introduction

Oriented two-dimensional percolation has been extensively studied. Durrett's 1984 paper [4] reviews the results known at that time and also proves some new results; for more recent work on that subject we refer the reader to [1],[2],[3],[8] and [9].The main goal of this paper is to prove a very intuitive inequality which is missing in the literature. We then derive a similar inequality for the contact process. For

background on the contact process we refer the reader to [6] and [7]. In the case of the contact process the inequality is stated in Gray's paper ([5]); but the proof given there depends on Theorem 1 of that paper whose proof has not been published. Some of the ideas of [5], including the notion of forbidden zones, are also present in this paper but we believe that there are some differences with Gray's approach. To state our results, let

$$\Lambda = \{(x, y) : x, y \in \mathbb{Z}, y \geq 0, x + y \in 2\mathbb{Z}\}$$

and for $n \in \mathbb{N}$ let

$$\Lambda_n = \{(x, y) \in \Lambda : y \leq n\}$$

Draw oriented bonds from each point (m, n) in Λ to $(m+1, n+1)$ and to $(m-1, n+1)$. In this paper we suppose that bonds are open independently of each other and that each bond is open with probability $p \in (0, 1)$.

A path in Λ is a sequence $(x_0, y_0), \ldots, (x_n, y_n)$ of points in Λ such that for all $0 \le i < n$, $|x_{i+1} - x_i| = 1$ and $y_{i+1} - y_i = 1$. We say that a path is open if all the bonds joining successive points of that path are open.

Given $C \subset 2\mathbb{Z} \times \{2i\}$ we say that $B \subset \Lambda$ can be generated from C if for all $(a, b) \in B$ there is a path $(x_0, y_0), \ldots, (x_n, y_n) = (a, b)$ such that $(x_0, y_0) \in C$ and $(x_i, y_i) \in B$ for all $0 \le i \le n$. In other words, each point in B can be attained from a point in C by means of an oriented path entirely contained in B. We say that a point (a, b) is to the left (right) of a path $\gamma = (x_0, y_0), \ldots, (x_n, y_n)$ if $a < x_i$ $(a > x_i)$ whenever $b = y_i$. Note that with this definition, any point (a, b) with $b < y_0$ or $b > y_n$ is to the left and to the right of γ.

We say that a subset B of Λ (or a path γ_2) is to the left (right) of a path γ_1 if all the points of B (or of γ_2) are to the left (right) of γ_1.

Suppose two paths $\gamma_1 = (x_0, y_0), \ldots, (x_n, y_n)$ and $\gamma_2 = (u_0, v_0), \ldots, (u_m, v_m)$ (or a path γ_1 and a set B) have a nonempty intersection. Then, we will call the last intersection of these paths (or the last intersection of γ_1 and B) the point in the intersection with the biggest second coordinate. In the case of two paths as above, if their last intersection is the point $(x_i, y_i) = (u_j, v_j)$ then, we define a new path:

$$\gamma_1\gamma_2 = (x_0, y_0), \ldots, (x_i, y_i), (u_{j+1}, v_{j+1}), \ldots, (u_m, v_m).$$

In our proofs it is important to consider open paths that do not use bonds in a forbidden zone of our graph. With this in mind we introduce the following notation: For any pair of arbitrary subsets B and D of Λ and any point $(m, n) \in \Lambda$,

$$\{D \xrightarrow{\Lambda \backslash B} (m, n)\}$$

will denote the event that there exists an open path from some point in D to (m, n) whose bonds do not have a point in B. When $B = \emptyset$ we will simply write $\{D \longrightarrow (m, n)\}$ and if D reduces to a point (a, b) we will write $\{(a, b) \longrightarrow (m, n)\}$

In Section 2 we prove:

Theorem 1.1. *Let $A \subset -2\mathbb{N}$. Then for any $C \subset 2\mathbb{N} \times \{0\}$, any $n \in \mathbb{N}$ and any set B which can be generated from C, we have*

$$P((A \cup \{0\}) \times \{0\} \xrightarrow{\Lambda \backslash B} (-1, 2n+1)) \ge P((A \cup \{0\}) \times \{0\} \xrightarrow{\Lambda \backslash B} (1, 2n+1)).$$

Moreover the inequality is strict if $A \cap \{-2n-2, -2n, \ldots, -2\} \ne \emptyset$ and $B = \emptyset$.

In Section 3 we derive from this theorem:

Corollary 1.1.

a) *For all $n > 0$ and all $i \in \{0, 1, \ldots, n\}$ we have*

$$P((0,0) \longrightarrow (2i, 2n)) > P((0,0) \longrightarrow (2i+2, 2n)).$$

b) *For all $n > 0$ and all $i \in \{1, \ldots, n+1\}$ we have*

$$P((0,0) \longrightarrow (2i-1, 2n+1)) > P((0,0) \longrightarrow (2i+1, 2n+1)).$$

Corollary 1.2. *Let ξ_t be the nearest neighbor contact process on $\mathbb{Z}$ with infection parameter $\lambda > 0$. Then, for all $t > 0$ and $i \ge 0$ $P(\xi_t^0(i) = 1) > P(\xi_t^0(i+1) = 1)$.*

To prove these corollaries we apply Theorem 1.1 in the special case $B = C = \emptyset$. It is nevertheless necessary to include these sets to carry out the inductive step in the proof of the theorem.

2. Proof of Theorem 1.1

Given $B \subseteq \Lambda$, we define $-B = \{(a,b) : (-a,b) \in B\}$. Consider now the following events:

$$D_1 = \{(0,0) \overset{\Lambda\backslash(B\cup -B)}{\longrightarrow} (-1,2n+1)\}^c \cap \{(0,0) \overset{\Lambda\backslash(B\cup -B)}{\longrightarrow} (1,2n+1)\}^c,$$

$$D_2 = \{(0,0) \overset{\Lambda\backslash(B\cup -B)}{\longrightarrow} (-1,2n+1)\} \cap \{(0,0) \overset{\Lambda\backslash(B\cup -B)}{\longrightarrow} (1,2n+1)\}^c,$$

$$D_3 = \{(0,0) \overset{\Lambda\backslash(B\cup -B)}{\longrightarrow} (-1,2n+1)\}^c \cap \{(0,0) \overset{\Lambda\backslash(B\cup -B)}{\longrightarrow} (1,2n+1)\},$$

$$D_4 = \{(0,0) \overset{\Lambda\backslash(B\cup -B)}{\longrightarrow} (-1,2n+1)\} \cap \{(0,0) \overset{\Lambda\backslash(B\cup -B)}{\longrightarrow} (1,2n+1)\},$$

$$F_1 = \{(A\cup\{0\}) \times \{0\} \overset{\Lambda\backslash B}{\longrightarrow} (1,2n+1)\}$$

and

$$F_{-1} = \{(A\cup\{0\}) \times \{0\} \overset{\Lambda\backslash B}{\longrightarrow} (-1,2n+1)\}.$$

Observe that if B is as in Theorem 1.1, then any path originated in $(0,0)$ and contained in $\Lambda \setminus (B \cup -B)$ is necessarily to the left of B and to the right of $-B$. Observe also that any point in $\{(x,y) : x \geq 0\}$ which is to the left of B must be to the right of $-B$. These properties will be used frequently in the sequel.

We now prove:

Proposition 2.1. *If A and B are as in Theorem 1.1, then*

a) $D_2 \cap F_1 = \emptyset$ *and*

b) $P(D_1 \cap F_{-1}) \geq P(D_1 \cap F_1)$.

Proof of Proposition 2.1. To show part a) note that if D_2 occurs there is an open path γ_1 from $(0,0)$ to $(-1,2n+1)$ contained in $\Lambda \setminus (B \cup -B)$. If F_1 occurs, there is an open path γ_2 from $(A \cup \{0\}) \times \{0\}$ to $(1,2n+1)$ contained in $\Lambda \setminus B$. But, if we have these two paths they must intersect and $\gamma_1\gamma_2$ is an open path from $(0,0)$ to $(1,2n+1)$. The portion of that path above the last intersection of γ_1 and γ_2 is to right of γ_1, hence to the right of $-B$. Therefore the whole path is contained in $\Lambda \setminus (B \cup -B)$ which contradicts the definition of D_2. Hence $D_2 \cap F_1 = \emptyset$.

To prove part b) we introduce several random objects: let

$$k = \sup\{r : 0 \leq r \leq n \text{ and } (0,0) \overset{\Lambda\backslash(B\cup -B)}{\longrightarrow} (0,2r)\}$$

where, by convention, $(0,0) \overset{\Lambda\backslash(B\cup -B)}{\longrightarrow} (0,0)$, let

$$I = \{m : k < m \leq n \text{ and } (A\cup\{0\}) \times \{0\} \overset{\Lambda\backslash B}{\longrightarrow} (0,2m)\}$$

and let

$$m_0 = \inf I$$

where, by convention $m_0 = \infty$ if $I = \emptyset$. Let

$$H = \begin{cases} \emptyset \text{ if } \; m_0 = \infty \\ \{x < 0 : (A \cup \{0\}) \times \{0\} \overset{\Lambda \setminus B}{\longrightarrow} (x, 2m_0)\} \text{ if } m_0 < \infty, \end{cases}$$

and let

$$K = \{(x,y) : (0,0) \overset{\Lambda \setminus (B \cup -B)}{\longrightarrow} (x,y) \text{ and } y \leq 2n+1\}.$$

The next figure shows each of the objects we have defined. In this example $A = \{-2,-4,-6,-8,-10\}$ and $C = \{(4,0)\}$. Moreover only the open bonds are drawn.

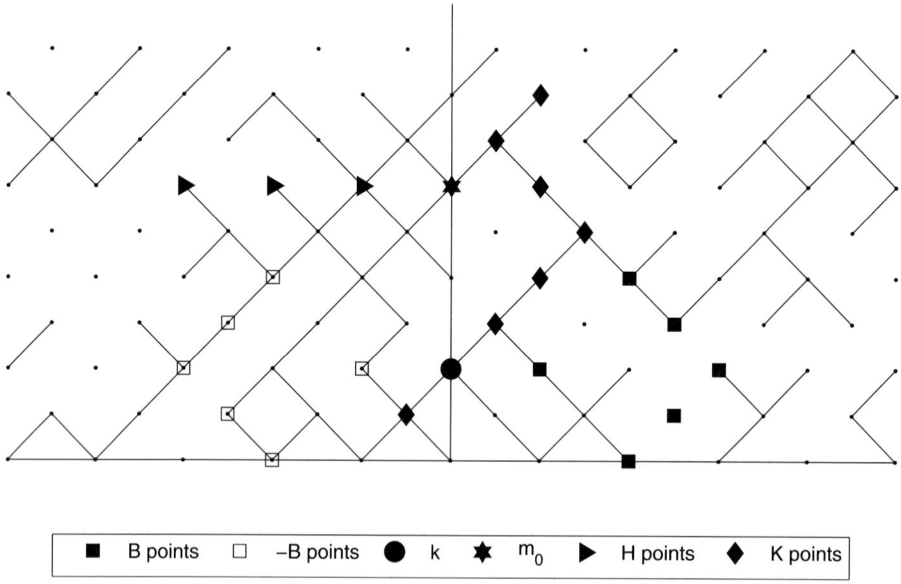

The H points in the drawing are really the points whose first coordinate is in that set and whose second coordinate is $2m_0$. Moreover, $(0, 2k)$ also belongs to K although it is not marked as a K point.

We now prove:

Lemma 2.1. *On $F_1 \cap D_1$ the following statements hold.*

a) $m_0 < \infty$.

b) *Any open path γ in $\Lambda_{2n+1} \setminus B$ starting from $(A \cup \{0\}) \times \{0\}$ and containing a point in the set $\{(x,y) : x > 0, y = 2m_0\}$ does not contain the point $(1, 2n+1)$ nor any point in the region $\{(x,y) : x \leq 0, y \geq 2m_0\}$. Moreover all the points in γ with second coordinate $y \geq 2m_0$ belong to K.*

c) $B \cap \{(x,y) : x \leq 0, y = 2m_0\} = \emptyset$.
d) $K \cap \{(x,y) : x \leq 0, y = 2m_0\} = \emptyset$.
e) *Any open path* $\gamma \in \Lambda_{2n+1} \setminus B$ *starting from the set* $\{(x,y) : x \leq 0, y = 2m_0\}$ *and containing a point in* K *does not contain the point* $(1, 2n+1)$ *nor the point* $(-1, 2n+1)$.

Remark. Statements b)–e) also hold on $\{m_0 < \infty\} \cap D_1$. By part a) $\{m_0 < \infty\} \cap D_1 \supset F_1 \cap D_1$. But, the reader can check that the proofs of parts b)–e) are also valid in this larger set.

To follow the proof of this lemma we recommend the reader to draw the paths that intervene in it.

Proof of Lemma 2.1. We start proving part a). Suppose F_1 occurs and let γ_1 be an open path from $(A \cup \{0\}) \times \{0\}$ to $(1, 2n+1)$ contained in $\Lambda \setminus B$. Let $(0, 2\ell)$ be the last intersection of this path with the set $\{(x,y) : x = 0\}$. It now suffices to show that on D_1 we have $\ell > k$. To do so we argue by contradiction: suppose $\ell \leq k$ and let γ_2 be an open path from $(0,0)$ to $(0, 2k)$ contained in $\Lambda \setminus (B \cup -B)$. Then γ_1 and γ_2 must intersect. The portion of $\gamma_2\gamma_1$ between the last intersection of γ_1 and γ_2 and the set $\{(x,y) : y = 2k\}$ is to the right of the corresponding portion of γ_2, hence to the right of $-B$ and therefore contained in $\Lambda \setminus (B \cup -B)$. Its end point must be $(2a, 2k)$ for some $a \geq 0$. Since $(0, 2\ell)$ is the last intersection of γ_1 and the set $\{(x,y) : x = 0\}$, and γ_1 reaches the point $(1, 2n+1)$, the portion of γ_1 strictly above the set $\{(x,y) : y = 2k\}$ must consist of points whose first coordinate is positive. Hence, this portion is also contained in $\Lambda \setminus (B \cup -B)$ and $\gamma_2\gamma_1$ contradicts the definition of D_1.

To prove part b), let γ_1 be an open path in $\Lambda_{2n+1} \setminus B$ starting from $(A \cup \{0\}) \times \{0\}$ and containing the point $(2a, 2m_0)$ for some $a > 0$ and let γ_2 be an open path from $(0,0)$ to $(0, 2k)$ contained in $\Lambda \setminus (B \cup -B)$. It now follows from the definition of m_0 that these paths must intersect. The portion of γ_1 between the last intersection of these paths and the set $\{(x,y) : y = 2k\}$ is to the right of γ_2, therefore it is contained in $\Lambda \setminus (B \cup -B)$. Recalling the definition of m_0 and the fact that $a > 0$, we see that the portion of γ_1 strictly above the set $\{(x,y) : y = 2k\}$ and below the point $(2a, 2m_0)$ must remain in the region $\{(x,y) : x > 0\}$. Since it is to left of B, it must be contained in $\Lambda \setminus (B \cup -B)$. We now show that the portion of γ_1 above the point $(2a, 2m_0)$ must remain in the region with strictly positive first coordinate. We argue by contradiction: let $(0, 2c)$ be the first point in γ_1 above $(2a, 2m_0)$ with first coordinate equal to 0. Then the portion of γ_1 between $(2a, 2m_0)$ and $(0, 2c)$ is in $\Lambda \setminus (B \cup -B)$ and the open path $\gamma_2\gamma_1$ contradicts the definition of k. Since γ_1 is in $\Lambda \setminus B$, this implies that the portion of γ_1 above the point $(2a, 2m_0)$ must be in $\Lambda \setminus (B \cup -B)$. Hence, the following two assertions hold:

1) The portion of $\gamma_2\gamma_1$ above the point $(2a, 2m_0)$ must remain in the region $\{(x,y); x > 0\}$.
2) The whole path $\gamma_2\gamma_1$ us contained in $\Lambda \setminus (B \cup -B)$.

All the statements of b) now follow easily from these two assertions and the definition of D_1.

Part c) follows from the definition of m_0 and the fact that B is to the right of any path starting from $(A \cup \{0\}) \times \{0\}$ and contained in $\Lambda \setminus B$.

To prove part d) we argue by contradiction. Consider an open path γ_1 in $\Lambda \setminus B$ from $(A \cup \{0\}) \times \{0\}$ to $(0, 2m_0)$ and suppose there is an open path γ_2 in $\Lambda \setminus (B \cup -B)$ from $(0,0)$ to $(x, 2m_0)$ for some $x \leq 0$. Then $\gamma_2\gamma_1$ is an open path in $\Lambda \setminus (B \cup -B)$ from $(0,0)$ to $(0, 2m_0)$ which contradicts the definitions of m_0 and k.

To prove part e), first note the following: part d) of this lemma and the definition of k imply that K cannot contain points in the region $\{(x,y) : x \leq 0, 2m_0 \leq y \leq 2n+1\}$. Therefore, if γ contains a point (x,y) in K, its first coordinate must be strictly positive and we can argue as in part b) to conclude that the portion of γ above this point is in $\Lambda \setminus (B \cup -B)$. This implies that this part of γ is in K. Hence, by definition of D_1, it cannot contain the point $(1, 2n+1)$ nor the point $(-1, 2n+1)$. □

We now return to the proof of part b) of the Proposition 2.1. We will proceed by induction on n. The result is obvious for $n = 0$. To prove the inductive step, let $L(i, R, A') = \{m_0 = i\} \cap \{K = R\} \cap \{H = A'\}$. It follows from part a) of Lemma 2.1 that:

$$F_1 \cap D_1 \subset \cup_{1 \leq i \leq n} \cup_{R,A'} L(i, R, A') \subset D_1$$

where A' takes all the possible values of H and R takes all the possible values of K not containing the points $(1, 2n+1)$ and $(-1, 2n+1)$. Now, using the convention $P(A|\emptyset) = 0$, write:

$$\begin{aligned} & P(F_1 \cap D_1) \\ & \quad = \sum_{1 \leq i \leq n} \sum_{R,A'} P(F_1 | D_1 \cap L(i,R,A')) P(D_1 \cap L(i,R,A')) \\ & \quad = \sum_{1 \leq i \leq n} \sum_{R,A'} P(F_1 | L(i,R,A')) P(L(i,R,A')) \qquad (*) \end{aligned}$$

where A' and R take the same values as above. From part d) of Lemma 2.1 we see that for each i the sum on R can be restricted to R's such that

$$R \cap \{(x, 2i) : x \leq 0\} = \emptyset, \qquad (**)$$

and from part c) of Lemma 2.1 we see that the sum on i can be restricted to values such that $B \cap \{(x, 2i) : x \leq 0\} = \emptyset$. Hence we may assume that all the terms in the right-hand side of $(*)$ are such that $\Lambda \cap T_{-2i}(B \cup R)$ can be generated from some subset of $2\mathbb{N} \times \{0\}$, where $T_{-2i}(x,y) = (x, y - 2i)$. It follows from the first statement of part b) of Lemma 2.1 that

$$F_1 \cap L(i,R,A') = \{(A' \cup \{0\}) \times \{2i\} \xrightarrow{\Lambda \setminus B} (1, 2n+1)\} \cap L(i,R,A').$$

But, since we are summing on R's not containing the points $(1, 2n+1)$ and $(-1, 2n+1)$ we can deduce from part e) of Lemma 2.1 that we also have:

$$F_1 \cap L(i, R, A') = \{(A' \cup \{0\}) \times \{2i\} \stackrel{\Lambda\backslash(B\cup R)}{\longrightarrow} (1, 2n+1)\} \cap L(i, R, A').$$

Hence

$$P(F_1 | L(i, R, A')) = P\Big(\{(A' \cup \{0\}) \times \{2i\} \stackrel{\Lambda\backslash(B\cup R)}{\longrightarrow} (1, 2n+1)\} | L(i, R, A')\Big).$$

The event $L(i, R, A')$ is independent of the state of the bonds whose initial points (x, y) are not in R and are such that $y \geq 2i$. Therefore,

$$\begin{aligned} &P\Big(\{(A' \cup \{0\}) \times \{2i\} \stackrel{\Lambda\backslash(B\cup R)}{\longrightarrow} (1, 2n+1)\} | L(i, R, A')\Big) \\ &\quad = P((A' \cup \{0\}) \times \{2i\} \stackrel{\Lambda\backslash(B\cup R)}{\longrightarrow} (1, 2n+1)), \end{aligned}$$

which by translation invariance is equal to

$$P((A' \cup \{0\}) \times \{0\} \stackrel{\Lambda\backslash T_{-2i}(B\cup R)}{\longrightarrow} (1, 2(n-i)+1)).$$

Since $\Lambda \cap T_{-2i}(B \cup R)$ can be generated from some subset of $2\mathbb{N} \times \{0\}$, we can apply the inductive hypothesis to conclude that this last probability is bounded above by:

$$P((A' \cup \{0\}) \times \{0\} \stackrel{\Lambda\backslash T_{-2i}(B\cup R)}{\longrightarrow} (-1, 2(n-i)+1)).$$

Hence, from $(*)$ we get:

$$\begin{aligned} &P(F_1 \cap D_1) \\ &\quad \leq \sum_{1\leq i\leq n} \sum_{R,A'} P((A' \cup \{0\}) \times \{0\} \stackrel{\Lambda\backslash T_{-2i}(B\cup R)}{\longrightarrow} (-1, 2(n-i)+1)) P(L(i, R, A')) \\ &\quad = \sum_{1\leq i\leq n} \sum_{R,A'} P((A' \cup \{0\}) \times \{2i\} \stackrel{\Lambda\backslash(B\cup R)}{\longrightarrow} (-1, 2n+1)) P(L(i, R, A')). \end{aligned}$$

Thanks to the remark following Lemma 2.1, we can argue as above to show that the last sum is equal to:

$$\begin{aligned} \sum_{1\leq i\leq n} \sum_{R,A'} P(F_{-1} \cap L(i, R, A')) &= P(F_{-1} \cap D_1, m_0 < \infty) \\ &\leq P(F_{-1} \cap D_1). \end{aligned}$$

□

Proof of Theorem 1.1. Since the sets D_i $(i = 1, \ldots, 4)$ form a partition of our probability space and F_1 contains both D_3 and D_4 we have:

$$P(F_1) = P(D_4) + P(D_1 \cap F_1) + P(D_2 \cap F_1) + P(D_3).$$

Similarly we get:

$$P(F_{-1}) = P(D_4) + P(D_1 \cap F_{-1}) + P(D_3 \cap F_{-1}) + P(D_2).$$

Since by symmetry $P(D_2) = P(D_3)$, the first statement of the theorem follows from parts a) and b) of Proposition 2.1. For the second statement just note that under its additional hypothesis we have that $P(D_3 \cap F_{-1}) > 0$. □

3. Proofs of Corollaries 1.1 and 1.2

Proof of Corollary 1.1. To prove part a) let

$$m_0 = \inf\{m : (0,0) \longrightarrow (2i+1, 2m+1)\}$$

and let

$$H = \begin{cases} \emptyset \text{ if } \ m_0 \geq n \\ \{x < 2i+1 : (0,0) \longrightarrow (x, 2m_0+1)\} \text{ if } m_0 < n, \end{cases}$$

Since

$$P((0,0) \longrightarrow (2i+2, 2n)) | m_0 \geq n) = 0,$$

and $P(m_0 < i) = 0$ we have

$$\begin{aligned} &P((0,0) \longrightarrow (2i+2, 2n)) \\ &= \sum_A \sum_{j=i}^{n-1} P((0,0) \longrightarrow (2i+2, 2n) | H = A, m_0 = j) P(H = A, m_0 = j) \\ &= \sum_A \sum_{j=i}^{n-1} P((A \cup \{2i+1\}) \times \{2j+1\} \longrightarrow (2i+2, 2n)) P(H = A, m_0 = j), \end{aligned}$$

where A ranges over all possible values of H. Using Theorem 1.1 with $B = C = \emptyset$ and translation invariance we get

$$\begin{aligned} &P((0,0) \longrightarrow (2i+2, 2n)) \\ &\leq \sum_A \sum_{j=i}^{n-1} P((A \cup \{2i+1\}) \times \{2j+1\} \longrightarrow (2i, 2n)) P(H = A, m_0 = j) \\ &= P((0,0) \longrightarrow (2i, 2n), m_0 < n). \end{aligned}$$

Part a) now follows from $P((0,0) \longrightarrow (2i, 2n), m_0 > n) > 0$.

The proof of part b) is similar and we omit it. □

Proof of Corollary 1.2. Corollary 1.2 can be deduced from our results for oriented percolation by means of a discrete time approximation of the contact process. Since the ideas are standard we will only sketch this proof. First modify our percolation model as follows: From each point $(2m, 2n)$ add an oriented bond from that point to $(2m, 2n+2)$. Then, let N be sufficiently large to satisfy: $\sqrt{\frac{\lambda}{N}} < 1$. Now suppose that each of the new bonds is open with probability $1 - 1/N$ and each of the diagonal bonds is open with probability $\sqrt{\frac{\lambda}{N}}$. All bonds being open independently of each other. Then, observe that in this percolation model we still have the following property: If a path γ_1 begins to the right (left) and ends to the left (right) of another path γ_2 then, these two paths must intersect. Thanks to this, Theorem 1.1 and its proof remain valid in this new context. Then, define ${}^N\xi_t(x)$ as 1 if there is an open path in our percolation model from $(0,0)$ to $(2x, 2[Nt])$ and as 0 if there is no such path ($[Nt]$ being the integer part of Nt). Letting N go to infinity ${}^N\xi_t$ converges to ξ_t^0, the contact process on $\mathbb{Z}$ starting from $\{0\}$ and we can transfer our results on oriented percolation to the contact process. To show that strict inequality is preserved at the limit, let

$$I = \{(0,0) \to (2i+1, 2k+1) \text{ for some } k < [Nt]\},$$

and note that it follows from Theorem 1.1 that

$$\begin{aligned} &P((0,0) \to (2i, 2[Nt])) - P((0,0) \to (2i+2, 2[Nt])) \\ &\quad \geq P(I^c \cap \{(0,0) \to (2i, 2[Nt]\}). \end{aligned}$$

Since the right-hand side is easily seen to be bounded away from 0 as N goes to infinity, the result follows. □

References

[1] P. Balister, B. Bollobás, A. Stacey (1994) Improved upper bounds for the critical probability of oriented percolation in two dimensions. *Random Structure Algorithms.* **5**, 573–589.

[2] V. Belitsky, P.A. Ferrari, N. Konno, T.M. Liggett (1997) A strong correlation inequality for contact processes and oriented percolation. *Stochastic Process. Appli.* **67**, 213–225.

[3] J. van den Berg, O. Haggstrom, J. Kahn (2006) Some conditional correlation inequalities for percolation and related processes. *Random Structure algorithms.* **4**, 417–435.

[4] R. Durrett (1984) Oriented percolation in two dimensions. *Ann. Probab.* **12**, 999–1040.

[5] L.F. Gray. (1991) Is the contact process dead? *Lectures in Appli. Math.* **27**, 19–29.

[6] T.M. Liggett (1989) Interacting particle systems. Springer-Verlag, New York.

[7] T.M. Liggett (1999) Stochastic interacting systems: contact, voter and exclusion processes. Springer-Verlag, Berlin.

[8] T. Kuczek (1989) The central limit theorem for the right edge of supercritical oriented percolation. *Ann. Probab.* **17**, 1322–1332.

[9] M.V. Menshikov, S.Yu. Popov, V.V. Sisko (2002) On the connection between oriented percolation and contact process. *J. Theoret. Probab.* **15**, 207–221.

Enrique D. Andjel
LATP-CMI
Université de Provence
39 Rue Joliot-Curie
F-13453 Marseille cedex 13
France
e-mail: andjel@cmi.univ-mrs.fr

Mariela Sued
Instituto de Cálculo
UBA-Conicet Argentina
Intendente Guiraldes 2670
1428 Buenos Aires
Argentina
e-mail: marielasued@gmail.com

Progress in Probability, Vol. 60, 31–58

Is Critical 2D Percolation Universal?

Vincent Beffara

Abstract. The aim of these notes is to explore possible ways of extending Smirnov's proof of Cardy's formula for critical site-percolation on the triangular lattice to other cases (such as bond-percolation on the square lattice); the main question we address is that of the choice of the lattice embedding into the plane which gives rise to conformal invariance in the scaling limit. Even though we were not able to produce a complete proof, we believe that the ideas presented here go in the right direction.

Mathematics Subject Classification (2000). 82B43, 32G15; 82B20, 82B27.

Keywords. Percolation, Conformal invariance, Complex structure.

Introduction

It is a strongly supported conjectured that many discrete models of random media, such as *e.g.* percolation and the Ising model, when taken in dimension 2 at their critical point, exhibit conformal invariance in the scaling limit. Indeed, the *universality* principle implies that the asymptotic behavior of a critical system after rescaling should not depend on the specific details of the underlying lattice, and in particular it should be invariant under rotations (at least under suitable symmetry conditions on the underlying lattice). Since by construction a scaling limit is also invariant under rescaling, it is natural to expect conformal invariance, as the local behavior of a conformal map is the composition of a rotation and a rescaling.

On the other hand, conformally invariant *continuous* models have been thoroughly studied by physicists, using tools such as conformal field theories. In 2000, Oded Schramm ([14]) introduced a one-parameter family of continuous bidimensional random processes which he called SLE processes, as the only possible scaling limits in this situation, under the assumption of conformal invariance; connections between SLE and CFT are now quite well understood (see, *e.g.*, [7, 1]).

However, actual convergence of discrete models to SLE in the scaling limit is known for only a few models. The case on which we focus in this paper is that of percolation. The topic of conformal invariance for percolation has a long history – see [13] and references therein for an in-depth discussion of it.

In the case of site-percolation on the triangular lattice, it is a celebrated result of Smirnov ([16]) that indeed the limit exists and is conformally invariant. While the proof is quite simple and extremely elegant (see Section 3 below and references therein), it is very specific to that particular lattice, to the point of being almost magical; it is a very natural question to ask how it can be generalized to other cases, and in particular to bond-percolation on the square lattice. Universality and conformal invariance have indeed been tested numerically for percolation in various geometries (see, e.g., [12]), and a partial version of conformal invariance (assuming the existence of the limit) is known in the case of Voronoi percolation (see [4]).

In fact, it seems that the question of convergence itself has hardly been addressed by physicists, at least in the CFT community – a continuous, conformally invariant object is usually the starting point of their work rather than its outcome. Techniques such as the renormalization group do give reason to expect the existence of a scaling limit and of critical exponents, but they seem to not give much insight into the emergence of rotational invariance.

This is not surprising in itself, for a very trivial reason: Take any discrete model for which you know that there is a conformally invariant scaling limit, say a simple random walk on $\mathbb{Z}^2$, and deform the underlying lattice, in a linear way, so as to change the aspect ratio of its faces. Then the scaling limit still exists (it is the image of the previous one by the same transformation); but obviously it is not rotationally invariant. Since all the rescaling techniques apply exactly the same way before and after deformation, they cannot be sufficient to derive rotational invariance. A trace of this appears in the most general statement of the universality hypothesis (see, *e.g.*, [13, Section 2.4]): To paraphrase it, given any two periodic planar graphs, the scaling limits of critical percolation on them are conjugated by *some* linear map g.

The main question we address in these notes is the following: Given a discrete model on a doubly periodic planar graph, how to embed this graph into the plane so as to make the scaling limit isotropic? If the graph has additional symmetry (as for instance in the case of the square or triangular lattices), the embedding has to preserve this symmetry; so a restating of the same question in the terms of the universality hypothesis would be, absent any additional symmetry for one of the two graphs involved, can one determine the map g?

The most surprising thing (to me at least) about the question, besides the fact that it appears to actually be orthogonal to the interests of physicists in that domain, is that its answer turns out to depend on the model considered. In other words, there is no absolute notion of a "conformal embedding" of a general graph. In the case of the simple random walk, the answer is quite easy to obtain, though it does not seem to have appeared in the literature in the form we present it here; in the case of percolation, I could find no reference whatsoever, the closest being the discussion and numerical study of *striated models* in [13] where, instead of looking at a different graph, the parameter p in the model is chosen to depend on the site in $\mathbb{Z}^2$ in a periodic fashion – which admittedly is a very related question.

The paper is roughly divided into two parts. In the first one, comprised of the firs two sections, we introduce some notation and the general framework of the approach, and we treat the case of the simple random walk. This is enough to prove that the correct embedding is not the same for it as for percolation; we then argue that circle packings might give a way to answer the question in the latter case. In the second part, which is of a more speculative nature, we investigate Smirnov's proof in some detail, and rephrase it in such a way that its general strategy can be applied to general triangulations. We then describe the two main steps of a strategy that could lead to its generalization, though we were able to perform none of the two.

1. Notation and setup

1.1. The graph

We first define the class of triangulations of the plane we are interested in. Let T be a 3-regular finite graph of genus 1 (*i.e.*, a graph that is embeddable in the torus $\mathbb{T}^2 := \mathbb{R}^2/\mathbb{Z}^2$ but not in the plane, and having only vertices of degree 3). For ease of notation, we assume that T is equipped with a fixed embedding in $\mathbb{T}^2$, which we also denote by T. The dual T^* of T (which we also assume to be embedded in the torus once and for all) is then a triangulation of $\mathbb{T}^2$.

Let $\hat{T}$ (resp. $\hat{T}^*$) be the universal cover of T (resp. T^*): Then $\hat{T}$ and $\hat{T}^*$ are mutually dual, infinite, locally finite planar graphs, on which $\mathbb{Z}^2$ acts by translation. We are interested in natural ways of embedding $\hat{T}$ into the complex plane $\mathbb{C}$. Let T_i (the meaning of the notation will become clear in a minute) be the embedding obtained by pulling T back using the canonical projection from $\mathbb{R}^2$ to $\mathbb{T}^2$ – we will call T_i the *square embedding* of T.

For every $\alpha \in \mathbb{C} \setminus \mathbb{R}$, let $\varphi_\alpha : \mathbb{C} \to \mathbb{C}$ be the $\mathbb{R}$-linear map defined by $\varphi_\alpha(x+iy) = x + \alpha y$ (*i.e.*, it sends 1 to itself and i to α) and let T_α be the image of T_i by φ_α. For lack of a better term, we will call T_α the *embedding of modulus* α of T into the complex plane.

Notice that the notation T_α depends on the *a priori* choice of the embedding of T in the flat torus; but, up to rotation and scaling, the set of proper embeddings of $\hat{T}$ obtained starting from two different embeddings of T is the same, so no generality is lost (as far as our purpose in these notes is concerned).

One very useful restriction on embeddings is the following:

Definition 1. We say that an embedding T_α of $\hat{T}$ in the complex plane is *balanced* if each of its vertices is the barycenter (with equal weights) of its neighbors; or, equivalently, if the simple random walk on it is a martingale.

Proposition 2. *Let T be a 3-regular graph of genus 1: Then, for every $\alpha \in \mathbb{H}$, there is a balanced embedding of $\hat{T}$ in the complex plane with modulus α. Moreover, this embedding is unique up to translations of the plane.*

Proof. We only give a sketch of the proof, because expanding it to a full proof is both straightforward and tedious. The main remark is that any periodic embedding which minimizes the sum S_2, over a period, of the squared lengths of its edges is balanced: Indeed, the gradient, with respect to the position of a given vertex, of S_2 is exactly the difference between this point and the barycenter of its neighbors. (This would be true in any Euclidean space.) It is easy to use a compactness argument to prove the existence of such a minimizer.

To prove uniqueness up to translation is a little trickier, but since it is not necessary for the rest of this paper, we allow ourselves to give an even sketchier argument. First, one can get rid of translations by assuming that a fixed vertex of $\hat{T}$ is put at the origin by the embedding; the set of all possible embeddings of modulus α is then parameterized by $2(|V(T)| - 1)$ real-valued parameters, which are the coordinates of the locations of the other vertices in one period of $\hat{T}$. In terms of these variables, S_2 is polynomial of degree 2. It is bounded below by the squared length of the longest edge in the embedding, which itself is bounded below, up to a constant depending only on the combinatorics of the graph, by the square of the largest of the $2(|V(T)| - 1)$ parameters; so it goes to infinity uniformly at infinity. This implies that its Hessian (which is constant) is positive definite, so S_2 is strictly convex as a function of those variables. This immediately implies the uniqueness of the minimizer. □

An essential point is that, even though our proof uses Euclidean geometry, the fact that the embedding is balanced is a linear condition. In particular, if the embedding T_i is balanced, then so are all the other T_α. The corresponding *a priori* embedding of T itself into the flat torus $\mathbb{T}^2$ (which is also unique up to translations) will be freely referred to as *the balanced embedding of T into the torus.*

1.2. The probabilistic model

We will be interested in critical site-percolation on the triangulation T_α^*; more specifically, the question we are interested is the following. Let Ω be a simply connected, smooth domain in the complex plane, and let A, B, C and D be four points on its boundary, in that order. For every $\delta > 0$, let Ω_δ be the largest connected component (in terms of graph connectivity) of the intersection of Ω with δT_α, and Ω_δ^* be its dual graph. Ω_δ should be seen as a discretization of Ω at scale δ. Let A_δ, B_δ, C_δ and D_δ be the vertices of Ω_δ that are closest to A, B, C and D respectively.

The model we are most interested in is critical site-percolation on Ω_δ^*; however, most of the following considerations remain valid for other lattice models. Let $C_\delta(\Omega, A, B, C, D)$ be the event that there is an open crossing in Ω_δ^*, between the intervals $A_\delta B_\delta$ and $C_\delta D_\delta$ of its boundary. Under some symmetry conditions on T, Russo-Seymour-Welsh theory ensures that at criticality, the probability of

$C_\delta(\Omega, A, B, C, D)$ is bounded away from both 0 and 1 as δ goes to 0. Its limit was conjectured by Cardy (see [6]) using non-rigorous arguments from conformal field theory; actual convergence was proved, in the case of the triangular lattice (embedded in such a way that its faces are equilateral triangles), by Smirnov (see [15, 3]). We defer the statement of the convergence to a later time. The following definition has become standard:

Definition 3. Assume that, for every choice of (Ω, A, B, C, D), the probability of the event $C_\delta(\Omega, A, B, C, D)$ has a limit $f_\alpha(\Omega, A, B, C, D)$ as $\delta \to 0$ – we will refer to this by saying that the model *has a scaling limit.* We say that the model is *conformally invariant in the scaling limit* if, for every conformal map Φ from Ω to $\Phi(\Omega)$, one has

$$f_\alpha(\Omega, A, B, C, D) = f_\alpha(\Phi(\Omega), \Phi(A), \Phi(B), \Phi(C), \Phi(D)).$$

This is equivalent to saying that $f_\alpha(\Omega, A, B, C, D)$ only depends on the modulus of the conformal rectangle (Ω, A, B, C, D).

(Notice that the extension of Φ to the boundary of Ω, which is necessary for the above definition to make sense, is ensured as soon as Ω is assumed to be regular enough.)

2. Periodic embeddings

2.1. Uniqueness of the modulus

Given T, it is natural to ask whether it is possible to choose a value for α which provides conformal invariance in the scaling limit. There are two possible strategies: Either give an explicit value for which "a miracle occurs" (in physical terms, for which the model is *integrable* – this is what Smirnov did in the case of the triangular lattice), or obtain its existence in a non-constructive way – which is what we are trying to do here.

A reassuring fact is that, whenever such an α exists, it is essentially unique:

Proposition 4. *For every graph T, there are either zero or two values of α such that critical site-percolation on T_α^* is conformally invariant in the scaling limit. In the latter case, the two values are complex conjugates of each other.*

Proof. The key remark is the following: Let β be a non-real complex number. Since the event C_δ is defined using purely combinatorial features, one can push the whole picture forward through φ_β without changing its probability. Let $\alpha' = \varphi_\beta(\alpha)$: φ_β then transforms Ω into $\varphi_\beta(\Omega)$ and the lattice T_α into $T_{\alpha'}$. So, assuming convergence on both sides, one always has

$$f_\alpha(\Omega, A, B, C, D) = f_{\alpha'}(\varphi_\beta(\Omega), \varphi_\beta(A), \varphi_\beta(B), \varphi_\beta(C), \varphi_\beta(D)).$$

In the case $\beta = -i$, φ_β is simply the map $z \mapsto \bar{z}$. In that case, the modulus of the conformal rectangle $(\varphi_{-i}(\Omega), \bar{D}, \bar{C}, \bar{B}, \bar{A})$ is the same as that of (Ω, A, B, C, D), and clearly the event C_δ is invariant when the order of the corners is reversed. So, conformal invariance for T_α and the previous remark implies that $f_{\bar\alpha}(\Omega, A, B, C, D)$ still only depends on the modulus of the conformal rectangle – in other words, if critical percolation T_α is conformally invariant in the scaling limit, that is also the case on $T_{\bar\alpha}$.

Now assume conformal invariance in the scaling limit for two choices of the modulus in the upper-half plane; these moduli can always be written as α and $\alpha' = \varphi_\beta(\alpha)$ for an appropriate choice of $\beta \in \mathbb{H} \setminus \{i\}$. Still using the above remark, all that is needed to arrive to a contradiction is to show that f_α does actually depend on the modulus of the rectangle (*i.e.*, that it is not constant), and that there exist two conformal rectangles with the same modulus and whose images by φ_β have different moduli.

For the former point, it is enough to prove that for every choice of $\rho, \rho' > 0$, the probability of crossing the rectangle $[0, \rho] \times [0, 1]$ horizontally is strictly larger than that of crossing $[0, \rho + \rho'] \times [0, 1]$. This is obvious by Russo-Seymour-Welsh: The event that there is a vertical dual crossing in $\delta T^*_\alpha \cap [\rho + \delta, \rho + \rho'] \times [0, 1]$ is independent of $C_\delta([0, \rho] \times [0, 1], \rho, \rho + i, i, 0)$ and its probability is bounded below, uniformly in $\delta < \rho'/10$, by some positive ε depending only on ρ and ρ'. Hence, still assuming that the limits all exist as $\delta \to 0$,

$$f_\alpha([0, \rho + \rho'] \times [0, 1], \rho + \rho', \rho + \rho' + i, i, 0) \leqslant (1 - \varepsilon) f_\alpha([0, \rho] \times [0, 1], \rho, \rho + i, i, 0).$$

For the latter point, assume that φ_β preserves the equality of moduli of conformal rectangles. Let $Q = [0, 1]^2$ be the unit square. By symmetry, the conformal rectangles $(Q, 0, 1, 1 + i, i)$ and $(Q, 1, 1 + i, i, 0)$ have the same modulus; on the other hand $\varphi_\beta(Q)$ is a parallelogram, and by our hypothesis on φ_β it has the same modulus in both directions. This easily implies that it is in fact a rhombus. If now Q' is the square with vertices $1/2$, $1 + i/2$, $1/2 + i$, $i/2$, $\varphi_\beta(Q')$ is both a rhombus (by the same argument) and a rectangle (because its vertices are the midpoints of the edges of $\varphi_\beta(Q)$ which is a rhombus). Hence $\varphi_\beta(Q')$ is a square, and so is $\varphi_\beta(Q)$, and in particular $\beta = \varphi_\beta(i) = i$, which is in contradiction with our hypothesis. □

When such a pair of moduli exists, we will denote by α_T^{perc} the one with positive imaginary part. The same reasoning can be done for various models, and in each case where the scaling limit exists and is non-trivial, there will be a pair of moduli making it conformally invariant; we will distinguish them from each other by using the name of the model as a superscript (so that for instance α_T^{RW} makes the simple random walk conformally invariant in the scaling limit – cf. below).

When an argument does not depend on the specific model (as is the case in the next subsection), we will use the generic notation α_T as a placeholder.

2.2. Obtaining α_T by symmetry arguments

It should be noted that, because the value of α_T (when it exists) is uniquely defined by the combinatorics of T, there are cases where additional symmetry specifies its value uniquely. Indeed, assume Ψ is a graph isomorphism of $\hat{T}$ which is neither a translation nor a central symmetry; for every α, it induces a topological isomorphism of T_α. Assume without loss of generality that the origin of the plane is chosen to be one of the vertices of T_α; let $z_0 = \Psi(0)$, $z_1 = \Psi(1)$ and $z_\alpha = \Psi(\alpha)$ (notice that both 1 and α are also vertices of T_α).

Assume $\alpha = \alpha_T^{\text{perc}}$. Because Ψ is an isomorphism, it preserves site-percolation; so, in particular, critical site-percolation on $\Psi(T_\alpha)$ is conformally invariant in the scaling limit. By Proposition 4, this implies that

$$\frac{z_\alpha - z_0}{z_1 - z_0} = \frac{\Psi(\alpha) - \Psi(0)}{\Psi(1) - \Psi(0)} \in \{\alpha, \bar{\alpha}\}. \tag{2.1}$$

This condition is then enough to obtain the value of α_T. There are two natural examples of that (illustrated in Figures 1 and 2), which we now describe.

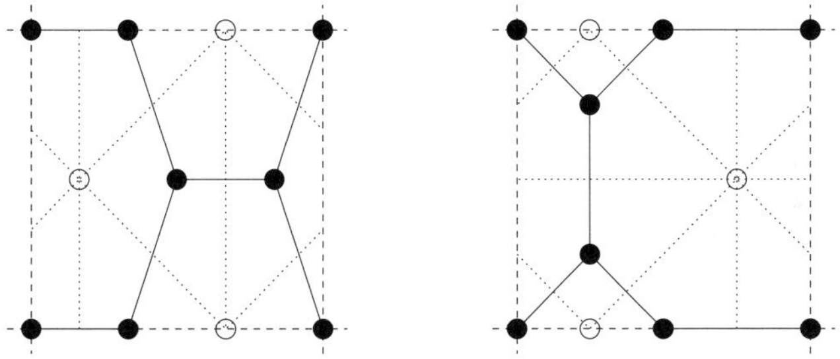

FIGURE 1. The graphs T_h (left) and T_s (right), embedded into $\mathbb{T}^2$ in a balanced way with a vertex at the origin; empty circles and dotted lines represent the dual graphs. Both are represented using their square embedding, so the triangles in T_h are not equilateral.

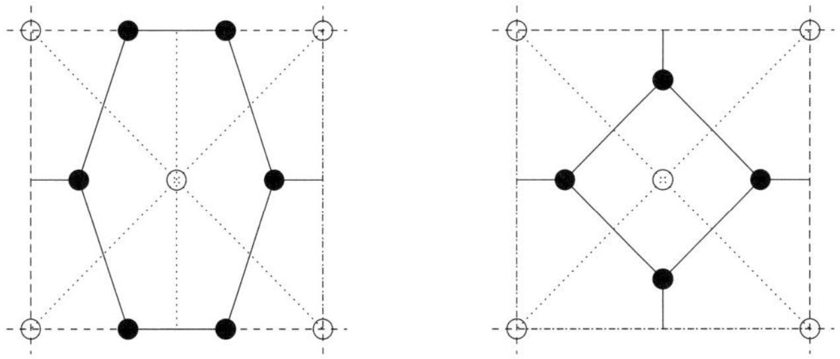

FIGURE 2. The same graphs as in Figure 1, with the origin on a vertex of the dual.

- Let T_h be one period of the honeycomb lattice, embedded into $\mathbb{T}^2$ in such a way that every vertex is the barycenter of its neighbors (we will call such an embedding *balanced*); since we take $\mathbb{T}^2$ to be a square, the coordinates of the vertices of T_h are $(0,0)$, $(1/3,0)$, $(1/2,1/2)$ and $(5/6,1/2)$. There is an isomorphism Ψ_h of order 3 of $\hat{T}_h$, corresponding to rotation around $(0,0)$; on T_α, it sends 0 to $z_0 = 0$, 1 to $z_1 = (3\alpha - 1)/2$ and α to $z_\alpha = -(1+\alpha)/2$. Since Ψ_h preserves orientation, Equation (2.1) leads to

$$\frac{z_\alpha - z_0}{z_1 - z_0} = \frac{1+\alpha}{1-3\alpha} = \alpha \quad \Longrightarrow \quad \alpha = \pm i\frac{\sqrt{3}}{3},$$

in other words $\alpha_{T_h} = i\sqrt{3}/3$. Not surprisingly, this corresponds to embedding the faces of $\hat{T}_h^*$ as equilateral triangles, and those of $\hat{T}_h$ as regular hexagons.

- Let T_s be chosen in such a way that $\hat{T}_s^*$ has the topology of the centered square lattice; if again the embedding is balanced, the coordinates of the vertices of T_s are $(0,0)$, $(1/2,0)$, $(1/4,1/4)$ and $(1/4,3/4)$. There is an isomorphism Ψ_s of order 4 of $\hat{T}_s$, corresponding to a rotation rotation around the vertex $(1/4,0)$ of T_s^*. In that case

$$\frac{z_\alpha - z_0}{z_1 - z_0} = \frac{(-3/4 - \alpha/4) - (1/4 - \alpha/4)}{(1/4 + 3\alpha/4) - (1/4 - \alpha/4)} = \frac{-1}{\alpha} = \alpha \quad \Longrightarrow \quad \alpha = \pm i,$$

so $\alpha_{T_s} = i$. Again not surprisingly, this corresponds to the usual embedding of the square lattice using – well, squares.

Of course, identifying α_T in those cases is a long way from a proof of conformal invariance; but it would seem that understanding, in the general case, what α_T^{perc} is would be a significant progress in our understanding of the process.

2.3. Embedding using random walks

As an aside, in this subsection and the next we describe two natural ways of embedding a doubly periodic graph into the complex plane, which both have something to do with conformal invariance.

Let T be a finite 3-regular graph of genus 1, embedded in $\mathbb{T}^2$ in a balanced way, and let $(X_n)_{n\geqslant 0}$ be a simple random walk on it. For simplicity, assume that (X_n) is irreducible as a Markov chain. (Both 3-regularity and irreducibility are completely unnecessary as far as the results presented here are concerned, and the same reasoning would work in the general case, but notation would be a little tedious.) Since T is finite, (X_n) converges in distribution to the unique invariant measure, which, because T is 3-regular, is the uniform measure on $V(T)$; moreover the convergence is exponentially fast.

Now pick $\alpha \in \mathbb{H}$, and lift (X_n) to a simple random walk (Z_n) on T_α. By the balance condition on the embedding, it is easy to check that (Z_n) is a martingale; exponential decay of correlations between its increments is enough to obtain a central limit theorem (cf. for instance [9] and references therein). To write the covariance matrix in a convenient form, we need some notation. For each (oriented) edge e of T, choose $z_1(e)$ and $z_2(e)$ in T_α in such a way that they are neighbors and the edge $(z_1(e), z_2(e))$ is a pre-image of e by the natural projection from T_α to T; let $e_\alpha := z_2(e) - z_1(e)$ – obviously it does not depend on the choice of $z_1(e)$

and $z_2(e)$. Define

$$\Sigma_\alpha^{xx}(T) := \frac{1}{|E(T)|} \sum_{e \in E(T)} (\Re e_\alpha)^2,$$

$$\Sigma_\alpha^{yy}(T) := \frac{1}{|E(T)|} \sum_{e \in E(T)} (\Im e_\alpha)^2,$$

$$\Sigma_\alpha^{xy}(T) := \frac{1}{|E(T)|} \sum_{e \in E(T)} (\Re e_\alpha)(\Im e_\alpha).$$

It is not difficult to compute the covariance matrix of the scaling limit of the walk:

Proposition 5. *As n goes to infinity, $n^{-1/2} Z_n$ converges in distribution to a Gaussian variable with covariance matrix*

$$\Sigma_\alpha(T) := \begin{bmatrix} \Sigma_\alpha^{xx}(T) & \Sigma_\alpha^{xy}(T) \\ \Sigma_\alpha^{xy}(T) & \Sigma_\alpha^{yy}(T) \end{bmatrix}.$$

Proof. The walk is centered by definition; the existence of a Gaussian limit is a direct consequence of the exponential decay of step correlations. All that remains to be done is to compute the covariance matrix. We focus on the first matrix entry, the others being similar. We have

$$\mathrm{Var}(n^{-1/2}\Re Z_n) = E\left[\frac{1}{n}\left(\sum_{k=0}^{n-1} \Re(Z_{k+1} - Z_k)\right)^2\right] = \frac{1}{n}\sum_{k=0}^{n-1} E\left[(\Re(Z_{k+1} - Z_k))^2\right]$$

(the other terms disappear by the martingale property). We know that $Z_{k+1} - Z_k$ converges in distribution, because the walk on T converges in distribution; its limit is the distribution of e_α where e is an edge of T chosen uniformly. By Cesàro's Lemma, the expression above then converges to Σ_α^{xx}; the computation of the other entries in $\Sigma_\alpha(T)$ it exactly similar. □

Even though the previous definition of conformal invariance in the scaling limit does not apply directly in this case, its natural counterpart is to ask for the scaling limit of the walk to be rotationally invariant (*i.e.*, to be standard two-dimensional Brownian motion); this is equivalent to saying that the covariance matrix $\Sigma_\alpha(T)$ is scalar, and since its entries are real, yet another equivalent formulation is

$$[\Sigma_\alpha^{xx}(T) - \Sigma_\alpha^{yy}(T)] + i\,[\Sigma_\alpha^{xy}(T)] = 0 \quad \Longleftrightarrow \quad \sum_{e \in E(T)} (e_\alpha)^2 = 0.$$

The last equation is a second-degree equation in α with real-valued coefficients. If $\alpha \in \mathbb{R}$, all the terms are non-negative and at least one is positive, so the equation has no solution in $\mathbb{R}$; letting α go to $+\infty$ along the real line leads to $|E(T)|$ positive terms, at least one of which is of order α^2, so the coefficient in α^2 in the equation is not zero. Hence the equation has exactly two solutions which are complex conjugate of each other – the situation is very similar to the one in Proposition 4. For further reference, we let α_T^{RW} be the one with positive imaginary part. One advantage of this choice (besides the fact that it exists for every doubly periodic graph) is that the value of α_T^{RW} is very easy to compute.

Remark 6. In the more general case of a doubly periodic graph but without the assumptions of 3-regularity and irreducibility (but still assuming that the embedding is balanced), the condition $\sum e_\alpha^2$ is still necessary and sufficient for the walk to be isotropic in the scaling limit – and the proof is essentially the same, so we do not delve into more detail.

Remark 7. Of course, in the cases where T has some additional symmetry, α_T^{RW} is the same as that obtained in the previous subsection using symmetry ...

Remark 8. One can also look at a simple random walk on the dual graph T_α^*, and ask for which values of α this dual walk is isotropic in the scaling limit. As it turns out, the modulus one obtains this way is the same as on the initial graph, in other words

$$\alpha_T^{\mathrm{RW}} = \alpha_{T^*}^{\mathrm{RW}}.$$

This is a very weak version of universality, and unfortunately there doesn't seem to be a purely discrete proof of it – say, using a coupling of the two walks.

There is another natural way to obtain the same condition. We are planning on studying convergence of discrete objects to conformally invariant limits, so it is a good idea to look for discrete-harmonic functions on T_α (with respect to the natural Laplacian, which is the same as the generator of the simple random walk on T_α). The condition of balanced embedding is exactly equivalent to saying that the identity map is harmonic on T_α; it is a linear condition, so it does not depend on the value of α.

The main difficulty when looking at discrete holomorphic maps is that the product of two such maps is not holomorphic in general. But we are interested in scaling limits, so maybe imposing that such a product is in fact "almost discrete holomorphic" (in the sense that it satisfies the Cauchy-Riemann equations up to an error term which vanishes in the scaling limit) would be sufficient.

Whether the previous paragraph makes sense or not – let us investigate whether the map $\zeta : z \mapsto z^2$ is discrete-harmonic. For every $z \in T_\alpha$, we can write

$$\Delta\zeta(z) = \frac{1}{3} \sum_{z' \sim z} (z'^2 - z^2) = \frac{1}{3} \sum_{e \in E_z(T)} (z + e_\alpha)^2 - z^2 = \frac{1}{3} \sum_{e \in E_z(T)} e_\alpha^2$$

(the term in $\sum z e_\alpha$ vanishes because the embedding is balanced). So, if ζ is discrete-harmonic, summing the above relation over $z \in T$ gives the very same condition $\sum e_\alpha^2 = 0$ as before; in other words, α_T^{RW} is the embedding for which $z \mapsto z^2$ is *discrete-harmonic on average*.

As a last remark, let us investigate how strong the condition of exact harmonicity of ζ is; so assume that α is chosen in such a way that $\Delta\zeta$ is identically

0. Let e be any oriented edge of T; let $e' := \tau.e$ and $e'' := \tau^2.e$ be the two other edges sharing the same source as e. The balance condition on the embedding plus harmonicity of ζ imply the following system:

$$\begin{cases} e_\alpha + e'_\alpha + e''_\alpha & = & 0 \\ e_\alpha^2 + (e'_\alpha)^2 + (e''_\alpha)^2 & = & 0. \end{cases} \tag{2.2}$$

Up to rotation and scaling, one can always assume that $e_\alpha = 1$, so the system reduces to $e'_\alpha + e''_\alpha = -1$ and $(e'_\alpha)^2 + (e''_\alpha)^2 = -1$. Squaring the first of these two relations and subtracting the second, one obtains $e'_\alpha e''_\alpha = 1$, so e'_α and e''_α are the two solutions of the equation

$$X^2 + X + 1 = 0$$

which implies that $\{e'_\alpha, e''_\alpha\} = \{e^{\pm 2\pi i/3}\}$. To sum it up:

Proposition 9. *The only* 3*-regular graph on which the map* $\zeta : z \mapsto z^2$ *is discrete-harmonic is the honeycomb lattice, embedded in such a way that its faces are regular hexagons.*

So, imposing ζ to be harmonic not only determines the embedding, it also restricts T to essentially one graph; but in terms of scaling limits, the condition that ζ is harmonic on the average makes as much sense as the exact condition.

2.4. Embedding using circle packings

There is another way to specify essentially unique embeddings of triangulations, which is very strongly related to conformal geometry, using the theory of circle packings. It is a fascinating subject in itself and a detailed treatment would be outside of the purpose of these notes, so the interested reader is advised to consult the book of Stephenson [18] and the references therein for the proofs of the claims in this subsection and much more.

We first give a version of a theorem of Köbe, Andreev and Thurston, specialized to our case. It is a statement about triangulations, which is why we actually apply it to T^* instead of directly to T. Notice that we *do not* assume T to be already embedded into the torus $\mathbb{T}^2$.

Theorem 10 (Discrete uniformization theorem [18, p. 51]**).** *Let* T^* *be a finite triangulation of the torus, and let* $\hat{T}^*$ *be its universal cover. There exists a locally finite family* $(C_v)_{v \in V(\hat{T}^*)}$ *of disks of positive radii and disjoint interiors, satisfying the following compatibility condition:* C_v *and* $C_{v'}$ *are tangent if, and only if,* v *and* v' *are neighbors in* $\hat{T}^*$.

Such a family is called a circle packing *associated to the graph* $\hat{T}^*$. *It is* essentially unique, *in the following sense: If* (C'_v) *is another circle packing associated to* $\hat{T}^*$, *then there is a map* $\varphi : \mathbb{C} \to \mathbb{C}$, *either of the form* $z \mapsto az + b$ *or of the form* $z \mapsto a\bar{z} + b$, *such that for every* $v \in V(\hat{T}^*)$, $C'_v = \varphi(C_v)$.

Remark 11. The "existence" part of the above theorem remains true in a much broader class of graphs; essentially all that is necessary is bounded degree and recurrence of the simple random walk on it. (One can see that a packing exists by completing the graph into a triangulation.) The "uniqueness" part however fails in general, as is made clear as soon as one tries to construct a circle packing associated to the square lattice ...

A consequence of the uniqueness part of the theorem is the following: Let $\theta : \hat{T}^* \to \hat{T}^*$ be a translation along one of the periods of $\hat{T}^*$, and let $\mathcal{C}'_v := \mathcal{C}_{\theta(v)}$; according to the theorem, let φ be such that $\mathcal{C}'_v = \varphi(\mathcal{C}_v)$ for all v. Up to composition of φ by itself, one can always assume that it is of the form $\varphi(z) = az + b$. By the assumption of local finiteness of the circle packing, one has $|a| = 1$; besides, the orbits of θ are unbounded, so those of φ are too, and in particular it does not have a fixed point, which implies that $a = 1$ and $b \neq 0$. In other words, φ is a translation, *i.e.*, the circle packing associated to $\hat{T}^*$ is itself doubly periodic.

As soon as one is given a circle packing associated to a planar graph, it comes with a natural embedding: Every vertex $v \in V(\hat{T}^*)$ will be represented by the center of $\mathcal{C}_v$, and if v' is a neighbor of v, the edge (v, v') will be embedded as a segment – which is the union of a radius of $\mathcal{C}_v$ and a radius of $\mathcal{C}_{v'}$, because those two disks are tangent. One can then specify an embedding of $\hat{T}$ by putting each of its vertex at the center of the disk inscribed in the corresponding triangular face of (the embedding of) $\hat{T}^*$; the collection of all those inscribed disks is in fact a circle packing associated with the graph $\hat{T}$ (see Figure 3).

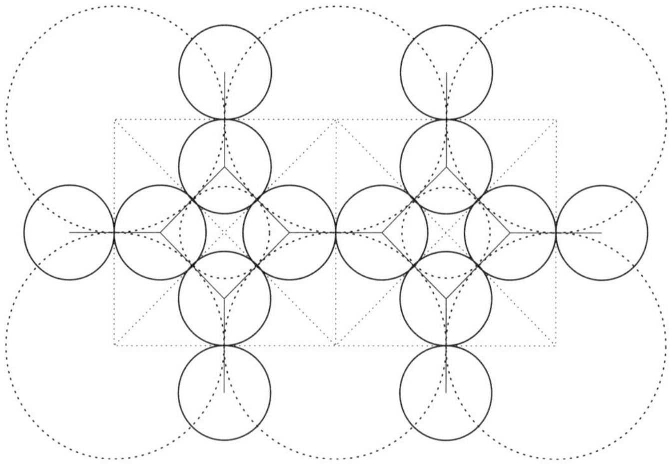

FIGURE 3. The circle packings associated to the graph T_s (solid lines) and its dual (dotted lines).

Of more interest to us is the fact that the embedding is itself doubly periodic, by the previous remarks. Up to rotation, and scaling and maybe complex conjugation, one can assume that the period corresponding to the translation by

1 (resp. i) in $\mathbb{T}^2$ is equal to 1 (resp. $\alpha \in \mathbb{H}$). Once again, the value of the modulus α is uniquely determined; for further reference, we will denote it by α_T^{CP}.

Yet again, as soon as one additional symmetry is present in $\hat{T}$, the value of α_T^{CP} is the same as that obtained using the symmetry; this is again a direct consequence of the essential uniqueness of the circle packing.

2.5. "Exotic" embeddings

Looking closely at Smirnov's proof, one notices that essentially the only place where the specifics of the graph are used is in the proof of "integrability" or exact cancellation; we will come back to this in the next section, let us just mention that the key ingredient in the phenomenon can be seen to be the fact that $\psi(e)$ (as will be introduced below) is identically 0. This is equivalent to saying that all the triangles of the triangular lattice are equilateral.

A way to try and generalize the proof is to demand that all the faces of T_α^* be equilateral triangles. Of course this cannot be done by embedding it in the plane, even locally – the total angle around a vertex would be equal to 2π only if the degree of the vertex is 6. But one can build a 2-dimensional manifold M_T with conic singularities by gluing together equilateral triangles according to the combinatorics of $\hat{T}^*$; since the average degree of a vertex of T^* is equal to 6, the average curvature of the manifold (defined *e.g.* as the limit of the normalized total curvature in large discs) is 0.

The manifold M_T is not flat in general (the only case where it is being the triangular lattice), but it is homeomorphic to the complex plane, and one can hope to see it as a perturbation of it on which some of the standard tools of complex analysis could have counterparts – the optimal being to be able to perform Smirnov's proof within it. This is no easy task, and is probably not doable anyway.

To relate M_T to the topic of this section, one can try to define a module out of it. A good candidate for that is the following: Assume that M_T can be realized as a sub-manifold of $\mathbb{R}^3$ (or in $\mathbb{R}^d$ for $d > 2$ large enough), in such a way that the (combinatorial) translations on $\hat{T}$ act by global translations of the ambient space, thus forming a *periodic sub-manifold.* Then there is a copy of $\mathbb{Z}^2$ acting on it, and the affine plane containing a given point of M_T and spanned by the directions of the two generators of that group is at finite Hausdorff distance from it; in other words, this realization of M_T looks like a bounded perturbation of a Euclidean plane.

One can then look at the orthogonal projections of the vertices of $\hat{T}^*$ (seen as points of M_T) onto that plane; this creates a doubly periodic, locally finite family of points of the Euclidean plane. It is not always possible to form an embedding of $\hat{T}^*$ in the plane from it (with disjoint edges); but it does define a value of α as above.

Unfortunately, there are cases when this value of α is not well defined, in the sense that it depends on the choice of M_T; this happens if the (infinite) polyhedron associated to $\hat{T}^*$, with equilateral faces, is *flexible.* The simplest example of this phenomenon is to take T to be two periods of the honeycomb lattice in each direction.

2.6. Comparing different methods of embedding

We now have at least two (forgetting about the last one) ways of giving a conformal structure to a torus equipped with a triangulation – which is but another way of referring to the choice of α. Assuming that critical percolation does have a scaling limit, it leads to a third choice α_T^{Perc} of it.

It would be a natural intuition that all these moduli are the same, and correspond to a notion of *conformal embedding* of a triangulation (or a 3-regular graph) in the plane; and they all have a claim to that name. But this is not true in general: We detail the construction of a counterexample. Start with the graph T_s and its dual T_s^*; and refine one of the "vertical" triangular faces of T_s^* by adding a vertex in the interior of it, connected to its three vertices. In terms of the primal graph, this correspond to replacing one of its vertices by a triangle – see Figure 4. Let T_s' be the graph obtained that way; we will refer to such a splitting as a *refinement*, and to the added vertex as a *new vertex*.

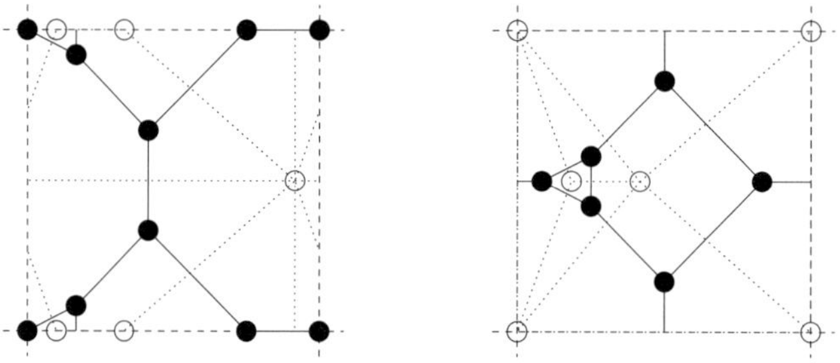

FIGURE 4. Square (but not balanced) embeddings of T_s' (solid) and its dual (dotted); the origin is taken as a point of T_s' on the left, and as a point of the dual on the right, corresponding to the ones chosen for Figures 1 and 2.

In terms of circle packings, this changes essentially nothing; the new vertex of $(T_s')^*$ can be realized as a new disc without modifying the rest of the configuration (cf. Figure 5). In terms of random walks, however, adding edges will modify the covariance matrix in the central limit theorem. The computation can be done easily, as explained above, and one gets the following values:

$$\alpha_{T_s'}^{\mathrm{CP}} = i = \alpha_{T_s}^{\mathrm{CP}} \;; \qquad \alpha_{T_s'}^{\mathrm{RW}} = i\sqrt{\frac{6}{7}} \neq i = \alpha_T^{\mathrm{RW}}.$$

In this particular case, the value of $\alpha_{(T_s')^*}^{\mathrm{RW}}$ is also $i\sqrt{6/7}$.

So, α^{RW} and α^{CP} are different in general. Is α^{Perc} (provided it exists) one of them? An easy fact to notice is the following: Let T^* be a triangulation of the torus and let $(T')^*$ be obtained from it by splitting a triangle into 3 as in the construction of T_s'. Then, consider two realizations of site-percolation at $p_c = 1/2$ on both universal covers, coupled in such a way that the common vertices are in the

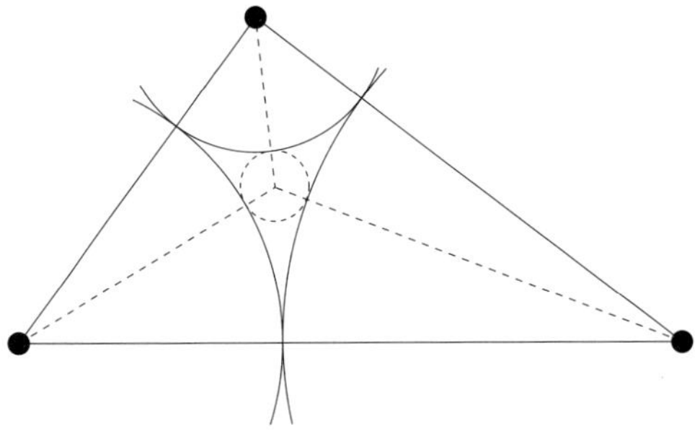

FIGURE 5. Splitting a face of a triangulation into 3 triangles, and the corresponding modification of its circle packing (added features represented by dashed lines).

same state for both models. In other words, start with a realization of percolation on $\hat{T}^*$ and without changing site states, refine a periodic family of triangles of it into 3, choosing the state of each new vertex independently of the others and of the configuration on $\hat{T}^*$.

If there is a chain of open vertices in $\hat{T}^*$, this chain is also a chain of open vertices in the refined graph – because all the edges are preserved in the refinement. Conversely, starting from a chain of open vertices in the refinement and removing each occurrence of a new vertex on it, one obtains a chain of open vertices in $\hat{T}^*$; the reason for that being that the triangle is a complete graph. Another way of stating the same fact is to say that opening (resp. closing) one of the new vertices cannot join two previously disjoint open clusters (resp. split a cluster into two disjoint components); they cannot be *pivotal* for a crossing event.

Hence, the probability that a large conformal rectangle is crossed is the same in both cases (at least if the choice of the discrete approximation of its boundary is the same for both graphs, which in particular implies that it contains no new vertex), and so is f_α for every choice of α (still assuming that it exists, of course). If one is conformally invariant in the scaling limit, the other also has to be. In short,

$$\alpha_T^{\text{Perc}} = \alpha_{T'}^{\text{Perc}}.$$

Looking at circle packings instead of percolation, we get the same identity (as was mentioned in the particular case of T_s), with a very similar proof: Adding a vertex does not change anything to the rest of the picture, and we readily obtain

$$\alpha_T^{\text{CP}} = \alpha_{T'}^{\text{CP}}.$$

This leads us to the following hope, which we state as a conjecture even though it is much closer to being wishful thinking:

Conjecture 12. *Let T^* be a triangulation of the torus. Then, the critical parameter for site-percolation on its universal cover $\hat{T}^*$ is equal to $1/2$, and for every $\alpha \in \mathbb{H}$,* *critical site-percolation on* $\hat{T}^*_\alpha$ *has a scaling limit. The value of the modulus* α *for which the model is conformally invariant in the scaling limit is that obtained from the circle packing associated to* $\hat{T}^*$*:*

$$\alpha_T^{\text{Perc}} = \alpha_T^{\text{CP}}.$$

3. Critical percolation on the triangular lattice

For reference, and as a way of introducing our general strategy, we give in this section a very shortened version of Smirnov's proof of the existence and conformal invariance of the scaling limit for critical site-percolation on the triangular lattice T_h. The interested reader is advised to consult our previous note [3] for an "extended shortening", or Smirnov's article [16] for the original proof; see the book of Bollobás and Riordan [5] for a more detailed treatment. Up to cosmetic changes, we follow the notation of [3].

Remark 13. Up to the last paragraph of the section, we are not assuming that the lattice we are working with is the honeycomb lattice; our only assumption is that we have an *a priori* bound for crossing probabilities of large rectangles which depends on their aspect ratio but not on their size (we "assume Russo-Seymour-Welsh conditions"). It is not actually clear how general those are; all the standard proofs require at least some symmetry in the lattice in addition to periodicity, but it is a natural conjecture that periodicity is enough.

Here and in the remainder of this paper, $\tau := e^{2\pi i/3}$ will be the third root of unity with positive imaginary part. Let T be a finite graph of genus 1, T_α an embedding of modulus α of T in the complex plane; let $V(T_\alpha)$ (resp. $E(T_\alpha)$) be the set of vertices (resp. *oriented* edges) of T_α. Each vertex $z \in V(T_\alpha)$ has three neighbours; let $E_z(T_\alpha)$ be the set of the three oriented edges in $E(T_\alpha)$ having their source at z. That set can be cyclically ordered counterclockwise; if $e \in E_z(T_\alpha)$ is one of the three edges starting at z, we will denote by $\tau.e$ (resp. $\tau^2.e$) the next (resp. second to next) edge in the ordering.

Remark 14. In the particular case of the honeycomb lattice, seeing each edge as a complex number (being the difference between its target and its source), the notation $\tau.e$ corresponds to complex multiplication by $e^{2\pi i/3}$ – in other words, $\tau.e = \tau e$ as a product of complex numbers. That is of course not the case in general, but we keep the formal notation for clarity. In what follows, whenever an algebraic expression involves the product of a complex number by an edge of T_α or T^*_α, as above the edge will be understood as the difference, as a complex number, between its target and its source; we will never use formal linear combinations of edges. The notation $\tau.e$ (with a dot) will be reserved for the "topological" rotation within $E_z(T_\alpha)$.

Let again Ω be a smooth Jordan domain in the complex plane, and let A, B, C and D be three points on its boundary, in that order when following $\partial\Omega$

counterclockwise. Let Ω_δ be the largest connected component of $\Omega \cap \delta T_\alpha$, and let A_δ (resp. B_δ, C_δ, D_δ) be the point of Ω_δ that is closest to A (resp. B, C, D). The main result in Smirnov's paper ([16]) is the following:

Theorem 15 (Smirnov). *In the case where T_α is the honeycomb lattice, embedded so as to make its faces regular hexagons* (i.e., *when* $\alpha = i\sqrt{3}/3$), *critical site-percolation has a conformally invariant scaling limit. If Ω is an equilateral triangle with vertices A, B and C, then*

$$f_\alpha(\Omega, A, B, C, D) = \frac{|CD|}{|CA|}.$$

Knowing this particular family of values of f_α is enough, together with conformal invariance, to compute it for any conformal rectangle. The formula obtained for a rectangle is known as *Cardy's formula.*

To each edge $e \in E(T_\alpha)$ corresponds its dual oriented edge $e^* \in E(T_\alpha^*)$, oriented in such a way that the angle (e, e^*) is in $(0, \pi)$. If $-e$ denotes the edge with the same endpoints as e but the reverse orientation, then we have $e^{**} = -e$. Define

$$\psi(e) := e^* + \tau(\tau.e)^* + \tau^2(\tau^2.e)^*$$

(where as above we interpret the edges e^*, $(\tau.e)^*$ and $(\tau^2.e)^*$ as complex numbers). It is easy to check that $\psi(e) = 0$ if, and only if, the face of T_α^* corresponding to the source of e is an equilateral triangle; so, $\psi(e)$ can be seen as a measure of the local deviation between T_α and the honeycomb lattice. An identity which will be useful later is the following:

$$\forall z \in V(T_\alpha), \qquad \sum_{e \in E_z(T_\alpha)} \psi(e) = 0. \tag{3.1}$$

For every $z \in \Omega_\delta$, let $E_{A,\delta}(z)$ be the event that there is a simple path of open vertices of Ω_δ^*, joining two points of the boundary of the domain, which separates z and A from B and C; let $H_{A,\delta} := P[E_{A,\delta}(z)]$. Define similar events for points B and C by a circular permutation of the letters, and let

$$S_\delta(z) := H_{A,\delta}(z) + H_{B,\delta}(z) + H_{C,\delta}(z),$$
$$H_\delta(z) := H_{A,\delta}(z) + \tau H_{B,\delta}(z) + \tau^2 H_{C,\delta}(z).$$

It is a direct consequence of Russo-Seymour-Welsh estimates that these functions are all Hölder with some universal positive exponent, with a norm which does not depend on δ, so by Ascoli's theorem they form a relatively compact family, and as $\delta \to 0$ they have subsequential limits which are Hölder maps from Ω to $\mathbb{C}$; all that is needed is prove that only one such limit is possible.

The key argument is to show that if h (resp. s) is any subsequential limit of (H_δ) (resp. (S_δ)) as $\delta \to 0$, then h and s are holomorphic; indeed, assume for a moment that they are. Since s is also real-valued, it has to be constant, and its value is 1 by boundary conditions (*e.g.* at point A). On the other hand, along the boundary arc $(A_\delta B_\delta)$ of $\partial\Omega_\delta$, $H_{C,\delta}$ is identically 0, so the image of the arc (AB)

by h is contained in the segment $[1, \tau]$ of $\mathbb{C}$; and similar statements hold *mutatis mutandis* for the arcs (BC) and (CA). By basic index theory, this implies that h is the unique conformal map sending Ω to the (equilateral) triangle of vertices 1, τ and τ^2, and that is enough to characterize it and to finish the proof of Theorem 15.

So, the crux of the matter, as expected, is to prove that the map h has to be holomorphic. The most convenient way to do that is to use Morera's theorem, which states that h is indeed holomorphic on Ω if, and only if, its integral along any closed, smooth curve contained in Ω is equal to 0.

Let γ be such a curve, and let $\gamma_\delta = (z_0, z_1, \dots, z_{L_\delta} = z_0)$ be a closed chain of vertices of Ω_δ which approximates it within Hausdorff distance δ and has $\mathcal{O}(\delta^{-1})$ points. Because the functions H_δ are uniformly Hölder, it follows that

$$\oint_{\gamma_\delta} H_\delta(z)\mathrm{d}z := \sum_{k=0}^{L_\delta - 1} H_\delta(z_k)(z_{k+1} - z_k) \to \oint_\gamma h(z)\mathrm{d}z.$$

We want to prove that, for a suitable choice of α, the discrete integral on the left-hand side of that equation vanishes in the scaling limit.

If $e = (z, z')$ is an oriented edge of Ω_δ, define $P_{A,\delta}(e) := P[E_{A,\delta}(z') \setminus E_{A,\delta}(z)]$; define $P_{B,\delta}$ and $P_{C,\delta}$ similarly. A very clever remark due to Smirnov, which is actually the only place in his proof where specifics of the model (as opposed to the lattice) are used, is that one can use color-swapping arguments to prove that, for every oriented edge,

$$P_{A,\delta}(e) = P_{B,\delta}(\tau.e) = P_{C,\delta}(\tau^2.e). \tag{3.2}$$

On the other hand, since differences of values of H_δ between points of Ω_δ can be computed in terms of these functions $P_{\cdot,\delta}$, the discrete integral above can be rewritten using them: Letting $E(\gamma_\delta)$ be the set of edges contained in the domain surrounded by γ_δ and using (3.2), one gets

$$\oint_{\gamma_\delta} H_\delta(z)\mathrm{d}z = \sum_{e \in E(\gamma_\delta)} \psi(e) P_{A,\delta}(e) + o(1). \tag{3.3}$$

A similar computation, together with the fact that $e^* + (\tau.e)^* + (\tau^2.e)^*$ is identically equal to 0, leads to

$$\oint_{\gamma_\delta} S_\delta(z)\mathrm{d}z = o(1). \tag{3.4}$$

We again refer the reader to [3] for the details of this construction.

Notice that it already implies that s is holomorphic, hence constant equal to 1, independently of the value of α; so, whether h is holomorphic or not, it will send $\bar{\Omega}$ to the triangle of vertices 1, τ and τ^2 anyway. In the case of the triangular lattice embedded in the usual way, $\psi(e)$ is also identically equal to 0, as was mentioned above, so h is itself holomorphic, and the proof is complete.

The remainder of these notes is devoted to some ideas about how to extend the general framework of the proof to more general cases; it is not clear how close one is to a proof, but it is likely that at least one fundamentally new idea will be required. However, we do believe that the overall strategy which we will now describe is the right angle of attack of the problem. Do not expect to find any formal proof in what follows, though.

4. Other triangulations

4.1. Using local shifts

The first natural idea when trying to generalize the construction of Smirnov is to try an apply it to more general periodic triangulations of the plane. Indeed, in all that precedes, up to and including Equation (3.3), nothing is specific to the regular triangular lattice, only Russo-Seymour-Welsh conditions (and their corollary that $p_c = 1/2$) are needed. It is only at the very last step, noticing that ψ was identically equal to 0, that the precise geometry was needed.

The key fact that makes hope possible is the following (and it is actually similar to one of the points we made earlier): In the expression of the discrete integral as a sum over interior edges, each term is the product of two contributions:

- $\psi(e)$ which depends on the geometry of the embedding, and through that on the value of α;
- $P_{A,\delta}(e)$ which is only a function of the combinatorics of Ω_δ.

Even though Ω_δ as a graph does depend on the choice of α, one can make the following remark: Applying the transformation φ_β (for some $\beta \in \mathbb{H}$) to both the domain Ω and the lattice δT_α does not change Ω_α as a graph. In particular it does not change the value of $P_{A,\delta}(e)$.

One can then see the whole sum as a function β, say $S_{\Omega,\delta}(\beta)$. Because $\varphi_\beta(z)$ is a real-affine function of β, so is $S_{\Omega,\delta}$; one can then try to solve the equation $S_{\Omega,\delta}(\beta) = 0$ in β. Using the corresponding φ_β, one gets a joint choice of a domain, a lattice modulus and mesh, and a curve γ making the discrete contour integral vanish.

If the modulus thus obtained actually did not depend on Ω, δ or γ, we would be done – call it α_T^{Perc} and there is only bookkeeping left to do. However we do not even know whether it has a limit as $\delta \downarrow 0 \dots$ An alternative is as follows. Because the lattice is periodic, it makes sense to first look at the sum $\sum \psi(e) P_{A,\delta}(e)$ over one fundamental domain. If that is small, then over the copy of the fundamental domain immediately to the right of the previous one, the terms $\psi(e)$ are exactly the same, and one is lead to compare $P_{A,\delta}$ for two neighboring pre-images of a given edge of T.

So, let e be an edge of Ω_δ, and let $e + \delta$ be its image by a translation of one period to the right. Making the dependency on the shape of the domain explicit in the notation, one can replace the translation of e by a translation of the domain itself and the boundary points in the opposite direction, to obtain

$$P_{A,\delta}^{\Omega}(e + \delta) = P_{A,\delta}^{\Omega-\delta}(e). \tag{4.1}$$

To estimate the difference between this term and the corresponding one in Ω, one can consider coupling two realizations of percolation, one on Ω_δ and the other in $\Omega_\delta - \delta$, so that they coincide on the intersection between the two.

The event corresponding to $P^{\Omega}_{A,\delta}(e)$ is that there is an open simple path separating the target of e and A from B, and C, and that no open simple path separates the source of e and A from B and C; this is equivalent to the existence of 3 disjoint paths from the 3 vertices of the face at the source of e to the 3 "sides" of the conformal triangle (Ω, A, B, C), two of them being formed of open vertices and the third being formed of closed vertices – cf. Figure 6. For this to happen in Ω but not in $\Omega - \delta$, one of these arms needs to go up to $\partial\Omega$ but not to $\partial(\Omega - \delta)$, and the only way for this to be realized is for a path of the opposite color to prevent it; this can be done in finitely many ways, Figure 6 being one of them; $P^{\Omega}_{A,\delta} - P^{\Omega-\delta}_{A,\delta}$ can then be written as the linear combination of the probabilities of finitely many terms of that form – half of these actually corresponding to the reversed situation, where arms go up to $\partial(\Omega - \delta)$ but not up to $\partial\Omega$.

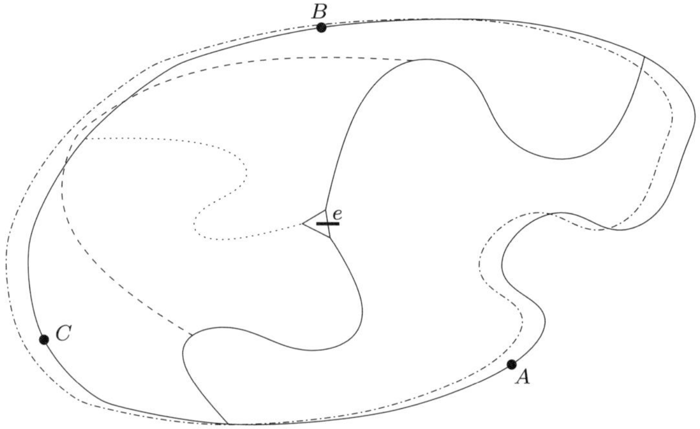

FIGURE 6. A typical case contributing to $P^{\Omega}_{A,\delta}(e) - P^{\Omega-\delta}_{A,\delta}(e)$. The original domain boundary is represented by a solid line, that of the shifted domain by a dashed-and-dotted line; open (resp. closed) arms from the source of e are represented as solid (resp. dotted) lines, and the additional open path preventing the closed arm from connecting to the boundary of $\Omega_\delta - \delta$ is represented as a dashed curve.

In the case corresponding to Figure 6, and all the similar ones, one sees that 3 arms connect the source of e to the boundary of $\Omega \cap (\Omega - \delta)$, and on at least one point of that boundaries there have to be 3 disjoint arms of diameter of order 1. There are $\mathcal{O}(\delta^{-1})$ points on the boundary, and the probability that 3 such arms exist from one of them is known – at least in the case of a polygon, which is enough for our purposes – to behave like δ^2, see, *e.g.*, [17].

Another possible reason for the non-existence of 3 arms from the source of e to the correct portions of the boundary of $\Omega - \delta$ (say) is that one of the corresponding arms in Ω actually lands very close to either A, B or C: Preventing it from touching the relevant part of $\partial(\Omega - \delta)$ requires only *one* additional arm from a δ-neighborhood of that vertex – *i.e.*, a total of 2 arms of diameter of order 1. The probability for that (see [17] also), still in the case when Ω is a polygon with none of A, B or C as a vertex, behaves like δ. Fortunately, there are only 3 corners on a conformal triangle, so the contribution of these cases is of the same order as previously

Putting everything together, one gets an estimate of the form

$$P_{A,\delta}^{\Omega-\delta}(e) = P_{A,\delta}^{\Omega}(e)\left[1 + \mathcal{O}(\delta)\right]. \tag{4.2}$$

Coming back to our current goal, let $\mathcal{E}$ be the set of oriented edges in a given period of Ω_δ, and let $\mathcal{E} + \delta$ be its image by the translation of vector δ. Then,

$$\begin{aligned}\sum_{e \in \mathcal{E}+\delta} \psi(e) P_{A,\delta}^{\Omega}(e) &= \sum_{e \in \mathcal{E}} \psi(e) P_{A,\delta}^{\Omega}(e)\left[1 + \mathcal{O}(\delta)\right] \\ &= \sum_{e \in \mathcal{E}} \psi(e) P_{A,\delta}^{\Omega}(e) + \mathcal{O}(\delta^{2+\eta})\end{aligned}$$

with $\eta > 0$; the existence of such an η is ensured by Russo-Seymour-Welsh type arguments again, which ensure that, uniformly in e and δ, for every edge e, $P_{A,\delta}^{\Omega}(e) = \mathcal{O}(\delta^\eta)$.

Now, if that is the way the proof starts, what needs to be done is quite clear:

- Fix a period $\mathcal{E}$ of the graph.
- Choose α so that the previous sum, over this period, of $\psi(e)P_{A,\delta}^{\Omega}(e)$ is equal to 0.
- Use the above estimate to give an upper bound for the same sum on neighboring periods.
- Try to somehow propagate the estimate up to the boundary.

The last part of the plan is the one that does not work directly, because one needs of the order of δ^{-1} steps to go from $\mathcal{E}$ to $\partial\Omega$, and the previous bound is not small enough to achieve that; one would need a term of the order of $\mathcal{O}(\delta^{3+\eta})$. It is however quite possible that a more careful decomposition of the events would lead to additional cancellation, though we were not able to perform it.

4.2. Using incipient infinite clusters

Another idea which might have a better chance of working out is based on the idea of incipient infinite clusters. We are trying to ensure that $\sum \psi(e) P_{A,\delta}(e)$ is equal to $o(\delta^2)$ over a period for a suitable choice of α; but for it to be exactly equal to 0 depends only on the ratios $P_A(e)/P_A(e')$ within the period considered, and not on their individual values. One can then let δ go to 0, or equivalently let Ω increase to cover the whole space, and look at this ratio.

Proposition 16. *There is a map* $\pi : E(\hat{T}) \to (0,+\infty)$ *such that the following happens. Let* e, e' *be two edges of* T_α, *which we identify with* $\hat{T}$ *for easier notation, and let* $\delta = 1$. *Then, as* Ω *increases to cover the whole plane,*

$$\frac{P^{\Omega}_{A,1}(e)}{P^{\Omega}_{A,1}(e')} \to \frac{\pi(e)}{\pi(e')},$$

uniformly in the choices of A, B *and* C *on* $\partial\Omega$. *The map* π *is periodic and does not depend on the choice of* α.

Proof. The argument is very similar to Kesten's proof of existence of the incipient infinite cluster (see [11]); it is based on Russo-Seymour-Welsh estimates. It will appear in an upcoming paper [2]. Notice that there is no requirement for A, B and C to remain separated from each other; this is similar to the fact that the incipient infinite cluster is also the limit, as $n \to \infty$, of critical percolation conditioned to the event that the origin is connected to the point $(n, 0)$ – which in turn is again a consequence of Russo-Seymour-Welsh theory. The speed of convergence is certainly different with and without such restrictions on the positions of A, B and C, though. □

Seeing this proposition, one is tempted to define α by solving the equation

$$\sum_{e\in\mathcal{E}} \psi(e)\pi(e) = 0, \tag{4.3}$$

where again the sum is taken over one period of the lattice. Indeed, all that remains in the sum, over the same period of the lattice, of $\psi(e)P_A(e)$ is composed of terms of a smaller order. However, because the limit taken to define π is uniform in the choices of A, B and C, in particular it is invariant by re-labelling of the corners of the conformal triangle; equivalently, taking P_B instead of P_A leads to the same limit. Combining this remark with Equation (3.2), one gets the following identities:

$$\forall e \in E(\hat{T}), \quad \pi(e) = \pi(\tau.e) = \pi(\tau^2.e). \tag{4.4}$$

In other words, $\pi(e)$ only depends on the source of e. For every edge $e = (z, z')$, let $\pi(z) := \pi(e)$: If $\mathcal{V}$ is a period of $V(\hat{T})$, one has

$$\sum_{e\in\mathcal{E}} \psi(e)\pi(e) = \sum_{z\in\mathcal{V}} \pi(z) \sum_{z'\sim z} \psi((z,z')) = 0$$

by using the remark in Equation (3.1). So, the equation (4.3) is actually always true, and does not help in finding the value of α ...

This is actually good news, because it is the sign of emerging cancellations in the scaling limit, which were not at first apparent; that means that the relevant terms in (3.3) are actually smaller than they look at first sight, which in turn means that making the leading term equal to 0 by the correct choice of α leads to even smaller terms.

Whether the overall strategy can be made to work actually depends on the speed of convergence in the statement of Proposition 16. In the case of the triangular lattice, one can actually use SLE to give an explicit expansion of the ratio $P_A(e)/P_A(e')$ as Ω increases, at least in some cases; this is the subject of an upcoming paper [2].

5. Other lattices

5.1. Mixed percolation

We conclude the speculative part of these notes by some considerations about bond-percolation on the planar square lattice. The combinatorial construction we perform here does apply to more general cases, but the probabilistic arguments which follow do not, so we restrict ourselves to the case of $\mathbb{Z}^2$.

The general idea it to map the problem of bond-percolation on $\mathbb{Z}^2$ to one of site-percolation on a suitable triangulation of the plane. Then, if the arguments in the previous section can be made to work, one could potentially prove the existence and conformal invariance of a scaling limit of critical percolation on the square lattice.

The key remark was already present in the book of Kesten [10]: For any bond-percolation model on a graph, one can construct the so-called *covering graph* on which it corresponds to site-percolation. More specifically, let G_1 be a connected graph with bounded degree; as usual, let $E(G_1)$ be the set of its edges and $V(G_1)$ be the set of its vertices. We construct a graph G_2 as follows: The set $V(G_2)$ of its vertices is chosen to be $E(G_1)$, and we put an edge between two vertices of G_2 if, and only if, the corresponding edges of G_1 share an endpoint. Notice that even if G_1 is assumed to be planar, G_2 does not have to be – see Figure 7 for the case of $\mathbb{Z}^2$.

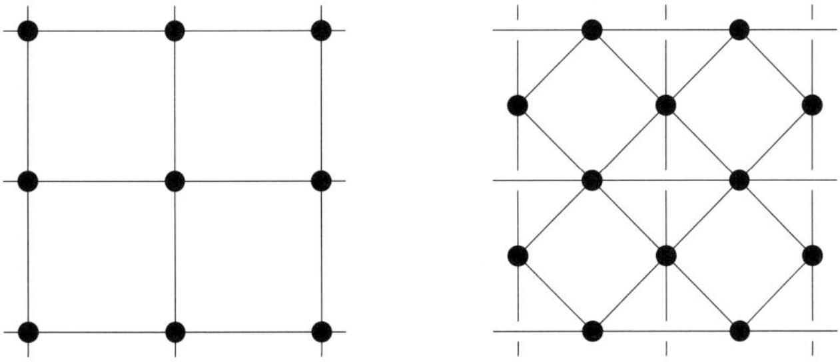

FIGURE 7. The square lattice $\mathbb{Z}^2$ and its covering graph.

The graph thus obtained from the square lattice is isomorphic to a copy of the square lattice where every second face, in a checkerboard disposition, is completed into a complete graph with 4 vertices. The next remark is the following: in terms of site-percolation, a complete graph with 4 vertices behaves the same way as a square with an additional vertex at the center, which is open with probability 1 – with the same meaning as when we looked at refinement of triangles in triangulations, *i.e.*, taking a chain of open vertices in the partially centered square lattice and removing from it the vertices which are face centers leads to a chain of open vertices in the covering graph of $\mathbb{Z}^2$.

So, let again G_s be the centered square lattice, as was introduced above, and let $q \in [0,1]$; split the vertices of G_s into three classes, to defined a non-homogeneous site-percolation model, as follows. Each vertex is either open or closed, independently of the others, and:

- The sites of $\mathbb{Z}^2$ are open with probability $p = 1/2$; we will call them *vertices of type I*, or p-sites for short, and denote by V_1 the set of such vertices.
- The vertices of coordinates $(k+1/2, l+1/2)$ with $k+l$ even are open with probability q; we will call them *vertices of type II*, or q-sites for short, and denote by V_2 the set of such vertices.
- The vertices of coordinates $(k+1/2, l+1/2)$ with $k+l$ odd are open with probability $1-q$; we will call them *vertices of type III*, or $(1-q)$-sites for short, and denote by V_3 the set of such vertices.

We will refer to that model as *mixed percolation* with parameters $p = 1/2$ and q, and denote by $P_{1/2,q}$ the associated probability measure. Two cases are of particular interest:

- If $q = 1/2$, the model is exactly critical site-percolation on the centered square lattice G_s.
- If $q = 0$ or $q = 1$ (the situation is the same in both cases up to a translation), from the previous remarks mixed percolation then corresponds to critical bond-percolation on the square lattice.

Besides, all the models obtained for $p = 1/2$ are critical and satisfy Russo-Seymour-Welsh estimates.

5.2. Model interpolation

We are now equipped to perform an interpolation between the models at $q = 0$ and $q = 1/2$. Let (Ω, A, B, C, D) be a simply connected subset of $\mathbb{Z}^2$ equipped with 4 boundary points – say, a rectangle; let $U = U_{\Omega,A,B,C,D}$ be the event, under mixed percolation with parameters $p = 1/2$ and q, that there is a chain of open vertices of Ω joining the boundary arcs (AB) and (CD). To estimate the difference between the probabilities of U for the two models we are most interested in, simply write

$$P_{1/2,1/2}[U] - P_{1/2,0}[U] = \int_0^{1/2} \frac{\partial}{\partial q} P_{1/2,q}[U] \, dq. \tag{5.1}$$

If percolation is indeed universal, then one would expect cancellation to occur, hopefully for each value of q; the optimal statement being of the form

$$\lim_{\Omega \uparrow \mathbb{Z}^2} \sup_{A,B,C,D \in \partial\Omega} \sup_{q \in (0,1)} \frac{\partial}{\partial q} P_{1/2,q}[U] = 0. \tag{5.2}$$

The main ingredient in the estimation of the derivative in q is, as one might expect, a slight generalization of Russo's formula; to state it, we need a definition:

Definition 17. Consider mixed percolation on G_s, and let E be a cylindrical increasing event for it (*i.e.*, an event which depends on the state of finitely many vertices). Given a realization ω of the model, we say that a vertex v is *pivotal for the event* E if E is realized for the configuration ω^v where v is made open, and not realized for the configuration ω_v where v is made closed. We will denote by $\mathrm{Piv}(E)$ the (random) set of pivotal vertices for E.

Proposition 18. *With the above notation, one has*

$$\frac{\partial}{\partial q} P_{1/2,q}[U] = E_{1/2,q}\left[|\mathrm{Piv}(U) \cap \Omega \cap V_2| - |\mathrm{Piv}(U) \cap \Omega \cap V_3|\right].$$

Proof. The argument is the same as in the proof of the usual formula (in the case of homogeneous percolation); we refer the reader to the book of Grimmett [8]. □

As was the case in the previous section, one can relate the event that a given site is pivotal to the presence of disjoint arms in the realization of the model, with appropriate color. More precisely, a q-site (say) $v \in \Omega$ is pivotal if, and only if, the following happens: v is at the center of a face of $\mathbb{Z}^2$; two opposite vertices of that face are connected respectively to the boundary arcs (AB) and (CD) by disjoint chains of open vertices; the other two vertices of the face are connected respectively to the boundary arcs (BC) and (AD) by disjoint chains of closed vertices; and none of the chains involved contains the vertex v. To state the previous description more quickly, there is a 4-arm configuration with alternating colors at vertex v, and the endpoints of the arms are appropriately located on $\partial\Omega$ – see Figure 8.

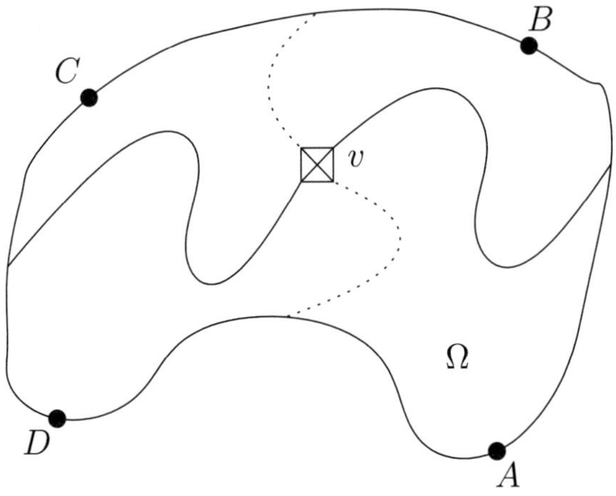

FIGURE 8. A four-arm configuration at vertex v making it pivotal for the event $U(\Omega, A, B, C, D)$.

The main feature of mixed percolation in the case of the centered square lattice is the following: Starting from a configuration sampled according to $P_{1/2,q}$ and shifting the state of all vertices by one lattice mesh to the right, or equivalently flipping the state of all vertices, or rotating the whole configuration by an angle of $\pi/2$ around a site of type I, one gets a configuration sampled according to $P_{1/2,1-q}$; on the other hand, rotating the picture by $\pi/2$ around a vertex of type II or III leaves the measure invariant.

Notice that the existence of 4 arms of alternating colors from a given vertex v is invariant by color-swapping; the configuration in Figure 8 is not though, because the arms obtained after the color change connect the neighbors of v to the wrong parts of the boundary. Nevertheless, one can try to apply the same reasoning as in the previous section, as follows: Let v' be the vertex that is one lattice step to the right of v. If v is a vertex of type II, then v' is a vertex of type III, and up to boundary terms, one can pair all the q-sites in Ω to corresponding $(1-q)$-sites.

To estimate the right-hand term in the statement of Proposition 18, let

$$\Delta(v) := P[v \in \mathrm{Piv}(U)] - P[v' \in \mathrm{Piv}(U)].$$

Our goal will be achieved if one is able to show that $\Delta(v) = o(|\Omega|^{-1})$; or equivalently, if Ω_δ is obtained from a fixed continuous domain by discretization with mesh δ, if one has

$$\Delta(v) = o(\delta^2).$$

In the case of critical site-percolation on the triangular lattice, arguments using SLE processes give an estimate to the probability that a vertex is pivotal, and from universality conjectures it is natural to expect that they extend to the case of mixed percolation on T_s. They involve the 4-arm exponent of percolation, and would read (still in the case of a fixed domain discretized at mesh δ) as

$$P[v \in \mathrm{Piv}(U)] \approx \delta^{5/4}.$$

So, shifting the domain instead of the point as we did in the last section, one would expect an estimate on $\Delta(v)$ of the order

$$\Delta(v) \approx \delta^{9/4}$$

(where the addition of 1 in the exponent corresponds to the presence of a 3-arm configuration at some point on the boundary on either the original domain or its image by the shift). Since $9/4 > 2$, that would be enough to conclude.

However, this approach does not work directly, because of the previous remark that the shift by one lattice step does change the measure, replacing q by $1-q$. If one is interested in the mere existence of the 4 arms around a vertex, combining the shift with color-flipping is enough to cancel the effect; but the estimate one obtains that way is of the form

$$P[v \in \mathrm{Piv}(U_{\Omega,A,B,C,D})] - P[v' \in \mathrm{Piv}(U_{\Omega,B,C,D,A})] \approx \delta P[v \in \mathrm{Piv}(U_{\Omega,A,B,C,D})] \tag{5.3}$$

(and Russo-Seymour-Welsh estimates are actually enough to obtain a formal proof of this estimate).

So, once again, what is missing is a way to estimate how much $P[v \in \mathrm{Piv}(U_{\Omega,A,B,C,D})]$ depends on the location of A, B, C and D along $\partial\Omega$; if the dependency is very weak, then the estimate in Equation (5.3) might actually be of the right order of magnitude. Once again, it is likely that the way to proceed is to use a modified version of the incipient infinite cluster conditioned to have 4 arms of alternating colors from the boundary, and that the order of magnitude of $\Delta(v)$ will be related to the speed of convergence of conditioned percolation to the incipient clusters; but we were not able to conclude the proof that way. It would seem that this part of the argument is easier to formalize than that of the previous section, though, and hopefully a clever reader of these notes will be able to do just that ...

References

[1] Michel Bauer and Denis Bernard, *Conformal field theories of stochastic Loewner evolutions*, Comm. Math. Phys. **239** (2003), no. 3, 493–521.

[2] Vincent Beffara, *Quantitative estimates for the incipient infinite cluster of 2D percolation*, in preparation.

[3] ———, *Cardy's formula on the triangular lattice, the easy way*, Universality and Renormalization (Ilia Binder and Dirk Kreimer, eds.), Fields Institute Communications, vol. 50, The Fields Institute, 2007, pp. 39–45.

[4] Itai Benjamini and Oded Schramm, *Conformal invariance of Voronoi percolation*, Comm. Math. Phys. **197** (1998), no. 1, 75–107.

[5] Béla Bollobás and Oliver Riordan, *Percolation*, Cambridge University Press, 2006.

[6] John Cardy, *Critical percolation in finite geometries*, J. Phys. A **25** (1992), L201–L206.

[7] ———, *Conformal invariance in percolation, self-avoiding walks, and related problems*, Ann. Henri Poincaré **4** (2003), no. suppl. 1, S371–S384.

[8] Geoffrey Grimmett, *Percolation*, second ed., Grundlehren der Mathematischen Wissenschaften, vol. 321, Springer-Verlag, Berlin, 1999.

[9] Galin L. Jones, *On the Markov chain central limit theorem*, Probability Surveys **1** (2004), 299–320.

[10] Harry Kesten, *Percolation theory for mathematicians*, Progress in Probability and Statistics, vol. 2, Birkhäuser, Boston, Mass., 1982.

[11] ———, *The incipient infinite cluster in two-dimensional percolation*, Probab. Theory Related Fields **73** (1986), no. 3, 369–394.

[12] R.P. Langlands, C. Pichet, Ph. Pouliot, and Y. Saint-Aubin, *On the universality of crossing probabilities in two-dimensional percolation*, J. Statist. Phys. **67** (1992), no. 3-4, 553–574.

[13] Robert Langlands, Yves Pouillot, and Yves Saint-Aubin, *Conformal invariance in two-dimensional percolation*, Bulletin of the A.M.S. **30** (1994), 1–61.

[14] Oded Schramm, *Scaling limits of loop-erased random walks and uniform spanning trees*, Israel Journal of Mathematics **118** (2000), 221–288.

[15] Stanislav Smirnov, *Critical percolation in the plane: Conformal invariance, Cardy's formula, scaling limits*, C.R. Acad. Sci. Paris Sér. I Math. **333** (2001), no. 3, 239–244.

[16] ———, *Critical percolation in the plane. I. Conformal invariance and Cardy's formula. II. Continuum scaling limit*, `http://www.math.kth.se/ stas/papers/percol.ps`, 2001.

[17] Stanislav Smirnov and Wendelin Werner, *Critical exponents for two-dimensional percolation*, Mathematical Research Letters **8** (2001), 729–744.

[18] Kenneth Stephenson, *Introduction to circle packing*, Cambridge University Press, 2005.

Vincent Beffara
CNRS – UMPA
École normale supérieure de Lyon
46 allée d'Italie
F-69364 Lyon cedex 07
France
e-mail: `vbeffara@ens-lyon.fr`

Progress in Probability, Vol. 60, 59–96

A New REM Conjecture

Gérard Ben Arous, Véronique Gayrard and Alexey Kuptsov

Abstract. We introduce here a new universality conjecture for levels of random Hamiltonians, in the same spirit as the local REM conjecture made by S. Mertens and H. Bauke. We establish our conjecture for a wide class of Gaussian and non-Gaussian Hamiltonians, which include the p-spin models, the Sherrington-Kirkpatrick model and the number partitioning problem. We prove that our universality result is optimal for the last two models by showing when this universality breaks down.

Mathematics Subject Classification (2000). 82B44, 60F99.

Keywords. Statistical mechanics, disordered media, spin-glasses.

1. Introduction

S. Mertens and H. Bauke recently observed ([Mer00], [BM04], see also [BFM04]) that the statistics of energy levels for very general random Hamiltonians are Poissonian, when observed micro-canonically, i.e., in a small window in the bulk. They are universal and identical to those of the simplest spin-glass model, the Random Energy Model or REM, hence the name of this (numerical) observation: the REM conjecture or more precisely the *local REM conjecture*.

This local REM conjecture was made for a wide class of random Hamiltonians of statistical mechanics of disordered systems, mainly spin-glasses (mean field or not), and for various combinatorial optimization problems (like number partitioning). Recently, two groups of mathematicians have established this conjecture in different contexts: C. Borgs, J. Chayes, C. Nair, S. Mertens, B. Pittel for the number partitioning question (see [BCP01], [BCMN05a], [BCMN05b]), and A. Bovier, I. Kurkova for general spin-glass Hamiltonians (see [BK06a], [BK06b]).

We introduce here a new kind of universality for the energy levels of disordered systems. We believe that one should find universal statistics for the energy levels of a wide class of random Hamiltonians if one re-samples the energy levels, i.e., draws a random subset of these energies. Put otherwise, our conjecture is thus that the level statistics should also be universal, i.e., Poissonian, when observed on a large random subset of the configuration space rather than in a micro-canonical window. We establish this new universality result (which could be called *the re-sampling REM universality* or *the REM universality by dilution*) for general (mean-field) spin-glass models, including the case of number partitioning and for large but sparse enough subsets. This approach has the following interesting property: the range of energies involved is not reduced to a small window as in the local REM conjecture. Thus we can study the extreme value distribution on the random subset, by normalizing the energies properly. Doing so we establish that the Gibbs measure restricted to a sparse enough random subset of configuration space has a universal distribution which is thus the same as for the REM, i.e., a Poisson-Dirichlet measure.

To be more specific, we specialize our setting to the case of random Hamiltonians $(H_N(\sigma))_{\sigma\in S_N}$ defined on the hypercube $S_N = \{-1,1\}^N$. We want to consider a sparse random subset of the hypercube, say X, and the restriction of the function H_N to X. We introduce the random point process

$$\mathcal{P}_N = \sum_{\sigma\in X} \delta_{H'_N(\sigma)}, \tag{1.1}$$

with a normalization:

$$H'_N(\sigma) = \frac{H_N(\sigma) - a_N}{b_N} \tag{1.2}$$

to be chosen. Our conjecture specializes to the following: the asymptotic behavior of the random point process $\mathcal{P}_N$ is universal, for a large class of Hamiltonians H_N, for appropriate sparse random subsets X, and appropriate normalization.

We will only study here the simplest possible random subset, i.e., a site percolation cluster $X = \{\sigma \in S_N : X_\sigma = 1\}$, where the random variables $(X_\sigma)_{\sigma\in S_N}$ are i.i.d. and Bernoulli:

$$\mathrm{P}(X_\sigma = 1) = 1 - \mathrm{P}(X_\sigma = 0) =: p_N = \frac{2^M}{2^N}. \tag{1.3}$$

Thus the mean size of X is 2^M and we will always assume that X is not too small, e.g., that $\log N = o(2^M)$. We will sometimes call X a *random cloud*.

In order to understand what the universal behavior should be, let us examine the trivial case where the $H_N(\sigma)$ are i.i.d. centered standard Gaussian random variables, i.e., the case of the Random Energy Model. Then, standard extreme value theory proves that if

$$a_N = \sqrt{2M\log 2 + 2\log b_N - \log 2} \text{ and } b_N = \sqrt{\frac{1}{M}}, \tag{1.4}$$

then $\mathcal{P}_N$ converges to a Poisson point process with intensity measure

$$\mu(dt) = \frac{1}{\sqrt{\pi}} e^{-t\sqrt{2\log 2}}\, dt. \tag{1.5}$$

We will now fix the normalization needed in (1.2) by choosing a_N and b_N as in (1.4). The basic mechanism of the REM universality we propose is that the influence of correlations between the random variables $H'_N(\sigma)$ should be negligible when the two-point correlation (the covariance) is a decreasing function of the Hamming distance

$$d_H(\sigma, \sigma') = \#\{i \leq N : \sigma_i \neq \sigma'_i\} \tag{1.6}$$

and when the random cloud is sparse enough. We first establish this universality conjecture for a large class of Gaussian Hamiltonians. This class contains the Sherrington-Kirkpatrick (SK) model as well as the more general p-spin models. It also contains the Gaussian version of the number partitioning problem.

Consider a Gaussian Hamiltonian H_N on the hypercube $S_N = \{-1, 1\}^N$ such that the random variables $(H_N(\sigma))_{\sigma \in S_N}$ are centered and whose covariance is a smooth decreasing function of the Hamming distance or, equivalently, a smooth increasing function ν of the overlap $R(\sigma, \sigma') = \frac{1}{N} \sum_{i=1}^{N} \sigma_i \sigma'_i$:

$$\operatorname{cov}(H_N(\sigma), H_N(\sigma')) = \nu(R(\sigma, \sigma')) = \nu\Big(1 - \frac{2d_H(\sigma, \sigma')}{N}\Big). \tag{1.7}$$

We will always assume that $\nu(0) = 0$ and that $\nu(1) = 1$. The first assumption is crucial, since it means that the correlation of the Hamiltonian vanishes for pairs of points on the hypercube which are at a typical distance. The second assumption simply normalizes the variance of $H_N(\sigma)$ to 1.

This type of covariance structure can easily be realized for ν real analytic, of the form:

$$\nu(r) = \sum_{p \geq 1} a_p^2 r^p. \tag{1.8}$$

Indeed such a covariance structure can be realized by taking mixtures of p-spin models. Let $H_{N,p}$ be the Hamiltonian of the p-spin model given by

$$H_{N,p}(\sigma) = \frac{1}{N^{(p-1)/2}} \sum_{1 \leq i_1, i_2, \ldots, i_p \leq N} g_{i_1, i_2, \ldots, i_p} \sigma_{i_1} \sigma_{i_2} \ldots \sigma_{i_p}, \tag{1.9}$$

where the random variables $(g_{i_1, i_2, \ldots, i_p})_{1 \leq i_1, i_2, \ldots, i_p \leq N}$ are independent standard Gaussians defined on a common probability space $(\Omega^g, \mathcal{F}^g, \mathbb{P})$. Then

$$H_N(\sigma) = \frac{1}{\sqrt{N}} \sum_{p \geq 1} a_p H_{N,p}(\sigma) \tag{1.10}$$

has the covariance structure given in (1.7)–(1.8) Let us recall that the case where $\nu(r) = r$ is the Gaussian version of the number partitioning problem ([BCP01]), the case where $\nu(r) = r^2$ is the SK model, and more generally, $\nu(r) = r^p$ defines the pure p-spin model.

Let us normalize H_N as above (see (1.4)):

$$H'_N(\sigma) = \frac{H_N(\sigma) - a_N}{b_N}, \tag{1.11}$$

and consider the sequence of point processes

$$\mathcal{P}_N = \sum_{\sigma \in X} \delta_{H'_N(\sigma)}. \tag{1.12}$$

Theorem 1.1 (Universality in the Gaussian case). *Assume that $M = o(\sqrt{N})$ if $\nu'(0) \neq 0$, and that $M = o(N)$ if $\nu'(0) = 0$. Then,* P*-almost surely, the distribution of the point process $\mathcal{P}_N$ converges weakly to the distribution of a Poisson point process $\mathcal{P}$ on $\mathbb{R}$ with intensity given by*

$$\mu(dt) = \frac{1}{\sqrt{\pi}} e^{-t\sqrt{2\log 2}} dt. \tag{1.13}$$

Remark 1.2. The condition $\log N = o(2^M)$ is needed in order to get P-almost sure results.

We extend this result, in Section 5, to a wide class of non-Gaussian Hamiltonians (introduced in [BCMN05a] for the case of number partitioning).
The theorem has the following immediate corollary. Let us fix the realization of the random cloud X. For configurations σ belonging to the cloud we consider the Gibbs' weights $G_{N,\beta}(\sigma)$ of the re-scaled Hamiltonian $H'_N(\sigma)$

$$G_{N,\beta}(\sigma) = \frac{e^{-\beta H'_N(\sigma)}}{\sum_{\varrho\in X} e^{-\beta H'_N(\varrho)}} = \frac{e^{-\beta\sqrt{M}H_N(\sigma)}}{\sum_{\varrho\in X} e^{-\beta\sqrt{M}H_N(\varrho)}}. \tag{1.14}$$

Reordering the Gibbs' weights $G_{N,\beta}(\sigma)$ of the configurations $\sigma \in X$ as a non-increasing sequence $(w_\alpha)_{\alpha\leq|X|}$ and defining $w_\alpha = 0$ for $\alpha > |X|$ we get a random element w of the space $\mathcal{S}$ of non increasing sequences of non negative real numbers with sum less than one.

Corollary 1.3 (Convergence to Poisson-Dirichlet). *If $\beta > \sqrt{2\log 2}$ then* P*-almost surely under the assumptions of Theorem* 1.1 *the law of the sequence $w = (w_\alpha)_{\alpha\geq 1}$ converges to the Poisson-Dirichlet distribution with parameter $m = \frac{\sqrt{2\log 2}}{\beta}$ on $\mathcal{S}$.*

The fact that Theorem 1.1 implies Corollary 1.3 is well known, see for instance [Tal03] (pp. 13–19) for a good exposition.
It is then a natural question to know if our sparseness assumption is optimal. When the random cloud is denser can this universality survive? We show that our sparseness condition is indeed optimal for the number partitioning problem and for the SK model, and that the universality does break down.

Theorem 1.4. [*Breakdown of Universality for the number partitioning problem*]

(i) *Let $\nu(r) = r$. Suppose that $\limsup \frac{M(N)}{\sqrt{N}} < \infty$. Then* P*-almost surely, the distribution of the point process $\mathcal{P}_N$ converges to the distribution of a Poisson point process if and only if $M = o(\sqrt{N})$.*

[*Breakdown of Universality for the Sherrington-Kirkpatrick model*]

(ii) *Let $\nu(r) = r^2$. Suppose that $\limsup \frac{M(N)}{N} < \frac{1}{8\log 2}$. Then* P*-almost surely, the distribution of the point process $\mathcal{P}_N$ converges to the distribution of a Poisson point process if and only if $M = o(N)$.*

We prove this theorem in Section 4 by showing that the second factorial moment of the point process does not converge to the proper value. The case of pure p-spin models, with $p \geq 3$, or more generally the case where $\nu'(0) = \nu''(0) = 0$ differs strongly (see Theorem 4.7). The asymptotic behavior of the first three moments is compatible with a Poissonian convergence. Proving or disproving Poissonian convergence (or REM universality) in this case is still open, as it is for the local REM conjecture.

The paper is organized as follows. In Section 2 we establish important combinatorial estimates about maximal overlaps of ℓ-tuples of points on the random cloud. We give a particular care to the case of pairs ($\ell = 2$) and triples ($\ell = 3$) which are important for the breakdown of universality results. In Section 3 we establish the universality in the Gaussian case (Theorem 1.1). We then prove, in Section 4, the breakdown of universality given in Theorem 1.4. Finally we extend the former results to a wide non-Gaussian setting in Section 5.

2. Combinatorial estimates

In this section we fix an integer $\ell \geq 1$ and study the maximal overlap

$$R_{\max}(\sigma^1, \dots, \sigma^\ell) = \max_{1 \leq i < j \leq \ell} |R(\sigma^i, \sigma^j)|. \tag{2.1}$$

For fixed $N \in \mathbb{N}$ and $R \in [0, 1)$ let us define the following subsets of S_N^ℓ

$$U_\ell(R) = \{(\sigma^1, \dots, \sigma^\ell) : R_{\max}(\sigma^1, \dots, \sigma^\ell) \leq R\} \tag{2.2}$$

and

$$V_\ell(R) = \{(\sigma^1, \dots, \sigma^\ell) : R_{\max}(\sigma^1, \dots, \sigma^\ell) = R\}. \tag{2.3}$$

More generally, let the sequence $R_N \in [0, 1)$ be given (respectively, the corresponding sequence of Hamming distances $d_N = \frac{N}{2}(1 - R_N))$ and introduce the sequence of sets $U_\ell(R_N)$ and $V_\ell(R_N)$ which we denote for simplicity of notation by $U_{N,\ell}$ and $V_{N,\ell}$ respectively.

For a set $Y \subset S_N^\ell$ we denote by Y^X its intersection with X^ℓ. In the following theorem we study the properties of the sets

$$U_{N,\ell}^X = \{(\sigma^1, \dots, \sigma^\ell) \in X^\ell : R_{\max}(\sigma^1, \dots, \sigma^\ell) \leq R_N\} \tag{2.4}$$

and

$$V_{N,\ell}^X = \{(\sigma^1, \dots, \sigma^\ell) \in X^\ell : R_{\max}(\sigma^1, \dots, \sigma^\ell) = R_N\}. \tag{2.5}$$

In order to state the main result of this section (Theorem 2.1) we define the function

$$\mathcal{J}(x) = \begin{cases} \frac{1-x}{2} \log(1-x) + \frac{1+x}{2} \log(1+x) & \text{if } x \in [-1, 1], \\ +\infty & \text{otherwise.} \end{cases} \tag{2.6}$$

Theorem 2.1. *Let the sequence R_N be such that $NR_N^2 \to \infty$.*

(i) *Then* P*-almost surely*

$$|U_{N,\ell}^X| = \mathrm{E}|U_{N,\ell}^X|(1+o(1)). \tag{2.7}$$

(ii) *If $R_N = o(1)$ and $M(N) \geq \log N$ then there exists $\alpha \in [0,1)$ and $C > 0$, depending only on ℓ, such that* P*-almost surely*

$$|V_{N,\ell}^X| \leq C\, e^{\alpha N \mathcal{J}(R_N)} \mathrm{E}|V_{N,\ell}^X|. \tag{2.8}$$

Proof. The proof is based on standard inequalities for i.i.d. random variables that result from exponential Chebychev inequality. We formulate them without proof: let $(X_i)_{1\leq i \leq n}$ be i.i.d. Bernoulli rv's with $\mathbb{P}(X_i = 1) = 1 - \mathbb{P}(X_i = 0) = p$ and let $Z = \sum_{i=1}^{n} X_i$. Then, for $t > 0$,

$$\mathbb{P}(Z - \mathbb{E}Z \geq t\mathbb{E}Z) \leq e^{-n\left(p(1+t)\log(1+t) + (1-p(1+t))\log\frac{1-p(1+t)}{1-p}\right)}, \tag{2.9}$$

$$\mathbb{P}(Z - \mathbb{E}Z \leq -t\mathbb{E}Z) \leq e^{-n\left(p(1-t)\log(1-t) + (1-p(1-t))\log\frac{1-p(1-t)}{1-p}\right)}. \tag{2.10}$$

If $p = p(n) \to 0$ and $t = t(n) \to 0$ as $n \to \infty$, the above inequalities imply that, for large enough n,

$$\mathbb{P}\big(|Z - \mathbb{E}Z| \geq t\mathbb{E}Z\big) \leq 2e^{-npt^2/4}, \tag{2.11}$$

whereas if $p(n) \to 0, t(n) \to \infty$, and $p(n)t(n) \to 0$ we get from (2.9) that for large enough n

$$\mathbb{P}(Z - \mathbb{E}Z \geq t\mathbb{E}Z) \leq e^{-\frac{1}{2}np(1+t)\log(1+t)}. \tag{2.12}$$

The proof of Theorem 2.1 relies on the following elementary lemma that again we state without proof.

Lemma 2.2.

(i) *For any sequence $R_N \in [0,1)$*

$$|U_{N,\ell}| \geq 2^{N\ell}\left(1 - 2\binom{\ell}{2} e^{-\frac{1}{8}NR_N^2}\right). \tag{2.13}$$

(ii) *Suppose R_N satisfies $NR_N^2 \to \infty$ and $R_N = o(1)$. Then for some $C > 0$ depending only on ℓ,*

$$|V_{N,\ell}| = 2^{N\ell}\frac{C}{\sqrt{N}} e^{-N\mathcal{J}(R_N)}(1+o(1)). \tag{2.14}$$

As an elementary consequence of part (i) of Lemma 2.2 one can prove that:

Corollary 2.3. P*-a.s.* $\max_{\sigma,\sigma' \in X} |R(\sigma,\sigma')| \leq \delta_N$, *where* $\delta_N \equiv 4\sqrt{\frac{M(N)\log 2}{N} + \frac{\log N}{N}}$.

The proof of part (i) of Theorem 2.1 then proceeds as follows. Let us first express the size of the random cloud $|X|$ as a sum of i.i.d. random variables

$$|X| = \sum_{\sigma \in S_N} \mathbb{1}_{X_\sigma = 1}. \tag{2.15}$$

Using (2.11) and the assumption that $\log N = o(2^M)$, we see that P-almost surely $|X|$ is given by its expected value, i.e., $|X| = 2^M(1 + o(1))$. Therefore $|X^\ell| = 2^{M\ell}(1 + o(1))$.

Since $U^X_{N,\ell} \subset X^\ell$ and $\mathrm{E}|U^X_{N,\ell}| = p^\ell_N |U_{N,\ell}| = 2^{M\ell}(1 + o(1))$, then proving part (i) of Theorem 2.1 is equivalent to proving that the set $U^X_{N,\ell}$ coincides, up to an error of magnitude $o(2^{M\ell})$, with the set X^ℓ. Let us rewrite $U^X_{N,\ell}$ as

$$U^X_{N,\ell} = \bigcap_{1\leq j<j'\leq \ell} (U^X_{N,\ell})_{jj'}, \tag{2.16}$$

where we defined

$$(U^X_N)_{jj'} = \Big\{(\sigma^1, \dots, \sigma^\ell) \in X^\ell : \ |R(\sigma^j, \sigma^{j'})| \leq R_N\Big\}. \tag{2.17}$$

If we prove that every set $(U^X_{N,\ell})_{jj'}$ coincides, up to an error of order $o(2^{M\ell})$, with the set X^ℓ, then the representation (2.16) implies part (i) of the theorem. We therefore concentrate on proving that

$$|(U^X_{N,\ell})_{jj'}| = \mathrm{E}|(U_{N,\ell})^X_{jj'}|(1 + o(1)), \quad \text{P-a.s.} \tag{2.18}$$

Without loss of generality we can consider the case of $j = 1$ and $j' = 2$. By definition of $(U^X_{N,\ell})_{jj'}$ we get

$$\begin{aligned} |(U^X_{N,\ell})_{12}| &= \Bigg(\sum_{\substack{\sigma^1,\sigma^2\in S_N: \\ |R(\sigma^1,\sigma^2)|\leq R_N}} \mathbf{1}_{X_{\sigma^1}=1}\mathbf{1}_{X_{\sigma^2}=1}\Bigg)\Bigg(\sum_{\sigma\in S_N} \mathbf{1}_{X_\sigma=1}\Bigg)^{\ell-2} \\ &= \Bigg(\sum_{\sigma^1\in S_N} \mathbf{1}_{X_{\sigma^1}=1} \sum_{\sigma^2:|R(\sigma^1,\sigma^2)|\leq R_N} \mathbf{1}_{X_{\sigma^2}=1}\Bigg)\Bigg(\sum_{\sigma\in S_N} \mathbf{1}_{X_\sigma=1}\Bigg)^{\ell-2}. \end{aligned} \tag{2.19}$$

As we already noted, the sum in the second factor of (2.19) concentrates on its expected value and is equal to $2^{M(\ell-2)}(1+o(1))$. Let us thus turn to the first factor in (2.19).

Introduce the set $(U^X_{N,\ell})_{\sigma^1} = \{\sigma^2 \in X : |R(\sigma^1, \sigma^2)| \leq R_N\}$. Then

$$|(U^X_{N,\ell})_{\sigma^1}| = \sum_{\sigma^2:|R(\sigma^1,\sigma^2)|\leq R_N} \mathbf{1}_{X_{\sigma^2}=1} \tag{2.20}$$

The summands in this sum are i.i.d. and it follows from part (i) of Lemma 2.2 that their number is at least $2^N(1-2e^{-\frac{1}{8}NR_N^2}) = 2^N(1+o(1))$. Applying (2.11) together with the assumption that $\log N = o(2^M)$ we obtain from the Borel-Cantelli Lemma that P-a.s., for any $\sigma^1 \in S_N$,

$$|(U^X_{N,\ell})_{\sigma^1}| = (1 + o(1))\mathrm{E}|(U^X_{N,\ell})_{\sigma^1}|. \tag{2.21}$$

From (2.19), (2.20) and (2.21) we immediately conclude that

$$|(U^X_{N,\ell})_{12}| = (1 + o(1))\mathrm{E}|(U^X_{N,\ell})_{12}|, \quad \text{P-a.s.} \tag{2.22}$$

This finishes the proof of part (i) of Theorem 2.1.

The proof of part (ii) is quite similar to the proof of part (i). By definition of $V^X_{N,\ell}$ we get

$$V^X_{N,\ell} \subseteq \bigcup_{1\le j<j'\le \ell} (V^X_{N,\ell})_{jj'}, \tag{2.23}$$

where

$$(V^X_{N,\ell})_{jj'} = \Big\{(\sigma^1,\dots,\sigma^\ell) \in V^X_{N,\ell} : \ |R(\sigma^j,\sigma^{j'})| = R_N\Big\}. \tag{2.24}$$

We claim that it suffices to prove that P-almost surely

$$|(V^X_{N,\ell})_{jj'}| \le e^{\alpha N \mathcal{J}(R_N)} \mathrm{E}|(V^X_{N,\ell})_{jj'}|. \tag{2.25}$$

Indeed, from (2.23) and from the above inequality we obtain that

$$|V^X_{N,\ell}| \le \sum_{1\le j<j'\le \ell} |(V^X_{N,\ell})_{jj'}| \le \sum_{1\le j<j'\le \ell} e^{\alpha N \mathcal{J}(R_N)} \mathrm{E}|(V^X_{N,\ell})_{jj'}|. \tag{2.26}$$

Using part (i) of Lemma 2.2 it is easy to establish that for all $1 \le j < j' \le \ell$

$$|V_{N,\ell}| = |(V_{N,\ell})_{jj'}|\big(1+o(1)\big), \tag{2.27}$$

and therefore

$$\mathrm{E}|V^X_{N,\ell}| = \mathrm{E}|(V^X_{N,\ell})_{jj'}|(1+o(1)). \tag{2.28}$$

Since (2.26) and (2.28) imply the result we concentrate on the proof of (2.25).

Without loss of generality we can take $j = 1$ and $j' = 2$. Then, by definition of $(V^X_{N,\ell})_{jj'}$, we get

$$\begin{aligned} |(V^X_{N,\ell})_{12}| &= \Bigg(\sum_{\substack{\sigma^1,\sigma^2\in S_N:\\ |R(\sigma^1,\sigma^2)|=R_N}} \mathbf{1}_{X_{\sigma^1}=1}\mathbf{1}_{X_{\sigma^2}=1}\Bigg)\Bigg(\sum_{\sigma\in S_N}\mathbf{1}_{X_\sigma=1}\Bigg)^{\ell-2} \\ &= \Bigg(\sum_{\sigma^1\in S_N}\mathbf{1}_{X_{\sigma^1}=1}\sum_{\sigma^2:|R(\sigma^1,\sigma^2)|=R_N}\mathbf{1}_{X_{\sigma^2}=1}\Bigg)\Bigg(\sum_{\sigma\in S_N}\mathbf{1}_{X_\sigma=1}\Bigg)^{\ell-2}. \end{aligned} \tag{2.29}$$

As in the proof of part (i) we see that the second part of (2.29) concentrates on its expected value and equals to $2^{M(\ell-2)}(1+o(1))$. We are thus left to treat the first part. Introducing the set

$$(V^X_{N,\ell})_{\sigma^1} = \{\sigma^2 \in X : |R(\sigma^1,\sigma^2)| = R_N\} \tag{2.30}$$

it is clear that

$$|(V^X_{N,\ell})_{\sigma^1}| = \sum_{\sigma^2:|R(\sigma^1,\sigma^2)|=R_N} \mathbf{1}_{X_{\sigma^2}=1} \tag{2.31}$$

There are $2\binom{N}{d_N}$ i.i.d. terms in the above sum. Applying (2.12) with $t+1 = e^{\alpha N \mathcal{J}(R_N)}$ where $\alpha \in [0,1)$ will be chosen later, we obtain

$$\mathrm{P}(|(V^X_{N,\ell})_{\sigma^1}| \ge e^{\alpha N \mathcal{J}(R_N)}\mathrm{E}|(V^X_{N,\ell})_{\sigma^1}|) \le e^{-\frac{1}{2}\mathrm{E}|(V^X_{N,\ell})_{\sigma^1}|(1+t)\log(1+t)}. \tag{2.32}$$

From Stirling's formula we see that for some $C > 0$,

$$\mathrm{E}|(V_{N,\ell}^X)_{\sigma^1}| = 2p_N \binom{N}{d_N} = 2^M \frac{C}{\sqrt{N}} e^{-N\mathcal{J}(R_N)}(1+o(1)), \tag{2.33}$$

and thus the exponent in (2.32) is

$$\begin{aligned}\frac{1}{2}\mathrm{E}|(V_{N,\ell}^X)_{\sigma^1}|(1+t)\log(1+t) &= \frac{C}{2\sqrt{N}} 2^M e^{-(1-\alpha)N\mathcal{J}(R_N)} \alpha N\mathcal{J}(R_N)(1+o(1)) \\ &= \frac{\alpha C}{2} e^{M\log 2-(1-\alpha)N\mathcal{J}(R_N)-\frac{1}{2}\log N} N\mathcal{J}(R_N)(1+o(1)). \end{aligned} \tag{2.34}$$

If for every $\sigma^1 \in S_N$ the set $(V_{N,\ell}^X)_{\sigma^1}$ is empty then there is nothing to prove. Otherwise, by Corollary 2.3 we obtain that P-almost surely $R_N < \delta_N$, and since $\mathcal{J}(x) = \frac{1}{2}x^2(1+O(x^2))$ near the origin we can choose $\alpha \in [0,1)$ in such a way that

$$M\log 2 - (1-\alpha)N\mathcal{J}(R_N) - \frac{1}{2}\log N > \gamma \log N \tag{2.35}$$

for some $\gamma > 0$. Hence $\sum_N e^{-\frac{1}{2}\mathrm{E}|(V_{N,\ell}^X)_{\sigma^1}|(1+t)\log(1+t)} < \infty$, and we obtain from (2.32) and the Borel-Cantelli Lemma that P-almost surely

$$|(V_{N,\ell}^X)_{\sigma^1}| \le e^{\alpha N\mathcal{J}(R_N)}\mathrm{E}|(V_{N,\ell}^X)_{\sigma^1}|. \tag{2.36}$$

It is easy to see that (2.36) and (2.29) imply (2.25). This concludes the proof of Theorem 2.1. □

In part (ii) of Theorem 2.1 we studied the properties of the sets $V_{N,\ell}^X \subset X^\ell$ for arbitrary $\ell \ge 2$. For $\ell = 2$ we can improve on Theorem 2.1.

Theorem 2.4. *Suppose that* $\limsup \frac{M}{N} < 1$.

(i) *If for some* $c_1 > \frac{1}{2}$

$$\mathcal{J}(R_N) \le \frac{M\log 2}{N} - \frac{c_1 \log N}{N}, \tag{2.37}$$

then P*-almost surely*

$$|V_{N,2}^X| = (1+o(1))\mathrm{E}|V_{N,2}^X|. \tag{2.38}$$

(ii) *If for positive constants* c_1, c_2

$$\frac{M\log 2}{N} - \frac{c_1\log N}{N} \le \mathcal{J}(R_N) \le \frac{M\log 2}{N} + \frac{c_2 \log N}{N} \tag{2.39}$$

then there is a constant c *such that* P*-almost surely*

$$|V_{N,2}^X| \le N^c \ \mathrm{E}|V_{N,2}^X|. \tag{2.40}$$

(iii) *If for some* $c_2 > \frac{3}{2}$

$$\mathcal{J}(R_N) > \frac{M\log 2}{N} + \frac{c_2\log 2}{N}, \tag{2.41}$$

then the set $V_{N,2}^X$ *is* P*-almost surely empty.*

Proof. (i) By definition of $V_{N,2}^X$,

$$|V_{N,2}^X| = \sum_{\sigma^1 \in S_N} \mathbf{1}_{X_{\sigma^1}=1} \sum_{\sigma^2 \in S_N : |R(\sigma^1,\sigma^2)|=R_N} \mathbf{1}_{X_{\sigma^2}=1}. \tag{2.42}$$

Using (2.30) the inner sum is $|(V_{N,2}^X)_{\sigma^1}|$. Since it is a sum of i.i.d. random variables then for all $t = t(N) = o(1)$ we get from (2.11) that

$$\mathrm{P}\left(||(V_{N,2}^X)_{\sigma^1}| - \mathrm{E}|(V_{N,2}^X)_{\sigma^1}|| \geq t\,\mathrm{E}|(V_{N,2}^X)_{\sigma^1}|\right) \leq 2e^{-t^2 \mathrm{E}|(V_{N,2}^X)_{\sigma^1}|/4}. \tag{2.43}$$

Using Stirling's approximation

$$\sqrt{2\pi} n^{n+\frac{1}{2}} e^{-n+\frac{1}{12n+1}} < n! < \sqrt{2\pi} n^{n+\frac{1}{2}} e^{-n+\frac{1}{12n}} \tag{2.44}$$

we obtain that

$$\mathrm{E}|(V_{N,2}^X)_{\sigma^1}| = 2p_N \binom{N}{d_N} = O\Big(\frac{e^{M\log 2 - N\mathcal{J}(R_N)}}{\sqrt{N(1-R_N^2)}}\Big). \tag{2.45}$$

Further, from (2.37) and Corollary 2.3 we obtain that $\mathrm{E}|(V_{N,2}^X)_{\sigma^1}| \geq CN^{c_1-1/2}$ for some positive constant $C > 0$. Choosing $t = N^{-\gamma}$ for some small enough $\gamma > 0$ in (2.43), we conclude from the Borel-Cantelli Lemma that

$$|(V_{N,2}^X)_{\sigma^1}| = \mathrm{E}|(V_{N,2}^X)_{\sigma^1}|(1+o(1)), \quad \text{P-a.s.} \tag{2.46}$$

Using (2.42), (2.46), and the fact that P-almost surely there are $2^M(1+o(1))$ configurations in the random cloud X proves (i).

(ii) The proof is similar to the proof of part (i). In particular, from the representation (2.42) it is easy to see that it suffices to prove that

$$|(V_{N,2}^X)_{\sigma^1}| \leq N^c\, \mathrm{E}|(V_{N,2}^X)_{\sigma^1}|, \quad \text{P-a.s.} \tag{2.47}$$

Choosing $1 + t = N^c$ we get from (2.45) that for some positive constant C

$$\frac{1}{2} E|(V_{N,2}^X)_{\sigma^1}|(1+t)\log(1+t) \geq CN^{c-c_2-1/2}\log N. \tag{2.48}$$

Choosing c large enough and applying (2.12) together with the Borel-Cantelli Lemma proves part (ii).

(iii) We again use the representation (2.42). Clearly it is enough to prove that for all $\sigma^1 \in X$ the set $(V_{N,2}^X)_{\sigma^1}$ is P-almost surely empty. By (2.45) $\mathrm{E}|(V_{N,2}^X)_{\sigma^1}| \leq CN^{-c_2-1/2}$ for some positive constant C. Thus, choosing $1 + t = \frac{1}{E|(V_{N,2}^X)_{\sigma^1}|} \to \infty$ we obtain from (2.12)

$$\mathrm{P}\left(|(V_{N,2}^X)_{\sigma^1}| \geq 1\right) = \mathrm{P}\left(|(V_{N,2}^X)_{\sigma^1}| \geq (1+t)\mathrm{E}|(V_{N,2}^X)_{\sigma^1}|\right) \leq (t+1)^{-1/2}. \tag{2.49}$$

By definition of t and by condition (2.41) we get

$$(t+1)^{-1/2} \leq C^{1/2} N^{-c_2/2-1/4}. \tag{2.50}$$

Applying the Borel-Cantelli Lemma proves (iii). □

In order to estimate the third moment in Theorems 4.1, 4.2, and 4.7 we give a result similar to Theorem 2.4 but for $\ell = 3$, i.e., for a given sequence of vectors $(R_{12}^N, R_{23}^N, R_{31}^N)$ we estimate the cardinal of the set

$$W_{N,3}^X = \Big\{(\sigma^1,\sigma^2,\sigma^3) \in X^3 : R(\sigma^1,\sigma^2) = R_{12}^N, R(\sigma^2,\sigma^3) = R_{23}^N, R(\sigma^3,\sigma^1) = R_{31}^N\Big\}. \tag{2.51}$$

Below we omit the explicit dependence of the sequence $(R_{12}^N, R_{23}^N, R_{31}^N)$ on N and instead of R_{12}^N, R_{23}^N and R_{31}^N will write R_{12}, R_{23} and R_{31} respectively. In order to formulate the theorem we introduce the following function on $\mathbb{R}^3$:

$$\begin{aligned}\mathcal{J}^{(2)}(x,y,z) = {} & \frac{1+x+y+z}{4}\log(1+x+y+z) + \frac{1+x-y-z}{4}\log(1+x-y-z) \\ & + \frac{1-x+y-z}{4}\log(1-x+y-z) + \frac{1-x-y+z}{4}\log(1-x-y+z) \end{aligned} \tag{2.52}$$

if $1 + x \geq |y+z|$ and $1 - x \geq |y-z|$ and $\mathcal{J}^{(2)}(x,y,z) = +\infty$ otherwise.

Theorem 2.5. *Suppose* $\limsup \frac{M}{N} < 1$.

(i) *If for some* $c_1 > \frac{1}{2}$ *the sequence* R_{12} *satisfies* (2.37), *and if for some* $c_1^{(2)} > 1$

$$\mathcal{J}^{(2)}(R_{12}, R_{23}, R_{31}) \leq \frac{M \log 2}{N} + \mathcal{J}(R_{12}) - \frac{c_1^{(2)} \log N}{N}, \tag{2.53}$$

then P*-almost surely*

$$|W_{N,3}^X| = (1 + o(1))\mathrm{E}|W_{N,3}^X|. \tag{2.54}$$

(ii) *If for positive constants* $c_1^{(2)}, c_2^{(2)}$,

$$\begin{aligned}\frac{M \log 2}{N} + \mathcal{J}(R_{12}) - \frac{c_1^{(2)} \log N}{N} & \leq \mathcal{J}^{(2)}(R_{12}, R_{23}, R_{31}) \\ & \leq \frac{M \log 2}{N} + \mathcal{J}(R_{12}) + \frac{c_2^{(2)} \log N}{N},\end{aligned} \tag{2.55}$$

then there is a constant c *such that* P*-almost surely*

$$|W_{N,3}^X| \leq N^c \mathrm{E}|W_{N,3}^X|. \tag{2.56}$$

(iii) *If for some* $c_2^{(2)} > \frac{3}{2}$

$$\mathcal{J}^{(2)}(R_{12}, R_{23}, R_{31}) > \frac{M \log 2}{N} + \mathcal{J}(R_{12}) + \frac{c_2^{(2)} \log 2}{N}, \tag{2.57}$$

then P*-almost surely the set* $W_{N,3}^X$ *is empty.*

Proof. From Lemma 2.6 stated below it follows that, for arbitrary configurations $\sigma^1, \sigma^2, \sigma^3 \in S_N$, the function $\mathcal{J}^{(2)}(R_{12}, R_{23}, R_{31})$ is well defined. This lemma, whose proof we will omit, is a direct consequence of the fact that the Hamming distance on S_N satisfies triangle inequality.

Lemma 2.6. *For arbitrary configurations* σ^1, σ^2 *and* $\sigma^3 \in S_N$ *we have* $1 + R_{12} \geq |R_{23} + R_{31}|$ *and* $1 - R_{12} \geq |R_{23} - R_{31}|$.

The proof of the Theorem 2.5 is similar to that of Theorem 2.4. We begin by writing the size of the set $W^X_{N,3}$ as

$$|W^X_{N,3}| = \sum_{\sigma^1\in S_N} \mathbf{1}_{X_{\sigma^1}=1} \sum_{\substack{\sigma^2\in S_N:\\ R(\sigma^1,\sigma^2)=R_{12}}} \mathbf{1}_{X_{\sigma^2}=1} \sum_{\substack{\sigma^3\in S_N:\\ R(\sigma^2,\sigma^3)=R_{23}\\ R(\sigma^1,\sigma^3)=R_{31}}} \mathbf{1}_{X_{\sigma^3}=1} \tag{2.58}$$

Let us first estimate the number of terms in the last sum. This means that, given $\sigma^1, \sigma^2 \in S_N$ with overlap $R(\sigma^1,\sigma^2) = R_{12}$, we have to calculate the number of configurations σ^3 with $R(\sigma^2,\sigma^3) = R_{23}$ and $R(\sigma^3,\sigma^1) = R_{31}$. Without loss of generality we can assume that all the spins of σ^1 are equal to 1. Further, let $C(\sigma^1,\sigma^2,\sigma^3)$ be a $3\times N$ matrix with rows $\sigma^1,\sigma^2,\sigma^3$. For a column vector $\boldsymbol{\delta} \in \{-1,1\}^3$ we let $n_{\boldsymbol{\delta}}$ be the number of columns of the matrix C that are equal to $\boldsymbol{\delta}$, i.e.,

$$n_{\boldsymbol{\delta}} = \left|\{j \le N : (\sigma^1_j, \sigma^2_j, \sigma^3_j) = \boldsymbol{\delta}\}\right|. \tag{2.59}$$

Then the overlaps can be written in terms of $n_{\boldsymbol{\delta}}$, namely

$$\begin{cases} n_{(1,1,1)} + n_{(1,1,-1)} - n_{(1,-1,1)} - n_{(1,-1,-1)} = NR_{12},\\ n_{(1,1,1)} - n_{(1,1,-1)} - n_{(1,-1,1)} + n_{(1,-1,-1)} = NR_{23},\\ n_{(1,1,1)} - n_{(1,1,-1)} + n_{(1,-1,1)} - n_{(1,-1,-1)} = NR_{31},\\ n_{(1,1,1)} + n_{(1,1,-1)} + n_{(1,-1,1)} + n_{(1,-1,-1)} = N. \end{cases} \tag{2.60}$$

Solving this system of linear equations we find

$$\begin{cases} n_{(1,1,1)} &= \frac14 N(1+R_{12}+R_{23}+R_{31}),\\ n_{(1,1,-1)} &= \frac14 N(1+R_{12}-R_{23}-R_{31}),\\ n_{(1,-1,1)} &= \frac14 N(1-R_{12}-R_{23}+R_{31}),\\ n_{(1,-1,-1)} &= \frac14 N(1-R_{12}+R_{23}-R_{31}). \end{cases} \tag{2.61}$$

Specifying the configuration σ^3 is equivalent to specifying a partition of the index set $\{1,2,\dots,N\}$ into subsets of seizes $n_{(1,1,1)}, n_{(1,1,-1)}, n_{(1,-1,1)}$, and $n_{1,-1,-1}$. Therefore the number of configurations $\sigma^3 \in S_N$ with overlaps $R(\sigma^2,\sigma^3) = R_{23}$, and $R(\sigma^3,\sigma^1) = R_{31}$ is

$$\begin{aligned} &\binom{n_{(1,1,1)} + n_{(1,1,-1)}}{n_{(1,1,1)}}\binom{n_{(1,-1,1)} + n_{(1,-1,-1)}}{n_{(1,-1,1)}}\\ &\quad = \frac{(n_{(1,1,1)} + n_{(1,1,-1)})!\,(n_{(1,-1,1)} + n_{(1,-1,-1)})!}{n_{(1,1,1)}!\,n_{(1,1,-1)}! \quad n_{(1,-1,1)}!\,n_{(1,-1,-1)}!}. \end{aligned} \tag{2.62}$$

Applying Stirling's approximation to (2.62) one obtains that the number of terms in the last summation in (2.58) is of order

$$\frac{2^N e^{N\mathcal{J}(R_{12}) - N\mathcal{J}^{(2)}(R_{12},R_{23},R_{31})}\sqrt{1-R_{12}^2}}{N\sqrt{P(R_{12},R_{23},R_{31})}},$$

where $P(x,y,z) = (1+x+y+z)(1+x-y-z)(1-x+y-z)(1-x-y+z)$. The rest of the proof essentially is a rerun of the proof of Theorem 2.4. We skip the details. □

3. Proof of Theorem 1.1

As in [BK06a] and [BCMN05a], [BCMN05b], the proof of the Poisson convergence is based on the analysis of factorial moments of the point processes $\mathcal{P}_N$ defined in (1.12).

In general, let ξ_N be a sequence of point processes defined on a common probability space $(\Omega, \mathcal{F}, \mathbb{Q})$ and let ξ be a Poisson point process with intensity measure μ. Define the ℓth factorial moment $\mathbb{E}(Z)_\ell$ of the random variable Z to be $\mathbb{E}Z(Z-1)\dots(Z-\ell+1)$. The following is a classical lemma that is a direct consequence of Theorem 4.7 in [Kal83].

Lemma 3.1. *If for every $\ell \geq 1$ and every Borel set A*

$$\lim_{N\to\infty} \mathbb{E}_{\mathbb{Q}}(\xi_N(A))_\ell = (\mu(A))^\ell, \tag{3.1}$$

then the distribution of $(\xi_N)_{N\geq 1}$ converges weakly to the distribution of ξ.

Applying Lemma 3.1 to the sequence of point processes $\mathcal{P}_N = \sum_{\sigma\in X} \delta_{H'_N(\sigma)}$ the following result proves Theorem 1.1.

Theorem 3.2. *Under the assumptions of Theorem 1.1, for every $\ell \in \mathbb{N}$ and every bounded Borel set A*

$$\lim_{N\to\infty} \mathbb{E}(\mathcal{P}_N(A))_\ell = (\mu(A))^\ell, \quad \text{P-}a.s., \tag{3.2}$$

where μ is defined in (1.13).

Proof. We start with the computation of the first moment of $\mathcal{P}_N(A)$:

$$\mathbb{E}\,\mathcal{P}_N(A) = \sum_{\sigma\in X} \mathbb{P}(H'_N(\sigma) \in A). \tag{3.3}$$

As we saw in the proof of Theorem 2.1 the size of the random cloud $|X|$ is P-almost surely $2^M(1+o(1))$. Since $H'_N(\sigma), \sigma \in S_N$, are identically distributed normal random variables with mean $-a_N/b_N$ and variance $1/b_N$, the sum in (3.3) can be written as

$$2^M \frac{e^{-a_N^2/2}}{\sqrt{2\pi}} b_N \int_A e^{-x^2 b_N^2/2 - a_N b_N x} dx\,(1+o(1)), \quad P\text{-a.s.} \tag{3.4}$$

By the dominated convergence theorem and the definition of a_N it follows from (3.4) that the limit of the first moment is $\mu(A)$.

To calculate factorial moments of higher order we follow [BCMN05a] and rewrite the ℓth factorial moment of $\mathcal{P}_N(A)$ as

$$\mathbb{E}(\mathcal{P}_N(A))_\ell = \sum_{\sigma^1,\dots,\sigma^\ell} \mathbb{P}\big(H'_N(\sigma^1) \in A, \dots, H'_N(\sigma^\ell) \in A\big) \tag{3.5}$$

where the sum runs over all ordered sequences of distinct configurations $\sigma^1,\dots,\sigma^\ell$ belonging to X. To analyze it we decompose the set S_N^ℓ into three non-intersecting subsets

$$S_N^\ell = U_\ell(R_N) \cup \big(U_\ell(\delta_N)\backslash U_\ell(R_N)\big) \cup (U_\ell(\delta_N))^c, \tag{3.6}$$

where U_ℓ is defined in (2.2), δ_N is defined in Corollary 2.3 and the sequence R_N is chosen such that:

$$R_N \to 0, M\nu(R_N) \to 0, NR_N^2 \to \infty. \tag{3.7}$$

This is possible since we have assumed that $M = o(\sqrt{N})$ if $\nu'(0) \neq 0$ and $M = o(N)$ if $\nu'(0) = 0$. We recall here that the function ν is defined in (1.7)–(1.8).

Having specified R_N, let us analyze the contribution to the sum (3.5) coming from the intersection of X^ℓ with the sets $U_\ell(R_N), U_\ell(\delta_N)\backslash U_\ell(R_N)$, and $(U_\ell(\delta_N))^c$. Firstly, by Corollary 2.3 the intersection of the set $(U_\ell(\delta_N))^c$ with X^ℓ is P-a.s. empty and therefore its contribution to the sum (3.5) is zero. Next, let us show that P-a.s.

$$\lim_{N\to\infty} \sum_{\sigma^1,\dots,\sigma^\ell} \mathbb{P}\left(H'_N(\sigma^1) \in A, \dots, H'_N(\sigma^\ell) \in A\right) = (\mu(A))^\ell, \tag{3.8}$$

where the sum is over all the sequences $(\sigma^1, \dots, \sigma^\ell) \in U_\ell(R_N) \cap X^\ell$.

For every $\ell \in \mathbb{N}$ let $B(\sigma^1, \dots, \sigma^\ell)$ denote the covariance matrix of random variables $H_N(\sigma^1), \dots, H_N(\sigma^\ell)$. By (1.7) its elements, b_{ij}, are given by

$$b_{ij} = \nu(R_{ij}) \tag{3.9}$$

where we wrote $R(\sigma^i, \sigma^j) = R_{ij}$. Since $R_N = o(1)$ the matrix $B(\sigma^1, \dots, \sigma^\ell)$ is non-degenerate. We therefore get for $(\sigma^1, \dots, \sigma^\ell) \in U_\ell(R_N) \cap X^\ell$ that

$$\begin{aligned}\mathbb{P}\big(H'_N(\sigma^1) \in A, \dots, H'_N(\sigma^\ell) \in A\big) &= \frac{b_N^\ell}{(2\pi)^{\ell/2}\sqrt{\det B}} \\ &\times \int_A \cdots \int_A e^{-(\vec{x}, B^{-1}\vec{x})b_N^2/2 - a_N b_N(\vec{x}, B^{-1}\vec{1}) - a_N^2(\vec{1}, B^{-1}\vec{1})/2} d\vec{x},\end{aligned} \tag{3.10}$$

where $B = B(\sigma^1, \dots, \sigma^\ell), \vec{x} = (x_1, \dots, x_\ell)^t$ and $\vec{1} = (1, \dots, 1)^t$. From the definition of a_N we further get from (3.10) that

$$\begin{aligned}\mathbb{P}\big(H'_N(\sigma^1) \in A, \dots, H'_N(\sigma^\ell) \in A\big) &= \frac{1}{2^{M\ell}} \frac{e^{(\ell - (\vec{1}, B^{-1}\vec{1}))(M \log 2 - 1/2 \log M)}}{\sqrt{\det B}} \\ &\times \frac{1}{(\sqrt{\pi})^\ell} \int_A \cdots \int_A e^{-(\vec{x}, B^{-1}\vec{x})b_N^2/2 - a_N b_N(\vec{x}, B^{-1}\vec{1})} d\vec{x}.\end{aligned} \tag{3.11}$$

Since the matrix B^{-1} is positive definite and since $a_N b_N \to \sqrt{2 \log 2}$ we conclude from the dominated convergence theorem that for all bounded Borel sets

A, uniformly in $(\sigma^1,\dots,\sigma^\ell) \in U_\ell(R_N) \cap X^\ell$,

$$\frac{1}{(\sqrt{\pi})^\ell} \int_A \cdots \int_A e^{-(\vec{x},B^{-1}\vec{x})b_N^2/2 - a_N b_N (\vec{x},B^{-1}\vec{1})} d\vec{x}$$
$$\to \frac{1}{(\sqrt{\pi})^\ell} \int_A \cdots \int_A e^{-\sqrt{2\log 2}(\vec{x},\vec{1})} d\vec{x} = (\mu(A))^\ell. \tag{3.12}$$

To evaluate term $e^{(\ell-(\vec{1},B^{-1}\vec{1}))(M\log 2 - 1/2 \log M)}$ in (3.11) we look at $(\ell - (\vec{1},B^{-1}\vec{1})) \det B$ as a multivariate function of $(b_{ij})_{1\le i<j\le \ell}$. It is a polynomial of degree ℓ with coefficients depending only on ℓ and without constant term. It implies that $|\ell - (\vec{1},B^{-1}\vec{1})| = O(\nu(R_N))$ and therefore, by (3.7),

$$e^{(\ell-(\vec{1},B^{-1}\vec{1}))(M\log 2 - 1/2 \log M)} = 1 + o(1). \tag{3.13}$$

Combining (3.11), (3.12), and (3.13) we may rewrite the sum (3.8) as

$$\sum_{\sigma^1,\dots,\sigma^\ell} \frac{1}{2^{M\ell}} (\mu(A))^\ell (1+o(1)) = \frac{|U_\ell(R_N) \cap X^\ell|}{2^{M\ell}} (\mu(A))^\ell (1+o(1)). \tag{3.14}$$

Now it follows from Theorem 2.1 (i) and Lemma 2.2 (i) that $|U_\ell(R_N) \cap X^\ell|$ concentrates around its expected value, namely $2^{M\ell}(1+o(1))$, so that (3.14) implies (3.8).

Next, let us establish that the contribution from the second set in (3.6) is negligible, i.e., let us prove that

$$\lim_{N\to\infty} \sum_{\sigma^1,\dots,\sigma^\ell} \mathbb{P}\left(H'_N(\sigma^1) \in A, \dots, H'_N(\sigma^\ell) \in A\right) = 0, \tag{3.15}$$

where the sum runs over all the sequences $(\sigma^1,\dots,\sigma^\ell) \in \big(U_\ell(\delta_N)\backslash U_\ell(R_N)\big) \cap X^\ell$. To do this we first bound the right-hand side of (3.11). By definition (2.1) and Corollary 2.3, $R_{\max}(\sigma^1,\dots,\sigma^\ell) \le \delta_N = o(1)$. Therefore following the same reasoning as above we obtain from (3.11) that for some constant $C > 0$ and for all $(\sigma^1,\dots,\sigma^\ell) \in \big(U_\ell(\delta_N)\backslash U_\ell(R_N)\big) \cap X^\ell$

$$\mathbb{P}\big(H'_N(\sigma^1) \in A, \dots, H'_N(\sigma^\ell) \in A\big) \le \frac{e^{CM\nu(R_{\max})}}{2^{M\ell}} (\mu(A))^\ell (1+o(1)). \tag{3.16}$$

For fixed N the overlap takes only a discrete set of values

$$K_N = \Big\{1 - \frac{2k}{N} : k = 0, 1, \dots, N\Big\}. \tag{3.17}$$

We represent the set $U_\ell(\delta_N)\backslash U_\ell(R_N)$ as a union of sets $V_\ell(R_{N,k})$, where we denoted $R_{N,k} = 1 - \frac{2k}{N} \in K_N \cap (R_N, \delta_N)$. Let us fix k and bound the contribution from the set $V_\ell(R_{N,k}) \cap X^\ell$, i.e.,

$$\sum_{\substack{(\sigma^1,\dots,\sigma^\ell)\in \\ V_\ell(R_{N,k})\cap X^\ell}} \mathbb{P}\left(H'_N(\sigma^1) \in A, \dots, H'_N(\sigma^\ell) \in A\right). \tag{3.18}$$

We obtain from Lemma 2.2 (ii) that

$$|V_\ell(R_{N,k})| = 2^{N\ell}\frac{C}{\sqrt{N}}e^{-N\mathcal{J}(R_{N,k})}(1+o(1)). \tag{3.19}$$

In the case $M(N) \leq \log N$ we can choose the sequence R_N in such a way that the set $U_\ell(\delta_N)\backslash U_\ell(R_N)$ is empty. Therefore we can assume without loss of generality that $M(N) \geq \log N$. Applying part (ii) of Theorem 2.1 we further conclude that

$$|V_\ell(R_{N,k}) \cap X^\ell| \leq 2^{M\ell}\frac{C}{\sqrt{N}}e^{-(1-\alpha)N\mathcal{J}(R_{N,k})}, \quad \text{P-a.s.} \tag{3.20}$$

Using (3.16) and the last inequality we bound the sum (3.18) by

$$\frac{C}{\sqrt{N}}e^{-(1-\alpha)N\mathcal{J}(R_{N,k})}e^{CM\nu(R_{N,k})}. \tag{3.21}$$

One can easily check that $M\nu(R_{N,k}) = o(NR_{N,k}^2)$ for $R_{N,k} \in (R_N, \delta_N)$. Together with $\mathcal{J}(x) \geq x^2/2$ it implies that for some positive constants C_1 and C_2 we can further bound (3.21) by

$$\frac{C_1}{\sqrt{N}}e^{-C_2NR_{N,k}^2}. \tag{3.22}$$

As a consequence, we obtain an almost sure bound

$$\begin{aligned} \sum_{\substack{R_{N,k}\in\\ K_N\cap(R_N,\delta_N)}} \sum_{\substack{(\sigma^1,\ldots,\sigma^\ell)\in\\ V_\ell(R_{N,k})\cap X^\ell}} \mathbb{P}\left(H_N'(\sigma^1)\in A,\ldots,H_N'(\sigma^\ell)\in A\right) \\ \leq \frac{C_1}{\sqrt{N}} \sum_{\substack{R_{N,k}\in\\ K_N\cap(R_N,\delta_N)}} e^{-C_2NR_{N,k}^2}. \end{aligned} \tag{3.23}$$

Introducing new variables $y_{N,k} = \sqrt{N}R_{N,k}$ we rewrite the above sum as

$$\frac{C_1}{2}\sum\nolimits_{y_{N,k}} e^{-C_2y_{N,k}^2}\frac{2}{\sqrt{N}}, \tag{3.24}$$

where the summation is over the discrete set $\sqrt{N}K_N \cap (\sqrt{N}R_N, \sqrt{N}\delta_N)$. Since $NR_N^2 \to \infty$, then for arbitrary $C>0$ and for large N we can further bound this sum by

$$\frac{C_1}{2}\sum\nolimits_{y_{N,k}\geq C} e^{-C_2y_{N,k}^2}\frac{2}{\sqrt{N}}. \tag{3.25}$$

Interpreting the last sum as a sum of areas of nonintersecting rectangles with one side equal $e^{-C_2y_{N,k}}$ and the other $\frac{2}{\sqrt{N}}$ we bound it with the integral

$$\int_{C-\frac{2}{\sqrt{N}}}^{\infty} e^{-C_2y^2}dy \tag{3.26}$$

Since the constant C is arbitrary we get that (3.15) is $o(1)$. This finishes the proof of Theorem 3.2 and therefore of Theorem 1.1. □

4. Proof of Theorem 1.4

In order to prove the breakdown of universality in Theorem 1.4 we use a strategy similar to that used in [BCMN05b] to disprove the local REM conjecture for the number partitioning problem and for the Sherrington-Kirkpatrick model when the energy scales are too large. We prove that P-a.s. for every bounded Borel set A

1. the limit of the first factorial moment exists and equals $\mu(A)$;
2. the second factorial moment $\mathbb{E}(\mathcal{P}_N(A))_2$ does not converge to $(\mu(A))^2$;
3. the third moment is bounded.

These three facts immediately imply that the sequence of random variables $\mathcal{P}_N(A)$ does not converge weakly to a Poisson random variable and so the sequence of point processes $\mathcal{P}_N$ does not converge weakly to a Poisson point process. Part (i) of Theorem 1.4 is thus obviously implied by the following

Theorem 4.1 (Breakdown of Universality for the number partitioning problem). *Let $\nu(r) = r$. For every bounded Borel set A*

$$\lim \mathbb{E}(\mathcal{P}_N(A))_1 = \mu(A), \quad \text{P-}a.s. \tag{4.1}$$

Moreover, if $\limsup \frac{M(N)}{\sqrt{N}} = \varepsilon < \infty$ *then* P-*a.s.*

(i) $\limsup \mathbb{E}(\mathcal{P}_N(A))_2 = e^{2\varepsilon^2 \log^2 2}(\mu(A))^2$,
(ii) $\limsup \mathbb{E}(\mathcal{P}_N(A))_3 < \infty$.

Similarly, part (ii) of Theorem 1.4 is implied by the following

Theorem 4.2 (Breakdown of Universality for the Sherrington-Kirkpatrick model). *Let $\nu(r) = r^2$. For every bounded Borel set A*

$$\lim \mathbb{E}(\mathcal{P}_N(A))_1 = \mu(A), \quad \text{P-}a.s. \tag{4.2}$$

Moreover, if $\limsup \frac{M(N)}{N} = \varepsilon < \frac{1}{8\log 2}$ *then* P-*a.s.*

(i) $\limsup \mathbb{E}(\mathcal{P}_N(A))_2 = \frac{(\mu(A))^2}{\sqrt{1-4\varepsilon \log 2}}$;
(ii) $\limsup \mathbb{E}(\mathcal{P}_N(A))_3 < \infty$.

Remark 4.3. Condition $\varepsilon < \frac{1}{8\log 2}$ in not optimal and could be improved. The reason for such a choice is that for $\varepsilon < \frac{1}{8\log 2}$ the third moment estimate is quite simple.

We will prove in detail Theorem 4.2 but omit the proof of Theorem 4.1 as it is very similar and much simpler.

Proof of Theorem 4.2. We successively prove the statement on the first, second, and third moment.

1. **First moment estimate.** Since all the random variables $H'_N(\sigma), \sigma \in S_N$, are identically distributed then

$$\mathbb{E}(\mathcal{P}_N(A))_1 = |X|\mathbb{P}(H'_N(\sigma) \in A). \tag{4.3}$$

We saw in the proof of Theorem 2.1 that for $M(N)$ satisfying $\log N = o(2^M)$, $|X| = 2^M(1+o(1))$ P-a.s. Combined with (3.11), the definition of a_N, and the dominated convergence theorem this fact implies that P-a.s.

$$\mathbb{E}(\mathcal{P}_N(A))_1 = 2^M(1+o(1))\frac{1}{2^M}\frac{1}{\sqrt{\pi}}\int\limits_A e^{-x^2b_N^2/2-a_Nb_Nx}dx = \mu(A)(1+o(1)). \tag{4.4}$$

Hence (4.2) is proven.

2. **Second moment estimate.** Next assume that $\limsup \frac{M(N)}{N} = \varepsilon$. We now want to calculate $\limsup \mathbb{E}(\mathcal{P}_N(A))_2$. For this we rewrite the second factorial moment as

$$\mathbb{E}(\mathcal{P}_N(A))_2 = \sum_{\sigma^1,\sigma^2} \mathbb{P}\left(H'_N(\sigma^1)\in A, H'_N(\sigma^2)\in A\right), \tag{4.5}$$

where the summation is over all pairs of distinct configurations $(\sigma^1,\sigma^2)\in X^2$. We then split the set S_N^2 into four non-intersecting subsets and calculate the contributions from these subsets separately (these calculations are similar to those of Theorem 3.2).

(1) We begin by calculating the contribution from the set $\mathcal{S}_1^X$, where

$$\mathcal{S}_1 = \left\{(\sigma^1,\sigma^2)\in S_N^2 : |R(\sigma^1,\sigma^2)|\le \tau_N\right\} \tag{4.6}$$

and where the sequence τ_N is chosen in such a way that $N\tau_N^4 \to 0$ and $e^{-N\tau_N^2}$ decays faster than any polynomial: this can be achieved by choosing, e.g., $\tau_N = \frac{1}{N^{1/4}\log N}$.

First, we get from (3.11) for $\ell = 2$

$$\mathbb{P}\big(H'_N(\sigma^1)\in A, H'_N(\sigma^2)\in A\big) = \frac{1}{2^{2M}}\frac{e^{\frac{2b_{12}}{1+b_{12}}(M\log 2 - 1/2\log M)}}{\sqrt{1-b_{12}^2}}$$
$$\times \frac{1}{(\sqrt{\pi})^2}\int\limits_A\int\limits_A e^{-(\vec{x},B^{-1}\vec{x})b_N^2/2-a_Nb_N(\vec{x},B^{-1}\vec{1})}dx_1dx_2, \tag{4.7}$$

where $b_{12} = \mathrm{cov}(H'_N(\sigma^1), H'_N(\sigma^2)) = R_{12}^2$. If $(\sigma^1,\sigma^2)\in\mathcal{S}_1$ then $b_{12} = o(1)$ and we get from (3.12) that uniformly in $(\sigma^1,\sigma^2)\in\mathcal{S}_1$ the second line in (4.7) is just $(\mu(A))^2(1+o(1))$.

Next, we represent the set $\mathcal{S}_1$ as a union of sets $V_2(R_{N,k})$ with $R_{N,k} = 1-\frac{2k}{N}\in K_N\cap[0,\tau_N]$. Applying Theorem 2.4 (i) we get that

$$|V_2(R_{N,k})\cap X^2| = 2\sqrt{\frac{2}{\pi N}}2^{2M}e^{-N\mathcal{J}(R_{N,k})}(1+o(1)). \tag{4.8}$$

Therefore, up to a multiplicative term of the form $1 + o(1)$, the contribution from the set $\mathcal{S}_1^X$ to the sum (4.5) is

$$\sum_{R_{N,k}} \frac{1}{2^{2M}} e^{\frac{2b_{12}}{1+b_{12}}(M\log 2 - 1/2\log M)} (\mu(A))^2 2\sqrt{\frac{2}{\pi N}} 2^{2M} e^{-N\mathcal{J}(R_{N,k})}$$

$$= 2\sqrt{\frac{2}{\pi N}}(\mu(A))^2 \sum_{R_{N,k}} e^{-N\mathcal{J}(R_{N,k})} e^{\frac{2b_{12}}{1+b_{12}}(M\log 2 - 1/2\log M)}. \tag{4.9}$$

Since $b_{12}(R_{N,k}) = R_{N,k}^2 > 0$ the last sum is monotone in M. As we will see below due to this fact it is sufficient to calculate the upper limit of $\mathbb{E}(\mathcal{P}(A))_2$ for sequences of the form $M(N) = \varepsilon N$ with $\varepsilon \in (0, \frac{1}{8\log 2})$. Thus let $M = \varepsilon N, \varepsilon \in (0, \frac{1}{8\log 2})$. We obtain

$$\frac{2b_{12}}{1+b_{12}}(M\log 2 - 1/2\log M) = 2\varepsilon N R_{N,k}^2 \log 2 + O(NR_{N,k}^4). \tag{4.10}$$

The choice of τ_N guarantees that for $R_{N,k} \le \tau_N$ the last term in the right-hand side of (4.10) is of order $o(1)$. Moreover, for $x < 1$ we observe that $\mathcal{J}(x) = \frac{1}{2}x^2 + O(x^4)$. Thus

$$N\mathcal{J}(R_{N,k}) = \frac{NR_{N,k}^2}{2} + o(1). \tag{4.11}$$

Using (4.10) and (4.11) the sum (4.9) becomes

$$2\sqrt{\frac{2}{\pi N}}(\mu(A))^2 \sum_{R_{N,k}} e^{-\frac{1}{2}NR_{N,k}^2(1-4\varepsilon\log 2)}(1+o(1)), \tag{4.12}$$

where the summation is over $R_{N,k} \in K_N \cap [0, \tau_N]$. Introducing new variables $y_{N,k} = \sqrt{N} R_{N,k}$ we further rewrite (4.12) as

$$\sqrt{\frac{2}{\pi}} \sum_{y_{N,k}} \frac{2}{\sqrt{N}} e^{-\frac{y_N^2}{2}(1-4\varepsilon\log 2)}(1+o(1)). \tag{4.13}$$

It is not difficult to see that for $\varepsilon < \frac{1}{4\log 2}$ the sum in (4.13) converges to the integral

$$\int_0^\infty e^{-\frac{y^2}{2}(1-4\varepsilon\log 2)} dy = \frac{1}{\sqrt{1-4\varepsilon\log 2}}\sqrt{\frac{\pi}{2}}. \tag{4.14}$$

Therefore the contribution from the set S_1^X is $\frac{(\mu(A))^2}{\sqrt{1-4\varepsilon\log 2}}(1+o(1))$.

We can extend this result to the case when $M(N) = \varepsilon N + o(N)$ with $\varepsilon < \frac{1}{8\log 2}$. Indeed, assume that $\varepsilon_n \uparrow \varepsilon$ as $n \to \infty$. Then using the above calculation for $M = \varepsilon_n N$ together with the monotonicity argument we have that for all $n \ge 1$

$$\mathbb{E}(\mathcal{P}_N(A))_2 \ge \frac{(\mu(A))^2}{\sqrt{1-4\varepsilon_n \log 2}}. \tag{4.15}$$

Taking the limit in n we obtain a lower bound. With exactly the same argument we prove the corresponding upper bound.

(2) Next, we estimate the contribution from the set $\mathcal{S}_2^X$, where

$$\mathcal{S}_2 = \{(\sigma^1, \sigma^2) \in S_N^2 : |R(\sigma^1, \sigma^2)| > \tau_N \text{ and } R(\sigma^1, \sigma^2) \text{ satisfies (2.37)}\}. \quad (4.16)$$

Since the set A is bounded an elementary computation yields that for a constant $C = C(A)$, uniformly in $(\sigma^1, \sigma^2) \in \mathcal{S}_2$,

$$\int_A \int_A e^{-(\vec{x}, B^{-1}\vec{x})\, b_N^2/2 - a_N b_N (\vec{x}, B^{-1}\vec{1})} dx_1 dx_2 \leq C. \quad (4.17)$$

For fixed N we let R_N^1 to be the largest value of the overlap satisfying condition (2.37). Then representing $\mathcal{S}_2$ as a union of sets $V_2(R_{N,k})$, where $R_{N,k} = 1 - \frac{2k}{N} \in K_N \cap [\tau_N, R_N^1]$, and using Theorem 2.4 (i), we conclude that the contribution from the set $\mathcal{S}_2^X$ is, up to a multiplicative term of the form $1 + o(1)$, bounded by

$$\sum_{R_{N,k}} 2\sqrt{\frac{2}{\pi N(1 - R_{N,k}^2)}} \frac{C}{\sqrt{1 - b_{12}^2}} e^{-N\mathcal{J}(R_{N,k}) + \frac{2b_{12}}{1+b_{12}} M \log 2}. \quad (4.18)$$

This quantity is monotone in M. As a consequence, to show that it is negligible in the limit $N \to \infty$ it suffices to show this fact under the assumption $M(N) = \varepsilon N$. Thus, letting $M = \varepsilon N$ and using that $\mathcal{J}(x) \geq \frac{1}{2}x^2$, we can bound the exponent in (4.18) by

$$-N\left(\mathcal{J}(R_{N,k}) - \frac{2b_{12}}{1 + b_{12}}\varepsilon \log 2\right) \leq -\frac{1}{2}Nx^2(1 - 4\varepsilon \log 2). \quad (4.19)$$

Since $1 - b_{12}^2 \geq 1/N$ and since the number of terms in (4.18) is at most N, we can further bound it, for some positive constant $C > 0$, by

$$CN\, 2\sqrt{\frac{2}{\pi}} e^{-\frac{1}{2}N\tau_N^2(1-4\varepsilon \log 2)}. \quad (4.20)$$

Since $\varepsilon < \frac{1}{8 \log 2}$ the contribution from the set $\mathcal{S}_2^X$ is negligible by definition of τ_N.

(3) We now analyze the contribution from the set $\mathcal{S}_3^X$, where

$$\mathcal{S}_3 = \{(\sigma^1, \sigma^2) \in S_N^2 : R(\sigma^1, \sigma^2) \text{ satisfies (2.39)}\}. \quad (4.21)$$

Let R_N^2 be the largest overlap value satisfying condition (2.39). To bound the contribution from the set $\mathcal{S}_3$ we represent it as a union of sets $V_2(R_{N,k})$, where $R_{N,k} = 1 - \frac{2k}{N}$ runs over the set $K_N \cap [R_N^1, R_N^2]$. Proceeding as in (2) and using Theorem 2.4 (ii), we bound it by

$$\sum_{R_{N,k}} 2\sqrt{\frac{2}{\pi N}} N^c \frac{C}{\sqrt{1 - b_{12}^2}} e^{\frac{2b_{12}}{1+b_{12}} M \log 2 - N\mathcal{J}(R_{N,k})}. \quad (4.22)$$

Again, the sum is monotone in M and thus it is enough to bound it for $M = \varepsilon N$.

For $R_{N,k}$ satisfying condition (2.39) we can bound the exponent in (4.22) as follows:

$$\begin{aligned}\frac{2b_{12}}{1+b_{12}}\varepsilon N\log 2 - N\mathcal{J}(R_{N,k}) &\leq \frac{2b_{12}}{1+b_{12}}\varepsilon N\log 2 - N\varepsilon\log 2 - c_1\log N\\ &= -\frac{1-b_{12}}{1+b_{12}}\varepsilon N\log 2 - c_1\log N \leq -CN\varepsilon\log 2 - c_1\log N, \qquad (4.23)\end{aligned}$$

where C is some positive constant.

Since $1-b_{12}^2 \geq 1/N$ and since there are at most N terms in the sum (4.22), we can bound the latter by $N^{c+1-c_1}\exp\bigl(-CN\varepsilon\log 2\bigr)$. Therefore, P-almost surely the contribution from the set $\mathcal{S}_3^X$ is negligible as $N\to\infty$.

(4) To finish the second moment estimate it remains to treat the set

$$\mathcal{S}_4 = \bigl\{(\sigma^1,\sigma^2)\in S_N^2 : R(\sigma^1,\sigma^2) \text{ satisfies (2.41)}\bigr\}. \qquad (4.24)$$

But by Theorem 2.4 (iii), the set $\mathcal{S}_4^X$ is P-almost surely empty. This finishes the proof of assertion (i) of Theorem 4.2.

3. **Third moment estimate.** To analyze the third factorial moment we use that, by formula (3.5) and definition (2.51), it can be written as

$$\mathbb{E}\bigl(\mathcal{P}_N(A)\bigr)_3 = \sum_{R_{12},R_{23},R_{13}} \frac{|W_{N,3}^X|}{2^{3M}\sqrt{\det B}} e^{(3-(\vec{1},B^{-1}\vec{1}))(M\log 2-1/2\log M)}$$
$$\times \int_A\int_A\int_A e^{-(\vec{x},B^{-1}\vec{x})b_N^2/2-a_Nb_N(\vec{x},B^{-1}\vec{1})}d\vec{x}, \qquad (4.25)$$

where the summation runs over all triplets of overlaps $(R_{12},R_{23},R_{31})\in K_N^3$ and $B=B(\sigma^1,\sigma^2,\sigma^3)$, the covariance matrix of the vector $(H_N(\sigma^1),H_N(\sigma^2),H_N(\sigma^3))$, is the matrix with elements $b_{ij}=R_{ij}^2, 1\leq i,j\leq 3$. To estimate this sum we rely on three auxiliary lemmas whose proofs we skip since they are simple.

Lemma 4.4. *If* $\varepsilon<\frac{1}{8\log 2}$ *then*

$$\limsup_{N\to\infty}\max_{\sigma^1,\sigma^2,\sigma^3\in X}\int_A\int_A\int_A e^{-(\vec{x},B^{-1}\vec{x})b_N^2/2-a_Nb_N(\vec{x},B^{-1}\vec{1})}d\vec{x}<\infty, \quad \text{P-}a.s. \quad (4.26)$$

Lemma 4.5. P-*almost surely for all configurations* $\sigma^1,\sigma^2,\sigma^3\in X$

$$3-(\vec{1},B^{-1}\vec{1}) \leq 2(R_{12}^2+R_{23}^2+R_{31}^2). \qquad (4.27)$$

Lemma 4.6. *For all* $\sigma^1,\sigma^2,\sigma^3\in S_N$

$$\mathcal{J}^{(2)}(R_{12},R_{23},R_{31}) \geq \frac{1}{4}\Bigl(R_{12}^2+R_{23}^2+R_{31}^2\Bigr). \qquad (4.28)$$

We are now ready to estimate sum (4.25). In the same spirit as for the second moment calculation, we will split (4.25) into four parts and show that every each of them is bounded. Moreover, by monotonicity argument similar to that used in calculation of the second moment we can restrict our attention to the case when $M=\varepsilon N$.

(1) We first calculate the contribution to (4.25) coming from the set

$$\mathcal{S}_1 = \{(\sigma^1, \sigma^2, \sigma^3) \in S_N^3 : \max\{|R_{12}|, |R_{23}|, |R_{31}|\} \leq \tau_N\}, \tag{4.29}$$

where $\tau_N = \frac{1}{N^{1/4} \log N}$. Using Theorem 2.5 (i) and Lemma 4.5, we obtain that the contribution from $\mathcal{S}_1$ is at most of order

$$\frac{1}{N^{3/2}} \sum_{R_{12}, R_{23}, R_{31}} e^{-N\mathcal{J}^{(2)}(R_{12}, R_{23}, R_{31}) + 2(R_{12}^2 + R_{23}^2 + R_{31}^2)M \log 2}, \tag{4.30}$$

where the summation is over $R_{12}, R_{23}, R_{31} \in K_N \cap [-\tau_N, \tau_N]$. Expanding in Taylor series we obtain that for $|R_{12}|, |R_{23}|, |R_{31}| \leq \tau_N$

$$\mathcal{J}^{(2)}(R_{12}, R_{23}, R_{31}) = \frac{1}{2}(R_{12}^2 + R_{23}^2 + R_{31}^2) + O(\tau_N^4), \tag{4.31}$$

and thus (4.30) is bounded by

$$\begin{aligned} &\frac{1}{N^{3/2}} \sum_{R_{12}, R_{23}, R_{31}} e^{-\frac{1}{2}N(R_{12}^2 + R_{23}^2 + R_{31}^2) + 2\varepsilon N(R_{12}^2 + R_{23}^2 + R_{31}^2) \log 2} \\ &\qquad = \left(\frac{1}{\sqrt{N}} \sum_{R_N \in K_N \cap [-\tau_N, \tau_N]} e^{-\frac{1}{2}NR^2(1 - 4\varepsilon \log 2)} \right)^3 < \infty. \end{aligned} \tag{4.32}$$

(2) We next calculate the contribution from the set

$$\begin{aligned} \mathcal{S}_2 = \{(\sigma^1, \sigma^2, \sigma^3) \in S_N^3 : &R_{12}, R_{23}, R_{21} \text{ satisfy } (2.53) \\ &\text{and } \max\{|R_{12}|, |R_{23}|, |R_{31}|\} > \tau_N\}. \end{aligned} \tag{4.33}$$

Without loss of generality we can assume that $|R_{12}| > \tau_N$. Then, using Theorem 2.5 (i) and Lemma 4.6, the contribution from this set is at most of order

$$\sum_{R_{12}, R_{23}, R_{31}} \frac{e^{-N\mathcal{J}^{(2)}(R_{12}, R_{23}, R_{31})}}{N^{3/2} P(R_{12}, R_{23}, R_{31})} e^{2(R_{12}^2 + R_{23}^2 + R_{31}^2)M \log 2}, \tag{4.34}$$

where $P(x, y, z) = (1+x+y+z)(1+x-y-z)(1-x+y-z)(1-x-y+z)$ and where the sum is over the triplets $(R_{12}, R_{23}, R_{31}) \in K_N^3$ satisfying (2.53) and $|R_{12}| > \tau_N$. Then, from Lemma 4.5, we further get that the sum (4.34) is bounded by

$$\begin{aligned} &\sum_{|R_{12}| > \tau_N} \frac{e^{-\frac{1}{4}N(R_{12}^2 + R_{23}^2 + R_{31}^2)}}{N^{3/2} P(R_{12}, R_{23}, R_{31})} e^{2(R_{12}^2 + R_{23}^2 + R_{31}^2)M \log 2} \\ &\qquad = \sum_{|R_{12}| > \tau_N} \frac{e^{-\frac{1}{4}N(1 - 8\varepsilon \log 2)(R_{12}^2 + R_{23}^2 + R_{31}^2)}}{N^{3/2} P(R_{12}, R_{23}, R_{31})} \end{aligned} \tag{4.35}$$

Since the number of terms in (4.35) is at most $(N+1)^3$, and since $e^{-\frac{1}{4}(1-8\varepsilon \log 2)N\tau_N^2}$ decreases faster than any polynomial it follows that sum (4.35) is of order $o(1)$.

(3) We now turn to the contribution from the set

$$\mathcal{S}_3 = \{(\sigma^1,\sigma^2,\sigma^3) \in S_N^3 : R_{12}, R_{23}, R_{31} \text{ satisfy (2.55)}\}. \tag{4.36}$$

By Lemma 2.5 (ii) the contribution from this set is at most of order

$$\sum_{R_{12},R_{23},R_{31}} \frac{N^c \, \mathrm{E}|W_{N,3}^X|}{2^{3M}\sqrt{\det B}} e^{(3-(\vec{1},B^{-1}\vec{1}))(M\log 2-1/2\log M)}, \tag{4.37}$$

where the summation is over the triplets $(R_{12}, R_{23}, R_{31}) \in K_N$ satisfying (2.55). Since $\mathrm{E}|W_{N,3}^X|$ is of order

$$\frac{2^{3M}e^{-N\mathcal{J}^{(2)}(R_{12},R_{23},R_{31})}}{N^{3/2}P(R_{12}, R_{23}, R_{31})} \tag{4.38}$$

we obtain, using Lemmas 4.5 and 4.6, that the sum (4.37) is bounded by

$$\sum_{R_{12},R_{23},R_{31}} \frac{N^c e^{-\frac{1}{4}N(R_{12}^2+R_{23}^2+R_{31}^2)}}{N^{3/2}P(R_{12}, R_{23}, R_{31})} e^{2\varepsilon N\log 2(R_{12}^2+R_{23}^2+R_{31}^2)}. \tag{4.39}$$

It is easy to show that the triplet (R_{12}, R_{23}, R_{31}) that satisfy (2.55) must satisfy either $|R_{23}| > \tau_N$ or $|R_{31}| > \tau_N$. Therefore we can further bound the contribution from the set $\mathcal{S}_3$ by the sum

$$\sum_{|R_{23}|>\tau_N \text{ or } |R_{31}|>\tau_N} \frac{N^c e^{-\frac{1}{4}N(R_{12}^2+R_{23}^2+R_{31}^2)}}{N^{3/2}P(R_{12}, R_{23}, R_{31})} e^{2\varepsilon N\log 2(R_{12}^2+R_{23}^2+R_{31}^2)}, \tag{4.40}$$

which is $o(1)$ by the same argument as in part (2).

(4) To finish the estimate of the third factorial moment we have to estimate the contribution to (4.25) coming from the set

$$\mathcal{S}_4 = \{(\sigma^1,\sigma^2,\sigma^3) \in S_N^3 : R_{12}, R_{23}, R_{21} \text{ satisfy (2.57)}\}. \tag{4.41}$$

By Theorem 2.5 (iii) the set $\mathcal{S}_4^X$ is P-a.s. empty and therefore its contribution is P-a.s. zero. This finishes the proof of assertion (ii) of Theorem 4.2. The proof of Theorem 4.2 is now complete. □

For comparison with the cases $\nu(r) = r^p$ for $p = 1$ and $p = 2$, we give here the asymptotic behavior of the first three factorial moments for the case $\nu(0) = \nu'(0) = 0$, i.e., for instance for the case $\nu(r) = r^p$ of pure p-spins when $p \geq 3$. This behavior is compatible with a Poisson convergence theorem.

Theorem 4.7. *Assume that $\nu(0) = \nu'(0) = 0$. For every bounded Borel set A*

$$\lim \mathbb{E}(\mathcal{P}_N(A))_1 = \mu(A), \quad \text{P-}a.s. \tag{4.42}$$

Moreover, if $\limsup \frac{M(N)}{N} < \frac{1}{8\log 2}$ *then* P-*a.s.*

(i) $\lim \mathbb{E}(\mathcal{P}_N(A))_2 = (\mu(A))^2$;

(ii) $\lim \mathbb{E}(\mathcal{P}_N(A))_3 = (\mu(A))^3 < \infty$.

We do not include a proof of this last statement, which again follows the same strategy as the proof of Theorem 4.2.

5. Universality for non-Gaussian Hamiltonians

In this section we extend the results of the previous sections to the case of non-Gaussian Hamiltonians. We are able to make this extension only for the pure p-spin models, i.e., $\nu(r) = r^p$. In this case we recall that the Hamiltonian is defined as

$$H_N(\sigma) = \frac{1}{\sqrt{N}} H_{N,p} = \frac{1}{\sqrt{N^p}} \sum_{1 \le i_1, \dots, i_p \le N} g_{i_1,\dots,i_p} \sigma_{i_1} \dots \sigma_{i_p}. \tag{5.1}$$

Our assumptions on the random variables $(g_{i_1,\dots,i_p})_{1 \le i_1,\dots,i_p \le N}$ in (5.1) are the same as were made in [BCMN05a] and [BCMN05b] for the number partitioning problem. That is, we assume that their distribution function admits a density $\rho(x)$ that satisfies the following conditions:

1. $\rho(x)$ is even;
2. $\int x^2 \rho(x) dx = 1$;
3. for some $\epsilon > 0$
$$\int_{-\infty}^{\infty} \rho(x)^{1+\epsilon} dx < \infty; \tag{5.2}$$
4. $\rho(x)$ has a Fourier transform that is analytic in some neighborhood of zero. We write
$$-\log \hat{\rho}(z) = \frac{1}{2}(2\pi)^2 z^2 + c_4 (2\pi)^4 z^4 + O(|z|^6). \tag{5.3}$$

Note that the inequality $\mathbb{E}(X^4) \ge \mathbb{E}(X^2)^2$ implies that necessarily $c_4 < \frac{1}{12}$.

Under these assumptions we will show, using the method introduced by C. Borgs, J. Chayes, S. Mertens and C. Nair in [BCMN05b], that Theorems 1.1 and 1.4 still hold.

5.1. Proof of universality

In this subsection we fix $p \ge 1$ and prove the analog of Theorem 1.1 in the non-Gaussian case assuming that the Hamiltonian is given by (5.1), and that the random variables $(g_{i_1,\dots,i_p})_{1 \le i_1,\dots,i_p \le N}$ satisfy conditions (1)–(4) above.

Theorem 5.1 (Universality in the Non-Gaussian case). *Assume $M(N) = o(\sqrt{N})$ for $p = 1$ and $M = o(N)$ for $p \ge 2$. Then* P*-almost surely the sequence of point processes $\mathcal{P}_N$ converges weakly to a Poisson point process $\mathcal{P}$ on $\mathbb{R}$ with intensity given by*

$$\mu(dt) = \frac{1}{\sqrt{\pi}} e^{-t\sqrt{2\log 2}} dt. \tag{5.4}$$

To prove Theorem 5.1 we essentially prove a local limit theorem. More precisely, for any fixed N let us introduce the Gaussian process Z_N on S_N that has the same mean and covariance matrix as the process $H'_N(\sigma)$ defined in (1.2). We will prove in Theorem 5.2 that P-a.s., for all sequences $(\sigma^1, \dots, \sigma^\ell) \in X^\ell$, the joint density of the random variables $H'_N(\sigma^1), \dots, H'_N(\sigma^\ell)$ is well approximated by the joint density of $Z_N(\sigma^1), \dots, Z_N(\sigma^\ell)$.

Theorem 5.2. *Assume $M(N) = o(\sqrt{N})$ for $p = 1$ and $M = o(N)$ for $p \geq 2$. Then* P*-almost surely for every $\ell \geq 1$ and every bounded Borel set A there exists $c > 0$ such that uniformly in $(\sigma^1, \ldots, \sigma^\ell) \in X^\ell$*

$$\mathbb{P}(H'_N(\sigma^j) \in A, j = 1, \ldots, \ell) = \mathbb{P}(Z_N(\sigma^j) \in A, j = 1, \ldots, \ell) \times \Big(1 + O(R_{\max}(\sigma^1, \ldots, \sigma^\ell)) + O\Big(\frac{M^2}{N^p}\Big)\Big) + O(e^{-cN^p}). \tag{5.5}$$

Applying Theorem 5.2 together with Theorem 1.1 we get from formula (3.5) that

$$\mathbb{E}(\mathcal{P}_N(A))_\ell = (\mu(A))^\ell(1 + o(1)) + O(2^{M\ell} e^{-cN^p}) \to (\mu(A))^\ell, \tag{5.6}$$

which, by Lemma 3.1, implies weak convergence of the sequence of point processes $\mathcal{P}_N$ to a Poisson point process with intensity measure μ, thus implying Theorem 5.1. We therefore focus on the proof of Theorem 5.2.

Proof. First, we obtain from the definition of $H'_N(\sigma)$ that

$$\Big\{H'_N(\sigma) \in (x, x + \Delta x)\Big\} = \Big\{ \sum_{1 \leq i_1, \ldots, i_p \leq N} g_{i_1, \ldots, i_p} \sigma_{i_1} \ldots \sigma_{i_p} \in (a_N + x b_N, a_N + (x + \Delta x) b_N) \sqrt{N^p} \Big\}. \tag{5.7}$$

Following [BCMN05b] we get an integral representation of the indicator function

$$\mathbf{1}_{H'_N(\sigma) \in (x, x+\Delta x)} = \Delta x\, b_N \sqrt{n} \int_{-\infty}^{\infty} \operatorname{sinc}\big(f \Delta x b_N \sqrt{n}\big)\, e^{2\pi i f \sum g_{i_1, \ldots, i_p} \sigma_{i_1} \ldots \sigma_{i_p} - 2\pi i f \alpha_N \sqrt{n}} df, \tag{5.8}$$

where, for brevity, we wrote $n = N^p, \alpha_N = a_N + b_N(x + \Delta x/2), \operatorname{sinc}(x) = \frac{\sin(\pi x)}{\pi x}$, and the sum in the exponent runs over all possible sequences $1 \leq i_1, \ldots, i_p \leq N$.

Changing the integration variable in (5.8) from f to $-f$ and applying the resulting formula to the product of indicator functions we arrive at the following representation

$$\prod_{j=1}^{\ell} \mathbf{1}_{H'_N(\sigma^j) \in (x_j, x_j + \Delta x_j)} = \prod_{j=1}^{\ell} \Delta x_j\, b_N \sqrt{n} \times \iiint_{-\infty}^{\infty} \prod_{j=1}^{\ell} \operatorname{sinc}\big(f_j \Delta x_j b_N \sqrt{n}\big)\, e^{-2\pi i f_j \sum g_{i_1 \ldots i_p} \sigma^j_{i_1} \ldots \sigma^j_{i_p} + 2\pi i f_j \alpha_N^{(j)} \sqrt{n}} df_j, \tag{5.9}$$

where $\alpha_N^{(j)} = a_N + b_N(x_j + \Delta x_j/2)$. Introducing the variables

$$v_{i_1, \ldots, i_p} = \sum_{j=1}^{\ell} f_j \sigma^j_{i_1} \ldots \sigma^j_{i_p}, \tag{5.10}$$

we rewrite the integral in the above formula as

$$\iiint_{-\infty}^{\infty} \prod_{1\leq i_1,\ldots,i_p\leq N} e^{-2\pi i g_{i_1\ldots i_p} v_{i_1,\ldots,i_p}} \prod_{j=1}^{\ell} \operatorname{sinc}\big(f_j \Delta x_j \sqrt{n}\big) e^{2\pi i f_j \alpha_N^{(j)} \sqrt{n}} df_j. \quad (5.11)$$

To get an integral representation of the joint density

$$\mathbb{P}\big(H'_N(\sigma^1) \in (x_1, x_1 + dx_1), \ldots, H'_N(\sigma^\ell) \in (x_\ell, x_\ell + dx_\ell)\big) \quad (5.12)$$

we have to take the expectation of (5.9) and then let $\Delta x_j \to 0$ for all $j = 1, 2, \ldots \ell$. As was proved in [BCMN05a] (see Lemma 3.4), the exchange of expectation and integration for $p = 1$ is justified when the rank of the matrix formed by the row vectors $\sigma^1, \ldots, \sigma^\ell$, is ℓ. To justify the exchange in our case we introduce an ℓ by N^p matrix, $C_p(\sigma^1, \ldots, \sigma^\ell)$, defined as follows: for any given set of configurations $\sigma^1, \ldots, \sigma^\ell$, the jth row is composed of all N^p products, $\sigma^j_{i_1}\sigma^j_{i_2}\ldots\sigma^j_{i_p}$ over all subsets $1 \leq i_1, \ldots, i_p \leq N$. By generalizing the arguments from [BCMN05a] the exchange can then be justified provided that the rank of the matrix $C_p(\sigma^1, \ldots, \sigma^\ell)$ is ℓ. As we will see in Lemma 5.3 below this holds true P-almost surely when $M = o(N)$.

Given a vector $\boldsymbol{\delta} \in \{-1, 1\}^\ell$ let $n_{\boldsymbol{\delta}}$ be the number of times the column vector $\boldsymbol{\delta}$ appears in the matrix C_p :

$$n_{\boldsymbol{\delta}} = n_{\boldsymbol{\delta}}(\sigma^1, \ldots, \sigma^\ell) = \big|\{j \leq N^p : (\sigma^1_j, \ldots, \sigma^\ell_j) = \boldsymbol{\delta}\}\big|. \quad (5.13)$$

With this notation we have:

Lemma 5.3. *Suppose $M = o(N)$. Then there exists a sequence $\lambda_N = o(1)$ such that* P*-almost surely for all collections $(\sigma^1, \ldots, \sigma^\ell) \in X^\ell$*

$$\max_{\boldsymbol{\delta}\in\{-1,1\}^\ell} \left| n_{\boldsymbol{\delta}} - \frac{n}{2^\ell} \right| \leq n\lambda_N. \quad (5.14)$$

Proof. We first prove by induction that the following simple fact holds true: if for a given sequence of configurations $(\sigma^1, \ldots, \sigma^\ell) \in S_N^\ell$ the matrix $C_1(\sigma^1, \ldots, \sigma^\ell)$ satisfies condition (5.14) then necessarily the matrix $C_p(\sigma^1, \ldots, \sigma^\ell)$ satisfies (5.14) for all $p \geq 1$.

For $p = 1$ there is nothing to prove. We now assume that the statement is true for the matrix $C_{p-1}(\sigma^1, \ldots, \sigma^\ell)$ and prove it for $C_p(\sigma^1, \ldots, \sigma^\ell)$.

Let $\sigma_\mu, 1 \leq \mu \leq N$ denote the columns of the matrix $C_1(\sigma^1, \ldots, \sigma^\ell)$. For every column vector σ_μ let us construct a matrix $C^\mu_{p-1} = C^\mu_{p-1}(\sigma^1, \ldots, \sigma^\ell)$ with entries

$$(C^\mu_{p-1})_{ij} = (\sigma_\mu)_j \, (C_{p-1})_{ij}. \quad (5.15)$$

For future convenience let $n^\mu_{\boldsymbol{\delta}}$ denote the variable $n_{\boldsymbol{\delta}}$ for the matrix C^μ_{p-1}. From the inductive assumption it follows that for all $1 \leq \mu \leq N$

$$\max_{\boldsymbol{\delta}} \left| n^\mu_{\boldsymbol{\delta}} - \frac{N^{p-1}}{2^\ell} \right| \leq N^{p-1}\lambda_N. \quad (5.16)$$

Now note that the $\ell \times N^p$ matrix $C_p(\sigma^1, \dots, \sigma^\ell)$ can be obtained by concatenating N matrices $C^\mu_{p-1}(\sigma^1, \dots, \sigma^\ell)$ each of size $\ell \times N^{p-1}$. Therefore for any sequence of configurations $(\sigma^1, \dots, \sigma^\ell)$ with matrix $C_1(\sigma^1, \dots, \sigma^\ell)$ satisfying (5.14) we have

$$\left|n_{\boldsymbol{\delta}} - \frac{N^p}{2^\ell}\right| \le \left|n^{\mu_1}_{\boldsymbol{\delta}} - \frac{N^{p-1}}{2^\ell}\right| + \dots + \left|n^{\mu_N}_{\boldsymbol{\delta}} - \frac{N^{p-1}}{2^\ell}\right| \le$$
$$\le N^{p-1}\lambda_N + \dots + N^{p-1}\lambda_N = N^p\lambda_N \tag{5.17}$$

and the induction is complete.

To prove Lemma 5.3 it is thus enough to demonstrate that P-almost surely there are no sequences $(\sigma^1, \dots, \sigma^\ell) \in X^\ell$ such that $C_1(\sigma^1, \dots, \sigma^\ell)$ violates condition (5.14). Let us prove this by induction in ℓ.

For $\ell = 1$ let us introduce the sets

$$\mathcal{T}_N = \left\{\sigma \in S : \max_{\boldsymbol{\delta} \in \{-1,1\}} \left|n_{\boldsymbol{\delta}} - \frac{N}{2}\right| \le N\lambda_N\right\}. \tag{5.18}$$

Then by Chernoff bound

$$|\mathcal{T}^c_N| = 2 \sum_{i \ge N\lambda_N} \binom{N}{\frac{N}{2} + i} \le 2^{N+1} e^{-\frac{1}{2}N((1+\lambda_N)\log(1+\lambda_N) - \lambda_N)}. \tag{5.19}$$

Let us choose $\lambda_N = o(1)$ in such a way that $M = o(N\lambda_N^2)$ and $\log N = o(N\lambda_N^2)$. Then using (2.12) we obtain that

$$\sum_{\sigma \in \mathcal{T}^c_N} \mathbf{1}_\sigma = 0, \quad \text{P-a.s.} \tag{5.20}$$

which proves the statement for $\ell = 1$. Now assume that P-a.s. for all sequences $(\sigma^1, \dots, \sigma^{\ell-1}) \in X^{(\ell-1)}$ the matrix $C_1(\sigma^1, \dots, \sigma^{\ell-1})$ satisfies condition (5.14). Since there is only a countable number of sequences $(\sigma^1, \dots, \sigma^{\ell-1})$ we fix a sequence $(\sigma^1, \dots, \sigma^{\ell-1}) \in X^{(\ell-1)}$ and prove that P-almost surely there are no configurations σ^ℓ such that $C_1(\sigma^1, \dots, \sigma^\ell)$ violates (5.14).

Let $\boldsymbol{\delta} \in \{-1,1\}^\ell$ be given. Define $\boldsymbol{\delta}_1(\boldsymbol{\delta}) \in \{-1,1\}^{\ell-1}$ as $(\boldsymbol{\delta}_1)_i = (\boldsymbol{\delta})_i$ for $1 \le i \le \ell - 1$ and also define $\boldsymbol{\delta}_2(\boldsymbol{\delta}) \in \{-1,1\}$ as $\boldsymbol{\delta}_2 = (\boldsymbol{\delta})_\ell$. Let us also introduce, for given $\boldsymbol{\delta}_1 \in \{-1,1\}^{\ell-1}$, the set

$$N_{\boldsymbol{\delta}_1} = \{j \le N : (\sigma^1_j, \dots, \sigma^{\ell-1}_j) = \boldsymbol{\delta}_1\}. \tag{5.21}$$

By the inductive assumption we conclude that for all $\boldsymbol{\delta}_1 \in \{-1,1\}^{\ell-1}$

$$\left||N_{\boldsymbol{\delta}_1}| - \frac{N}{2^{\ell-1}}\right| \le N\lambda_N. \tag{5.22}$$

From the definition of $N_{\boldsymbol{\delta}}$ it is not hard to see that for all $\boldsymbol{\delta} \in \{-1,1\}^\ell$

$$n_{\boldsymbol{\delta}} = \left|\{j \in N_{\boldsymbol{\delta}_1(\boldsymbol{\delta})} : \sigma^\ell_j = \boldsymbol{\delta}_2(\boldsymbol{\delta})\}\right|. \tag{5.23}$$

Using the above relation together with (2.11) and the assumptions on λ_N we get that P-almost surely

$$\left|n_{\boldsymbol{\delta}} - \frac{|N_{\boldsymbol{\delta}_1(\boldsymbol{\delta})}|}{2}\right| \le \frac{N\lambda_N}{2}. \tag{5.24}$$

Therefore from (5.22) and (5.24)

$$\left|n_{\boldsymbol{\delta}} - \frac{N}{2^\ell}\right| \le \left|n_{\boldsymbol{\delta}} - \frac{|N_{\boldsymbol{\delta}_1(\boldsymbol{\delta})}|}{2}\right| + \left|\frac{|N_{\boldsymbol{\delta}_1(\boldsymbol{\delta})}|}{2} - \frac{N}{2^\ell}\right| \le \frac{N\lambda_N}{2} + \frac{N\lambda_N}{2} = N\lambda_N. \quad (5.25)$$

The induction is now complete and the lemma is proved. □

Lemma 5.3 implies that P-almost surely, for all $(\sigma^1, \ldots, \sigma^\ell) \in X^\ell$,

$$n_{\min}(\sigma^1, \ldots, \sigma^\ell) = \min_{\boldsymbol{\delta} \in \{-1,1\}^\ell} n_{\boldsymbol{\delta}} = \frac{n}{2^\ell}(1 + O(\lambda_N)), \quad (5.26)$$

and hence, for sufficiently large N, the rank of the matrix $C_p(\sigma^1, \ldots, \sigma^\ell)$ is ℓ. The exchange of integration and expectation is thus justified.

Using once again Lemma 5.3, condition (5.2), and the dominated convergence theorem we obtain that the joint density is

$$\mathbb{P}\big(H'_N(\sigma^j) \in (x_j, x_j + dx_j) \text{ for } j = 1, \ldots, \ell\big) = \prod_{j=1}^{\ell} b_N \sqrt{n}\, dx_j$$
$$\times \iiint_{-\infty}^{\infty} \prod_{1 \le i_1, \ldots, i_p \le N} \hat{\rho}(v_{i_1,\ldots,i_p}) \prod_{j=1}^{\ell} e^{2\pi i f_j \alpha_N^{(j)} \sqrt{n}} df_j, \quad (5.27)$$

where we redefined $\alpha_N^{(j)} = a_N + b_N x_j$. We remark for future use that $\alpha_N^{(j)} = O(a_N)$ for all $1 \le j \le \ell$.

It is straightforward at this point to adapt the saddle point analysis used in [BCMN05b] to calculate the integrals of such type. The only difference is that instead of the matrix $C_1(\sigma^1, \ldots, \sigma^\ell)$ with rows formed by row vectors $\sigma^1, \ldots, \sigma^\ell$, we use the matrix $C_p(\sigma^1, \ldots, \sigma^\ell)$. By analogy with Lemma 5.3 from [BCMN05b] we first approximate the integral in (5.27) by an integral over a bounded domain, i.e., for some $c_1 > 0$ depending on $\mu_1 > 0$

$$\iiint_{-\infty}^{\infty} \prod_{i_1,\ldots,i_p} \hat{\rho}(v_{i_1,\ldots,i_p}) \prod_{j=1}^{\ell} e^{2\pi i f_j \alpha_N^{(j)} \sqrt{n}} df_j =$$
$$\iiint_{-\mu_1}^{\mu_1} \prod_{i_1,\ldots,i_p} \hat{\rho}(v_{i_1,\ldots,i_p}) \prod_{j=1}^{\ell} e^{2\pi i f_j \alpha_N^{(j)} \sqrt{n}} df_j + O(e^{-c_1 n_{\min}}). \quad (5.28)$$

We next rewrite the integral in the right-hand side of (5.28) as

$$\iiint_{-\mu_1}^{\mu_1} e^{2\pi i n \mathbf{f} \cdot \boldsymbol{\alpha}} \prod_{\boldsymbol{\delta} \in \{-1,1\}^\ell} \hat{\rho}(\mathbf{f} \cdot \boldsymbol{\delta})^{n_{\boldsymbol{\delta}}} \prod_{j=1}^{\ell} df_j, \quad (5.29)$$

where $\boldsymbol{\alpha} = \big(\frac{\alpha_N^{(1)}}{\sqrt{n}}, \ldots, \frac{\alpha_N^{(\ell)}}{\sqrt{n}}\big)$, $\boldsymbol{f} = (f_1, \ldots, f_\ell)$, and where $\boldsymbol{\alpha} \cdot \boldsymbol{f} = \alpha_1 f_1 + \cdots + \alpha_\ell f_\ell$ is the standard scalar product.

Using Lemma 5.3 again we can apply Lemma 5.4 from [BCMN05b] to conclude that given μ_1 there are constants $c_1(\mu_1) > 0$ and $\mu_2 > 0$ such that the following equality holds whenever $\eta_1, \ldots, \eta_\ell$ is a sequence of real numbers with $\sum_j |\eta_j| \leq \mu_2$ and $\eta_j \alpha_N^{(j)} \geq 0$ for all $j = 1, \ldots, \ell$

$$\iiint_{-\mu_1}^{\mu_1} \prod_{\boldsymbol{\delta}} \hat{\rho}(\mathbf{f} \cdot \boldsymbol{\delta})^{n_{\boldsymbol{\delta}}} \prod_{j=1}^{\ell} e^{2\pi i n f_j \alpha_N^{(j)}} df_j \tag{5.30}$$

$$= \iiint_{-\mu_1}^{\mu_1} e^{2\pi n (i\mathbf{f}\cdot\boldsymbol{\alpha} - \boldsymbol{\eta}\cdot\boldsymbol{\alpha})} \prod_{\boldsymbol{\delta}} \hat{\rho}(\mathbf{f} \cdot \boldsymbol{\delta} + i\boldsymbol{\eta}\cdot\boldsymbol{\delta})^{n_{\boldsymbol{\delta}}} \prod_{j=1}^{\ell} df_j + O(e^{-\frac{1}{2}c_1 n_{\min}}).$$

The values of the shifts $\eta_1, \ldots, \eta_\ell$ are determined by the following system:

$$\sum_{\boldsymbol{\delta}} \frac{n_{\boldsymbol{\delta}}}{n} \delta_j F'(i\boldsymbol{\delta}\cdot\boldsymbol{\eta}) = 2\pi i \frac{\alpha_N^{(j)}}{\sqrt{n}}, \quad j = 1, \ldots, \ell, \tag{5.31}$$

where we wrote $F = -\log \hat{\rho}$.

Since $\max_{\sigma,\sigma' \in X} |R(\sigma, \sigma')|$ is P-almost surely of order $o(1)$ when $M = o(N)$ we can apply Lemma 5.5 from [BCMN05b] and obtain that this system has a unique solution

$$\boldsymbol{\eta}(\boldsymbol{\alpha}) = \frac{1}{2\pi} B^{-1} \boldsymbol{\alpha} \left(1 + O(||\boldsymbol{\alpha}||_2^2)\right). \tag{5.32}$$

Moreover, for sufficiently small μ_1,

$$\iiint_{-\mu_1}^{\mu_1} e^{2\pi n (i\mathbf{f}\cdot\boldsymbol{\alpha} - \boldsymbol{\eta}\cdot\boldsymbol{\alpha})} \prod_{\boldsymbol{\delta}} \hat{\rho}(\mathbf{f} \cdot \boldsymbol{\delta} + i\boldsymbol{\eta}\cdot\boldsymbol{\delta})^{n_{\boldsymbol{\delta}}} \prod_{j=1}^{\ell} df_j$$

$$= e^{-nG_{n,\ell}(\boldsymbol{\alpha})} \left(\frac{1}{2\pi n}\right)^{\ell/2} \left(1 + O(n^{-1/2}) + O(a_N^2/n) + O(R_{\max})\right), \tag{5.33}$$

where

$$G_{n,\ell}(\boldsymbol{\alpha}) = \sum_{\boldsymbol{\delta}} \frac{n_{\boldsymbol{\delta}}}{n} F(i\boldsymbol{\delta}\cdot\boldsymbol{\eta}(\boldsymbol{\alpha})) + 2\pi \boldsymbol{\eta}(\boldsymbol{\alpha}) \cdot \boldsymbol{\alpha}. \tag{5.34}$$

Therefore

$$\mathbb{P}\big(H_N'(\sigma^j) \in (x_j, x_j + dx_j) \text{ for } j = 1, \ldots, \ell\big) = \left(\frac{b_N}{\sqrt{2\pi}}\right)^{\ell} e^{-nG_{n,\ell}(\boldsymbol{\alpha})} \prod_{j=1}^{\ell} dx_j$$

$$\times \left(1 + O(n^{-1/2}) + O(a_N^2/n) + O(R_{\max})\right) + O(b_N^\ell n^{\ell/2} e^{-\frac{1}{2}c_1 n_{\min}}). \tag{5.35}$$

Expanding $G_{n,\ell}$ we get the approximation

$$nG_{n,\ell}(\boldsymbol{\alpha}) = \frac{n}{2}(\boldsymbol{\alpha}, B^{-1}\boldsymbol{\alpha}) + O\Big(\frac{a_N^4}{n}\Big). \tag{5.36}$$

By definition $a_N = O(\sqrt{M})$. Under the assumptions of Theorem 5.2 we get that $a_N = o(\sqrt[4]{N})$ for $p = 1$, that $a_N = o(\sqrt{N})$ for $p \geq 2$, and thus that $a_N^4 = o(n)$. It implies that asymptotically the joint density (5.12) is Gaussian.

More precisely, it follows from the equations (5.35) and (5.36) that P-a.s., for all collections $(\sigma^1, \ldots, \sigma^\ell) \in X^\ell$,

$$\mathbb{P}\big(H'_N(\sigma^j) \in (x_j, x_j + dx_j) \text{ for } j = 1, \ldots, \ell\big) = \left(\frac{b_N}{\sqrt{2\pi}}\right)^\ell e^{-n(\boldsymbol{\alpha}, B^{-1}\boldsymbol{\alpha})/2}$$
$$\times \prod_{j=1}^{\ell} dx_j \left(1 + O(R_{\max}) + O\left(\frac{a_N^4}{n}\right)\right) + O(b_N^\ell n^{\ell/2} e^{-\frac{1}{2}c_1 n_{\min}}). \quad (5.37)$$

By Lemma 5.3 the term $O(b_N^\ell n^{\ell/2} e^{-\frac{1}{2}c_1 n_{\min}})$ is of order $o(e^{-cN^p})$ as $N \to \infty$. This finishes the proof of Theorem 5.2. □

5.2. Breakdown of universality

In this subsection we follow the same strategy as we used in the proof of Theorem 1.4 – we fix a bounded set A and study the first three factorial moments of the random variable $\mathcal{P}_N(A)$. In the case of the number partitioning problem the following theorem implies that the Poisson convergence fails as soon as $\limsup M/\sqrt{N} > 0$.

Theorem 5.4 (Number partitioning problem). *Fix* $p = 1$ *and let the Hamiltonian be given by* (5.1). *If* $\limsup \frac{M(N)}{N} = \varepsilon < \infty$ *then* P*-a.s. for every bounded Borel set* A *we have*

(i) $\lim \mathbb{E}(\mathcal{P}_N(A))_1 = \mu(A) e^{-4c_4\varepsilon^2 \log^2 2}$;
(ii) $\limsup \mathbb{E}(\mathcal{P}_N(A))_2 = e^{2\varepsilon^2 \log 2 - 32c_4\varepsilon^2 \log^2 2}(\mu(A))^2$;
(iii) $\limsup \mathbb{E}(\mathcal{P}_N(A))_3 < \infty$.

Therefore the limit of the ratio of the second factorial moment to the square of the first is

$$\frac{\mathbb{E}(\mathcal{P}_N(A))_2}{\mathbb{E}(\mathcal{P}_N(A))_1^2} \to e^{2\varepsilon^2 \log^2 2 - 24c_4\varepsilon^2 \log^2 2} = e^{2\varepsilon^2 \log^2 2(1-12c_4)}. \quad (5.38)$$

Taking into account that $c_4 < \frac{1}{12}$ we conclude that the ratio is strictly larger than one and thus there is no Poisson convergence for $\varepsilon > 0$.

And in the case of the Sherrington-Kirkpatrick model the failure of Poisson convergence follows from

Theorem 5.5 (Sherrington-Kirkpatrick model). *Fix* $p = 2$ *and let the Hamiltonian be given by* (5.1). *If* $\limsup \frac{M(N)}{N} = \varepsilon < \frac{1}{8\log 2}$ *then* P*-a.s. for every bounded Borel set* A *we have*

(i) $\lim \mathbb{E}(\mathcal{P}_N(A))_1 = \mu(A) e^{-4c_4\varepsilon^2 \log^2 2}$;
(ii) $\limsup \mathbb{E}(\mathcal{P}_N(A))_2 = \frac{e^{-32c_4\varepsilon^2 \log^2 2}}{\sqrt{1-4\varepsilon \log 2}}(\mu(A))^2$;
(iii) $\limsup \mathbb{E}(\mathcal{P}_N(A))_3 < \infty$.

The ratio of the second factorial moment to the square of the first moment is

$$\frac{\mathbb{E}(\mathcal{P}_N(A))_2}{\mathbb{E}(\mathcal{P}_N(A))_1^2} \to \frac{e^{-24c_4\varepsilon^2 \log^2 2}}{\sqrt{1-4\varepsilon \log 2}}. \tag{5.39}$$

For $\varepsilon > 0$ the above ratio is strictly larger than one and thus convergence to a Poisson point process fails.

We will give the proof of Theorem 5.5 only since the case $p = 1$ is based on essentially the same computations.

Proof of Theorem 5.5. As in the proof of Theorem 4.2 we successively prove the statement on the first, second, and third moment. To simplify our computations we will assume that $M = \varepsilon N$. (Using the monotonicity argument, the case of general sequences $M(N)$ can be analyzed just as in Theorem 4.2 of Section 4.)

1. **First moment estimate.** Following the same steps as in Subsection 5.1 we approximate the density of $H'_N(\sigma)$ by

$$e^{-nG_{n,1}(\alpha_N)}\frac{1}{\sqrt{2\pi}}\left(1+O\Big(\frac{1}{\sqrt{N}}\Big)+O\Big(\frac{\alpha_N^2}{n}\Big)\right)+O(e^{-c_1 n}), \tag{5.40}$$

where according to the notation introduced above $n = N^2$.

In the case $M = \varepsilon N$ we need a more precise approximation of the function $G_{n,\ell}$ than given by formula (5.36). Expanding the solution of the system (5.31) as

$$\boldsymbol{\eta}(\boldsymbol{\alpha}) = \frac{1}{2\pi}B^{-1}\boldsymbol{\alpha} + \frac{4c_4}{2\pi}\sum_{\boldsymbol{\delta}}\frac{n_{\boldsymbol{\delta}}}{n}(\boldsymbol{\delta}, B^{-1}\boldsymbol{\alpha})^3 B^{-1}\boldsymbol{\delta} + O(||\alpha||^5) \tag{5.41}$$

and applying (5.41) we obtain from (5.34) and (5.3) that

$$\begin{aligned} G_{n,\ell}(\boldsymbol{\alpha}) &= -\frac{(2\pi)^2}{2}\sum_{\boldsymbol{\delta}}\frac{n_{\boldsymbol{\delta}}}{n}(\boldsymbol{\delta}\cdot\boldsymbol{\eta})^2 \\ &\quad + c_4(2\pi)^4\sum_{\boldsymbol{\delta}}\frac{n_{\boldsymbol{\delta}}}{n}(\boldsymbol{\delta}\cdot\boldsymbol{\eta})^4 + 2\pi(\boldsymbol{\eta}\cdot\boldsymbol{\alpha}) + O(||\boldsymbol{\alpha}||^6) \\ &= \frac{1}{2}(\boldsymbol{\alpha}, B^{-1}\boldsymbol{\alpha}) + c_4\sum_{\boldsymbol{\delta}}\frac{n_{\boldsymbol{\delta}}}{n}(\boldsymbol{\delta}, B^{-1}\boldsymbol{\alpha})^4 + O(||\boldsymbol{\alpha}||^6). \end{aligned} \tag{5.42}$$

Using the approximation for $G_{n,1}$ given by formula (5.42), we obtain

$$nG_{n,1} = \frac{\alpha_N^2}{2} + c_4\frac{\alpha_N^4}{n} + O\Big(\frac{\alpha_N^6}{n^2}\Big), \tag{5.43}$$

where $\alpha_N = a_N + b_N x$. Since $\alpha_N^4/n = 4c_4\varepsilon^2\log^2 2(1+o(1))$, we see that up to an ε-dependent multiplier the density of $H'_N(\sigma)$ is given by the normal density, more precisely, the density of $H'_N(\sigma)$ is

$$\frac{1}{\sqrt{2\pi}}e^{-\alpha_N^2/2}\, e^{-4c_4\varepsilon^2\log^2 2}\left(1+O\Big(\frac{1}{\sqrt{N}}\Big)+O\Big(\frac{\alpha_N^2}{n}\Big)\right)+O(e^{-c_1 n}). \tag{5.44}$$

Therefore the first factorial moment of $\mathcal{P}_N(A)$ is

$$\mu(A)e^{-4c_4\varepsilon^2\log^2 2}(1+o(1))+o(1) \tag{5.45}$$

and part (i) is proven.

2. **Second moment estimate.** When analyzing the second moment

$$\mathbb{E}(\mathcal{P}_N(A))_2 = \sum_{(\sigma^1,\sigma^2)\in X^2} \mathbb{P}\left(H'_N(\sigma^1)\in A, H'_N(\sigma^2)\in A\right), \tag{5.46}$$

it is useful to distinguish between "typical" and "atypical" sets of configurations $(\sigma^1,\dots,\sigma^\ell)\in X^\ell$, a notion introduced in [BCMN05a].

Take some sequence $\theta_N \to 0$ such that $N\theta_N^2 \to \infty$ and consider the $\ell\times n$ matrix $C_p(\sigma^1,\dots,\sigma^\ell)$, introduced in Subsection 5.1. Then all but a vanishing fraction of the configurations $(\sigma^1,\dots,\sigma^\ell)\in S_N^\ell$ obey the condition

$$\max_{\delta\in\{-1,1\}^\ell}\left| n_\delta - \frac{n}{2^\ell}\right| \le n\theta_N. \tag{5.47}$$

When $M=o(N)$, Lemma 5.3 guarantees that for a properly chosen sequences θ_N, P-almost surely, all the sampled sets $(\sigma^1,\dots,\sigma^\ell)\in X^\ell$ obey condition (5.47). It is no longer the case when $M=\varepsilon N$ and thus we have to consider the contribution from the sets violating (5.47).

Fix $p=2$ and $\ell=2$. Let a sequence $\theta_N\to 0$ be given, and define

$$I=\sum_{\sigma^1,\sigma^2}\mathbb{P}\big(H'_N(\sigma^1)\in A, H'_N(\sigma^2)\in A\big), \tag{5.48}$$

where the sum runs over all pairs of distinct configurations $(\sigma^1,\sigma^2)\in X^2$ satisfying condition (5.47) (the so-called "typical" configurations). Also define

$$II=\sum_{\sigma^1,\sigma^2}\mathbb{P}\big(H'_N(\sigma^1)\in A, H'_N(\sigma^2)\in A\big), \tag{5.49}$$

where the sum is over all pairs of distinct configurations $(\sigma^1,\sigma^2)\in X^2$ violating (5.47) (the "atypical" configurations). For later use we introduce the quantities I_g and II_g – the analogs of the variables I and II in the case where the random variables $(g_{i_1,i_2})_{1\le i_1,i_2\le N}$ are i.i.d. standard normals.

Lemma 5.6. *Let $\theta_N=\frac{1}{\sqrt{N}\log^2 N}$. If $\varepsilon\in(0,\frac{1}{4\log 2})$ then* P*-almost surely $II=o(1)$ and*

$$I=\frac{e^{-32c_4\varepsilon^2\log^2 2}(\mu(A))^2}{\sqrt{1-4\varepsilon\log 2}}(1+o(1)). \tag{5.50}$$

Proof. To prove that II is $o(1)$ let us bound the quantity II by II_g and show that II_g is almost surely negligible for $\theta_N=\frac{1}{\sqrt{N}\log^2 N}$. We start with the proof of the second statement, for which we will need the following simple observation.

Consider a sequence $\lambda_N \to 0$ such that $N\lambda_N \to \infty$ and define the set of configurations $\sigma \in \{-1,1\}^N$ with almost equal number of spins equal to 1 and to -1 :

$$\mathcal{T}_N = \Big\{\sigma \in S_N : \big|\#\{\sigma_i = 1\} - \#\{\sigma_i = -1\}\big| \leq N\lambda_N\Big\}. \tag{5.51}$$

It is not hard to prove that configurations $\sigma^1, \sigma^2 \in \mathcal{T}_N$ with overlap $|R(\sigma^1,\sigma^2)| \leq \lambda_N$ must satisfy (5.47) with $\theta_N = \lambda_N^2$. Therefore the set of pairs $(\sigma^1,\sigma^2) \in X^2$ violating condition (5.47) with $\theta_N = \lambda_N^2$ is contained in the set

$$\Big\{(\sigma^1,\sigma^2) \in X^2 : |R(\sigma^1,\sigma^2)| > \lambda_N \text{ or } \sigma^1 \in \mathcal{T}_N^c \text{ or } \sigma^2 >\in \mathcal{T}_N^c\Big\}. \tag{5.52}$$

Thus to prove that II_g is $o(1)$ for $\theta_N = \frac{1}{\sqrt{N}\log^2 N}$ it suffices to prove that

$$\sum_{\sigma^1,\sigma^2} \mathbb{P}(H'_N(\sigma^1) \in A, H'_N(\sigma^2) \in A) = o(1), \tag{5.53}$$

where the summation is over all pairs of distinct configurations contained in the set (5.52) with $\lambda_N = \sqrt{\theta_N} = \frac{1}{N^{1/4}\log N}$.

Let us prove (5.53). Since we already proved in Section 4 that the contribution from the set

$$\{(\sigma^1,\sigma^2) \in X^2 : |R(\sigma^1,\sigma^2)| > \lambda_N\} \tag{5.54}$$

to the sum (5.46) is negligible, it is enough to consider the sum (5.46) restricted to the set

$$\Big\{(\sigma^1,\sigma^2) \in X^2 : \sigma^1 \text{ or } \sigma^2 \in (\mathcal{T}_N^c)^X \text{ and } |R(\sigma^1,\sigma^2)| < \lambda_N\Big\}. \tag{5.55}$$

By Stirling's formula we obtain that $|\mathcal{T}_N^c| = \sqrt{\frac{2}{\pi N}}2^N e^{-N\mathcal{J}(\lambda_N)}(1+O(\lambda_N^2))$ and using this fact one can prove, proceeding as in part (i) of Theorem 2.4, that

$$|(\mathcal{T}_N^c)^X| = \mathrm{E}|(\mathcal{T}_N^c)^X|(1+o(1)) = \sqrt{\frac{2}{\pi N}}\, 2^M e^{-N\mathcal{J}(\lambda_N)}(1+o(1)). \tag{5.56}$$

Thus for large enough N the contribution from the set (5.55) to the sum (5.46) is bounded by

$$2\sqrt{\frac{2}{\pi N}} 2^{2M} e^{-N\mathcal{J}(\lambda_N)} \frac{C}{2^{2M}} e^{\frac{2b12}{1+b_{12}}(M\log 2 - 1/2\log M)}, \tag{5.57}$$

where the constant C is from (4.17). Using that $N\mathcal{J}(\lambda_N) = \frac{1}{2}N\lambda_N^2 + O\Big(\frac{1}{\log^4 N}\Big)$ we can further bound (5.57) by

$$2C\sqrt{\frac{2}{\pi N}} e^{-\frac{1}{2}N\lambda_N^2(1-4\varepsilon\log 2)}, \tag{5.58}$$

which is $o(1)$ by the choice of λ_N and ε.

Our next step is to bound the sum II by II_g. For this purpose we need to give an estimate of the joint density of $H'_N(\sigma^1), H'_N(\sigma^2)$ that would be valid also for pairs (σ^1,σ^2) violating condition (5.47). As we already noted for such (σ^1,σ^2) the results of Subsection 5.1 cannot be applied directly since it is no longer true that $\max_{\sigma,\sigma'\in X}|R(\sigma,\sigma')|$ is $o(1)$. Fortunately, we have only 2×2 covariance matrix $B(\sigma^1,\sigma^2)$ and using this fact we can easily adapt the results of Subsection 5.1 to the case where $\max_{\sigma,\sigma'\in X}|R(\sigma,\sigma')|$ is not $o(1)$. We start with formula (5.27) which, in the case $\ell=2$, can be rewritten as

$$\mathbb{P}\big(H'_N(\sigma^j)\in(x_j,x_j+dx_j)\text{ for }j=1,2\big)$$
$$=b_N^2 n\,dx_1dx_2\iint_{-\infty}^{\infty}\prod_{1\le i_1,i_2\le N}\hat\rho(v_{i_1,i_2})e^{2\pi i\sqrt{n}\left(f_1\alpha_N^{(1)}+f_2\alpha_N^{(2)}\right)}df_1df_2. \tag{5.59}$$

We can rewrite the integral in the above expression as

$$\iint_{-\infty}^{\infty}\prod_{\boldsymbol\delta}\hat\rho(\mathbf{f}\cdot\boldsymbol\delta)^{n_\delta}\,e^{2\pi in\mathbf{f}\cdot\boldsymbol\alpha}df_1df_2, \tag{5.60}$$

where $\boldsymbol\delta\in\{-1,1\}^2$. Since the function $\hat\rho$ is even we obtain

$$\prod_{\boldsymbol\delta}\hat\rho(\mathbf{f}\cdot\boldsymbol\delta)^{n_\delta}=\hat\rho(f_1+f_2)^{n_{(1,1)}+n_{(-1,-1)}}\hat\rho(f_1-f_2)^{n_{(1,-1)}+n_{(-1,1)}}. \tag{5.61}$$

One obvious relation between $n_{(1,1)},n_{(-1,-1)},n_{(1,-1)}$ and $n_{(-1,1)}$ is

$$n_{(1,1)}+n_{(-1,-1)}+n_{(1,-1)}+n_{(-1,1)}=n. \tag{5.62}$$

The other one we obtain by noting that

$$n_{(1,1)}+n_{(-1,-1)}-n_{(-1,1)}-n_{(1,-1)}=\sum_{i,j}\sigma_i^1\sigma_j^1\sigma_i^2\sigma_j^2=nR_{12}^2. \tag{5.63}$$

Therefore

$$\begin{cases} n_{(1,1)}+n_{(-1,-1)}=\frac12 n(1+R_{12}^2),\\ n_{(1,-1)}+n_{(-1,1)}=\frac12 n(1-R_{12}^2).\end{cases} \tag{5.64}$$

By Theorem 2.4 we conclude that P-a.s. $\max_{\sigma,\sigma'\in X}|R(\sigma,\sigma')|<1$ and therefore, for some positive constant c,

$$n_{(1,1)}+n_{(-1,-1)}\ge n_{(1,-1)}+n_{(-1,1)}\ge cn. \tag{5.65}$$

The above inequality allows us to approximate (5.60) by

$$\iint_{-\mu_1}^{\mu_1}e^{2\pi n(i\mathbf{f}\cdot\boldsymbol\alpha-\boldsymbol\eta\cdot\boldsymbol\alpha)}\prod_{\boldsymbol\delta}\hat\rho(\mathbf{f}\cdot\boldsymbol\delta+i\boldsymbol\eta\cdot\boldsymbol\delta)^{n_\delta}\,df_1df_2+O\big(e^{-c_1n}\big). \tag{5.66}$$

Adapting the proof of Lemma 5.5 from [BCMN05b] we get

$$\iint_{-\mu_1}^{\mu_1}e^{2\pi n(i\mathbf{f}\cdot\boldsymbol\alpha-\boldsymbol\eta\cdot\boldsymbol\alpha)}\prod_{\boldsymbol\delta}\hat\rho(\mathbf{f}\cdot\boldsymbol\delta+i\boldsymbol\eta\cdot\boldsymbol\delta)^{n_\delta}\,df_1df_2$$
$$=e^{-nG_{n,2}(\boldsymbol\alpha)}\frac{\sqrt{\det B(\sigma^1,\sigma^2)}}{2\pi n}\Big(1+O\Big(\frac{1}{\sqrt n}\Big)+O\Big(\frac{a_N^2}{n}\Big)\Big). \tag{5.67}$$

Finally, we obtain that

$$\mathbb{P}\big(H'_N(\sigma^j) \in (x_j, x_j + dx_j) \text{ for } j = 1,2\big) = dx_1 dx_2\, b_n e^{-nG_{n,2}(\boldsymbol{\alpha})} \\ \times \frac{\sqrt{\det B}}{2\pi}\Big(1 + O\Big(\frac{1}{\sqrt{n}}\Big) + O\Big(\frac{a_N^2}{n}\Big)\Big) + O\big(b_N^2 n e^{-c_1 n}\big). \tag{5.68}$$

From (5.42) we see that for some constant C

$$nG_{n,2} = \frac{n}{2}(\boldsymbol{\alpha}, B^{-1}\boldsymbol{\alpha}) + \frac{16c_4}{(1+R_{12}^2)^3}\frac{a_N^4}{n} + O\Big(\frac{a_N^6}{n^2}\Big) \geq \frac{n}{2}(\boldsymbol{\alpha}, B^{-1}\boldsymbol{\alpha}) + C, \tag{5.69}$$

and thus the joint density of $H'_N(\sigma^1), H'_N(\sigma^2)$ is bounded by

$$\frac{b_N^2\sqrt{\det B}}{2\pi}\, e^{-n(\boldsymbol{\alpha},B^{-1}\boldsymbol{\alpha})/2-C}\Big(1 + O\Big(\frac{1}{\sqrt{n}}\Big) + O\Big(\frac{a_N^2}{n}\Big)\Big) + O\big(b_N^2 n e^{-c_1 n}\big). \tag{5.70}$$

This last bound for the joint density clearly implies that the sum II could be bounded by II_g plus an error resulting from the second term. But the cumulative error coming from the second term is of order $O\big(2^{2M} b_N^2 n e^{-c_1 n}\big)$ which is negligible even in the case $\limsup M/N > 0$.

To prove the second statement of the lemma we will approximate the sum I by I_g, which was already calculated in Section 4. We first notice that for sequences $(\sigma^1, \dots, \sigma^\ell)$ satisfying condition (5.47) $R_{\max}(\sigma^1, \dots, \sigma^\ell) = O(\theta_N)$. Furthermore, for configurations $(\sigma^1, \dots, \sigma^\ell)$ obeying condition (5.47) it is possible to derive from (5.42) that

$$G_{n,\ell}(\boldsymbol{\alpha}) = \frac{1}{2}(\boldsymbol{\alpha}, B^{-1}\boldsymbol{\alpha}) + c_4\ell(1 + 3(\ell-1))\frac{a_N^4}{n} + O\Big(\frac{a_N^6}{n^3}\Big) + O\Big(\frac{a_N^2}{n^2}\theta_N\Big). \tag{5.71}$$

For the details of the derivation we refer to Subsection 5.4 of [BCMN05b] and in particular to formula (5.57) in there. Using formula (5.71) with $\ell = 2$ and substituting it into (5.35) we obtain that

$$I = I_g\, e^{-32c_4\varepsilon^2\log^2 2}\Big(1 + O\Big(\frac{1}{\sqrt{n}}\Big) + O\Big(\frac{a_N^2}{n}\Big)\Big). \tag{5.72}$$

This finishes the proof of Lemma 5.6. □

To conclude the calculation of the second moment we notice that summing I and II we get the second factorial moment

$$\mathbb{E}(\mathcal{P}_N(A))_2 = \frac{e^{-32c_4\varepsilon^2\log^2 2}(\mu(A))^2}{\sqrt{1-4\varepsilon\log 2}}(1 + o(1)). \tag{5.73}$$

Assertion (i) of Theorem 5.5 is thus proven.

3. **Third moment estimate.** To deal with the third moment

$$\mathbb{E}(\mathcal{P}_N(A))_3 = \sum_{(\sigma^1,\sigma^2,\sigma^3)\in X^3} \mathbb{P}\left(H'_N(\sigma^1) \in A, H'_N(\sigma^2) \in A, H'_N(\sigma^3) \in A\right), \tag{5.74}$$

we use the same strategy as we used to calculate the second moment. In particular, we fix $\ell = 3$ and split the sum (5.74) in two parts:

$$I = \sum_{\sigma^1,\sigma^2,\sigma^3} \mathbb{P}\big(H'_N(\sigma^1) \in A, H'_N(\sigma^2) \in A, H'_N(\sigma^3) \in A\big), \tag{5.75}$$

where the sum runs over all sequences of distinct configurations $(\sigma^1, \sigma^2, \sigma^3) \in X^3$ satisfying condition (5.47), and

$$II = \sum_{\sigma^1,\sigma^2,\sigma^3} \mathbb{P}\big(H'_N(\sigma^1) \in A, H'_N(\sigma^2) \in A, H'_N(\sigma^3) \in A\big), \tag{5.76}$$

where the sum is over all sequences of distinct configurations $(\sigma^1, \sigma^2, \sigma^3) \in X^3$ violating (5.47).

By exactly the same argument as in the calculation of the second moment the contribution from the "typical" collections, I, is bounded. We therefore concentrate on the analysis of the contribution from the "atypical" collections, II. Since we are interested only in the estimate of the third moment from above it suffices to bound the joint density $\mathbb{P}\big(H'_N(\sigma^j) \in (x_j, x_j + dx_j) \text{ for } j = 1, 2, 3\big)$. We start with formula (5.27) which, in the case $\ell = 3$, can be rewritten as

$$\begin{aligned} \mathbb{P}\big(H'_N(\sigma^j) \in (x_j, x_j + dx_j) \text{ for } j = 1, 2, 3\big) &= b_N^3 n \, dx_1 dx_2 dx_3 \\ \times \iiint_{-\infty}^{\infty} \prod_{i_1,i_2,i_3} \hat{\rho}(v_{i_1,i_2,i_3}) &e^{2\pi i\sqrt{n}\left(f_1\alpha_N^{(1)} + f_2\alpha_N^{(2)} + f_3\alpha_N^{(3)}\right)} df_1 df_2 df_3. \end{aligned} \tag{5.77}$$

We can rewrite the integral in the above expression as

$$\iiint_{-\infty}^{\infty} \prod_{\boldsymbol{\delta}} \hat{\rho}(\mathbf{f} \cdot \boldsymbol{\delta})^{n_{\boldsymbol{\delta}}} \, e^{2\pi i n \mathbf{f}\cdot\boldsymbol{\alpha}} df_1 df_2 df_3, \tag{5.78}$$

where $\boldsymbol{\delta} \in \{-1, 1\}^3$. Since the function $\hat{\rho}$ is even we obtain

$$\begin{aligned} \prod_{\boldsymbol{\delta}} \hat{\rho}(\mathbf{f} \cdot \boldsymbol{\delta})^{n_{\boldsymbol{\delta}}} &= \hat{\rho}(f_1 + f_2 + f_3)^{n_1} \hat{\rho}(f_1 + f_2 - f_3)^{n_2} \\ &\times \hat{\rho}(f_1 - f_2 + f_3)^{n_3} \hat{\rho}(f_1 - f_2 - f_3)^{n_4}, \end{aligned} \tag{5.79}$$

where

$$\begin{cases} n_1 = n_{(1,1,1)} + n_{(-1,-1,-1)}, \\ n_2 = n_{(1,1,-1)} + n_{(-1,-1,1)}, \\ n_3 = n_{(1,-1,1)} + n_{(-1,1,-1)}, \\ n_4 = n_{(-1,1,1)} + n_{(1,-1,-1)}. \end{cases} \tag{5.80}$$

By definition of the matrix $C_2(\sigma^1, \sigma^2, \sigma^3)$ we have

$$\begin{cases} n_1 + n_2 - n_3 - n_4 = nR_{12}^2, \\ n_1 - n_2 - n_3 + n_4 = nR_{23}^2, \\ n_1 - n_2 + n_3 - n_4 = nR_{31}^2, \\ n_1 + n_2 + n_3 + n_4 = n. \end{cases} \tag{5.81}$$

Solving the system

$$\begin{cases} n_1 = \frac{1}{4}n(1 + R_{12}^2 + R_{23}^2 + R_{31}^2), \\ n_2 = \frac{1}{4}n(1 + R_{12}^2 - R_{23}^2 - R_{31}^2), \\ n_3 = \frac{1}{4}n(1 - R_{12}^2 - R_{23}^2 + R_{31}^2), \\ n_4 = \frac{1}{4}n(1 - R_{12}^2 + R_{23}^2 - R_{31}^2). \end{cases} \tag{5.82}$$

From Theorem 2.4 we obtain that P-almost surely

$$\limsup_{N\to\infty} \max_{\sigma^1,\sigma^2\in X} \mathcal{J}\big(R(\sigma^1,\sigma^2)\big) \le \varepsilon \log 2. \tag{5.83}$$

Since the function $\mathcal{J}$ is monotone we obtain from (5.83) and from assumption $\varepsilon < \frac{1}{8\log 2}$ that P-a.s.

$$\limsup_{N\to\infty} \max_{\sigma^1,\sigma^2\in X} |R(\sigma^1,\sigma^2)| < \frac{1}{2}. \tag{5.84}$$

It implies that $\min\{n_1,n_2,n_3,n_4\} \ge cn$ for some positive constant c. It allows us to approximate the integral in (5.77) by

$$\iiint_{-\mu_1}^{\mu_1} e^{2\pi n(i\mathbf{f}\cdot\boldsymbol{\alpha}-\boldsymbol{\eta}\cdot\boldsymbol{\alpha})} \prod_{\boldsymbol{\delta}} \hat{\rho}(\mathbf{f}\cdot\boldsymbol{\delta} + i\boldsymbol{\eta}\cdot\boldsymbol{\delta})^{n_\delta}\, df_1 df_2 df_3 + O\big(e^{-cn}\big). \tag{5.85}$$

Adapting the proof of Lemma 5.5 from [BCMN05b] we get

$$\iiint_{-\mu_1}^{\mu_1} e^{2\pi n(i\mathbf{f}\cdot\boldsymbol{\alpha}-\boldsymbol{\eta}\cdot\boldsymbol{\alpha})} \prod_{\boldsymbol{\delta}} \hat{\rho}(\mathbf{f}\cdot\boldsymbol{\delta} + i\boldsymbol{\eta}\cdot\boldsymbol{\delta})^{n_\delta}\, df_1 df_2 df_3$$

$$= e^{-nG_{n,3}(\boldsymbol{\alpha})} \frac{\sqrt{\det B(\sigma^1,\sigma^2)}}{(2\pi n)^{3/2}} \Big(1 + O\Big(\frac{1}{\sqrt{n}}\Big) + O\Big(\frac{a_N^2}{n}\Big)\Big). \tag{5.86}$$

Next, after a little algebra, one can derive from (5.42) that for some constant C

$$nG_{n,3} \ge \frac{n}{2}(\boldsymbol{\alpha}, B^{-1}\boldsymbol{\alpha}) + C \tag{5.87}$$

and this estimate is enough to bound the joint density of $H'_N(\sigma^1), H'_N(\sigma^2), H'_N(\sigma^3)$ by the joint density of $Z_N(\sigma^1), Z_N(\sigma^2), Z_N(\sigma^3)$. Thus Theorem 5.5 is proved. □

Acknowledgment

We thank J.Černý for a helpful reading of the first versions of the manuscript. V. Gayrard thanks the Chair of Stochastic Modeling of the École Polytechnique Fédérale of Lausanne for financial support. A. Kuptsov gratefully acknowledges the support of NSF grant DMS-0102541 "Collaborative Research: Mathematical Studies of Short-Ranged Spin Glasses."

References

[BFM04] H. Bauke, S. Franz, and S. Mertens, *Number partitioning random energy model.* Journal of Statistical Mechanics: Theory and Experiment, page P04003, 2004

[BM04] H. Bauke and S. Mertens, *Universality in the level statistics of disordered systems.* Physical Review E **70**, 025102(R), 2004

[BCMN05a] C. Borgs, J.T. Chayes, S. Mertens, and Ch. Nair, *Proof of the local REM conjecture for number partitioning I: Constant energy scales.* preprint, 2005. http://arxiv.org/abs/cond-mat/0501760, to appear in Random Structures & Algorithms.

[BCMN05b] C. Borgs, J.T. Chayes, S. Mertens, and Ch. Nair, *Proof of the local REM conjecture for number partitioning II: Growing energy scales.* preprint, 2005. http://arxiv.org/abs/cond-mat/0508600

[BCP01] C. Borgs, J.T. Chayes and B. Pittel, *Phase transition and finite-size scaling for the integer partitioning problem* Random Structures & Algorithms, **19** (2001), no. 3-4, 247–288.

[B06] A. Bovier, *Statistical mechanics of disordered systems.* Cambridge Series in Statistical and Probabilistic Mathematics 18, Cambridge University Press, 2006.

[BK06a] A. Bovier and I. Kurkova, *Local Energy Statistics in Disordered Systems: A Proof of the Local REM Conjecture.* Commun. Math. Phys. 263, 513–533 (2006).

[BK06b] A. Bovier and I. Kurkova, *Local energy statistics in spin glasses,* Journal of Statistical Physics, (Online).

[BK04] A. Bovier and I. Kurkova, *Poisson convergence in the restricted k-partitioning problem.* WIAS preprint 964, to appear in Random Structures & Algorithms, 2006.

[BKL02] A. Bovier, I. Kurkova, and M. Löwe, *Fluctuations of the free energy in the REM and the p-spin SK models.* Ann. Probab. 30, 605–651 (2002)

[Kal83] O. Kallenberg, *Random measures.* 3rd, rev. and enl. ed. New York : Academic Press, 1983

[Mer00] S. Mertens, *Random costs in combinatorial optimization.* Phys. Rev. Lett. **84** (2000), no. 6, 1347–1350.

[Tal03] M. Talagrand, *Spin glasses: a challenge for mathematicians: cavity and mean field models.* Berlin; New York: Springer, © 2003

Gérard Ben Arous
Swiss Federal Institute of Technology (EPFL), CH-1015 Lausanne, Switzerland
and
Courant Institute of Mathematical Sciences, New York University,
251 Mercer Street, New York, NY 10012
e-mail: benarous@cims.nyu.edu

Véronique Gayrard
Laboratoire d'Analyse, Topologie, Probabilités
CMI, 39 rue Joliot-Curie, F-13453 Marseille Cedex
e-mail: gayrard@latp.univ-mrs.fr, veronique@gayrard.net

Alexey Kuptsov
Swiss Federal Institute of Technology (EPFL), CH-1015 Lausanne, Switzerland
and
Courant Institute of Mathematical Sciences, New York University,
251 Mercer Street, New York, NY 10012
e-mail: kuptsov@cims.nyu.edu

Progress in Probability, Vol. 60, 97–116

The Biham-Middleton-Levine Traffic Model for a Single Junction

Itai Benjamini, Ori Gurel-Gurevich and Roey Izkovsky

Abstract. In the Biham-Middleton-Levine traffic model cars are placed in some density p on a two dimensional torus, and move according to a (simple) set of predefined rules. Computer simulations show this system exhibits many interesting phenomena: for low densities the system self organizes such that cars flow freely while for densities higher than some critical density the system gets stuck in an endless traffic jam. However, apart from the simulation results very few properties of the system were proven rigorously to date. We introduce a simplified version of this model in which cars are placed in a single row and column (a junction) and show that similar phenomena of self-organization of the system and phase transition still occur.

Mathematics Subject Classification (2000). 60K35.

Keywords. Traffic, phase transition, cellular automata.

1. The BML traffic model

The Biham-Middleton-Levine (BML) traffic models was first introduced in [3] published 1992. The model involves two types of cars: "red" and "blue". Initially the cars are placed in random with a given density p on the $N \times N$ torus. The system dynamics are as follows: at each turn, first all the red cars try to move simultaneously a single step to the right in the torus. Afterwards all blue cars try to move a single step upwards. A car succeeds in moving as long as the relevant space above/beside it (according to whether it is blue/red) is vacant.

The basic properties of this model are described in [3] and some recent more subtle observations due to larger simulations are described in [4]. The main and most interesting property of the system originally observed in simulations is a *phase transition*: for some critical density p_c one observes, that while filling the torus randomly with cars in density $p < p_c$ the system self organizes such that after some time all cars flow freely with no car ever running into another car (see Figure 1), by slightly changing the density to some $p > p_c$ not only does system

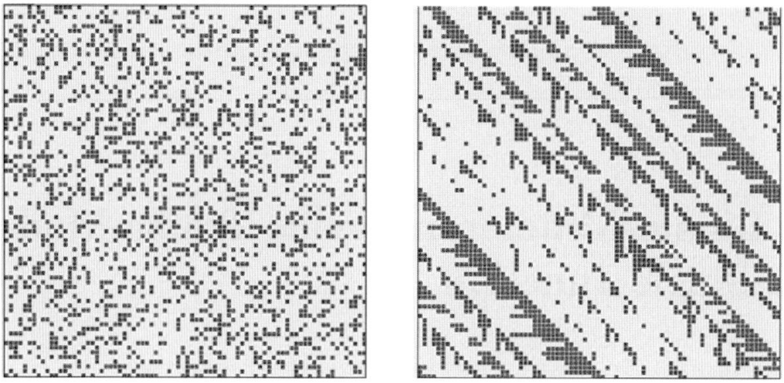

FIGURE 1. Self organization for $p = 0.3$: the initial configuration in the left organizes after 520 steps to the "diagonal" configuration on the right which flows freely.

not reach a free flow, but it will eventually get stuck in a configuration in which no car can ever move (see Figure 2).

Very little of the above behaviour is rigorously proven for the BML model. The main rigorous result is that of Angel, Holroyd and Martin [2], showing that for some fixed density $p < 1$, very close to 1, the probability of the system getting stuck tends to 1 as $N \to \infty$. In [1] one can find a study of the very low density regime (when $p = O(\frac{1}{N})$).

First, we introduce a slight variant of the original BML model by allowing a car (say, red) to move not only if there is a vacant place right next to it but also if there is a red car next to it that moves. Thus, for sequence of red cars placed in a row with a single vacant place to its right – all cars will move together (as oppose to only the rightmost car in the sequence for the original BML model). Not only does this new variant exhibits the same phenomena of self-organization and phase transition, they even seem to appear more quickly (i.e., it takes less time for the system to reach a stable state). Actually the demonstrated simulations (Figures 1, 2) were performed using the variant model. Note that the results of [2] appear to apply equally well to the variant model.

In the following, we will analyze a simplified version of the BML model: BML on a single junction. Meaning, we place red cars in some density p on a single row of the torus, and blue cars are placed in density p on a single column. We will show that for all p the system reaches *optimal speed*[1], depending only on p. For $p < 0.5$ we will show the system reaches speed 1, while for $p > 0.5$ the speed cannot be 1, but the system will reach the same speed, regardless of the initial configuration. Moreover, at $p = 0.5$ the system's behaviour undergoes a phase transition: we will prove that while for $p < 0.5$ the stable configuration will have linearly many

[1]The speed of the system is the asymptotic average rate in which a car moves – i.e., number of actual moves per turn.

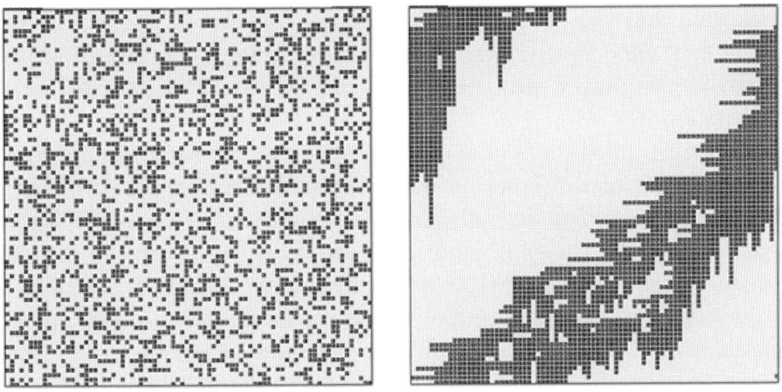

FIGURE 2. Traffic jam for $p = 0.35$: the initial configuration on the left gets to the stuck configuration on the right after 400 steps, in which no car can ever move.

sequences of cars, for $p > 0.5$ we will have only $O(1)$ different sequences after some time. We will also examine what happens at a small window around $p = 0.5$.

Note that in the variant BML model (and unlike the original BML model) car sequences are never split. Therefore, the simplified version of the variant BML model can be viewed as some kind of 1-dimensional coalescent process.

Much of the proofs below rely on the fact that we model BML on a symmetric torus. Indeed, the time-normalization of Section 2.2 would take an entirely different form if the height and width were not equal. We suspect that the model would exhibit similar properties if the height and width had a large common denominator, e.g., if they had a fixed proportion. More importantly, we believe that for height and width with low common denominator (e.g., relatively prime), we would see a clearly different behaviour. As a simple example, note that in the relatively prime case, a speed of precisely 1 cannot be attained, no matter how low is p, in contrast with Corollary 3.14. This dependence on the arithmetic properties of the dimensions is also apparent in [4].

2. The junction model

2.1. Basic model

We start with the exact definition of our simplified model. On a cross shape, containing a single horizontal segment and a single vertical segment, both of identical length N, red cars are placed in exactly pN randomly (and uniformly) chosen locations along the row, and blue cars are similarly placed in pN locations along the column. p will be called the density of the configuration. For simplicity and consistency with the general BML model, we refer to the cars placed on the horizontal segment as red and those on the vertical segment as blue. For simplicity we may assume that the *junction*, i.e., the single location in the intersection of the segments, is left unoccupied. The segments are cyclic – e.g., a red car located at the rightmost position ($N-1$) that moves one step to the right re-emerges at the leftmost position (0).

At each turn, all the red cars move one step to the right, except that the red car that is just left of the junction will not move if a blue car is in the junction (i.e., blocking it), in which case also all the red cars immediately following it will stay still. Afterwards, the blue cars move similarly, with the red and blue roles switched. As in the original BML, we look at the asymptotic speed of the system, i.e., the (asymptotic) average number of steps a car moves in one turn. It is easily seen that this speed is the same for all blue cars and all red cars, since the number of steps any two cars of the same color have taken cannot differ by more than N. It is somewhat surprising, perhaps, that these two speed must also be the same. For instance, there is no configuration for which a blue car completes 2 cycles for every 1 a red car completes.

2.2. Time-normalized model

Though less natural, it will be sometimes useful to consider the equivalent description, in which two rows of cars – red and blue – are placed one beneath the other, with a "special place" – the junction – where at most one car can be present at any time. In every step first the red line shifts one to the right (except cars immediately to the left of the junction, if it contains a blue car) and then the blue line does the same. Furthermore, instead of having the cars move to the right, we can have the junction move to the left, and when a blue car is in the junction, the (possibly empty) sequence of red cars immediately to the left of the junction moves to the left, and vice verse. Figure 3 illustrates the correspondence between these models.

From the discussion above we get the following equivalent system, which we will call the time-normalized junction:

1. Fix $S = N-1 \in \mathbb{Z}_N$ and fix some initial configuration $\{R_i\}_{i=0}^{N-1}, \{B_i\}_{i=0}^{N-1} \in \{0,1\}^N$ representing the red and blue cars respectively, i.e., $R_i = 1$ iff there's a red car at the ith place. We require that $\sum R_i = \sum B_i = p \cdot N$, and at place S $(= N-1)$ there is at most one car in both rows[2].
2. In each turn:
   - If place S contains a blue car, and place $S-1$ contains a red car (if $B_S = R_{S-1} = 1$), push this car one step to the left. By pushing the car we mean also moving all red cars that immediately followed it one step to the left, i.e., set $R_{S-1} = 0$, $R_{S-i} = 1$ for $i = \min_{j\geq 1}[R_{S-j} = 0]$.
   - If place S does not contain a blue car and place $S-1$ contains both a red and a blue car (if $B_S = 0$ and $R_{S-1} = B_{S-1} = 1$), push the blue car at $S-1$ one step to the left (set $B_{S-1} = 0$ and $B_{S-i} = 1$ for $i = \min_{j\geq 1}[B_{S-j} = 0]$).
   - set $S = S-1$

[2]The last requirement is that the junction itself can contain only one car at the beginning.

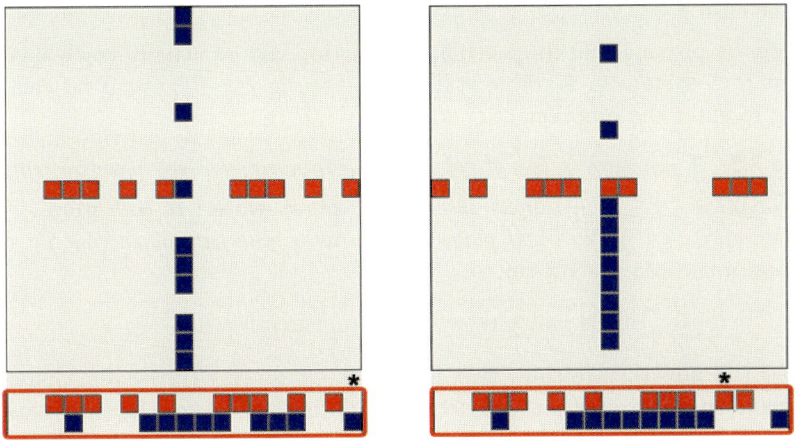

FIGURE 3. On the left: a junction configuration and the analogous configuration beneath it. The junction is marked with an asterisk. On the right: same configuration after 3 turns in both views.

Note that indeed the dynamics above guarantee that after a turn place $(S-1)$ – the new junction – contains at most one car. Generally, as long as cars flow freely, the time-normalized system configuration does not change (except for the location of the junction). Cars in the time-normalized system configuration actually move only when some cars are waiting for the junction to clear (in the non-time-normalized system).

3. Analysis of an (N, p) junction

Our analysis of the junction will argue that for any p, regardless of the initial configuration, the system will reach some optimal speed, depending only on p. First let us state what is this optimal speed:

Theorem 3.1. *For a junction with density p, the maximal speed for any configuration is* $\min(1, \frac{1}{2p})$.

Proof. Obviously, the speed cannot exceed 1. Observe the system when it reaches its stable state, and denote the speed by s. Thus, at time t, a car has advanced $ts(1 + o(ts))$ steps (on average) which means that it has passed the junction $ts/N(1 + o(1))$ times. As only one car can pass the junction at any time, the total number of cars passing the junction at time t is bounded by t. Therefore, we have $2pN \times ts/N \leq t$ which implies $s \leq 1/2p$. □

We will now show that the system necessarily reaches these speeds from any starting configuration.

3.1. The case $p < 0.5$

We begin by proving that for $p < 0.5$ the junction will eventually reach speed very close to 1. A system in a stable state is said to be *free-flowing* if no car is ever waiting to enter the junction.

Lemma 3.2. *A junction is free-flowing iff the time-normalized junction satisfies:*

(1) *For all* $0 \leq i \leq N-1$ *there is only one car in place* i *in both rows.*
(2) *For all* $0 \leq i \leq N-1$ *if place* i *contains a blue car, place* $(i-1) \mod N$ *does not contain a red car.*

Proof. Obviously, this is just a reformulation of free-flowing. □

We will now turn to show that for $p < 0.5$, following the system dynamics of the time-normalized junction will necessarily bring us to a free flowing state, or at worst an "almost" free flowing state, meaning a state for which the system speed will be arbitrarily close to 1 for large enough N.

For this let us consider some configuration and look at the set of "violations" to Lemma 3.2, i.e., places that either contain both a blue and a red car, or that contain a blue car and a red car is in the place one to the left. As the following lemma will show, the size of the set of violations is very closely related to the system speed, and posses a very important property: it is non-increasing in time.

More formally, for a configuration R, B we define two disjoint sets for the two types of violations:

$$V_B = \{0 \leq i \leq N-1 : R_{i-1} = B_{i-1} = 1, B_i = 0\},$$

$$V_R = \{0 \leq i \leq N-1 : R_{i-1} = B_i = 1\}.$$

Also, let $V = V_B \cup V_R$ be the set of all violations. It will be sometimes useful to refer to a set of indicators $X = \{X(i)\}_{i=0}^{N-1}$, where each $X(i)$ is 1 if $i \in V$ and 0 otherwise, thus $|V| = \sum_{i=0}^{N-1} X(i)$.

For a junction with some initial configuration R, B, let R^t, B^t be the system configuration at time t, and let $V_t = V_{B^t} \cup V_{R^t}$ be the set of violations for this configuration, and X^t be the corresponding indicator vector. Similarly, let S^t denote the junction's position at time t.

Lemma 3.3.

1. $|V_{t+1}| \leq |V_t|$.
2. *For any* t, *the system speed is at least* $(1 + \frac{|V_t|}{N})^{-1} \geq 1 - \frac{|V_t|}{N}$.

Proof. Property (1) follows from the system dynamics. To see this, examine the three possible cases for what happens at time t:

1. If in turn t place S^t does not contain a violation then the configurations do not change during the next turn, i.e., $R^{t+1} = R^t$, $B^{t+1} = B^t$ hence clearly $|V_{t+1}| = |V_t|$.
2. If $S^t \in V_{B^t}$, then the configuration B^{t+1} changes in two places:

(a) $B^t_{S^t-1}$ is changed from 1 to 0. Thus, place S^t is no longer in V_{B^t}, i.e., $X^t(S^t)$ changes from 1 to 0.

(b) $B^t_{S^t-i}$ is changed from 0 to 1 for $i = \min_{j\geq 1}(B^t_{S^t-j} = 0)$. This may affect $X^{t+1}(S^t - i)$, and $X^{t+1}(S^t - i + 1)$. However, for place $S^t - i + 1$, by changing $B^t_{S^t-i}$ from 0 to 1 no new violation can be created (since by definition of i, $B^{t+1}_{S^t-i+1} = 1$, so $X^{t+1}(S^t - i + 1) = 1$ iff $R^{t+1}_{S^t-i} = 1$ regardless of $B^{t+1}_{S^t-i}$).

For other indices $X^{t+1}(i) = X^t(i)$ since R, B do not change, so between times t and $t+1$, we have that $X^t(S^t)$ changes from 1 to 0, and at worst only one other place - $X^t(S^t - i + 1)$ changes from 0 to 1, so $|V_{t+1}| = \sum_{i=0}^{N-1} X^{t+1}(i) \leq \sum_{i=0}^{N-1} X^t(i) = |V_t|$.

3. Similarly, if place $S^t \in V_{R^t}$ then the configuration R^{t+1} changes in two places:
   (a) $R^{t+1}_{S^t-1}$ is changed from 1 to 0. Thus, $X^t(S^t)$ changes from 0 to 1.
   (b) R_{S^t-i} is changed from 0 to 1 for $i = \min_{j\geq 1}(R^t_{S^t-j} = 0)$, affecting $X^{t+1}(S^t - i)$, $X^{t+1}(S^t - i + 1)$. However for place $S^t - i$ changing R_{S^t-i} does not affect whether this place is a violation or not, so at worst $X^t(S^t - i + 1)$ changed from 0 to 1.

   By the same argument we get $|V_{t+1}| \leq |V_t|$.

Therefore, $|V_{t+1}| \leq |V_t|$.

For property (2) we note that in the time-normalized system, following a specific car in the system, its "current speed" is $\frac{N}{N+k} = (1 + \frac{k}{N})^{-1}$ where k is the number of times the car was pushed to the left during the last N system turns. We note that if a car at place j is pushed to the left at some time t, by some violation at place S^t, this violation can reappear only to the left of j, so it can push the car again only after S^t passes j. Hence any violation can push a car to the left only once in a car's cycle (i.e., N moves). Since by (1) at any time from t onwards the number of violations in the system is at most $|V_t|$, then each car is pushed left only $|V_t|$ times in N turns, so its speed from time t onwards is at least $(1 + \frac{|V_t|}{N})^{-1} > 1 - \frac{|V_t|}{N}$ as asserted. □

With Lemma 3.3 at hand we are now ready to prove system self organization for $p < 0.5$. We will show that for $p < 0.5$, after $2N$ system turns $|V_t| = O(1)$, and hence deduce by part (2) of Lemma 3.3 the system reaches speed $1 - O(\frac{1}{N}) \to 1$ (as $N \to \infty$).

As the junction advances to the left it pushes some car sequences, thus affecting the configuration to its left. The next lemma will show that when $p < 0.5$, for some $T < N$, the number of cars affected to the left of the junction is only a constant, independent of N.

Lemma 3.4. *Consider a junction with density* $p < 0.5$. *There exists some constant* $C = C(p) = \frac{p}{1-2p}$, *independent of* N, *for which:*

From any configuration R, B *with junction at place* S *there exist some* $0 < T < N$ *such that after* T *turns:*

(1) *For* $i \in \{S-T,\ldots,S\}$, $X^T(i)=0$ (*i.e., there are no violations there*).
(2) *For* $i \in \{S+1,\ldots,N-1,0,\ldots,S-T-C\}$, $R_i^T = R_i^0$ *and* $B_i^T = B_i^0$ (R, B *are unchanged there*).

Proof. First let us consider $T=1$. For $T=1$ to not satisfy the lemma conclusions, there need to be a car sequence (either red or blue) of length exceeding C, which is pushed left by the junction as it moves. Progressing through the process, if, for some $T<N-C$, the sequence currently pushed by the junction is shorter then C, then this is the T we seek. Therefore, for the conclusions *not* to hold, the length of the car sequence pushed by the junction must exceed C for all $0<T<N$. If this is the case, then leaving the junction we see alternating red and blue sequences, all of lengths exceeding C and one vacant place after any blue sequence and before any red one.

However, if this is the case for all $0<T\le T'$ then the average number of cars per location in $\{S-T',\ldots,S\}$ at time T' must be at least $\frac{2C}{2C+1}$ (at least $2C$ cars between vacant places). Therefore, the total number of cars in $\{S-T',\ldots,S\}$ at time T' is more than $\frac{2C}{2C+1}T' = 2pT'$ (recall that $C=\frac{2p}{1-2p}$).

Since there are only $2pN$ cars in the system, this cannot hold for all T up to N. Thus, there must be some time for which the conclusions of the lemma are satisfied. □

We are now ready to easily prove the main result for this section.

Theorem 3.5. *A junction of size* N *with density* $p<0.5$ *reaches speed of* $1-\frac{C(p)}{N}$ *from any initial configuration.*

Proof. Let R,B be some initial configuration, with $S=N-1$ and let V the corresponding set of violations and X the matching indicators vector. By Lemma 3.4 there exist $T_0>0$ for which $X^{T_0}(i)=0$ for $i\in[N-1-T_0,N-1]$. Now starting at R^{T_0},B^{T_0} and $S=N-1-T_0$ reusing the lemma there exist $T_1>0$ s.t. $X^{T_0+T_1}(i)=0$ for $i\in[N-1-T_1,N-1-T_0]$, and also, as long as $N-1-T_0-T_1 > C(p)=\frac{2p}{1-2p}$, $X^{T_0+T_1}(i)=X^{T_0}(i)=0$ for $i\in[N-1-T_0,N-1]$ as well.

Proceeding in this manner until $T=\sum T_i \ge N$ we will get that after T turns, $X^T(i)=0$ for all but at most $C(p)$ places, hence by Lemma 3.3 the system speed from this time onward is at least $1-\frac{C(p)}{N}$. □

We remark one cannot prove an actual speed 1 (rather than $1-o(1)$) for the case $p<\frac{1}{2}$ since this is definitely not the case. Figure 4 (a) demonstrates a general construction of a junction with speed $1-\frac{1}{N}$ for all N. Finally we remark that a sharper constant $C'(p)$ can be provided, that will also meet a provable lower bound for the speed (by constructions similar to the above). This $C'(p)$ is exactly half of $C(p)$. We will see proof of this later in this paper as we obtain Theorem 3.5 as a corollary of a different theorem, using a different approach.

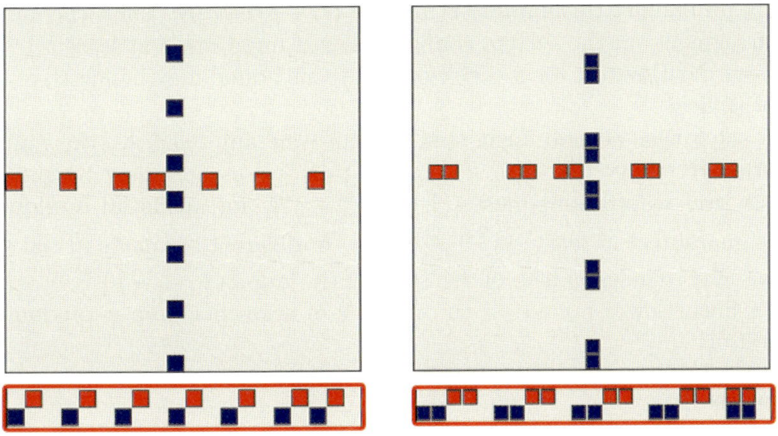

FIGURE 4. (a) A junction configuration with density $p = \frac{1}{3}$ ($\frac{p}{1-2p} = 1$) not reaching speed 1 (speed is $1 - \frac{1}{N}$). (b) A similar construction for $p = 0.4$ ($\frac{p}{1-2p} = 2$) reaching speed of only $1 - \frac{2}{N}$.

3.2. Number of segments for $p < 0.5$

The proof in the previous section were combinatorial in nature and showed that for a density $p < 0.5$ the system can slow-down only by some constant number independent of N regardless of the configuration. As we turn to examine the properties of the stable configuration of course we can not say much if we allow any initial configuration. There are configurations in which the number of different segments in each row will be $\Theta(N)$ while clearly if the cars are arranged in a single red sequence and a single blue sequence in the first place, we will have only one sequence of each color at any time.

However, we will show that for a random initial configuration, the system will have linearly many different segments of cars with high probability.

Theorem 3.6. *A junction of size N with density $p < 0.5$, started from a random initial configuration, will have $\Theta(N)$ different segments at all times with high probability (w.r.t. N).*

Proof. As we have already seen in the proof of 3.4, as the system completes a full round, since every place (but a constant number of places) contains at most a single car there must be $(1-2p)N$ places in which no car is present. Each two such places that are not adjacent must correspond to a segment in the cars configuration.

It is evident by the system dynamics, that the number of places for which $R_i = B_i = R_{i-1} = B_{i-1} = 0$ is non-increasing. More precisely, only places for which $R_i = B_i = R_{i-1} = B_{i-1} = 0$ in the initial configuration can satisfy this in the future. In a random initial configuration with density p, the initial number of these places is expected to be $(1-p)^4 N$, and by standard CLT, we get that with high probability this number is at most $((1-p)^4+\varepsilon)N$. Thus, the number of different segments in the system configuration at any time is at least $((1-2p)-(1-p)^4-\varepsilon)N$. However for p very close to $\frac{1}{2}$ this bound may be negative, so this does not suffice.

To solve this we note that similarly also for any fixed K, the number of consecutive K empty places in a configuration is non-increasing by the system dynamics, and w.h.p. is at most $((1-p)^{2K}+\varepsilon)N$ for an initial random state. But this guarantees at least $\frac{(1-2p)-(1-p)^{2K}-\varepsilon}{K-1}N$ different segments in the system. Choosing $K(p)$ (independent of N) sufficiently large s.t. $(1-p)^{2K} < (1-2p)$ we get a linear lower bound on the number of segments from a random initial configuration. □

3.3. The case $p > 0.5$

The proofs of speed optimality and segment structure for $p > 0.5$ will rely mainly of a the combinatorial properties of a *stable configuration.* A *stable configuration* for the system is a configuration that re-appears after running the system for some M turns. Since for a fixed N the number of possible configurations of the system is finite, and the state-transitions (traffic rules) are time independent, the system will necessarily reach a stable configuration at some time regardless of the starting configuration.

We will use mainly two simple facts that must hold after a system reached a stable configurations: (a) $|V_t|$ cannot change – i.e., no violation can disappear. This is clear from Lemma 3.3; (b) Two disjoint segments of car cannot merge to one (i.e., one pushed until it meets the other), since clearly the number of segments in the system is also non-increasing in time.

These two facts alone already provide plenty of information about the stable configuration for $p > 0.5$. We begin with the following twin lemmas on the stable state.

Lemma 3.7. *Let R, B be a stable configuration with junction at $S = 0$ and $B_0 = 0$. Assume that there is a sequence of exactly s_R consecutive red cars at places $[N-s_R, N-1]$ and s_B blue cars at places $[N-s_B, N-1]$, $s_R, s_B \geq 1$. Then:*

1. $B_i = 0$ *for* $i \in [N-s_R-s_B-1, N-s_B-1]$.
2. $R_i = 1$ *for* $i \in [N-s_R-s_B-1, N-\max(s_R, s_B)-2]$.
3. $R_i = 0$ *for* $i \in [N-\max(s_R, s_B)-1, N-s_R-1]$.

Proof. Since it is easy to lose the idea in all the notations, a visual sketch of the proof is provided in Figure 5.

By the assumptions:

- $B_i = 1$ for $i \in [N-s_B, N-1]$, $B_{N-s_B-1} = 0$.
- $R_i = 1$ for $i \in [N-s_R, N-1]$, $R_{N-s_R-1} = 0$.

To get (1), we note that with the assumption $B_0 = 0$, the blue sequence will be pushed to the left in the next s_R turns, so by restriction (b), $B_i = 0$ for

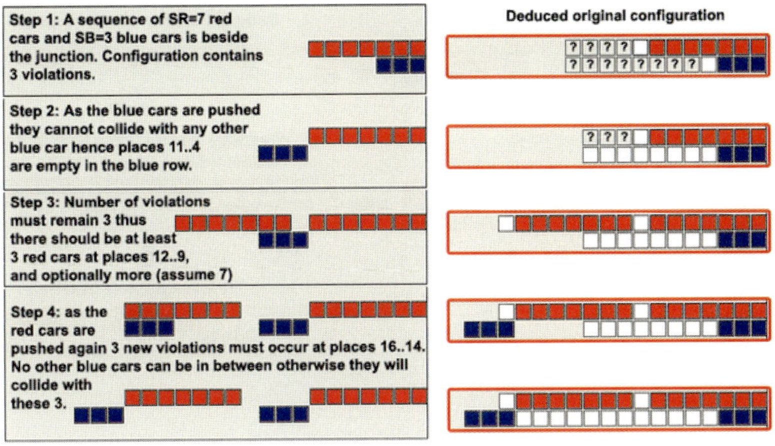

FIGURE 5. Sketch of proof ideas for Lemmas 3.7 (steps 1–3) and 3.8 (step 4).

$i \in [N - s_B - s_R - 1, N - s_B - 1]$ since otherwise 2 disjoint blue segments will merge while the sequence is pushed.

Thus following the system, after s_R turns we will get: $B_i = 1$ for $i \in [N - s_R - s_B, N - s_R - 1]$ and $B_i = 0$ for $i \in [N - s_R, N - 1]$, and B_i not changed left to $N - s_R - s_B$, and R unchanged.

Note that originally R, B contained $\min(s_R, s_B)$ consecutive violations in places $[N - \min(s_R, s_B), 0]$ which all vanished after s_R turns. Possible violations at places $[N - \max(s_R, s_B), N - \min(s_R, s_B)]$ remained as they were. From here we get that we must have $R_i = 1$ for $\min(s_R, s_B)$ places within $[N - s_R - s_B - 1, N - \max(s_R, s_B) - 1]$. Since for place $N - \max(s_R, s_B) - 1$ either R or B are empty by the assumption, we must therefore have $R_i = 1$ for $i \in [N - s_B - s_R - 1, N - \max(s_R, s_B) - 2]$, giving (2).

If we follow the system for s_B more steps we note that any red car in $[N - \max(s_R, s_B) - 1, N - s_R - 1]$ will be pushed left until eventually hitting the red car already proven to be present at $N - \max(s_R, s_B) - 2]$, thus $R_i = 0$ for $i \in [N - \max(s_R, s_B) - 1, N - s_R - 1]$ giving (3). □

The following lemma is completely analogous when reversing the roles of R, B, and can be proven the same way.

Lemma 3.8. *Let R, B be a stable configuration with junction at $S = 0$ and $B_0 = 1$. Assume that there is a sequence of exactly s_R consecutive red cars at places $[N - s_R, N - 1]$ and s_B blue cars at places $[N - s_B + 1, 0]$, $s_R, s_B \geq 1$. Then:*

1. $R_i = 0$ *for* $i \in [N - s_B - s_R - 1, N - s_R - 1]$.
2. $B_i = 1$ *for* $i \in [N - s_B - s_R, N - \max(s_B, s_R) - 1]$.
3. $B_i = 0$ *for* $i \in [N - \max(s_B, s_R), N - s_B]$.

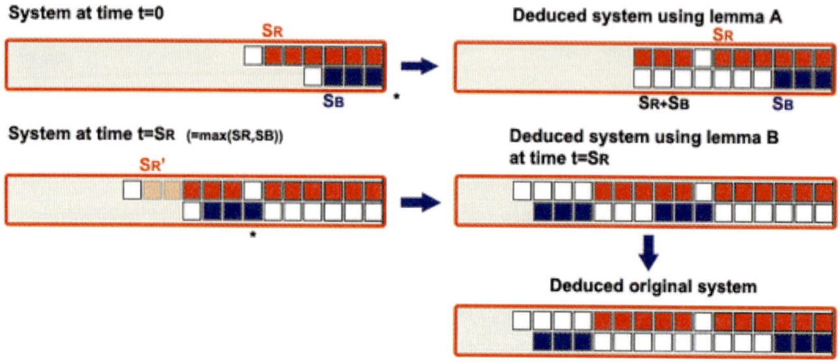

FIGURE 6. Lemmas 3.7 (A) and 3.8 (B) combined together yield Lemma 3.9.

Putting Lemmas 3.7, 3.8 together we get the following characterization for stable configurations:

Lemma 3.9. *Let R, B be a stable configuration with junction at $S = 0$ and $B_0 = 0$. Assume that there is a sequence of exactly s_R consecutive red cars at places $[N - s_R, N - 1]$ and s_B blue cars at places $[N - s_B, N - 1]$. Denote $M = \max(s_B, s_R)$ Then*

1. *There are no additional cars are at $[N - M, N - 1]$.*
2. *Place $i = N - M - 1$ is empty – i.e., $R_i = B_i = 0$.*
3. *Starting at $N - M - 2$ there is a sequence of $K_1 \geq \min(s_B, s_R)$ places for which $R_i = 1; B_i = 0; (i \in [N - M - K_1 - 1, N - M - 2])$.*
4. *Starting $N - M - K_1 - 2$ (i.e., right after the red sequence) there is a sequence of $K_2 \geq \min(s_B, s_R)$ places for which $B_i = 1; R_i = 0; (i \in [N - M - K_1 - K_2 - 1, N - M - K_1 - 2])$.*

Proof. Figure 6 outlines the proof, without the risk of getting lost in the indices. For the full proof, first get from Lemma 3.7 that:

1. $B_i = 0$ for $i \in [N - s_R - s_B - 1, N - s_B - 1]$.
2. $R_i = 1$ for $i \in [N - s_R - s_B - 1, N - \max(s_R, s_B) - 2]$.
3. $R_i = 0$ for $i \in [N - \max(s_R, s_B) - 1, N - s_R - 1]$.

From (1), (3) we get in particular, $B_i = 0$ for $i \in [N - \max(s_R, s_B) - 1, N - s_B - 1]$, and $R_i = 0$ for $i \in [N - \max(s_R, s_B) - 1, N - s_R - 1]$, so indeed no additional cars are at $[N - \max(s_B, s_R), N - 1]$, and place $N - \max(s_R, s_B) - 1$ is empty, proving claims 1, 2 in the lemma.

From (2), (3) we get $B_i = 0$ for $i \in [N - s_R - s_B - 1, N - \max(s_R, s_B) - 2]$ and $R_i = 1$ for $i \in [N - s_R - s_B - 1, N - \max(s_R, s_B) - 2]$, thus places $[N - s_R - s_B - 1, N - \max(s_R, s_B) - 2]$ contain a sequence of length $\min(s_R, s_B)$ of red cars with no blue cars in parallel to it. This sequence is possibly a part of a larger sequence of length $s'_R \geq \min(s_R, s_B)$, located at $[N - \max(s_R, s_B) - s'_R - 1, N - \max(s_R, s_B) - 2]$.

Now running the system for $\max(s_R, s_B)$ turns, we will have the junction at place $S = N - \max(s_R, s_B) - 1$, $B_S = 1$, followed by sequences of s'_R reds and $s'_B = \min(s_B, s_R)(\leq s'_R)$ blues. Applying Lemma 3.8 for the system (rotated by $\max(s_R, s_B)$, i.e., for $N' = N - \max(s_R, s_B) - 1$):

1. $R_i = 0$ for $i \in [N' - s'_B - s'_R - 1, N' - s'_R - 1] = [N - \max(s_R, s_B) - \min(s_R, s_B) - s'_R - 2, N - \max(s_R, s_B) - s'_R - 2] = [N - s_R - s_B - s'_R - 2, N - \max(s_R, s_B) - s'_R - 2]$.
2. $B_i = 1$ for $i \in [N' - s'_B - s'_R, N' - \max(s'_B, s'_R) - 1] = [N - s_R - s_B - s'_R - 1, N - \max(s_R, s_B) - \max(s'_B, s'_R) - 2] = [N - s_R - s_B - s'_R - 1, N - \max(s_R, s_B) - s'_R - 2]$.
3. $B_i = 0$ for $i \in [N' - \max(s'_B, s'_R), N' - s'_B] = [N - \max(s_R, s_B) - s'_R - 1, N - \max(s_R, s_B) - \min(s_R, s_B) - 1] = [N - \max(s_R, s_B) - s'_R - 1, N - s_R - s_B - 1]$.

In particular from (3) we get that no blue cars are in parallel to the entire red segment in $[N - \max(s_R, s_B) - s'_R - 1, N - \max(s_R, s_B) - 2]$: We were previously assured this is true up to place $N - s_R - s_B - 1$, and for places $[N - \max(s_R, s_B) - s'_R - 1, N - s_R - s_B - 2] \subseteq [N - \max(s_R, s_B) - s'_R - 1, N - \max(s_R, s_B) - s'_B - 1]$ this holds by (3).

Furthermore by (2) we get that a sequence of blue cars which begins from place $N - \max(s_R, s_B) - s'_R - 2$ with no red cars in parallel to it by (1). Note that $N - \max(s_R, s_B) - s'_R - 2$ is exactly to the left of $N - \max(s_R, s_B) - s'_R - 1$ where the red sequence ended. Now clearly choosing $K_1 = s'_R; K_2 = \min(s_R, s_B)$ we get claims 3, 4, 5 in the lemma. □

Putting it all together we can now get a very good description of a stable state.

Theorem 3.10. *Let R, B be a stable configuration with junction at $S = 0$ and $B_0 = 0$. Assume that there is a sequence of exactly s_R consecutive red cars at places $[N - s_R, N - 1]$ and s_B blue cars at places $[N - s_B, N - 1]$. Denote $M = \max(s_B, s_R)$.*

Then no additional cars are at $[N - M, N - 1]$, and at places $[0, N - M - 1]$ the configurations R, B satisfies:

1. *Each place contains at most one type of car, red or blue.*
2. *Place $N - M - 1$ is empty. Each empty place, is followed by a sequence of places containing red cars immediately left to it, which is followed by a sequence of places containing blue cars immediately left to it.*
3. *Any sequence of red or blue cars is of length at least $\min(s_R, s_B)$.*

Proof. This is merely applying Lemma 3.9 repeatedly. Applying Lemma 3.9 we know that there exist $K_1, K_2 \geq \min(s_R, s_B)$ such that: Place $N - M - 1$ is empty, followed by K_1 consecutive places with only red cars and K_2 consecutive places with only blue cars left to it, thus the assertion holds for the segment $[T, N - M - 1]$ for $T = N - M - 1 - K_1$.

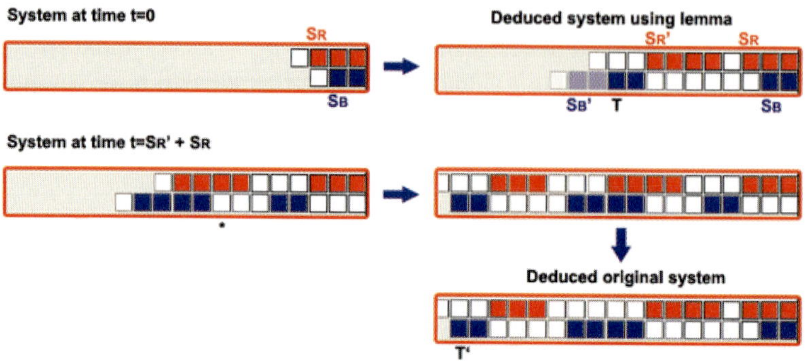

FIGURE 7. Repetitively applying Lemma 3.9, we unveil a longer segment in the configuration, $[T, N-1]$, for which properties of Theorem 3.10 hold.

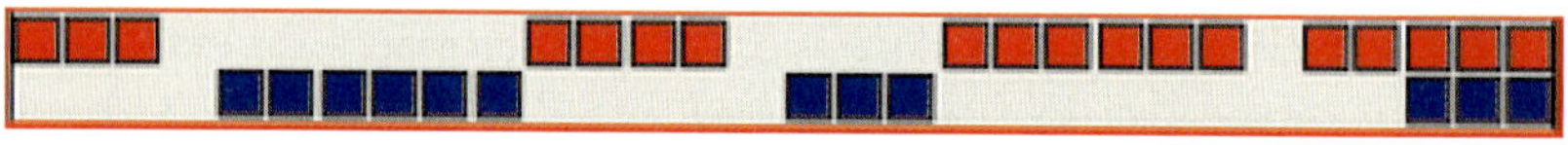

FIGURE 8. A typical stable configuration.

Now we completely know R, B in $[N-M-1-K_1, N-M-1]$, and this is enough to advance the system for K_1+M turns. The s_B blue segment is pushed left s_R places, further pushing the K_1 red sequence $\min(s_R, s_B)$ places to the left such that its last $\min(s_R, s_B)$ cars now overlap with the K_2 blue sequence.

So after K_1+M turns the system evolves to a state where $S = N - K_1 - M$, $B_S = 0$, and left to S there are $K_2' > K_2$ consecutive blue cars and exactly $\min(s_R, s_B)$ consecutive red cars. Noting that this time

$$M' = \max(K_2', \min(s_R, s_B)) = K_2',$$

once again we can deduce from Lemma 3.9 that: there are no additional cars in $[N-M-K_1-K_2', N-M-K_1-1]$ (thus we are assured that the entire blue segment of length K_2' does not have red cars parallel to it). Place $N-M-K_1-K_2'-1$ is empty, followed by some K_3 consecutive places with only red cars and K_4 consecutive places with only blue cars left to it, for $K_3, K_4 \geq \min(K_1, K_2) = \min(s_R, s_B)$ thus the assertion holds for the segment $[T', N-M-1]$ for $T' = N-M-K_1-K_2'-K_3-1 < T$.

Repeatedly applying Lemma 3.9 as long as $T > 0$, we repeatedly get that the assertion holds for some $[T, N-M-1]$ for T strictly decreasing, so the assertion holds in $[0, N-M-1]$. □

We have worked hard for Theorem 3.10, but it will soon turn to be worthwhile. Let us obtain some useful corollaries.

Corollary 3.11. *Let R, B be a stable configuration with junction at $S = 0$ and $B_0 = 0$. Assume that there is a sequence of exactly s_R consecutive red cars at places $[N - s_R, N - 1]$ and s_B blue cars at places $[N - s_B, N - 1]$. Denote $m = \min(s_B, s_R)$ Then:*

1. *The number of blue segments in the system equals the number of red segments.*
2. *System speed is at least $(1 + \frac{m}{N})^{-1}$.*
3. *Total number of cars in the system is at most $N + m$.*
4. *Total number of cars in the system is at least $\frac{2m}{2m+1}N + m$.*
5. *Total number of segments in the system is at most $\frac{N}{m} + 1$.*

Proof. (1) Follows by the structure described in 3.10, since there is exactly one red and one blue sequence in $[N - M, N - 1]$ (as always $M = \max(s_R, s_B)$) and an equal number of reds and blues in $[0, N - M - 1]$ since any red sequence is immediately proceeded by a blue sequence.

For (2) we note that by 3.10 the configuration R, B contains exactly m violations: the m overlapping places of the segments s_R, s_B in $[N - M, N - 1]$ are the only violations in the configuration since any places in $[0, N - M - 1]$ contains a single car, and no red car can be immediately to the left of a blue car. By 3.3 the system speed is hence at least $(1 + \frac{m}{N})^{-1}$.

By 3.10 any place in $[0, N - m - 1]$ contains at most one car, or it is empty, and places $[N - m, N - 1]$ contain both a red car and a blue car. Thus the total number of cars in the system is at most $N - m + 2m = N + m$ giving (3).

On the other hand, Since any sequence of a red car or a blue car is of length at least m, an empty place in $[0, N - m]$ can occur only once in $2m + 1$ places, and other places contain one car. Thus the number of cars is lower bounded by $\frac{2m}{2m+1}(N - m) + 2m = \frac{2m}{2m+1}N + \frac{2m^2+2m}{2m+1} \geq \frac{2m}{2m+1}N + m$ giving (4).

The last property follows the fact that any sequence of cars is of length at least m, and by (3) total number of cars is at most $N + m$, thus number of different sequences is at most $\frac{N+m}{m} = \frac{N}{m} + 1$. □

Theorem 3.12. *All cars in the system have the same asymptotic speed.*

Proof. As we have seen, when it reaches a stable state, the system consists of alternating red and blue sequences of cars. Obviously, the order of the sequences cannot change. Therefore, the difference between the number of steps two different cars have taken cannot be more then the length of the longest sequence, which is less then N. Thus, the asymptotic (w.r.t. t) speed is the same for all cars. □

With these corollaries we can now completely characterize the stable state of a junction with $p > 0.5$, just by adding the final simple observation, that since the number of cars in the model is greater than N, there are violations at all times, including after reaching a stable state. Now let us look at some time when the junction reaches a violation when the system is in stable state. At this point the conditions of Theorem 3.10 are satisfied, thus:

Theorem 3.13. *A junction of size N and density $p > 0.5$ reaches speed of $\frac{1}{2p} - O(\frac{1}{N})$ (i.e., arbitrarily close to the optimal speed of $\frac{1}{2p}$, for large enough N), and contains at most a bounded number (depending only on p) of car sequences.*

Proof. We look at the system after it reached a stable state. Since $2pN > N$ at some time after that conditions of Theorem 3.10 are satisfied for some $s_R, s_B \geq 1$. Let $m = \min(s_R, s_B)$ at this time. Using claims (3), (4) in Corollary 3.11 we get:

$$\frac{2m}{2m+1}N + m \leq 2pN \leq N + m.$$

From here we get

$$(*) \quad (2p-1)N \leq m \leq (2p - \frac{2m}{2m+1})N = (2p-1)N + \frac{N}{2m+1}$$

and reusing $m \geq (2p-1)N$ on the left-hand size we get:

$$(2p-1)N \leq m \leq (2p-1)N + \frac{1}{4p-1}.$$

For $C = \frac{1}{4p-1}$ a constant independent of N ($C = O(1)$). So $m = (2p-1)N + K$, for $K \leq C$. Now by claim (2) in 3.11, system speed is at least

$$(1 + \frac{m}{N})^{-1} = (1 + \frac{(2p-1)N + K}{N})^{-1} = (2p + \frac{K}{N})^{-1} \geq_{(2p>1)} \frac{1}{2p} - \frac{K}{N}.$$

But by Theorem 3.1 system speed is at most $\frac{1}{2p}$, thus system speed is exactly $\frac{1}{2p} - \frac{K'}{N}$ for some $0 \leq K' \leq K \leq C = O(1)$, thus $K' = O(1)$ proving the first part of the theorem.

By (5) we get total number of segments in the system is at most $\frac{N}{m} + 1$, applying $m \geq (2p-1)N$ we get the number of segments is bounded by:

$$\frac{N}{m} + 1 \leq \frac{1}{2p-1} + 1 = \frac{2p}{2p-1} = O(1).$$

Thus the second part proven. □

3.4. $p < 0.5$ revisited

The characterization in 3.10 can be also proven useful to handle $p < 0.5$. Actually the main result for $p < 0.5$, Theorem 3.5, can be shown using similar technique, and even sharpened.[3]

Corollary 3.14. *For $p < 0.5$ the junction reaches speed of at least $1 - \frac{C(p)}{N}$, for $C(p) = \left\lfloor \frac{p}{1-2p} \right\rfloor$. In particular, for $p < \frac{1}{3}$ the junction reaches speed 1, for any initial configuration.*

Proof. Let R, B be any initial configuration. Looking at the configuration after it reached the stable state, if the system reached speed 1 we have nothing to

[3] To Theorem 3.5 defence, we should note that key components of its proof, such as Lemma 3.3, were also used in the proof of 3.10.

prove. Assume the speed is less than 1. Since in this case violations still occur, at some time the stable configuration will satisfy Theorem 3.10. As before, letting $m = \min(s_R, s_B)$ at this time, by claim (4) in Corollary 3.11 we have:

$$\frac{2m}{2m+1}N + m \leq 2pN \Rightarrow m \leq (2p - 1 + \frac{1}{2m+1})N.$$

In particular $2p - 1 + \frac{1}{2m+1} > 0$, rearranging we get $m < \frac{p}{1-2p}$, and since $m \in \mathbb{Z}$, $m \leq \left\lfloor \frac{p}{1-2p} \right\rfloor = C(p)$. m must be positive, thus for $p < \frac{1}{3}$, having $C(p) = \left\lfloor \frac{p}{1-2p} \right\rfloor = 0$ we get a contradiction, thus the assumption (speed < 1) cannot hold. For $p \geq \frac{1}{3}$ by claim (2) in Corollary 3.11 the system speed is at least $(1 + \frac{C(p)}{N})^{-1} \geq 1 - \frac{C(p)}{N}$. □

3.5. The critical $p = 0.5$

Gathering the results so far we get an almost-complete description of the behaviour of a junction. For junction of size N and density p:

- If $p < 0.5$ the junction will reach speed $1 - o(1)$ (asymptotically optimal), and contain linearly many different segments in the stable state.
- If $p > 0.5$ the junction will reach speed $\frac{1}{2p} - o(1)$ (asymptotically optimal), and contain constant many segments in the stable state.

From the description above one sees that the junction system goes through a sharp phase transition at $p = 0.5$, as the number of segments of cars as the system stabilizes drops from being linear to merely constant. The last curiosity, is what happens at $p = 0.5$. Once again by using the powerful Theorem 3.10 we can deduce:

Theorem 3.15. *A junction of size N with $p = 0.5$ reaches speed of at least $1 - \frac{1}{\sqrt{N}}$ and contains at most $\sqrt{N}$ different segments.*

Proof. For $p = 0.5$ we have exactly N cars in the system. As we reach stable state, violations must still occur (since a system with exactly N cars must contain at least one violation), thus at some time Theorem 3.10 is satisfied. For $m = \min(s_R, s_B)$ at this time, by claim (4) in Corollary 3.11 we have:

$$\frac{2m}{2m+1}N + m \leq N \Rightarrow m(2m+1) \leq N.$$

Thus $m < \sqrt{N}$. From here by claim (2) the system speed is at least $(1 + \frac{\sqrt{N}}{N})^{-1} \geq 1 - \frac{1}{\sqrt{N}}$.

If S is the number of segments, then by Theorem 3.10 we can deduce that the total number of cars, N, is: $2m$ cars in places $[N - m, N - 1]$, and $N - m - S$ cars in places $[0, N - M - 1]$ (since each place contains one car, except transitions between segments that are empty).

$$N = (N - m - S) + 2m = N + m - S \Rightarrow S = m \leq \sqrt{N}.$$

Thus the configuration contains at most $\sqrt{N}$ segments. □

Simulation results show that these bounds are not only tight, but typical, meaning that a junction with a random initial configuration with density p indeed has $O(\sqrt{N})$ segments in the stable state, with the largest segment of size near $n^{1/2}$. This suggests that the system undergoes a second order phase transition.

4. Simulation results

Following are computer simulation results for the junction for critical and near critical p, demonstrating the phase transition. The columns below consist of N, the average asymptotic speed, the average number of car segments in the stable state, the average longest segment in the stable state, and the average number of segments divided by N and $\sqrt{N}$.

4.1. $p = 0.48$

For large N, system reaches speed 1 and the average number of segments is linear (approx. $0.037N$).

N	Speed	No. segs	Longest	No. segs/N	No. segs/$\sqrt{N}$
1000	0.99970	38.7	6.8	0.0387	1.2238
5000	1.00000	186.4	8.5	0.0373	2.6361
10000	1.00000	369.8	7.5	0.0370	3.6980
50000	1.00000	1850.6	8.2	0.0370	8.2761

4.2. $p = 0.52$

For large N, system reaches speed $0.961 = \frac{1}{2p}$ and the average number of segments is about constant.

N	Speed	No. segs	Longest	No. segs/N	No. segs/$\sqrt{N}$
1000	0.95703	5.7	76.7	0.0057	0.1802
5000	0.96041	6.9	330.0	0.0014	0.0976
10000	0.96091	7.3	416.1	0.0007	0.0730
50000	0.96142	7.2	3207.1	0.0001	0.0322

4.3. $p = 0.5$

At criticality, the speed is approaching 1 like $1 - \frac{C}{\sqrt{N}}$ and the average number of segments is around $0.43 \cdot \sqrt{N}$.

N	Speed	No. segs	Longest	No. segs/N	No. segs/$\sqrt{N}$
1000	0.98741	13.4	38.4	0.0134	0.4237
5000	0.99414	30.0	82.8	0.0060	0.4243
10000	0.99570	43.8	142.1	0.0044	0.4375
50000	0.99812	95.0	248.4	0.0019	0.4251

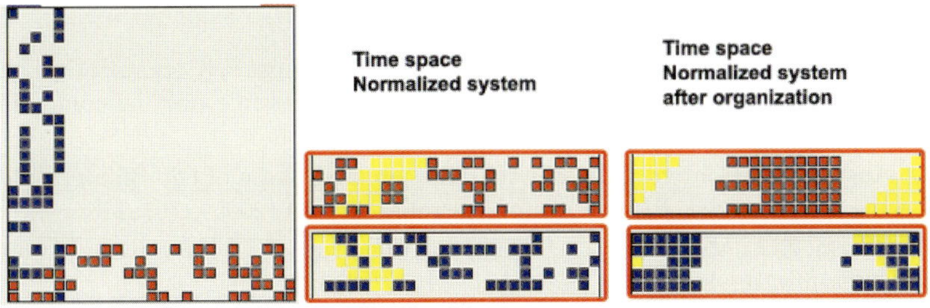

FIGURE 9. K-Lanes junction in a time-space normalized view, before and after organization. Junction is marked in yellow and advances left each turn

5. Summary

The fascinating phenomena observed in the BML traffic model are still far from being completely understood. In this paper we showed a very simplified version of this model, which, despite its relative simplicity, displayed very similar phenomena of phase transition at some critical density and of self-organization, which in our case both can be proven and well understood.

We used two approaches in this paper: The first one was to use some sort of "independence" in the way the system evolves, as in the way we handle $p < 0.5$. We showed that the system self-organizes "locally" and with a bounded affect on the rest of the configuration, thus it will eventually organize globally. The second approach is the notion of the stable configuration, i.e., we characterize the combinatorial structure of any state that the system can "preserve" and use it to show it is optimal (in a way saying, that as long as we are not optimal the system must continue to evolve).

Can these results be extended to handle more complicated settings than the junction? Possibly yes. For example, considering a k-lanes junction (i.e., k consecutive red lines meeting k consecutive blue rows), one can look at a time-space-normalized version of the system, as shown in Figure 9, with the junction now being a $k \times k$ parallelogram travelling along the red and blue lines, and "propagating violations" (which now have a more complicated structure depending on the $k \times k$ configuration within the junction). Stable states of this configurations seem to have the same characteristics of a single junction, with a red (blue) car equivalent to some red (blue) car in any of the k red (blue) lines. Thus a zero-effort corollary for a k-lanes junction is that it reaches speed 1 for $p < \frac{1}{2k}$, but for k nearing $O(N)$ this bound is clearly non-significant. It is not surprising though, since combinatorics alone cannot bring us too far, at least for the complete BML model – even as little as $2N$ cars in an N^2 size torus can be put in a stuck configuration – i.e., reach speed 0.

References

[1] T. Austin, I. Benjamini, For what number of cars must self organization occur in the Biham-Middleton-Levin traffic model from any possible starting configuration?, 2006.

[2] O. Angel, A.E. Holroyd, J.B. Martin, The jammed phase of the Biham-Middleton-Levine traffic model, Elect. Comm. in Probability 10 (2005), 167–178.

[3] O. Biham, A.A. Middleton, D. Levine, Self organization and a dynamical transition in traffic flow models, Phys. Rev. A 46 (1992), R6124.

[4] R.M. D'Souza, Geometric structure of coexisting phases found in the Biham-Middleton-Levine traffic model, Phys. Rev. E 71 (2005), 0066112.

Itai Benjamini
Department of Mathematics,
Weizmann Institute of Science,
Rehovot, 76100, Israel
e-mail: `itai.benjamini@weizmann.ac.il`

Ori Gurel-Gurevich
Department of Mathematics,
Weizmann Institute of Science,
Rehovot, 76100, Israel
e-mail: `ori.gurel-gurevich@weizmann.ac.il`

Roey Izkovsky
Department of Mathematics,
Weizmann Institute of Science,
Rehovot, 76100, Israel
e-mail: `roey.izkovsky@weizmann.ac.il`

Progress in Probability, Vol. 60, 117–135

Box-Crossings and Continuity Results for Self-Destructive Percolation in the Plane

J. van den Berg, R. Brouwer and B. Vágvölgyi

Abstract. A few years ago (see [1]) two of us introduced, motivated by the study of certain forest-fire processes, the self-destructive percolation model (abbreviated as sdp model). A typical configuration for the sdp model with parameters p and δ is generated in three steps: First we generate a typical configuration for the ordinary site percolation model with parameter p. Next, we make all sites in the infinite occupied cluster vacant. Finally, each site that was already vacant in the beginning or made vacant by the above action, becomes occupied with probability δ (independent of the other sites).

Let $\theta(p, \delta)$ be the probability that some specified vertex belongs, in the final configuration, to an infinite occupied cluster. In our earlier paper we stated the conjecture that, for the square lattice and other planar lattices, the function $\theta(\cdot, \cdot)$ has a discontinuity at points of the form (p_c, δ), with δ sufficiently small. We also showed (see [2]) remarkable consequences for the forest-fire models.

The conjecture naturally raises the question whether the function $\theta(\cdot, \cdot)$ is continuous outside some region of the above-mentioned form. We prove that this is indeed the case. An important ingredient in our proof is a somewhat modified (improved) form of a recent RSW-like (box-crossing) result of Bollobás and Riordan ([4]). We believe that this modification is also useful for many other percolation models.

Mathematics Subject Classification (2000). 60K35, 82B43.

Keywords. Percolation, RSW results, box crossings.

1. Introduction and outline of results

1.1. Background and motivation

The self-destructive percolation model (abbreviated as sdp model) on the square lattice is described as follows: First we perform independent site percolation on this lattice: we declare each site *occupied* with probability p, and *vacant* with

Part of J. van den Berg's research has been funded by the Dutch BSIK/BRICKS project.

probability $1-p$, independent of the other sites. We will use the notation $\{V \leftrightarrow W\}$ for the event that there is an occupied path from the set of sites V to the set of sites W. We write $\{V \leftrightarrow \infty\}$ for the event that there is an infinite occupied path starting at V.
Let, as usual, $\theta(p)$ denote the probability that a given site, say $O = (0,0)$, belongs to an infinite occupied cluster. It is known that there is a critical value $0 < p_c < 1$ such that $\theta(p) > 0$ for all $p > p_c$, and $\theta(p) = 0$ for all $p \leq p_c$. Now suppose that, by some catastrophe, the infinite occupied cluster (if present) is destroyed; that is, each site in this cluster becomes vacant. Further suppose that after this catastrophe we give the sites independent 'enhancements', as follows: Each site that was already vacant in the beginning, or was *made* vacant by the catastrophe, becomes occupied with probability δ, independent of the others. Let $\mathcal{P}_{p,\delta}$ be the distribution of the final configuration.

A more formal, and often very convenient description of the model is as follows: Let X_i, $i \in \mathbb{Z}^2$ be independent $0-1$-valued random variables, each X_i being 1 with probability p and 0 with probability $1-p$. Further, let Y_i, $i \in \mathbb{Z}^2$, be independent $0-1$-valued random variables, each Y_i being 1 with probability δ and 0 with probability $1-\delta$. Moreover, we take the collection of Y_i's independent of that of the X_i's. Let $X_i^*, i \in \mathbb{Z}^2$ be defined by

$$X_i^* = \begin{cases} 1 \text{ if } X_i = 1 \text{ and there is no } X\text{-occupied path from } i \text{ to } \infty \\ 0 \text{ otherwise,} \end{cases} \tag{1}$$

where by 'X-occupied path' we mean a path on which each site j has $X_j = 1$. Finally, define $Z_i = X_i^* \vee Y_i$. This collection $(Z_i, i \in \mathbb{Z}^2)$ is (with 0 meaning 'vacant' and 1 'occupied') what we called 'the final configuration', and the above-mentioned $\mathcal{P}_{p,\delta}$ is its distribution.

We use the notation $\theta(p,\delta)$ for the probability that, in the final configuration, O is in an infinite occupied cluster:

$$\theta(p,\delta) := \mathcal{P}_{p,\delta}(O \leftrightarrow \infty).$$

Note that O is occupied in the final configuration if and only if the above-mentioned enhancement was successful, or O belonged initially (before the catastrophe) to a non-empty but finite occupied cluster. This gives

$$\mathcal{P}_{p,\delta}(O \text{ is occupied }) = \delta + (1-\delta)(p - \theta(p)).$$

Also note that, in the case that $p \leq p_c$, nothing happens in the above catastrophe, so that in the final configuration the sites are independently occupied with probability $p + (1-p)\delta$. Formally, if $p \leq p_c$, then

$$\mathcal{P}_{p,\delta} = \mathcal{P}_{p+(1-p)\delta}, \tag{2}$$

where we use the notation $\mathcal{P}_p$ for the product measure with parameter p. In particular,

$$\theta(p_c,\delta) = \theta(p_c + (1-p_c)\delta) > 0, \tag{3}$$

for each $\delta > 0$.

Remark 1.1. *Most of what we said above has straightforward analogs for arbitrary countable graphs, but there are subtle differences. For instance, on the cubic lattice it has not yet been proved that* $\theta(p_c) = 0$ *(although this is generally believed to be true). So, for that lattice,* (2) *with* $p = p_c$*, and hence* (3)*, are not rigorously known.*

It is also clear from the construction that $\mathcal{P}_{p,\delta}$ stochastically dominates $\mathcal{P}_{\delta}$. Hence, if $\delta > p_c$ then $\theta(p, \delta) \geq \theta(\delta) > 0$ for all p.

It turns out (see Proposition 3.1 of [1]) that, if $p > p_c$, a 'non-negligible' enhancement is needed after the catastrophe to create again an infinite occupied cluster. More precisely, for each $p > p_c$ there is a $\delta > 0$ with $\theta(p, \delta) = 0$. A much more difficult question is whether the needed enhancement goes to 0 as $p \downarrow p_c$. By (3) one might be tempted to reason intuitively that this is indeed the case. In [1] it was shown that for the analogous model on the binary tree this is correct. However, in [1] a conjecture is presented which says, in particular, that for the square lattice (and other planar lattices) there is a $\delta > 0$ for which $\theta(p, \delta) = 0$ for all $p > p_c$. In Section 4 of [1] and in [2] we showed remarkable consequences for certain forest-fire models.

Note that, since $\theta(p_c, \delta) > 0$, the above conjecture says that the function $\theta(\cdot, \cdot)$ has discontinuities at points of the form (p_c, δ) with δ sufficiently small. This naturally raises the question whether this function is continuous in the complement of a region of such form: is there a $\delta > 0$ such that $\theta(\cdot, \cdot)$ is continuous outside the set $\{p_c\} \times [0, \delta]$? In the next subsection we state that this is indeed the case, and give a summary of the methods and intermediate results used in the proof. At the end of Section 6 we point out why our proof does not work at points (p_c, δ) with small δ. We hope our arguments provide a better understanding of the earlier mentioned conjecture and will trigger new attempts to prove (or disprove) it.

1.2. Outline of results

The conjecture mentioned in the previous subsection raises the natural question whether $\theta(\cdot, \cdot)$ *is* continuous outside the indicated 'suspected' region. The following theorem states that this is indeed the case.

Theorem 1.2. *There is a* $\delta \in (0, 1)$ *such that the function* $\theta(\cdot, \cdot)$ *is continuous outside the segment* $\{p_c\} \times (0, \delta)$.

As could be expected, the proof widely uses tools and results from ordinary percolation. However, the dependencies introduced by the self-destructive mechanism cause complications. Until recently, a serious obstacle was the absence of a suitable RSW-like theorem. This obstacle could be removed by the use of (a modified and somewhat stronger form of) a recent theorem of Bollobás and Riordan ([4]).

A rough outline of the proof of Theorem 1.2, and the needed intermediate results that are interesting in themselves, is as follows: In Section 2 we list some basic properties of our model, which will be used later. The results in Section 3, which are also contained in the recent PhD thesis [6] of one of us, show that if $\theta(\cdot,\cdot)$ is strictly positive in some open region, then it is continuous on this region. It is also shown that if $\theta(p,\delta) = 0$, then $\theta(\cdot,\cdot)$ is continuous at (p,δ). These two results reduce the proof of Theorem 1.2 to showing that if $\theta(p,\delta) > 0$ and $p \neq p_c$, then $\theta(p,\delta) > 0$ in an open neighborhood of (p,δ). This in turn requires a suitable finite-size criterion (see below) for sdp. In Section 4 we give the modified form of the Bollobás-Riordan theorem. This is used in Section 5 to obtain the above-mentioned finite-size criterion. Finally, in Section 6 we combine these results and prove the main theorem.

We end this section with the following remark: When we say that a function f is 'increasing' ('decreasing') this should, unless this is preceded by the word 'strictly', be interpreted in the weak sense: $x < y$ implies $f(x) \leq f(y)$.

2. Basic properties

In this section we state some basic properties which will be used later.

First some more terminology and notation: If $v = (v_1, v_2)$ and $w = (w_1, w_2)$ are two vertices, we let $|v - w|$ denote their (graph) distance $|v_1 - w_1| + |v_2 - w_2|$. By $B(v,k)$ and $\partial B(v,k)$ we denote the set of vertices w for which $|v - w|$ is at most k, respectively equal to k. For $V, W \subset \mathbb{Z}^2$, we define the distance between V and W as $\min\{|v - w| : v \in V, w \in W\}$.

Recall that $\mathcal{P}_{p,\delta}$ denotes the sdp distribution (that is, the distribution of the collection $(Z_i,\, i \in \mathbb{Z}^2)$ defined in Subsection 1.1). This is a distribution on $\Omega := \{0,1\}^{\mathbb{Z}^2}$ (with the usual σ-field). Elements of Ω are typically denoted by $\omega(= (\omega_i, i \in \mathbb{Z}^2))$, σ etc. We write $\omega \leq \sigma$ if $\omega_i \leq \sigma_i$ for all i.
Let V be a set of vertices and A an event. We say that A lives on V if $\omega \in A$ and $\sigma_i = \omega_i$ for all $i \in V$, implies $\sigma \in A$. And we say that A is a cylinder event if A lives on some finite set of vertices. As usual, we say that A is increasing if $\omega \in A$ and $\omega_i \leq \sigma_i$ for all i, implies $\sigma \in A$. The first two lemma's below come from Section 2.2 and 2.4 respectively in [1].

Lemma 2.1. *Let A and B be two increasing cylinder events. We have*

$$\mathcal{P}_{p,\delta}(A \cap B) \geq \mathcal{P}_{p,\delta}(A)\mathcal{P}_{p,\delta}(B).$$

As to monotonicity, it is obvious that the sdp model has monotonicity in δ: If $\delta_1 \geq \delta_2$, then $\mathcal{P}_{p,\delta_1}$ stochastically dominates $\mathcal{P}_{p,\delta_2}$. Although there seems to be no 'nice' monotonicity in p we have the following property.

Lemma 2.2. *If $p_2 \geq p_1$ and $p_2 + (1 - p_2)\delta_2 \leq p_1 + (1 - p_1)\delta_1$, then*

$$\mathcal{P}_{p_1,\delta_1} \text{ dominates } \mathcal{P}_{p_2,\delta_2}.$$

The next result is about 'almost independence' of cylinder events which live on widely separated sets. As usual, the lattice which has the same vertices as the square lattice but where each vertex has, besides the four edges to its nearest neighbours, also four 'diagonal edges' is called the matching lattice (of the square lattice). To distinguish paths and circuits in the matching lattice from those in the square lattice, we use the terminology *-paths and *-circuits.

Lemma 2.3. *Let k be a positive integer and let V and W be subsets of $\mathbb{Z}^2$ that have distance larger than $2k$. Further, let A and B be events which live on V and W respectively. Then*

$$|\mathcal{P}_{p,\delta}(A \cap B) - \mathcal{P}_{p,\delta}(A)\mathcal{P}_{p,\delta}(B)| \leq 2(|V| + |W|) \\ \mathcal{P}_p\left(\exists \textit{ vacant *-circuit surrounding } O \textit{ and some vertex in } \partial B(O,k)\right). \tag{4}$$

Proof. Recall how we formally defined the sdp model in terms of random variables X, Y and Z. We use a modification of those variables: Let X and Y be as before, but in addition to X^* and Z we now define $X^{*(f)}$ and $Z^{(f)}$ by

$$X_i^{*(f)} = \begin{cases} 1 \text{ if } X_i = 1 \text{ and } \nexists\ X\text{-occupied path from } i \text{ to } \partial B(i,k) \\ 0 \text{ otherwise;} \end{cases}$$

$$Z_i^{(f)} = X_i^{*(f)} \vee Y_i. \tag{5}$$

Let $\mathcal{P}_{p,\delta}^{(f)}$ denote the distribution of $Z^{(f)}$. It is clear that the random variables $Z_i^{(f)}, i \in V$ are independent of the random variables $Z_i^{(f)}, i \in W$, and hence

$$\mathcal{P}_{p,\delta}^{(f)}(A \cap B) = \mathcal{P}_{p,\delta}^{(f)}(A)\mathcal{P}_{p,\delta}^{(f)}(B). \tag{6}$$

Also note that if $Z_i \neq Z_i^{(f)}$, then the X-occupied cluster of i intersects $\partial B(i,k)$ but is finite. Hence there is an X-vacant circuit in the matching lattice that surrounds i and some site in $\partial B(i,k)$. Hence, since the X-variables are Bernoulli random variables with parameter p, we have for any finite set K of vertices and any event E living on K,

$$|\mathcal{P}_{p,\delta}(E) - \mathcal{P}_{p,\delta}^{(f)}(E)| \leq P(Z_K \neq Z_K^{(f)}) \leq \\ |K|\mathcal{P}_p\left(\exists \text{ a vacant *-circuit surrounding } O \text{ and some vertex in } \partial B(O,k)\right). \tag{7}$$

The lemma now follows easily from (6) and (7) □

Our last result in this section is on the uniqueness of the infinite cluster.

Lemma 2.4. *If $\theta(p,\delta) > 0$, then*

$$\mathcal{P}_{p,\delta}(\exists \textit{ a unique infinite occupied cluster}) = 1.$$

Proof. From the earlier construction of the sdp model in terms of the X- and Y-variables, it is clear that $\mathcal{P}_{p,\delta}$ is stationary and ergodic. It is also clear that in the sdp model the conditional probability that a given site is occupied given the configuration at all other sites, is at least δ. So this model has the so-called positive finite energy property. The result now follows from an extension in [9] of the well-known Burton-Keane ([7]) uniqueness result. □

3. Partial continuity results

In this section we first prove that in the sdp model the probabilities of cylinder events are continuous functions of (p, δ). Next we prove that the function $\theta(\cdot, \cdot)$ is continuous at (p, δ) if $\theta(p, \delta) = 0$ or there is an open neighborhood of (p, δ) on which θ is strictly positive. Note that, once we have this, the proof of Theorem 1.2 is basically reduced to showing that if $p \neq p_c$ and $\theta(p_c, \delta) > 0$, then $\theta(\cdot, \cdot)$ is strictly positive on an open neighborhood of (p, δ).

Lemma 3.1. *Let A be a cylinder event. The function $(p, \delta) \to \mathcal{P}_{p,\delta}(A)$ is continuous on $[0, 1]^2$.*

Remark 3.2. *The proof (see below) uses the well-known fact that $\theta(p_c) = 0$. For many lattices (e.g., the cubic lattice) this fact has not been proved. For those lattices the arguments below show that the function in the statement of* 3.1 *is continuous on $[0, 1]^2 \setminus (\{p_c\} \times [0, 1])$. Related to this, Proposition* 3.3 *below would also need some modification for those lattices.*

Proof. Let A be an event which lives on some finite set V. Recall the construction of the sdp model in terms of random variables X, Y and Z. Let, for $\sigma \in \Omega$, σ_V denote the tuple $(\sigma_i, i \in V)$. It is clear that the distribution of X^*_V is a function of p only, and that, conditioned on X^*_V, the probability that $Z_V \in A$ is a polynomial (of degree $|V|$) in δ. Therefore it is sufficient to prove that, for each $\alpha \in \{0, 1\}^V$, the function $f : p \to \mathcal{P}(X^*_V = \alpha)$ is continuous. Recall that the X-variables are Bernoulli random variables (with parameter p). Now let $0 < p_1 < p_2$. In a standard way, by introducing independent, uniformly on the interval $(0, 1)$ distributed random variables $U_i, i \in \mathbb{Z}^2$, we can suitably couple two collections of Bernoulli random variables with parameters p_1, respectively p_2. Such argument easily gives that $|f(p_2) - f(p_1)|$ is less than or equal to the sum over $i \in V$ of $\mathcal{P}(U_i \in (p_1, p_2)) + \mathcal{P}(i$ is in an infinite p_2-open but not in an infinite p_1-open cluster), which equals

$$|V|(p_2 - p_1 + \theta(p_2) - \theta(p_1)).$$

The lemma now follows from the continuity of $\theta(.)$. □

Proposition 3.3. *Let $(p, \delta) \in [0, 1]^2$. If* (a) *or* (b) *below holds, the function $\theta(\cdot, \cdot)$ is continuous at (p, δ).*

(a) *$\theta(\cdot, \cdot) > 0$ on an open neighborhood of (p, δ).*
(b) *$\theta(p, \delta) = 0$,*

Proof. For this (and some other) results it is convenient to describe the sdp model in terms of Poisson processes: Assign to each site, independently of the other sites, a Poisson clock with rate 1. These clocks govern the following time evolution: Initially each site is vacant. Whenever the clock of a site rings, the site becomes occupied. (If it was already occupied, the ring is ignored.) Note that if occupied sites would always remain occupied, then for each time t, the configuration at time t would be a collection of independent Bernoulli random variables with parameter $1 - \exp(-t)$. In particular, before and at time t_c, defined by the relation $p_c = 1 - \exp(-t_c)$, there would be no infinite occupied cluster, but after t_c there would be a (unique) infinite cluster. However, we do allow occupied sites to become vacant, although only once, as follows: Fix a time τ, a parameter of the time evolution. At time τ all sites in the infinite occupied cluster become vacant. (If there is no infinite occupied cluster, which is a.s. the case if $\tau \leq t_c$, nothing happens.) After time τ we let the evolution behave as before; that is, each vacant site becomes occupied when its Poisson clock rings. Let, for this time evolution with parameter τ, $\hat{\mathcal{P}}_{\tau,t}$ denote the distribution of the configuration at time t , and let

$$\hat{\theta}(\tau, t) = \hat{\mathcal{P}}_{\tau,t}(O \text{ is in an infinite occupied cluster}). \tag{8}$$

It is easy to see that

$$\hat{\mathcal{P}}_{\tau,t} = \mathcal{P}_{p,\delta}, \tag{9}$$

where $p = 1 - \exp(-\tau)$ and $\delta = 1 - \exp(-(t - \tau))$. It is also easy to see that $\hat{\mathcal{P}}_{\tau,t}$ is stochastically decreasing in τ and stochastically increasing in t. In fact this is the key behind Lemma 2.2.

Now we come back to the proof of Proposition 3.3. From (8) and (9) we get (since the map between pairs (p, δ) and (τ, t) in (9) is continuous) that this proposition is equivalent to saying that if $\hat{\theta}(\tau, t) = 0$ or $\hat{\theta}$ is strictly positive on an open neighborhood of (τ, t_c), then $\hat{\theta}$ is continuous at (τ, t). To prove this equivalent form of Proposition 3.3 we use ideas from [3]. The introduction of pairs (τ, t) as replacement of (p, δ) not only has the advantage that, as we already saw, we now have a more suitable form of monotonicity, but, more importantly, that we now have a more 'detailed' structure (the Poisson processes) in the background which gives the appropriate 'room' needed to get a suitable modification of the arguments in [3].

Let (τ, t) be as above. Divide the parameter space in four 'quadrants', numbered I to IV:

$$\begin{aligned} I &:= [0, \tau] \times [t, \infty), \\ II &:= [\tau, \infty) \times [t, \infty), \\ III &:= [\tau, \infty) \times [0, t], \\ IV &:= [0, \tau] \times [0, t]. \end{aligned}$$

Note that it is sufficient to prove that for each monotone sequence $(\tau_i, t_i)_{i \geq 0}$ that lies in one of the above quadrants and converges to (τ, t), one has

$$\lim_{i \to \infty} \hat{\theta}(\tau_i, t_i) = \hat{\theta}(\tau, t).$$

We handle each of the quadrants separately.

Quadrant I) This is easy and corresponds to the (easy) proof of right continuity of ordinary percolation: Let (τ_i) be a monotone sequence which converges from below to τ and let (t_i) be a monotone sequence which converges from above to t. Let A_n denote the event that there is an occupied path from O to $\partial B(O, n)$. By monotonicity and Lemma 3.1 we have that

$$\text{For each } i, \hat{\mathcal{P}}_{\tau_i,t_i}(A_n) \downarrow \hat{\theta}(\tau_i, t_i) \text{ as } n \to \infty; \tag{10}$$

$$\hat{\mathcal{P}}_{\tau,t}(A_n) \downarrow \hat{\theta}(\tau, t) \text{ as } n \to \infty; \tag{11}$$

$$\text{For each } n, \hat{\mathcal{P}}_{\tau_i,t_i}(A_n) \downarrow \hat{\mathcal{P}}_{\tau,t}(A_n) \text{ as } i \to \infty, \tag{12}$$

From these three statements it is easy to see that $\hat{\theta}(\tau_i, t_i)$ tends to $\hat{\theta}(\tau, t)$ as $i \to \infty$.

Quadrant III) Let (τ_i) be a monotone sequence which converges from above to τ and (t_i) a monotone sequence which converges from below to t. By the earlier monotonicity arguments, the sequence $\hat{\theta}(\tau_i, t_i)$ is increasing in i, and has a limit smaller than or equal to $\hat{\theta}(\tau, t)$. So for the situation where $\hat{\theta}(\tau, t) = 0$, the proof is done. Now we handle the other situation: we assume $\hat{\theta}$ is positive in an open neighborhood of (τ, t). For this situation considerable work has to be done. Note that in the dynamic description given earlier in this section, the underlying Poisson processes were the same for each choice of the model parameter τ. This allows us (and we already used this to derive some monotonicity properties) to couple the models with the different τ_i's and τ.

Let, for $s < u$, $C_{s,u}$ denote the occupied cluster of site O at time u in the process with parameter s (that is, under the time evolution where the infinite occupied cluster is destroyed at time s). Further, we use the notation $\omega(s, u)$ for the configuration at time u in that model. It is also convenient to consider $\omega(u)$, the configuration at time u in the model where *no* destruction takes place. (So, $\omega_v(u), v \in \mathbb{Z}^2$, are independent $0-1$-valued random variables, each being 1 with probability $1 - \exp(-u)$). Again we emphasize that all these models are defined in terms of the same Poisson processes. From monotonicity (note that $C_{\tau_i,t_i} \subset C_{\tau_{i+1},t_{i+1}}$ for all i) it is clear that

$$\lim_{i\to\infty} \hat{\theta}(\tau_i, t_i) = \mathcal{P}(\exists i \, |C_{\tau_i,t_i}| = \infty),$$

and

$$\hat{\theta}(\tau, t) - \lim_{i\to\infty} \hat{\theta}(\tau_i, t_i) = \mathcal{P}(|C_{\tau,t}| = \infty, \forall i \, |C_{\tau_i,t_i}| < \infty). \tag{13}$$

So we have to show that the r.h.s. of (13) is 0. Fix a j with the property that $\hat{\theta}(\tau_j, t_j) > 0$. Such j exists by the condition we assumed for (τ, t). To show that the r.h.s. of (13) is 0, it is sufficient (and necessary) to prove the following claim:

Claim

Apart from an event of probability 0, *the event* $\{|C_{\tau,t}| = \infty\}$ *is contained in the event that there is a* $k > j$ *for which* $|C_{\tau_k,t_k}| = \infty$.

So suppose $|C_{\tau,t}| = \infty$. By our choice of j we may assume that $\omega(\tau_j, t_j)$ has an infinite occupied cluster, and by Lemma 2.4 that this cluster is unique. We denote it by I_j. If $O \in I_j$ we are done. From monotonicity and the uniqueness of the infinite cluster (see Lemma 2.4), we have $I_j \subset C_{\tau,t}$. Hence there is a finite path π from O to some site in I_j such that $\omega(\tau, t) \equiv 1$ on π. Since, a.s. there are no vertices whose clock rings exactly at time t or τ, we may assume that for every site v on π, (a) or (b) below holds:

(a) The clock of v rings in the interval (τ, t).
(b) $\omega_v(\tau) = 1$ but the occupied cluster of v in $\omega(\tau)$ is finite.

If (a) occurs we define:

$$i_v := \min\{i : i \geq j \text{ and the clock of } v \text{ rings in } (\tau_i, t_i)\}.$$

Note that then, by the monotonicity of the sequence (τ_i, t_i), the clock of v rings in the interval (τ_l, t_l) for all $l \geq i_v$. If (a) does not occur, (b) occurs, and hence there is a finite set K_v of sites on which $\omega(\tau) = 0$ and which separates v from ∞. Then we use the following alternative definition of i_v:

$$i_v := \min\{i : i \geq j \text{ and } \omega(\tau_i) \equiv 0 \text{ on } K_v\}.$$

This minimum exists since K_v is finite and (again) we assume that no Poisson clock rings exactly at time τ. Now let

$$k := \max_{v \in \pi} i_v,$$

which exists since π is finite.

From the above procedure it is clear that $\omega_v(\tau_k, t_k) = 1$ for all v on π. Further, since $k \geq j$ and by monotonicity, also $\omega_v(\tau_k, t_k) = 1$ for all $v \in I_j$. Since π is a path from O to I_j this implies that I_j is contained in C_{τ_k, t_k} and hence that $|C_{\tau_k, t_k}| = \infty$. This proves the Claim above.

Quadrants II) and IV)
The required results for these quadrants follow very easily from monotonicity and the above results for quadrants I and III: Let (τ_i, t_i) be a sequence in quadrant II that converges to (τ, t). We have, by earlier stated monotonicity properties,

$$\hat{\theta}(\tau_i, t) \leq \hat{\theta}(\tau_i, t_i) \leq \hat{\theta}(\tau, t_i).$$

Since the sequence $(\tau_i, t))$ lies in quadrant III and the sequence (τ, t_i) lies in quadrant I, the upper and lower bound both converge to $\hat{\theta}(\tau, t)$. This completes the treatment of quadrant II. Quadrant IV is treated in the same way. This completes the proof of Proposition 3.3 □

4. An RSW-type result

For our main result we need to prove that if the crossing probability of an n by n square goes to 1 as $n \to \infty$, then also the crossing probability of a (say) $3n \times n$ rectangle in the 'difficult direction' goes to 1 as $n \to \infty$. Such (and stronger) results were proved for ordinary percolation in the late nineteen seventies by Russo, and by Seymour and Welsh, and therefore became known as RSW theorems. Their proofs used careful conditioning on the lowest horizontal crossing in a rectangle, after which the area above that crossing was treated, and a new, vertical crossing in that area was 'constructed'. Such arguments work for ordinary percolation because there the above-mentioned area can be treated as 'fresh' territory. However, they usually break down in situations where we have dependencies, as in the sdp model.

Recently, Bollobás and Riordan ([4]) made significant progress on these matters. For the so-called Voronoi percolation model they proved an RSW type result. That result is one of the main ingredients in their proof that the critical probability for Voronoi percolation equals 1/2 (which had been conjectured but stayed open for a long time). Although they explicitly proved their RSW type result only for the Voronoi model, their proof works (as they remark in their paper) for a large class of models. The result we needed is a little stronger than that of [4]. We think that this improvement is useful for many other models as well. The rest of this section is organised as follows. First we give a short introduction to Voronoi percolation. Then we state the above-mentioned RSW-like theorem of [4], and point out where and how its proof needs to be modified to obtain the stronger version. Finally we state the analog for the sdp model and explain why the proof for the Voronoi model works for this model as well.

4.1. The Voronoi percolation model

We start with a brief description of the Voronoi percolation model. The (random) Voronoi percolation model is as follows: Let Z denote the (random) set of points in a Poisson point process with density 1 in the plane. This set gives rise to a random Voronoi tessellation of the plane: Assign to each $z \in Z$ the set of all $x \in \mathbb{R}^2$ for which z is the nearest point in Z. The closure of this set is called the (Voronoi) cell of z. It is known that (with probability 1) each Voronoi cell is a convex polygon, and that two cells are either disjoint or share an entire edge. In the latter case the two cells are said to be neighbours or adjacent. This notion of adjacency gives, in a natural way, rise to the notion of paths, clusters etc.

Now consider the percolation model where each cell, independently of everything else, is coloured black with probability p and white with probability $1-p$. Based on analogies with ordinary percolation (in particular with the self-matching property of the usual triangular lattice) it has been conjectured for a long time that the critical value for this percolation model is 1/2: for $p < 1/2$ there is (a.s.) no infinite black cluster, but for $p > 1/2$ there is an infinite black cluster (a.s.). As we said before, this was recently proved rigorously by Bollobás and Riordan ([4]), and a key ingredient in their proof is an ingenious RSW-like result.

4.2. The RSW-like result for Voronoi percolation

As in [4] we define, for the Voronoi percolation model with parameter p, $f_p(\rho, s)$ as the probability that there is a horizontal black crossing of the rectangle $[0, \rho s] \times [0, s]$. The following is Theorem 4.1 in [4]

Theorem 4.1. (*Bollobás and Riordan*) *Let* $0 < p < 1$ *be fixed.*

$$\text{If} \qquad \liminf_{s\to\infty} f_p(1,s) > 0, \tag{14}$$

$$\text{then} \qquad \limsup_{s\to\infty} f_p(\rho,s) > 0 \text{ for all } \rho > 0.$$

Studying the proof we realised that the condition can be weakened, so that the following theorem is obtained:

Theorem 4.2. *Let $0 < p < 1$ be fixed.*

$$\text{If} \qquad \limsup_{s\to\infty} f_p(\rho, s) > 0 \text{ for some } \rho > 0,$$

$$\text{then} \qquad \limsup_{s\to\infty} f_p(\rho, s) > 0 \text{ for all } \rho > 0.$$

As we shall point out, this somewhat stronger Theorem 4.2 can be proved in almost the same way as Theorem 4.1. But see Remark 4.4 about the global structure of the proof. First note that Theorem 4.2 is (trivially) equivalent to the following:

Theorem 4.3. *Let $0 < p < 1$ be fixed.*

$$\text{If } \limsup_{s\to\infty} f_p(\rho, s) = 0 \text{ for some } \rho > 0, \tag{15}$$

then

$$\limsup_{s\to\infty} f_p(\rho, s) = 0 \text{ for all } \rho > 0. \tag{16}$$

This is the form we will prove, following (with some small changes) the steps in [4].

Proof. (Theorem 4.2 and 4.3). Since p is fixed we will omit it from our notation. In particular we will write f instead of f_p.

First we rewrite the condition (15) in Theorem 4.3: If $\limsup_{s\to\infty} f(\rho, s) = 0$ for some $\rho \le 1$ then, since $f(\rho, s)$ is decreasing in ρ, this lim sup is 0 for all $\rho > 1$. Moreover, the well-known pasting techniques from ordinary percolation show easily that if $\limsup_{s\to\infty} f(\rho, s) > 0$ for some $\rho > 1$, then this lim sup is positive for all $\rho' > \rho$, and hence (using again monotonicity of f in ρ) for all $\rho > 1$. Equivalently, if $\limsup_{s\to\infty} f(\rho, s) = 0$ for some $\rho > 1$, then this limit equals 0 for all $\rho > 1$. Hence, the condition in Theorem 4.3 is equivalent to

$$\limsup_{s\to\infty} f_p(\rho, s) = 0 \text{ for all } \rho > 1, \tag{17}$$

and this is also equivalent to condition (4.2) in Section 4 of [4]:

$$\limsup_{s\to\infty} f_p(\rho, s) = 0 \text{ for some } \rho > 1\,. \tag{18}$$

We will assume (17) (or its equivalent form (18)) and show how, following basically the proof of Theorem 4.1, the equation in (16) can be derived from it for all $\rho > 1/2$. Then we make clear that, for each k, very similar arguments work for $1/k$ instead of $1/2$, which completes the proof of Theorem 4.3.

Remark 4.4. *Bollobás and Riordan prove their theorem by contradiction: They assume (as we do here) (18) above, and, moreover they assume condition (14) (see equation (4.1) in their paper). Then, after a number of steps (claims), they reach a contradiction, which completes the proof. However, most of these steps do not use the 'additional' assumption (14) at all. We found a 'direct' (that is, not by* *contradiction) proof, as sketched below, more clarifying since it leads more easily to further improvements. For our goal most of the steps (claims) in the proof in [4] remain practically unchanged. Therefore we (re)write only some of them in more detail (Claim 1 is stated to give an impression of the start of the proof, and Claim 4 because that already gives a good indication of the strong consequences of (18)). For the other claims we only describe which changes have to be made for our purpose.*

First some notation and terminology: T_s is defined as the strip $[0, s] \times \mathbb{R}$. An event is said to hold *with high probability*, abbreviated *whp* if its probability goes to 1 as $s \to \infty$ (and all other parameters, e.g., p and ϵ are fixed).

Claim 1 (Claim 4.3 in [4]).
Let $\varepsilon > 0$ be fixed, and let L be the line-segment $\{0\} \times [-\varepsilon s, \varepsilon s]$. Assuming that (18) holds, the probability that there is a black path P in T_s starting from L and going outside $S' = [0, s] \times [-(1/2 + 2\varepsilon)s, (1/2 + 2\varepsilon)s]$ tends to 0 as $s \to \infty$.

This claim is exactly the same as in [4], except that in their formulation not only (18) but also (14) is assumed. However, their proof of this claim does not use the latter assumption.

The above, quite innocent looking claim, leads step by step to stronger and eventually very strong claims. We will not rewrite **Claim 2** and **Claim 3**; like Claim 1, they are exactly the same as their corresponding Claims (4.4 and 4.5 respectively) in [4], except that we do not assume (14). And, again, the proof remains as in [4].

Claim 4 (Claim 4.6 in [4]) *Let $C > 0$ be fixed, and let $R = R_s$ be the s by $2Cs$ rectangle $[0, s] \times [-Cs, Cs]$. For $0 \le j \le 4$, set $R_j = [js/100, (j+96)s/100] \times [-Cs, Cs]$. Assuming that (18) holds, whp every black path P crossing R horizontally contains 16 disjoint black paths P_i, $1 \le i \le 16$, where each P_i crosses some R_j horizontally.*

Again, in the formulation in [4] also (14) is assumed, but this is not used in the proof.

Following [4] we now define, for a rectangle R, the random variable $L(R)$ as the minimum length of a black path crossing R horizontally. (More precisely, it is the minimum length of a piecewise-linear black curve that crosses R horizontally.) If there is no horizontal black crossing of R we take $L(R) = \infty$. A complicating property of L is that if R_1 and R_2 are two disjoint rectangles, $L(R_1)$ and $L(R_2)$ are not independent (no matter how large the distance between the two rectangles). Therefore, below Claim 12.4 in their paper, Bollobás and Riordan introduce a suitable modification $\tilde{L}$. The key idea is that whp the colours inside a rectangle with length and width of order s, are completely determined by the Poisson points within distance of order $o(s)$ of the rectangle.

There are many suitable choices of $\tilde{L}$, and we will not rewrite the precise definition given in [4], but only highlight the following three key properties (which neither use (18) nor (14)), where R_s denotes an s by $2s$ rectangle:

$$\tilde{L}(R_s) = L(R_s), \text{ whp}, \tag{19}$$

$$\tilde{L}(R_s) \geq s, \text{ whp}, \tag{20}$$

and:

Claim 5. (Claim 4.7 in [4]). *Let R_1 and R_2 be two s by $2s$ rectangles, separated by a distance of at least $s/100$. If s is large enough, then the random variables $\tilde{L}(R_1)$ and $\tilde{L}(R_2)$ are independent.*

Remark 4.5. *In fact, the independence property of $\tilde{L}$ is only used in the proof of Claim 4.8 in* [4], *and there it could be replaced by the following property (which follows from* (19) *and Claim* 5)*:*

For each $\varepsilon > 0$ there is a u such that for all $s > u$ and all s by $2s$ rectangles R_1 and R_2 that are separated by a distance of at least $s/100$, we have

$$\sup_{x,y>0} |P(L(R_1) < x,\, L(R_2) < y) - P(L(R_1) < x)\, P(L(R_2) < y)| < \varepsilon.$$

Now choose an arbitrary number $\hat{\eta}$ smaller than 10^{-4}. This deviates from the choice of η by Bollobás and Riordan, who add an extra condition, related to their assumption of (14). Define

$$\hat{t}(s) = \sup\{x \,:\, \mathcal{P}(\tilde{L}(R_s) < x) \leq \hat{\eta}\}, \tag{21}$$

where R_s is an s by $2s$ rectangle.

This definition of $\hat{t}$ is the same in form as that of t in [4] (see two lines below (4.15) in [4]); however our way of choosing $\hat{\eta}$ was different. A consequence of this difference is that, in our setup, $\hat{t}(s)$ can be ∞. As in [4], we do have that

$$\hat{t}(s) \geq s, \text{ for all sufficiently large } s. \tag{22}$$

Claim 6. (Claim 4.8 in [4]).
Let R_s be a fixed 0.96 s by $2s$ rectangle. If (18) *holds, then*

$$\mathcal{P}\left(L(R_s) < \hat{t}(0.47s)\right) \leq 200\hat{\eta}^2.$$

This statement is the same as in [4], except that in [4] also (14) is assumed, and that we use $\hat{t}$ and $\hat{\eta}$ instead of t, respectively η. The proof is the same as in [4].

From the above (in particular Claim 4, Claim 6 and (19)), the following quite startling Proposition (which, essentially is equation (4.17) in [4]) now follows quite easily.

Proposition 4.6. (Corresponds with (4.17) in [4]). *If* (18) *holds, then, for all sufficiently large s,*

$$\hat{t}(s) \geq 16\hat{t}(0.47s). \tag{23}$$

Proof. Practically the same as the proof of equation (4.17) in [4]. □

Now Theorem 4.3 follows in a few lines from this proposition and the definition of $\hat{t}(s)$: It is easy to show (see the arguments below (4.17) in [4]) that (22) and Proposition 4.6 together imply that $\hat{t}(s) > s^3$, for all sufficiently large s. Hence, by the definition of $\hat{t}(s)$ (and by (19)) we get that

$$\limsup_{s\to\infty} \mathcal{P}(L(R_s) < s^3) \leq \hat{\eta}, \tag{24}$$

where R_s is an s by $2s$ rectangle.

It is also easy to show (see the arguments below equation (4.18) in [4]) that

$$\mathcal{P}(s^3 \leq L(R_s) < \infty) \to 0 \text{ as } s \to \infty. \tag{25}$$

Combining (24) and (25), and recalling that $L(R_s) = \infty$ iff there is no horizontal black crossing of R_s, immediately gives

$$\limsup_{s\to\infty} \mathcal{P}(\exists \text{ a horizontal black crossing of } R_s) \leq \hat{\eta}.$$

Now, since $\hat{\eta}$ was an arbitrary number between 0 and 10^{-4}, we get $\lim_{s\to\infty} \mathcal{P}(\exists \text{ horizontal black crossing of } R_s) = 0$, that is,

$$f(1/2, s) \to 0, \text{ as } s \to \infty. \tag{26}$$

Note that in the last part of the above arguments (after Claim 4) we worked in particular with s by $2s$ rectangles. A careful look at the arguments shows that the choice of this factor 2 is, in fact, immaterial: if we would take s by $3s$ rectangles or, more generally, fix an $N \geq 2$ and take s by Ns rectangles, the arguments remain practically the same. To see this, one can easily check that in Claims 1–4 (Claims 4.3–4.6 in [4]) the factor 2 plays no role at all: here the rectangles under consideration are s by $2Cs$, where C is a fixed but arbitrary positive number. Further, the proof of Claim 5 remains the same when, for some fixed positive number C, we replace the factor 2 by $2C$. And, the definition of $\hat{t}(s)$ (see (21)), which was given in terms of s by $2s$ rectangles, has, for each $C > 0$, an obvious analog for s by $2Cs$ rectangles:

$$t_C(s) := \sup\{x : \mathcal{P}(\tilde{L}(R_s^C) < x) \leq \hat{\eta}\}, \tag{27}$$

where, for each s, R_s^C is some fixed s by $2Cs$ rectangle.

In the generalization of Claim 6 (Claim 4.8 in [4]) we now fix $C \geq 1$, and take $R_s^C := [0, 0.96s] \times [-Cs, Cs]$. In the proof of this claim we have to replace, on the vertical scale, s by Cs. For instance, the segments L_i, which in the original proof in [4] have length $0.02s$, will now have length $0.02Cs$, and R_0 and R_1 which in the original proof are $0.47s$ by $2 \times 0.47s$ rectangles, are now $0.47s$ by $2C0.47s$ rectangles. In this way we get if (18) holds, for each fixed $C > 1$ the following analog of (26):

$$f(1/(2C), s) \to 0, \text{ as } s \to \infty. \tag{28}$$

This proves Theorem 4.3 and hence Theorem 4.2. □

In the above we were dealing with black horizontal crossings. Obviously, completely analogous results hold for white horizontal crossings: If we denote (for a fixed value of the parameter p of the Voronoi percolation model), the probability of a vertical white crossing of a given ρs by s rectangle by $g(\rho, s)$, we have that if $\lim_{s \to \infty} g(\rho, s) = 0$ for *some* $\rho > 0$, then this limit is 0 for *all* $\rho > 0$. Since a rectangle has either a horizontal black crossing or a vertical white crossing (and hence $g(\rho, s) = 1 - f(\rho, s)$) this gives:

Corollary 4.7.

$$\text{If} \qquad \lim_{s \to \infty} f(\rho, s) = 1 \text{ for some } \rho > 0, \tag{29}$$
$$\text{then} \qquad \lim_{s \to \infty} f(\rho, s) = 1 \text{ for all } \rho > 0.$$

4.3. An RSW analog for self-destructive percolation

In the previous subsection we considered (and somewhat strengthened) an RSW-like result of Bollobás and Riordan ([4]) for the Voronoi percolation model. Only a few properties of the model are used in its proof. As remarked in [4] (at the end of Section 4; see also [5], Section 5.1), these properties are basically the following: First of all, crossings of rectangles are defined in terms of 'geometric paths' in such a way that, for example, horizontal and vertical black crossings meet, which enables to form longer paths by pasting together several small paths. Further, a form of FKG is used (e.g., that events of the form 'there is a black path from A to B' are positively correlated. Also some symmetry is needed. Bollobás and Riordan say that "invariance of the model under the symmetries of $\mathbb{Z}^2$ suffices, as we need only consider rectangles with integer coordinates". Finally, some form of asymptotic independence is needed (see Remark 4.5). Similar considerations hold wrt the somewhat stronger Theorem 4.2.
Using the results in Section 2, is not difficult to see that the sdp model has the above-mentioned properties:

- The indicated geometric properties are just the well-known intersection properties of paths in the square lattice (and in its matching lattice).
- Lemma 2.1 gives the needed FKG-like properties.
- Asymptotic independence: Note that for $p \leq p_c$ the sdp model is an ordinary percolation model, where this property is trivial. If $p > p_c$, then $1 - p$ is smaller than the critical probability of the matching lattice. In that case the needed asymptotic independence (of the form described in Lemma 4.5) comes from Lemma 2.3 and the well-known exponential decay theorems for ordinary subcritical percolation.
- The sdp model on the square lattice clearly has all the symmetries of $\mathbb{Z}^2$.

Further, to carry out for the sdp model the analog of the arguments that led from Theorem 4.3 to Corollary 4.7, we note that the random collection of vacant sites on the matching lattice clearly also has the above-mentioned properties. So we get the following theorem for the sdp model:

Theorem 4.8. *The analogs of Theorems 4.2 and 4.3 and Corollary 4.7 hold for the self-destructive percolation model. In particular, let for the sdp model with parameters p and δ, $f(\rho, s) = f_{p,\delta}(\rho, s)$ denote the probability that there is an occupied horizontal crossing of a given $\rho s \times s$ rectangle. We have*

$$\begin{aligned} &\textit{If} && \lim_{s\to\infty} f(\rho, s) = 1 \textit{ for some } \rho > 0, \\ &\textit{then} && \lim_{s\to\infty} f(\rho, s) = 1 \textit{ for all } \rho > 0. \end{aligned} \tag{30}$$

In the next sections this result will play an important role in the completion of the proof of Theorem 1.2. In particular, in Section 5 it will be used to prove a finite-size criterion for supercriticality of the sdp model.

5. A finite-size criterion

The main result of this section is a suitable finite-size criterion for supercriticality of the sdp model. The overall structure of the argument is similar to that in ordinary percolation (see [8]), but the dependencies in the model require extra attention. One of the main ingredients, a suitable RSW-like theorem for this model, was obtained in the previous section.

Theorem 5.1. *Let $f = f_{p,\delta}$ as in Theorem 4.8. There is a universal constant $\alpha > 0$ and there is a decreasing function $\hat{N} : (p_c, 1) \to \mathbb{N}$ such that for all $p > p_c$ and all $\delta > 0$ the following two assertions,* (i) *and* (ii) *below, are equivalent.*

$$\begin{aligned} &\text{(i)} && \theta(p, \delta) > 0. \\ &\text{(ii)} && \exists n \geq \hat{N}(p) \textit{ such that } f_{p,\delta}(3, n) > 1 - \alpha. \end{aligned} \tag{31}$$

Remark 5.2. *In ordinary percolation $\hat{N}(p)$ can be taken constant* 1. *Remark* 6.1 *below explains the impact of this difference.*

Proof. Consider for each $n \in \mathbb{N}$ the events

$$\begin{aligned} A &= \{\exists \text{ a vertical vacant *-crossing of } [0, 9n] \times [0, 3n]\}; \\ B &= \{\exists \text{ a vertical vacant *-crossing of } [0, 9n] \times [0, n]\}; \\ C &= \{\exists \text{ a vertical vacant *-crossing of } [0, 9n] \times [2n, 3n]\}. \end{aligned} \tag{32}$$

Let $h(\rho, n)$ denote the probability of a vertical vacant *-crossing (in the matching lattice) of a ρn by n box. So, $h(\rho, n) = 1 - f(\rho, n)$. Clearly, $\mathcal{P}_{p,\delta}(B) = \mathcal{P}_{p,\delta}(C) = h_{p,\delta}(9, n)$ and $\mathcal{P}_{p,\delta}(A) = h_{p,\delta}(3, 3n)$. It is also clear that $A \subset B \cap C$. From this, Lemma 2.3, the fact that the r.h.s. of (4) is decreasing, and the well-known exponential decay results for ordinary subcritical percolation applied to (4), it follows that there is an increasing, function $\phi : (p_c, 1) \to (0, \infty)$ such that for all $p > p_c$,

$$h(3, 3n) \leq h(9, n)^2 + \exp(-n\phi(p)). \tag{33}$$

Further note that if the event B occurs, there must be a vacant vertical *-crossing of one of the rectangles $[0, 3n] \times [0, n]$, $[2n, 5n] \times [0, n]$, $[4n, 7n] \times [0, n]$,

$[6n, 9n] \times [0, n]$, or a vacant horizontal *-crossing of one of the rectangles $[2n, 3n] \times [0, n]$, $[4n, 5n] \times [0, n]$, $[6n, 7n] \times [0, n]$.

Hence

$$h(9, n) \leq 4h(3, n) + 3h(1, n) \leq 7h(3, n), \tag{34}$$

which combined with (33) gives

$$h(3, 3n) \leq 49h(3, n)^2 + \exp(-n\phi(p)). \tag{35}$$

Take α so small that $49\alpha^2 < \alpha/4$. Let, for each $p > p_c$, $\hat{N}(p)$ be the smallest positive integer for which

$$\exp(-\hat{N}(p)\phi(p)) < \alpha/4.$$

Sine ϕ is increasing, $\hat{N}$ is decreasing in p.

Now suppose $p > p_c$ and $\delta \in (0, 1)$ are given and suppose that (ii) holds. So there exists an n that satisfies:

$$\exp(-n\phi(p)) < \alpha/4 \text{ and } h(3, n) < \alpha. \tag{36}$$

From (35), (36) and the choice of α we get

$$h(3, 3n) \leq 49\alpha^2 + \alpha/4 < \alpha/4 + \alpha/4 = \alpha/2, \tag{37}$$

and

$$\exp(-3n\phi(p)) < (\alpha/4)^3 < (\alpha/2)/4.$$

Hence, (36) with n replaced by $3n$, and α replaced by $\alpha/2$ holds. So we can iterate (37) and conclude that, for all integers $k \geq 0$,
$h(3,\, 3^k n) < \alpha/(2^k)$.

The last part of the argument is exactly as for ordinary percolation: Note that if none of the rectangles $[0, 3^{2k+1}n] \times [0, 3^{2k}n]$ and $[0, 3^{2k+1}n] \times [0, 3^{2k+2}n]$, $k = 0, 1, 2, \ldots$ has a white *-crossing in the 'easy' (short) direction, then each of these rectangles has a black crossing in the long direction. Moreover, all these black crossings together form an infinite occupied path. Hence,

$$\theta(p, \delta) \geq 1 - \sum_{k=0}^{\infty} h(3, n3^k) \geq 1 - \alpha \sum_{k=0}^{\infty} (1/2)^k = 1 - 2\alpha > 0.$$

This proves that (ii) implies (i).

Now we show that (i) implies (ii): Suppose $\theta(p, \delta) > 0$. Then there is (a.s.) an infinite occupied cluster, and by Lemma 2.4 this cluster is unique. From the usual spatial symmetries, positive association, and the above-mentioned uniqueness one can, in exactly the same way as for ordinary percolation (see [10], Theorem 8.97) show that $f(1, n) \to 1$ as $n \to \infty$. By Theorem 4.8 it follows that also $f(3, n) \to 1$ as $n \to \infty$; so (ii) holds. □

6. Proof of Theorem 1.2

We are now ready to prove Theorem 1.2:

Proof. For $p < p_c$, we have (see Section 1) $\theta(p,\delta) = \theta(p + (1-p)\delta)$, so that continuity follows from continuity for ordinary percolation. If $p = p_c$ and $\delta > p_c + \varepsilon$ for some $\varepsilon > 0$, then (trivially) there is a neighborhood of (p,δ) where the sdp model dominates ordinary percolation with parameter $p_c + \varepsilon/2 > p_c$; hence $\theta(\cdot,\cdot) > 0$ on this neighborhood, and Proposition 3.3 implies continuity of $\theta(\cdot,\cdot)$ at (p_c,δ). (In fact, by combining this argument with an Aizenman-Grimmett type argument, one can extend this result and show that there is an $\varepsilon > 0$ such that $\theta(\cdot,\cdot)$ is continuous at (p_c,δ) if $\delta > p_c - \varepsilon$).

Finally, we consider the case where $p > p_c$. If $\theta(p,\delta) = 0$, continuity at (p,δ) follows from part (b) of Proposition 3.3. So suppose $\theta(p,\delta) > 0$. Let α as in Theorem 5.1. By that theorem there is an $n \geq \hat{N}(p)$ with

$$f_{p,\delta}(3,n) > 1 - \alpha.$$

Hence, by Lemma 3.1 there is an open neighborhood W of (p,δ) such that

$$f_{p',\delta'}(3,n) > 1 - \alpha, \tag{38}$$

for all $(p',\delta') \in W$. Since $n \geq \hat{N}(p)$ and $\hat{N}(.)$ is decreasing, it follows from (38) and Theorem 5.1 that $\theta(\cdot,\cdot) > 0$ on S, where S is the set of all $(p',\delta') \in W$ with $p' \geq p$. From this and Lemma 2.2 we conclude that $\theta(\cdot,\cdot)$ is also strictly positive on the set

$$U := \{(p',\delta') : p' < r \text{ and } p' + (1-p')\delta' > r + (1-r)\beta \text{ for some } (r,\beta) \in S\}.$$

It is easy to see that $S \cup U$ contains an open neighborhood of (p,δ). Now it follows from part (a) of Proposition 3.3 that $\theta(\cdot)$ is continuous at (p,δ). This completes the proof of the main theorem. □

Remark 6.1. *A crucial role in the proof is the finite-size criterion, Theorem* 5.1. *That theorem has been formulated for* $p > p_c$. *When* $p = p_c$ *(or* $< p_c$*) the sdp model is an ordinary percolation model, for which a similar criterion is known. In fact, for ordinary percolation we do not have the dependency problems which led to the introduction of* $\hat{N}$. *Consequently, for* $p = p_c$ *we can take* $\hat{N} = 1$. *But, on the other hand, if we let* p *tend to* p_c *from above, the upper bound on* $\hat{N}(p)$ *obtained from our arguments in Section* 5 *tends to* ∞. *And that, in turn, comes from the fact that our bound on dependencies, Lemma* 2.3, *is in terms of path probabilities for an ordinary percolation model (on the matching lattice, with parameter* $1-p$*) which is subcritical but approaches criticality (which makes these bounds worse and worse) as* p *approaches* p_c *from above. This is essentially why the proof of Theorem* 1.2 *does not work at* p_c. *Of course, if it would work, the conjecture referred to in Section* 1 *would be false. We hope that attempts to stretch the arguments in our paper as far as possible will substantially increase insight in the conjecture and help to obtain a solution.*

References

[1] van den Berg, J. and Brouwer, R. Self-destructive percolation, *Random Structures and Algorithms* **24** Issue 4, 480–501 (2004).

[2] van den Berg, J. and Brouwer, R. Self-organized forest-fires near the critical time, *Comm. Math. Phys.* **267**, 265–277 (2006.)

[3] van den Berg, J. and Keane, M. On the continuity of the percolation probability function, *Particle Systems, Random Media and Large Deviations* (R.T. Durrett, ed.), Contemporary Mathematics series, vol. 26, AMS, Providence, R.I., 61–65 (1984).

[4] Bollobás, B. and Riordan, O., The critical probability for random Voronoi percolation in the plane is 1/2, *Prob. Th. Rel. Fields* **136**, 417–468 (2006).

[5] Bollobás, B. and Riordan, O., Sharp thresholds and percolation in the plane, *Random Structures and Algorithms* **29**, 524–548 (2006).

[6] Brouwer, R., *Percolation, forest-fires and monomer-dimers*, PhD thesis, VUA, October 2005.

[7] Burton, R.M. and Keane, M. (1989), Density and uniqueness in percolation, *Comm. Math. Phys.* **121**(3), 501–505 (1989)

[8] Chayes, J.T. and Chayes, L. Percolation and random media, *Critical Phenomena, Random Systems and Gauge Theories* (K. Oswalder and R. Stora, eds.), Les Houches, Session XLIII, 1984, Elsevier, Amsterdam, pp. 1001–1142.

[9] Gandolfi, A., Keane, M.S. and Newman, C.M., Uniqueness of the infinite component in a random graph with applications to percolation and spin glasses, *Prob. Th. Rel. Fields* **92**, 511–527 (1992).

[10] Grimmett, G.R. *Percolation*, 2nd ed. Springer-Verlag (1999).

J. van den Berg
Centrum voor Wiskunde en Informatica
Kruislaan 413
NL-1098 SJ Amsterdam, The Netherlands
e-mail: `J.van.den.Berg@cwi.nl`

R. Brouwer
University Medical Center Utrecht
Division of Neuroscience
House A.01.126
P.O. Box 85500
NL-3508 GA Utrecht, The Netherlands
e-mail: `r.m.brouwer-4@umcutrecht.nl`

B. Vágvölgyi
Vrije Universiteit Amsterdam
De Boelelaan 1081
NL-1081 HV Amsterdam, The Netherlands
e-mail: `bvagvol@few.vu.nl`

Progress in Probability, Vol. 60, 137–160

A Quenched Invariance Principle for Certain Ballistic Random Walks in i.i.d. Environments

Noam Berger and Ofer Zeitouni

Abstract. We prove that every random walk in i.i.d. environment in dimension greater than or equal to 2 that has an almost sure positive speed in a certain direction, an annealed invariance principle and some mild integrability condition for regeneration times also satisfies a quenched invariance principle. The argument is based on intersection estimates and a theorem of Bolthausen and Sznitman.

Mathematics Subject Classification (2000). Primary 60K37; secondary 60F05.

Keywords. Random walk in random environment, quenched invariance principle.

1. Introduction

Let $d \geq 1$. A Random Walk in Random Environment (RWRE) on $\mathbb{Z}^d$ is defined as follows. Let $\mathcal{M}^d$ denote the space of all probability measures on $\mathcal{E}_d = \{\pm e_i\}_{i=1}^d$ and let $\Omega = \left(\mathcal{M}^d\right)^{\mathbb{Z}^d}$. An *environment* is a point $\omega = \{\omega(x,e)\}_{x\in\mathbb{Z}^d,\, e\in\mathcal{E}_d} \in \Omega$. Let P be a probability measure on Ω. For the purposes of this paper, we assume that P is an i.i.d. measure, i.e.,

$$P = Q^{\mathbb{Z}^d}$$

for some distribution Q on $\mathcal{M}^d$ and that Q is *uniformly elliptic*, i.e., there exists a $\kappa > 0$ such that for every $e \in \mathcal{E}_d$,

$$Q(\{\omega(0,\cdot) : \omega(0,e) < \kappa\}) = 0.$$

For an environment $\omega \in \Omega$, the *Random Walk* on ω is a time-homogeneous Markov chain with transition kernel

$$P_\omega\left(X_{n+1} = x + e \,|\, X_n = x\right) = \omega(x,e).$$

O.Z. was partially supported by NSF grant DMS-0503775. N.B. was partially supported by NSF grant DMS-0707226.

The **quenched law** P_ω^x is defined to be the law on $\left(\mathbb{Z}^d\right)^{\mathbb{N}}$ induced by the transition kernel P_ω and $P_\omega^x(X_0 = x) = 1$. With some abuse of notation, we write P_ω also for P_ω^0. We let $\mathcal{P}^x = P \otimes P_\omega^x$ be the joint law of the environment and the walk, and the **annealed** law is defined to be its marginal

$$\mathbb{P}^x(\cdot) = \int_\Omega P_\omega^x(\cdot)dP(\omega).$$

We use $\mathbb{E}^x$ to denote expectations with respect to $\mathbb{P}^x$. We consistently omit the superscript x if $x = 0$.

We say that the RWRE $\{X(n)\}_{n\geq 0}$ satisfies the law of large numbers with deterministic speed v if $X_n/n \to v$, $\mathbb{P}$-a.s. For $x \geq 0$, let $[x]$ denote the largest integer less than or equal to x. We say that the RWRE $\{X(n)\}_{n\geq 0}$ satisfies the *annealed* invariance principle with deterministic, positive definite covariance matrix $\sigma_{\mathbb{P}}^2$ if the linear interpolations of the processes

$$B^n(t) = \frac{X([nt]) - [nvt]}{\sqrt{n}}\,, t \geq 0 \tag{1.1}$$

converge in distribution (with respect to the supremum topology on the space of continuous function on $[0, 1]$) as $n \to \infty$, under the measure $\mathbb{P}$, to a Brownian motion of covariance $\sigma_{\mathbb{P}}^2$. We say the process $\{X(n)\}_{n\geq 0}$ satisfies the *quenched* invariance principle with variance $\sigma_{\mathbb{P}}^2$ if for P-a.e. ω, the above convergence holds under the measure P_ω^0. Our focus in this paper are conditions ensuring that when an annealed invariance principle holds, so does a quenched one.

To state our results, we need to recall the regeneration structure for random walk in i.i.d. environment, developed by Sznitman and Zerner in [SZ99]. We say that t is a regeneration time (in direction e_1) for $\{X(\cdot)\}$ if

$$\langle X(s), e_1\rangle < \langle X(t), e_1\rangle \text{ whenever } s < t$$

and

$$\langle X(s), e_1\rangle \geq \langle X(t), e_1\rangle \text{ whenever } s > t\,.$$

When ω is distributed according to an i.i.d. P such that the process $\{\langle X(n), e_1\rangle\}_{n\geq 0}$ is $\mathbb{P}$-almost surely transient to $+\infty$, it holds by [SZ99] that, $\mathbb{P}$-almost surely, there exist infinitely many regeneration times for $\{X(\cdot)\}$. Let

$$t^{(1)} < t^{(2)} < \cdots,$$

be all of the regeneration times for $\{X(\cdot)\}$. Then, the sequence

$$\{(t^{(k+1)} - t^{(k)}), (X(t^{(k+1)}) - X(t^{(k)}))\}_{k\geq 1}$$

is an i.i.d. sequence under $\mathbb{P}$. Further, if $\lim_{n\to\infty} n^{-1}\langle X(n), e_1\rangle > 0$, $\mathbb{P}$-a.s., then we get, see [SZ99], that

$$\mathbb{E}(t^{(2)} - t^{(1)}) < \infty. \tag{1.2}$$

The main result of this paper is the following:

Theorem 1.1. *Let $d \geq 4$ and let Q be a uniformly elliptic distribution on $\mathcal{M}^d$. Set $P = Q^{\mathbb{Z}^d}$. Assume that the random walk $\{X(n)\}_{n\geq 0}$ satisfies the law of large numbers with a positive speed in the direction e_1, that is*

$$\lim_{n\to\infty} \frac{X(n)}{n} = v\,, \mathbb{P} - a.s \quad \text{with } v \text{ deterministic such that } \langle v, e_1\rangle > 0\,. \tag{1.3}$$

Assume further that the process $\{X(n)\}_{n\geq 0}$ satisfies an annealed invariance principle with variance $\sigma_{\mathbb{P}}^2$.

Assume that there exists an $\epsilon > 0$ such that $\mathbb{E}(t^{(1)})^\epsilon < \infty$ and, with some $r \geq 2$,

$$\mathbb{E}[(t^{(2)} - t^{(1)})^r] < \infty\,. \tag{1.4}$$

If $d = 4$, assume further that (1.4) holds with $r > 8$. Then, the process $\{X(\cdot)\}$ satisfies a quenched invariance principle with variance $\sigma_{\mathbb{P}}^2$.

(The condition $r \geq 2$ for $d \geq 5$ can be weakened to $r > 1 + 4/(d+4)$ by choosing in (3.7) below $r' = r$ and modifying appropriately the value of K_d in Proposition 3.1 and Corollary 3.2.) We suspect, in line with Sznitman's conjecture concerning condition T', see [Szn02], that (1.4) holds for $d \geq 2$ and all $r > 0$ as soon as (1.3) holds.

A version of Theorem 1.1 for $d = 2, 3$ is presented in Section 4. For $d = 1$, the conclusion of Theorem 1.1 does not hold, and a quenched invariance principle, or even a CLT, requires a different centering [Zei04, Gol07, Pet08]. (This phenomenon is typical of dimension $d = 1$, as demonstrated in [RAS06] in the context of the totally asymmetric, non-nearest neighbor, RWRE.) Thus, some restriction on the dimension is needed.

Our proof of Theorem 1.1 is based on a criterion from [BS02], which uses two independent RWRE's in the same environment ω. This approach seems limited, in principle, to $d \geq 3$ (for technical reasons, we restrict attention to $d \geq 4$ in the main body of the paper), regardless of how good tail estimates on regeneration times hold. An alternative approach to quenched CLT's, based on martingale methods but still using the existence of regeneration times with good tails, was developed by Rassoul-Agha and Seppäläinen in [RAS05], [RAS07a], and some further ongoing work of these authors. While their approach has the potential of reducing the critical dimension to $d = 2$, at the time this paper was written, it had not been successful in obtaining statements like in Theorem 1.1 without additional structural assumptions on the RWRE. [1]

[1] After the first version of this paper was completed and posted, Rassoul-Agha and Seppäläinen posted a preprint [RAS07b] in which they prove a statement similar to Theorem 1.1, for all dimensions $d \geq 2$, under somewhat stronger assumptions on moments of regeneration times. While their approach differs significantly from ours, and is somewhat more complicated, we learnt from their work an extra ingredient that allowed us to extend our approach and prove Theorem 1.1 in all dimensions $d \geq 2$. For the convenience of the reader, we sketch the argument in Section 4 below.

Since we will consider both the case of two independent RWRE's in different environments and the case of two RWRE's evolving in the same environment, we introduce some notation. For $\omega_i \in \Omega$, we let $\{X_i(n)\}_{n\geq 0}$ denote the path of the RWRE in environment ω_i, with law $P^0_{\omega_i}$. We write P_{ω_1,ω_2} for the law $P^0_{\omega_1} \times P^0_{\omega_2}$ on the pair $(\{X_1(\cdot), X_2(\cdot)\})$. In particular,

$$E_{P\times P}[P_{\omega_1,\omega_2}(\{X_1(\cdot)\} \in A_1, \{X_2(\cdot)\} \in A_2)] = \mathbb{P}(\{X_1(\cdot)\} \in A_1) \cdot \mathbb{P}(\{X_2(\cdot)\} \in A_2)$$

represents the annealed probability that two walks $\{X_i(\cdot)\}$, $i = 1, 2$, in independent environments belong to sets A_i, while

$$\begin{aligned} &E_P[P_{\omega,\omega}(\{X_1(\cdot)\} \in A_1, \{X_2(\cdot)\} \in A_2)] \\ &\quad = \int P_\omega(\{X_1(\cdot)\} \in A_1) \cdot P_\omega(\{X_2(\cdot)\} \in A_2) dP(\omega) \end{aligned}$$

is the annealed probability for the two walks in the *same* environment.

We use throughout the notation

$$t_i^{(1)} < t_i^{(2)} < \cdots, \quad i = 1, 2$$

for the sequence of regeneration times of the process $\{X_i(\cdot)\}$. Note that whenever P satisfies the assumptions in Theorem 1.1, the estimate (1.4) holds for $(t_i^{(2)} - t_i^{(1)})$, as well.

Notation Throughout, C denotes a constant whose value may change from line to line, and that may depend on d and κ only. Constants that may depend on additional parameters will carry this dependence in the notation. Thus, if F is a fixed function then C_F denotes a constant that may change from line to line, but that depends on F, d and κ only. For $p \geq 1$, $\|\cdot\|_p$ denotes the L^p norm on $\mathbb{R}^d$ or $\mathbb{Z}^d$, while $\|\cdot\|$ denotes the supremum norm on these spaces.

2. An intersection estimate and proof of the quenched CLT

As mentioned in the introduction, the proof of the quenched CLT involves considering a pair of RWRE's $(X_1(\cdot), X_2(\cdot))$ in the same environment. The main technical tool needed is the following proposition, whose proof will be provided in Section 3. Let $H_K := \{x \in \mathbb{Z}^d : \langle x, e_1 \rangle > K\}$.

Proposition 2.1. *We continue under the assumptions of Theorem* 1.1. *Let*

$$W_K := \{\{X_1(\cdot)\} \cap \{X_2(\cdot)\} \cap H_K \neq \emptyset\}.$$

Then

$$E_P[P_{\omega,\omega}(W_K)] < CK^{-\kappa_d} \tag{2.1}$$

where $\kappa_d = \kappa_d(\epsilon, r) > 0$ *for* $d \geq 4$.

We can now bring the

Proof of Theorem 1.1 (assuming Proposition 2.1). For $i = 1, 2$, define $B_i^n(t) = n^{-1/2}(X_i([nt]) - [nvt]))$, where the processes $\{X_i\}$ are RWRE's in the same environment ω, whose law is P. We introduce the space $C(\mathbb{R}_+, \mathbb{R}^d)$ of continuous $\mathbb{R}^d$-valued functions on $\mathbb{R}_+$, and the $C(\mathbb{R}_+, \mathbb{R}^d)$-valued variable

$$\beta_i^n(\cdot) = \text{the polygonal interpolation of } \tfrac{k}{n} \to B_i^n(\tfrac{k}{n}),\ k \geq 0\ . \tag{2.2}$$

It will also be useful to consider the analogously defined space $C([0,T], \mathbb{R}^d)$, of continuous $\mathbb{R}^d$-valued functions on $[0,T]$, for $T > 0$, which we endow with the distance

$$d_T(v, v') = \sup_{s \leq T} |v(s) - v'(s)| \wedge 1\,. \tag{2.3}$$

With some abuse of notation, we continue to write $\mathbb{P}$ for the law of the pair (β_1^n, β_2^n). By Lemma 4.1 of [BS02], the claim will follow once we show that for all $T > 0$, for all bounded Lipschitz functions F on $C([0,T], \mathbb{R}^d)$ and $b \in (1, 2]$:

$$\sum_m (E_P[E_\omega(F(\beta_1([b^m]))E_\omega(F(\beta_2([b^m]))] - \mathbb{E}[F(\beta_1([b^m]))]\mathbb{E}[F(\beta_2([b^m])]) < \infty\,. \tag{2.4}$$

When proving (2.4), we may and will assume that F is bounded by 1 with Lipschitz constant 1.

Fix constants $1/2 > \theta > \theta'$. Write $N = [b^m]$. Let

$$s_i^m := \min\{t > N^\theta/2 :\ X_i(t) \in H_{N^{\theta'}},\ t \text{ is a regeneration time for } X_i(\cdot)\}\,, i = 1, 2.$$

Define the events

$$A_i^m := \{s_i^m \leq N^\theta\}\,, i = 1, 2,$$

and

$$C_m := \{\{X_1(n + s_1^m)\}_{n\geq 0} \cap \{X_2(n + s_2^m)\}_{n\geq 0} = \emptyset\}\,, B_m := A_1^m \cap A_2^m \cap C_m\,.$$

Finally, write $\mathcal{F}_i := \sigma(X_i(t), t \geq 0)$ and

$$\mathcal{F}_i^\Omega := \sigma\{\omega_z : \text{there exists a } t \text{ such that } X_i(t) = z\} \vee \mathcal{F}_i, i = 1, 2.$$

Note that, for $i = 1, 2$,

$$\begin{aligned}
&\mathbb{P}\left((A_i^m)^c\right) \\
&\leq \mathbb{P}(\max_{j=1}^{N^\theta}[t_i^{(j+1)} - t_i^{(j)}] \geq N^\theta/4) + \mathbb{P}(t_i^{(1)} > N^\theta/4) + \mathbb{P}(X_i(N^\theta/2) \notin H_{N^{\theta'}}) \\
&\leq \frac{4^r N^\theta \mathbb{E}[(t_i^{(2)} - t_i^{(1)})^r]}{N^{r\theta}} + \frac{4^\epsilon \mathbb{E}\left([t_i^{(1)}]^\epsilon\right)}{N^{\theta\epsilon}} + \mathbb{P}\left(\sum_{j=1}^{N^{\theta'}} (t_i^{(j+1)} - t_i^{(j)}) > \frac{N^\theta}{4}\right) \qquad (2.5)\\
&\leq N^{-\delta'} + 4N^{\theta' - \theta}\mathbb{E}\left[t_i^{(2)} - t_i^{(1)}\right] \leq 2N^{-\delta'}\,,
\end{aligned}$$

with $\delta' = \delta'(\epsilon, \theta) > 0$ independent of N. Using the last estimate and Proposition 2.1, one concludes that

$$\sum_m E_P[P_{\omega,\omega}(B_m^c)] < \infty. \tag{2.6}$$

Now,

$$|\mathbb{E}[F(\beta_1^{[b^m]})F(\beta_2^{[b^m]})] - \mathbb{E}[\mathbf{1}_{B_m} F(\beta_1^{[b^m]})F(\beta_2^{[b^m]})]| \le \mathbb{P}(B_m^c)\,. \tag{2.7}$$

Let the process $\bar{\beta}_i^{[b^m]}(\cdot)$ be defined exactly as the process $\beta_i^{[b^m]}(\cdot)$, except that one replaces $X_i(\cdot)$ by $X_i(\cdot + s_i^m)$. On the event A_i^m, we have by construction that

$$\sup_t \left|\beta_i^{[b^m]}(t) - \bar{\beta}_i^{[b^m]}(t)\right| \le 2N^{\theta-1/2}\,,$$

and therefore, on the event $A_1^m \cap A_2^m$,

$$\left|[F(\beta_1^{[b^m]})F(\beta_2^{[b^m]})] - [F(\bar{\beta}_1^{[b^m]})F(\bar{\beta}_2^{[b^m]})]\right| \le CN^{\theta-1/2}\,, \tag{2.8}$$

for some constant C (we used here that F is Lipschitz (with constant 1) and bounded by 1).

On the other hand, writing ω' for an independent copy of ω with the same distribution P,

$$\begin{aligned}
&\mathbb{E}[\mathbf{1}_{B_m} F(\bar{\beta}_1^{[b^m]})F(\bar{\beta}_2^{[b^m]})] = E_P(E_\omega[\mathbf{1}_{B_m} F(\bar{\beta}_1^{[b^m]})F(\bar{\beta}_2^{[b^m]}]) \\
= \;& \mathbb{E}\left(\mathbf{1}_{A_m^1} F(\bar{\beta}_1^{[b^m]})E_\omega[\mathbf{1}_{A_m^2 \cap C_m} F(\bar{\beta}_2^{[b^m]}]) \mid \mathcal{F}_1^\Omega]\right) \\
= \;& E_P\left(E_\omega\left[\mathbf{1}_{A_m^1} F(\bar{\beta}_1^{[b^m]})E_\omega[\mathbf{1}_{A_m^2 \cap C_m} F(\bar{\beta}_2^{[b^m]}]) \mid \mathcal{F}_1^\Omega]\right]\right) \\
= \;& E_P\left(E_\omega\left[\mathbf{1}_{A_m^1} F(\bar{\beta}_1^{[b^m]})E_{\omega'}[\mathbf{1}_{A_m^2 \cap C_m} F(\bar{\beta}_2^{[b^m]}]) \mid \mathcal{F}_1^\Omega]\right]\right) \\
= \;& E_P\left(E_{\omega,\omega'}\left[\mathbf{1}_{A_m^1} F(\bar{\beta}_1^{[b^m]})\mathbf{1}_{A_m^2 \cap C_m} F(\bar{\beta}_2^{[b^m]}])\right]\right) \\
= \;& E_P\left(E_{\omega,\omega'}\left[\mathbf{1}_{B_m} F(\bar{\beta}_1^{[b^m]})F(\bar{\beta}_2^{[b^m]}])\right]\right)\,.
\end{aligned} \tag{2.9}$$

The third equality follows from the fact that we multiply by the indicator of the event of non-intersection. Since

$$\left|E_P\left(E_{\omega,\omega'}\left[\mathbf{1}_{B_m} F(\beta_1^{[b^m]})F(\beta_2^{[b^m]}])\right]\right) - E_P\left(E_{\omega,\omega'}\left[F(\beta_1^{[b^m]})F(\beta_2^{[b^m]}])\right]\right)\right| \le \mathbb{P}(B_m^c),$$

and

$$E_P\left(E_{\omega,\omega'}\left[F(\beta_1^{[b^m]})F(\beta_2^{[b^m]}])\right]\right) = \mathbb{E}\left[F(\beta_1^{[b^m]})\right]\mathbb{E}\left[F(\beta_2^{[b^m]}])\right]\,,$$

we conclude from the last two displays, (2.9) and (2.8) that

$$\left|\mathbb{E}[F(\beta_1^{[b^m]})F(\beta_2^{[b^m]})] - \mathbb{E}\left[F(\beta_1^{[b^m]})\right]\mathbb{E}\left[F(\beta_2^{[b^m]}])\right]\right| \le 2\mathbb{P}(B_m^c) + 2CN^{\theta-1/2}\,.$$

Together with (2.6), we conclude that (2.4) holds, and complete the proof of Theorem 1.1. □

3. Intersection structure

In this section we prove Proposition 2.1, that is we establish estimates on the probability that two independent walks in the same environment intersect each other in the half space $H_K = \{x \in \mathbb{Z}^d : \langle x, e_1 \rangle > K\}$. It is much easier to obtain such estimates for walks in different environments, and the result for different environments will be useful for the case of walks in the same environment.

3.1. The conditional random walk

Under the assumptions of Theorem 1.1, the process $\{\langle X(\cdot), e_1\rangle\}$ is $\mathbb{P}$-a.s. transient to $+\infty$. Let

$$D := \{\forall_{n\geq 0}, \langle X(n), e_1\rangle \geq \langle X(0), e_1\rangle\}\,.$$

By, e.g., [SZ99], we have that

$$\mathbb{P}(D) > 0\,. \tag{3.1}$$

3.2. Intersection of paths in independent environments

In this subsection, we let $\omega^{(1)}$ and $\omega^{(2)}$ be independent environments, each distributed according to P. Let $\{Y_1(n)\}$ and $\{Y_2(n)\}$ be random walks in the environments (respectively) $\omega^{(1)}$ and $\omega^{(2)}$, with starting points $U_i = Y_i(0)$. In other words, $\{Y_1(n)\}$ and $\{Y_2(n)\}$ are independent samples taken from the annealed measures $\mathbb{P}^{U_i}(\cdot)$. For $i = 1, 2$ set

$$D_i^{U_i} = \{\langle Y_i(n), e_1\rangle \geq \langle U_i, e_1\rangle \text{ for } \; n \geq 0\}\,, \quad i = 1, 2.$$

For brevity, we drop U_i from the notation and use $\mathbf{P}$ for $\mathbb{P}^{U_1} \times \mathbb{P}^{U_2}$ and $\mathbf{P}^D$ for $\mathbb{P}^{U_1}(\cdot|D_1^{U_1}) \times \mathbb{P}^{U_2}(\cdot|D_2^{U_2})$.

First we prove some basic estimates. While the estimates are similar for $d = 4$ and $d \geq 5$, we will need to prove them separately for the two cases.

3.2.1. Basic estimates for $d \geq 5$.

Proposition 3.1. ($d \geq 5$) *With notation as above and assumptions as in Theorem 1.1,*

$$\mathbf{P}^D\left(\{Y_1(\cdot)\} \cap \{Y_2(\cdot)\} \neq \emptyset\right) < C\|U_1 - U_2\|^{-K_d}$$

where $K_d = \frac{d-4}{4+d}$

The proof is very similar to the proof of Lemma 5.1 of [Ber06], except that here we need a quantitative estimate that is not needed in [Ber06].

Proof of Proposition 3.1. We first note that the (annealed) law of $\{Y_i(\cdot) - U_i\}$ does not depend on i, and is identical to the law of $\{X(\cdot)\}$. We also note that on the event $D_i^{U_i}$, $t_i^{(1)} = 0$.

For $z \in \mathbb{Z}^d$, let

$$F_i(z) = \mathbf{P}^D(\exists_k Y_i(k) = z)$$

and let

$$F_i^{(R)}(z) = F(z) \cdot \mathbf{1}_{\|z - U_i\| > R}.$$

We are interested in $\|F_i\|_2$ and in $\|F_i^{(R)}\|_2$, noting that none of the two depends on i or U_i. We have that

$$F_i(z) = \sum_{n=1}^{\infty} G_i(z,n) \quad \text{and} \quad F_i^{(R)}(z) = \sum_{n=1}^{\infty} G_i^{(R)}(z,n) \tag{3.2}$$

where

$$G_i(z,n) = \mathbf{P}^D(\exists_{t_i^{(n)} \le k < t_i^{(n+1)}} Y_i(k) = z).$$

and

$$G_i^{(R)}(z,n) = \mathbf{P}^D(\exists_{t_i^{(n)} \le k < t_i^{(n+1)}} Y_i(k) = z) \cdot \mathbf{1}_{\|z-U_i\|>R}.$$

are the occupation functions of $\{Y_i(\cdot)\}$.

By the triangle inequality,

$$\|F_i\|_2 \le \sum_{n=1}^{\infty} \|G_i(\cdot,n)\|_2 \tag{3.3}$$

and

$$\|F_i^{(R)}\|_2 \le \sum_{n=1}^{\infty} \|G_i^{(R)}(\cdot,n)\|_2. \tag{3.4}$$

Thus we want to bound the norm of $G_i(\cdot,n)$ and $G_i^{(R)}(\cdot,n)$. We start with $G_i(\cdot,n)$. Thanks to the i.i.d. structure of the regeneration slabs (see [SZ99]),

$$G_i(\cdot,n) = Q_i^n \star J\,,$$

where Q_i^n is the distribution function of $Y_i(t_i^{(n)})$ under $\mathbb{P}(\cdot|D_i^{U_i})$,

$$J(z) = \mathbf{P}^D(\exists_{0=t_i^{(1)} \le k < t_i^{(2)}} Y_i(k) - U_i = z),$$

and $\star$ denotes (discrete) convolution. Positive speed ($\langle v, e_1\rangle > 0$) tells us that

$$\Gamma := \|J\|_1 \le \mathbb{E}(t^{(2)} - t^{(1)}|D) < \infty$$

and thus

$$\|G_i(\cdot,n)\|_2 \le \Gamma \|Q_i^n\|_2$$

Under the law $\mathbf{P}^D$, Q_i^n is the law of a sum of integrable i.i.d. random vectors $\Delta Y_i^k = Y_i(t_i^{k+1}) - Y_i(t_i^k)$, that due to the uniform ellipticity condition are non-degenerate. By the same computation as in [Ber06, Proof of claim 5.2], we get

$$\|Q_i^n\|_2 \le Cn^{-d/4},$$

and thus

$$\|G_i^{(R)}(\cdot,n)\|_2 \le \|G_i(\cdot,n)\|_2 \le Cn^{-d/4}. \tag{3.5}$$

(We note in passing that these estimates can also be obtained from a local limit theorem applied to a truncated version of the variables ΔY_i^k.) It follows from the last two displays and (3.3) that for $d \ge 5$,

$$\|F_i\|_2 < C. \tag{3.6}$$

For $F_i^{(R)}$ we have a fairly primitive bound: by Markov's inequality and the fact that the walk is a nearest neighbor walk, for any $r' > 1$,

$$\begin{aligned} \|G_i^{(R)}(\cdot,n)\|_2 &\le \|G_i^{(R)}(\cdot,n)\|_1 \le \mathbb{E}(\mathbf{1}_{t_i^{(n+1)}>R}(t_i^{(n+1)}-t_i^{(n)})|D) \qquad (3.7)\\ &\le \mathbb{E}[\mathbf{1}_{t_i^{(n)}>\frac{R}{2}}(t_i^{(n+1)}-t_i^{(n)})\,|D] + \mathbb{E}[\mathbf{1}_{t_i^{(n+1)}-t_i^{(n)}>\frac{R}{2}}(t_i^{(n+1)}-t_i^{(n)})\,|D]\\ &\le \frac{2n\mathbb{E}\left[t_i^{(n+1)}-t_i^{(n)}\right]}{R} + C\mathbb{E}\left(\frac{(t_i^{(n+1)}-t_i^{(n)})^{r'}}{R^{(r'-1)}}\right) \le C\frac{n\mathbb{E}((t^{(2)}-t^{(1)})^2|D)}{R}, \end{aligned}$$

where the choice $r' = 2$ was made in deriving the last inequality. Together with (3.5), we get, with $K = \lceil R^{4/(d+4)} \rceil$,

$$\begin{aligned} \|F_i^{(R)}\|_2 \le C\left[\sum_{n=1}^{K}\frac{n}{R} + \sum_{n=K+1}^{\infty} n^{-d/4}\right] &\le C\left[K^2/R + K^{1-d/4}\right] \\ &\le CR^{(4-d)/(d+4)}. \end{aligned} \qquad (3.8)$$

Let $R := \|U_2 - U_1\|/2$. An application of the Cauchy-Schwarz inequality yields

$$\mathbf{P}^D\left(\{Y_1(\cdot)\}\cap\{Y_2(\cdot)\}\neq\emptyset\right) \le \|F_1^{(R)}\|_2^2 + 2\|F_1^{(R)}\|_2\|F_1\|_2 = O\left(R^{(4-d)/(d+4)}\right)$$

for $d \ge 5$. □

Now assume that the two walks do intersect. How far from the starting points could this happen? From (3.8) we immediately get the following corollary.

Corollary 3.2. ($d \ge 5$) *Fix R, $Y_1(\cdot)$ and $Y_2(\cdot)$ as before. Let A_i be the event that $Y_1(\cdot)$ and $Y_2(\cdot)$ intersect, but the intersection point closest to $U_i = Y_i(0)$ is at distance $\ge R$ from $Y_i(0)$. Then*

$$\mathbf{P}^D(A_1 \cap A_2) < CR^{(4-d)/(d+4)}. \qquad (3.9)$$

3.2.2. Basic estimates for $d = 4$. We will now see how to derive the same estimates for dimension 4 in the presence of bounds on higher moments of the regeneration times. The crucial observation is contained in the following lemma.

Lemma 3.3. *Let $d \ge 3$ and let v_i be i.i.d., $\mathbb{Z}^d$-valued random variables satisfying, for some $r \in [2, d-1]$,*

$$\langle v_1, e_1\rangle \ge 1 \text{ a.s.}, \quad \text{and } E\|v_1\|^r < \infty. \qquad (3.10)$$

Assume that, for some $\delta > 0$,

$$P(\langle v_1, e_1\rangle = 1) > \delta, \qquad (3.11)$$

and

$$P(v_1 = z|\langle v_1, e_1\rangle = 1) > \delta, \text{ for all } z \in \mathbb{Z}^d \text{ with } \|z-e_1\|_2 = 1 \text{ and } \langle z, e_1\rangle = 1. \qquad (3.12)$$

Then, with $W_n = \sum_{i=1}^n v_i$*, there exists a constant* $c > 0$ *such that for any* $z \in \mathbb{Z}^d$,

$$P(\exists_i : W_i = z) \le c|\langle z, e_1\rangle|^{-r(d-1)/(r+d-1)}, \tag{3.13}$$

and, for all integer K,

$$\sum_{z:\langle z,e_1\rangle = K} P(\exists_i : W_i = z) \le 1. \tag{3.14}$$

Proof. We set $T_K = \min\{n : \langle W_n, e_1\rangle \ge K\}$. We note first that because of (3.11), for some constant $c_1 = c_1(\delta) > 0$ and all $t > 1$,

$$P(A_t) \le e^{-c_1 t}. \tag{3.15}$$

where

$$A_t = \{\#\{i \le t : \langle v_i, e_1\rangle = 1\} < c_1 t\}.$$

Set $\bar{v} = Ev_1$ and $v = E\langle v_1, e_1\rangle$. Then, for any $\alpha \le 1$, we get from (3.10) and the Marcinkiewicz-Zygmund inequality (see, e.g., [Sh84, Pg. 469] or, for Burkholder's generalization, [St93, Pg. 341]) that for some $c_2 = c_2(r, v, \alpha)$, and all $K > 0$,

$$P(T_K < K^\alpha/2v) \le c_2 K^{-r(2-\alpha)/2}. \tag{3.16}$$

Let $\mathcal{F}_n := \sigma(\langle W_i, e_1\rangle, i \le n)$ denote the filtration generated by the e_1-projection of the random walk $\{W_n\}$. Denote by $W_n^\perp$ the projection of W_n on the hyperplane perpendicular to e_1. Conditioned on the filtration $\mathcal{F}_n$, $\{W_n^\perp\}$ is a random walk with independent (not identically distributed) increments, and the assumption (3.12) together with standard estimates shows that, for some constant $c_3 = c_3(\delta, d)$,

$$\sup_{y\in\mathbb{Z}^{d-1}} \mathbf{1}_{A_t^c} P(W_t = y|\mathcal{F}_t) \le c_3 t^{-(d-1)/2}, \ a.s. \tag{3.17}$$

Therefore, writing $z_1 = \langle z, e_1\rangle$, we get for any $\alpha \le 1$,

$$\begin{aligned}
&P(\exists_i : W_i = z) \le P(T_{z_1} < z_1^\alpha/2v) + P(W_i = z \text{ for some } i \ge z_1^\alpha/2v)\\
&\le c_2 z_1^{-r(2-\alpha)/2} + \sum_{i=z_1^\alpha/2v}^{z_1} P(W_i = z) \le c_2 z_1^{-r(2-\alpha)/2} + \sum_{i=z_1^\alpha/2v}^{z_1} E(P(W_i = z|\mathcal{F}_i))\\
&\le c_2 z_1^{-r(2-\alpha)/2} + \sum_{i=z_1^\alpha/2v}^{z_1} P(A_i) + \sum_{i=z_1^\alpha/2v}^{z_1} E(\mathbf{1}_{T_{z_1}=i} \sup_{y\in\mathbb{Z}^{d-1}} \mathbf{1}_{A_i^c} P(W_i^\perp = y|\mathcal{F}_i))\\
&\le c_2 z_1^{-r(2-\alpha)/2} + z_1 e^{-c_1 z_1^\alpha/2v} + c_3(z_1/2v)^{-\alpha(d-1)/2} \sum_{i=z_1^\alpha/2v}^{z_1} P(T_{z_1} = i),
\end{aligned} \tag{3.18}$$

where the second inequality uses (3.16), and the fifth uses (3.15) and (3.17). The estimate (3.18) yields (3.13) by choosing $\alpha = 2r/(r + d - 1) \le 1$.

To see (3.14), note that the sum of probabilities is exactly the expected number of visits to $\{z : \langle z, e_1\rangle = K\}$, which is bounded by 1. □

We are now ready to state and prove the following analogue of Proposition 3.1.

Proposition 3.4. ($d = 4$) *With notation as in Proposition* 3.1, $d = 4$ *and* r *in* (1.4) *satisfying* $r > 8$, *we have*

$$\mathbf{P}^D(\{Y_1(\cdot)\} \cap \{Y_2(\cdot)\} \neq \emptyset) < C\|U_1 - U_2\|^{-K_4}$$

where $K_4 > 0$.

Proof. Fix $\nu > 0$ and write $U = |U_1 - U_2|$. Let $\{v_i\}_{i\geq 1}$ denote an i.i.d. sequence of random variables, with v_1 distributed like $Y_1(t^{(2)}) - Y_1(t^{(1)})$ under $\mathbf{P}^D$. This sequence clearly satisfies the assumptions of Lemma 3.3, with $\delta = \kappa^2\mathbb{P}(D)$.

Let $T := \mathbb{E}(t^{(2)} - t^{(1)})$. By our assumption on the tails of regeneration times, for $\nu \in (0,1)$ with $\nu r > 1$,

$$\mathbf{P}^D\left(\exists_{i\geq \frac{U}{8T}} : t_1^{(i+1)} - t_1^{(i)} > i^\nu\right) \leq \sum_{i=U/8T}^{\infty} \frac{C}{i^{\nu r}} \leq CU^{1-\nu r}. \tag{3.19}$$

By Doob's maximal inequality, and our assumption on the tails of regeneration times,

$$\begin{aligned}
\mathbf{P}^D\left(\exists_{i\geq\frac{U}{8T}} : t_1^{(i)} > 2Ti\right) &\leq \mathbf{P}^D\left(\exists_{i\geq\frac{U}{8T}} : (t_1^{(i)} - \mathbb{E}t_1^{(i)}) > Ti\right)\\
&\leq \sum_{j=0}^{\infty} \mathbf{P}^D\left(\exists_{i\in[\frac{2^jU}{8T},\frac{2^{j+1}U}{8T})} : (t_1^{(i)} - \mathbb{E}t_1^{(i)}) > Ti\right)\\
&\leq \sum_{j=0}^{\infty} \mathbf{P}^D\left(\exists_{i\leq\frac{2^{j+1}U}{8T}} : (t_1^{(i)} - \mathbb{E}t_1^{(i)}) > 2^jU/8\right)\\
&\leq C\sum_{j=0}^{\infty} \frac{1}{(2^jU)^{r/2}} \leq \frac{C}{U^{r/2}}.
\end{aligned} \tag{3.20}$$

For integer k and $i = 1, 2$, let $s_{k,i} = \max\{n : \langle Y_i(t_i^{(n)}), e_1\rangle \leq k\}$. Let

$$\mathcal{A}_{i,U,\nu} := \cap_{k\geq U/8T}\{t_i^{(s_{k,i}+1)} - t_i^{(s_{k,i})} \leq (2Tk)^\nu\}.$$

Combining (3.20) and (3.19), we get

$$\mathbf{P}^D((\mathcal{A}_{i,U,\nu})^c) \leq C[U^{1-\nu r} + U^{-r/2}]. \tag{3.21}$$

For an integer K, set $\mathcal{C}_K = \{z \in \mathbb{Z}^d : \langle z, e_1\rangle = K\}$. Note that on the event $\mathcal{A}_{1,U,\nu} \cap \mathcal{A}_{2,U,\nu}$, if the paths $Y_1(\cdot)$ and $Y_2(\cdot)$ intersect at a point $z \in \mathcal{C}_K$, then there exist integers α, β such that $|Y_1(t_1^{(\alpha)}) - Y_2(t_2^{(\beta)})| \leq 2(2TK)^\nu$. Therefore, with

$W_n = \sum_{i=1}^n v_i$, we get from (3.20) and (3.21) that, with $r_0 = r \wedge 3$,

$$\begin{aligned}
&\mathbf{P}^D\left(\{Y_1(\cdot)\} \cap \{Y_2(\cdot)\} \neq \emptyset\right) \\
\leq\ & 2\mathbf{P}^D\left(t_1^{(U/8T)} \geq U/2\right) + 2\mathbf{P}^D\left((\mathcal{A}_{i,U,\nu})^c\right) \\
&+ \sum_{K>U/8T} \sum_{z \in \mathcal{C}_K} \sum_{z':|z-z'|<2(2TK)^\nu} \mathbf{P}(\exists i : W_i = z)\mathbf{P}(\exists j : W_j = z') \\
\leq\ & C\left[U^{-\epsilon} + U^{1-\nu r} + U^{-r/2} + U^{\left[1-\frac{3r_0}{r_0+3}+4\nu\right]}\right], \qquad (3.22)
\end{aligned}$$

as long as $1 - 3r_0/(r_0+3) + 4\nu < 0$, where Lemma 3.3 and (3.21) were used in the last inequality. With $r > 8$ (and hence $r_0 = 3$), one can chose $\nu > 1/r$ such that all exponents of U in the last expression are negative, yielding the conclusion. □

Equivalently to Corollary 3.2, the following is an immediate consequence of the last line of (3.22)

Corollary 3.5. *With notation as in Corollary* 3.2, $d = 4$ *and* r *in* (1.4) *satisfying* $r > 8$, *we have*

$$\mathbf{P}^D(A_1 \cap A_2) < CR^{-K_4'}.$$

with $K_4' = K_4'(r) > 0$.

3.2.3. Main estimate for random walks in independent environments. Let $R > 0$ and let $T_i^Y(R) := \min\{n : Y_i(n) \in H_R\}$.

Proposition 3.6. ($d \geq 4$) *Let* $Y_1(\cdot)$ *and* $Y_2(\cdot)$ *be random walks in independent environments satisfying the assumptions of Theorem* 1.1, *with starting points* U_1, U_2 *satisfying* $\langle U_1, e_1\rangle = \langle U_2, e_1\rangle = 0$. *Let*

$$\begin{aligned}
A(R) := &\{\forall_{n<T_1^Y(R)} \langle Y_1(n), e_1\rangle \geq 0\} \\
&\cap \{\forall_{m<T_2^Y(R)} \langle Y_2(m), e_1\rangle \geq 0\} \qquad (3.23)\\
&\cap \{\forall_{n<T_1^Y(R), m<T_2^Y(R)} Y_1(n) \neq Y_2(m)\}.
\end{aligned}$$

Then,

1. *There exists* $\rho > 0$ *such that for every choice of* R *and* U_1, U_2 *as above,*

$$\mathbf{P}(A(R)) > \rho. \qquad (3.24)$$

2. *Let* $\hat{B}_i(n)$ *be the event that* $Y_i(\cdot)$ *has a regeneration time at* $T_i^Y(n)$, *and let*

$$B_i(R) := \bigcup_{n=R/2}^{R} \hat{B}_i(n). \qquad (3.25)$$

Then

$$\mathbf{P}\left(\{\{Y_1(n)\}_{n=1}^\infty \cap \{Y_2(m)\}_{m=1}^\infty \neq \emptyset\} \cap A(R) \cap B_1(R) \cap B_2(R)\right) < CR^{-\beta_d} \qquad (3.26)$$

with $\beta_d = \beta_d(r, \epsilon) > 0$ *for* $d \geq 4$.

Proof. To see (3.24), note first that due to uniform ellipticity, we may and will assume that $|U_1 - U_2| > C$ for a fixed arbitrary large C. Since $\zeta := \mathbb{P}(D_1 \cap D_2) > 0$ does not depend on the value of C, the claim then follows from Propositions 3.1 and 3.4 by choosing C large enough such that $\mathbf{P}^D\left(A(R)^c\right) < \zeta/2$.

To see (3.26), note the event $A(R) \cap B_1(R) \cap B_2(R)$ implies the event $D_1^{U_1} \cap D_2^{U_2}$, and further if $\{Y_1(n)\}_{n=1}^{\infty} \cap \{Y_1(m)\}_{m=1}^{\infty} \neq \emptyset$ then for $i = 1, 2$ the closest intersection point to U_i is at distance greater than or equal to $R/2$ from U_i. Therefore (3.26) follows from Corollary 3.2 and Corollary 3.5. □

3.3. Intersection of paths in the same environment

In this subsection we take $\{X_1(n)\}$ and $\{X_2(n)\}$ to be random walks in the same environment ω, with $X_i(0) = U_i$, $i = 1, 2$, and ω distributed according to P. As in subsection 3.2, we also consider $\{Y_1(n)\}$ and $\{Y_2(n)\}$, two independent random walks evolving in independent environments, each distributed according to P. We continue to use $\mathbb{P}^{U_1,U_2}$ (or, for brevity, $\mathbb{P}$) for the annealed law of the pair $(X_1(\cdot), X_2(\cdot))$, and $\mathbf{P}$ for the annealed law of the pair $(Y_1(\cdot), Y_2(\cdot))$. Note that $\mathbf{P} \neq \mathbb{P}$. Our next proposition is a standard statement, based on coupling, that will allow us to use some of the results from Section 3.2, even when the walks evolve in the same environment and we consider the law $\mathbb{P}$.

In what follows, a stopping time T with respect to the filtration determined by a path X will be denoted $T(X)$.

Proposition 3.7. *With notation as above, let $T_i(\cdot)$, $i = 1, 2$ be stopping times such that $T_i(X_i)$, $i = 1, 2$ are $\mathbb{P}$-almost surely finite. Assume $X_1(0) = Y_1(0)$ and $X_2(0) = Y_2(0)$. Set*

$$I_X := \left\{ \{X_1(n)\}_{n=0}^{T_1(X_1)} \bigcap \{X_2(n)\}_{n=0}^{T_2(X_2)} = \emptyset \right\}$$

and

$$I_Y := \left\{ \{Y_1(n)\}_{n=0}^{T_1(Y_1)} \bigcap \{Y_2(n)\}_{n=0}^{T_2(Y_2)} = \emptyset \right\}.$$

Then, for any nearest neighbor deterministic paths $\{\lambda_i(n)\}_{n \geq 0}$, $i = 1, 2$,

$$\begin{aligned} &\mathbf{P}\left(Y_i(n) = \lambda_i(n), 0 \leq n \leq T_i(Y_i), i = 1, 2; I_Y\right) \\ &= \mathbb{P}\left(X_i(n) = \lambda_i(n), 0 \leq n \leq T_i(X_i), i = 1, 2; I_X\right). \end{aligned} \tag{3.27}$$

Proof. For every pair of non-intersecting paths $\{\lambda_i(n)\}_{n \geq 0}$, define three i.i.d. environments $\omega^{(1)}$, $\omega^{(2)}$ and $\omega^{(3)}$ as follows: Let $\{J(z)\}_{z \in \lambda_1 \cup \lambda_2}$ be a collection of i.i.d. variables, of marginal law Q. At the same time, let $\{\eta^j(z)\}_{z \in \mathbb{Z}^d}$, $j = 1, 2, 3$ be three independent i.i.d. environments, each P-distributed. Then define

$$\omega^{(1)}(z) = \begin{cases} J(z) & \text{if } z \in \lambda^{(1)} \\ \eta^{(1)}(z) & \text{otherwise,} \end{cases}$$

$$\omega^{(2)}(z) = \begin{cases} J(z) & \text{if } z \in \lambda^{(2)} \\ \eta^{(2)}(z) & \text{otherwise,} \end{cases}$$

and

$$\omega^{(3)}(z) = \begin{cases} J(z) & \text{if } z \in \lambda^{(1)} \cup \lambda^{(2)} \\ \eta^{(3)}(z) & \text{otherwise,} \end{cases}$$

and let Y_1 evolve in $\omega^{(1)}$, let Y_2 evolve in $\omega^{(2)}$ and let X_1 and X_2 evolve in $\omega^{(3)}$. Then by construction,

$$P_{\omega^{(1)},\omega^{(2)}}(Y_i(n)=\lambda_i(n), 0\le n\le T_i(Y_i)) = P_{\omega^{(3)}}(X_i(n)=\lambda_i(n), 0\le n\le T_i(X_i)).$$

Integrating and then summing we get (3.27). □

An immediate consequence of Proposition 3.7 is that the estimates of Proposition 3.6 carry over to the processes $(X_1(\cdot), X_2(\cdot))$. More precisely, let $R > 0$ be given and set $T_i^X(R) := \min\{n : X_i(n) \in H_R\}$. Define $A(R)$ and $B_i(R)$ as in (3.23) and (3.25), with the process X_i replacing Y_i.

Corollary 3.8. ($d \geq 4$) *Let $X_1(\cdot)$ and $X_2(\cdot)$ be random walks in the same environment satisfying the assumptions of Theorem* 1.1, *with starting points U_1, U_2 satisfying $\langle U_1, e_1\rangle = \langle U_2, e_1\rangle = 0$. Then,*

1. *There exists $\rho > 0$ such that for every choice of R and U_1, U_2 as above,*
$$\mathbb{P}(A(R)) > \rho. \tag{3.28}$$
2. *With $C < \infty$ and $\beta_d > 0$ as in* (3.26),
$$\mathbb{P}(\{\{X_1(n)\}_{n=1}^{\infty} \cap \{X_2(m)\}_{m=1}^{\infty} \neq \emptyset\} \cap A(R) \cap B_1(R) \cap B_2(R)) < CR^{-\beta_d}. \tag{3.29}$$

With β_d as in (3.29) and ϵ as in the statement of Theorem 1.1, fix $0 < \psi_d$ satisfying

$$\psi_d < \beta_d(1 - \psi_d) \text{ and } (1 + \epsilon)(1 - \psi_d) > 1. \tag{3.30}$$

For R integer, let

$$\mathcal{K}_k(R) = \{\exists_{(k+0.5)R^{1-\psi_d}<j<(k+1)R^{1-\psi_d}} \text{ s.t. } T_i(j) \text{ is a regeneration time for } X_i(\cdot)\},$$

and let

$$C_i(R) := \bigcap_{k=1}^{[2R^{\psi_d}]} \mathcal{K}_k(R). \tag{3.31}$$

Proposition 2.1 will follow from the following lemma:

Lemma 3.9. ($d \geq 4$) *Under the assumptions of Theorem* 1.1, *there exist constants C and $\gamma_d > 0$ such that for all integer K,*

$$\mathbb{P}(W_K \cap C_1(K) \cap C_2(K)) < CK^{-\gamma_d}.$$

Proof of Lemma 3.9. Let $w := [K^{1-\psi_d}]$ and for $k = 1, \ldots, [K^{\psi_d}/2]$ define the event

$$S_k = \left\{ \begin{array}{r} \forall_{T_1(kw)\le j<T_1((k+1)w)} X_1(j) > kw \quad \text{and} \\ \forall_{T_2(kw)\le j<T_2((k+1)w)} X_2(j) > kw \quad \text{and} \\ \{X_1(j)\}_{j=T_1(kw)}^{T_1((k+1)w)-1} \bigcap \{X_2(j)\}_{j=T_2(kw)}^{T_2((k+1)w)-1} = \emptyset \end{array} \right\} \tag{3.32}$$

By (3.28),

$$\mathbb{P}\left(S_k|S_1^c \cap S_2^c \cap \cdots \cap S_{k-1}^c\right) \geq \rho\,.$$

Therefore,

$$\mathbb{P}\left(\cup_k S_k\right) \geq 1 - (1-\rho)^{[K^{\psi_d}/2]}. \tag{3.33}$$

Now, by (3.29),

$$\mathbb{P}\left(S_k \cap C_1(K) \cap C_2(K) \cap W_K\right) < Cw^{-\beta_d} = CK^{-\beta_d(1-\psi_d)}.$$

We therefore get that

$$\mathbb{P}\left(\cup_k S_k \cap C_1(K) \cap C_2(K) \cap W_K\right) < CK^{-\beta_d(1-\psi_d)}K^{\psi_d} = CK^{\psi_d-\beta_d(1-\psi_d)}.$$

Combined with (3.33), we get that

$$\mathbb{P}\left(\{X_1(\cdot)\} \cap \{X_2(\cdot)\} \cap H_K \neq \emptyset \;\cap\; C_1(K) \cap C_2(K)\right) < CK^{-\gamma_d}$$

for every choice of $\gamma_d < \beta_d(1-\psi_d) - \psi_d$. □

Proof of Proposition 2.1. Note that by the moment conditions on the regeneration times,

$$\mathbb{P}\left(C_i(K)^c\right) \leq CK^{-\epsilon(1-\psi_d)} + CK \cdot K^{-(1+\epsilon)(1-\psi_d)} = CK^{-\epsilon(1-\psi_d)} + CK^{1-(1+\epsilon)(1-\psi_d)}.$$

By the choice of ψ_d, see (3.30), it follows that (2.1) holds for

$$\kappa_d < \min\left\{(1+\epsilon)(1-\psi_d) - 1, \gamma_d\right\}.$$

□

4. Addendum: $d = 2, 3$

After the first version of this work was completed and circulated, F. Rassoul-Agha and T. Seppäläinen have made significant progress in their approach to the CLT, and posted an article [RAS07b] in which they derive the quenched CLT for all dimensions $d \geq 2$, under a somewhat stronger assumption on the moments of regeneration times than (1.4). (In their work, they consider finite range, but not necessarily nearest neighbor, random walks, and relax the uniform ellipticity condition.) While their approach is quite different from ours, it incorporates a variance reduction step that, when coupled with the techniques of this paper, allows one to extend Theorem 1.1 to all dimensions $d \geq 2$, with a rather short proof. In this addendum, we present the result and sketch the proof.

Theorem 4.1. *Let $d = 2, 3$. Let Q and $\{X(n)\}$ be as in Theorem* 1.1, *with $\epsilon = r \geq 40$. Then, the conclusions of Theorem* 1.1 *still hold.*

Remark: The main contribution to the condition $r \geq 40$ comes from the fact that one needs to transfer estimates on regenerations times in the direction e_1 to regeneration times in the direction v, see Lemma 4.5 below. If $e_1 = v$, or if one is willing to assume moment bounds directly on the regeneration times in direction v, then the same proof works with $\epsilon > 0$ arbitrary and $r > 14$.

Proof of Theorem 4.1 (*sketch*). The main idea of the proof is that the condition "no late intersection of independent random walks in the same environment" may be replaced by the condition "intersections of independent random walks in the same environment are rare".

Recall, c.f. the notation and proof of Theorem 1.1, that we need to derive a polynomially decaying bound on $\mathrm{Var}(E_\omega F(\beta^N))$ for $F : C([0,1], \mathbb{R}^d) \to \mathbb{R}$ bounded Lipschitz and β^N the polygonal interpolation as in (2.2). In the sequel, we write $F^N(X) := F(\beta^N)$ if β^N is the polygonal interpolation of the scaling (as in (2.2)) of the path $\{X_n\}_{n=0,\ldots,N}$.

For any k, let $S_k = \min\{n : X_n \in H_k\}$. For two paths p_1, p_2 of length T_1, T_2 with $p_i(0) = 0$, let $p_1 \circ p_2$ denote the concatenation, i.e.,

$$p_1 \circ p_2(t) = \begin{cases} p_1(t)\,, & t \le T_1 \\ p_1(T_1) + p_2(t - T_1)\,, & t \in (T_1, T_2]\,. \end{cases}$$

Use the notation $X_i^j = \{X_i, \ldots, X_2, \ldots, X_j\}$. Then, we can write, for any k,

$$F^N(X_0^N) = F^N(X_0^{S_k \wedge N} \circ [X^N_{S^k \wedge N} - X_{S^k \wedge N}])\,.$$

Now comes the main variance reduction step, which is based on martingale differences. Order the vertices in an L^1 ball of radius N centered at 0 in $\mathbb{Z}^d$ in lexicographic order $\ell(\cdot)$. Thus, z is the predecessor of z', denoted $z = p(z')$, if $\ell(z') = \ell(z) + 1$. Note that (because of our choice of lexicographic order), if $z_1 < z_1'$ then $\ell(z) < \ell(z')$.

Let $\delta > 1/r$ be given such that $2\delta < 1$. Define the event

$$W_N := \{\exists i \in [0, N] : t^{(i+1)} - t^{(i)} > N^\delta/3 \text{ or } t^{(i+N^\delta)} - t^{(i)} > N^{3\delta/2}\}\,.$$

By our assumptions, we have that $\mathbb{P}(W_N) \le C(N^{-\epsilon\delta} + N^{1-\delta r})$, and hence decays polynomially.

$$\mathrm{Var}(E_\omega F(\beta^N)) \le \mathrm{Var}(E_\omega F(\beta^N) \mathbf{1}_{W_N^c}) + O(N^{-\delta'})\,,$$

for some $\delta' > 0$. In the sequel we write $\bar{F}^N(X) = F^N(X)\mathbf{1}_{W_N^c}$.

Set $\mathcal{G}_z^N := \sigma(\omega_x : \ell(x) \le \ell(z), \|x\|_1 \le N)$, and write $\hat{H}_k = \{z : \langle z, e_1 \rangle = k\}$. We have the following martingale difference representation:

$$\begin{aligned} E_\omega \bar{F}^N(X) - \mathbb{E}\bar{F}^N(X) &= \sum_{z:|z|_1 \le N} \left[\mathbb{E}\left(\bar{F}^N(X) | \mathcal{G}_z\right) - \mathbb{E}\left(\bar{F}^N(X) | \mathcal{G}_{p(z)}\right)\right] \\ &=: \sum_{k=-N}^{N} \sum_{z \in \hat{H}_k, |z|_1 \le N} \Delta_z^N\,. \end{aligned} \tag{4.1}$$

Because it is a martingale differences representation, we have

$$\mathrm{Var}(E_\omega \bar{F}^N(X)) = \sum_{k=-N}^{N} \sum_{z \in \hat{H}_k, |z|_1 \le N} \mathbb{E}\left(\Delta_z^N\right)^2\,. \tag{4.2}$$

Because of the estimate $\mathbb{E}[(t^{(1)})^\epsilon] < \infty$, the Lipschitz property of F, and our previous remarks concerning W_N, the contribution of the terms with $k \le 2N^\delta$ to the sum in (4.2) decays polynomially. To control the terms with $k > 2N^\delta$, for $z \in \hat{H}_k$ let τ_z denote the largest regeneration time $t^{(i)}$ smaller than S_{k-N^δ}, and write τ_z^+ for the first regeneration time larger than S_{k+N^δ}. Then,

$$F^N(X) = F^N(X_0^{\tau_z} \circ [X_{\tau_z}^{\tau_z^+} - X_{\tau_z}] \circ [X_{\tau_z^+}^N - X_{\tau_z^+}]) \,.$$

Because of the Lipschitz property of F, our rescaling, and the fact that we work on the event W_N^c, we have the bound

$$|\bar{F}^N(X_0^{\tau_z} \circ [X_{\tau_z}^{\tau_z^+} - X_{\tau_z}] \circ [X_{\tau_z^+}^N - X_{\tau_z^+}]) - \bar{F}^N(X_0^{\tau_z} \circ [X_{\tau_z^+}^N - X_{\tau_z^+}])| \le 4N^{(3\delta-1)/2} \,.$$

One then obtains by standard manipulations

$$\mathbb{E}\left(\Delta_z^N\right)^2 \le CN^{3\delta-1} E[(E_\omega[\mathbf{1}_{X \text{ visits z}}])^2] \,.$$

Let I_N denote the number of intersections, up to time N, of two independent copies of $\{X(n)\}_{n\ge 0}$ *in the same environment.* Then,

$$\sum_{z:\|z\|_1 \le N} [E(E_\omega[\mathbf{1}_{X \text{ visits z}}])^2] = E(E_{\omega\times\omega} I_N) \,. \tag{4.3}$$

Combining these estimates, we conclude that

$$\mathrm{Var}(E_\omega \bar{F}^N(X)) \le CN^{3\delta-1} E(E_{\omega\times\omega} I_N) + N^{-\delta'} \,. \tag{4.4}$$

The proof of Theorem 4.1 now follows from the following lemma.

Lemma 4.2. *Under the assumptions of Theorem* 4.1, *for $d \ge 2$ and $r > 40$, we have that for $r' < r/4 - 1/2$ and any $\epsilon' \in (0, 1/2 - 4/r' + 2/(r')^2)$,*

$$[E(E_{\omega\times\omega} I_N)] \le CN^{1-\epsilon'} \,, \tag{4.5}$$

where C depends only on ϵ'.

Indeed, equipped with Lemma 4.2, we deduce from (4.4) that

$$\mathrm{Var}(E_\omega F^N(X)) \le N^{-\delta'} + CN^{1-\epsilon'} N^{3\delta-1} \,.$$

Thus, whenever $\delta > 1/r$ is chosen such that $3\delta < \epsilon'$, (which is possible as soon as $r > 3/\epsilon'$, which in turn is possible for some $\epsilon' < 1/2 - 4/r' + 2/(r')^2$ if $r \ge 40$), $\mathrm{Var}(E_\omega F^N(X)) \le CN^{-\delta}$, for some $\delta > 0$. As mentioned above, this is enough to conclude. □

Before proving Lemma 4.2, we need the following estimate:

Lemma 4.3. *Let S_n be an i.i.d. random walk on $\mathbb{R}$ with $ES_1 = 0$ and $E|S_1|^r < \infty$ for $r > 3$. Let U_n be a sequence of events such that, for some constant $a_3 > 3/2$, and all n large,*

$$P(U_n) \ge 1 - \frac{1}{n^{a_3}} \,. \tag{4.6}$$

In addition we assume that $\{U_k\}_{k<n}$ is independent of $\{S_k - S_n\}_{k\ge n}$ for every n.

Let $a_1 \in (0,1)$ and $a_2 > 0$ be given. Suppose further that for any n finite,

$$P(\text{for all } t \le n,\, S_t \ge \lfloor t^{\frac{a_1}{2}} \rfloor \text{ and } U_t \text{ occurs}) > 0\,.$$

Then, there exists a constant $C = C(a_1, a_2, a_3) > 0$ such that for any T,

$$P(\text{for all } t \le T,\, S_t \ge \lfloor t^{\frac{a_1}{2}} \rfloor \text{ and } U_t \text{ occurs}) \ge \frac{C}{T^{1/2+a_2}}\,. \tag{4.7}$$

Proof. Fix constants $\bar{\epsilon} > 0$, $\alpha \in (0,1)$ and $\beta \in (1,2)$ (eventually, we will take $\alpha \to 1, \beta \to 2$ and $\bar{\epsilon} \to \infty$). Throughout the proof, C denote constants that may change from line to line but may depend only on these parameters. Define $b_i = \lfloor i^{\alpha\bar{\epsilon}} \rfloor$ and $c_i = \lceil i^{\bar{\epsilon}+1} \rceil$. Consider the sequence of stopping times $\tau_0 = 0$ and

$$\tau_{i+1} = \min\{n > \tau_i : S_n - S_{\tau_i} > c_{i+1} - c_i \text{ or } S_n - S_{\tau_i} < b_{i+1} - c_i\}.$$

Declare an index i good if $S_{\tau_i} - S_{\tau_{i-1}} = c_i - c_{i-1}$. Note that if the indices $i = 1, \ldots, K$ are all good, then $S_n \ge b_{i-1}$ for all $n \in (\tau_{i-1}, \tau_i]$, $i = 1, \ldots, K$.

Let the overshoot O_i of $\{S_n\}$ at time τ_i be defined as $S_{\tau_i} - S_{\tau_{i-1}} - (c_i - c_{i-1})$ if i is good and $S_{\tau_i} - S_{\tau_{i-1}} - (b_i - c_{i-1})$ if i is not good. By standard arguments (see, e.g., [RAS07b, Lemma 3.1]), $E(|O_i|^{r-1}) < \infty$. By considering the martingale S_n, we then get

$$P(i \text{ is good}) \sim (1 - \frac{1+\bar{\epsilon}}{i})\,, \tag{4.8}$$

as $i \to \infty$. By considering the martingale $S_n^2 - nES_1^2$, we get

$$E(\tau_{i+1} - \tau_i) = \Omega(i^{1+2\bar{\epsilon}})\,, \tag{4.9}$$

as $i \to \infty$. In particular,

$$P(\tau_{i+1} - \tau_i > i^{2+2\bar{\epsilon}+\delta}) \le \frac{C}{i^{1+\delta}}\,, \tag{4.10}$$

while, from our assumption on the moments of S_1 and Doob's inequality,

$$P(\tau_{i+1} - \tau_i \le i^{\bar{\epsilon}\beta}) \le \frac{C}{i^{r\bar{\epsilon}(2-\beta)/2}}\,. \tag{4.11}$$

We assume in the sequel that $r\bar{\epsilon}(2-\beta)/2 > 2$ and that $a_3(\bar{\epsilon}\beta+1) > 5+3\bar{\epsilon}+\delta$ (both these are possible by choosing any $\beta < 2$ so that $a_3\beta > 3$, and then taking $\bar{\epsilon}$ large). We say that $i+1$ is *very good* if it is good and in addition, $\tau_{i+1} - \tau_i \in [i^{\bar{\epsilon}\beta}, i^{2+2\bar{\epsilon}+\delta}]$. By (4.8), (4.10) and (4.11), we get

$$P(i \text{ is very good}) \sim (1 - \frac{1+\bar{\epsilon}}{i})\,. \tag{4.12}$$

Declare an index i *excellent* if i is very good and in addition, U_n occurs for all $n \in [\tau_{i-1}, \tau_i)$.

On the event that the first K i's are very good, we have that $\tau_{K-1} \ge CK^{\bar{\epsilon}\beta+1}$ and $\tau_K \le CK^{3+2\bar{\epsilon}+\delta} =: T_K$, and $S_n \ge K^{\bar{\epsilon}\alpha}$ for $n \in [\tau_{K-1}, \tau_K]$. Letting $\mathcal{M}_K$ denote the event that the first $K-1$ i's are excellent, and K is very good, we then have, for every n,

$$P(U_n^c \mathbf{1}_{n\in[\tau_{K-1},\tau_K)} | \mathcal{M}_K) \le K^{-a_3(\bar{\epsilon}\beta+1)}/P(\mathcal{M}_K)\,.$$

We now show inductively that $P(\mathcal{M}_K) \geq C/K^{1+\bar\epsilon}$. Indeed, under the above hypotheses, we get

$P(U_n^c \text{ for some } n \in [\tau_{K-1}, \tau_K) | \mathcal{M}_K) \leq K^{1+\bar\epsilon+3+2\bar\epsilon+\delta-a_3(\bar\epsilon\beta+1)} = K^{4+\delta-a_3+\bar\epsilon(3-a_3\beta)}$

and thus, with our choice of constants and (4.8), we conclude that under the above hypothesis,

$$P(\mathcal{M}_{K+1} | \mathcal{M}_K) \sim (1 - \frac{1+\bar\epsilon}{K+1}). \tag{4.13}$$

We thus get inductively that the hypothesis propagates and in particular we get

$$P(i \text{ is excellent for } i \leq K) \geq \frac{C}{K^{1+\bar\epsilon}}. \tag{4.14}$$

Further, if the first K i's are excellent (an event with probability bounded below by $C/K^{1+\bar\epsilon}$), we have that $\tau_K \leq T_K$. Note that if $t = CK^{\bar\epsilon\beta+1}$ then on the above event we have that by time t, at least $Ct^{1/(3+2\bar\epsilon+\delta)}$ of the τ_i's are smaller than t, and hence $S_t \geq Ct^{\bar\epsilon\alpha/(3+2\bar\epsilon+\delta)}$. We thus conclude that, for all T large,

$$P(\text{for all } t \leq T,\ S_t \geq t^{\bar\epsilon\alpha/(3+2\bar\epsilon+\delta)}, \text{ and } U_t \text{ occurs}) \geq \frac{C}{T^{(1+\bar\epsilon)/(1+\bar\epsilon\beta)}}.$$

Taking $\bar\epsilon$ large and β close to 2 (such that still $r\bar\epsilon(2-\beta) > 2$), and α close to 1, completes the proof. □

Proof of Lemma 4.2 (*sketch*). Let

$$v = \lim_{n\to\infty} \frac{X_n}{n} \neq 0$$

be the limiting direction of the random walk, and let u be a unit vector which is orthogonal to v.

In what follows we will switch from the regenerations in direction e_1 that we used until now, and instead use regenerations in the direction v, whose definition, given below, is slightly more general than the definition of regenerations in the direction e_1 given in Section 1.

Definition 4.4. We say that t is a regeneration time for $\{X_n\}_{n=1}^\infty$ in direction v if

- $\langle X_s, v\rangle \leq \langle X_{t-1}, v\rangle$ for every $s < t-1$.
- $\langle X_t, v\rangle > \langle X_{t-1}, v\rangle$.
- $\langle X_s, v\rangle \geq \langle X_t, v\rangle$ for every $s > t$.

We denote by $t^{v,(n)}$ the successive regeneration times of the RWRE X_n in direction v (when dealing with two RWRE's $X_i(n)$, we will use the notation $t_i^{v,(n)}$). The sequence $t^{v,(n+1)} - t^{v,(n)}$, $n \geq 1$, is still i.i.d., and with D^v defined in the obvious way, the law of $t^{v,(2)} - t^{v,(1)}$ is identical to the law of $t^{v,(1)}$ conditioned on the event D^v. The following lemma, of maybe independent interest, shows that, up to a fixed factor, the regeneration time $t^{v,(1)}$ (and hence, also $t^{v,(2)} - t^{v,(1)}$) inherits moment bounds from $t^{(1)}$.

Lemma 4.5. *Assume* $r > 10$ *and* $\mathbb{E}((t^{(1)})^r) < \infty$. *Then* $\mathbb{E}(\langle X_{t^{v,(1)}}, v\rangle)^{2r'} < \infty$ *and* $\mathbb{E}((t^{v,(1)})^{r'}) < \infty$ *with* $r' < r/4 - 1/2$.

Proof. On the event $(D^v)^c$, define $\tau_0 = \min\{n > 0 : \langle X_n, v\rangle \le 0\}$ and set $M = \max\{\langle X_n, v\rangle : n \in [0, \tau_0]\}$. By [Szn02, Lemma 1.2], $\langle X_{t^{v,(1)}}, v\rangle$ is (under the annealed law) stochastically dominated by the sum of a geometric number of independent copies of $M+1$. Hence, if $\mathbb{E}[M^p|(D^v)^c] < \infty$ for some p, then $\mathbb{E}|\langle X_{t^{v,(1)}}, v\rangle|^p < \infty$.

Fix a constant $\chi < 1/2(\mathbb{E}t^{(1)} \wedge \mathbb{E}(t^{(2)} - t^{(1)}))$ small enough so that $(2 + 2\|v\|_2)\chi < \|v\|_2^2$. Now fix some (large) number x. On the event $M > x$, either

- $t^{(\chi x)} \ge x$

or

- $t^{(k+1)} - t^{(k)} \ge \chi k \quad$ for some $k > \chi x$

or

- $\{|t^{(k)} - \mathbb{E}t^{(k)}| > \chi k\}$ or $\{\|X_{t^{(k)}} - \mathbb{E}X_{t^{(k)}}\| > \chi k\}$ for some $k > \chi x$.

(Indeed, on the event $M > x$ with x large, the RWRE has to satisfy that at some large time $t > x$, $\langle X_t, v\rangle$ is close to 0 instead of close to $\|v\|_2^2 t$.)

Due to the moment bounds on $t^{(1)}$ and $t^{(2)} - t^{(1)}$, and the chosen value of χ, we have $\mathbb{P}(t^{(\chi x)} \ge x) \le Cx^{-r/2}$. We also have

$$\mathbb{P}(t^{(k+1)} - t^{(k)} \ge \chi k, \quad \text{some } k > \chi x) \le C \sum_{k=\chi x}^{\infty} k^{-r} = Cx^{-r+1},$$

and

$$\begin{aligned}&\mathbb{P}(\{|t^{(k)} - \mathbb{E}t^{(k)}| > \chi k\} \text{ or } \{\|X_{t^{(k)}} - \mathbb{E}X_{t^{(k)}}\| > \chi k\}, \text{ some } k > \chi x)\\ &\quad \le C \sum_{k=\chi x}^{\infty} k^{-r/2} = Cx^{-r/2+1}.\end{aligned}$$

We conclude that $\mathbb{E}M^p \le C + C\int_1^\infty x^{p-1} x^{-r/2+1} dx < \infty$ if $p < r/2 - 1$. This proves that

$$\mathbb{E}|\langle X_{t^{v,(1)}}, v\rangle|^p < \infty \quad \text{if } p < r/2 - 1. \tag{4.15}$$

We can now derive moment bounds on $t^{v,(1)}$ (which imply also moment bounds on $\bar{t}^v := t^{v,(2)} - t^{v,(1)}$). Suppose $\mathbb{E}((t^{v,(1)})^{p'}) = \infty$. For any $\epsilon'' > 0$ we can then find a sequence of integers $x_m \to \infty$ such that $\mathbb{P}(t^{v,(1)} > x_m) \ge C/x_m^{p'+\epsilon''}$. Therefore, using (4.15) and the assumed moment bounds,

$$\begin{aligned}&\mathbb{P}(|t^{v,(x_m)} - \mathbb{E}(t^{v,(x_m)})| > x_m/2)\\ \ge\ & \mathbb{P}(t^{v,(1)} - \mathbb{E}(t^{v,(1)}) > x_m)\mathbb{P}(|t^{v,(x_m)} - t^{v,(1)} - (x_m - 1)\mathbb{E}(\bar{t}^v)| < \chi x_m)\\ \ge\ & Cx_m^{-(p'+\epsilon'')}.\end{aligned}$$

Therefore,

$$\begin{aligned}\mathbb{P}\left(\left|t^{v,(x_m)} - \mathbb{E}\left[t^{v,(x_m)}\right]\right| > x_m/2 \ ; \ |\langle X_{t^{v,x_m}}, v\rangle - \mathbb{E}\langle X_{t^{v,x_m}}, v\rangle| < \chi x_m\right)&\\ \ge \frac{C}{x_m^{p'+\epsilon''}} - \frac{C}{x_m^{p/2}} \ge \frac{C}{x_m^{p'+\epsilon''}},&\end{aligned} \tag{4.16}$$

if $p' < p/2 < r/4 - 1/2$. On the other hand, the event depicted in (4.16) implies that at some time t larger than x_m, the ratio $\langle X_t, v\rangle/t$ is not close to $\|v\|_2^2$, an event whose probability is bounded above (using the regeneration times $t^{(n)}$) by

$$Cx_m^{-r/2} + C\sum_{k=Cx_m}^{\infty} k^{-r/2} \leq Cx_m^{1-r/2}.$$

Since $1 - r/2 < -p'$, we achieved a contradiction. □

Consider temporarily the walks X_1 and X_2 as evolving in independent environments. We define the following i.i.d. one dimensional random walk:

$$S_n = \left\langle X_1\left(t_1^{v,(n)}\right) - X_2\left(t_2^{v,(n)}\right), u\right\rangle.$$

Set $r' < r/4 - 1/2$. For κ and η to be determined below, we define the events

$$B_n = \left\{t_i^{v,(n)} - t_i^{v,(n-1)} < n^\eta, i = 1, 2\right\},$$

$$C_n = \left\{\left|\langle X_i(t_i^{v,(n)}) - E(X_i(t_i^{v,(n)})), v\rangle\right| < n^\kappa, i = 1, 2\right\},$$

$$D_n = \left\{\max_{k\in[n-n^\kappa,n]} |\langle X_i(t_i^{v,(k)}) - X_i(t_i^{v,(n)}), u\rangle| < n^\eta, i = 1, 2\right\},$$

and $U_n = B_n \cap C_n \cap D_n$. By our assumptions, Lemma 4.5, and standard random walk estimates, $P(B_n^c) \leq n^{-\eta r'}$, $P(C_n^c) \leq n^{-r'(2\kappa-1)}$ and $P(D_n^c) \leq n^{-r'(2\eta-\kappa)/2}$. With $r' > 15/2$, choose $\kappa > 1/2, \eta < 1/2$ such that $r'\eta > 3/2$, $r'(2\kappa - 1) > 3/2$ and $r'(2\eta - \kappa) > 3$, to deduce that $P(U_n^c) \leq n^{-a_3}$ for some $a_3 > 3/2$. (This is possible with η close to $1/2$ and κ close to $2(\eta + 1)/5$.)

Fix $\eta' \in (0, 1/2 - \eta)$ and define the event

$$A(T) = \{\text{for all } n \leq T,\ S_n \geq \lfloor n^{\frac{1}{2}-\eta'}\rfloor\}.$$

Note that there exists k_0 such that on the event $A(T) \cap_{n=1}^T U_n$, $X_1[k_0, T/2] \cap X_2[k_0, T/2] = \emptyset$.

From Lemma 4.3 we have that $\mathbb{P}^D(A(T) \cap_{n=1}^T U_n) \geq C/T^{1/2+a_2}$, for some constant $a_2 > 0$. By ellipticity, this implies

$$\mathbb{P}^D(X_1[1, t_1^{(T)}] \cap X_2[1, t_2^{(T)}] = \emptyset) \geq C/T^{1/2+a_2}. \tag{4.17}$$

uniformly over the starting points. (This estimate, which was derived initially for walks in independent environments, obviously holds for walks in the same environment, i.e., under $\mathbb{P}$, too, because it involves a non-intersection event.)

Fix T, and let

$$G(T) = \sum_{i,j} \mathbf{1}_{X_1(i)=X_2(j)} \mathbf{1}_{\langle X_1(i),v\rangle\in[T-0.5,T+0.5)}.$$

We want to bound the sum of

$$F(T) = E\left[E_{\omega\times\omega}\left(G(T)\right)\right].$$

Claim 4.6.

$$\sum_{t=1}^{N} E\left[P_{\omega\times\omega}\left(G(t)\neq 0\right)\right] \le CN^{1/2+a_2+1/r'} . \tag{4.18}$$

Proof. We define variables $\{\psi_n\}_{n=1}^{\infty}$ and $\{\theta_n\}_{n=0}^{\infty}$ inductively as follows:

$$\psi_1 := \max\{t_1^{v,(1)}, t_2^{v,(1)}\}, \quad \theta_0 = 0$$

and then, for every $n \ge 1$,

$$\theta_n := \min\{k > \psi_n : G(k) \neq 0\}, \quad \psi_{n+1} := \max\{\tau_1(\theta_n), \tau_2(\theta_n)\},$$

with

$$\tau_i(k) := \min\{\langle X_{t_i^{v,(m)}}, v\rangle : \langle X_{t_i^{v,(m)}}, v\rangle > k+1\}.$$

We define $h_n = \psi_n - \theta_{n-1}$ and $j_n = \theta_n - \psi_n$. By (4.17), for every k

$$\mathbb{P}(j_n > k | j_1, \dots, j_{n-1}, h_1, \dots, h_n) \ge C/k^{1/2+a_2} . \tag{4.19}$$

Let

$$K := \min\left\{n : \sum_{i=1}^{n} j_i > N\right\}.$$

Let $Y_i^{(N)} = \max_{k=0}^{N}[t_i^{v,(k+1)} - t_i^{v,(k)}]$ be the length of the longest of the first N regenerations of X_i, $i = 1, 2$, in direction v, and set $Y_N = \max(Y_1^{(N)}, Y_2^{(N)})$. Then

$$\sum_{t=1}^{N} \mathbf{1}_{G(t)\neq 0} \le K \cdot Y_N .$$

We see below in (4.21) that $E(Y_N^p) \le CN^{p/r'}$ for $p < r'$. In addition, by the moment bound (4.19), for any t,

$$\mathbf{P}(K > t) = \mathbf{P}\left(\sum_{i=1}^{t} j_i < N\right) \le \exp\left(-C\frac{t}{N^{\frac{1}{2}+a_2}}\right).$$

From here we get

$$\sum_{t=1}^{N} E\left[P_{\omega\times\omega}\left(G(t)\neq 0\right)\right] \le CN^{1/2+a_2+1/r'+\epsilon''} \tag{4.20}$$

for every $\epsilon'' > 0$. The fact that a_2 was an arbitrary positive number allows the removal of ϵ'' from (4.20). □

In addition, $G(t)$ is bounded by the product of the length of the $\{X_1\}$ regeneration containing t and that of the $\{X_2\}$ regeneration containing t. So for all $t < N$,

$$G(t) \le Y_1^{(N)} \cdot Y_2^{(N)},$$

and therefore, for any $p < r'/2$,

$$E\left[G(t)^p\right] \le \sqrt{E((Y_1^{(N)})^{2p})E((Y_2^{(N)})^{2p})} \le CN^{2p/r'},$$

where in the last inequality we used the estimate

$$E((Y_i^{(N)})^{2p}) \leq A^{2p} + 2pN \int_A^\infty y^{2p-1} P(\tau_i^{(2)} - \tau_i^{(1)} > y) dy \leq A^{2p} + CNA^{2p-r'}, \tag{4.21}$$

with $A = N^{1/r'}$. Thus, with $1/q = (p-1)/p$,

$$\begin{aligned} E[G(t)] &= E[G(t) \cdot \mathbf{1}_{G(t) \neq 0}] \leq (EG(t)^p)^{1/p} \left(E\left[P_{\omega \times \omega}\left(G(t) \neq 0\right)\right]\right)^{1/q} \\ &\leq CN^{2/r'} \left(E\left[P_{\omega \times \omega}\left(G(t) \neq 0\right)\right]\right)^{1/q}. \end{aligned} \tag{4.22}$$

Thus,

$$E[E_{\omega,\omega} I_N] \leq \sum_{t=1}^N E[G(t)] \leq CN^{2/r'} N^{1/p} \left(\sum_{t=1}^N E\left[P_{\omega \times \omega}\left(G(t) \neq 0\right)\right] \right)^{1/q} \tag{4.23}$$

Using (4.18), we see that

$$E[E_{\omega,\omega} I_N] \leq CN^{\frac{2}{r'} + \frac{1}{p} + (\frac{p-1}{p})(\frac{1}{2} + \frac{1}{r}' + a_2)}.$$

By choosing $2p < r'$ close to r' and a_2 small, we can get in the last exponent any power strictly larger than $4/r' + 1/2 - 2/(r')^2$. □

Acknowledgment

We thank N. Zygouras for useful discussions. Section 4 was written following a very useful conversation with F. Rassoul-Agha and T. Seppäläinen, who described to one of us their work in [RAS07b]. We thank T. Seppäläinen for very useful comments on an earlier draft of this paper.

References

[Ber06] Noam Berger. On the limiting velocity of high dimensional random walk in random environment, *Ann. Probab* **36**(2):728–738, 2008.

[BS02] Erwin Bolthausen and Alain-Sol Sznitman. On the static and dynamic points of view for certain random walks in random environment. *Methods Appl. Anal.*, 9(3):345–375, 2002. Special issue dedicated to Daniel W. Stroock and Srinivasa S.R. Varadhan on the occasion of their 60th birthday.

[Gol07] Ilya Y. Goldsheid. Simple transient random walks in one-dimensional random environment: the central limit theorem. *Probab. Theory Rel. Fields* 139:41–64, 2007.

[Pet08] Jonathon Peterson. PhD thesis (forthcoming), University of Minnesota, 2008.

[RAS05] Firas Rassoul-Agha and Timo Seppäläinen. An almost sure invariance principle for random walks in a space-time random environment. *Probab. Theory Related Fields*, 133(3):299–314, 2005.

[RAS06] Firas Rassoul-Agha and Timo Seppäläinen. Ballistic random walk in a random environment with a forbidden direction. *Alea*, **1**:111–147, 2006.

[RAS07a] Firas Rassoul-Agha and Timo Seppäläinen. Quenched invariance principle for multidimensional ballistic random walk in a random environment with a forbidden direction. *Ann. Probab*, **35**:1–31, 2007.

[RAS07b] Firas Rassoul-Agha and Timo Seppäläinen. Almost sure functional limit theorem for ballistic random walk in random environment. *Arxiv:*0705.4116*v*2, 2007.

[Sh84] A.N. Shiryayev, *Probability.* Springer, 1984.

[Spi76] Frank Spitzer. *Principles of random walks.* Second edition. Graduate Texts in Mathematics, Vol. **34.** Springer-Verlag, New York-Heidelberg, 1976. xiii+408 pp.

[St93] Daniel W. Stroock. *Probability theory, an analytic view.* Cambridge University Press, 1993.

[SZ99] Alain-Sol Sznitman and Martin Zerner. A law of large numbers for random walks in random environment. *Ann. Probab.*, 27(4):1851–1869, 1999.

[Szn02] Alain-Sol Sznitman. An effective criterion for ballistic behavior of random walks in random environment. *Probab. Theory Related Fields*, 122(4):509–544, 2002.

[Zei04] Ofer Zeitouni. Random walks in random environment. In *Lectures on probability theory and statistics*, volume 1837 of *Lecture Notes in Math.*, pages 189–312. Springer, Berlin, 2004.

Noam Berger
Department of Mathematics
University of California at Los Angeles,
and Department of Mathematics,
Hebrew University, Jerusalem
e-mail: `berger@math.huji.ac.il`

Ofer Zeitouni
Department of Mathematics,
University of Minnesota,
and Department of Mathematics,
Weizmann Institute of Sciences, Rehovot
e-mail: `zeitouni@math.umn.edu`

Progress in Probability, Vol. 60, 161–183

Homogenenous Multitype Fragmentations

Jean Bertoin

Abstract. A homogeneous mass-fragmentation, as it has been defined in [6], describes the evolution of the collection of masses of fragments of an object which breaks down into pieces as time passes. Here, we show that this model can be enriched by considering also the types of the fragments, where a type may represent, for instance, a geometrical shape, and can take finitely many values. In this setting, the dynamics of a randomly tagged fragment play a crucial role in the analysis of the fragmentation. They are determined by a Markov additive process whose distribution depends explicitly on the characteristics of the fragmentation. As applications, we make explicit the connection with multitype branching random walks, and obtain multitype analogs of the pathwise central limit theorem and large deviation estimates for the empirical distribution of fragments.

Mathematics Subject Classification (2000). 60J80, 60G18.

Keywords. Multitype, fragmentation, branching process, Markov additive process.

1. Introduction

In recent years, there has been some interest for a class of stochastic processes which are meant to serve as models for the evolution of an object that breaks down into smaller pieces, randomly and repeatedly as time passes. We refer to the monograph [6] and the survey [5] for a detailed account and references. Several crucial hypotheses have to be made in order to deal with models that can be analyzed by standard probabilistic techniques. Typically, one assumes that the process enjoys the branching property, in the sense that the dynamics of a given fragment do not depend on the others. A further important assumption which is made in [6], is that each fragment is characterized by a real number which can be viewed as its size. The latter requirement does not allow us to consider geometrical properties like the shape of a fragment, although such notions could be relevant for describing how an object breaks down.

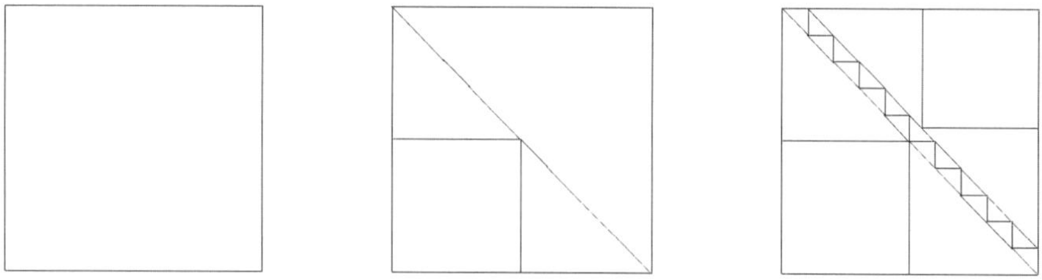

FIGURE 1. Example of a fragmentation of a square into squares and triangles.

In the simpler case when time is discrete, one can analyze a fragmentation chain using the framework of branching random walks or that of multiplicative cascades. In this setting, it is therefore natural to enrich the model by assigning to each fragment a *type*, which, for instance, may describe its shape, and let the evolution of each fragment depend on its initial type. The study of the latter can then be developed directly by translating the literature on multitype branching random walks or cascades (see, e.g., [2, 9, 10]).

However, we shall be interested here in the much more delicate case where time is continuous and each fragment may splits immediately, a situation which cannot be handled directly by discrete techniques based on branching processes. In the monotype setting, Kingman's theory of exchangeable random partitions provides the key for the construction and the study of fragmentation processes in continuous time; this was pointed out first by Pitman [15], see Chapter 3 in [6] for a complete account. In the first part of this work (Sections 2 and 3), we shall briefly explain how Kingman's theory can be extended to the multitype setting (for any finite family of types), and how this extension enables us to develop an adequate framework for multitype fragmentation processes. In short, the main result states that the dynamics of a homogeneous multitype fragmentation are characterized by a family of erosion coefficients and a family of dislocation measures. Each erosion coefficient describes the rate at which a fragment with a given type melts down as time passes, and each dislocation measure specifies the statistics of its sudden splits. Once the correct setting is found, statements are straightforward modifications of that in the monotype situation, and for the sake of avoiding what would be essentially a lengthy and boring duplication of existing material, our presentation will be rather sketchy and proofs will be omitted. Non-specialist readers may wish to consult first Chapters 2 and 3 of [6] for getting the flavor of the arguments.

The second part of this work (Section 4) is devoted to the study of the tagged fragment, i.e., the fragment which contains a point which has been tagged at random and independently of the fragmentation process, and its applications. It departs more significantly from the monotype situation, in the sense that the evolution of the tagged fragment is now given in terms of a Markov additive process (instead of a subordinator), which depends explicitly on the characteristics of the fragmentation. The central limit theorem for Markov additive processes then enables us to determine the asymptotic behavior of certain multitype fragmentation processes, extending an old result of Kolmogorov [14] in this area. We will also develop the natural connection with multitype branching random walks from which we derive some sharp large-deviation estimates based on the work of Biggins and Rahimzadeh Sani [10].

Throughout this text, we shall consider a finite family of types, say with cardinal $k+1 \geq 2$, which can thus be identified with $\{0,1,\ldots,k\}$. The type 0 is special and will only be used in peculiar situations. As it has been mentioned above, it may be convenient to think of a type as a geometrical shape (see Figure 1 above for an example), but the type can also be used, for instance, to distinguish between active and inactive fragments in a frozen fragmentation (see [11] for a closely related notion in the setting of coalescents). Last but not least, it was observed recently by Haas *et al.* [12] that homogeneous fragmentations bear close connections with certain continuum random trees, a class of random fractal spaces which has been introduced by Aldous. It is likely that more generally, multitype fragmentations can be used to construct some multifractal continuum random trees, following the analysis developed in [12].

2. Kingman's theory for partitions with types

The purpose of this section is to provide a brief presentation of an extension of Kingman's theory (see [13] or Section 2.3.2 in [6]) to partitions with types, which is a key step in the analysis of random fragmentations.

2.1. Partitions with types

We shall deal with two natural notions of partitions with types, which correspond to two different points of view. The first one focuses on the masses (and the types) of the components, whereas, roughly speaking, the second one corresponds to a discretization of the object which breaks down.

We call any numerical sequence $\mathbf{x} = (x_1, x_2, \ldots)$ with

$$x_1 \geq x_2 \geq \cdots \geq 0 \quad \text{and} \quad \sum_1^\infty x_n \leq 1$$

a *mass-partition*, and write $\mathcal{P}_{\mathrm{m}}$ for the space of mass-partitions. A mass-partition with types is a pair $\bar{\mathbf{x}} = (\mathbf{x}, \mathbf{i})$ with $\mathbf{x} = (x_1, x_2, \ldots) \in \mathcal{P}_{\mathrm{m}}$ and $\mathbf{i} = (i_1, i_2, \ldots)$ a sequence in $\{0,1,\ldots,k\}$, such that for every $n \in \mathbb{N} = \{1,2,\ldots\}$

$$x_n = 0 \Leftrightarrow i_n = 0\,, \tag{2.1}$$

and further

$$x_n = x_{n+1} \Rightarrow i_n \geq i_{n+1}\,,$$

i.e., the sequence $(x_1, i_1), (x_2, i_2), \ldots$ is non-increasing in the lexicographic order. We shall write indifferently

$$\bar{\mathbf{x}} = (\mathbf{x}, \mathbf{i}) = ((x_1, i_1), (x_2, i_2), \ldots)$$

by a slight abuse of notation.

We should think of x_n as the size of the nth largest component of some object with total mass 1 which has been split, and of i_n as its type. A component with size 0 means that it is absent or empty, and thus has the special type 0. Note that a mass-partition $\mathbf{x}$ can be *improper*, in the sense that $\sum_1^\infty x_n < 1$. Then the mass-defect $x_0 = 1 - \sum_1^\infty x_n$ is called the mass of *dust*, where the dust is viewed as a set of infinitesimal particles. It may be convenient to think that the special type 0 is also assigned to these infinitesimal particles.

We write $\bar{\mathcal{P}}_{\mathrm{m}}$ for the space of mass-partitions with types and endow it with the following distance. Let $(e_1, \ldots, e_k)$ denote the canonical basis of the Euclidean space $\mathbb{R}^k$, and associate to any mass-partition with types $\bar{\mathbf{x}} \in \bar{\mathcal{P}}_{\mathrm{m}}$ the probability measure on the axes of the unit cube

$$\varphi_{\bar{\mathbf{x}}} = x_0 \delta_0 + \sum_{n=1}^{\infty} x_n \delta_{x_n e_{i_n}},$$

where $x_0 = 1 - \sum_1^\infty x_n$ is the mass of dust. Then we define the distance $d(\bar{\mathbf{x}}, \bar{\mathbf{x}}')$ for every $\bar{\mathbf{x}}, \bar{\mathbf{x}}' \in \bar{\mathcal{P}}_{\mathrm{m}}$ as the Prohorov distance between the probability measures $\varphi_{\bar{\mathbf{x}}}$ and $\varphi_{\bar{\mathbf{x}}'}$, which makes $(\bar{\mathcal{P}}_{\mathrm{m}}, d)$ a compact space. We stress that the distance d is strictly weaker than other perhaps simpler distances on $\bar{\mathcal{P}}_{\mathrm{m}}$ such as, for instance,

$$\max\{|x_n - x'_n| 1\!\!1_{\{i_n = i'_n\}} + (x_n + x'_n) 1\!\!1_{\{i_n \neq i'_n\}} : n \in \mathbb{N}\}. \tag{2.2}$$

It may be also worthy to point out that our choice for the distance is well adapted to the requirement (2.1). Typically, denote for $n \in \mathbb{N}$ by $\bar{\mathbf{x}}^{(n,i)}$ the mass-partition with types which consists in n identical fragments with mass $1/n$ and fixed type $i \in \{1, \ldots, k\}$. Then $\bar{\mathbf{x}}^{(n,i)}$ converges as $n \to \infty$ to the degenerate partition of pure dust (and type 0). Such a natural convergence would fail if we had chosen a stronger distance on $\bar{\mathcal{P}}_{\mathrm{m}}$ like (2.2).

Next we turn our attention to the second notion of partition. We call any subset of $\mathbb{N} = \{1, 2, \ldots\}$ a block. A *partition* of a block $B \subseteq \mathbb{N}$ is a sequence $\pi = (\pi_1, \pi_2, \ldots)$ of pairwise disjoint blocks with $\cup \pi_n = B$, which is ranked according to the increasing order of the least elements, i.e., $\inf \pi_m \leq \inf \pi_n$ whenever $m \leq n$ (with the usual convention that $\inf \emptyset = \infty$). We write $\mathcal{P}_B$ for the space of partitions of B.

Given a partition $\pi = (\pi_1, \pi_2, \ldots) \in \mathcal{P}_B$, we can assign to each block π_n a type $i_n \in \{0, \ldots, k\}$, with the following convention which is related to (2.1):

$$i_n = 0 \Leftrightarrow \pi_n \text{ is either empty or a singleton.} \tag{2.3}$$

We write $\bar{\pi} = (\pi, \mathbf{i}) = ((\pi_1, i_1), (\pi_2, i_2), \ldots)$ and call $\bar{\pi}$ a partition with types of B. We denote by $\bar{\mathcal{P}}_B$ the space of partitions with types of some block B.

For every block $B \subseteq \mathbb{N}$ and every partition $\pi = (\pi_1, \ldots)$ of $\mathbb{N}$, we define $\pi_{|B}$, the restriction of π to B, as the partition of B whose blocks are given by $\pi_n \cap B$, $n \in \mathbb{N}$. If $\bar{\pi} = (\pi, \mathbf{i})$ is now a partition with types, we assign types to the blocks of the restricted partition $\pi_{|B}$ as follows. The type of $\pi_n \cap B$ coincides with the type i_n of the block π_n if $\pi_n \cap B$ is neither empty nor a singleton, and is 0 otherwise in order to agree with (2.3). We then write $\bar{\pi}_{|B}$ for the restriction to B of the partition with types $\bar{\pi}$.

For every pair $(\bar{\pi}, \bar{\pi}')$ of partitions with types, we define

$$d(\bar{\pi}, \bar{\pi}')1/\sup\{n \in \mathbb{N} : \bar{\pi}_{|[n]} = \bar{\pi}'_{|[n]}\},$$

where $[n] = \{1, \ldots, n\}$ and $1/\sup \mathbb{N} = 0$. Note that, as the type assigned to singletons is always 0, the identity $\bar{\pi}_{|[1]} = \bar{\pi}'_{|[1]}$ holds in all cases and thus $d(\bar{\pi}, \bar{\pi}') \leq 1$. It is easily checked that $d(\bar{\pi}, \bar{\pi}')$ defines a distance which makes $\bar{\mathcal{P}}_{\mathbb{N}}$ a compact set; see Lemma 2.6 in [6] on its page 96.

Finally, we say that a block $B \subseteq \mathbb{N}$ possesses an asymptotic frequency if and only if

$$|B| = \lim_{n \to \infty} n^{-1} \mathrm{Card}(B \cap [n])$$

exists. If all the blocks of $\bar{\pi} \in \bar{\mathcal{P}}_{\mathbb{N}}$ possess an asymptotic frequency, then we say that $\bar{\pi}$ has asymptotic frequencies, and we write $|\bar{\pi}|^{\downarrow} = (\mathbf{x}, \mathbf{i})$ for the sequence of the asymptotic frequencies and types of the blocks of $\bar{\pi}$ ranked in the non-increasing lexicographic order. Note from Fatou's lemma that $\sum_1^{\infty} |\pi_n| \leq 1$ is a mass-partition and thus $|\bar{\pi}|^{\downarrow} \in \bar{\mathcal{P}}_{\mathrm{m}}$.

2.2. Exchangeability and paintbox construction

A finite permutation is a bijection $\sigma : \mathbb{N} \to \mathbb{N}$ such that $\sigma(n) = n$ when n is sufficiently large. The group of finite permutations acts naturally on the space $\bar{\mathcal{P}}_{\mathbb{N}}$ of partitions with types. Specifically, we write σ^{-1} for the finite permutation obtained as the inverse σ. Given an arbitrary partition with types of $\mathbb{N}$, $\bar{\pi} = (\pi, \mathbf{i})$, σ^{-1} maps each block π_n of π into a block $\sigma^{-1}(\pi_n)$ of some partition denoted by $\sigma(\pi)$. We decide to assign the type i_n of the block π_n to the block $\sigma^{-1}(\pi_n)$. This way, we obtain a partition with types denoted by $\sigma(\bar{\pi})$.

A measure on $\bar{\mathcal{P}}_{\mathbb{N}}$ is called *exchangeable* if it is invariant under the action of finite permutations. Following Kingman [13], we can associate to every mass-partition with types $\bar{\mathbf{x}} = (\mathbf{x}, \mathbf{i}) \in \bar{\mathcal{P}}_{\mathrm{m}}$ an exchangeable probability measure on $\bar{\mathcal{P}}_{\mathbb{N}}$ by the paintbox construction. Specifically, introduce a pair of random variables (ξ, τ) with values in $\mathbb{Z}_+ \times \{0, 1, \ldots, k\}$ whose distribution is specified by the following:

$$\mathbb{P}((\xi, \tau) = (n, i_n)) = x_n \quad \text{for every } n \in \mathbb{N} \quad \text{and} \quad \mathbb{P}((\xi, \tau) = (0, 0)) = 1 - \sum_{n=1}^{\infty} x_n .$$

Then consider a sequence $(\xi_1, \tau_1), \ldots$ of i.i.d. copies of (ξ, τ) and define a random partition with types $\bar{\pi} = (\pi, \mathbf{i})$ by declaring that two distinct integers ℓ, m are in the same block of π if and only if $\xi_m = \xi_\ell \geq 1$, and then decide that the type of that block is $\tau_m = \tau_\ell$. Integers ℓ such that $\xi_\ell = 0$ form the class of singletons of π, and their type is of course 0. Similarly, if some block of π is empty, then its type is necessarily 0 by our convention. The distribution of $\bar{\pi}$ will be denoted by $\rho_{\bar{\mathbf{x}}}$ and called the paintbox based on $\bar{\mathbf{x}}$.

A slight variation of this paintbox construction can be illustrated as follows. Suppose for simplicity that the mass-partition with types $\bar{\mathbf{x}}$ can be represented by splitting some geometric object, for instance a rectangle with unit area, into smaller components, for instance squares, rectangles and triangles. Each component has an area and a shape which we called a type. Imagine that we pick at random a sequence of i.i.d. uniform points $U_1, U_2, \ldots$ in the initial object. A random partition with types is obtained by declaring that two distinct indices are in the same block of the partition if the corresponding random points belong to the same component of the object, and the type of this block is then the type of this component. See Figure 2 below.

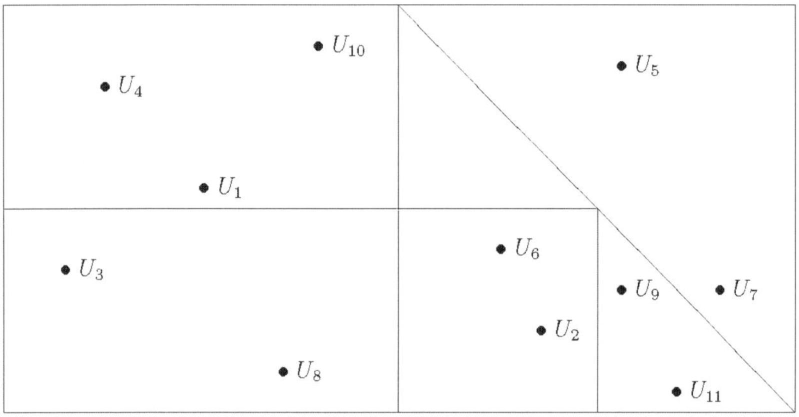

FIGURE 2. Paintbox with types for a partition of a rectangle; types: square = 1, rectangle = 2 and triangle = 3.

$$\bar{\pi}_{|[11]} = ((\{1,4,10\},2), (\{2,6\},1), (\{3,8\},2), (\{5,7\},3), (\{9,11\},3), (\emptyset,0), \ldots)$$

By the law of large numbers, for every positive integer ℓ, the block $B = \{m \in \mathbb{N} : \xi_m = \ell\}$ has an asymptotic frequency

$$|B| = \lim_{n\to\infty} n^{-1}\mathrm{Card}(B \cap [n]) = x_\ell\,,$$

and clearly the type i_ℓ. One can arrange the sequence of the pairs (asymptotic frequency, type) of the blocs of $\bar{\pi}$ in the non-increasing lexicographic order and then it coincides with $\bar{\mathbf{x}}$.

Another important observation is that $|\bar{\pi}_1|$, the asymptotic frequency and the type of the first block $\bar{\pi}_1$ of a paintbox based on a mass-partition with type $\bar{\mathbf{x}}$, has the distribution of a size-biased sample of $\bar{\mathbf{x}}$, that is

$$\rho_{\bar{\mathbf{x}}}(|\bar{\pi}_1| = (x_n, i_n)) = x_n \text{ for every } n \in \mathbb{N} \text{ and } \rho_{\bar{\mathbf{x}}}(|\bar{\pi}_1| = (0,0)) = x_0 = 1 - \sum_1^\infty x_n \,. \tag{2.4}$$

Plainly, if σ is a finite permutation, then $(\xi_{\sigma(1)}, \tau_{\sigma(1)}), (\xi_{\sigma(2)}, \tau_{\sigma(2)}), \ldots$ is again a sequence of i.i.d. copies of (ξ, τ) and the corresponding partition with types is given by $\sigma(\bar{\pi})$. Thus $\rho_{\bar{\mathbf{x}}}$ is an exchangeable probability measure on $\bar{\mathcal{P}}_{\mathbb{N}}$, and more generally any mixture of paintboxes produces an exchangeable probability measure on $\bar{\mathcal{P}}_{\mathrm{m}}$. The converse to the latter assertion is a slight variation of the fundamental theorem of Kingman [13], see, e.g., Theorem 2.1 in [6] on its page 100.

Theorem 2.1. *Let ρ be an exchangeable probability measure on $\bar{\mathcal{P}}_{\mathbb{N}}$. Then ρ-almost every $\bar{\pi} \in \bar{\mathcal{P}}_{\mathbb{N}}$ possesses asymptotic frequencies, and if ϱ stands for the distribution of $|\bar{\pi}|^{\downarrow}$ under ρ, then there is the following disintegration of the measure ρ :*

$$\rho(\mathrm{d}\bar{\pi}) = \int_{\bar{\mathcal{P}}_{\mathrm{m}}} \rho_{\bar{\mathbf{x}}}(\mathrm{d}\bar{\pi}) \varrho(\mathrm{d}\bar{\mathbf{x}}) \,, \qquad \bar{\pi} \in \bar{\mathcal{P}}_{\mathbb{N}} \,. \tag{2.5}$$

Conversely, for every probability measure ϱ on $\bar{\mathcal{P}}_{\mathrm{m}}$, (2.5) *defines an exchangeable probability measure on $\bar{\mathcal{P}}_{\mathbb{N}}$.*

We next turn our attention to an extension of Kingman's theorem to certain sigma-finite measures on $\bar{\mathcal{P}}_{\mathbb{N}}$. In this direction, it is convenient to denote for every type $i \in \{1, \ldots, k\}$ and every block B that is neither empty nor a singleton, by $\mathbf{1}_{B,i}$ the partition with type of B given by $((B,i), (\emptyset, 0), \ldots)$. We also write $\mathbf{1}_i = ((1,i), (0,0), \ldots) \in \bar{\mathcal{P}}_{\mathrm{m}}$ for a related mass-partition with types. For every $n \in \mathbb{N}$, we denote by $\epsilon_{(n,i)}$ for the partition with types of $\mathbb{N}$ which has exactly two non-empty blocks, $(\mathbb{N} \backslash \{n\}, i)$ and $(\{n\}, 0)$. The exchangeable measure on $\bar{\mathcal{P}}_{\mathbb{N}}$

$$\epsilon_i = \sum_{n \in \mathbb{N}} \delta_{\epsilon_{(n,i)}}$$

will be referred to as the erosion measure with type i.

The following extension of Theorem 2.1 to certain possibly infinite measures is the multitype version of Theorem 3.1 in [6] on its page 127. Recall the notation $\bar{\pi}_{|B}$ for the partition with types restricted to some block B, and that $[2] = \{1, 2\}$.

Theorem 2.2. *Fix a type $i \in \{1, \ldots, k\}$ and let μ_i be an exchangeable measure on $\bar{\mathcal{P}}_{\mathbb{N}}$ such that*

$$\mu_i(\{\mathbf{1}_{\mathbb{N},i}\}) = 0 \text{ and } \mu_i\left(\bar{\pi} \in \bar{\mathcal{P}}_{\mathbb{N}} : \bar{\pi}_{|[2]} \neq \mathbf{1}_{[2],i}\right) < \infty \,. \tag{2.6}$$

Then the following holds:

(i) *μ_i-almost every partition $\bar{\pi} \in \bar{\mathcal{P}}_{\mathbb{N}}$ possesses asymptotic frequencies.*
(ii) *Let $|\mu_i|^{\downarrow}$ be the image measure of μ_i by the mapping $\bar{\pi} \to |\bar{\pi}|^{\downarrow}$. The restriction*

$$\nu_i(\mathrm{d}\bar{\mathbf{x}}) = \mathbb{1}_{\{\bar{\mathbf{x}} \neq \mathbf{1}_i\}} |\mu_i|^{\downarrow}(\mathrm{d}\bar{\mathbf{x}}) \tag{2.7}$$

of $|\mu_i|^\downarrow$ *to* $\bar{\mathcal{P}}_{\rm m}\backslash\{\mathbf{1}_i\}$ *fulfills*

$$\int_{\bar{\mathcal{P}}_{\rm m}} (1 - x_1 1\!\!1_{\{i_1=i\}})\nu_i(\mathrm{d}\bar{\mathbf{x}}) < \infty \tag{2.8}$$

and there is the disintegration

$$1\!\!1_{\{|\bar\pi|^\downarrow \neq \mathbf{1}_i\}}\mu_i(\mathrm{d}\bar\pi) = \int_{\bar{\mathcal{P}}_{\rm m}} \rho_{\bar{\mathbf{x}}}(\mathrm{d}\bar\pi)\nu_i(\mathrm{d}\bar{\mathbf{x}})\,.$$

(iii) *There is a real number* $\mathtt{c}_i \geq 0$ *such that*

$$1\!\!1_{\{|\bar\pi|^\downarrow = \mathbf{1}_i\}}\mu_i(\mathrm{d}\bar\pi) = \mathtt{c}_i\epsilon_i(\mathrm{d}\bar\pi)\,. \tag{2.9}$$

Conversely, for every real number $\mathtt{c}_i \geq 0$ *and every measure* ν_i *on* $\bar{\mathcal{P}}_{\rm m}$ *without atom at* $\mathbf{1}_i$ *and that satisfies* (2.8)*, the measure on* $\bar{\mathcal{P}}_{\mathbb{N}}$

$$\mu_i(\mathrm{d}\bar\pi) = \mathtt{c}_i\epsilon_i(\mathrm{d}\bar\pi) + \int_{\bar{\mathcal{P}}_{\rm m}} \rho_{\bar{\mathbf{x}}}(d\bar\pi)\nu_i(\mathrm{d}\bar{\mathbf{x}})$$

is exchangeable and fulfills (2.6) *and* (2.7).

As the erosion measure ϵ_i has infinite total mass, we see that $\mathtt{c}_i$ must be zero whenever μ_i has a finite total mass. In this situation, Theorem 2.2 is an immediate consequence of Theorem 2.1.

3. The structure of multitype fragmentations

The purpose of this section is to describe the structure of multitype fragmentations. In the monotype case, dynamics of a homogeneous fragmentation are entirely determined by an erosion rate $\mathtt{c} \geq 0$, which accounts for the smooth evolution of the process, and a dislocation measure ν on the space $\mathcal{P}_{\rm m}$ of mass-partitions, which, as its name suggests, characterizes the statistics of the sudden dislocations. See Sections 3.1 and 3.2 in [6]. A similar description remains valid in the multitype situation, more precisely dynamics are then determined by a family $(\mathtt{c}_i)_{i\in\{1,\dots,k\}}$ of erosion rates and a family $(\nu_i)_{i\in\{1,\dots,k\}}$ of dislocation measures on $\bar{\mathcal{P}}_{\rm m}$, where the index i refers of course to the type of the fragment that is eroded or dislocated. This will be achieved first in the setting of partitions with types of $\mathbb{N}$, and then shifted to the more intuitive framework of mass-partitions.

3.1. Basic definitions

We first introduce the natural notion of homogeneous fragmentation for mass-partitions with types, which bears strong similarities with that of multitype branching process. Specifically, let $\bar X = (\bar X(t), t \geq 0)$ be a Markov process with values in $\bar{\mathcal{P}}_{\rm m}$ and càdlàg sample paths. For every $i \in \{1,\dots,k\}$, we write $\mathbb{P}_i$ for its distribution starting from $\bar X(0) = \mathbf{1}_i$, i.e., at the initial time, there is a single unit mass with type i.

For every mass-partition with types $\bar{\mathbf{x}} = (\mathbf{x}, \mathbf{i})$ and every real number $r \geq 0$, it will be convenient to write

$$r\bar{\mathbf{x}} = (r\mathbf{x}, \mathbf{i}) = ((rx_1, i_1), (rx_2, i_2), \ldots) .$$

We then introduce a sequence of independent processes $Y^{(1)}, Y^{(2)}, \ldots$ such that for every $n \in \mathbb{N}$, $Y^{(n)}$ is distributed as $x_n\bar{X}$ under $\mathbb{P}_{i_n}$. For every $t \geq 0$, we write $Y(t)$ for the rearrangement in the non-increasing lexicographic order of the terms of the random mass-partitions with types $Y^{(1)}(t), Y^{(2)}(t), \ldots$, and denote by $\mathbb{P}_{\bar{\mathbf{x}}}$ the distribution of the process $Y = (Y(t), t \geq 0)$. In particular $\mathbb{P}_{\mathbf{1}_i} = \mathbb{P}_i$.

Definition 3.1. The process $\bar{X}$ is called a **homogeneous multitype mass-fragmentation** if the conditional law of $(\bar{X}_{t+s}, s \geq 0)$ given $\bar{X} = \bar{\mathbf{x}} \in \bar{\mathcal{P}}_{\mathrm{m}}$ is $\mathbb{P}_{\bar{\mathbf{x}}}$.

The preceding section incites us to translate Definition 3.1 in the setting of partitions with types of $\mathbb{N}$. In this direction, the notion of fragmentation operator (see Definition 3.1 in [6] on its page 114) has a natural extension in the multitype setting.

Specifically, consider $\pi \in \mathcal{P}_B$ a partition of some block B and $\bar{\pi}^{(\cdot)} = (\bar{\pi}^{(n)}, n \in \mathbb{N})$ a sequence of partitions with types. We then write $\mathrm{Frag}(\pi, \bar{\pi}^{(\cdot)})$ for the partition with types which is obtained from the collection of blocks with types of the sequence of the restrictions $\bar{\pi}^{(n)}_{|\pi_n}$ of $\bar{\pi}^{(n)}$ to the nth block π_n of π for $n \in \mathbb{N}$, by rearrangement in the non-increasing lexicographic order. In other words, each block π_n of π is split using $\bar{\pi}^{(n)}$. Note that if the block π_n is either a singleton or empty, then the partition with types $\bar{\pi}^{(n)}_{|\pi_n}$ does not depend on $\bar{\pi}^{(n)}$; more precisely, it is always given by $((\pi_n, 0), (\emptyset, 0), (\emptyset, 0), \ldots)$.

Let $\bar{\Pi} = (\bar{\Pi}(t), t \geq 0)$ be a Markov process with values in $\bar{\mathcal{P}}_{\mathbb{N}}$ with càdlàg sample paths. By a slight abuse of notation, for every $i \in \{1, \ldots, k\}$, we write $\mathbb{P}_i$ for its distribution starting from $\bar{\Pi}(0) = \mathbf{1}_{\mathbb{N},i}$.

Definition 3.2. The process $\bar{\Pi}$ is called a **homogeneous multitype fragmentation** if for every time $t \geq 0$ and every type $i \in \{1, \ldots, k\}$ the distribution of $\bar{\Pi}(t)$ under $\mathbb{P}_i$ is exchangeable, and the semigroup of $\bar{\Pi}$ can be described as follows:

Fix $t, t' \geq 0$ and consider a partition with types $\bar{\pi} = (\pi, \mathbf{i})$ where $\mathbf{i} = (i_1, i_2, \ldots)$ is a sequence in $\{0, 1, \ldots, k\}$. Let $\bar{\pi}^{(\cdot)} = (\pi^{(1)}, \ldots)$ denote a sequence of independent exchangeable random partitions with types, such that for every $n \in \mathbb{N}$ with $i_n \neq 0$, $\bar{\pi}^{(n)}$ is distributed as $\bar{\Pi}(t')$ under $\mathbb{P}_{i_n}$. When $i_n = 0$, the block π_n is either empty or a singleton; the role of $\bar{\pi}^{(n)}$ has no importance and its law can be chosen arbitrarily. The conditional distribution of $\bar{\Pi}(t + t')$ given $\bar{\Pi}(t) = (\pi, \mathbf{i})$ is then the law of $\mathrm{Frag}(\pi, \bar{\pi}^{(\cdot)})$.

Let us now explain the connection between these two definitions. When $\bar{\Pi}$ is a homogeneous multitype fragmentation, we know from Kingman's Theorem 2.1 that for every $t \geq 0$, the exchangeable random partition with types $\bar{\Pi}(t)$ possesses asymptotic frequencies a.s. If we write $\bar{X}(t) = |\bar{\Pi}(t)|^{\downarrow}$ for the random multitype mass-partition obtained by reordering these asymptotic frequencies in the non-increasing lexicographic order, it can be proved that the process $\bar{X} = (\bar{X}(t), t \geq$

0) is then a homogeneous multitype mass-fragmentation. Technically, the main difficulty is to establish that the paths of $t \to |\bar{\Pi}(t)|^{\downarrow}$ are càdlàg; in this direction we stress that this could fail if we had equipped $\bar{\mathcal{P}}_{\mathrm{m}}$ with a stronger distance such as that given by (2.2). In the converse direction, one can rephrase the argument of Berestycki [3] and show that given a homogeneous multitype mass-fragmentation $\bar{X} = (\bar{X}(t), t \geq 0)$, there exists a homogeneous multitype fragmentation $\bar{\Pi}$ such that the process $(|\bar{\Pi}(t)|^{\downarrow}, t \geq 0)$ is distributed as $\bar{X}$. In short, there a bijective correspondence between the laws of homogeneous multitype mass-fragmentations and laws of homogeneous multitype fragmentations.

Of course, the fundamental difference with the monotype case (see for instance Definition 3.2 in [6] on its page 119) is that the distribution of the sequence $\bar{\pi}^{(\cdot)}$ which is used to split the partition with types $\bar{\Pi}(t)$ into finer the blocks depends on $\bar{\Pi}(t)$. However, this dependence only arises through the types of the blocks of $\bar{\Pi}(t)$, and does not involve directly the partition $\Pi(t)$. This preserves the possibility of adapting the approach developed in Chapter 3 of [6], provided that one can handle some technical issues.

In particular, it is easily seen that the fragmentation operation is compatible with the restriction of partitions with types, in the sense that for every integer n

$$\mathrm{Frag}(\pi, \bar{\pi}^{(\cdot)})_{|[n]} = \mathrm{Frag}(\pi_{|[n]}, \bar{\pi}^{(\cdot)}) . \tag{3.1}$$

This entails that the Markov property still holds for the restricted process $\bar{\Pi}_{|[n]} = (\bar{\Pi}_{|[n]}(t), t \geq 0)$, and as the latter can only take finitely many values, $\bar{\Pi}_{|[n]}$ is a Markov chain in continuous times. Note that $\bar{\Pi}_{|[n]}$ coincides with the restriction of $\bar{\Pi}_{|[n+1]}$ to $[n]$, and that the initial process $\bar{\Pi}$ can be recovered from the sequence of Markov chains $\bar{\Pi}_{|[n]}$, $n \in \mathbb{N}$.

Just as in Section 3.1.2 of [6], these observations enable us to characterize the law of $\bar{\Pi}$ by a finite family of measures $(\mu_i)_{i\in\{1,\ldots,k\}}$ on $\bar{\mathcal{P}}_{\mathbb{N}}$ as follows. For every fixed $i \in \{1, \ldots, k\}$, $n \geq 2$ and every partition with types $\bar{\gamma}$ of $[n]$ with $\bar{\gamma} \neq \mathbf{1}_{[n],i}$, we introduce the jump rate of $\bar{\Pi}_{|[n]}$ from $\mathbf{1}_{[n],i}$ to $\bar{\gamma}$,

$$q_{n,\bar{\gamma}} = \lim_{t\to 0+} \frac{1}{t} \mathbb{P}_i \left(\bar{\Pi}_{|[n]}(t) = \bar{\gamma} \right) .$$

By the very same arguments as in Section 3.1.2 of [6], one can check that the collection of those jump rates characterize the evolution of the restricted Markov chains $\bar{\Pi}_{|[n]}$, and thus of the process $\bar{\Pi}$. Further these jump rates can be represented as

$$q_{n,\bar{\gamma}} = \mu_i \left(\{ \bar{\pi} \in \bar{\mathcal{P}}_{\mathbb{N}} : \bar{\pi}_{|[n]} = \bar{\gamma} \} \right) , \tag{3.2}$$

where μ_i is an exchangeable measure on $\bar{\mathcal{P}}_{\mathbb{N}}$ with

$$\mu_i(\{\mathbf{1}_{\mathbb{N},i}\}) = 0 \text{ and } \mu_i \left(\{ \bar{\pi} \in \bar{\mathcal{P}}_{\mathbb{N}} : \bar{\pi}_{|[n]} \neq \mathbf{1}_{[n],i} \} \right) < \infty \text{ for any } n \geq 2 ,$$

and these requirements determine the measure μ_i uniquely. Note that when the condition above is fulfilled for $n = 2$, then, thanks to the exchangeability, it is fulfilled for every $n \geq 2$, therefore it is equivalent to (2.6). We shall refer to the family $(\mu_i)_{i\in\{1,\ldots,k\}}$ as the splitting rates of $\bar{\Pi}$.

3.2. Poissonian constructions

Our first goal in this section is to show that any family of exchangeable measures $(\mu_i)_{i\in\{1,\dots,k\}}$ which fulfill (2.6) can be viewed as the splitting rates of a homogeneous multitype fragmentation $\bar{\Pi}$. More precisely, we shall briefly present a Poissonian construction of $\bar{\Pi}$ which mimics that in Section 3.1.3 of [6] in the monotype case. For the sake of simplicity, we assume that the initial state has been chosen equal to $\mathbf{1}_{\mathbb{N},i_0}$ for some type $i_0 \in \{1,\dots,k\}$.

For every type $i \in \{1,\dots,k\}$, consider the atoms $(t_{i,m}, \bar{\pi}^{(i,m)}, \ell_{i,m})_{m\in\mathbb{N}}$ of a Poisson random measure in $\mathbb{R}_+ \times \bar{\mathcal{P}}_{\mathbb{N}} \times \mathbb{N}$ with intensity $dt \otimes \mu_i \otimes \#$, where $\#$ stands for the counting measure on $\mathbb{N}$. This means that for every measurable set $A \subseteq \mathbb{R}_+ \times \bar{\mathcal{P}}_{\mathbb{N}} \times \mathbb{N}$, the cardinal of the collection of indices m for which $(t_{i,m}, \bar{\pi}^{(i,m)}, \ell_{i,m}) \in A$ has the Poisson distribution with parameter $dt \otimes \mu_i \otimes \#(A)$ and to disjoint sets correspond independent Poisson variables. We assume that these Poisson random measures are independent for different types $i \in \{1,\dots,k\}$.

For every integer n, we can construct a Markov chain in continuous time $\bar{\Pi}^{[n]} = (\bar{\Pi}^{[n]}(t), t \geq 0)$ with values in $\bar{\mathcal{P}}_{[n]}$ as follows. The atoms $(t, \bar{\pi}, \ell)$ of the Poisson point measures such that $\bar{\pi}_{|[n]} = \mathbf{1}_{[n],i}$ or $\ell > n$ play no role in the construction and can thus be removed. Thanks to (2.6), the instants t at which an atom $(t, \bar{\pi}, \ell)$ that has not been removed arises, form a discrete set of $\mathbb{R}_+$, and the chain $\bar{\Pi}^{[n]}$ can only jump at such times. More precisely, if $(t_{i,m}, \bar{\pi}^{(i,m)}, \ell_{i,m})$ is an atom which has not been removed, then we look at the $\ell_{i,m}$th block of $\bar{\Pi}^{[n]}(t_{i,m}-)$, say B. If the type of this block is different from the type i of the atom (in particular if B is either empty or a singleton), then we decide that $\bar{\Pi}^{[n]}(t_{i,m}) = \bar{\Pi}^{[n]}(t_{i,m}-)$ and $t_{i,m}$ is not a jump time for the chain $\bar{\Pi}^{[n]}$. If the type of B is the same as the type i of the atom, then $\bar{\Pi}^{[n]}(t_{i,m})$ is the partition obtained from $\bar{\Pi}^{[n]}(t_{i,m}-)$ by replacing B, that is the $\ell_{i,m}$th block of $\bar{\Pi}^{[n]}(t_{i,m}-)$, by the restriction of $\bar{\pi}^{(i,m)}$ to this block, and leaving the other blocks and types unchanged.

To give an example, take for instance $n = 7$, $\ell_{i,m} = 2$,

$$\bar{\pi}^{(i,m)} = ((\{1,3,5,7,\dots\},3), (\{2,4,6,\dots\},1),\dots) ,$$

and set for simplicity $t = t_{i,m}$. Assume also that

$$\bar{\Pi}^{[n]}(t-) = ((\{1,2\},1), (\{3,4,5\},i), (\{6,7\},2), (\emptyset,0),\dots) .$$

As $\ell_{i,m} = 2$, we look at the 2nd block of $\bar{\Pi}^{[n]}(t-)$, which is $B = \{3,4,5\}$ and has type i, and thus coincides with the type the atom $(t_{i,m}, \bar{\pi}^{(i,m)}, \ell_{i,m})$. At time t, we split B using the partition with types $\bar{\pi}^{(i,m)}$. This produces two new blocks with types: $\{3,5\}$ which has type 3, and $\{4\}$ which is a singleton and thus has type 0. We conclude that

$$\bar{\Pi}^{[n]}(t) = ((\{1,2\},1), (\{3,5\},3), (\{4\},0), (\{6,7\},2), (\emptyset,0),\dots) .$$

It is easily seen that this construction is compatible with the restriction, in the sense that for every $n \in \mathbb{N}$, $\bar{\Pi}^{[n]}$ coincides with the restriction of $\bar{\Pi}^{[n+1]}$ to $[n]$. We refer to Lemma 3.3 in [6] on its page 118 for the argument in the monotype case. This implies that there exists a process $\bar{\Pi}$ with values in $\bar{\mathcal{P}}_{\mathbb{N}}$ such that the restriction of $\bar{\Pi}$ to $[n]$ is $\bar{\Pi}^{[n]}$; see Lemma 2.5 in [6] on its page 95 for a closely related argument.

A crucial step is to show that for every $t \geq 0$, the distribution of $\bar{\Pi}(t)$ is exchangeable. The proof relies on the following technical lemma, which is the multitype version of Lemma 3.2 in [6] on its page 116.

Lemma 3.3. *Let $\bar{\pi} = (\pi, \mathbf{i}) \in \bar{\mathcal{P}}_{\mathbb{N}}$ be an exchangeable random partition with types and $\bar{\pi}^{(\cdot)} = (\bar{\pi}^{(n)}, n \in \mathbb{N})$ a sequence of random partitions with types. Suppose that:*

- *π and $\bar{\pi}^{(\cdot)}$ are independent conditionally on* $\mathbf{i}$,
- *the sequence $\bar{\pi}^{(\cdot)}$ is* **doubly-exchangeable**, *in the sense that for every finite permutation σ of $\mathbb{N}$, the sequences*

$$\left(\sigma(\bar{\pi}^{(n)}), n \in \mathbb{N}\right) \quad \text{and} \quad \left(\bar{\pi}^{(\sigma(n))}, n \in \mathbb{N}\right)$$

both have the same law as $\bar{\pi}^{(\cdot)}$. Then the random partitions with types $\bar{\pi}$ and $\mathrm{Frag}(\pi, \bar{\pi}^{(\cdot)})$ *are jointly exchangeable, that is their joint distribution is invariant by the action of permutations.*

Sketch of the proof: One observes that with probability one, the conditional distribution of π given $\mathbf{i}$ is an exchangeable probability measure on $\mathcal{P}_{\mathbb{N}}$. One can then follow the argument of the proof of Lemma 3.2 in [6]. □

It is then easy to verify from standard properties of Poisson random measures that the process $\bar{\Pi}^{[n]}$ which has just been constructed is a Markov chain in continuous time, and that its jumps rates

$$q_{n,\bar{\gamma}} = \lim_{t \to 0+} \frac{1}{t} \mathbb{P}_i \left(\bar{\Pi}^{[n]}(t) = \bar{\gamma}\right)$$

for every $\bar{\gamma} \in \bar{\mathcal{P}}_{[n]}$ with $\bar{\gamma} \neq \mathbf{1}_{[n],i}$, are given by (3.2). This shows that the process $\bar{\Pi}$, which is specified by the requirement that its restriction to $[n]$ coincides with $\bar{\Pi}^{[n]}$, is a homogeneous multitype fragmentation with splitting rates $(\mu_i)_{i \in \{1,\ldots,k\}}$. Applying Theorem 2.2, we can summarize this analysis in the following statement.

Proposition 3.4. *Let $\bar{\Pi}$ be a homogeneous multitype fragmentation. There exists a unique family $(\mathtt{c}_i)_{i \in \{1,\ldots,k\}}$ of nonnegative real numbers and a unique family $(\nu_i)_{i \in \{1,\ldots,k\}}$ of measures on $\bar{\mathcal{P}}_{\mathrm{m}}$ which fulfill (2.8), such that the family $(\mu_i)_{i \in \{1,\ldots,k\}}$ of splitting rates of $\bar{\Pi}$ is given by (2.9).*

Conversely, for every family $(\mathtt{c}_i)_{i \in \{1,\ldots,k\}}$ of nonnegative real numbers and every family $(\nu_i)_{i \in \{1,\ldots,k\}}$ of measures on $\bar{\mathcal{P}}_{\mathrm{m}}$ which fulfill (2.8), if we define measures μ_i on $\bar{\mathcal{P}}_{\mathbb{N}}$ by (2.9), then the Poissonian construction above produces a homogeneous multitype fragmentation $\bar{\Pi}$ with splitting rates $(\mu_i)_{i \in \{1,\ldots,k\}}$.

Berestycki [3] established a related Poissonian construction for monotype mass-fragmentations. The latter can be extended to the multitype setting provided that the erosion coefficients $\mathtt{c}_i$ are all the same, which enlighten the probabilistic interpretation of the dislocation measures ν_i.

Specifically, for each type $i \in \{1, \dots, k\}$, consider the atoms

$$(t_{i,m}, \bar{\mathbf{x}}^{(i,m)}, \ell_{i,m})_{m \in \mathbb{N}}$$

of a Poisson random measure in $\mathbb{R}_+ \times \bar{\mathcal{P}}_{\mathrm{m}} \times \mathbb{N}$ with intensity $dt \otimes \nu_i \otimes \#$, where $\#$ stands for the counting measure on $\mathbb{N}$. Assume as usual that these Poisson measures are independent for different types. One can construct a pure jump process $\big(\bar{Y}(t), t \geq 0\big)$ in $\bar{\mathcal{P}}_{\mathrm{m}}$ which jumps only at times $t_{i,m}$ at which some atom $(t_{i,m}, \bar{\mathbf{x}}^{(i,m)}, \ell_{i,m})$ occurs. The jump (i.e., the dislocation) induced by such an atom can be described as follows.

We consider the mass-partition with types immediately before time $t_{i,m}$, that is $\bar{Y}(t_{i,m}-)$, and look at its $\ell_{i,m}$th term, say (y, j) for some $y \geq 0$ and $j \in \{0, \dots, k\}$ (recall that the terms of a mass-partition with types are ranked in the non-increasing lexicographic order). If the type j is different from the type i of the atom, then we simply set $\bar{Y}(t_{i,m}-) = \bar{Y}(t_{i,m})$. Otherwise, the $\ell_{i,m}$th term of $\bar{Y}(t_{i,m}-)$ is dislocated according to $\bar{\mathbf{x}}^{(i,m)}$, that is it is replaced by the mass-partition with types $y\bar{\mathbf{x}}^{(i,m)}$. The other terms of $\bar{Y}(t_{i,m}-)$ are left unchanged, and $Y(t_{i,m})$ then results from the rearrangement in the non-increasing lexicographic order of all the terms.

For instance, if

$$\ell_{i,m} = 2 \ , \ \bar{\mathbf{x}}^{(i,m)} = ((\frac{2}{3}, 2), (\frac{1}{3}, 1), (0, 0), \dots)$$

and

$$\bar{Y}(t_{i,m}-) = \left((\frac{1}{2}, 4), (\frac{1}{3}, i), (\frac{1}{6}, 1), (0, 0), \dots \right)$$

then at time $t_{i,m}$ the second term of $\bar{Y}(t_{i,m}-)$, i.e., $(\frac{1}{3}, i)$ is dislocated using $\bar{\mathbf{x}}^{(i,m)}$. This produces the sequence $((\frac{2}{9}, 2), (\frac{1}{9}, 1), (0, 0), \dots)$, and finally

$$\bar{Y}(t_{i,m}) = \left((\frac{1}{2}, 4), (\frac{2}{9}, 2), (\frac{1}{6}, 1), (\frac{1}{9}, 1), (0, 0), \dots \right) .$$

The process $\bar{Y}$ is then a homogeneous multitype fragmentation with zero erosion and dislocation measures $(\nu_i)_{i \in \{1, \dots, k\}}$. Following an argument in Berestycki [3], one can check that for every $\mathsf{c} \geq 0$, the exponentially discounted process $\bar{X}(t) = \mathrm{e}^{-\mathsf{c}u} \bar{Y}(t)$, $t \geq 0$, is then a homogeneous multitype mass-fragmentation with dislocation measures $(\nu_i)_{i \in \{1, \dots, k\}}$ and erosion coefficients $\mathsf{c}_i = \mathsf{c}$ for every $i \in \{1, \dots, k\}$. Unfortunately, this simple transformation cannot be extended to the case when the erosion coefficients are distinct. Informally, when the erosion coefficients depend on the type of the fragments, one would need information about the types of the ancestors of each fragment of $\bar{Y}(t)$ in order to determine the proportion of its mass that has been turned to dust at time t. This information is available for processes with values in $\bar{\mathcal{P}}_{\mathbb{N}}$, but not for those with values in $\bar{\mathcal{P}}_{\mathrm{m}}$.

4. The tagged fragment

Up to a few technical issues, the analysis of multitype fragmentations was so far an easy translation of that in the monotype situation. However more significant differences appear when dealing with finer aspects of these processes. Here, we shall focus on the evolution of the tagged fragment, that is the fragment which contains a point which has been picked uniformly at random and independently of the fragmentation process. The relevance of this study stems from the fact that, even though the tagged fragment alone does not characterize the evolution of the fragmentation, it captures some useful information. In particular, this will enable us to determine the asymptotic behavior of the fragmentation, by making explicit the connection with multitype branching random walks.

Let $\bar{\mathbf{X}} = (X, T)$ be a homogeneous multitype mass-fragmentation, where:

$X_n(t)$ stands for the mass of the n-th largest fragment at time t

and

$T_n(t)$ denotes the type of the n-th largest fragment at time t.

It will be convenient to think of $\bar{\mathbf{X}}$ as associated to a homogeneous multitype fragmentation $\bar{\Pi}$ by $\bar{\mathbf{X}} = |\bar{\Pi}|^{\downarrow}$. In order to avoid technical discussions, we shall assume throughout this section that the fragmentation process is *conservative*, i.e., for every type $i \in \{1, \ldots, k\}$, the erosion coefficient is $\mathsf{c}_i = 0$ and the dislocation measure satisfies

$$\nu_i \left(\left\{ \bar{\mathbf{x}} \in \bar{\mathcal{P}}_{\mathrm{m}} : x_0 = 1 - \sum_{1}^{\infty} x_n > 0 \right\} \right) = 0 \,. \tag{4.1}$$

The description of the evolution of the tagged fragment relies on the notion of *Markov additive processes*. We first provide some background in this area, referring to Section XI.2 of Asmussen [1] for details.

4.1. Background on Markov additive processes

The class of Markov additive processes that will be useful in this work is that formed by bivariate Markov processes $(S_t, J_t)_{t \geq 0}$, where $(J_t)_{t \geq 0}$ is a continuous time Markov chain with values in the finite space of types $\{1, \ldots, k\}$, and, roughly speaking, on every time-interval on which J stays constant, S evolves as a subordinator (i.e., an increasing process with independent and stationary increments) with characteristics specified by the value of J. More precisely, one requires that for every $t, t' \geq 0$,

$$\mathbb{E}((f(S_{t+t'}) - f(S_t))g(J_{t+t'}) \mid \sigma((S_u, J_u), 0 \leq u \leq t)) = \mathbb{E}_{0,J_t}(f(S_{t'})g(J_{t'})) \tag{4.2}$$

where $\mathbb{E}_{s,j}$ refers to the mathematical expectation when the process (S, J) starts from the state (s, j) and f, g denote two generic measurable nonnegative functions.

The law of the Markov chain $(J_t)_{t \geq 0}$ is specified by its intensity matrix $\boldsymbol{\Lambda} = (\lambda_{ij})_{i,j \in \{1,\ldots,k\}}$, i.e., for $i \neq j$, λ_{ij} is the jump rate of J from i to j and

$\sum_{j=1}^{k} \lambda_{ij} = 0$. On every time-interval $[t, t+t')$ on which $J \equiv i$, S evolves as a subordinator $S^{(i)}$ with Bernstein exponent $\Psi^{(i)}$, i.e.,

$$\mathbb{E}(\exp(-\theta S_t^{(i)})) = \exp(-t\Psi^{(i)}(\theta)) .$$

The Bernstein exponent is a concave increasing function which can take the value $-\infty$, and is nonnegative and finite on $[0, \infty)$. Note that, since we are dealing with subordinators, we shall work with the Bernstein exponent $\Psi^{(i)}$ whereas Asmussen [1] uses the cumulant $\theta \to -\Psi^{(i)}(-\theta)$.

Further, a jump of J from i to $j \neq i$ has a probability p_{ij} of inducing a jump of S at the same time, the distribution of which is denoted by B_{ij}, and we write $\hat{B}_{ij}(\theta) = \int \mathrm{e}^{-\theta x} B_{ij}(\mathrm{d}x)$. It is also convenient to agree that $p_{ii} = 0$.

For every types $i, j \in \{1, \ldots, k\}$ and every $\theta \geq 0$ and $t \geq 0$, there is the following identity between $k \times k$ matrices:

$$\mathbb{E}_{0,i}(\mathrm{e}^{-\theta S_t}, J_t = j) = \left(\mathrm{e}^{-t\mathbf{\Phi}(\theta)}\right)_{ij} ,$$

where

$$\mathbf{\Phi}(\theta) = -\mathbf{\Lambda} + \left(\Psi^{(i)}(\theta)\right)_{\mathrm{diag}} + (\lambda_{ij} p_{ij}(1 - \hat{B}_{ij}(\theta))) , \tag{4.3}$$

and the notation $(\lambda_i)_{\mathrm{diag}}$ refers to the diagonal matrix whose ith diagonal coefficient is λ_i. See Proposition 2.2 in [1] on its page 311. We shall refer to $\mathbf{\Phi}$ as the *Bernstein matrix* of (S, J).

4.2. Distribution of the tagged fragment

We are interested in the process of the asymptotic frequency and the type of the first block $(|\bar{\Pi}_1(t)|, t \geq 0)$ of a homogeneous multitype fragmentation $\bar{\Pi}$. Recall from the paintbox construction that $|\bar{\Pi}_1(t)|$ can be viewed as the mass and type of the fragment which contains some point that has been picked at random according to the mass-distribution and independently of the fragmentation.

The conditions which have been enforced at the beginning of this section ensure that for every $t \geq 0$, the first block $\Pi_1(t)$ is neither empty nor a singleton, hence its asymptotic frequency is strictly positive and its type is not 0. This allows us to introduce the process (S, J) with values in $\{1, \ldots, k\} \times \mathbb{R}_+$ by

$$|\bar{\Pi}_1(t)| = (\exp(-S_t), J_t) , \qquad t \geq 0 .$$

Theorem 4.1. *Suppose that the homogeneous multitype fragmentation $\bar{\Pi}$ has erosion coefficients $\mathsf{c}_i = 0$ and that its dislocation measures fulfill* (4.1). *Then (S, J) is a Markov additive process with Bernstein matrix given for every $\theta \geq 0$ by*

$$\mathbf{\Phi}(\theta) = \left(\int_{\bar{\mathcal{P}}_{\mathrm{m}}} \left(\mathbb{1}_{\{i=j\}} - \sum_{n=1}^{\infty} x_n^{1+\theta} \mathbb{1}_{\{i_n = j\}}\right) \nu_i(\mathrm{d}\bar{\mathbf{x}})\right)_{i,j \in \{1,\ldots,k\}} .$$

Proof. The fact that (S, J) is a Markov process that satisfies (4.2) can be seen from the Poissonian construction and the arguments in Section 3.2.2 of [6]. It may be interesting to stress that the evolution of the type J_t of the tagged fragment is Markovian and does not depend on the size $|\Pi_1(t)| = \exp(-S_t)$.

The determination of the Bernstein matrix also relies on the Poissonian construction. First, note that for $i \neq j$, the jump rate λ_{ij} of the type process J coincides with the rate of occurrence of atoms $(t_{i,m}, \bar{\pi}^{(i,m)}, 1)$ with $\bar{\pi}^{(i,m)} = (\pi^{(i,m)}, \mathbf{i})$ and $i_1 = j$. Using (2.4) and Theorem 2.2, this yields

$$\lambda_{ij} = \int_{\bar{\mathcal{P}}_{\mathrm{m}}} \left(\sum_{n=1}^{\infty} x_n \mathbb{1}_{\{i_n = j\}} \right) \nu_i(\mathrm{d}\bar{\mathbf{x}}) , \qquad \text{for } i \neq j.$$

As $\sum_{j=1}^{k} \lambda_{ij} = 0$, this entails

$$\lambda_{ii} = - \int_{\bar{\mathcal{P}}_{\mathrm{m}}} \left(\sum_{n=1}^{\infty} x_n \mathbb{1}_{\{i_n \neq i\}} \right) \nu_i(\mathrm{d}\bar{\mathbf{x}}) .$$

Thus, by (4.1), we obtain the general formula

$$\lambda_{ij} = \int_{\bar{\mathcal{P}}_{\mathrm{m}}} \left(\sum_{n=1}^{\infty} x_n \mathbb{1}_{\{i_n = j\}} - \mathbb{1}_{\{i = j\}} \right) \nu_i(\mathrm{d}\bar{\mathbf{x}}) , \qquad i, j \in \{1, \ldots, k\}. \tag{4.4}$$

A slight refinement of this argument enables us to compute the finite measure $\lambda_{ij} p_{ij} B_{ij}$. Specifically, one finds for $i \neq j$

$$\lambda_{ij} p_{ij} \int f(b) B_{ij}(\mathrm{d}b) = \int_{\bar{\mathcal{P}}_{\mathrm{m}}} \left(\sum_{n=1}^{\infty} x_n \mathbb{1}_{\{i_n = j\}} f(-\ln x_n) \right) \nu_i(\mathrm{d}\bar{\mathbf{x}}) .$$

This gives

$$\lambda_{ij} p_{ij} (1 - \hat{B}_{ij}(\theta)) = \int_{\bar{\mathcal{P}}_{\mathrm{m}}} \left(\sum_{n=1}^{\infty} \mathbb{1}_{\{i_n = j\}} (x_n - x_n^{1+\theta}) \right) \nu_i(\mathrm{d}\bar{\mathbf{x}}) .$$

Finally, the calculation of the Bernstein functions of the subordinators $S^{(i)}$ is made by reduction to the monotype situation. Specifically, we shall work under the law $\mathbb{P}_i$, and we denote by $\nu_i^{\dagger}$ the image of ν_i by the map $\dagger_i : \bar{\mathcal{P}}_{\mathrm{m}} \to \mathcal{P}_{\mathrm{m}}$ where $\dagger_i(\bar{\mathbf{x}})$ is the mass-partition given by rearrangement of the terms $\mathbb{1}_{\{i_n = i\}} x_n$. Informally, this means that all the components of $\bar{\mathbf{x}}$ which are not of type i are reduced to dust. Then $\nu_i^{\dagger}$ is a (monotype) dislocation measure. It should be plain from the Poissonian construction that if we denote by ζ the instant (i.e., the first coordinate) of the first atom $(t_{i,m}, \bar{\pi}^{(i,m)}, 1)$ with $\bar{\pi}^{(i,m)} = (\pi^{(i,m)}, \mathbf{i})$ and $i_1 \neq i$, then ζ is the first jump time of the type process J and the process killed at time ζ, $(|\Pi_1(t)|, u < \zeta)$, can be viewed as the process of the tagged fragment in a homogeneous monotype fragmentation with no erosion and dislocation measure $\nu_i^{\dagger}$. This yields

$$\Psi^{(i)}(\theta) = \Psi^{(i,\dagger)}(\theta) - \Psi^{(i,\dagger)}(0) ,$$

where, according to Theorem 3.2 in [6] on its page 135,

$$\Psi^{(i,\dagger)}(\theta) = \int_{\mathcal{P}_{\mathrm{m}}} \left(1 - \sum_{n=1}^{\infty} x_n^{1+\theta} \right) \nu_i^{\dagger}(\mathrm{d}\mathbf{x}) = \int_{\bar{\mathcal{P}}_{\mathrm{m}}} \left(1 - \sum_{n=1}^{\infty} \mathbb{1}_{\{i_n = i\}} x_n^{1+\theta} \right) \nu_i(\mathrm{d}\bar{\mathbf{x}}) .$$

Hence

$$\Psi^{(i)}(\theta) = \int_{\bar{\mathcal{P}}_{\mathrm{m}}} \left(\sum_{n=1}^{\infty} \mathbb{1}_{\{i_n=i\}} (x_n^{1+\theta} - x_n) \right) \nu_i(\mathrm{d}\bar{\mathbf{x}}) .$$

Putting the pieces together in (4.3), this establishes our claim. □

If we introduce

$$\underline{\theta} = \inf \left\{ \theta \in \mathbb{R} : \int_{\bar{\mathcal{P}}_{\mathrm{m}}} \left| \mathbb{1}_{\{i=j\}} - \sum_{n=1}^{\infty} x_n^{1+\theta} \mathbb{1}_{\{i_n=j\}} \right| \nu_i(\mathrm{d}\bar{\mathbf{x}}) < \infty \right.$$
$$\left. \text{for every } i,j \in \{1,\dots,k\} \right\},$$

then the Bernstein matrix function $\theta \to \mathbf{\Phi}(\theta)$ possesses an analytic extension to $(\underline{\theta}, \infty)$ and Theorem 4.1 still holds for $\theta \in (\underline{\theta}, \infty)$.

4.3. Connection with multitype branching random walks

The preceding analysis of the evolution of the tagged fragment provides us with the key to shift some deep results on multitype branching random walks to homogeneous fragmentations. The approach is quite similar to that in [7], so again we shall skip details.

Just as in the monotype case, we consider the logarithms of the masses of the fragments. Recall that $X_n(t)$ denotes the mass of the nth largest fragment at time t and $T_n(t)$ its type. We introduce for every $t \geq 0$ the empirical measure $\mathbf{Z}^{(t)} = (Z_1^{(t)}, \dots, Z_k^{(t)})$, where

$$Z_j^{(t)} = \sum_{n=1}^{\infty} \mathbb{1}_{\{T_n(t)=j\}} \delta_{-\ln X_n(t)} \, . \tag{4.5}$$

For every fixed step-parameter $a > 0$, the process in discrete time $(\mathbf{Z}^{(an)}, n \in \mathbb{Z}_+)$ is then a multitype branching random walk; see [10] for a precise definition. For the sake of simplicity, we shall focus on the case when $a = 1$ and compute first a quantity of fundamental importance in terms of the characteristics of the fragmentation.

The analysis of multitype branching random walks relies on the Laplace transform of the intensity

$$m_{ij}(\theta) = \mathbb{E}_i \left(\int_{\mathbb{R}} \mathrm{e}^{-\theta x} Z_j^{(1)}(\mathrm{d}x) \right) \mathbb{E}_i \left(\sum_{n=1}^{\infty} \mathbb{1}_{\{T_n(1)=j\}} X_n^{\theta}(1) \right) .$$

Recall now that $\bar{X}(t) = |\bar{\Pi}(t)|^{\downarrow}$ and, from (2.4), that conditionally on $\bar{X}(t)$, the tagged fragment $|\bar{\Pi}_1(t)| = (\exp -S_t, J_t)$ is distributed as a size-biased sample of $\bar{X}(t)$. Hence, for every $\theta > \underline{\theta} + 1$, we have

$$m_{ij}(\theta) = \mathbb{E}_{i,0}(\exp(-(\theta-1)S_1), J_1 = j) = \left(\mathrm{e}^{-\mathbf{\Phi}(\theta-1)} \right)_{ij} .$$

We shall now make a further assumption on the fragmentation $\bar{\mathbf{X}}$, which will be crucial to investigate its asymptotic behavior. Specifically, we assume henceforth that the process J of the type of the tagged fragment is ergodic, i.e., the intensity matrix $\boldsymbol{\Lambda} = (\lambda_{ij})$ given by (4.4) is irreducible. We recall from the Perron-Frobenius theory (see for instance Seneta [16] or Section I.6 and II.4 in Asmussen [1]) that for every $\theta > \underline{\theta}$, the matrix $\exp(-\boldsymbol{\Phi}(\theta))$ has a unique eigenvalue with maximal modulus which is real, has multiplicity one and can be expressed as $\mathrm{e}^{-\varphi(\theta)}$. In other words, $\varphi(\theta)$ is the eigenvalue of the Bernstein matrix $\boldsymbol{\Phi}$ with minimal real part. We also write $\mathbf{u}(\theta) = (u_1(\theta), \ldots, u_k(\theta))$ and $\mathbf{v}(\theta) = (v_1(\theta), \ldots, v_k(\theta))$ for the left and right eigenvectors associated with $\mathrm{e}^{-\varphi(\theta)}$, normalized so that $\sum_{i=1}^k u_i(\theta) = 1$ and $\sum_{i=1}^k u_i(\theta)v_i(\theta) = 1$.

It may be useful to compare our notation with that in [10], see in particular Theorem 1 there. The matrix $M(\theta)$, the eigenvalue $\rho(\theta)$ and the eigenvectors $u(\theta)$ and $v(\theta)$ there coincide respectively with $\exp(-\boldsymbol{\Phi}(\theta-1))$, $\exp(-\varphi(\theta-1))$, $\mathbf{u}(\theta-1)$ and $\mathbf{v}(\theta-1)$ here.

We are now able to turn our attention to a fundamental family of martingales, which have been introduced first by Biggins [8] in the monotype situation.

Theorem 4.2. *Assume that the erosion coefficients c_i are all zero, that the dislocation measures ν_i are conservative (in the sense that (4.1) holds), and that the intensity matrix (4.4) is irreducible.*

(i) *The equation*

$$\varphi(\theta) = (\theta+1)\varphi'(\theta)$$

possesses a unique solution $\bar{\theta} \geq 0$. The function $\theta \to \varphi(\theta)/(\theta+1)$ increases on $(\underline{\theta}, \bar{\theta})$ and decreases on $(\bar{\theta}, \infty)$, and thus reaches its unique maximum at $\bar{\theta}$.

(ii) *For every $\theta \in (\underline{\theta}, \bar{\theta})$, the process*

$$M(\theta, t) = \mathrm{e}^{t\varphi(\theta)} \sum_{n=1}^{\infty} v_{T_n(t)}(\theta) X_n^{\theta+1}(t)\,, \qquad t \geq 0$$

is a martingale which converges a.s. and in $L^1(\mathbb{P}_i)$ for every type $i \in \{1,\ldots,k\}$. Further, this convergence is uniform for θ in any compact set in $(\underline{\theta}, \bar{\theta})$, almost surely.

Proof. (i) It can be shown that the function

$$\varphi : (\underline{\theta}, \infty) \to \mathbb{R} \text{ is concave, increasing, and } \varphi(\theta) = o(\theta) \text{ as } \theta \to \infty. \tag{4.6}$$

See Theorem 3.7 in Seneta [16] for the concavity assertion, the fact that φ increases is similar. Finally observe from Theorem 4.1 that $\lim_{\theta\to\infty} \theta^{-1}\boldsymbol{\Phi}(\theta) = 0$, which entails $\varphi(\theta) = o(\theta)$. We can then follow the arguments of the proof of Lemma 1 in [4].

(ii) We start by recall from the size-biased sampling formula (2.4) that, if $(\mathcal{F}_t)_{t\geq 0}$ denotes the natural filtration of $\bar{\mathbf{X}}$, then

$$\begin{aligned} M(\theta,t) &= e^{t\varphi(\theta)}\sum_{n=1}^{\infty} v_{T_n(t)}(\theta)X_n^{\theta+1}(t) \\ &= e^{t\varphi(\theta)}\mathbb{E}\left(\exp(-\theta S_t)v_{J_t}(\theta)\mid \mathcal{F}_t\right). \end{aligned}$$

As $\mathbf{v}(\theta)$ is a right eigenvector of $\exp(-t\mathbf{\Phi}(\theta))$ corresponding to the eigenvalue $e^{-t\varphi(\theta)}$, we easily see from the Markov property of (S,J) that the process $e^{t\varphi(\theta)}\exp(-\theta S_t)v_{J_t}(\theta)$ is a martingale in its own filtration. By projection on $(\mathcal{F}_t)_{t\geq 0}$, we conclude that $M(\theta,t)$ is an $(\mathcal{F}_t)$-martingale.

By an argument of discretization analogous to that in [7], it suffices to establish the statement when t goes to infinity along the sequence an for some arbitrary $a>0$. For the sake of simplicity, we shall focus on the case $a=1$ without loss of generality and aim at applying Theorems 2 and 3 of [10] to the discrete time martingale

$$M(\theta,n)=e^{n\varphi(\theta)}\sum_{j=1}^{k}v_j(\theta)\int e^{-(\theta+1)x}Z_j^{(n)}(dx)\,,\qquad n\in\mathbb{Z}_+.$$

Recall from Theorem 1(ii) in [10] that $v_j(\theta)\neq 0$ for every $\theta>\underline{\theta}$ and $j=1,\ldots,k$. An application of the conditional form of Jensen's inequality to the identity

$$M(\theta,1)=e^{\varphi(\theta)}\mathbb{E}\left(\exp(-\theta S_1)v_{J_1}(\theta)\mid \mathcal{F}_1\right)$$

shows that for every $\theta>\underline{\theta}$, there is $\alpha>1$ such that $\mathbb{E}_i(M(\theta,1)^{\alpha})<\infty$ for all types $i=1,\ldots,k$. On the other hand, we deduce from (i) that whenever $\theta\in(\underline{\theta},\bar{\theta})$, we can find $\alpha>1$ close to 1 such that $\varphi(\theta)/(\theta+1)<\varphi(\theta')/(\theta'+1)$ where $\theta'=\alpha(\theta+1)-1$. This implies that

$$\exp(-\varphi(\alpha(\theta+1)-1)+\alpha\varphi(\theta))<1\,,$$

and Theorems 2 and 3 in [10] now entails our claim. □

Let us mention an interesting consequence of Theorem 4.2 to the rate of decay of the largest fragment as time goes to infinity, which follows readily from Theorem 4.2 (see, e.g., Corollary 1 in [5] for a related argument):

Corollary 4.3. *We have*

$$\lim_{t\to\infty}\frac{1}{t}\ln X_1(t)=-\varphi'(\bar{\theta})\qquad a.s.$$

More precisely, for every type $j\in\{1,\ldots,k\}$,

$$\lim_{t\to\infty}\frac{1}{t}\ln \xi_j(t)=-\varphi'(\bar{\theta})\qquad a.s.\,,$$

where $\xi_j(t)=\max\{X_n(t),T_n(t)=j\}$.

4.4. Asymptotic behavior of the empirical measure

We shall now conclude this section by presenting a couple of applications to the asymptotic behavior of homogeneous multitype mass-fragmentations. The first belongs to the same vein as Corollary 3.3 in [6] on its page 158. In the monotype case, a version of the result in discrete time goes back to Kolmogorov [14], in probably the first rigorous work ever on fragmentation processes. Roughly speaking, Kolmogorov provided an explanation to the fact which has been observed experimentally in mineralogy, that the logarithms of the masses of mineral grains are often normally distributed. We shall see that a similar feature holds for the more general model of multitype fragmentations.

Corollary 4.4. *Assume that the erosion coefficients* c_i *are all zero, that the dislocation measures* ν_i *are conservative (in the sense that* (4.1) *holds), and that the intensity matrix* (4.4) *is irreducible. Denote for simplicity by* $\mathbf{u}(0) = \mathbf{u} = (u_1, \dots, u_k)$ *the stationary distribution on* $\{1, \dots, k\}$ *of the Markov chain* J*, i.e.,* $\mathbf{u}$ *is the unique probability vector with*

$$\mathbf{u}\boldsymbol{\Lambda} = \mathbf{0}\,.$$

Suppose further that φ *is twice differentiable at* 0*. Then the following limits hold in* $L^2(\mathbb{P}_i)$ *for any initial type* $i \in \{1, \dots, k\}$ *and every continuous bounded function* $f : \mathbb{R} \times \{1, \dots, k\} \to \mathbb{R}$*:*

$$\lim_{t\to\infty} \sum_{n=1}^{\infty} X_n(t) f(t^{-1} \ln X_n(t), T_n(t)) = \sum_{j=1}^{k} u_j f(-\varphi'(0), j)$$

and

$$\lim_{t\to\infty} \sum_{n=1}^{\infty} X_n(t) f(t^{-1/2}(\ln X_n(t) + \varphi'(0)t), T_n(t)) = \sum_{j=1}^{k} u_j \mathbb{E}(f(\mathcal{N}(0, -\varphi''(0)), j))\,,$$

where $\mathcal{N}(0, -\varphi''(0))$ *denotes a centered Gaussian variable with variance* $-\varphi''(0)$*.*

Informally, the first limit means that the masses of most fragments decay exponentially fast with rate $\varphi'(0)$ and their types are distributed according to the stationary law $\mathbf{u}$ of the Markov chain J. The second limit is a refinement of the first and shows that, in a pathwise sense, fluctuations are Gaussian and independent of the type.

Proof. The two limits can be established by first and second moments estimates which rely respectively on the law of large numbers and the central limit theorem for the Markov additive process (S, J), and an argument of propagation of chaos. For the sake of conciseness, we shall focus on the second limit.

Under the present assumptions, we know from Corollary 2.8 in [1] on its page 313 that as $t \to \infty$,

$$\frac{S_t - \varphi'(0)t}{\sqrt{t}} \Rightarrow \mathcal{N}(0, -\varphi''(0))\,,$$

where $\Rightarrow$ is used as a symbol for convergence in distribution. On the other hand, we also have $J(t) \Rightarrow \tau$, where τ is a random type distributed according to the stationary law $\mathbf{u}$. Further an easy argument using the fact that the Markov chain J mixes exponentially fast shows the asymptotic independence, in the sense that

$$\left(\frac{S_t - \varphi'(0)t}{\sqrt{t}}, J_t\right) \Rightarrow (\mathcal{N}(0, -\varphi''(0)), \tau) ,$$

where in the right-hand side, the variables $\mathcal{N}(0, -\varphi''(0)$ and τ are assumed independent.

Recall now that $\bar{X}(t) = |\bar{\Pi}(t)|^{\downarrow}$ and, from (2.4), that conditionally on $\bar{X}(t)$, the tagged fragment $|\bar{\Pi}_1(t)| = (\exp -S_t, J_t)$ is distributed as a size-biased sample of $\bar{X}(t)$. Hence

$$\mathbb{E}_i\left(\sum_{n=1}^{\infty} X_n(t) f(t^{-1/2}(\ln X_n(t) + \varphi'(0)t), T_n(t))\right) = \mathbb{E}_i\left(f\left(\frac{-S_t + \varphi'(0)t}{\sqrt{t}}, J_t\right)\right) ,$$

and therefore

$$\begin{aligned} &\lim_{t\to\infty} \mathbb{E}_i\left(\sum_{n=1}^{\infty} X_n(t) f(t^{-1/2}(\ln X_n(t) + \varphi'(0)t), T_n(t))\right) \\ = \; & \mathbb{E}\left(f\left(\mathcal{N}(0, -\varphi''(0)), \tau\right)\right) \\ = \; & \sum_{j=1}^{k} u_j \mathbb{E}(f(\mathcal{N}(0, -\varphi''(0)), j)) . \end{aligned}$$

By an argument of propagation of chaos similar to that in the proof of Corollary 3.3 in [6] on its pages 159–160, we can estimate the second moment and get

$$\begin{aligned} &\lim_{t\to\infty} \mathbb{E}_i\left(\left(\sum_{n=1}^{\infty} X_n(t) f(t^{-1/2}(\ln X_n(t) + \varphi'(0)t), T_n(t))\right)^2\right) \\ = \; & \left(\sum_{j=1}^{k} u_j \mathbb{E}(f(\mathcal{N}(0, -\varphi''(0)), j))\right)^2 . \end{aligned}$$

This entails the convergence in $L^2(\mathbb{P}_i)$ which has been stated. □

Finally, using time discretization techniques similar to those in [7], we can translate Theorem 7 in [10] to multitype fragmentations. This yields a pathwise large deviation limit theorem for the empirical distribution of the fragmentation which refines considerably Corollary 4.4. In this direction, we shall further assume that the eigenvalue function φ is strictly concave and that the branching random walk $\mathbf{Z}^{(n)}$ is strongly non-lattice (see [10] for the terminology), which are very natural and mild conditions. Recall Theorem 4.2 and denote the terminal value of the martingale $M(\theta, t)$ by $M(\theta, \infty)$.

Corollary 4.5. *Let $h : \mathbb{R} \times \{1, \ldots, k\} \to \mathbb{R}$ be a continuous function with compact support. Under the preceding assumptions, we have*

$$\lim_{t\to\infty} \sqrt{t}\, \mathrm{e}^{-t((\theta+1)\varphi'(\theta)-\varphi(\theta))} \sum_{n=1}^{\infty} h(t\varphi'(\theta) + \ln X_n(t), T_n(t))$$

$$= \frac{M(\theta,\infty)}{\sqrt{2\pi|\varphi''(\theta)|}} \sum_{j=1}^{k} u_j(\theta) \int_{-\infty}^{\infty} h(y,j)\mathrm{e}^{-(\theta+1)y}\mathrm{d}y\,,$$

uniformly for θ in compact subsets of $(\underline{\theta}, \bar{\theta})$, almost surely.

In particular, this implies that for every $a < b \in \mathbb{R}$, $\theta \in (\underline{\theta}, \bar{\theta})$ and $j \in \{1, \ldots, k\}$, there is the estimate as $t \to \infty$

$$\# \left\{ n \in \mathbb{N} : a\mathrm{e}^{-t\varphi'(\theta)} \le X_n(t) \le b\mathrm{e}^{-t\varphi'(\theta)} \text{ and } T_n(t) = j \right\}$$

$$\sim \quad A_\theta u_j(\theta) t^{-1/2} \exp(t((\theta+1)\varphi'(\theta) - \varphi(\theta))) \left(\mathrm{e}^{-a(\theta+1)} - \mathrm{e}^{-b(\theta+1)} \right),$$

where A_θ is some strictly positive random variable with finite mean. See Corollary 2 in [10].

Agradecimentos

In August 2004, I had the pleasure to be invited in Ubatuba, to give a short course on self-similar fragmentation chains for the 8th Brazilian School of Probability. The present paper can be viewed, in some sense, as a natural prolongation of the material which I presented during that course. I would like to thank again Pablo Ferrari for his very kind invitation and the wonderful organization, and Vladas Sidoravicius and Maria-Eulalia Vares for making my stays so pleasant each time I have the chance to go to Rio.

References

[1] Asmussen, S. (2003). *Applied Probability and Queues.* Second edition. Applications of Mathematics. Stochastic Modelling and Applied Probability. Springer-Verlag, New York.

[2] Barral, J. (2001). Generalized vector multiplicative cascades. *Adv. in Appl. Probab.* **33**, 874–895.

[3] Berestycki, J. (2002). Ranked fragmentations. *ESAIM, Probabilités et Statistique* **6**, 157–176. Available via `http://www.edpsciences.org/ps/OnlinePSbis.html`

[4] Bertoin, J. (2003). The asymptotic behavior of fragmentation processes. *J. Euro. Math. Soc.* **5**, 395–416.

[5] Bertoin, J. (2006). Some aspects of a random fragmentation model. *Stochastic Process. Appl.* **116**, 345–369.

[6] Bertoin, J. (2006). *Random Fragmentation and Coagulation Processes.* Cambridge University Press, Cambridge.

[7] Bertoin, J. and Rouault, A. (2005). Discretization methods for homogeneous fragmentations. *J. London Math. Soc.* **72**, 91–109.

[8] Biggins, J.D. (1977). Martingale convergence in the branching random walk. *J. Appl. Probability* **14**, no. 1, 25–37.

[9] Biggins, J.D. and Kyprianou, A.E. (2004). Measure change in multitype branching. *Adv. Appl. Probab.* **36**, 544–581.

[10] Biggins, J.D. and Rahimzadeh Sani, A. (2005). Convergence results on multitype, multivariate branching random walks. *Adv. Appl. Probab.* **37**, 681–705.

[11] Dong, R., Gnedin, A., and Pitman, J. (2006). Exchangeable partitions derived from Markovian coalescents.
Preprint available via: `http://arxiv.org/abs/math.PR/0603745`

[12] Haas, B., Miermont, G., Pitman, J., and Winkel, M. (2006). Continuum tree asymptotics of discrete fragmentations and applications to phylogenetic models. Preprint available via `http://arxiv.org/abs/math.PR/0604350`

[13] Kingman, J.F.C. (1982). The coalescent. *Stochastic Process. Appl.* **13**, 235–248.

[14] Kolmogoroff, A.N. (1941). Über das logarithmisch normale Verteilungsgesetz der Dimensionen der Teilchen bei Zerstückelung. *C. R. (Doklady) Acad. Sci. URSS* **31**, 99–101.

[15] Pitman, J. (1999). Coalescents with multiple collisions. *Ann. Probab.* **27**, 1870–1902.

[16] Seneta, E. (1973). *Non-Negative Matrices. An Introduction to Theory and Applications.* Halsted Press, New York.

Jean Bertoin
Laboratoire de Probabilités et Modèles Aléatoires
Université Pierre et Marie Curie
and
DMA, Ecole Normale Supérieure
Paris, France
e-mail: `jean.bertoin@upmc.fr`

Progress in Probability, Vol. 60, 185–210

A Hierarchical Bayes Model for Combining Precipitation Measurements from Different Sources

Tamre P. Cardoso and Peter Guttorp

Abstract. Surface rain rate is an important climatic variable and many entities are interested in obtaining accurate rain rate estimates. Rain rate, however, cannot be measured directly by currently available instrumentation. A hierarchical Bayes model is used as the framework for estimating rain rate parameters through time, conditional on observations from multiple instruments such as rain gauges, ground radars, and distrometers. The hierarchical model incorporates relationships between physical rainfall processes and collected data. A key feature of this model is the evolution of drop-size distributions (DSD) as a hidden process. An unobserved DSD is modeled as two independent component processes: 1) an AR(1) time-varying mean with GARCH errors for the total number of drops evolving through time, and 2) a time-varying lognormal distribution for the size of drops. From the modeled DSDs, precipitation parameters of interest, including rain rate, are calculated along with associated uncertainty. This model formulation deviates from the common notion of rain gauges as "ground truth"; rather, information from the various precipitation measurements is incorporated into the parameter estimates and the estimate of the hidden process. The model is implemented using Markov chain Monte Carlo methods.

Mathematics Subject Classification (2000). 62P12, 62M09, 86A10.

Keywords. Hierarchical Bayes, MCMC, Precipitation, Rain rate, Drop-size distribution.

1. Introduction

Surface rainfall is an important environmental variable that is incorporated across many areas of study, including meteorology, climatology, agriculture, land use, and hydrology. These various fields of study require estimation of precipitation at a range of temporal and spatial scales. Hydrologists and land use planners may be interested in short-term rainfall in relatively small regional areas for studies involving flood and flash flood forecasting. Researchers in climatology or agriculture may be interested in climatic studies that focus on weekly, monthly, or annual totals over large spatial extents.

Although of considerable interest, surface precipitation rates and amounts are difficult to estimate. Unlike other atmospheric variables, rainfall can display extreme heterogeneity in space and time. Both the occurrence of precipitation and the rate at which it falls may be highly variable, even within a single rain event. Due to the importance of and difficultly with estimation, there has been much research in the area of precipitation measurement and estimation.

Precipitation is measured using many different instruments. Some instrumentation measures rainfall directly, while others make indirect measurements of quantities that can be related to rainfall. All of the instrumentation, however, make indirect measurements of the usual quantity of interest, namely, rain rate. The most common instruments include rain gauges and ground-based scanning radar. Distrometer in situ measurements and satellite deployed instrumentation are also used, although less commonly. Each of the instruments has inherent strengths and weaknesses, which affect the observed measurements and subsequent estimation of surface rainfall.

There exist many algorithms, each with its own set of assumptions, for calculating rainfall related parameters from the instrumental observations. Many times the observations obtained from one instrument are used in the calculation of estimates based on the observations from another instrument. Rain gauge data are often considered to be "ground truth" and are used to adjust the estimates based on other instruments. This is done despite the known fact that observations from rain gauges are inherently prone to errors ([12] and [8]) and that they are no more "true" than the observations from other instruments.

Surface precipitation can be defined by a population of falling drops. The distribution of the size diameters of the drops characterizes the population behavior; thus, the drop-size distribution (DSD) forms a basic descriptor in the modeling of rain microphysics. The DSD is used to compute a variety of "derived" properties of rainfall through mathematical relationships; common computed properties include water content, rain rate, and reflectivity. DSD data, however, are not collected on a routine basis due to the expense of distrometer instruments and the limited spatial coverage of an individual instrument.

The literature is full of modeling approaches for the estimation of surface rainfall, including empirical statistical models ([24] and [21], for example) and a variety of stochastic models ([3], [16], [6], [15], and [22]). More recently, hierarchical Bayesian models have been used in a number of environmental applications ([17], [29], [4], [5], [30], and [9]).

In this paper, using a hierarchical Bayesian approach, we develop a model for estimating the underlying distribution of drop sizes at a site. This then can be used to calculate different quantities related to precipitation (such as rain rate, the instantaneous rate of precipitation across a unit area), and to calibrate various measurement devices (such as tipping rain gauges and radar reflectivity). The advantage with this approach is that there is no need to think of a particular kind of measurement as "ground truth." Rather, the ground truth is the estimated distribution, and the methodology allows researchers to assess quantitatively the uncertainty associated with that distribution.

2. Hierarchical model

DSD spectra can be used to calculate many rainfall parameters. The availability of such spectra, however, is often limited and surface rainfall estimates usually rely on other data sources. Model development is focused on an unobserved process model for the evolution of drop-size distributions through time. Information contained in more widely accessible data is used to augment the parameter estimates for the hidden process. Specifically, these data include rain gauge and ground radar observations.

This section provides a summary of key characteristics of DSD spectra that are used in formulating the model for the unobserved process. This is followed by a conceptual description of the model and then model details. Model components are developed in the time domain only, assuming a single, fixed spatial location. The model is a flexible five-stage Bayesian hierarchical formulation that is based on a hidden model for generating rainfall drop-size distributions. Estimates of the unobserved process, denoted by $N(D)_t$, are the primary quantity of interest. The first stage of the model specifies a measurement error model for the observational data, radar data denoted by Z and rain gauge data denoted by G, both of which are observations of functions of $N(D)_t$ with error. The second stage of the model incorporates a time series formulation of the unobserved DSD process. Time series parameters and temporal dynamic terms of the DSD evolution process are modeled in the third stage. The Bayesian formulation is completed in stages four and five by specifying priors on model parameters. Figure 1 shows a schematic that summarizes the relationships among stages of the hierarchical model specification.

For readability, the model is presented in a different order than the component stages of the hierarchical framework. The complete model for the hidden process, including temporal dynamics and priors on related model parameters is presented first, followed by a description of the observational data components.

2.1. Empirical analysis of DSD data

Exploratory data analysis was conducted on a set of one-minute drop size spectra collected using a Josh-Waldvogel distrometer (JWD) from a site in Eureka, California from January through March 1999. Distrometers are direct measuring devices that collect observations of the ground rainfall drop-size distribution at a spatial point location with high temporal resolution. The JWD is a electro-mechanical spectrometer that separates raindrops entering a 50 cm^2 sampling area into 20 different drop-size classes using a 20-channel pulse-height analyzer. Drops with diameters between $0.3 - 5.0$ mm are partitioned into the first 19 bins. The last bin contains all drops with diameter >5 mm [28].

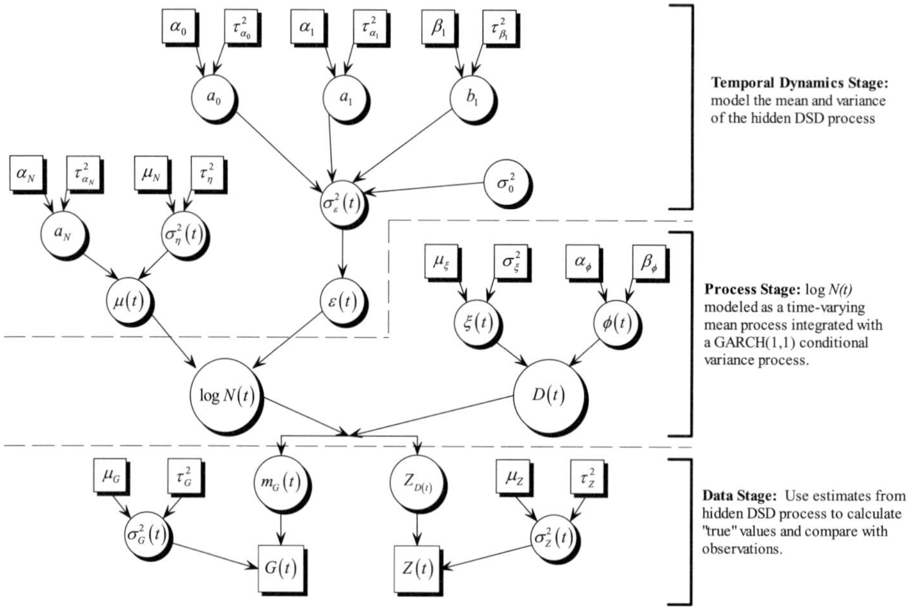

FIGURE 1. Summary of model stages showing the relationship among the data, process, and temporal dynamics stages. Rectangles represent observed data and fixed hyperparameters. Circles represent unobserved parameters.

Rain rates calculated for each of the one-minute spectra were used to identify rainfall events. Seven events spanning early February to mid-March, ranging from 12 to 24 hours in duration, were chosen for evaluation. Barplots of the binned drops and time series of the total number of drops/minute were evaluated for each event. Two general observations were made:

1. The distributions of drop-sizes were generally unimodal and right-skewed. Based on visual inspection and five number summaries, the shapes of the distributions of the one-minute DSD spectra were similar both within and across rainfall events. Figure 2 shows barplots for some representative one-minute DSDs.
2. The total number of drops from one-minute to the next was highly variable both within and across events. Evaluation of log-transformed time series data suggested an underlying auto-regressive process. Figure 3 shows some time series plots for total number of drops.

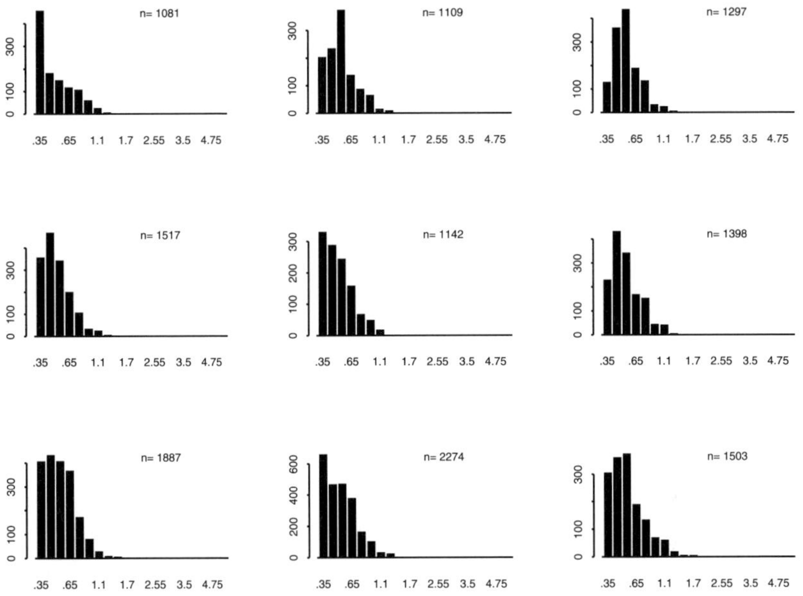

FIGURE 2. Barplots for nine consecutive minutes of distrometer DSD data from a Eureka rainfall event; n is the total number of drops in each one-minute interval.

2.2. Unobserved process

A DSD, denoting the number of drops per mm diameter bin interval and per m^3 of air, is well defined as

$$N(D_i)_t = \frac{n(D_i)_t}{AtV(D_i)\Delta D_i} \tag{1}$$

where $n(D_i)_t$ is the observed number of drops (raw counts) in each diameter bin interval D_i for sampling interval t, A is the sample surface area in m^2, t is the length of sample collection in seconds, $V(D_i)$ is the terminal fall speed in still air in m/s, and ΔD is the drop diameter interval in mm.

Now, consider defining the number of drops over all diameters D as

$$n(D)_t = n_t \times P(D)_t \tag{2}$$

where n_t is the total number of raw counts from a JWD instrument over all drop diameters for the time interval t and is assumed independent of $P(D)_t$, a probability density function defining the shape of the DSD over the interval t; $P(D)_t$ gives the probability of finding a drop within a diameter D to $D + \Delta D$.

Substituting (2) into (1) yields

$$N(D_i)_t = \frac{n_t \times P(D_i)_t}{AtV(D_i)\Delta D_i}. \tag{3}$$

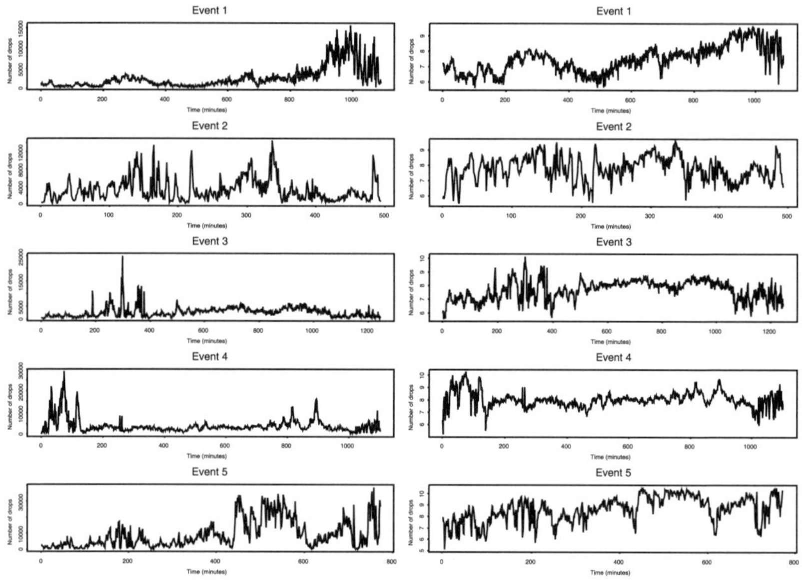

FIGURE 3. Time series of the one-minute total number of drops (left) and log transformed total number of drops (right) for each of five Eureka rainfall events.

Next, define N_t as the total number of drops per m^3 of air. Substitution into (3) gives the DSD formulated in terms of total counts as

$$N(D_i)_t = \frac{N_t \times P(D_i)_t}{\Delta D_i}. \tag{4}$$

For a rainfall event of duration T, the unobserved states over all drop diameters $N(D)_t$ will be estimated through models for N_t and $P(D)_t$; $t = 1, \ldots, T$. To be consistent with commonly collected distrometer data, each t will represent a one-minute interval.

The total number of drops is treated as independent of shape of the DSD spectra. Sauvageot and Lacaux [18] suggest the possibility that the total number of drops in a DSD varies with rain rate and this total can be thought of as independent of a generic shape function that reflects an equilibrium between coalescence and collisional breakup processes. Models for the two independent components of $N(D)_t$ are detailed below.

2.2.1. Model for total number of drops, $\mathbf{N_t}$. Let N_t denote the total number of drops at time t, $t = 0, 1, 2, \ldots, T$. To account for the serial correlation and burst variability observed in many of the time series of total number of drops in contiguous DSD spectra, we introduce a model composed of a time-varying mean process integrated with a generalized autoregressive conditional heteroscedasticity

(GARCH) conditional variance process. GARCH provides conditional models that allow for time-varying changes in the distributions that govern each data point, as opposed to assuming a constant variance model that applies equally anywhere in the series.

Let

$$TN_t = \ln N_t = \mu_{N_t} + \epsilon_t.$$

The ϵ_t's are an independent sequence of correlated errors from an unknown distribution. The ϵ_t values, conditional on all previous values are assumed to be normally distributed:

$$\epsilon_t | \epsilon_{t-1}, \epsilon_{t-2}, \ldots \sim N(0, \sigma^2_{\epsilon_t}).$$

Based on a analysis of data from several rainfall events, we selected a $GARCH(1,1)$ model, which provides a single lagged error square term and one autoregressive variance term to capture the "burst" variability observed in many of the total number of drop time series. Under a $GARCH(1,1)$ model

$$\sigma^2_{\epsilon_t} = a_0 + a_1 \epsilon^2_{t-1} + b_1 \sigma^2_{\epsilon_{t-1}}$$

where the parameters a_0, a_1, and b_1 are estimated, subject to the constraints

$$a_0 > 0,\ a_1, b_1 \geq 0, \quad \text{and} \quad a_1 + b_1 \leq 1 \quad [1].$$

By substitution,

$$\begin{aligned}
\epsilon_1 | \epsilon_0 &\sim N(0, a_0 + a_1 \epsilon_0^2 + b_1 \sigma^2_{\epsilon_0}) \\
\epsilon_2 | \epsilon_1, \epsilon_0 &\sim N(0, a_0 + a_1 \epsilon_1^2 + b_1 \sigma^2_{\epsilon_1}) \\
&\vdots \\
\epsilon_t | \epsilon_{t-1}, \epsilon_{t-2}, \ldots, \epsilon_1, \epsilon_0 &\sim N(0, a_0 + a_1 \epsilon^2_{t-1} + b_1 \sigma^2_{\epsilon_{t-1}}).
\end{aligned}$$

Given the conditional distribution of $\epsilon_t | \epsilon_{t-1}, \epsilon_{t-2}, \ldots, \epsilon_1, \epsilon_0$, the joint probability for ϵ is:

$$\begin{aligned}
P(\epsilon_t, \epsilon_{t-1}, \ldots, \epsilon_0) &= P(\epsilon_t | \epsilon_{t-1}, \epsilon_{t-2}, \ldots, \epsilon_0) P(\epsilon_{t-1} | \epsilon_{t-2}, \epsilon_{t-3}, \ldots, \epsilon_0) \\
&\quad \times \ldots \times P(\epsilon_0) \\
&= \left[\prod_{i=1}^{t} P(\epsilon_{t+1-i} | \epsilon_{t-i}, \epsilon_{t-i-1}, \epsilon_{t-i-2}, \ldots, \epsilon_0) \right] P(\epsilon_0) \\
&= \left[\prod_{i=1}^{t} \frac{1}{\sqrt{2\pi} \sigma_{\epsilon_{t+1-i}}} \exp\Big\{ -\frac{1}{2\sigma^2_{\epsilon_{t+1-i}}} (\epsilon^2_{t+1-i}) \Big\} \right] \\
&\quad \times \frac{1}{\sqrt{2\pi}} \exp\Big\{ -\frac{1}{2} (\epsilon_0^2) \Big\}.
\end{aligned}$$

Assume that conditional on μ_{N_t} and the parameters $\theta_N = \{a_0, a_1, b_1, \sigma^2_{\epsilon 0}\}$, the $\ln N_t$'s are independent and are distributed as $N(\mu_{N_t}, \sigma^2_{\epsilon_t})$. Then, the joint distribution of $\ln N$ is:

$$\begin{aligned} P(TN) &= \left[TN_1, \ldots, TN_T \,\middle|\, \mu_{N_1}, \ldots, \mu_{N_T}; \sigma^2_{\epsilon_1}, \ldots, \sigma^2_{\epsilon_T}\right] \\ &= \left[\prod_{t=1}^{T} \frac{1}{\sqrt{2\pi}\sigma_{\epsilon_t}} \exp\left\{ -\frac{1}{2\sigma^2_{\epsilon_t}} (TN_t - \mu_{N_t})^2 \right\}\right] \\ &\quad \times \frac{1}{\sqrt{2\pi}\sigma_{\epsilon_0}} \exp\left\{ -\frac{1}{2\sigma^2_{\epsilon_0}} (TN_0 - \mu_{N_0})^2 \right\}. \end{aligned}$$

Set $TN_0 = \mu_{N_0}$; this reduces $P(\epsilon_0)$ to the constant term $\frac{1}{\sqrt{2\pi}\sigma_{\epsilon_0}}$ and yields:

$$P(TN) = \left[\prod_{t=1}^{T} \frac{1}{\sqrt{2\pi}\sigma_{\epsilon_t}} \exp\left\{ -\frac{1}{2\sigma^2_{\epsilon_t}} (TN_t - \mu_{N_t})^2 \right\}\right] \times \frac{1}{\sqrt{2\pi}\sigma_{\epsilon_0}}.$$

Since the GARCH parameters are constrained to be greater than zero, each coefficient is assigned a Lognormal prior:

$$\begin{aligned} a_0 &\sim \text{Lognormal}(\alpha_0, \tau^2_{\alpha_0}) \\ a_1 &\sim \text{Lognormal}(\alpha_1, \tau^2_{\alpha_1}) \\ b_1 &\sim \text{Lognormal}(\beta_1, \tau^2_{\beta_1}). \end{aligned}$$

The hyperparameters α_0, α_1, β_1, $\tau^2_{\alpha_0}$, $\tau^2_{\alpha_1}$, and $\tau^2_{\beta_1}$ are taken as fixed, with values that may vary by application.

The mean for TN_t, μ_{N_t}, is modeled as a first-order autoregressive process. Let $\mu_{N_t} = a_N \mu_{N_{t-1}} + \eta_{N_t}$ where η_{N_t} is a random normal process with mean 0 and variance σ^2_η. The parameters are $\theta_\mu = \{a_N, \sigma^2_\eta\}$. Conditional on a_N and $\mu_{N_{t-1}}$, $\mu_{N_t} \sim N(a_N \mu_{N_{t-1}}, \sigma^2_\eta)$ with $\sigma^2_\eta \sim IG(\mu_\eta, \tau^2_\eta)$. The parameter a_N needs to be estimated. The prior distribution for a_N is defined as $a_N \sim N(\alpha_N, \tau^2_{\alpha N})$. The hyperparameters α_N, $\tau^2_{\alpha N}$, μ_η, and τ^2_η are taken as fixed, with values that may vary for different types of rainfall events (e.g., light rainfall versus heavy rainfall).

2.2.2. DSD shape, $\mathbf{P(D)_t}$. A number of distributions have been used to model DSDs including the gamma distribution proposed by Ulbrich [26], Ulbrich and Atlas [27], and Willis and Tattelman [32], in addition to the lognormal distribution ([18] and [22]). A three parameter normalized gamma distribution ([31], [25], and [10]), along with the Weibull distribution ([20], [11] and [19]) have also been used to model DSDs. We considered exponential, lognormal and gamma distributions for modeling the DSD shape. An exponential distribution tended to underestimate the number of drops in the 0.3 mm to about 1 mm diameters, while overestimating drops in the < 0.3 mm bin. The use of a gamma distribution was dismissed due to constraints on available computing resources and the prohibitive run time that was required due to excessive required calls to evaluate the Γ function. Thus, the DSD shape, is modeled as a lognormal distribution with time varying parameters

ξ_t and ϕ_t^2. Conditional on ξ_t and ϕ_t^2 the D_ts are assumed independent yielding the joint distribution for D as

$$[D|\xi_t, \phi_t^2] = \prod_{t=1}^{T} \frac{1}{\sqrt{2\pi} d_t \phi_t} \exp\left\{ - \frac{(\ln d_t - \xi_t)^2}{2\phi_t^2} \right\}$$

where $\xi_t \sim N(\mu_\xi, \sigma_\xi^2)$; $\phi_t^2 \sim IG(\alpha_\phi, \beta_\phi)$. The hyperparameters μ_ξ, σ_ξ^2, α_ϕ, and β_ϕ are taken as fixed, with values that may vary for different rainfall events.

2.3. Data components

The hierarchical model incorporates rain gauge and ground radar observations, which are two of the most common data sources for measuring surface rainfall. Since, given a DSD, the instantaneous (true) rain rate at a given time and the derived (true) radar reflectivity can be calculated, we develop a measurement error model for both the rain gauge and radar observations.

2.3.1. Gauge observations, G. Rain gauges measure the amount of rainfall accumulating over a fixed time at a point in space. If the amount of rainfall measured at a point is measured without error, the gauge measurement, G, can be equated to the appropriate time integral of rain rate. The predominant error in gauge measurements is a systematic bias induced by winds, which often results in an underestimate of the surface rainfall. Thus, define a measurement error model for gauges as a function of the true rain rate and wind speed (taking other systematic biases as negligible). Define a gauge observation as

$$G_t = \left\{ c(w_t) \int_{t-\Delta t}^{t} R_s ds \right\} + \epsilon_{G_t}$$

where R_t is the derived instantaneous rain rate at time t; $c(w_t)$ are gauge type-specific coefficients ($c \leq 1$) primarily based on wind speed, w; and ϵ_{G_t} are independent measurement errors modeled as $N(0, \sigma_G^2)$; $\sigma_G^2 \sim IG(\mu_G, \tau_G^2)$.

Given a DSD, the instantaneous rain rate for a given time t can be derived based on meteorological principles as a function of the third power of drop diameter [2]. The rain rate R, for a given t, is estimated as

$$R_t = c_R \frac{\pi}{6} \int_0^\infty D^3 V(D) N(D)_t dD$$

where D is the drop diameter in mm; $V(D)$ is a deterministic function for the terminal velocity for drops with diameter D; $N(D)_t$ is the DSD giving the number of drops per mm diameter D per m^3 at time t; and c_R is an additional constant that accounts for units.

Given the model for $N(D_i)$ (equation 4), summing over all ΔD, and substituting for $N(D)_t$ yields

$$R_t = c_R \frac{\pi}{6} \left[\int_0^\infty D^3 V(D) [N_t f(D)_t] dD \right] \tag{5}$$

where N_t is the total drop concentration at time t and $f(D)_t$ is the pdf for D_t defined through $P(D)_t$.

Assuming this independent Gaussian measurement error model, and conditional on rain rate as derived from a DSD and parameter $\theta_G = \{\sigma_G^2\}$, the elements of G are independent and in each case $G_t \sim N(m_{G_t}, \sigma_G^2)$, where m_{G_t} is the integral, over the time interval between gauge measurements, of the instantaneous rain rate defined above.

2.3.2. Ground radar observations, Z. Reflectivity values are related to the intensity of reflected power that is bounced off of raindrops in the atmosphere. Reflectivity will vary depending on the number of and the size of the drops. The equivalent radar reflectivity observations from ground radar, Z, are a function of the average returned power and range of the radar scan [2]. If there are no measurement errors, Z will conform to the reflectivity derived the DSD. To accommodate various factors that can lead to errors when estimating Z, we construct a simple measurement error model for equivalent radar reflectivity as a function of the derived reflectivity and random errors; $Z_t = Z_{D_t} + \epsilon_{Z_t}$, where Z_{D_t} is the meteorologically derived reflectivity at time t and ϵ_{Z_t} are independent measurement errors modeled as $N(0, \sigma_Z^2)$; $\sigma_Z^2 \sim IG(\mu_Z, \tau_Z^2)$.

The treatment of derived reflectivity, Z_D, is similar to that for rain rate. Given a DSD, Z_D can be derived as the sixth moment of the DSD [2]. Derived reflectivity at time t is estimated as:

$$Z_{D_t} = c_Z \int_0^\infty D^6 N(D)_t dD$$

where D is the drop diameter in mm; $N(D)_t$ is the current drop spectra giving the number of drops per mm per m^3, at time t, with diameter D; and c_Z is a constant related to the sample volume and equals one for a one-m^3 sample.

As for gauge observations, summing over all ΔD and substituting $N_t \times P(D)_t$ for $N(D)_t$ yields:

$$Z_{D_t} = c_Z \left[\int_0^\infty D^6 [N_t f(D)_t] dD \right]$$

where variables are as previously defined. In practice, Z_{D_t} is estimated as a sum over binned drop sizes. Assuming this independent Gaussian measurement error model and conditional on Z_D derived from the DSD and parameter $\theta_Z = \{\sigma_Z^2\}$, the elements of Z are independent and in each case $Z_t \sim N(Z_{D_t}, \sigma_Z^2)$.

2.4. Model summary

We use the components presented above to construct the hierarchical model formulation. At the highest level we assume, based on a set of conditional independence assumptions, that given a DSD ($N(D)$), the gauge (G) and radar (Z) observations are conditionally independent with the following factorization:

$$\begin{aligned}[N(D), G, Z] &= [N(D)][G, Z|N(D)] \\ &= [N(D)][G|N(D)][Z|N(D)].\end{aligned}$$

TABLE 1. Hierarchical Model Summary

Variables	Conditional Probabilities for Model Components[a]
Observational Data	$[G_t \mid N(D)_t, \theta_G]$[b] $[Z_t \mid N(D)_t, \theta_Z]$
Hidden Process	$[TN_t \mid \mu_{N_t}, \theta_N]$[c] $[D_t \mid \theta_D]$
Temporal Dynamics for mean of TN	$[\mu_{N_t} \mid \theta_\mu]$
Model Parameters	$[\theta_G, \theta_Z, \theta_N, \theta_D, \theta_\mu \mid \theta_H] = [\theta_G \mid \theta_{H_G}][\theta_Z \mid \theta_{H_Z}][\theta_N \mid \theta_{H_N}][\theta_\mu \mid \theta_{H_\mu}][\theta_D \mid \theta_{H_D}]$ where θ_H is the collection of hyperparameters and $\theta_G = \{\sigma_G^2\} = [\sigma_G^2 \mid \mu_G; \tau_G^2]$ $\theta_Z = \{\sigma_Z^2\} = [\sigma_Z^2 \mid \mu_Z; \tau_Z^2]$ $\theta_N = \{a_0, a_1, b_1, \sigma_0^2\} = [a_0 \mid \alpha_0; \tau_{\alpha_0}^2][a_1 \mid \alpha_1; \tau_{\alpha_1}^2][b_1 \mid \beta_1; \tau_{\beta_1}^2][\sigma_0^2]$ $\theta_\mu = \{a_N, \sigma_{\eta t}^2\} = [a_N \mid \alpha_N; \tau_{\alpha N}^2][\sigma_{\eta t}^2 \mid \mu_\eta; \tau_\eta^2]$ $\theta_D = \{\lambda_t\} = [\lambda_t \mid \mu_\lambda; \sigma_\lambda^2]$
Hyper-parameters	$\theta_H = \{\mu_G, \tau_G^2, \mu_Z, \tau_Z^2, \alpha_0, \alpha_1, \beta_1, \tau_{\alpha_0}^2, \tau_{\alpha_1}^2, \tau_{\beta_1}^2, \alpha_N, \tau_{\alpha_N}^2, \mu_\eta, \tau_\eta^2, \mu_\lambda, \sigma_\lambda^2\}$

[a] The bracket notation, $[\cdot|\cdot]$, is used as a shorthand for denoting conditional distributions.
[b] Recall that $N(D)_t$ is general notation for $N_t \times P(D)_t$.
[c] Recall that $TN = \ln(N_t)$.

The components $[G|N(D)]$ and $[Z|N(D)]$ represent likelihoods based on the gauge and radar data, respectively. $[N(D)]$, represents the prior probability for the DSD process. The conditional probabilities for the full model, including the parameters and hyperparameters, are summarized in Table 1.

Estimated model parameters are obtained by using a Markov Chain Monte Carlo (MCMC) implementation to sample from the posterior distribution of the hierarchical model. The appropriate Markov chain is constructed using a single-component Metropolis-Hastings algorithm based on the hierarchical structure of the model and on the conditional probabilities associated with the various model components.

Starting values, hyperparameter values, the length of burn-in, and the number of iterations must be determined for each application. Starting values are random, based on the prior distributions of the parameters; the distributions can be altered by the choice of the fixed hyperparameters. The required number of iterations for a given application depends on the length of the burn-in, autocorrelation of the parameters, and the number of samples needed for parameter estimation. Convergence is monitored using visual analysis of plotted parameters in two ways: 1) by performing multiple runs using different random starts; and 2) running multiple chains using the same starting values.

3. Application to Eureka rainfall

We applied the model developed in the previous sections to a site in Eureka, California. The Eureka site has three sources of available rainfall data: distrometer, rain gauge, and ground radar data. The distrometer data were collected using a JWD instrument. The distrometer was located near the Eureka radar site at $40°48'N/124°09'W$. Each data file contained one hours worth of raw bin counts collected at one-minute intervals, where raw counts are the actual number of drops passing through the instrument in each one-minute interval. The raw bin counts were converted to one-minute DSD spectra counts, N_{D_t} ($\mathrm{m}^{-3}\mathrm{mm}^{-1}$), by adjusting raw counts for the instrument collection area, time, and terminal fall speed, and drop diameter.

A set of processed 10-minute distrometer data was also available. The 10-minute data are drop spectra that had been averaged over 10 contiguous one-minute periods. The observed 10-minute drop spectra were used to calculate rain rates and reflectivity values that were used in place of gauge and radar observations in model verification runs.

Hourly accumulation data (TD-3240) were obtained from the National Climatic Data Center (NCDC) on-line service for a gauging station located close to the radar and distrometer instruments at $40°49'N/124°10'W$. Data were collected using Fischer-Porter precipitation gauges with automated readouts. Precipitation data from the NCDC database are quality checked and edited, as necessary, by an automated and a manual edit [7]. The gauge data were used in conjunction with the distrometer data to identify rainfall events.

Level II WSR-88D radar scans were ordered from NCDC for the Eureka radar site located at ($40°29'54''N/124°17'31''W$), for periods coinciding with rainfall events identified by gauge and distrometer data. There was a radar scan approximately every six minutes for several elevation increments. The spherical coordinate Level II data were interpolated to a 2 km $\times$ 2 km Cartesian coordinate system of equivalent reflectivity (dBZ) values using standard programs for NEXRAD radar. The interpolated data were only available for one of the identified events; the availability of radar data defined, and limited, the time period over which the model was implemented.

Appropriate reflectivity values needed to be extracted from the 2 km $\times$ 2 km gridded values to correspond with the fixed spatial location for the model runs. Reflectivity in the time domain was extracted from the spatial data by first identifying the closest pixel located vertically above the distrometer and nearby rain gauge. Given the uncertainty in the vertical flow of rainfall above the surface of the earth to a point on the ground, an equivalent reflectivity was calculated as the aerial average of first and second order neighbors using the identified pixel and the eight neighboring pixels. Issues of beam blockage along the path of the radar to the location of the distrometer were considered to be negligible due to the gradually increasing nature of the coastal terrain of Northern California. Therefore, the reflectivity values were extracted from the lowest radar beam (tilt elevation of 0.5°).

3.1. Hyperparameters, initial values, constraints

We used exploratory analysis of distrometer data from five February rainfall events to come up with informative means and variances for the fixed hyperparameters. The constant hyperparameter values are summarized in Table 2 with model results.

In addition to the hyperparameter values, several $t = 0$ starting values were specified as follows: $\epsilon_0 = 0$, $\sigma^2_{\epsilon_t} = 0.001$, and $\mu_{N_0} = 3.0$. We based initial conditions for the remaining parameters on random draws from the appropriate prior distributions.

We introduced cutoffs for constraints due to rainfall dynamics, instrumentation and statistical considerations. Constraints based on rainfall dynamics included upper bounds for three parameters:

1. We limited the total logged number of drops in a one-minute interval to 6.0, about 2.5% above the maximum value observed from the distrometer data. This corresponds to a total of $\exp(6.0) \approx 403$ raw count drops when summed across the 20 drop diameter bins. This cutoff was applied to the parameters TN_t and μ_{N_t}.
2. We limited the upper bound for drop diameter size such that $P(D \geq 5 \text{ mm}) \approx 0.000001$. This limitation was imposed since the rain event of interest represented light rainfall. The distrometer data had no raw counts beyond a mean drop diameter of 3.15 mm.

Proposals for parameter updating were drawn from uniform distributions centered at the current parameter values. This implementation made it possible to draw negative-valued proposals, depending on the current value of the parameter. Constraints based on statistical considerations thus included lower bounds on the variance parameters (> 0) and $\lambda_t > 0$. Required model constraints were also placed on the three GARCH coefficients.

3.2. Estimating m_G and Z_D

At each time period for which there were gauge or radar observations, the data components of the model required computation of the two quantities m_{G_t} and Z_{D_t}, the true equivalent gauge value and the true derived reflectivity, respectively. Both of these values are deterministic quantities based on the current state of the parameters that define the hidden DSD process.

We calculate the true equivalent gauge value as the sum over the previous 60 minutes of each of the instantaneous rain rates in mm/min derived from the state of the DSD for each point in time. The instantaneous rain rates defined by equation 5 are approximated by summing over the 20 drop diameter bins that would typically be observed using a JWD instrument as follows:

$$R_t = \frac{3.6\pi}{6000} \sum_{i=1}^{20} \left[\frac{\exp\{TN_t\} \times P(D_i)_t}{\Delta D_i} \right] D_i^3 V(D_i) \Delta D_i$$

where $P(D_i)_t$ is based on the cumulative distribution of a lognormal distribution with parameters ξ_t and ϕ_t^2. The terminal fall speed, $V(D_i)$, was calculated as

$3.78D_i^{0.67}$ [22], where D_i is the mean drop diameter for bin i. Then,

$$m_{G_t} = \sum_{i=t-60+1}^{t} \frac{R_i}{60}$$

True derived reflectivity is calculated as an average, over the previous 10 minutes, of the sixth moment of the DSD. As with the equivalent gauge calculation, we approximate derived reflectivity by summing over the 20 drop diameter bins using

$$Z_t = \sum_{i=1}^{20} \left[\frac{\exp\{TN_t\} \times P(D_i)_t}{\Delta D_i} \right] D_i^6 \Delta D_i.$$

Then,

$$Z_{D_t} = \frac{\sum_{i=t-10+1}^{t} Z_i}{60}$$

where Z_{D_t}, in units of $\mathrm{mm}^6\mathrm{m}^{-3}$, are converted to dBZ as $10\log_{10}(Z_{D_t})$.

3.3. Tuning, thinning, and convergence

Tuning the chain was optimized by making many small runs (5000 to 15000 iterations in length) and monitoring the acceptance rates for each of the parameters. Adjustments to the acceptance rates were made by either increasing or decreasing the upper and lower bounds on the uniformly distributed parameter proposals. After much experimentation, bounds on the uniform proposals were adjusted to yield acceptance rates ranging from about 0.42 on the low end to about 0.85 on the high end. Most parameters had acceptance rates in the range of 0.60 - 0.65. While these values exceed the commonly recommended values of 0.15 to 0.5 for updating multidimensional components [14], we found that mixing and convergence performed better at these higher levels.

We used thinning to obtain roughly independent samples from the joint posterior distribution. We used *gibbsit* [13] to get estimates of the ideal spacing between iterations and the number of burn-in iterations, as well as the number of iterations needed for a desired precision. In most cases, a thinning value of 150 was more than sufficient to obtain approximately independent samples although, on occasion, a parameter thinning value would hit 300 or so.

We monitored convergence by visual inspection of plotted output. The number of burn-in iterations recommended by *gibbsit* was generally low, and always < 5000. Convergence was also assessed by looking at the path of several runs using both the same and different starting values.

3.4. Results

Results for two types of model runs are summarized in the following sections. Since there is no way to specifically validate the model in terms of comparing outputs to "true" values, model verification was conducted by assessing the ability of the model to realistically reproduce targeted signals and integrated parameters associated with surface rainfall. For model verification we replaced NEXRAD radar observations with derived reflectivity calculated from 10-minute distrometer data. Rain gauge data were omitted for this assessment. The number of inputs were maximized by using a 10-minute time step, thus insuring a data value at each time step. The full model runs use no distrometer data, but rather the gauge and radar data are combined with the model for the hidden DSD process to calibrate the parameter estimates with all available information.

3.4.1. Model verification. The verification run includes 49 derived reflectivity observations computed from the one set of distrometer data. The MCMC run was ended after 280,000 iterations following 100,000 burn-in iterations. Based on graphical output, all parameters converged within the 100,000 burn-in period. Results for a typical model verification run showing time independent parameter estimates and time varying parameter estimates (using the 800 MCMC samples generated after applying a thinning value of 350) are shown in Tables 2 and 3, respectively. Further, using the posterior parameter estimates for the hidden process we can compare calculated integrated rainfall parameters with values obtained from the 10-minute distrometer data. Figures 4 and 5 show the comparisons for rain rate and reflectivity for some representative samples from the model run. Note that although we compare the model output with quantities derived from distrometer

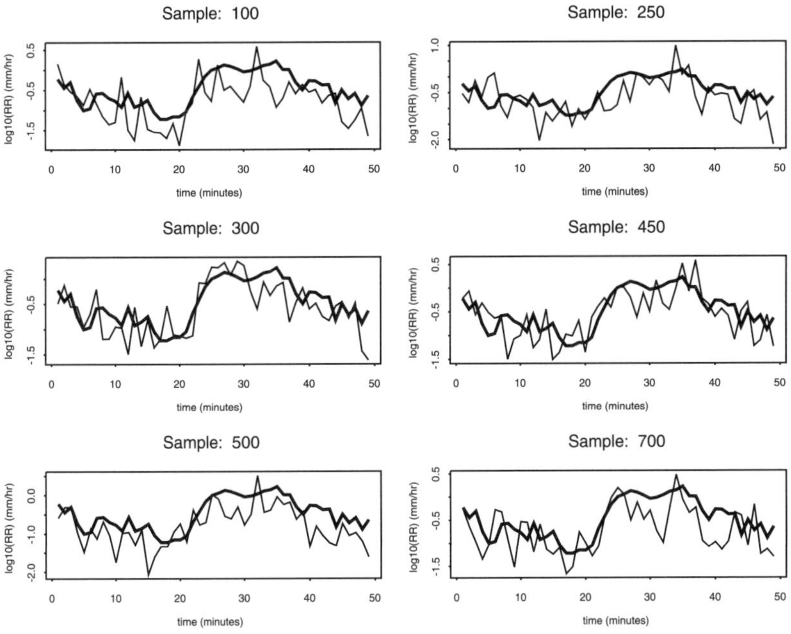

FIGURE 4. Time series posterior $\log_{10}$ rain rate estimates calculated for samples from the validation run (light lines) with rain rate estimates from the distrometer data (heavy lines).

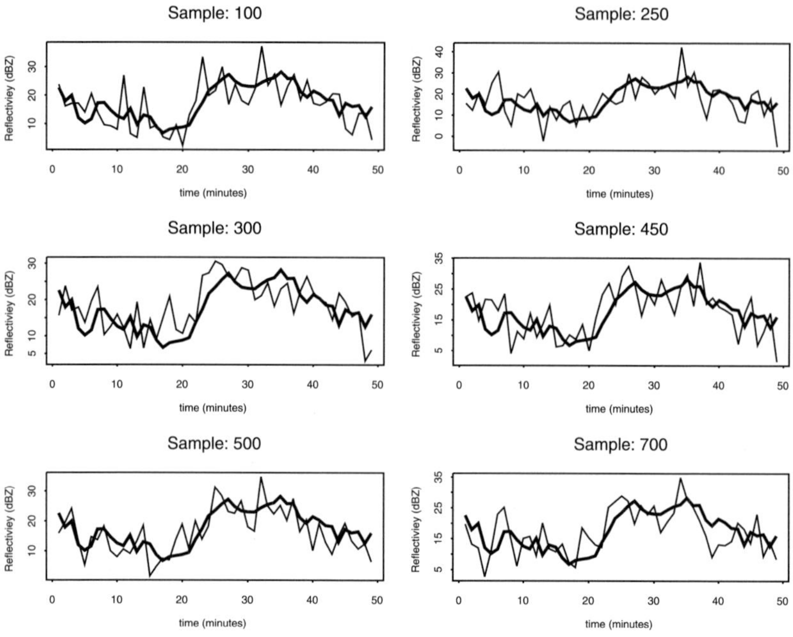

FIGURE 5. Time series posterior reflectivity estimates calculated for samples from the validation run (light lines) with reflectivity estimates from the distrometer data (heavy lines).

data, we do not consider the distrometer data as ground truth. Distrometer data are subject to measurement errors and our distrometer data set represents a single realization of the process.

The posterior means of the three GARCH coefficients are all lower than those specified for the prior distributions. The variability for the coefficients is slightly less than that for the priors. The mean of the AR(1) coefficient is slightly higher than that specified in the prior. The variance parameter for the AR(1) errors, σ_η^2, is considerably higher than that specified in the prior, as is the measurement error variance for radar (σ_Z^2). The mean values for TN and μ_N are about the same and posterior estimates of TN are close to expected counts although the model tends to underestimate the counts. In general, the DSD distribution parameters ξ_t and ϕ_t^2 are more variable than estimates based on the distrometer data. On average, the posterior estimates of ξ_t are similar to the values estimated from the distrometer data. The model tends to slightly underestimate ϕ_t^2.

The posterior estimates for both integrated parameters (rain rate and reflectivity) are more variable than values calculated from the distrometer data realization. This is expected since these estimates include the distrometer data realization with some uncertainty. Aside from the extra variability, the model estimates are a good match with the integrated quantities calculated from distrometer observations. The model does a good job of capturing the overall means and the signals present in the time series.

3.4.2. Full model run. The full model run includes 9 hourly rain gauge observations and 92 effective reflectivity observations from ground radar. The MCMC run was ended after 245,000 iterations following a burn-in period of 135,000 iterations. After applying a thinning value of 350, there were 700 MCMC samples for posterior estimates. Convergence for the full model run is much slower than that for the verification run, taking up to about 60,000 iterations for a few parameters. With the exception of the AR(1) variance term, σ_η^2, all parameters appear to converge by 60,000 iterations. The σ_η^2 variance hardly moves until 60,000 iterations, at which point the value starts bouncing around and then settles down at about 0.5, followed by more intermittent bouncing. This pattern could be due to slow mixing or identifiability issues.

Results for a typical full model run showing time independent parameter estimates and time varying parameter estimates are shown in Tables 2 and 3, respectively, along side the verification run results. Time series results for integrated

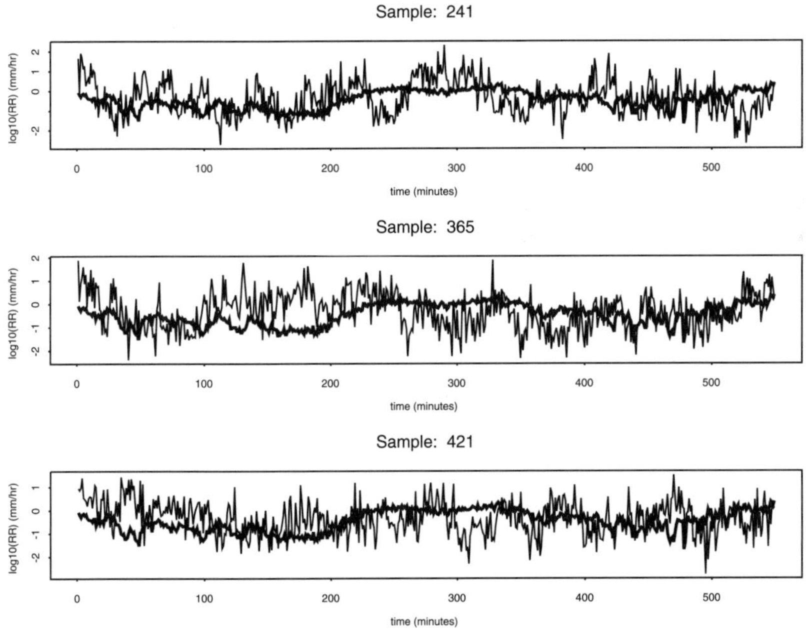

FIGURE 6. Time series plots of full model run posterior $\log_{10}$ rain rate estimates calculated for selected samples (light lines) with logged rain rate values based on distrometer data (heavy lines).

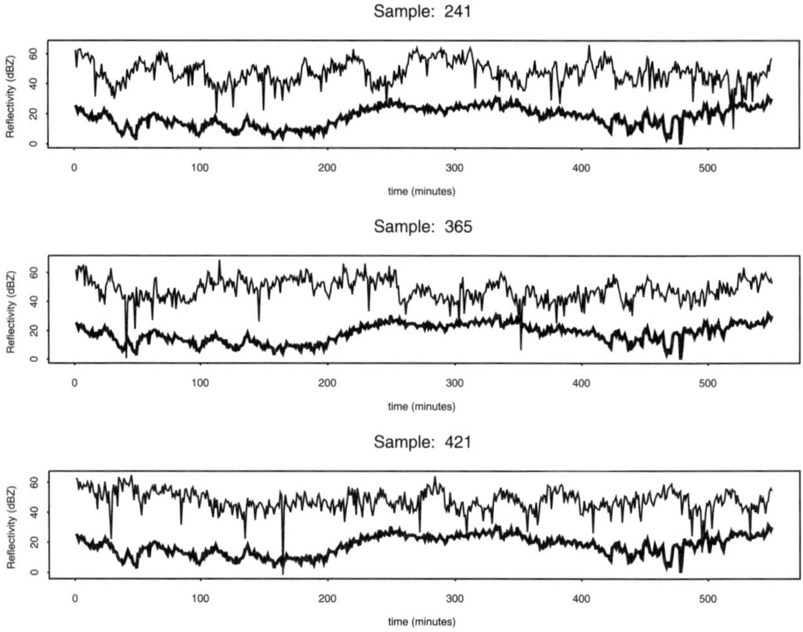

FIGURE 7. Time series full model run posterior reflectivity estimates calculated for selected samples (light lines) with reflectivity values based on distrometer data (heavy lines).

parameters calculated from the estimated parameters for the hidden DSD process are shown in Figures 6 through 9.

Except for a_0, the posterior estimates for the GARCH parameters are similar to those from the verification run. The generally larger standard errors reflect the greater uncertainty in the one-minute model, which does not have data at every time step. The mean error terms, ϵ_t, are not significantly different from zero based on 95% confidence intervals for the means. The posterior distributions are relatively symmetric and approximately normal, which is in line with the GARCH model for ϵ_t. The within time posterior means for TN_t and μ_{N_t} are essentially the same. The posterior distributions for TN_t are nearly symmetrical and appear to be approximately normally distributed. The representative ξ_t distributions all have means close to the mean for the prior distribution. The posterior distributions are fairly symmetric; values for ξ_t are variable, ranging from -1.82 to 1.01 and do contain the observed estimates based on the one realization of distrometer data. The representative ϕ_t^2 distributions all have means close to 0.18, the mean of the prior distribution. The posterior distributions are right-skewed; values for ϕ_t^2 are variable, ranging from 0.05 to 1.03 and do contain the observed estimates calculated from the distrometer data.

TABLE 2. MCMC Summary statistics for time independent parameters.

Model Parameter	Prior Mean	Prior SD	MCMC Posterior Mean (SE) Verification Run	Full Run
GARCH Coefficients a_0	0.023	0.0062	0.0080 (0.00012)	0.049 (0.0012)
a_1	0.242	0.1007	0.200 (0.00082)	0.164 (0.0020)
b_1	0.742	0.1119	0.677 (0.00099)	0.684 (0.0031)
AR(1) Process a_N	0.95	0.04	0.981 (0.00018)	0.990 (0.0002)
σ_η^2	0.005	0.0007	0.418 (0.0023)	0.478 (0.0033)
Measurement Error σ_G^2	5.0	1.02	—	5.82 (0.044)
σ_Z^2	2.86	1.429	31.68 (0.132)	104.01 (0.865)

The in-series variability for TN_t, ξ_t, and ϕ_t^2 is greater than that for the observed distrometer data. Except for ϕ_t^2, the time series values estimated from the distrometer data are nearly centered in the posterior model results. As with the verification run, the full model underestimates the DSD shape parameter, ϕ_t^2.

From the posterior parameter estimates for the hidden DSD process, we can estimate the integrated values for rain rate and derived reflectivity. The time series of calculated rain rates show higher variability than that produced from the one realization of distrometer data. The large variability makes it difficult to compare the signals, though there appears to be some amount of agreement in the gross signal features for rain rate (Figure 6).

The time series plots of derived reflectivity also show higher variability than that produced from one realization of distrometer data, though not to the same extent as for rain rate (Figure 7). The model output consistently overestimates distrometer derived values. While the cause of the bias is unknown, it's possible that the observed bias indicates poor calibration of the radar. A mean 1 normalized plot for reflectivity (Figure 8) removes the bias, but basically shows that the fitted reflectivity estimates are stationary. This is contrary to what is observed for reflectivity values derived from the distrometer data, and may reflect a lack of

TABLE 3. MCMC summary statistics for time varying parameters for three time points. For the verification run, times are $t = 9, 24, 39$. For the full run, $t = 90, 235, 391$.

	Verification Run		Full Model Run	
Model Parameter	**Observed Value**	**Posterior Mean (SE)**	**Observed Value**	**Posterior Mean (SE)**
Total # Drops				
TN_{t1}	6.41	6.00 (0.026)	4.37	5.00 (0.056)
TN_{t2}	7.59	6.36 (0.029)	5.90	5.28 (0.055)
TN_{t3}	6.64	6.29 (0.028)	5.08	5.04 (0.051)
Mean # Drops				
$\mu_{N_{t1}}$	—	6.00 (0.026)	—	5.01 (0.052)
$\mu_{N_{t2}}$	—	6.34 (0.028)	—	5.25 (0.049)
$\mu_{N_{t3}}$	—	6.30 (0.026)	—	5.05 (0.048)
Errors for TN_t				
ϵ_{t1}	—	−0.0001 (0.008)	—	−0.017 (0.0214)
ϵ_{t2}	—	0.005 (0.008)	—	−0.016 (0.0221)
ϵ_{t3}	—	0.012 (0.010)	—	−0.023 (0.0217)
Error Variance				
$\sigma^2_{\epsilon_{t1}}$	—	0.050 (0.0025)	—	0.320 (0.0077)
$\sigma^2_{\epsilon_{t2}}$	—	0.071 (0.0036)	—	0.329 (0.0084)
$\sigma^2_{\epsilon_{t3}}$	—	0.080 (0.0040)	—	0.319 (0.0075)
Lognormal Scale Parameter Prior mean: −0.50; SD: 0.400				
ξ_{t1}	−0.54[a]	−0.68 (0.009)	−0.56[b]	−0.49 (0.015)
ξ_{t2}	−0.59	−0.46 (0.009)	−0.61	−0.54 (0.015)
ξ_{t3}	−0.63	−0.49 (0.009)	−0.63	−0.53 (0.015)
Lognormal Shape Parameter Prior mean: 0.182; SD: 0.0857				
ϕ_{t1}	0.316[c]	0.155 (0.0018)	0.26[d]	0.18 (0.003)
ϕ_{t2}	0.338	0.176 (0.0023)	0.34	0.19 (0.003)
ϕ_{t3}	0.371	0.166 (0.0023)	0.37	0.19 (0.004)

[a] Estimated from 10-minute distrometer data
[b] Estimated from 1-minute distrometer data
[c] Estimated from 10-minute distrometer data
[d] Estimated from 1-minute distrometer data

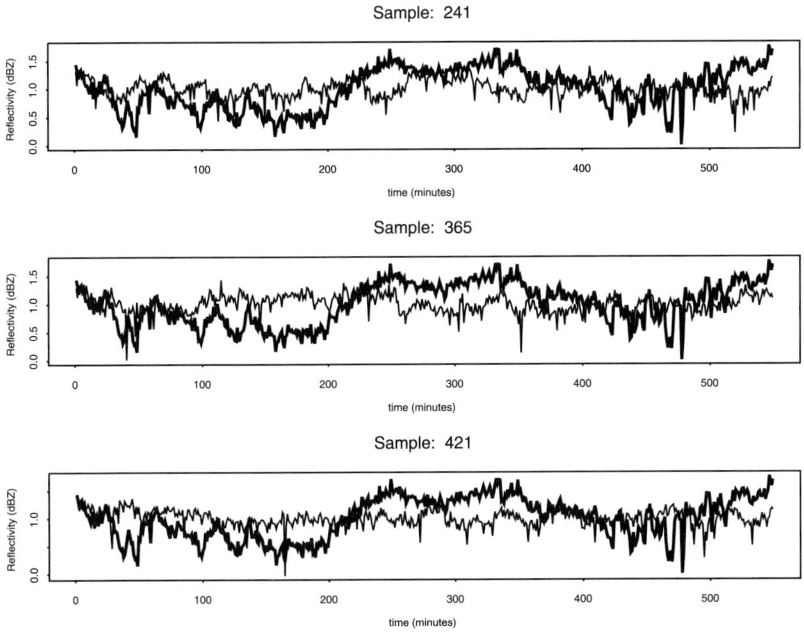

FIGURE 8. Time series full model run posterior reflectivity (normalized to mean 1) estimates calculated for selected samples (light lines) with reflectivity values based on distrometer data (heavy lines).

structure in parameters used to model the DSD shape. In general, a comparison of the signals between any two of the time series is difficult since we are trying to compare multiple realizations based on model output with one realization based on a set of distrometer data.

An alternate visualization of the calculated rain rate parameter based on model output displays histograms for rain rates from 12 points along the time series (Figure 9). Corresponding values calculated from the distrometer data are also shown. The individual histograms contain the distrometer-based rain rates, although the distrometer-based values are in the lower ends of the distributions for most time points.

4. Discussion

The model results for the Eureka rainfall are promising in that the use of hierarchical Bayes methods make it possible to integrate information from multiple data sources for purposes of estimating rain rate and related parameters. While we are unable to validate the posterior estimates of TN_t, ξ_t, and ϕ_t^2, the main parameters

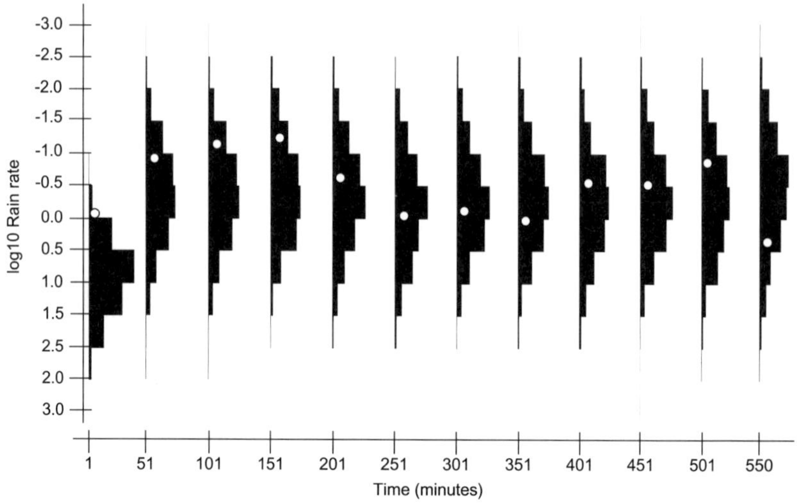

FIGURE 9. Frequency histograms of the full model run posterior distribution of $\log_{10}$ rain rate estimates for selected time points within the 550 minute modeled event. All histograms are scaled with equivalent y-axes. The white circles correspond to the values of rain rate calculated from one realization of distrometer data.

for the unobserved DSD process, we can compare their values with the one available distrometer realization. For TN_t, many of the MCMC realizations, though not all, produce time series signals similar to that observed for the distrometer realization with model output that is always more variable. This higher variability is expected, and is due in part to the higher uncertainty that is introduced when observations from two different data sources (gauge and radar) are considered.

The DSD scale and shape parameters capture the gross features of the drop spectra, but the time series for ξ_t and ϕ_t^2 are many times more variable than those produced from distrometer estimates. A comparison of autocorrelations for these parameters based on distrometer data versus model output suggests that there is more structure in the time evolution of the DSD shape than what is accounted for in the current model. Estimates of the parameters for the lognormal shape distribution based on distrometer data are highly autocorrelated out to about 25 minutes. The estimates based on the hierarchical model are essentially uncorrelated, corresponding with the implemented model parametrization for DSDs.

Ultimately the estimated DSD parameters are used to produce estimates of the integrated quantities rain rate and derived reflectivity. The verification runs do a good job of capturing the features of these quantities; the full model runs capture the gross features, but again produce higher variable time series output and overestimates of derived reflectivity. Aside from poor calibration of the radar, the bias seen in derived reflectivity could be due to a number of other factors;

most notably, differences may be related to sampling errors associated with distrometer data. Derived reflectivity is more adversely affected by small differences in drop-size distributions since Z_D is a function of D^6 while R is a function of D^3. Either overestimation of larger drops by the model, or under recording of larger drops by the distrometer could lead to these observed differences. Smith et al. [23] showed a resulting skewness in the sampling distributions for DSD spectra when the samples are small (size in raw counts). This skewness tends to result in an underestimate of the integrated quantities. Verification runs produced better consistency between model derived reflectivity and distrometer derived values; use of 10-minute distrometer data helps compensate for distrometer sampling errors that arise due to the small sample volume collected in a one-minute interval.

There is no measurable truth when it comes to determining rain rates at the earth's surface. For each instrument deployed in the field, there are claims and beliefs as to how well each performs their respective tasks, in light of uncertainty. Instead of using rain gauge data as "ground truth", the posterior model distributions capture multiple sources of uncertainty that reflect uncertainty associated with one or more sources such as: the inability to directly measure quantities of interest; measurement and sampling errors associated with the multiple instruments; sparse gauge data; and, incorrect or inadequate assumptions and specifications related to equations used to represent the underlying physical processes.

As a proof of concept, we were able to incorporate Eureka rainfall data into a hierarchical Bayes model to generate estimates for the modeled parameters. Most of the estimated parameter values seem to track to some extent the information contained in the single distrometer realization. Using existing model runs as baselines we can further characterize the observed uncertainties by making stepwise modifications to model stages or by incorporating additional data. In the short term, a temporal dynamic stage should be developed for the evolution of DSD through time. Also, we should run the model for other rainfall events. The particular Eureka event consisted of relatively light rainfall. Determining how well the model performs for heavier rainfall events will provide some idea as to the portability/usability of this model under a variety of rainfall scenarios. In the longer term, some ideas for expanding and using this modeling approach include:

1. If applicable, incorporate wind corrections for gauge observations in future model runs. Gauge catchments are inherently affected by winds; a wind effect would be one component of the overall uncertainty observed in the model output.
2. Incorporate a vertical correction model component for radar reflectivity observations, based on an underlying process model, to compensate for the fact that ground-based radar provides information about precipitation above the earth's surface. Include data from vertical pointing radar to augment the vertical correction model component.
3. Expand the time only model to a space-time model. Such a model could provide a tool that allows for the creation of spatial rainfall maps with un-

certainty estimates, while also providing the opportunity for more expanded analysis of the component uncertainties.

4. Use model output to estimate better Z-R relationships dynamically.
5. Use model output to assess the value added by using other types of precipitation measurements (or by removing some instruments).

Acknowledgment

We would like to thank Dr. Sandra Yuter, who provided the distrometer and NEXRAD radar data sets, for her help with data pre-processing and her invaluable insights related to instrumentation, meteorological properties of precipitation, and relationships between the two.

References

[1] C. Alexander. Volatility and correlation: measurement, models, and applications. In C. Alexander, editor, *Risk Management and Analysis. Vol.* 1. John Wiley Sons Ltd, 1998.

[2] Louis J. Battan. *Radar observation of the atmosphere.* University Chicago Press, Chicago, revised edition, 1973.

[3] Thomas L. Bell. A space-time stochastic model of rainfall for satellite remote-sensing studies. *Journal of Geophysical Research*, 92(D8):9631–9643, 1987.

[4] L. Mark Berliner. Hierarchical Bayesian modeling in the environmental sciences. *Allgemeines Statistisches Archiv*, 84:141–153, 2000.

[5] L. Mark Berliner, Richard A. Levine, and Dennis J. Shea. Bayesian climate change assessment. *Journal of Climate*, 13(21):3805–3820, 2000.

[6] D.R. Cox and Valerie Isham. A simple spatial-temporal model of rainfall. *Proceedings of the Royal Society of London*, A415:317–328, 1988.

[7] Greg Hammer and Pete Steurer. Data documentation for hourly precipitation data td-3240. Technical report, National Climatic Data Center, February 11, 1998.

[8] E. Hanna. How effective are tipping-bucket raingauges? A review. *Weather*, 50(10): 336–342, 1995.

[9] B. Hrafnkelsson. Hierarchical modeling of count data with application to nuclear fall-out. *Journal of Environmental and Ecological Statistics*, 10:179–200, 2003.

[10] A.J. Illingworth and M.T. Blackman. The need to represent raindrop size spectra as normalized gamma distributions for the interpretation of polarized radar observations. *Journal of Applied Meteorology*, 41:286–297, 2002.

[11] H. Jiang, M. Sano, and M. Sekine. Weibull raindrop-size distribution and its application to rain attenuation. *IEE Proceedings-Microwaves, Antennas and Propagation*, 144(3):197–200, 1997.

[12] E.L. Neff. How much rain does a rain gage gage? *Journal of Hydrology*, 35:213–220, 1977.

[13] A. Raftery and S. Lewis. Implementing mcmc. In Gilks W., S. Richardson, and D. Spiegelhalter, editors, *Markov Chain Monte Carlo in Practice.* Chapman Hall, London, 1996.

[14] G.O. Roberts. Markov chain concepts related to sampling algorithms. In Gilks W., S. Richardson, and D. Spiegelhalter, editors, *Markov Chain Monte Carlo in Practice.* Chapman Hall, London, 1996.

[15] Ignacio Rodriguez-Iturbe, D.R. Cox, and V. Isham. A point process model for rainfall: further developments. *Proceedings of the Royal Society London*, A417:283–298, 1988.

[16] Ignacio Rodriguez-Iturbe and Peter S. Eagleson. Mathematical models of rainstorm events in space and time. *Water Resources Research*, 23(1):181–190, 1987.

[17] J.A. Royle, L.M. Berliner, C.K. Wikle, and R. Milliff. A hierarchical spatial model for constructing wind fields from scatterometer data in the Labrador sea. In *Case Studies in Bayesian Statistics IV.* Springer-Verlag, 1998.

[18] Henri Sauvageot and Jean-Pierre Lacaux. The shape of averaged drop size distributions. *Journal of the Atmospheric Sciences*, 52(8):1070–1083, 1995.

[19] M. Sekine, S. Ishii, I. Hwang, and S. Sayama. Weibull raindrop-size distribution and its application to rain attenuation from 30 ghz to 1000 ghz. *International Journal of Infrared Millimeter Waves*, 28:383–392, 2007.

[20] Matsuo Sekine and Goran Lind. Rain attenuation of centimeter, millimeter and submillimeter radio waves. In *Proceedings of the* 12*th European Microwave conference*, pages 584–589, Helsinki, Finland, 1982.

[21] T. Skaugen, J.D. Creutin, and L. Gottschald. Reconstruction and frequency estimates of extreme daily areal precipitation. *Journal of Geophysical Research*, 101(D21):26287–25295, 1996.

[22] James A. Smith. Marked point process models of raindrop-size distributions. *Journal of Applied Meteorology*, 32:284–296, 1993.

[23] James A. Smith and Richard D. De Veaux. The temporal and spatial variability of rainfall power. *Environmetrics*, 3(1):29–53, 1992.

[24] R. Stern and R. Coe. A model fitting analysis of daily rainfall data. *Journal of the Royal Statistical Society A*, 147(Part 1); 1–34, 1984.

[25] J. Testud, E. Le Bouar, E. Obligis, and M. Ali-Mehenni. The rain profiling algorithm applied to polarimetric weather radar. *Journal of Atmospheric and Oceanic Technology*, 17:332–356, 2000.

[26] Carlton W. Ulbrich. Natural variations in the analytical form of the raindrop size distribution. *Journal of Climate and Applied Meteorology*, 22:1764–1775, 1983.

[27] Carlton W. Ulbrich and David Atlas. Rainfall microphysics and radar properties: analysis methods for drop size spectra. *Journal of Applied Meteorology*, 37:912–923, 1998.

[28] A. Waldvogel. The n0 jump of raindrop spectra. *Journal of the Atmospheric Sciences*, 31:1067–1078, 1974.

[29] Christopher K. Wikle. Hierarchical Bayesian space-time models. *Environmental and Ecological Statistics*, 5:117–154, 1998.

[30] Christopher K. Wikle, R.F. Milliff, D. Nychka, and M. Berliner. Spatiotemporal hierarchial Bayesian modeling: tropical ocean surface winds. *Journal of the American Statistical Association*, 96(454):382–397, 2001.

[31] Paul T. Willis. Functional fits to some observed drop size distributions and parameterization of rain. *Journal of Atmos. Sci.*, 41:1648–1661, 1984.

[32] Paul T. Willis and Paul Tattelman. Drop-size distributions associated with intense rainfall. *Journal of Applied Meteorology*, 28:3–15, 1989.

Tamre P. Cardoso and Peter Guttorp
Department of Statistics
Box 354322, University of Washington
Seattle, WA 98195-4322
e-mail: tamre@u.washington.edu

e-mail: peter@stat.washington.edu

Progress in Probability, Vol. 60, 211–227

A Lower Bound on the Disconnection Time of a Discrete Cylinder

Amir Dembo and Alain-Sol Sznitman

Abstract. We study the asymptotic behavior for large N of the disconnection time T_N of simple random walk on the discrete cylinder $(\mathbb{Z}/N\mathbb{Z})^d \times \mathbb{Z}$. When d is sufficiently large, we are able to substantially improve the lower bounds on T_N previously derived in [3], for $d \geq 2$. We show here that the laws of N^{2d}/T_N are tight.

Mathematics Subject Classification (2000). 60J10, 60K35, 82C41.

Keywords. Disconnection time, random walk, discrete cylinders.

1. Introduction

In this note we consider random walk on an infinite discrete cylinder having a base modelled on a d-dimensional discrete torus of side-length N. We investigate the asymptotic behavior for large N of the time needed by the walk to disconnect the cylinder, or in a more picturesque language, the problem of a "termite in a wooden beam", see [3], [6], for recent results on this question. Building up on recent progress concerning the presence of a well-defined giant component in the vacant set left by a random walk on a large discrete torus in high dimension at times that are small multiples of the number of sites of the torus, cf. [1], we are able to substantially sharpen the known lower bounds on the disconnection time when d is sufficiently large.

Before describing the results of this note, we present the model more precisely. For $N \geq 1$, we consider the discrete cylinder

$$E = (\mathbb{Z}/N\mathbb{Z})^d \times \mathbb{Z}\,, \tag{1.1}$$

endowed with its natural graph structure. A finite subset $S \subseteq E$ is said to *disconnect* E if for large M, $(\mathbb{Z}/N\mathbb{Z})^d \times [M, \infty)$ and $(\mathbb{Z}/N\mathbb{Z})^d \times (-\infty, -M]$ are contained in distinct connected components of $E \backslash S$.

We denote with P_x, $x \in E$, the canonical law on $E^{\mathbb{N}}$ of the simple random walk on E starting at x, and with $(X_n)_{n\geq 0}$ the canonical process. Our principal object of interest is the disconnection time:

$$T_N = \inf\{n \geq 0; X_{[0,n]} \text{ disconnects } E\}\,. \tag{1.2}$$

Under P_x, $x \in E$, the Markov chain $X_.$ is irreducible, recurrent, and it is plain that T_N is P_x-a.s. finite. Also T_N has the same distribution under all measures P_x, $x \in E$. Moreover it is known, cf. Theorem 1 of [3], Theorem 1.2 of [6] that for $d \geq 1$:

$$\lim_N P_0[N^{2d(1-\delta)} \leq T_N \leq N^{2d}(\log N)^{4+\epsilon}] = 1, \quad \text{for any } \delta > 0,\, \epsilon > 0\,. \tag{1.3}$$

The main object of this note is to sharpen the lower bound on T_N, (arguably the bound that requires a more delicate treatment in [3] and [6]), when d is sufficiently large. More precisely consider for $\nu \geq 1$, integer, see also (2.4),

$$q(\nu) = \text{the return probability to the origin of simple random walk on } \mathbb{Z}^\nu\,. \tag{1.4}$$

It is known that for large ν, $q(\nu) \sim (2\nu)^{-1}$, cf. (5.4) in [5]. Our main result is:

Theorem 1.1. *Assume that d is such that*

$$7\Big(\frac{2}{d+1} + \Big(1 - \frac{2}{d+1}\Big)\, q(d-1)\Big) < 1\,, \tag{1.5}$$

(*note that necessarily $d \geq 4$, and* (1.5) *holds for large d, see also Remark* 3.1), *then*

$$\lim_{\gamma\to 0}\, \liminf_N P_0[\gamma N^{2d} \leq T_N] = 1\,. \tag{1.6}$$

(*i.e., the laws of N^{2d}/T_N, $N \geq 2$, are tight*).

In fact we expect that the laws of $\frac{T_N}{N^{2d}}$ on $(0,\infty)$ are tight, when $d \geq 2$, but the present note does not contain any novel upper bound on T_N complementing (1.6), nor a treatment of the small values of d.

As previously mentioned we build up on some recent progress made in the study of the vacant set left at times of order N^{d+1} by a simple random walk on a $(d+1)$-dimensional discrete torus of side-length N, cf. [1]. It is shown there that when d is large enough and u chosen adequately small, the vacant set on the discrete torus left by the walk at time uN^d contains with overwhelming probability a well-characterized giant component. In the present work we, loosely speaking, benefit for large N from the presence in all blocks

$$C_j = (\mathbb{Z}/N\mathbb{Z})^d \times \Big(\Big[\Big(j - \frac{3}{4}\Big)N,\ \Big(j + \frac{3}{4}\Big)N\Big] \cap \mathbb{Z}\Big),\ j \in \mathbb{Z}\,, \tag{1.7}$$

of such a giant component left by the walk on E by time γN^{2d}, with a probability that can be made arbitrarily close to 1, by choosing γ suitably small. Due to their characterization, the giant components corresponding to neighboring blocks do meet and thus offer a route left free by the walk at time γN^{2d}, linking together "top" and "bottom" of E. This is in essence the argument we use when proving (1.6).

Let us mention that the strategy we employ strongly suggests that the disconnection of E takes place in a block C_j, which due to the presence of a sufficient number of excursions of the walk reaches criticality for the presence of a giant component. So far the study in [1] only pertains to small values of the factor u mentioned above, i.e., within the super-critical regime for the existence of a giant component in the vacant set. And the evidence of a critical and sub-critical regime, corresponding to the behavior when u gets close or beyond a threshold where all components in the vacant set are typically small, rests on simulations. So the above remark is of a conjectural nature and rather points out to possible avenues of research. Incidentally progress on some of these questions may come from the investigation in [8] of a translation invariant model of "random interlacements" on $\mathbb{Z}^{d+1}$, $d \geq 2$, which provides a microscopic picture of the trace left in the bulk by the walk on E for the type of time regimes we are interested in.

Let us now give more precise explanations on how we prove Theorem 1.1. This involves the following steps. We first analyze excursions of the walk corresponding to visits of the walk to a block of height $2N$, centered at level jN, thus containing C_j, and departures from a block with double height (centered at level jN as well). We show that when $d \geq 3$, with overwhelming probability for large N, C_j contains a wealth of segments along coordinate axes of length $c \log N$, which have not been visited by the walk up to the time of the $[uN^{d-1}]$th excursion described above, for a small enough u (see Proposition 2.1 for a precise statement).

We then derive in Theorem 2.2 an exponential estimate which is crucial in specifying the giant components in the various blocks C_j, and proving that giant components in overlapping blocks do meet. It shows that when $d \geq 4$, the probability that the walk by the time of the $[uN^{d-1}]$th excursion has visited all points of any given set A sitting in an affine-plane section of the block C_j, cf. (2.12) for the precise definition, decays exponentially in a uniform fashion with the cardinality of A, when N is large, if u is chosen small. Then a Peierl-type argument, related to the appearance of the constant 7 in (1.5), shows that when (1.5) holds, with overwhelming probability the segments in C_j of length $c_0 \log N$ not visited by a time n prior to the $[u_0\, N^{d-1}]$th excursion of the walk, all belong to the same connected component of the vacant set in C_j left by the walk at time n. Here both c_0 and u_0 are positive constants solely depending on d, cf. Corollary 2.6.

To prove Theorem 1.1, it simply remains to ascertain that with probability arbitrarily close to 1 for large N, when picking γ suitably small, no more than $u_0 N^{d-1}$ excursions have occurred in all blocks corresponding to the various levels jN, $j \in \mathbb{Z}$. This is performed with the help of a coupling between the local time of simple random walk on $\mathbb{Z}$ and that of Brownian motion, cf. [2], and a scaling argument.

Let us now describe the structure of this note. In Section 1 we first introduce some further notation. We then show in Proposition 2.1, Theorem 2.2, Corollary 2.6 the key ingredients for the proof of Theorem 1.1, i.e., the presence in a suitable regime of a multitude of segments of length $c \log N$ in the vacant set left in a block C_j, the above-mentioned exponential bound, and the interconnection in the vacant set left in a block C_j of the various segments of length $c_0 \log N$ it contains.

In Section 2 we complete the proof of Theorem 1.1 with the help of the key steps laid out in Section 1.

Finally throughout the text c or c' denote positive constants which solely depend on d, with value changing from place to place. The numbered constants $c_0, c_1, \dots,$ are fixed and refer to their first appearance in the text below. Dependence of constants on additional parameters appears in the notation. For instance $c(\gamma)$ denotes a positive constant depending on d and γ.

2. Preparation

In this section we first introduce some further notation and then provide with Proposition 2.1, Theorem 2.2, Corollary 2.6, the main ingredients for the proof of Theorem 1.1 as explained in the Introduction.

We write π_E for the canonical projection from $\mathbb{Z}^{d+1}$ onto E. We denote with $(e_i)_{1\le i\le d+1}$ the canonical basis of $\mathbb{R}^{d+1}$. For $x \in \mathbb{Z}^{d+1}$, resp. $x \in E$, we let x^{d+1} stand for the last component, resp. the projection on $\mathbb{Z}$, of x. We denote with $|\cdot|$ and $|\cdot|_\infty$ the Euclidean and ℓ^∞-distances on $\mathbb{Z}^{d+1}$, or the corresponding distances induced on E. We write $B(x,r)$ for the closed $|\cdot|_\infty$-ball with radius $r \ge 0$, and center $x \in \mathbb{Z}^{d+1}$, or $x \in E$. For A and B subsets of E or $\mathbb{Z}^{d+1}$, we write $A+B$ for the set of elements of the form $x+y$, with $x \in A$ and $y \in B$. We say that x, y in $\mathbb{Z}^{d+1}$ or E are neighbors, resp. $\star$-neighbors if $|x-y|=1$, resp. $|x-y|_\infty = 1$. The notions of connected or $\star$-connected subsets of $\mathbb{Z}^{d+1}$ or E are then defined accordingly, and so are the notions of nearest neighbor path or $\star$-nearest neighbor path on $\mathbb{Z}^{d+1}$ or E. For U a subset of $\mathbb{Z}^{d+1}$ or E, we denote with $|U|$ the cardinality of U and ∂U the boundary of U:

$$\partial U = \{x \in U^c; \exists y \in U,\, |x-y|=1\}\,. \tag{2.1}$$

We let $(\theta_n)_{n\ge 0}$ and $(\mathcal{F}_n)_{n\ge 0}$ respectively stand for the canonical shift and filtration for the process $(X_n)_{n\ge 0}$ on $E^{\mathbb{N}}$. For $U \subseteq E$, H_U and T_U denote the entrance time and exit time in or from U:

$$H_U = \inf\{n \ge 0;\ X_n \in U\},\ T_U = \inf\{n \ge 0;\ X_n \notin U\}\,, \tag{2.2}$$

and $\widetilde{H}_U$ the hitting time of U:

$$\widetilde{H}_U = \inf\{n \ge 1;\ X_n \in U\}\,. \tag{2.3}$$

When $U = \{x\}$, we write H_x or $\widetilde{H}_x$ in place of $H_{\{x\}}$, or $\widetilde{H}_{\{x\}}$. For $\nu \ge 1$ and $x \in \mathbb{Z}^\nu$, Q^ν_x denotes the canonical law on $(\mathbb{Z}^\nu)^{\mathbb{N}}$ of the simple random walk on $\mathbb{Z}^\nu$ starting from x. When this causes no confusion, we use the same notation as above for the corresponding canonical process, canonical shift, the entrance, exit, or hitting times. So for instance, cf. (1.4), we have

$$q(\nu) = Q_0^\nu[\widetilde{H}_0 < \infty], \text{ for } \nu \geq 1\,. \tag{2.4}$$

We are interested in certain excursions of the walk on E around the level jN, $j \in \mathbb{Z}$, in the discrete cylinder. For this purpose we introduce for $j \in \mathbb{Z}$, the blocks:

$$\begin{aligned} B_j &= (\mathbb{Z}/N\mathbb{Z})^d \times \big[(j-1)\,N, (j+1)N\big] \subseteq \widetilde{B}_j \\ &= (\mathbb{Z}/N\mathbb{Z})^d \times \big[(j-2)N+1, (j+2)N-1\big]\,, \end{aligned} \tag{2.5}$$

so that with the definition (1.7) we find

$$C_j \subseteq B_j \subseteq \widetilde{B}_j, \text{ for } j \in \mathbb{Z}\,. \tag{2.6}$$

When $j=0$, we simply drop the subscript from the notation. We are specifically interested in the successive returns R_k^j, $k \geq 1$, to B_j, and departures D_k^j, $k \geq 1$, from $\widetilde{B}_j$:

$$\begin{aligned} R_1^j &= H_{B_j},\; D_1^j = T_{\widetilde{B}_j} \circ \theta_{R_1^j} + R_1^j, \text{ and for } k \geq 1\,, \\ R_{k+1}^j &= H_{B_j} \circ \theta_{D_k^j} + D_k^j,\; D_{k+1}^j = D_1^j \circ \theta_{D_k^j} + D_k^j\,, \end{aligned} \tag{2.7}$$

so that

$$0 \leq R_1^j \leq D_1^j \leq \dots \leq R_k^j \leq D_k^j \leq \dots \leq \infty\,,$$

and for any $x \in E$, P_x-a.s. these inequalities are strict except maybe for the first one. We also use the convention $R_0^j = D_0^j = 0$, for $j \in \mathbb{Z}$, as well as $R_t^j = R_{[t]}^j$, $D_t^j = D_{[t]}^j$, for $t \geq 0$.

It will be convenient in what follows to "spread out" the distribution of the starting point of the walk on E, and to this end we define:

$$\begin{aligned} P = \;&\text{the law of the walk on } E \text{ with initial distribution,} \\ &\text{the uniform measure on } B, \end{aligned} \tag{2.8}$$

where we recall the convention stated below (2.6). We write $E[\cdot]$ for the corresponding expectation. As noted below (1.2) T_N has the same distribution under any P_x, $x \in E$, and it coincides with its distribution under P.

As we now see, when $d \geq 3$, for large N, with overwhelming P-probability, there is in C_j a wealth of segments along the coordinate axes of length $c \log N$ that have not been visited by the walk up to time $D_{uN^{d-1}}^j$, when u is small enough. With this in mind we introduce for $K>0$, $j \in \mathbb{Z}$, $t \geq 0$, the event:

$$\begin{aligned} \mathcal{V}_{K,j,t} = \big\{&\text{for all } x \in C_j,\; e \in \mathbb{Z}^{d+1} \text{ with } |e| = 1, \text{ for some} \\ &0 \leq i < \sqrt{N},\; X_{[0,D_t^j]} \cap \{x + (i + [0, K\log N])\,e\} = \emptyset\big\}\,. \end{aligned} \tag{2.9}$$

The first step on our route to the proof of Theorem 1.1 is:

Proposition 2.1. ($d \geq 3$)
For any $K > 0$,

$$\limsup_N N^{-\frac{1}{4}} \log \sup_{j \in \mathbb{Z}} P[\mathcal{V}^c_{K,j,uN^{d-1}}] < 0, \text{ for small } u > 0\,. \tag{2.10}$$

Proof. Theorem 1.2 of [1] adapted to the present context states that for small $u > 0$,

$$\sup_{j \in \mathbb{Z}} P[\mathcal{V}^c_{K,j,uN^{d-1}}] \to 0$$

as $N \to \infty$ (take there $\beta = 1/2$ and note that the dimension $d+1$ plays the role of d in [1]). Moreover, taking $\beta_2 = \frac{1}{4}$ and $\beta_1 = \frac{1}{4} + \frac{1}{16}$ in (2.27) of [1], it is not hard to verify that the proof of this theorem actually yields the exponential decay of probabilities as in (2.10). Indeed the probabilities in question are bounded (up to multiplicative factor N^{d+1}), by the sum of those in (1.49) and (1.56) of [1], each of whom is shown there to be of the stated exponential decay. The intuition behind the argument lies in a coupon-collector heuristics. Roughly speaking the strategy of the argument is the following. Given any C_j, x in C_j and coordinate direction, we consider a collection of $[N^{\beta_1}]$ segments of length $[K \log N]$ on the "half line" starting at x with the above-chosen coordinate direction, and regular interspacing of order $[N^{\beta_1 - \beta_2}]$. We introduce a decimation process of the above collection of segments. We consider the successive excursions between times R^j_k and D^j_k, $k \geq 1$, of the walk. At first all segments are active and we look at the first excursion visiting one of the above segments. We call it successful and take out from the list of active segments the first (active) segment, which this excursion visits. We then look for the next successful excursion visiting a segment of the list of remaining active segments. We then delete from the list of active segments the first active segment hit by this excursion. We then carry on the decimation procedure until there is no active segment left.

As in (1.49) of [1], one can show that when u is small for large N, uniformly in $x \in C_j$ and in the coordinate direction, no more than $[N^{\beta_1}] - [N^{\beta_2}]$ successful excursions can occur up to time $D^j_{uN^{d-1}}$ except on a set of probability decaying exponentially in $N^{\frac{\beta_1+\beta_2}{2}}$.

Then as in (1.56) of [1], one shows that during the first $[N^{\beta_1}] - [N^{\beta_2}]$ successful excursions the total number of additional segments visited after the first hit of active segments does not exceed $\frac{1}{2}\,[N^{\beta_2}]$, except on a set of probability decaying exponentially in N^{β_2}.

This enables to bound the total number of segments visited by the walk up to $D^j_{uN^{d-1}}$ by $[N^{\beta_1}] - \frac{1}{2}\,[N^{\beta_2}]$, except on a set of probability decaying exponentially in N^{β_2}. Taking into account the polynomial growth in N due to the various possible choices of x in C_j and coordinate direction we obtain (2.10). We refer the reader to [1] for more details. □

Our next step is an exponential bound for which we need some additional notation. For $1 \le m \le d+1$, we write $\mathcal{L}_m$ for the collection of subsets of E that are image under the projection π_E of affine lattices of $\mathbb{Z}^{d+1}$ generated by m distinct vectors of the canonical basis $(e_i)_{1 \le i \le d+1}$:

$$\begin{aligned}\mathcal{L}_m = \Big\{ & F \subseteq E;\ \text{for some } I \subseteq \{1,\dots,d+1\}, \text{ with } |I| = m \text{ and some} \\ & y \in \mathbb{Z}^{d+1},\ F = \pi_E\Big(y + \sum_{i \in I} \mathbb{Z}\, e_i\Big)\Big\},\ 1 \le m \le d+1\,.\end{aligned} \tag{2.11}$$

For $j \in \mathbb{Z}$, $1 \le m \le d+1$, we consider

$$\begin{aligned}\mathcal{A}^j_m = &\text{ the collection of non-empty subsets } A \text{ of } C_j \\ &\text{ such that } A \subseteq F \text{ for some } F \in \mathcal{L}_m\,.\end{aligned} \tag{2.12}$$

It is plain that $\mathcal{A}^j_m$ increases with m, and $\mathcal{A}^j_{d+1}$ is the collection of non-empty subsets of C_j. Very much in the spirit of Theorem 2.1 of [1], we have the exponential bound:

Theorem 2.2. $(d \ge 3, 1 \le m \le d-2)$

Assume that $\lambda > 0$ *satisfies*

$$\chi(\lambda) \stackrel{\text{def}}{=} e^{\lambda}\left(\frac{m}{d+1} + \Big(1 - \frac{m}{d+1}\Big)\, q(d+1-m)\right) < 1\,, \tag{2.13}$$

then for $u > 0$,

$$\limsup_N \sup_{j \in \mathbb{Z}, A \in \mathcal{A}^j_m} |A|^{-1} \log E\big[e^{\lambda \sum_{x \in A} 1\{H_x \le D^j_{uN^{d-1}}\}}\big] \le cu\, \frac{e^{\lambda} - 1}{1 - \chi(\lambda)}\,, \tag{2.14}$$

and there exist $N_1(d,m,\lambda) > 0$, $u_1(d,m,\lambda) > 0$, *such that for* $N \ge N_1$:

$$P\big[X_{[0, D^j_{u_1 N^{d-1}}]} \supseteq A\big] \le \exp\{-\lambda |A|\},\ \textit{for all } j \in \mathbb{Z},\ A \in \mathcal{A}^j_m\,. \tag{2.15}$$

Proof. We use a variation on the ideas used in the proof of Theorem 2.2 of [1]. We consider for $j \in \mathbb{Z}$, $A \in \mathcal{A}^j_m$, $1 \le m \le d-2$, and $\lambda > 0$, the function

$$\phi_j(z) = E_z\big[e^{\lambda \sum_{x \in A} 1\{H_x < T_{\tilde{B}^j}\}}\big]\ (\ge 1),\ \text{for } z \in E\,. \tag{2.16}$$

It follows from the application of the strong Markov property at time H_A, that:

$$\begin{aligned}\phi_j(z) &= P_z[H_A \ge T_{\tilde{B}^j}] + E_z[H_A < T_{\tilde{B}^j}, \phi_j(X_{H_A})] \\ &= 1 + E_z[H_A < T_{\tilde{B}^j}, (\phi_j(X_{H_A}) - 1)]\,.\end{aligned} \tag{2.17}$$

Now for $z \in \partial(B^c_j)\ (= (\mathbb{Z}/N\mathbb{Z})^d \times \{(j-1)N, (j+1)N\})$, we have

$$\begin{aligned}\phi_j(z) &= 1 + E_z[H_A < T_{\tilde{B}^j}, \phi_j(X_{H_A}) - 1] \\ &\le 1 + P_z[H_A < T_{\tilde{B}^j}](\|\phi_j\|_\infty - 1) \\ &\le 1 + c\,\frac{|A|}{N^{d-1}}\,(\|\phi_j\|_\infty - 1) \le \exp\Big\{c\,\frac{|A|}{N^{d-1}}\,(\|\phi_j\|_\infty - 1)\Big\}\,,\end{aligned} \tag{2.18}$$

where in the first inequality of the last line we have used estimates on the Green function of simple random walk killed outside a strip, cf. (2.14) of [7], to bound $P_z[H_A < T_{\widetilde{B}^j}]$ from above. We thus see that for $k \geq 1$,

$$E\big[e^{\lambda \sum_{x\in A} 1\{H_x < D^j_{k+1}\}}\big] \leq E\big[e^{\lambda \sum_{x\in A} 1\{H_x < D^j_k\}} \\ E_{X_{R^j_{k+1}}}\big[e^{\lambda \sum_{x\in A} 1\{H_x < T_{\widetilde{B}^j}\}}\big]\big] \overset{(2.18)}{\leq} E\big[e^{\lambda \sum_{x\in A} 1\{H_x < D^j_k\}}\big] \\ e^{c\frac{|A|}{N^{d-1}}(\|\phi_j\|_\infty - 1)}\,. \tag{2.19}$$

With the help of the symmetry of the Green function of the walk killed outside $\widetilde{B}^j$ we show at the end of the proof of Lemma 2.3 of [3] that $P[H_A < D^j_1] \leq c|A|N^{-(d-1)}$ in case A is a sub-block of side length $[N^\gamma]$ for fixed $0 < \gamma < 1$. Since exactly the same argument (and bound) applies for all subsets A of C_j, we also have:

$$E\big[e^{\lambda \sum_{x\in A} 1\{H_x < D^j_1\}}\big] \leq 1 + P[H_A < D^j_1](\|\phi_j\|_\infty - 1) \\ \leq 1 + c\,\frac{|A|}{N^{d-1}}\,(\|\phi_j\|_\infty - 1) \leq e^{c\,\frac{|A|}{N^{d-1}}\,(\|\phi_j\|_\infty - 1)}\,. \tag{2.20}$$

Combining (2.19) and (2.20), using induction as well as (2.20) for the last term, we obtain:

$$E\big[e^{\lambda \sum_{x\in A} 1\{H_x < D^j_{uN^{d-1}}\}}\big] \leq e^{c\,\frac{|A|}{N^{d-1}}\,uN^{d-1}(\|\phi_j\|_\infty - 1)} \\ = \exp\{cu\,|A|(\|\phi_j\|_\infty - 1)\}\,. \tag{2.21}$$

We will now bound $\|\phi_j\|_\infty$.

Lemma 2.3. ($d \geq 3$, $1 \leq m \leq d-2$, $e^\lambda m < d+1$, $N \geq 2$)

$$\|\phi_j\|_\infty \leq \frac{e^\lambda}{1 - e^\lambda \frac{m}{d+1}}\left(1 - \frac{m}{d+1}\right)\big(1 + (\|\phi_j\|_\infty - 1)q_N\big)\,, \tag{2.22}$$

where we use the notation

$$q_N = \sup_{F\in\mathcal{L}_m, z\in\partial F} P_z[H_F < T_{\widetilde{B}^j}]\,, \tag{2.23}$$

and this quantity does not depend on j due to translation invariance of $\mathcal{L}_m$ and the walk.

Proof. We consider $F \in \mathcal{L}_m$, $A \subseteq F \cap (C_j \cup \partial C_j)$, and introduce the return time to F:

$$R_F = H_F \circ \theta_{T_F} + T_F\,. \tag{2.24}$$

For $z \in E$, we find:

$$\begin{aligned}
\phi_j(z) &= E_z\big[e^{\lambda \sum_{x\in A} 1\{H_x < T_{\widetilde{B}j}\}}\big] \\
&\le E_z\big[e^{\lambda (T_F + 1\{R_F < T_{\widetilde{B}j}\}\{(\sum_{x\in A} 1\{H_x < T_{\widetilde{B}j}\})\circ\theta_{R_F}\})}\big] \\
&= E_z\big[e^{\lambda T_F}\big(1\{R_F \ge T_{\widetilde{B}j}\} + 1\{R_F < T_{\widetilde{B}j}\}\, e^{\lambda \sum_{x\in A} 1\{H_x < T_{\widetilde{B}j}\}} \circ \theta_{R_F}\big)\big] \\
&= E_z\big[e^{\lambda T_F}\big(1 + 1\{R_F < T_{\widetilde{B}j}\}(\phi_j(X_{R_F}) - 1)\big)\big] \\
&\le E_z[e^{\lambda T_F}] + E_z\big[e^{\lambda T_F}\, P_{X_{T_F}}[H_F < T_{\widetilde{B}j}]\big](\|\phi_j\|_\infty - 1) \\
&\overset{(2.23)}{\le} E_z[e^{\lambda T_F}]\,\big(1 + (\|\phi_j\|_\infty - 1) q_N\big)\,,
\end{aligned} \tag{2.25}$$

where we have used the strong Markov property respectively at time R_F and T_F in the fourth and fifth line. Then observe that when $z \notin F$, $T_F = 0$, P_z-a.s., whereas when $z \in F$, T_F has geometric distribution with success probability $1 - \frac{m}{d+1}$, so that with λ satisfying the hypothesis of the lemma,

$$\begin{aligned}
E_z[e^{\lambda T_F}] &= \sum_{k\ge 1} \Big(1 - \frac{m}{d+1}\Big)\Big(\frac{m}{d+1}\Big)^{k-1} e^{\lambda k} \\
&= e^{\lambda}\Big(1 - \frac{m}{d+1}\Big)\Big(1 - \frac{e^{\lambda} m}{d+1}\Big)^{-1}.
\end{aligned} \tag{2.26}$$

Our claim (2.22) follows from the last line of (2.25). □

We now relate q_N to $q(d+1-m)$, cf. (1.4) and (2.4). Note that our assumptions ensure that $d+1-m \ge 3$.

Lemma 2.4. $(d \ge 3, 1 \le m \le d-2)$

$$\limsup_N\, q_N \le q(d+1-m)\,. \tag{2.27}$$

Proof. Without loss of generality we set $j = 0$ in (2.23). Then for $M \ge 1$, $F \in \mathcal{L}_m$, $z \in \partial F$, we have

$$P_z[H_F < T_{\widetilde{B}}] \le P_z[H_F < MN^2] + P_z[T_{\widetilde{B}} > MN^2]\,. \tag{2.28}$$

Using the fact that, cf. (2.19) of [3]:

$$\sup_{x\in\widetilde{B}} E_x\Big[\exp\Big\{\frac{c}{N^2}\, T_{\widetilde{B}}\Big\}\Big] \le c'\,, \tag{2.29}$$

to bound the last term of (2.28), we obtain:

$$\begin{aligned}
P_z[H_F < T_{\widetilde{B}}] &\le P_z[H_F < MN^2] + c'\, e^{-cM} \\
&\le P_{\widetilde{z}}^{(\mathbb{Z}/N\mathbb{Z})^{d+1}}[H_{\widetilde{F}} < MN^2] + c'\, e^{-cM}\,,
\end{aligned} \tag{2.30}$$

where $\widetilde{z}, \widetilde{F}$ are the respective images of z and F under the canonical projection from E onto $(\mathbb{Z}/N\mathbb{Z})^{d+1}$, and $P_{\widetilde{z}}^{(\mathbb{Z}/N\mathbb{Z})^{d+1}}$ denotes the canonical law of simple random walk on $(\mathbb{Z}/N\mathbb{Z})^{d+1}$ starting from $\widetilde{z}$. If we now consider the motion of the walk "transversal to $\widetilde{F}$", we find that for $N \geq 2$,

$$P_{\widetilde{z}}^{(\mathbb{Z}/N\mathbb{Z})^{d+1}}[H_{\widetilde{F}} < MN^2] \leq P_{e_1}^{(\mathbb{Z}/N\mathbb{Z})^{d+1-m}}[H_0 < MN^2], \tag{2.31}$$

with hopefully obvious notation. The right-hand side is the probability that simple random walk on $\mathbb{Z}^{d+1-m}$ starting at e_1 reaches $N\mathbb{Z}^{d+1-m}$ before time MN^2. The proof of Lemma 2.3 of [1] shows that the contribution of points of $N\mathbb{Z}^{d+1-m}$ other than 0 becomes negligible as N tends to infinity, so that

$$\limsup_N P_{e_1}^{(\mathbb{Z}/N\mathbb{Z})^{d+1-m}}[H_0 < MN^2] \leq q(d+1-m), \tag{2.32}$$

so that with (2.30) we find

$$\limsup_N q_N \leq q(d+1-m) + c' e^{-cM}, \text{ for } M \geq 1 \text{ arbitrary}. \tag{2.33}$$

Letting M tend to infinity, we obtain (2.27). □

With (2.22), (2.27), it is straightforward to deduce that when $\chi(\lambda) < 1$, cf. (2.13),

$$\limsup_N \sup_{j \in \mathbb{Z}, A \in \mathcal{A}_m^j} (\|\phi_j\|_\infty - 1) \leq \frac{e^\lambda - 1}{1 - \chi(\lambda)}. \tag{2.34}$$

Coming back to (2.21), taking logarithms and dividing by $|A|$, the claim (2.14) follows. As for (2.15), we pick $\widetilde{\lambda}(d,m,\lambda) > \lambda$, $\widetilde{q}(d,m,\lambda) > q(d+1-m)$, so that

$$1 - e^{\widetilde{\lambda}}\Big(\frac{m}{d+1} + \Big(1 - \frac{m}{d+1}\Big)\widetilde{q}\Big) = \tfrac{1}{2}\,(1 - \chi(\lambda)). \tag{2.35}$$

Applying (2.14) with $\widetilde{\lambda}$ (which satisfies $\chi(\widetilde{\lambda}) < 1$), we see that for $u > 0$, $N \geq N_2(d,m,\lambda,u)$, and any $j \in \mathbb{Z}$, $A \in \mathcal{A}_m^j$, one has with (2.35):

$$\begin{aligned} P[X_{[0, D^j_{uN^{d-1}}]} \supseteq A] &\leq P\Big[\sum_{x \in A} 1\{H_x < H_{D^j_{uN^{d-1}}}\} \geq |A|\Big] \\ &\leq \exp\Big\{ -\widetilde{\lambda}\,|A| + cu\,\frac{e^{\widetilde{\lambda}} - 1}{1 - \chi(\widetilde{\lambda})}\,|A|\Big\}. \end{aligned} \tag{2.36}$$

Choosing $u = u_1(d,m,\lambda)$ small enough, and setting $N_1(d,m,\lambda) = N_2(d,m,\lambda,u_1)$, we obtain (2.15). □

We will now use the above exponential control combined with a Peierl-type argument to ensure the typical presence for large N of a well-specified giant component in the vacant set left by the walk in a block C_j, as long as the number of excursions between B_j and $\widetilde{B}_j$ does not exceed a small multiple of N^{d-1}. This construction will force giant components corresponding to neighboring blocks to have non-empty intersection. We recall that $\star$-nearest neighbor paths have been defined at the beginning of this section, and introduce

$$a(n) = \text{the cardinality of the collection of } \star\text{-nearest neighbor self-avoiding paths on } \mathbb{Z}^2, \text{ starting at the origin, with } n \text{ steps}. \tag{2.37}$$

One has the straightforward upper bound

$$a(n) \leq 8\, 7^{n-1}, \text{ for } n \geq 1. \tag{2.38}$$

Given $N \geq 1$, $K > 0$, $j \in \mathbb{Z}$, $t \geq 0$, we introduce the event

$$\begin{aligned} \mathcal{U}_{K,j,t} = {} & \text{for any } F \in \mathcal{L}_2,\ n \leq D_t^j, \text{ any connected subsets } O_1, O_2 \\ & \text{of } F \cap C_j \backslash X_{[0,n]}, \text{ with } |\cdot|_\infty\text{-diameter at least } [K \log N] \\ & \text{are in the same connected component of } F \cap C_j \backslash X_{[0,n]}. \end{aligned} \tag{2.39}$$

This event will be helpful in specifying the above-mentioned giant components. We recall the notation (2.4).

Corollary 2.5. *If $d \geq 4$ is such that*

$$\rho \overset{\text{def}}{=} 7 \Big(\frac{2}{d+1} + \Big(1 - \frac{2}{d+1}\Big)\, q(d-1)\Big) < 1, \tag{2.40}$$

as this happens for any large d, then there are constants $c_0 > 0$, cf. (2.45), and u_0, cf. (2.43), such that

$$\limsup_N N^{2d} \sup_j P[\mathcal{U}^c_{c_{0,j,u_0 N^{d-1}}}] = 0. \tag{2.41}$$

Proof. Note that $q(\nu) \sim (2\nu)^{-1}$, cf. (5.4) of [5], and (2.40) holds for any large enough d. Assume that (2.40) holds and choose $\lambda_0(d)$ such that

$$e^{\lambda_0} = 7\rho^{-\frac{1}{2}} > 7, \text{ so that } e^{\lambda_0}\Big(\frac{2}{d+1} + \Big(1 - \frac{2}{d+1}\Big)\, q(d-1)\Big) = \rho^{\frac{1}{2}} < 1. \tag{2.42}$$

When N is large, on $\mathcal{U}^c_{K,j,t}$, one can find $F \in \mathcal{L}_2$, $n \leq D_t^j$, $O_1, O_2 \subseteq F \cap C_j \backslash X_{[0,n]}$, distinct connected components of $F \cap C_j \backslash X_{[0,n]}$ with $|\cdot|_\infty$-diameter at least $[K \log N]$. If the last vector e_{d+1} of the canonical basis does not enter the definition of F, cf. (2.11), then $F \subseteq C_j$, and we can introduce an affine projection of $\mathbb{Z}^2$ onto F, and define $\widehat{O}_i$, $i = 1, 2$, the inverse images of O_i under this affine projection. Considering separately the case when at least one of the $\widehat{O}_i$, $i = 1, 2$, has bounded components, (necessarily of $|\cdot|_\infty$-diameter at least $[K \log N]$), or both of the $\widehat{O}_i$ have unbounded components, one can construct a $\star$-nearest neighbor self-avoiding path π with $[K \log N]$ steps in $\partial O_1 \cap F$ or $\partial O_2 \cap F \subseteq F \cap X_{[0,n]} \subseteq F \cap C_j \cap X_{[0,D_t^j]}$, see also Proposition 2.1, p. 387, in [4]. On the other hand if the last vector e_{d+1} of the canonical basis enters the definition of F, we introduce an affine projection of $\mathbb{Z}^2$ onto F so that the inverse image of $F \cap C_j$ coincides with the strip $\mathbb{Z} \times ([-\frac{3}{4}N, \frac{3}{4}N] \cap \mathbb{Z})$. Defining as above $\widehat{O}_i$, $i = 1, 2$, the inverse images of O_i under this affine projection, we can separately consider the case when at least one of the $\widehat{O}_i$, $i = 1, 2$ has bounded components, (necessarily of $|\cdot|_\infty$-diameter at least

$[K \log N]$), or both of the $\widehat{O}_i$ have unbounded components. We can then construct a $\star$-nearest neighbor self-according path π with $[K \log N]$ steps in $\partial O_1 \cap F \cap C_j$ or $\partial O_2 \cap F \cap C_j \subseteq F \cap C_j \cap X_{[0,n]} \subseteq F \cap C_j \cap X_{[0,D_t^j]}$.

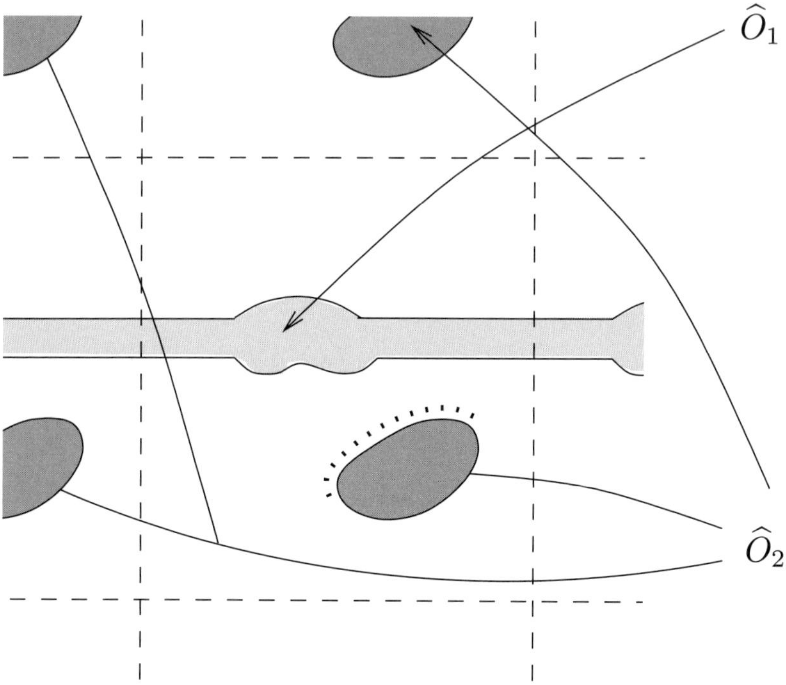

FIGURE 1. An example of possible $\widehat{O}_1$, $\widehat{O}_2$ is depicted in the case when e_{d+1} does not enter the definition of F. The square delimited by dashed lines is a "fundamental domain" for the affine projection. The dotted line is $*$-connected and projects on a subset of $F \cap C_j \cap X_{[0,n]}$.

As a result setting, cf. above (1.15),

$$u_0(d) = u_1\big(d, m = 2, \lambda = \lambda_0(d)\big)\,, \tag{2.43}$$

we see that for large enough N, for any $j \in \mathbb{Z}$,

$$\begin{aligned} P[\mathcal{U}^c_{K,j,u_0 N^{d-1}}] &\le \sum_F \sum_\pi P\big[X_{[0,D^j_{u_0 N^{d-1}}]} \supseteq A\big] \\ &\overset{(2.15)}{\le} \sup_j \sum_F \sum_\pi e^{-\lambda_0 |A|} \\ &\overset{(2.38)}{\le} \sup_j \sum_F cN^2\, 7^{[K \log N]-1}\, e^{-\lambda_0 [K \log N]} \\ &\overset{(2.42)}{\le} c\, N^{d+1} \rho^{\frac{1}{2}[K \log N]}\,, \end{aligned} \tag{2.44}$$

where the sum over F pertains to $F \in \mathcal{L}_2$ with $F \cap C_j \neq \phi$, the sum over π pertains to the collection of $\star$-nearest neighbor self-avoiding paths with values in $F \cap C_j$ with $[K \log N]$ steps, and A stands for the set of points visited by π. If we now specify K to take the value

$$c_0 = 8d\Big(\log\frac{1}{\rho}\Big)^{-1}, \tag{2.45}$$

the claim (2.41) follows from the last line of (2.44). □

We now introduce for $j \in \mathbb{Z}, t \geq 0$, the event, cf. (2.9), (2.39):

$$\mathcal{G}_{j,t} = \mathcal{V}_{c_0,j,t} \cap \mathcal{U}_{c_0,j,t}\,. \tag{2.46}$$

Corollary 2.6. *Assume $d \geq 3$, and $N \geq 2$ large enough so that $(\mathbb{Z}/N\mathbb{Z})^d$ has $|\cdot|_\infty$-diameter bigger than $c_0 \log N$. Then for $j \in \mathbb{Z}$, $t \geq 0$, on $\mathcal{G}_{j,t}$, for any $0 \leq n \leq D_t^j$,*

all segments in $C_j \backslash X_{[0,n]}$ of length $L_0 \overset{\text{def}}{=} [c_0 \log N]$ belong to the same connected component of $C_j \backslash X_{[0,n]}$, (2.47)

any $F \in \mathcal{L}_1$ intersecting C_j contains a segment of length L_0 included in $C_j \backslash X_{[0,n]}$. (2.48)

Moreover when $d \geq 4$ satisfies (2.40), with the notation (2.43), one has:

$$\lim_N N^{2d} \sup_{j\in\mathbb{Z}} P[\mathcal{G}^c_{j,u_0 N^{d-1}}] = 0\,. \tag{2.49}$$

Proof. The claim (2.48) readily follows from the inclusion $\mathcal{G}_{j,t} \subseteq \mathcal{V}_{c_0,j,t}$, and (2.9). To prove (2.47), consider $n \leq D_t^j$ and note that $\mathcal{G}_{j,t} \subseteq \mathcal{U}_{c_0,j,t}$. Hence any two segments of length L_0 contained in $\widetilde{F} \cap C_j \backslash X_{[0,n]}$, with $\widetilde{F} \in \mathcal{L}_2$, belong to the same connected component of $F \cap C_j \backslash X_{[0,n]}$. Then consider $\widetilde{F}, \widetilde{F}_2 \in \mathcal{L}_2$, we see that with (2.48) and the above,

when $\widetilde{F}_1 \cap \widetilde{F}_2 \in \mathcal{L}_1$ and intersects C_j, all segments of length L_0 in $(\widetilde{F}_1 \cup \widetilde{F}_2) \cap C_j \backslash X_{[0,n]}$ are in the the same connected component of $C_j \backslash X_{[0,n]}$. (2.50)

Next, given y_0 and y in C_j, we can construct a nearest neighbor path $(y_i)_{0 \leq i \leq m}$ in C_j, with $y_m = u$. Given $\widetilde{F} \ni y_0$, with $\widetilde{F} \in \mathcal{L}_2$, we can construct a sequence $\widetilde{F}_i \in \mathcal{L}_2$, $0 \leq i \leq m$, such that

$\widetilde{F}_0 = \widetilde{F}$, $y_i \in \widetilde{F}_i$, for $0 \leq i \leq m$, and either $\widetilde{F}_{i-1} = \widetilde{F}_i$, or $\widetilde{F}_{i-1} \cap \widetilde{F}_i \in \mathcal{L}_1$ and intersects C_j, for $1 \leq i \leq m$, (2.51)

(see for instance below (2.60) of [1] for a similar argument).

Analogously when $\widetilde{F}, \widetilde{F}' \in \mathcal{L}_2$ have a common point $y \in C_j$, we can define $\widetilde{F}_i \in \mathcal{L}_2$, $0 \le i \le 2$, such that

$$\begin{aligned}&\widetilde{F}_0 = \widetilde{F},\ \widetilde{F}_2 = \widetilde{F}', \text{ with } y \in \widetilde{F}_i,\ 0 \le i \le 2, \text{ and either } \widetilde{F}_i = \widetilde{F}_{i-1} \text{ or}\\ &\widetilde{F}_i \cap \widetilde{F}_{i-1} \in \mathcal{L}_1 \text{ (and intersects } C_j\text{), for } i = 1, 2.\end{aligned} \tag{2.52}$$

Combining (2.48) and (2.50)–(2.52), we obtain (2.47). Finally (2.49) is a direct consequence of (2.10) and (2.41). □

3. Denouement

We now use the results of the previous section to prove Theorem 1.1. As mentioned in the Introduction, the rough idea is that by making γ small one can ensure that with arbitrarily high probability, when N is large enough, for all j in $\mathbb{Z}$ the times $D^j_{u_0 N^{d-1}}$ are bigger than γN^{2d}. Then on "most" of the event $\{\inf_{j\in\mathbb{Z}} D^j_{u_0 N^{d-1}} > \gamma N^{2d}\}$, there is a profusion of segments of length L_0 in each $C_j \backslash X_{[0,[\gamma N^{2d}]]}$, and they all lie in the same connected component of $X^c_{[0,[\gamma N^{2d}]]}$. This now forces the disconnection time T_N to be bigger than γN^{2d}.

Proof of Theorem 1.1. As note below (1.2), we can with no loss of generality replace P_0 with P in the claim (1.6) to be proved. We introduce for $0 < \gamma < 1$, $t \ge 0$, the events:

$$\mathcal{C}_{\gamma,t} = \mathcal{D}_{\gamma,t} \cap \Big(\bigcap_{|j| \le 2N^{2d-1}} \mathcal{G}_{j,t} \Big), \text{ where} \tag{3.1}$$

$$\mathcal{D}_{\gamma,t} = \bigcap_{|j| \le 2N^{2d-1}} \{D^j_t > \gamma N^{2d}\}\,. \tag{3.2}$$

Note that P-a.s., the vertical component of the walk up to time N^{2d} remains in $[-N^{2d} - N, N^{2d} + N]$. Hence for large N, with (2.47), (2.48), P-a.s. on $\mathcal{C}_{\gamma,t}$, there is a nearest neighbor path in $X^c_{[0,[\gamma N^{2d}]]}$ starting in $(\mathbb{Z}/N\mathbb{Z})^d \times (-\infty, -M]$ and ending in $(\mathbb{Z}/N\mathbb{Z})^d \times [M, +\infty)$ for any $M \ge 1$, and therefore $T_N > \gamma N^{2d}$. As a result Theorem 1.1 will be proved once we show that with the notation of (2.43)

$$\liminf_N P\Big[\bigcap_{|j| \le 2N^{2d-1}} \mathcal{G}_{j,u_0 N^{d-1}} \Big] = 1, \text{ and} \tag{3.3}$$

$$\lim_{\gamma \to 0} \liminf_N P[\mathcal{D}_{\gamma,uN^{d-1}}] = 1, \text{ for any } u > 0\,. \tag{3.4}$$

The claim (3.3) readily follows from (2.49), so we only need to prove (3.4). To this end we introduce the following sequence of $(\mathcal{F}_n)$-stopping times that are P_x-a.s. finite for any $x \in E$:

$$\begin{aligned}\tau_0 &= H_{(\mathbb{Z}/N\mathbb{Z})^d \times N\mathbb{Z}}, \text{ cf. (2.2) for the notation, and for } k \ge 0,\\ \tau_{k+1} &= \inf\{n > \tau_k;\ |X^{d+1}_n - X^{d+1}_{\tau_k}| = N\}\,,\end{aligned} \tag{3.5}$$

where according to the notation of the beginning of Section 1, $X^{d+1}_{\cdot}$ denotes the $\mathbb{Z}$-component of $X_{\cdot}$. We also write $\tau_t = \tau_{[t]}$, for $t \geq 0$. The count of visits of X_{τ_n}, $n \geq 0$, to level ℓN with n at most t, is then expressed by

$$L_N(\ell, t) = \sum_{0 \leq n \leq t} 1\{X^{d+1}_{\tau_n} = \ell N\}, \text{ for } \ell \in \mathbb{Z}, t \geq 0\,. \tag{3.6}$$

Pick $M \geq 1$, and note that as soon as $\tau_{M\gamma N^{2d-2}} > \gamma N^{2d}$ and $L_N(\ell, M\gamma N^{2d-2}) < \frac{1}{2}\,[uN^{d-1}]$, for all $|\ell| \leq 2N^{2d-1}+1$, then $D^j_{uN^{d-1}} > \gamma N^{2d}$ for all $|j| \leq 2N^{2d-1}$, and hence $\mathcal{D}_{\gamma, uN^{d-1}}$ occurs. As a result we find that

$$\begin{aligned} &P[\mathcal{D}_{\gamma,uN^{d-1}}] \\ &\geq P\Big[\theta^{-1}_{\tau_0}\Big(\Big\{\sup_{|\ell| \leq 2N^{2d-1}+1} L_N(\ell, M\gamma N^{2d-2}) < \tfrac{1}{2}\,[uN^{d-1}]\Big\}\Big) \\ &\qquad \cap \big\{\tau_{M\gamma N^{2d-2}} \circ \theta_{\tau_0} > \gamma N^{2d}\big\}\Big] \\ &\geq P_0\Big[\Big\{\sup_{|\ell| \leq 2N^{2d-1}+2} L_N(\ell, M\gamma N^{2d-2}) < \tfrac{1}{2}\,[uN^{d-1}]\Big\} \\ &\qquad \cap \big\{\tau_{M\gamma N^{2d-2}} > \gamma N^{2d}\big\}\Big] \geq a_1 - a_2, \text{ with the notation} \end{aligned} \tag{3.7}$$

$$\begin{aligned} a_1 &= P_0\Big[\sup_{|\ell| \leq 2N^{2d-1}+2} L_N(\ell, M\gamma N^{2d-2}) < \tfrac{1}{2}\,[uN^{d-1}]\Big]\,, \\ a_2 &= P_0\big[\tau_{M\gamma N^{2d-2}} \leq \gamma N^{2d}\big]\,, \end{aligned} \tag{3.8}$$

and where we have used the strong Markov property at time τ_0, as well as translation invariance when going from the second to the third line of (3.7).

Note that under P_0, τ_k, $k \geq 0$, has stationary independent increments and applying the invariance principle to the $\mathbb{Z}$-component of X, we also have

$$E_0\Big[\exp\Big\{-\frac{c_1}{N^2}\,\tau_1\Big\}\Big] \leq e^{-c_2}\,. \tag{3.9}$$

With Cheybyshev's inequality we thus find

$$\begin{aligned} P_0[\tau_{M\gamma N^{2d-2}} \leq \gamma N^{2d}] &\leq e^{c_1\gamma N^{2d-2}} E_0\big[e^{-\frac{c_1}{N^2}\,\tau_{M\gamma N^{2d-2}}}\big] \\ &\leq c\,e^{(c_1 - c_2 M)\gamma N^{2d-2}}\,. \end{aligned}$$

Choosing from now on $M > \frac{c_1}{c_2}$, we find that

$$\lim_N a_2 = 0\,. \tag{3.10}$$

Coming back to a_1, we observe that under P_0, $L_N(\cdot,\cdot)$ has the same distribution as the local time process of simple random walk on $\mathbb{Z}$ starting at the origin.

In fact, cf. (1.20) of [2], we can construct on some auxiliary probability space $(\widetilde{\Sigma}, \widetilde{\mathcal{A}}, \widetilde{P})$ a one-dimensional Brownian motion $(\widetilde{B}_t)_{t \geq 0}$, and a simple random walk on $\mathbb{Z}$ starting at the origin, $Z_k, k \geq 0$, so that setting $\widetilde{L}(x,t)$, $x \in \mathbb{R}$, $t \geq 0$, be a jointly continuous version of the local time of $\widetilde{B}_{\cdot}$ and

$$L(x,k) = \sum_{n=0}^{k} 1\{Z_n = x\}\,, x \in \mathbb{Z},\, k \geq 0\,, \tag{3.11}$$

be the local time of the simple random walk $Z_.$, one has

$$\widetilde{P}\text{-a.s., for all } \rho > 0, \lim_{n\to\infty} n^{-\frac{1}{4}-\rho} \sup_{x\in\mathbb{Z}} |\widetilde{L}(x,n) - L(x,n)| = 0\,. \tag{3.12}$$

With this we find that for any $\gamma > 0, u > 0$,

$$\begin{aligned}
\liminf_N a_1 &\geq \liminf_N \widetilde{P}\Big[\sup_{|\ell|\leq 3N^{2d-1}} L(\ell,[M\gamma N^{2d-2}]) < \frac{1}{2}\,[uN^{d-1}]\Big] \\
&\overset{(3.12)}{\geq} \liminf_N \widetilde{P}\Big[\sup_{w\in\mathbb{R}} \widetilde{L}(w, M\gamma N^{2d-2}) < \frac{1}{4}\,uN^{d-1}\Big] \\
&\overset{\text{scaling}}{=} \widetilde{P}\Big[\sup_{v\in\mathbb{R}} \widetilde{L}(v, M\gamma) < \frac{1}{4}\,u\Big]\,.
\end{aligned}$$

It thus follows from the $\widetilde{P}$-a.s. joint continuity of $\widetilde{L}(\cdot,\cdot)$ that

$$\lim_{\gamma\to 0} \liminf_N a_1 = 1\,. \tag{3.13}$$

Together with (3.7) and (3.10), this concludes the proof of Theorem 1.1. □

Remark 3.1. It is plain that the condition (1.5) under which Theorem 1.1 holds requires $d \geq 14$. Using a numerical evaluation of $q(\nu)$ in (1.4) based on the formula (5.1) of [5], which was kindly provided to us by Wesley P. Petersen, one can see that (1.5) holds when $d \geq 17$ and fails when $d < 17$. On the other hand the numerical simulations performed in the context of the investigation of the vacant set left by random walk on the discrete torus in [1] make it plausible that the conclusion of Theorem 1.1 should hold for all $d \geq 1$ (the case $d = 1$ being trivial). □

Acknowledgments

We wish to thank Wesley P. Petersen for kindly communicating to us a table of numerical values of the return probability of simple random walk to the origin.

Amir Dembo would like to thank the FIM for hospitality and financial support during his visit to ETH. His research was also partially supported by the NSF grants DMS-0406042, DMS-FRG-0244323.

References

[1] I. Benjamini and A.S. Sznitman. Giant component and vacant set for random walk on a discrete torus. *J. Eur. Math. Soc.*, **10**(1) (2008), 133–172.

[2] E. Csáki and P. Revesz. Strong invariance for local times. *Z. für Wahrsch. verw. Geb.*, **62** (1983), 263–278.

[3] A. Dembo and A.S. Sznitman. On the disconnection of a discrete cylinder by a random walk. *Probab. Theory Relat. Fields*, **136**(2) (2006), 321–340.

[4] H. Kesten. *Percolation theory for Mathematicians.* Birkhäuser, Basel, 1982.

[5] E.W. Montroll. Random walks in multidimensional spaces, especially on periodic lattices. *J. Soc. Industr. Appl. Math.*, **4**(4) (1956), 241–260.

[6] A.S. Sznitman. How universal are asymptotics of disconnection times in discrete cylinders? *Ann. Probab.*, **36**(1) (2008), 1–53.

[7] A.S. Sznitman. On new examples of ballistic random walks in random environment. *Ann. Probab.*, **31**(1) (2003), 285–322.

[8] A.S. Sznitman. Vacant set of random interlacements and percolation. Preprint available at: http://www.math.ethz.ch/u/sznitman/preprints.

Amir Dembo
Department of Statistics and
Department of Mathematics
Stanford University
Stanford, CA 94305, USA
e-mail: `amir@math.stanford.edu`

Alain-Sol Sznitman
Departement Mathematik
ETH Zürich
CH-8092 Zürich, Switzerland
e-mail: `sznitman@math.ethz.ch`

Progress in Probability, Vol. 60, 229–256

On the Large Deviations Properties of the Weighted-Serve-the-Longest-Queue Policy

Paul Dupuis, Kevin Leder and Hui Wang

Abstract. We identify the large deviation rate function for a single server with multi-class arrivals in which the service priority is determined according to the weighted-serve-the-longest-queue policy. The problem setup falls into the general category of systems with discontinuous statistics. Our analysis, which is largely based on a weak convergence approach, does not require any symmetry or dimensional restrictions.

Mathematics Subject Classification (2000). 60F10, 60K25, 90B22.

Keywords. Large deviations, serve-the-longer, service policy, discontinuous statistics.

1. Introduction

Consider a single server that must serve multiple queues of customers from different classes. A common service discipline in this situation is the serve-the-longest-queue policy, in which the longest queue is given priority. In this paper we will consider a natural generalization of this discipline, namely, the *weighted-serve-the-longest-queue* (WSLQ) policy. Under WSLQ, each queue length is multiplied by a constant to determine a "score" for that queue, and the queue with the largest score is granted priority. Such service policies are more appropriate than serve-the-longest-queue policy when the different arrival queues or customer classes have different requirements or statistical properties. For example, if there is a finite queueing capacity to be split among the different classes, one may want to choose the partition and the weighting constants in order to optimize a certain performance measure.

Because WSLQ is a frequently proposed discipline for queueing models in communication problems, a large deviations analysis of this protocol is useful [12].

Paul Dupuis was supported in part by the National Science Foundation (NSF-DMS-0306070 and NSF-DMS-0404806) and the Army Research Office (W911NF-05-1-0289).
Kevin Leder was supported in part by the National Science Foundation (NSF-DMS-0404806).
Hui Wang was supported in part by the National Science Foundation (NSF-DMS-0404806).

However, service policies such as WSLQ are not smooth functions of the system state and lead to multidimensional stochastic processes with *discontinuous statistics*. In general, large deviation properties of processes with discontinuous statistics are hard to analyze [1, 7, 8, 9]. This is especially true when the discontinuities appear on the interior of the state space, rather than the boundary. In fact, very general results with an explicit identification of the rate function only exist for the case where two regions of smooth statistical behavior are separated by an interface of codimension one [3]. For the WSLQ policy, the large deviation analysis has been limited to special, two-dimensional cases [11]. Large deviations for a weighted-serve-the-largest-workload policy are treated in [10], but [10] considers a particular event rather than the sample path large deviations principle.

The purpose of the present work is to show that a complete large deviation analysis of WSLQ is possible without any symmetry or dimensional assumptions. Given the intrinsic difficulties in models with discontinuous statistics, it is worthwhile to explain what makes such an analysis possible for WSLQ. To this end, we recall the main difficulty in the large deviation analysis of systems with discontinuous statistics. A large deviation upper bound can often be established using the results in [6], which assumes little regularity on the statistical behavior of the underlying processes. However, this upper bound is generally *not* tight, even for the very simple situation of two regions of constant statistical behavior separated by a hyperplane of codimension one [5].

The reason for this gap is most easily identified by considering the corresponding lower bound. When proving a large deviation lower bound, it is necessary to analyze the probability that the process closely follows or tracks a constant velocity trajectory that lies on the interface of two or more regions of smooth statistical behavior. For this one has to consider all changes of measure in these different regions that lead to the desired tracking behavior. The thorny issue is how to characterize such changes of measure. In the case of two regions [5], this can be done in a satisfactory fashion and it turns out that the large deviation rate function is a modified version of the upper bound in [6]. The modification is made to explicitly include certain "stability about the interface" conditions, and part of the reason that everything works out nicely in the setup of [5] is that these stability conditions can be easily characterized. However, the analogous characterization of stability is not known for more elaborate settings such as WSLQ, where the regions of constant statistical behavior are defined according to the partition of the state space by a finite number of hyperplanes of codimension one. Therefore, at present there is no general theory that subsumes WSLQ as a special case.

However, a key observation that makes possible a large deviations analysis for WSLQ is that for this model, the required stability conditions are *implicitly* and *automatically* built into the upper bound rate function of [6]. More precisely, it can be shown that in the lower bound analysis one can restrict, a priori, to a class of changes of measure for which the stability conditions are automatically implied. Thus while it is true that the upper bound of [6] is not tight in general, it is so in this case due to the structural properties of WSLQ policy.

The study of the large deviation properties of WSLQ is partly motivated by the problem of estimating buffer overflow (rare event) probabilities for stable WSLQ systems using importance sampling. It turns out that the simple form of the large deviation local rate function as exhibited in (3.1) and (3.2) is essential toward constructing simple and asymptotically optimal importance sampling schemes using a game theoretic approach. These results will be reported elsewhere.

The paper is organized as follows. In Section 2, we introduce the single server system with WSLQ policy. In Section 3, we state the main result, whose proof is presented in Section 4. Collected in the appendices are a lengthy technical part of the proof that involves the approximation of continuous trajectories, as well as miscellaneous results.

2. Problem Setup

Consider a server with d customer classes, where customers of class i arrive according to a Poisson process with rate λ_i and are buffered at queue i for $i = 1, \ldots, d$. The service time for a customer of class i is exponentially distributed with rate μ_i.

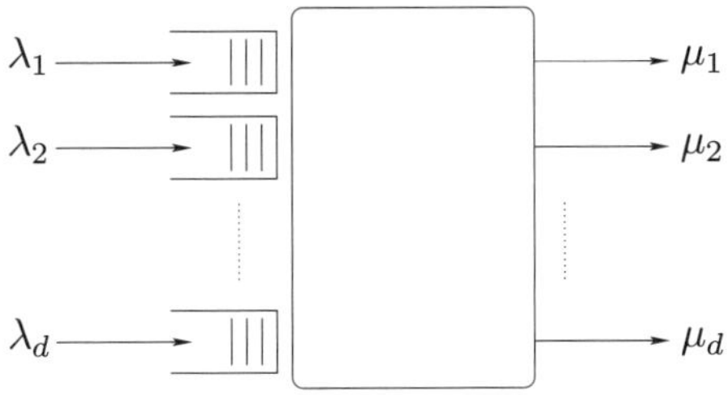

FIGURE 1. WSLQ system.

The service policy is determined according to the WSLQ discipline that can be described as follows. Let c_i be the weight associated with class i. If the size of queue i is q_i, then the "score" of queue i is defined as $c_i q_i$ and service priority will be given to the queue with the maximal score. When there are multiple queues with the maximal score, the assignment of priority among these queues can be arbitrary – the choice is indeed non-essential and will lead to the same rate function. We adopt the convention that when there are ties, the priority will be given to the queue with the largest index.

The system state at time t is the vector of queue lengths and is denoted by $Q(t) \doteq (Q_1(t), \ldots, Q_d(t))$. Then Q is a continuous time pure jump Markov process whose possible jumps belong to the set

$$\Theta = \{\pm e_1, \pm e_2, \ldots, \pm e_d\}.$$

For $v \in \Theta$, let $r(x; v)$ denote the jump intensity of process Q from state x to state $x + v$. Under the WSLQ discipline, these jump intensities are as follows. For

$x = (x_1, \ldots, x_d) \in \mathbb{R}^d_+$ and $x \neq 0$, let $\pi(x)$ denote the indices of queues that have the maximal score, that is,

$$\pi(x) \doteq \left\{ 1 \leq i \leq d : c_i x_i = \max_j c_j x_j \right\}.$$

Then

$$r(x; v) = \begin{cases} \lambda_i, & \text{if } v = e_i \text{ and } i = 1, \ldots, d, \\ \mu_i, & \text{if } v = -e_i \text{ where } i = \max \pi(x), \\ 0\,, & \text{otherwise.} \end{cases}$$

For $x = 0$, there is no service and the jump intensities are

$$r(0; v) = \begin{cases} \lambda_i, & \text{if } v = e_i \text{ and } i = 1, \ldots, d, \\ 0\,, & \text{otherwise.} \end{cases}$$

We also set

$$\pi(0) \doteq \{0, 1, 2, \ldots, d\}.$$

An illustrative figure for the case of two queues in given below.

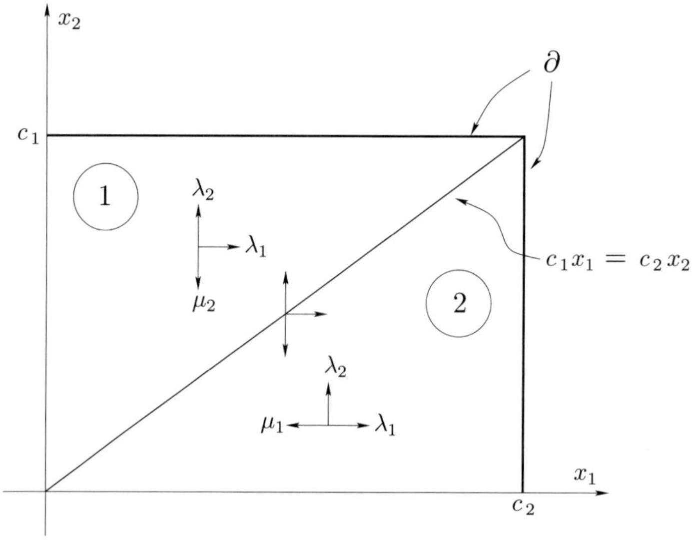

FIGURE 2. System dynamics for $d = 2$.

Remark 2.1. It is not difficult to see that the system dynamics have constant statistical behavior in the regions where $\pi(\cdot)$ is constant. Discontinuity occurs when $\pi(\cdot)$ changes, and every x with $|\pi(x)| \geq 2$ (i.e., when there is a tie) is indeed a discontinuous point. Therefore, we have various discontinuity interfaces with different dimensions. For example, for every subset $A \subset \{1, 2, \ldots, d\}$ with $|A| \geq 2$ or $A = \{0, 1, 2, \ldots, d\}$, the set $\{x \in \mathbb{R}^d_+ : \pi(x) = A\}$ defines an interface with dimension $d - |A| + 1$.

Remark 2.2. The definition of $\pi(0)$ is introduced to cope with the discontinuous dynamics at the origin. Note that with this definition, $\pi(x)$ can only be $\{0,1,2,\ldots,d\}$ if $x=0$ and a subset of $\{1,2,\ldots,d\}$ if $x \neq 0$.

Remark 2.3. A useful observation is that π is *upper semicontinuous* as a set-valued function. That is, for any $x \in \mathbb{R}^d_+$, $\pi(y) \subset \pi(x)$ for all y in a small neighborhood of x.

3. The main result

In order to state a large deviation principle on path space, we fix arbitrarily $T>0$, and for each $n \in \mathbb{N}$ let $\{X^n(t) : t \in [0,T]\}$ be the scaled process defined by

$$X^n(t) \doteq \frac{1}{n} Q(nt).$$

Then X^n is a continuous time Markov process with generator

$$\mathcal{L}^n f(x) = n \sum_{v \in \Theta} r(x;v) \left[f\left(x + v/n\right) - f(x) \right].$$

The processes $\{X^n\}$ live in the space of cadlag functions $\mathcal{D}([0,T] : \mathbb{R}^d)$, which is endowed with the Skorohod metric and thus a Polish space.

3.1. The rate function

For each $i = 1,\ldots,d$, let $H^{(i)}$ be the convex function given by

$$H^{(i)}(\alpha) \doteq \mu_i(e^{-\alpha_i} - 1) + \sum_{j=1}^{d} \lambda_j (e^{\alpha_j} - 1),$$

for all $\alpha = (\alpha_1,\ldots,\alpha_d) \in \mathbb{R}^d$. We also define for $i = 0$,

$$H^{(0)}(\alpha) \doteq \sum_{j=1}^{d} \lambda_j (e^{\alpha_j} - 1)$$

for $\alpha \in \mathbb{R}^d$.

For each non-empty subset $A \subset \{1,\ldots,d\}$ or $A = \{0,1,\ldots,d\}$, let L^A be the Legendre transform of $\max_{i \in A} H^{(i)}$, that is,

$$L^A(\beta) \doteq \sup_{\alpha \in \mathbb{R}^d} \left[\langle \alpha, \beta \rangle - \max_{i \in A} H^{(i)}(\alpha) \right]$$

for each $\beta \in \mathbb{R}^d$. Clearly, L^A is convex and non-negative. When A is a singleton $\{i\}$, we simply write L^A as $L^{(i)}$. A useful representation of L^A [4, Corollary D.4.3] is

$$L^A(\beta) = \inf \sum_{i \in A} \rho^{(i)} L^{(i)}(\beta^{(i)}), \tag{3.1}$$

where the infimum is taken over all $\{(\rho^{(i)}, \beta^{(i)}) : i \in A\}$ such that

$$\rho^{(i)} \geq 0, \sum_{i \in A} \rho^{(i)} = 1, \sum_{i \in A} \rho^{(i)} \beta^{(i)} = \beta. \tag{3.2}$$

Furthermore, for $x \in \mathbb{R}^d_+$ let $L(x, \beta) \doteq L^{\pi(x)}(\beta)$.

Now we can define the process level rate function. For $x \in \mathbb{R}^d_+$, define $I_x : \mathcal{D}([0,T] : \mathbb{R}^d) \to [0, \infty]$ by

$$I_x(\psi) \doteq \int_0^T L(\psi(t), \dot{\psi}(t))dt \tag{3.3}$$

if $\psi(0) = x$, ψ is absolutely continuous, and $\psi(t) \in \mathbb{R}^d_+$ for all $t \in [0, T]$. Otherwise set $I_x(\psi) \doteq \infty$. The family of rate functions $\{I_x : x \in \mathbb{R}^d_+\}$ has compact level sets on compacts in the sense that, for every $M \geq 0$ and every compact set $C \in \mathbb{R}^d_+$, the set

$$\cup_{x \in C} \left\{\psi \in \mathcal{D}([0,T] : \mathbb{R}^d) : I_x(\psi) \leq M\right\}$$

is compact [6, Theorem 1.1].

3.2. The main theorem

The main result of this paper can be stated as follows. Let E_{x_n} denote the expectation conditioned on $X^n(0) = x_n$.

Theorem 3.1. *The process $\{X^n(t) : t \in [0,T]\}$ satisfies the uniform Laplace principle with rate functions $\{I_x : x \in \mathbb{R}^d_+\}$. That is, for any sequence $\{x_n\} \subset \mathbb{R}^d_+$ such that $x_n \to x$ and any bounded continuous function $h : \mathcal{D}([0,T] : \mathbb{R}^d) \to \mathbb{R}$, we have*

$$\lim_{n\to\infty} -\frac{1}{n} \log E_{x_n} \{\exp[-nh(X^n)]\} = \inf_{\psi \in \mathcal{D}([0,T]:\mathbb{R}^d)} \{I_x(\psi) + h(\psi)\}.$$

In particular, $\{X^n(t) : t \in [0,T]\}$ satisfies the large deviation principle with rate function $\{I_x : x \in \mathbb{R}^d_+\}$.

4. Proof of the main theorem

Throughout the rest of the paper, we will assume without loss of generality that $T = 1$. We only need to show the uniform Laplace principle which automatically implies the large deviation principle [4, Theorem 1.2.3]. The uniform Laplace principle upper bound is implied by the large deviation upper bound [6, Theorem 1.1] and an argument similar to [4, Theorem 1.2.1]. Therefore, it is only necessary to prove the uniform Laplace principle lower bound. That is,

$$\limsup_{n\to\infty} -\frac{1}{n} \log E_{x_n} \{\exp[-nh(X^n)]\} \leq \inf_{\psi \in \mathcal{D}([0,1]:\mathbb{R}^d)} \{I_x(\psi) + h(\psi)\}. \tag{4.1}$$

One can a priori restrict to ψ such that $I_x(\psi) < \infty$ since the inequality holds trivially otherwise. Note that such ψ's are necessarily absolutely continuous by the definition of I_x.

4.1. An approximation lemma

A very important step in the proof of (4.1) is the following approximation lemma. A function ψ^* has property $\mathcal{P}$ if there exists a $k \in \mathbb{N}$ and $0 = t_0 < t_1 < \cdots < t_{k-1} < t_k = 1$ such that on each interval $(t_{i-1}, t_i), i = 1, \ldots, k$, $\dot{\psi}^*$ and $\pi(\psi^*)$ are both constant. Then define $\mathcal{N}$ be the collection of functions $\psi^* \in \mathcal{C}([0,1] : \mathbb{R}^d_+)$ that have property $\mathcal{P}$.

Lemma 4.1. *Given any $\psi \in \mathcal{D}([0,1] : \mathbb{R}^d)$ such that $I_x(\psi) < \infty$ and any $\delta > 0$, there exists $\psi^* \in \mathcal{N}$ such that $\|\psi - \psi^*\|_\infty < \delta$ and $I_x(\psi^*) \le I_x(\psi) + \delta$.*

The proof of Lemma 4.1 is lengthy and technical, and is deferred to Appendix A.

We claim that, in order to show inequality (4.1), it suffices to show that for any $x_n \to x$ and $\psi^* \in \mathcal{N}$,

$$\limsup_{n\to\infty} -\frac{1}{n} \log E_{x_n} \{\exp[-nh(X^n)]\} \le I_x(\psi^*) + h(\psi^*). \tag{4.2}$$

This reduction follows easily from Lemma 4.1 and the continuity of h. We omit the details.

4.2. A stochastic control representation

The proof of (4.2) uses the weak convergence approach, and is based on the following formula. Define the function ℓ by

$$\ell(x) = \begin{cases} x \log x - x + 1, & \text{if } x \ge 0, \\ \infty, & \text{if } x < 0, \end{cases}$$

with the convention $0\ell(0/0) \doteq 0$. Let $\bar{r}(x,t;v)$ be non-negative, uniformly bounded, and piecewise constant in t, and also satisfy $\bar{r}(x,t;v) = 0$ whenever $r(x;v) = 0$. Let $\bar{X}^n$ be the non-stationary jump Markov process with generator

$$\bar{\mathcal{L}}^n f(x,t) = n \sum_{v\in\Theta} \bar{r}(x,t;v) \left[f(x+v/n) - f(x)\right].$$

Then

$$\begin{aligned} & -\frac{1}{n} \log E_x \{\exp[-nh(X^n)]\} \\ \le\ & \inf_{\bar{r}} E_x \left[\int_0^1 \sum_{v\in\Theta} r(\bar{X}^n(t);v) \ell\left(\frac{\bar{r}(\bar{X}^n(t),t;v)}{r(\bar{X}^n(t);v)}\right) dt + h(\bar{X}^n)\right]. \end{aligned} \tag{4.3}$$

In this inequality $\bar{r}$ can be viewed as a control and $\bar{X}^n$ a controlled process.

The proof of (4.3) follows from the relative entropy representation formula for exponential integrals. Let P be a probability measure and let h be a bounded continuous function on $\mathcal{D}([0,1] : \mathbb{R}^d)$. For another probability measure Q on

$\mathcal{D}([0,1]:\mathbb{R}^d)$ let $R(Q\,\|P)$ denote the relative entropy of Q with respect to P. Then [4]

$$-\frac{1}{n}\log\int_{\mathcal{D}([0,1]:\mathbb{R}^d)} e^{-nh}dP = \inf\left[\frac{1}{n}R(Q\,\|P) + \int_{\mathcal{D}([0,1]:\mathbb{R}^d)} hdQ\right],$$

where the infimum is over all Q. If one restricts Q to the probability measures induced by the class of Markov processes described above, one obtains an inequality. Finally, we substitute the explicit form of the Radon-Nikodym derivative dQ/dP (as in [11, Theorem B.6]) into the definition of $R(Q\,\|P)$ to arrive at the inequality in (4.3).

Thanks to the control representation (4.3), the Laplace principle lower bound (4.2) for $\psi^* \in \mathcal{N}$ follows if one can, for an arbitrarily fixed $\varepsilon > 0$, construct a control (abusing the notation) $\bar{r} = \bar{r}^\varepsilon$ such that

$$\begin{aligned}\limsup_{n\to\infty} E_{x_n}&\left[\int_0^1 \sum_{v\in\Theta} r(\bar{X}^n(t);v)\ell\left(\frac{\bar{r}(\bar{X}^n(t),t;v)}{r(\bar{X}^n(t);v)}\right)dt + h(\bar{X}^n)\right] \\ &\leq\; I_x(\psi^*) + h(\psi^*) + \varepsilon.\end{aligned} \tag{4.4}$$

The details of the construction will be carried out in the next section.

4.3. Properties of the rate function

We give some useful representation formulae for the local rate functions L^A.

Lemma 4.2. *Given $\beta \in \mathbb{R}^d$, the following representation for $L^{(i)}(\beta)$ holds.*

1. *For each $i = 1, 2, \ldots, d$,*

$$L^{(i)}(\beta) = \inf\left\{\mu_i\ell\left(\frac{\bar{\mu}_i}{\mu_i}\right) + \sum_{j=1}^d \lambda_j\ell\left(\frac{\bar{\lambda}_j}{\lambda_j}\right) : -\bar{\mu}_i e_i + \sum_{j=1}^d \bar{\lambda}_j e_j = \beta\right\}.$$

2. *For $i = 0$,*

$$L^{(0)}(\beta) = \inf\left\{\sum_{j=1}^d \lambda_j\ell\left(\frac{\bar{\lambda}_j}{\lambda_j}\right) : \sum_{j=1}^d \bar{\lambda}_j e_j = \beta\right\}.$$

In every case the infimum is attained.

Proof. Note that for every $\lambda > 0$ and $v \in \mathbb{R}^d$, the Legendre transform of the convex function

$$h(\alpha) \doteq \lambda\left(e^{\langle\alpha,v\rangle} - 1\right)$$

is

$$h^*(\beta) \doteq \begin{cases} \lambda\ell(\bar{\lambda}/\lambda), & \text{if } \beta = \bar{\lambda}v \text{ for some } \bar{\lambda}\in\mathbb{R}, \\ \infty, & \text{otherwise}\end{cases}$$

for every $\beta \in \mathbb{R}^d$. The proof of this claim is straightforward computation and we omit the details. Now the representation for $L^{(i)}$ follows directly from [4, Corollary D.4.2]. The attainability of the infimum is elementary. □

Lemma 4.3. *Given $\beta \in \mathbb{R}^d$, we have the following representation for $L^A(\beta)$.*

1. *Assume $A \in \{1, 2, \ldots, d\}$ is non-empty. Then*

$$L^A(\beta) = \inf \left[\sum_{i \in A} \rho^{(i)} \mu_i \ell \left(\frac{\bar{\mu}_i}{\mu_i} \right) + \sum_{j=1}^{d} \lambda_j \ell \left(\frac{\bar{\lambda}_j}{\lambda_j} \right) \right],$$

where the infimum is taken over all collections of $(\rho^{(i)}, \bar{\mu}_i, \bar{\lambda}_j)$ such that

$$\rho^{(i)} \geq 0, \ \sum_{i \in A} \rho^{(i)} = 1, \ -\sum_{i \in A} \rho^{(i)} \bar{\mu}_i e_i + \sum_{j=1}^{d} \bar{\lambda}_j e_j = \beta. \tag{4.5}$$

2. *For $A = \{0, 1, 2, \ldots, d\}$, we have*

$$L^A(\beta) = \inf \left[\sum_{i=1}^{d} \rho^{(i)} \mu_i \ell \left(\frac{\bar{\mu}_i}{\mu_i} \right) + \sum_{j=1}^{d} \lambda_j \ell \left(\frac{\bar{\lambda}_j}{\lambda_j} \right) \right],$$

where the infimum is taken over all collections of $(\rho^{(i)}, \bar{\mu}_i, \bar{\lambda}_j)$ such that

$$\rho^{(i)} \geq 0, \ \sum_{i=0}^{d} \rho^{(i)} = 1, \ -\sum_{i=1}^{d} \rho^{(i)} \bar{\mu}_i e_i + \sum_{j=1}^{d} \bar{\lambda}_j e_j = \beta.$$

Proof. We only present the proof for Part 1. The proof for Part 2 is similar and thus omitted. Thanks to Lemma 4.2 and equations (3.1)–(3.2), we have

$$L^A(\beta) = \inf \sum_{i \in A} \rho^{(i)} \left\{ \mu_i \ell \left(\frac{\bar{\mu}_i^{(i)}}{\mu_i} \right) + \sum_{j=1}^{d} \lambda_j \ell \left(\frac{\bar{\lambda}_j^{(i)}}{\lambda_j} \right) \right\},$$

where the infimum is taken over all $(\rho^{(i)}, \bar{\mu}_i^{(i)}, \bar{\lambda}_j^{(i)})$ such that

$$\rho^{(i)} \geq 0, \ \sum_{i \in A} \rho^{(i)} = 1, \ \sum_{i \in A} \rho^{(i)} \left[-\bar{\mu}_i^{(i)} e_i + \sum_{j=1}^{d} \bar{\lambda}_j^{(i)} e_j \right] = \beta. \tag{4.6}$$

Abusing the notation a bit, write $\bar{\mu}_i = \bar{\mu}_i^{(i)}$ for $i \in A$, and let $\bar{\lambda}_j \doteq \sum_{i \in A} \rho^{(i)} \bar{\lambda}_j^{(i)}$ for $j = 1, 2, \ldots, d$. Thanks to (4.6), the collection $(\rho^{(i)}, \bar{\mu}_i, \bar{\lambda}_j)$ satisfies the constraints (4.5). Observing that, by convexity of ℓ,

$$\sum_{i \in A} \rho^{(i)} \sum_{j=1}^{d} \lambda_j \ell \left(\frac{\bar{\lambda}_j^{(i)}}{\lambda_j} \right) = \sum_{j=1}^{d} \lambda_j \sum_{i \in A} \rho^{(i)} \ell \left(\frac{\bar{\lambda}_j^{(i)}}{\lambda_j} \right) \geq \sum_{j=1}^{d} \lambda_j \ell \left(\frac{\bar{\lambda}_j}{\lambda_j} \right),$$

with equality if $\lambda_j^{(i)} = \lambda_k^{(i)}$ for every j, k. The first part of Lemma 4.3 now follows readily. □

Remark 4.4. The representation of L^A in Lemma 4.3 remains valid if we further constrain $\rho^{(i)}$, $\bar{\mu}_i$, and $\bar{\lambda}_i$ to be strictly positive for every $i \in A$. This is an easy consequence of the fact that ℓ is finite and continuous on the interval $[0, \infty)$. We omit the details.

Remark 4.5. Given a non-empty subset $A \subsetneq \{1,2,\ldots,d\}$ and $\beta = (\beta_1,\ldots,\beta_d) \in \mathbb{R}^d$, $L^A(\beta)$ is finite if and only if $\beta_j \geq 0$ for all $j \notin A$. For $A = \{1,2,\ldots,d\}$ or $\{0,1,2,\ldots,d\}$, $L^A(\beta)$ is finite for every $\beta \in \mathbb{R}^d$.

4.4. The construction of controls and the cost

Fix $\varepsilon > 0$. We will use ψ^* to construct a control $\bar{r} = \bar{r}^\varepsilon$ based on the representation of the local rate function as in Lemma 4.3. For notational simplicity, we drop the superscript ε.

Since $\psi^* \in \mathcal{N}$, there exist $0 = t_0 < t_1 < \cdots < t_K = 1$ such that for every k there are β_k and A_k such that $\dot{\psi}^*(t) \equiv \beta_k$ and $\pi(\psi^*(t)) \equiv A_k$ for all $t \in (t_k, t_{k+1})$. We start by defining a suitable collection of $\{(\rho_k^{(i)}, \bar{\mu}_{i,k}, \bar{\lambda}_{j,k}) : 0 \leq i \leq d, 1 \leq j \leq d\}$. We consider the following two cases.

Case 1. Suppose $A_k = \{0,1,2,\ldots,d\}$. Lemma 4.3 and Remark 4.4 imply the existence of a collection $\{(\rho_k^{(i)}, \bar{\mu}_{i,k}, \bar{\lambda}_{j,k}) : 0 \leq i \leq d, 1 \leq j \leq d\}$ such that $\bar{\mu}_{i,k} > 0$, $\bar{\lambda}_{i,k} > 0$ for all i and

$$\rho_k^{(i)} > 0, \; \sum_{i=0}^{d} \rho_k^{(i)} = 1, \; -\sum_{i=1}^{d} \rho_k^{(i)} \bar{\mu}_{i,k} e_i + \sum_{j=1}^{d} \bar{\lambda}_{j,k} e_j = \beta_k, \tag{4.7}$$

$$\sum_{i=1}^{d} \rho_k^{(i)} \mu_i \ell\left(\frac{\bar{\mu}_{i,k}}{\mu_i}\right) + \sum_{j=1}^{d} \lambda_j \ell\left(\frac{\bar{\lambda}_{j,k}}{\lambda_j}\right) \leq L^{A_k}(\beta_k) + \varepsilon. \tag{4.8}$$

Case 2. Suppose $A_k \subset \{1,2,\ldots,d\}$. According to Lemma 4.3 and Remark 4.4, for each k there exist a collection $\{(\rho_k^{(i)}, \bar{\mu}_{i,k}, \bar{\lambda}_{j,k}) : i \in A_k, 1 \leq j \leq d\}$ such that $\bar{\mu}_{i,k} > 0$, $\bar{\lambda}_{i,k} > 0$ for all $i \in A_k$ and

$$\rho_k^{(i)} > 0, \; \sum_{i \in A_k} \rho_k^{(i)} = 1, \; -\sum_{i \in A_k} \rho_k^{(i)} \bar{\mu}_{i,k} e_i + \sum_{j=1}^{d} \bar{\lambda}_{j,k} e_j = \beta_k, \tag{4.9}$$

$$\sum_{i \in A_k} \rho_k^{(i)} \mu_i \ell\left(\frac{\bar{\mu}_{i,k}}{\mu_i}\right) + \sum_{j=1}^{d} \lambda_j \ell\left(\frac{\bar{\lambda}_{j,k}}{\lambda_j}\right) \leq L^{A_k}(\beta_k) + \varepsilon. \tag{4.10}$$

We extend the definition by letting $\rho_k^{(i)} \doteq 0$, $\bar{\mu}_{i,k} \doteq \mu_i$ for $i \notin A_k$ and $i \neq 0$, and letting $\rho_k^{(0)} \doteq 0$.

The control $\bar{r}$ is defined as follows. For $t \in [t_k, t_{k+1})$, let

$$\bar{r}(x,t;v) = \begin{cases} \bar{\lambda}_{j,k}, & \text{if } v = e_j \text{ and } j = 1,\ldots,d, \\ \bar{\mu}_{j,k}, & \text{if } v = -e_j \text{ where } j = \max \pi(x) \text{ and } x \neq 0, \\ 0\;, & \text{otherwise.} \end{cases}$$

Thus, on time interval $[t_k, t_{k+1})$, the system has arrival rates $\{\bar{\lambda}_{1,k}, \ldots, \bar{\lambda}_{d,k}\}$ and service rates $\{\bar{\mu}_{1,k}, \ldots, \bar{\mu}_{d,k}\}$ under this control $\bar{r}$. The corresponding running cost for $t \in [t_k, t_{k+1})$ is

$$\begin{aligned}
&\sum_{v\in\Theta} r(\bar{X}^n(t); v)\ell\left(\frac{\bar{r}(\bar{X}^n(t), t; v)}{r(\bar{X}^n(t); v)}\right) \qquad (4.11)\\
&= \sum_{j=1}^{d} \mu_j \ell\left(\frac{\bar{\mu}_{j,k}}{\mu_j}\right) 1_{\{\max \pi(\bar{X}^n(t))=j, \bar{X}^n(t)\neq 0\}} + \sum_{j=1}^{d} \lambda_j \ell\left(\frac{\bar{\lambda}_{j,k}}{\lambda_j}\right)\\
&= \sum_{i\in A_k, i\neq 0} \mu_i \ell\left(\frac{\bar{\mu}_{i,k}}{\mu_i}\right) 1_{\{\max \pi(\bar{X}^n(t))=i, \bar{X}^n(t)\neq 0\}} + \sum_{j=1}^{d} \lambda_j \ell\left(\frac{\bar{\lambda}_{j,k}}{\lambda_j}\right),
\end{aligned}$$

here the last equality holds since $\bar{\mu}_{j,k} = \mu_j$ for $j \notin A_k$ and $\ell(1) = 0$.

For future use, we also define for each $i = 1, \ldots, d$,

$$\beta_k^{(i)} \doteq -\bar{\mu}_{i,k} e_i + \sum_{j=1}^{d} \bar{\lambda}_{j,k} e_j, \qquad (4.12)$$

which is the law of large number limit of the velocity of the controlled process if queue of class i is served. Analogously, we also define (when none of the queues are being served)

$$\beta_k^{(0)} \doteq \sum_{j=1}^{d} \bar{\lambda}_{j,k} e_j. \qquad (4.13)$$

4.5. Weak convergence analysis

In this section we characterize the limit processes. Below are a few definitions. For each j, define random measures $\{\gamma_j^n\}$ on $[0, 1]$ by

$$\begin{aligned}
\gamma_j^n\{B\} &\doteq \int_B 1_{\{\max \pi(\bar{X}^n(t))=j, \bar{X}^n(t)\neq 0\}} dt, \ j = 1, 2, \ldots, d,\\
\gamma_0^n\{B\} &\doteq \int_B 1_{\{\bar{X}^n(t)=0\}} dt,
\end{aligned}$$

for Borel subsets $B \subset [0, 1]$, and denote $\gamma^n \doteq (\gamma_0^n, \gamma_1^n, \ldots, \gamma_d^n)$. We also define the stochastic processes

$$S^n(t) \doteq x_n + \sum_{j=0}^{d} \left[\sum_{k=0}^{\kappa(t)-1} \beta_k^{(j)} \gamma_j^n\{[t_k, t_{k+1})\} + \beta_{\kappa(t)}^{(j)} \gamma_j^n\{[t_{\kappa(t)}, t)\} \right],$$

where $\kappa(t) = \max\{0 \le k \le K : t_k \le t\}$.

Proposition 4.6. *Given any subsequence of $(\gamma^n, S^n, \bar{X}^n)$, there exist a subsubsequence, a collection of random measures $\gamma \doteq (\gamma_0, \gamma_1, \ldots, \gamma_d)$ on $[0, 1]$, and a continuous process $\bar{X}$ such that*

(a) *The subsubsequence converges in distribution to $(\gamma, \bar{X}, \bar{X})$.*

(b) *With probability one, γ_j is absolutely continuous with respect to the Lebesgue measure on $[0,1]$, and its density, denoted by h_j, satisfies*

$$\sum_{j=0}^{d} h_j(t) = \sum_{j\in\pi(\bar{X}(t))} h_j(t) = 1$$

for almost every t.

(c) *With probability one, the process $\bar{X}$ satisfies*

$$\bar{X}(t) = x + \sum_{j=0}^{d}\left[\sum_{k=0}^{\kappa(t)-1} \beta_k^{(j)}\gamma_j\{[t_k,t_{k+1})\} + \beta_{\kappa(t)}^{(j)}\gamma_j\{[t_{\kappa(t)},t)\}\right]$$

for every t. Therefore, $\bar{X}$ is absolutely continuous with derivative

$$\frac{d\bar{X}(t)}{dt} = \sum_{j=0}^{d}\beta_{\kappa(t)}^{(j)}h_j(t).$$

Proof. The family of random measures $\{\gamma_j^n\}$ is contained in the compact set of sub-probability measures on $[0,1]$ and therefore tight. Furthermore, since $\{S^n\}$ is uniformly Lipschitz continuous, it takes values in a compact subset of $\mathcal{C}([0,1]:\mathbb{R}^d)$, and therefore is also tight. We also observe that for every $\varepsilon > 0$,

$$\lim_{n\to\infty} P(\|\bar{X}^n - S^n\|_\infty > \varepsilon) = 0, \tag{4.14}$$

which in turn implies that $\{\bar{X}^n\}$ is tight. Equation (4.14) is trivial since $\bar{X}^n - S^n$ is a martingale (e.g., [11, Appendix B.2]), whence the process $\|\bar{X}^n - S^n\|^2$ is a submartingale. Therefore, by the submartingale inequality and the uniform boundedness of the jump intensity $\bar{r}$

$$P\left(\sup_{0\le t\le 1}\|\bar{X}^n(t) - S^n(t)\| > \varepsilon\right) \le \frac{2}{\varepsilon^2}E\left[\|\bar{X}^n(1) - S^n(1)\|^2\right] \to 0.$$

It follows that there exists a subsubsequence that converges weakly to say (γ, S, S), with $\gamma = (\gamma_0,\gamma_1,\ldots,\gamma_d)$. By the Skorohod representation theorem, we assume without loss of generality that the convergence is almost sure convergence, and everything is defined on some probability space, say $(\bar{\Omega},\bar{\mathcal{F}},\bar{P})$.

Since the $\{\gamma_j^n\}$ are absolutely continuous with respect to Lebesgue measure on $[0,1]$ with uniformly bounded densities (i.e., Radon-Nikodým derivatives) in both n and t, the limit γ_j is also absolutely continuous. Furthermore, if we define the process $\bar{X}$ as in (c), the above consideration yields that, for every $t\in[0,1]$, $S^n(t)$ converges to $\bar{X}(t)$ almost surely. Therefore, with probability one, $S(t) = \bar{X}(t)$ for all these rational $t\in[0,1]$. Since both S and $\bar{X}$ are continuous, $S = \bar{X}$ with probability one.

It remains to show the two equalities of (b). Since $\sum_{j=0}^{d}\gamma_j^n$ equals Lebesgue measure for every n, we have $\sum_{j=0}^{d} h_j(t) = 1$ for almost every t. The proof of the second equality is similar to that of [4, Theorem 7.4.4(c)]. Consider an $\omega\in\bar{\Omega}$ such that $\bar{X}(t,\omega)$ is a continuous function of $t \in [0,1]$, $\gamma^n(\omega) \Rightarrow \gamma(\omega)$, and such that $\bar{X}^n(\cdot,\omega)$ converges to $\bar{X}(\cdot,\omega)$ in the Skorohod metric (whence also in sup-norm since $\bar{X}(\cdot,\omega)$ is continuous [4, Theorem A.6.5]). By the upper semicontinuity of $\pi(\cdot)$ [Remark 2.3], it follows that for any $t \in (0,1)$ and $A \subset \{0,1,\ldots,d\}$ such that $\pi(\bar{X}(t,\omega)) \subset A$, there exist an open interval (a,b) containing t and $N \in \mathbb{N}$ such that $\pi(\bar{X}^n(s,\omega)) \subset A$ for all $n \geq \mathbb{N}$ and $s \in (a,b)$. Therefore $\sum_{j\notin A} \gamma_j^n(\omega)\{(a,b)\} = 0$ for all $n \geq \mathbb{N}$. Taking the limit as $n \to \infty$ we have $\sum_{j\notin A} \gamma_j(\omega)\{(a,b)\} = 0$, which in turn implies that

$$\sum_{j\notin A} \gamma_j(\omega)\{t \in (0,1) : \pi(\bar{X}(t,\omega)) \subset A\} = 0,$$

or equivalently,

$$\int_0^1 \sum_{j\notin A} h_j(t,\omega) 1_{\{\pi(\bar{X}(t,\omega))\subset A\}}\, dt = 0.$$

We claim that this implies the desired equality at ω. Otherwise, there exists a subset $D \subset (0,1)$ with positive Lebesgue measure such that for every $t \in D$,

$$\sum_{j\notin\pi(\bar{X}(t,\omega))} h_j(t,\omega) > 0.$$

Since $\pi(\bar{X}(t,\omega))$ can only take finitely many possible values, there exists a subset (abusing the notation) $A \subset \{0,1,\ldots,d\}$ such that the set

$$\bar{D} \doteq \{t \in D : \pi(\bar{X}(t,\omega)) = A\}$$

has positive Lebesgue measure. It follows that

$$\begin{aligned}
\int_0^1 \sum_{j\notin A} h_j(t,\omega) 1_{\{\pi(\bar{X}(t,\omega))\subset A\}}\, dt &\geq \int_{\bar{D}} \sum_{j\notin A} h_j(t,\omega) 1_{\{\pi(\bar{X}(t,\omega))\subset A\}}\, dt \\
&= \int_{\bar{D}} \sum_{j\notin\pi(\bar{X}(t,\omega))} h_j(t,\omega)\, dt \\
&> 0,
\end{aligned}$$

a contradiction. This completes the proof. □

4.6. Stability analysis

In this section we prove a key lemma in the analysis that identifies the weak limit $\bar{X}$ as ψ^*. The proof uses the implied "stability about the interface" in a crucial way.

We discuss the main idea behind this stability property before giving the detailed proof. For the large deviation analysis, it is important to analyze the probability that the process tracks a segment of trajectory that lies on an interface, say $\{x : \pi(x) = A\}$, with a constant velocity, say β. To this end, it is natural to use the change of measure induced by β through the local rate function L as described in Section 4.4. However, for general systems, this very natural construction does *not* guarantee that $\bar{X}$ will follow or track ψ, and certain "stability about the interface" conditions have to be added for this to happen.

For the WSLQ system, a stability condition is not explicitly needed since it is implicitly and automatically built into the upper bound local rate function L. To be more precise, denote the arrival and service rates under the change of measure by $\bar{\lambda}_i$ and $\bar{\mu}_i$ respectively. Then for the tracking behavior to take place it is required that the proportion of time that the process $\bar{X}$ spent in the region $\{x : \max \pi(x) = i\}$ equals $\rho^{(i)}$, where $\{\rho^{(i)}\}$ is the (strictly positive) solution to the system of equations

$$c_i(\bar{\lambda}_i - \rho^{(i)}\bar{\mu}_i) = c_j(\bar{\lambda}_j - \rho^{(j)}\bar{\mu}_j), \ i, j \in A, \text{ and } \sum_{i \in A} \rho^{(i)} = 1.$$

Thanks to Proposition 4.6, such proportions are characterized by $\{h_i(t)\}$ where

$$\sum_{i \in \pi(\bar{X}(t))} h_i(t) = 1.$$

Therefore, we would like to show that $\pi(\bar{X}(t)) \equiv A$ and $h_i(t) \equiv \rho^{(i)}$ for $i \in A$. This can be shown, and the argument is based on the simple fact that for any non-empty subset $B \subset \{1, 2, \ldots, d\}$ and any b, the solution $\{x_i : i \in B\}$ to the system of equations

$$c_i(\bar{\lambda}_i - x_i\bar{\mu}_i) = c_j(\bar{\lambda}_j - x_j\bar{\mu}_j), \ i, j \in B, \text{ and } \sum_{i \in B} x_i = b, \tag{4.15}$$

is unique and component-wise strictly increasing with respect to b.

For example, suppose on some time interval (a, b) that $\pi(\bar{X}(t)) \equiv B$ where B is a strict subset of A. It is not difficult to see that $\{h_i(t) : i \in B\}$ is a solution to equation (4.15) because, thanks to Proposition 4.6,

$$\frac{d}{dt}(\bar{X}(t))_i = \bar{\lambda}_i - h_i(t)\bar{\mu}_i, \text{ for } i \in B.$$

Since

$$\sum_{i \in B} \rho^{(i)} = 1 - \sum_{i \in A \setminus B} \rho^{(i)} < 1,$$

the monotonicity implies $h_i(t) > \rho^{(i)}$ for all $i \in B$. This in turn yields [see the proof of (4.25) for details] that for $i \in B$ and $j \in A \setminus B$,

$$\frac{d}{dt}\left[c_i((\bar{X}(t))_i - c_j((\bar{X}(t))_j\right] < 0. \tag{4.16}$$

Thus the differences between weighted queue lengths grows smaller, and the state is "pushed" towards the interface A. This derivation can be used to prove by contradiction that $A \subset \pi(\bar{X}(t))$. The other direction $\pi(\bar{X}(t)) \subset A$ can be shown similarly. Once $\pi(\bar{X}(t)) = A$ is shown, $h_i(t) = \rho^{(i)}$ follows immediately from the uniqueness of the solution to equation (4.15).

We want to point out that in general such stability conditions are equivalent to the existence of certain Lyapunov functions, and for the case of WSLQ, inequality (4.16) indicates that the difference of weighted queue lengths is a Lyapunov function.

Lemma 4.7. *Let $(\gamma, \bar{X}, \bar{X})$ be a limit of any weakly converging subsubsequence $(\gamma^n, S^n, \bar{X}^n)$ as in Proposition 4.6. Then with probability one, $\bar{X}(t) = \psi^*(t)$ for every $t \in [0,1]$, and for each j,*

$$h_j(t) = \sum_{k=0}^{K-1} \rho_k^{(j)} 1_{(t_k, t_{k+1})}(t)$$

for almost every $t \in [0,1]$.

Proof. The proof is by induction. By definition $\bar{X}(0) = x = \psi^*(0)$. Assume that $\bar{X}(t) = \psi^*(t)$ for all $t \in [0, t_k]$. The goal is to show that $\bar{X}(t) = \psi^*(t)$ for all $t \in [0, t_{k+1}]$. Define $A_k = \pi(\psi^*(t))$, $t \in (t_k, t_{k+1})$ and $A = \pi(\psi^*(t_k))$. Note that $A_k \subset A$ thanks to the continuity of ψ^* and the upper semicontinuity of π. Define the random time

$$\tau_k = \inf \left\{ t > t_k : \pi(\bar{X}(t)) \not\subset A \right\}.$$

Since $\pi(\bar{X}(t_k)) = \pi(\psi^*(t_k)) = A$ and $\bar{X}$ is continuous, the upper semicontinuity of π implies that $\tau_k > t_k$ and $\pi(\bar{X}(\tau_k)) \not\subset A$. We claim that it suffices to show $\bar{X}(t) = \psi^*(t)$ and $h_j(t) = \rho_k^{(j)}$ for all $t \in (t_k, t_{k+1} \wedge \tau_k)$. Indeed, if this is the case, we must have $\tau_k \geq t_{k+1}$ with probability one, since otherwise by continuity $\bar{X}(\tau_k) = \psi^*(\tau_k)$ and thus $\pi(\bar{X}(\tau_k)) = A_k \subset A$, a contradiction.

The proof of $\bar{X}(t) = \psi^*(t)$ and $h_j(t) = \rho_k^{(j)}$ for all $t \in (t_k, t_{k+1} \wedge \tau_k)$ proceeds in three steps.

Step 1. For every $t \in (t_k, t_{k+1} \wedge \tau_k)$, either $A_k \subset \pi(\bar{X}(t))$ or $\pi(\bar{X}(t))$ is a strict subset of A_k.

Step 2. For every $t \in (t_k, t_{k+1} \wedge \tau_k)$, $\pi(\bar{X}(t)) = A_k$.

Step 3. For every $t \in (t_k, t_{k+1} \wedge \tau_k)$, $\bar{X}(t) = \psi^*(t)$ and $h_j(t) = \rho_k^{(j)}$.

Note that, for $t \in (t_k, t_{k+1} \wedge \tau_k)$, the definition of τ_k implies $\pi(\bar{X}(t)) \subset A$. It follows from (4.12), (4.13), and Proposition 4.6 that,

$$\frac{d}{dt}\bar{X}(t) = \sum_{i \in \pi(\bar{X}(t))} \beta_k^{(i)} h_i(t) = \sum_{j=1}^{d} \bar{\lambda}_{j,k} e_j - \sum_{i \in \pi(\bar{X}(t)),\ i \neq 0} \bar{\mu}_{i,k} h_i(t) e_i. \tag{4.17}$$

To verify Step 1, we assume that A_k is a strict subset of A and a strict subset of $\{1, 2, \ldots, d\}$ as well, since otherwise the claim is trivial. It follows from the definitions of π, A, and A_k that for every $i \in A_k$ and $j \in A \setminus A_k$ such that $j \neq 0$,

$$c_i \left(\psi^*(t_k)\right)_i - c_j \left(\psi^*(t_k)\right)_j = 0$$

and for $t \in (t_k, t_{k+1})$

$$c_i \left(\psi^*(t)\right)_i - c_j \left(\psi^*(t)\right)_j > 0.$$

Since $\dot{\psi}^*(t) \equiv \beta_k$ for $t \in (t_k, t_{k+1})$, the last display and (4.9) yield

$$0 < c_i(\beta_k)_i - c_j(\beta_k)_j = c_i(\bar{\lambda}_{i,k} - \rho_k^{(i)}\bar{\mu}_{i,k}) - c_j\bar{\lambda}_{j,k}. \tag{4.18}$$

Note that Step 1 amounts to claiming that

$$\begin{aligned}\pi(\bar{X}(t)) \text{ cannot be written as } B \cup C, \text{ where } B \text{ is a strict}\\ \text{subset of } A_k \text{ and } C \subset A \setminus A_k \text{ is non-empty.}\end{aligned} \tag{4.19}$$

Indeed, note that $\pi(\bar{X}(t))$ can always be written as $\pi(\bar{X}(t)) = \bar{B} \cup \bar{C}$ where $\bar{B} \subset A_k$ and $\bar{C} \cap A_k = \emptyset$. The sets $\bar{B}, \bar{C}$ are uniquely determined by

$$\bar{B} = \pi(\bar{X}(t)) \cap A_k, \ \bar{C} = \pi(\bar{X}(t)) \cap (A \setminus A_k).$$

Then Step 1 amounts to claiming that either $\bar{B} = A_k$, or $\bar{B}$ is a strict subset of A_k and $\bar{C} = \emptyset$. This is clearly equivalent to (4.19).

We will prove (4.19) by contradiction and assume that there exists an $s \in (t_k, t_{k+1} \wedge \tau_k)$ such that $\pi(\bar{X}(s))$ can be written as such a union $B \cup C$. Note that C must contain at least one non-zero element, since otherwise $C = \{0\}$ and by Remark 2.2 $B \cup C = \{0, 1, 2, \ldots, d\}$ or $B = \{1, 2, \ldots, d\}$, which contradicts the assumption that B is a strict subset of A_k. Let

$$\bar{t} \doteq \sup\left\{t \leq s : \pi(\bar{X}(t)) \cap (A_k \setminus B) \neq \emptyset\right\}.$$

We claim that $\bar{t} \in [t_k, s)$ and $\pi(\bar{X}(\bar{t})) \cap (A_k \setminus B) \neq \emptyset$. Indeed, $\bar{t} \geq t_k$ is trivial since $\pi(\bar{X}(t_k)) = A \supset A_k$ and $A_k \setminus B$ is non-empty. Thanks to the upper semicontinuity of π and the continuity of $\bar{X}$, there exists a small neighborhood of s such that for any t in this small neighborhood $\pi(\bar{X}(t)) \subset \pi(\bar{X}(s)) = B \cup C$. It follows readily that $\bar{t} < s$. An analogous use of upper semicontinuity shows that $\pi(\bar{X}(\bar{t})) \cap (A_k \setminus B) \neq \emptyset$.

Fix $i \in \pi(\bar{X}(\bar{t})) \cap (A_k \setminus B)$. Note that $i \notin \pi(\bar{X}(t))$ for $t \in (\bar{t}, s)$. Thus, for every $j \in C$ such that $j \neq 0$ and $t \in (\bar{t}, s)$, it follows from equations (4.17) and (4.18) that

$$\begin{aligned}\frac{d}{dt}\left[c_i((\bar{X}(t))_i - c_j((\bar{X}(t))_j\right] &= c_i\bar{\lambda}_{i,k} - c_j(\bar{\lambda}_{j,k} - \bar{\mu}_{j,k}h_j(t))\\ &= \left[c_i(\bar{\lambda}_{i,k} - \rho_k^{(i)}\bar{\mu}_{i,k}) - c_j\bar{\lambda}_{j,k}\right]\\ &\quad + \left[c_i\rho_k^{(i)}\bar{\mu}_{i,k} + c_j\bar{\mu}_{j,k}h_j(t)\right]\\ &> 0.\end{aligned}$$

Therefore,

$$0 \geq c_i((\bar{X}(s))_i - c_j((\bar{X}(s))_j > c_i((\bar{X}(\bar{t}))_i - c_j((\bar{X}(\bar{t}))_j \geq 0,$$

here the first inequality holds since $j \in C \subset \pi(\bar{X}(s))$ and the last inequality is due to $i \in \pi(\bar{X}(\bar{t}))$. This is a contradiction. Thus (4.19) holds and Step 1 is completed.

We now show Step 2. Thanks to Step 1, it suffices to show that $|\pi(\bar{X}(t))| \geq |A_k|$ for $t \in (t_k, t_{k+1} \wedge \tau_k)$. Let $t^* \in (t_k, t_{k+1} \wedge \tau_k)$ be such that

$$|\pi(\bar{X}(t^*))| = \min\{|\pi(\bar{X}(t))| : t \in (t_k, t_{k+1} \wedge \tau_k)\},$$

and assume that $|\pi(\bar{X}(t^*))| < |A_k|$ for the purpose of getting a contradiction. It follows from Step 1 that $B \doteq \pi(\bar{X}(t^*))$ is a strict subset of A_k. Note that $B \subset \{1, 2, \ldots, d\}$ (otherwise $B = \{0, 1, \ldots, d\}$ by Remark 2.2, a clear contradiction). Thanks to the upper semicontinuity of π and the continuity of $\bar{X}$, there exists an open interval containing t^* such that $\pi(\bar{X}(s)) \subset \pi(\bar{X}(t^*))$ for all s in this interval. We will assume (a, b) to be the largest of such intervals. By the definition of t^*, $\pi(\bar{X}(s)) = \pi(\bar{X}(t^*)) = B$ for every $s \in (a, b) \cap (t_k, t_{k+1} \wedge \tau_k)$. Obviously $a \geq t_k$ and $\pi(\bar{X}(a))$ contains B as a strict subset by the upper semicontinuity of π. Therefore, for $s \in (a, b)$, $c_i(\bar{X}(s))_i = c_j(\bar{X}(s))_j$ if $i, j \in B$. However, (4.17) yields

$$\frac{d}{ds}(\bar{X}(s))_j = \begin{cases} \bar{\lambda}_{j,k} & \text{if } j \in A_k \setminus B, \ j \neq 0, \\ \bar{\lambda}_{j,k} - h_j(s)\bar{\mu}_{j,k} & \text{if } j \in B. \end{cases} \tag{4.20}$$

It follows that

$$c_i(\bar{\lambda}_{i,k} - h_i(s)\bar{\mu}_{i,k}) = c_j(\bar{\lambda}_{j,k} - h_j(s)\bar{\mu}_{j,k}) \text{ for } i, \, j \in B. \tag{4.21}$$

In addition, thanks to Proposition 4.6(b),

$$\sum_{i \in B} h_i(s) = 1. \tag{4.22}$$

Solving the system of equations (4.21)–(4.22), we obtain the unique solution

$$h_i(s) = \frac{\bar{\lambda}_{i,k}}{\bar{\mu}_{i,k}} - \frac{1}{c_i \bar{\mu}_{i,k}} \cdot \left(\sum_{j \in B} \frac{1}{c_j \bar{\mu}_{j,k}} \right)^{-1} \cdot \left(-1 + \sum_{j \in B} \frac{\bar{\lambda}_{j,k}}{\bar{\mu}_{j,k}} \right).$$

On the other hand, since $\pi(\psi^*(s)) \equiv A_k \supset B$ and $\dot{\psi}^*(s) \equiv \beta_k$ for $s \in (a, b)$, we have $c_i(\beta_k)_i = c_j(\beta_k)_j$ for every $i, j \in B$. Invoking (4.7) and (4.9), we arrive at

$$c_i(\bar{\lambda}_{i,k} - \rho_k^{(i)} \bar{\mu}_{i,k}) = c_j(\bar{\lambda}_{j,k} - \rho_k^{(j)} \bar{\mu}_{j,k}) \text{ for } i, \, j \in B, \tag{4.23}$$

and that

$$b \doteq \sum_{i \in B} \rho_k^{(i)} = 1 - \sum_{i \in A_k \setminus B} \rho_k^{(i)} < 1. \tag{4.24}$$

Similarly, one can solve (4.23)-(4.24) to uniquely determine $\{\rho_k^{(i)}\}$ for $i \in B$:

$$\rho_k^{(i)} = \frac{\bar{\lambda}_{i,k}}{\bar{\mu}_{i,k}} - \frac{1}{c_i \bar{\mu}_{i,k}} \cdot \left(\sum_{j \in B} \frac{1}{c_j \bar{\mu}_{j,k}} \right)^{-1} \cdot \left(-b + \sum_{j \in B} \frac{\bar{\lambda}_{j,k}}{\bar{\mu}_{j,k}} \right).$$

Since $b < 1$ it follows that $h_i(s) > \rho_k^{(i)}$ for all $i \in B$ and $s \in (a, b)$.

Assume for now that there exists $j \in A_k \setminus B$ such that $j \neq 0$. Then for $i \in B$ and $s \in (a,b)$ we have, thanks to (4.20) and (4.23),

$$\begin{aligned}\frac{d}{ds}\left[c_i(\bar{X}(s))_i - c_j(\bar{X}(s))_j\right] &= c_i\left[\bar{\lambda}_{i,k} - h_i(s)\bar{\mu}_{i,k}\right] - c_j\bar{\lambda}_{j,k} \qquad (4.25)\\ &< c_i\left[\bar{\lambda}_{i,k} - \rho_k^{(i)}\bar{\mu}_{i,k}\right] - c_j\bar{\lambda}_{j,k}\\ &= -c_j\rho_k^{(j)}\bar{\mu}_{j,k}\\ &< 0.\end{aligned}$$

However, since $\pi(\bar{X}(a)) \neq B$, Step 1 implies that $\pi(\bar{X}(a)) \cap (A_k \setminus B) \neq \emptyset$. Pick any j in this set. It follows from (4.25) that for any $i \in B$,

$$0 \geq c_i(\bar{X}(a))_i - c_j(\bar{X}(a))_j > c_i(\bar{X}(t^*))_i - c_j(\bar{X}(t^*))_j.$$

This contradicts that $i \in B = \pi(\bar{X}(t^*))$. Therefore, we have $|\pi(\bar{X}(t))| \geq |A_k|$. It follows from Step 1 that $A_k \subset \pi(\bar{X}(t))$, for $t \in (t_k, t_{k+1} \wedge \tau_k)$.

It remains to consider the case where $A_k \setminus B = \{0\}$. In this case we necessarily have $B = \{1,2,\ldots,d\}$ and $A_k = \{0,1,\ldots,d\} = A$. It follows that $\psi^*(t) \equiv 0$ for $t \in (t_k, t_{k+1})$ and thus $\beta_k = 0$. Then for any $i \in B$ and $s \in (a,b)$ we have, thanks to (4.20) and (4.7),

$$\frac{d}{ds}\left[c_i(\bar{X}(s))_i\right] = c_i\left[\bar{\lambda}_{i,k} - h_i(s)\bar{\mu}_{i,k}\right] < c_i\left[\bar{\lambda}_{i,k} - \rho_k^{(i)}\bar{\mu}_{i,k}\right] = (\beta_k)_i = 0.$$

However, since $\pi(X(a))$ contains B as a strict subset, we must have $\pi(X(a)) = \{0,1,\ldots,d\}$ or $X(a) = 0$. It follows then

$$0 = c_i(\bar{X}(a))_i > c_i(\bar{X}(t^*))_i.$$

This contradicts the non-negativity of $\bar{X}$, and again we have $|\pi(\bar{X}(t))| \geq |A_k|$. It follows from Step 1 that $A_k \subset \pi(\bar{X}(t))$, for $t \in (t_k, t_{k+1} \wedge \tau_k)$.

If $A = A_k$ then we finish the proof of Step 2 since $\pi(\bar{X}(t)) \subset A$. Consider the case when A_k is a strict subset of A, which in turn implies that $A_k \subset \{1,2,\ldots,d\}$. Following an argument analogous to that leading to equations (4.21)–(4.22) and (4.23)–(4.24), we have, for every $i,j \in A_k$ and almost every $t \in (t_k, t_{k+1} \wedge \tau_k)$,

$$c_i(\bar{\lambda}_{i,k} - h_i(t)\bar{\mu}_{i,k}) = c_j(\bar{\lambda}_{j,k} - h_j(t)\bar{\mu}_{j,k}), \ \sum_{j \in A_k} h_j(t) \leq 1. \qquad (4.26)$$

$$c_i(\bar{\lambda}_{i,k} - \rho_k^{(i)}\bar{\mu}_{i,k}) = c_j(\bar{\lambda}_{j,k} - \rho_k^{(j)}\bar{\mu}_{j,k}), \ \sum_{i \in A_k} \rho_k^{(i)} = 1. \qquad (4.27)$$

One can solve these two equations like before to obtain $h_i(t) \leq \rho_k^{(i)}$ for every $i \in A_k$ and $t \in (t_k, t_{k+1} \wedge \tau_k)$, whence

$$\frac{d}{dt}\left[c_i(\bar{X}(t))_i - c_i(\psi^*(t))_i\right] = c_i(\bar{\lambda}_{i,k} - h_i(t)\bar{\mu}_{i,k}) - c_i(\bar{\lambda}_{i,k} - \rho_k^{(i)}\bar{\mu}_{i,k}) \geq 0.$$

It follows that for any $i \in A_k$,

$$c_i(\bar{X}(t))_i - c_i(\psi^*(t))_i \geq c_i(\bar{X}(t_k))_i - c_i(\psi^*(t_k))_i = 0, \qquad (4.28)$$

or $(\bar{X}(t))_i \geq (\psi^*(t))_i$. However, since $\pi((\psi^*(t)) \equiv A_k \subset \{1, 2, \ldots, d\}$, we must have $\bar{X}(t) \neq 0$ and thus $\pi(\bar{X}(t)) \subset \{1, 2, \ldots, d\}$.

On the other hand, for any $j \in A \backslash A_k$ and $j \neq 0$, since $(\dot{\psi}^*(t))_j = (\beta_k)_j = \bar{\lambda}_{j,k}$ thanks to (4.9), we have

$$\frac{d}{dt}\left[c_j(\bar{X}(t))_j - c_j(\psi^*(t))_j\right] = c_j(\bar{\lambda}_{j,k} - h_j(t)\bar{\mu}_{j,k}) - c_j\bar{\lambda}_{j,k} \leq 0.$$

It follows, since $\bar{X}(t_k) = \psi^*(t_k)$ and (4.28), that for every $t \in (t_k, t_{k+1} \wedge \tau_k)$ and $i \in A_k$,

$$c_j(\bar{X}(t))_j - c_j(\psi^*(t))_j \leq 0 \leq c_i(\bar{X}(t))_i - c_i(\psi^*(t))_i.$$

But $(\psi^*(t))_i > (\psi^*(t))_j$ by definition of A_k, and therefore $(\bar{X}(t))_i > (\bar{X}(t))_j$. It follows that $j \notin \pi(\bar{X}(t))$. Therefore $\pi(\bar{X}(t)) \equiv A_k$ and we finish Step 2.

The proof of Step 3 is simple. Assume first $A_k \subset \{1, 2, \ldots, d\}$. Since $\pi(\bar{X}(t)) = A_k$ we know that $\{h_i(t)\}$ satisfies the equation (4.26) except now $\sum_{j \in A_k} h_j(t) = 1$. Compared with equation (4.27), it follows easily that $h_j(t) = \rho_k^{(j)}$. Note that equations (4.9) and (4.12) imply

$$\sum_{i \in A_k} \beta_k^{(i)} \rho_k^{(i)} = \beta_k,$$

whence

$$\frac{d}{dt}\bar{X}(t) = \sum_{i \in A_k} \beta_k^{(i)} h_i(t) = \sum_{i \in A_k} \beta_k^{(i)} \rho_k^{(i)} = \beta_k = \dot{\psi}^*(t),$$

which implies $\bar{X}(t) = \dot{\psi}^*(t)$ for all $t \in (t_k, t_{k+1} \wedge \tau_k)$.

For the case where $A_k = \{0, 1, \ldots, d\}$, we must have $\bar{X}(t) = \psi^*(t) = 0$. It is not difficult to see that equations (4.26) and (4.27) reduce to

$$c_i(\bar{\lambda}_{i,k} - h_i(t)\bar{\mu}_{i,k}) = 0 = c_i(\bar{\lambda}_{i,k} - \rho_k^{(i)} \bar{\mu}_{i,k}), \; i = 1, 2, \ldots, d.$$

Thus $h_i(t) = \rho_k^{(i)}$ for all $i = 1, 2, \ldots, d$, and also $h_0(t) = \rho_k^{(0)}$ since $\sum_{i=0}^d h_i(t) = \sum_{i=0}^d \rho_k^{(i)} = 1$. □

4.7. Analysis of the cost

In this section, we prove the Laplace lower bound, or inequality (4.4).

Proof. Thanks to (4.11) and (4.10),

$$\lim_{n \to \infty} E_{x_n}\left[\int_{t_k}^{t_{k+1}} \sum_{v \in \Theta} r(\bar{X}^n(t); v)\ell\left(\frac{\bar{r}(\bar{X}^n(t), t; v)}{r(\bar{X}^n(t); v)}\right) dt\right]$$

$$= \lim_{n \to \infty} E_{x_n}\left[\int_{t_k}^{t_{k+1}} \sum_{i \in A_k, i \neq 0} \mu_i \ell\left(\frac{\bar{\mu}_{i,k}}{\mu_i}\right) \gamma_i^n(dt) + \sum_{j=1}^d \lambda_j \ell\left(\frac{\bar{\lambda}_{j,k}}{\lambda_j}\right) dt\right]$$

$$= \int_{t_k}^{t_{k+1}} \left[\sum_{i \in A_k, i \neq 0} \mu_i \ell\left(\frac{\bar{\mu}_{i,k}}{\mu_i}\right) h_i(t) + \sum_{j=1}^d \lambda_j \ell\left(\frac{\bar{\lambda}_{j,k}}{\lambda_j}\right)\right] dt$$

$$
\begin{aligned}
&= (t_{k+1} - t_k) \left[\sum_{i \in A_k, i \neq 0} \mu_i \ell \left(\frac{\bar{\mu}_{i,k}}{\mu_i} \right) \rho_k^{(i)} + \sum_{j=1}^{d} \lambda_j \ell \left(\frac{\bar{\lambda}_{j,k}}{\lambda_j} \right) \right] \\
&\leq (t_{k+1} - t_k) \cdot [L^{A_k}(\beta_k) + \varepsilon] \\
&= \int_{t_k}^{t_{k+1}} L(\psi^*(t), \dot{\psi}^*(t)) dt + (t_{k+1} - t_k)\varepsilon.
\end{aligned}
$$

Summing over k and observing

$$\lim_{n \to \infty} E_{x_n} h(\bar{X}^n) = E_x h(\bar{X}) = h(\psi^*),$$

(4.4) follows readily. This completes the proof of inequality (4.4) and also the proof of Theorem 3.1. □

5. Summary

In this paper we analyze the large deviation properties for a class of systems with discontinuous dynamics, namely, the WSLQ policy for a network with multiple queues and a single server. A key observation that allows us to derive explicitly the large deviation rate function on path space is that the stability condition about the interface is automatically implied in the formulation of a general large deviation upper bound.

This is not an unprecedented situation, and indeed analogous results are proved in [2] for processes that model Jackson networks and head-of-the-line processor sharing. These problems also feature discontinuous statistics, though the discontinuities appear on the boundary of the state space rather than the interior, and whence alternative techniques based on Skorohod mapping can be used. With the inclusion of the WSLQ policy there are now a number of physically meaningful models with discontinuous statistics for which the upper bound of [6] is tight. An intriguing question that will be investigated elsewhere is whether there is a common property of these models that can be easily identified and recognized.

Appendix A. Proof of Lemma 4.1

Before getting into the details of the construction, it is worth pointing out that the main difficulty in the proof is dealing with the situation where the function ψ may spiral into (or away from) a low dimensional interface while hitting higher dimensional interfaces infinitely many times in the process. More precisely, there can be a time t and a sequence $t_n \uparrow t$ (or $t_n \downarrow t$) such that $|\pi(\psi(t_n))| < |\pi(\psi(t))|$, but with the sets $\{\pi(\psi(t_n))\}$ different for successive n. The problem is how to approximate ψ on a small neighborhood of t with a small cost, which is made difficult by the fact that the domain of finiteness of $L(x, \cdot)$ is not continuous in x.

Proof. Throughout this section we assume that ψ is Lipschitz continuous. This is without loss of generality, since for a given continuous function ψ and any $\delta > 0$, there exists a Lipschitz continuous function ζ such that $\|\zeta - \psi\|_\infty \leq \delta$ and $I_x(\zeta) \leq I_x(\psi) + \delta$. The proof of this claim is based on a time-rescaling argument very much analogous to that of [4, Lemma 6.5.3], and we omit the details.

A.1. Dividing the time interval

The approximation requires a suitable division of the time interval $[0, 1]$. The following results are useful in proving the main result of this section, namely, Lemma A.4. Note that whenever we say two intervals are "non-overlapping", it means that the two intervals cannot have common interiors but may have a same endpoint.

Lemma A.1. *Consider a non-empty closed interval* $[a, b]$ *and assume*

$$k \doteq \max\{|\pi(\psi(t))| : t \in [a, b]\} \geq 2.$$

Then for arbitrary $\sigma > 0$*, there exists a finite collection of non-overlapping intervals* $\{[\alpha_j, \beta_j]\}$ *such that*

1. $\alpha_j, \beta_j \in [a, b]$ *and* $0 \leq \beta_j - \alpha_j \leq \sigma$ *for every* j.
2. $|\pi(\psi(t))| \leq k - 1$ *for every* $t \in [a, b] \setminus \cup_j [\alpha_j, \beta_j]$.

Lemma A.2. *Consider a non-empty interval* (a, b) *such that for some* $k \geq 2$, $|\pi(\psi(t))| \leq k$ *for* $t \in (a, b)$ *and* $|\pi(\psi(a))| \wedge |\pi(\psi(b))| \geq k + 1$. *Then for arbitrary* $\sigma > 0$ *and* $\varepsilon > 0$*, there exist a finite collection of non-overlapping intervals* $\{[\alpha_j, \beta_j]\}$ *and* $\varepsilon_1, \varepsilon_2 \in [0, \varepsilon)$ *such that*

1. $\alpha_j, \beta_j \in [a + \varepsilon_1, b - \varepsilon_2]$ *and* $0 \leq \beta_j - \alpha_j \leq \sigma$ *for each* j,
2. *for* $t \in (\alpha_j, \beta_j)$, $\pi(\psi(t)) \subset \pi(\psi(\alpha_j)) = \pi(\psi(\beta_j))$,
3. $|\pi(\psi(\alpha_j))| = |\pi(\psi(\beta_j))| = k$ *for each* j,
4. *for any* $t \in (a, a + \varepsilon_1]$, $\pi(\psi(t))$ *is a strict subset of* $\pi(\psi(a))$ *and* $|\pi(\psi(t))| \leq k \leq |\pi(\psi(a + \varepsilon_1))|$,
5. *for any* $t \in [b - \varepsilon_2, b)$, $\pi(\psi(t))$ *is a strict subset of* $\pi(\psi(b))$ *and* $|\pi(\psi(t))| \leq k \leq |\pi(\psi(b - \varepsilon_2))|$,
6. $|\pi(\psi(t))| \leq k - 1$ *for every* $t \notin \cup_j [\alpha_j, \beta_j] \cup (a, a + \varepsilon_1] \cup [b - \varepsilon_2, b)$.

Lemma A.3. *Consider a non-empty interval* $[a, b)$ *such that for some* $2 \leq k \leq d$, $|\pi(\psi(t))| \leq k$ *for* $t \in [a, b)$ *and* $|\pi(\psi(b))| \geq k + 1$. *Then for arbitrary* $\sigma > 0$ *and* $\varepsilon > 0$*, there exist a finite collection of non-overlapping intervals* $\{[\alpha_j, \beta_j]\}$ *and* $\varepsilon_2 \in [0, \varepsilon)$ *such that*

1. $\alpha_j, \beta_j \in [a, b - \varepsilon_2]$ *and* $0 \leq \beta_j - \alpha_j \leq \sigma$ *for each* j,
2. *for* $t \in (\alpha_j, \beta_j)$, $\pi(\psi(t)) \subset \pi(\psi(\alpha_j)) = \pi(\psi(\beta_j))$,
3. $|\pi(\psi(\alpha_j))| = |\pi(\psi(\beta_j))| = k$ *for each* j,
4. *for every* $t \in [b - \varepsilon_2, b)$, $\pi(\psi(t))$ *is a strict subset of* $\pi(\psi(b))$ *and* $|\pi(\psi(t))| \leq k \leq |\pi(\psi(b - \varepsilon_2))|$,
5. $|\pi(\psi(t))| \leq k - 1$ *for every* $t \notin \cup_j [\alpha_j, \beta_j] \cup [b - \varepsilon_2, b)$.

Symmetric results hold for a non-empty interval $(a, b]$ *such that for some* $k \geq 2$, $|\pi(\psi(t))| \leq k$ *for* $t \in (a, b]$ *and* $|\pi(\psi(a))| \geq k + 1$.

We will only provide the details of the proof for Lemma A.2. The proofs for Lemma A.1 and Lemma A.3 are very similar (indeed, simpler versions of the proof for Lemma A.2), and thus omitted.

Proof. If the set $\{t \in (a, b) : |\pi(\psi(t))| = k\}$ is empty, then the claim holds trivially since we can let $\varepsilon_1 = \varepsilon_2 = 0$ and $\{[\alpha_j, \beta_j]\} \doteq \emptyset$. Assume from now on that this set is non-empty. We first define ε_1 and ε_2. Let

$$\bar{a} \doteq \inf\{t \in (a, b) : |\pi(\psi(t))| = k\}, \; \bar{b} \doteq \sup\{t \in (a, b) : |\pi(\psi(t))| = k\}.$$

If $\bar{a} = a$, then the upper semicontinuity of π implies that there exists $0 < \varepsilon_1 < \varepsilon$ such that $|\pi(\psi(a + \varepsilon_1))| = k$ and $\pi(\psi(t))$ is a subset (whence a strict subset) of $\pi(\psi(a))$ for every $t \in (a, a + \varepsilon_1]$. If $\bar{a} > a$ then we let $\varepsilon_1 = 0$. ε_2 is defined in a completely analogous fashion. It is easy to see that parts 4 and 5 of the claim are satisfied. We will define α_j and β_j recursively.

1. Let $\alpha_1 \doteq a + \varepsilon_1$ if $\bar{a} = a$ and $\alpha_1 \doteq \bar{a}$ if $\bar{a} > a$. Clearly, in either case $\alpha_1 \in [a+\varepsilon_1, b-\varepsilon_2]$. Moreover, we have $|\pi(\psi(\alpha_1))| = k$. Indeed, when $\bar{a} = a$ it follows from the definition, and when $\bar{a} > a$, it follows from a simple argument by contradiction, thanks to the upper semicontinuity of π and the assumption that $|\pi(\psi(t))| \leq k$ for every $t \in (a, b)$.
2. Suppose now $\alpha_j \in [a + \varepsilon_1, b - \varepsilon_2]$ is given such that $|\pi(\psi(\alpha_j))| = k$. We wish to define β_j. Let
$$\begin{aligned} t_j &\doteq \inf\{t \in (\alpha_j, b) : \pi(\psi(t)) \not\subset \pi(\psi(\alpha_j))\}, \\ s_j &\doteq \sup\{t \in [\alpha_j, t_j \wedge (\alpha_j + \sigma)) : \pi(\psi(t)) = \pi(\psi(\alpha_j))\}, \\ \beta_j &\doteq s_j \wedge (b - \varepsilon_2). \end{aligned}$$
It is easy to check that $\pi(\psi(s_j)) = \pi(\psi(\alpha_j))$. We claim that $\pi(\psi(\beta_j)) = \pi(\psi(\alpha_j))$. This is trivial when $\beta_j = s_j$. Assume for now that $\beta_j < s_j$. Then we must have $s_j > b - \varepsilon_2$ and $\beta_j = b - \varepsilon_2$. This could happen only if $\varepsilon_2 > 0$, in which case $|\pi(\psi(b-\varepsilon_2))| = k$, or $|\pi(\psi(\beta_j))| = k$. But $\pi(\psi(\beta_j)) \subset \pi(\psi(\alpha_j))$ since $\beta_j < s_j \leq t_j$. Thus we must have $\pi(\psi(\alpha_j)) = \pi(\psi(\beta_j))$. It is clear now that parts 1, 2, and 3 of the lemma holds.
3. If $\beta_j = \bar{b}$ (when $\bar{b} < b$) or $\beta_j = b - \varepsilon_2$ (when $\bar{b} = b$), we stop. Otherwise, define $\alpha_{j+1} \doteq \inf\{t \in (\beta_j, b) : |\pi(\psi(t))| = k\}$. Then $\alpha_{j+1} \in [a + \varepsilon_1, b - \varepsilon_2]$ and $|\pi(\psi(\alpha_{j+1}))| = k$. Now repeat step 2.

Note that part 6 holds by the construction. It only remains to show that the construction will terminate in finitely many steps. Observe that $\alpha_{j+1} \geq t_j \wedge (\alpha_j + \sigma)$, since by definition, for every $t \in (\beta_j, t_j \wedge (\alpha_j + \sigma))$, $\pi(\psi(t))$ is a strict subset of $\pi(\psi(\alpha_j))$ and whence $|\pi(\psi(t))| < k$. Therefore, $\alpha_{j+1} - \alpha_j \geq (t_j - \alpha_j) \wedge \sigma$, and it suffices to show that $t_j - \alpha_j$ is uniformly bounded away from 0.

We should first consider the case $k \le d-1$. Observe that in the construction we indeed have $\alpha_j, \beta_j \in \bar{I} \doteq [(a+\varepsilon_1)\vee \bar{a}, (b-\varepsilon_2)\wedge \bar{b}] \subset (a,b)$. Define

$$c \doteq \inf\left\{ \max_{i=1,\ldots,d} c_i(\psi(t))_i - \max_{i\notin\pi(\psi(t))} c_i(\psi(t))_i : t\in \bar{I}, |\pi(\psi(t))| = k \right\}.$$

We claim that $c>0$. If this is not the case, there exists a sequence of $\{t_n\}\subset \bar{I}$ such that

$$\max_{i=1,\ldots,d} c_i(\psi(t_n))_i - \max_{i\notin\pi(\psi(t_n))} c_i(\psi(t_n))_i \downarrow 0.$$

One can find a subsequence of $\{t_n\}$, still denoted by $\{t_n\}$, such that $t_n \to t^* \in I$ and $\pi(\psi(t_n)) \equiv B$ for some $B\subset\{1,2,\ldots,d\}$ with $|B| = k$. Thanks to the upper semicontinuity of π, $B\subset \pi(\psi(t^*))$. But $|\pi(\psi(t^*))| \le k$, thus $B = \pi(\psi(t^*))$. However, for every $j\in B$,

$$\begin{aligned}
0 &= \lim_n \left[\max_{i=1,\ldots,d} c_i(\psi(t_n))_i - \max_{i\notin\pi(\psi(t_n))} c_i(\psi(t_n))_i \right] \\
&= \lim_n \left[c_j(\psi(t_n))_j - \max_{i\notin B} c_i(\psi(t_n))_i \right] \\
&= c_j(\psi(t^*))_j - \max_{i\notin B} c_i(\psi(t^*))_i \\
&> 0,
\end{aligned}$$

a contradiction. Therefore $c>0$.

Now, by the definition of t_j, there exists $l\notin \pi(\psi(\alpha_j))$ such that $l\in\pi(\psi(t_j))$. Therefore, for every $i\in\pi(\psi(\alpha_j))$,

$$c_i(\psi(t_j))_i - c_l(\psi(t_j))_l \le 0.$$

But $c_i(\psi(\alpha_j))_i - c_l(\psi(\alpha_j))_l \ge c$ and ψ is Lipschitz continuous. It follows readily that $t_j - \alpha_j$ is uniformly bounded away from 0.

The case $k=d$ can be treated in a completely analogous fashion with

$$c \doteq \inf\left\{ \max_{i=1,\ldots,d} c_i(\psi(t))_i : t\in\bar{I}, |\pi(\psi(t))| = d \right\}.$$

We omit the details. □

The next lemma is the main result of this section, from which one can construct the approximating function in $\mathcal{N}$. The intervals $\{E_i\}$ in the lemma are indeed introduced to take care of the "spiraling" problem.

Lemma A.4. *Consider the interval* $[0,1]$. *Given* $\sigma>0$, *there exist two finite collections of non-overlapping intervals* $\{I_j\}$ *and* $\{E_i\}$ *such that*

1. $I_j\subset[0,1]$ *is of the form* $[a_j,b_j]$ *with* $0\le b_j-a_j\le\sigma$ *for each* j,
2. $E_i\subset[0,1]$ *is either of the form* $[z_i,d_i)$ *or* $(z_i,d_i]$, *and* $\sum_i(d_i-z_i)\le\sigma$,
3. $|\pi(\psi(t))| = 1$ *for all* $t\notin(\cup_j I_j)\cup(\cup_i E_i)$,
4. *for every* j *and every* $t\in(a_j,b_j)$, $\pi(\psi(t))\subset\pi(\psi(a_j)) = \pi(\psi(b_j))$,
5. *for every* j, $|\pi(\psi(a_j))| = |\pi(\psi(b_j))| \ge 2$,

6. *for all* $t \in E_i = (z_i, d_i]$, $\pi(\psi(t))$ *is a strict subset of* $\pi(\psi(z_i))$ *and* $|\pi(\psi(t))| \leq |\pi(\psi(d_i))|$,
7. *for all* $t \in E_i = [z_i, d_i)$, $\pi(\psi(t))$ *is a strict subset of* $\pi(\psi(d_i))$ *and* $|\pi(\psi(t))| \leq |\pi(\psi(z_i))|$.

Proof. Let $N \doteq \max\{|\pi(\psi(t))| : t \in [0,1]\}$. If $N = 1$ then the claim holds trivially. Assume from now on that $N \geq 2$. The construction can be done in N steps.

Step 1*:* Apply Lemma A.1 to the interval $[0,1]$ with $k = N$. This will produce a collection of closed intervals $\{I_j^{(1)}\}$. Let $U_1 \doteq [0,1] \setminus \cup_j I_j^{(1)}$.

Step 2*:* Note that U_1 is the union of finitely many non-overlapping intervals. These intervals are of two types. They are either open intervals that satisfy the conditions of Lemma A.2 with $k = N - 1$, or intervals of type $[0, a)$ or $(b, 1]$ that satisfy the conditions of Lemma A.3. Apply the corresponding lemma to these intervals to generate a collection of closed intervals $\{I_j^{(2)}\}$ and intervals $\{E_i^{(2)}\}$ of type $[a, b)$ or $(b, a]$. Since the number of intervals in $\{E_i^{(2)}\}$ is at most twice the number of intervals in U_1, one can choose ε in Lemma A.2 and Lemma A.3 so small that the total Lebesgue measure of $\cup_i E_i^{(2)}$ is bounded from above by σ/N. Define $U_2 \doteq U_1 \setminus ((\cup_j I_j^{(2)}) \cup (\cup_i E_i^{(2)}))$.

Step m*:* For $3 \leq m \leq N - 1$, the procedure of Step m is just like Step 2, except U_1 is replaced by U_{m-1}.

Step N*:* Note that U_{N-1} consists of intervals on which $\pi(\psi(\cdot))$ is a singleton. We stop the construction.

Let $\{I_j\} \doteq \cup_l \{I_j^{(l)}\}$ and $\{E_i\} \doteq \cup_l \{E_j^{(l)}\}$, and it is not difficult to see that $\{I_j\}$ and $\{E_i\}$ have the required properties. □

A.2. Construction of the approximating function

In this section we construct an approximating function of ψ parameterized by σ. Note that the set $[0,1] \setminus ((\cup_j I_j) \cup (\cup_i E_i))$ is the union of finitely many non-overlapping intervals, and we will denote these intervals by $\{G_k\}$. Without loss of generality assume that the length of each interval G_k is bounded above by σ (if necessary, we can divide G_k into the union of several smaller intervals each with a length less than σ).

Define a piece-wise constant function u^σ on interval $[0,1]$ as follows. Given any interval $D \in \{I_j\} \cup \{E_i\} \cup \{G_k\}$ with positive length, define for every $t \in D$,

$$u^\sigma(t) \doteq \frac{1}{\text{length of } D} \int_D \dot{\psi}(s)\, ds.$$

Then the candidate approximating function will be

$$\psi^\sigma(t) \doteq \psi(0) + \int_0^t u^\sigma(s) ds.$$

Note that $\psi^\sigma(t)$ coincides with $\psi(t)$ at those t that are end points of intervals $\{I_j\}$, $\{E_i\}$, and $\{G_k\}$. Furthermore, ψ^σ is affine on each of these intervals.

We claim that $\psi^\sigma \in \mathcal{N}$. All we need is to show is that $\pi(\psi^\sigma(t))$ remain unchanged in the interior of each interval. This is immediate from the following result, whose proof is simple and straightforward, and thus omitted.

Lemma A.5. *Let ϕ be an affine function on interval $[a,b]$. If $\pi(\phi(a)) \cap \pi(\phi(b)) \neq \emptyset$, then for every $t \in (a,b)$ we have $\pi(\phi(t)) = \pi(\phi(a)) \cap \pi(\phi(b))$.*

A.3. The analysis of the rate function

Since the length of any interval in $\{I_j\}$, $\{E_i\}$, and $\{G_k\}$ is bounded from above by σ, $\lim_{\sigma\to 0} \|\psi^\sigma - \psi\|_\infty = 0$. Therefore it only remains to show that given $\delta > 0$ we have $I_x(\psi^\sigma) \leq I_x(\psi) + \delta$ for σ small enough. Clearly we can assume $I_x(\psi) < \infty$ hereafter.

The analysis for intervals in $\{I_j\}$ and $\{G_k\}$ is simple. Consider an interval $D \in \{I_j\} \cup \{G_k\}$, and denote the interior of D by (a,b) [D itself could be $[a,b]$, (a,b), $(a,b]$, or $[a,b)$]. Since $\psi(a) = \psi^\sigma(a)$ and $\psi(b) = \psi^\sigma(b)$ by construction, it follows from Lemma A.4 and Lemma A.5 that

$$\pi(\psi(t)) \subset \pi(\psi(a)) \cap \pi(\psi(b)) = \pi(\psi^\sigma(a)) \cap \pi(\psi^\sigma(b)) = \pi(\psi^\sigma(t)) \doteq B.$$

for every $t \in (a,b)$. Note that $L^B(\beta) \leq L^A(\beta)$ for every β whenever $A \subset B$, thanks to Lemma 4.3. Therefore,

$$\int_a^b L(\psi(t), \dot{\psi}(t))\, dt = \int_a^b L^{\pi(\psi(t))}(\dot{\psi}(t))\, dt \geq \int_a^b L^B(\dot{\psi}(t))\, dt.$$

Furthermore, for $t \in (a,b)$, the construction of ψ^σ implies that

$$\dot{\psi}^\sigma(t) \equiv \frac{\psi(b) - \psi(a)}{b-a} \doteq v.$$

Thanks to the convexity of L^B and Jensen's inequality, we arrive at

$$\int_a^b L^B(\dot{\psi}(t))\, dt \geq (b-a)L^B(v) = \int_a^b L(\psi^\sigma(t), \dot{\psi}^\sigma(t))dt,$$

whence

$$\int_a^b L(\psi(t), \dot{\psi}(t))\, dt \geq \int_a^b L(\psi^\sigma(t), \dot{\psi}^\sigma(t))dt. \tag{A.1}$$

for any σ.

Now we turn to analyze the intervals in $\{E_i\}$. The next two lemmas claim that over any interval, say $D \in \{E_i\}$, ψ^σ will always be in the domain of finiteness of the local rate function L. The first lemma considers the case where $D = (a,b]$ and the second considers the case where D is of form $[a,b)$.

Lemma A.6. *Suppose that $I_x(\psi) < \infty$ and there exists an interval $(a,b] \subset [0,1]$ such that $\pi(\psi(t))$ is a strict subset of $\pi(\psi(a))$ for $t \in (a,b]$. Then $(\psi(a))_j \leq (\psi(b))_j$ for all $j \in \{1,\ldots,d\}$.*

Proof. For notational simplicity, we assume $c_1 = c_2 = \cdots = c_d$. The case of general $(c_1, \ldots, c_d)$ is shown in exactly the same fashion with $(\psi(t))_j$ replaced by $c_j(\psi(t))_j$, and whence omitted.

Let $B \doteq \pi(\psi(a))$. For $j \notin B$, we have $j \notin \pi(\psi(t))$ and thus $\dot{\psi}(t) \geq 0$ for almost every $t \in (a, b]$, thanks to Remark 4.5 and the assumption that $I_x(\psi)$ is finite. Therefore for $j \notin B$ the claim holds. Now we consider those $j \in B$. Fix arbitrarily $\varepsilon > 0$ and let

$$t^* \doteq \inf\{t > a : (\psi(t))_j < (\psi(a))_j - \varepsilon \text{ for some } j \in B\} \wedge b.$$

It suffices to show that $t^* = b$ always holds. Indeed, if this is the case, we have $(\psi(b))_j \geq (\psi(a))_j - \varepsilon$ for every $j \in B$. Since ε is arbitrary, we arrive at the desired inequality for $j \in B$.

We will argue by contradiction and assume $t^* < b$. It follows that $(\psi(t^*))_j \geq (\psi(a))_j - \varepsilon$ for every $j \in B$, and there exist $j^* \in B$ and a sequence $t_n \downarrow t^*$ such that

$$(\psi(t_n))_{j^*} < (\psi(a))_{j^*} - \varepsilon. \tag{A.2}$$

In particular, $(\psi(t^*))_{j^*} = (\psi(a))_{j^*} - \varepsilon$.

We claim that $(\psi(t^*))_j = (\psi(a))_j - \varepsilon$ holds for every $j \in B$. If this is not the case, then there exists (abusing the notation) $j \in B$ such that

$$(\psi(t^*))_j > (\psi(a))_j - \varepsilon.$$

Since $j, j^* \in B = \pi(\psi(a))$ we have $(\psi(a))_j = (\psi(a))_{j^*}$, and whence $(\psi(t^*))_j > (\psi(t^*))_{j^*}$. Therefore we can find a small interval, say $[t^*, t^* + \delta)$ such that for any t in this interval we have $(\psi(t))_j > (\psi(t))_{j^*}$, which in turn implies $j^* \notin \pi(\psi(t))$. Thanks to the finiteness of $I_x(\psi)$ and Remark 4.5, $(\dot{\psi}(t))_{j^*} \geq 0$ for almost every $t \in [t^*, t^* + \delta)$. In particular, $(\psi(t))_{j^*} \geq (\psi(t^*))_{j^*} = (\psi(a))_{j^*} - \varepsilon$ for all $t \in [t^*, t^* + \delta)$, which contradicts equation (A.2) for large n. Therefore $(\psi(t^*))_j = (\psi(a))_j - \varepsilon$ holds for every $j \in B$.

Since $B = \pi(\psi(a))$, it follows that $(\psi(t^*))_j$ takes the same value for every $j \in B$. This contradicts the assumption that $\pi(\psi(t^*))$ is a strict subset of B. We complete the proof. ☐

Lemma A.7. *Suppose that $I_x(\psi) < \infty$ and there exists an interval $[a, b) \subset [0, 1]$ such that $\pi(\psi(t))$ is a strict subset of $\pi(\psi(b))$ for every $t \in [a, b)$. Let $B \doteq \pi(\psi(a))$. If $|\pi(\psi(t))| \leq |B|$ for every $t \in [a, b)$, then $(\psi(a))_j \leq (\psi(b))_j$ for all $j \notin B$ and $j \neq 0$.*

Proof. As in the proof of Lemma A.6 we assume without loss of generality that $c_1 = c_2 = \cdots = c_d = 1$. Let $A \doteq \pi(\psi(b))$. For $j \notin A$, $(\dot{\psi}(t))_j \geq 0$ for almost every $t \in [a, b)$ since $j \notin \pi(\psi(t))$. Thus the claim holds for $j \notin A$. It remains to show for those $j \in A \setminus B$ such that $j \neq 0$. We will argue by contradiction and assume that there is a $j^* \in A \setminus B$ such that $(\psi(a))_{j^*} > (\psi(b))_{j^*}$.

For each $i \in B$ let $t_i \doteq \inf\{t > a : (\psi(t))_i \leq (\psi(a))_{j^*}\}$. Note that $(\psi(a))_i > (\psi(a))_{j^*}$ since $i \in B = \pi(\psi(a))$ and $j^* \notin B$, whence $t_i > a$. Similarly, since

$j^* \in A = \pi(\psi(b))$, $(\psi(b))_i \leq (\psi(b))_{j^*} < (\psi(a))_{j^*}$, whence $t_i < b$. It follows that $(\psi(t_i))_i = (\psi(a))_{j^*}$.

We claim that $i \in \pi(\psi(t_i))$ for all $i \in B$. Otherwise, there exists a small positive number δ such that $i \notin \pi(\psi(t))$ for $t \in (t_i - \delta, t_i + \delta)$. Thanks to the assumption that $I_x(\psi)$ is finite, $(\dot{\psi}(t))_i \geq 0$ for almost every $t \in (t_i - \delta, t_i + \delta)$. In particular, $(\psi(t_i - \delta))_i \leq (\psi(t_i))_i = (\psi(a))_{j^*}$, which contradicts the definition of t_i. Therefore, $i \in \pi(\psi(t_i))$ for all $i \in B$. An immediate consequence is that for any $i, k \in B$, $(\psi(t_i))_k \leq (\psi(t_i))_i = (\psi(a))_{j^*}$, whence $t_k \leq t_i$ by definition. Therefore, for all $i, k \in B$, $t_i = t_k \doteq t^*$ and $B \subset \pi(\psi(t^*))$. However, since $|\pi(\psi(t^*))| \leq |B|$ by assumption, we have necessarily $B = \pi(\psi(t^*))$.

Since $j^* \notin B = \pi(\psi(t^*))$, we have $(\psi(t^*))_{j^*} < (\psi(a))_{j^*}$. Define

$$s \doteq \sup\{t \in [a, t^*) : (\psi(t))_{j^*} \geq (\psi(a))_{j^*}\} \wedge t^*.$$

Clearly $s \in [a, t^*)$ and $(\psi(s))_{j^*} = (\psi(a))_{j^*}$. By the definitions of s and t_i and that $t^* \equiv t_i$ for all $i \in B$, we have, for any $t \in (s, t^*)$ and any $i \in B$,

$$(\psi(t))_{j^*} < (\psi(a))_{j^*} < (\psi(t))_i.$$

Therefore, $j^* \notin \pi(\psi(t))$ and whence $(\dot{\psi}(t))_{j^*} \geq 0$ for almost every $t \in (s, t^*)$. In particular, $(\psi(a))_{j^*} = (\psi(s))_{j^*} \leq (\psi(t^*))_{j^*} < (\psi(a))_{j^*}$, a contradiction. We complete the proof. □

We are now ready to show that $I_x(\psi^\sigma) \leq I_x(\psi) + \delta$ for σ small enough. Let M be the Lipschitz constant for ψ, and define

$$C \doteq \max_{A \subset \{1,\ldots,d\}} \sup\{L^A(\beta) : \beta \in \mathrm{dom}(L^A), \|\beta\| \leq M\}.$$

That C is finite follows easily from Lemma 4.3 and the definition of ℓ.

We now analyze the intervals in $\{E_i\}$. If $E_i = [z_i, d_i)$, then by Lemma A.5 and Lemma A.4, $\pi(\psi^\sigma(t)) = \pi(\psi(z_i)) \doteq B$ for all $t \in (z_i, d_i)$. Since

$$\dot{\psi}^\sigma(t) \equiv \frac{\psi(d_i) - \psi(z_i)}{d_i - z_i} \doteq v,$$

it follows from Lemma A.7 and Remark 4.5 that $v \in \mathrm{dom}(L^B)$. Therefore

$$\int_{z_i}^{d_i} L(\psi^\sigma(t), \dot{\psi}^\sigma(t))\, dt = (d_i - z_i) L^B(v) \leq C(d_i - z_i). \tag{A.3}$$

The same inequality holds for the case $E_i = (z_i, d_i]$, where Lemma A.6 is invoked in place of Lemma A.7.

By (A.1), (A.3), the non-negativity of L, and Lemma A.4, we have

$$\begin{aligned} \int_0^1 L(\psi^\sigma(t), \dot{\psi}^\sigma(t))\, dt &\leq \int_0^1 L(\psi(t), \dot{\psi}(t))\, dt + C \sum_i (d_i - z_i) \\ &\leq \int_0^1 L(\psi(t), \dot{\psi}(t))\, dt + C\sigma. \end{aligned}$$

Choose $\sigma < \delta/C$ and we complete the proof. □

Acknowledgment

We thank the referee for correcting an error in the first version of this paper and for helpful suggestions.

References

[1] M. Alanyali and B. Hajek. On large deviations of Markov processes with discontinuous statistics. *Ann. Appl. Probab.*, 8:45–66, 1998.

[2] R. Atar and P. Dupuis. Large deviations and queueing networks: methods for rate function identification. *Stoch. Proc. and Their Appl.*, 84:255–296, 1999.

[3] P. Dupuis and R.S. Ellis. The large deviation principle for a general class of queueing systems, I. *Trans. Amer. Math. Soc.*, 347:2689–2751, 1996.

[4] P. Dupuis and R.S. Ellis. *A Weak Convergence Approach to the Theory of Large Deviations*. John Wiley & Sons, New York, 1997.

[5] P. Dupuis and R.S. Ellis. Large deviations for Markov processes with discontinuous statistics, II: Random walks. *Probab. Th. Rel. Fields*, 91:153–194, 1992.

[6] P. Dupuis, R.S. Ellis, and A. Weiss. Large deviations for Markov processes with discontinuous statistics, I: General upper bounds. *Annals of Probability*, 19:1280–1297, 1991.

[7] R. Foley and D. McDonald. Join the shortest queue: stability and exact asymptotics. *Ann. Appl. Probab.*, 11:569–607, 2001.

[8] I. Ignatiouk-Robert. Large deviations for processes with discontinuous statistics. *Ann. Probab.*, 33:1479–1508, 2005.

[9] K. Majewski. Large deviations of the steady-state distribution of reflected processes with applications to queueing systems. *Queueing Systems*, 29:351–381, 1998.

[10] K. Ramanan and S. Stolyar. Largest weighted delay first scheduling: Large deviations and optimality. *The Annals of Applied Probab.*, 11:1–49, 2001.

[11] A. Shwartz and A. Weiss. *Large Deviations for Performance Analysis: Queues, Communication and Computing*. Chapman and Hall, New York, 1995.

[12] L. Ying, R. Srikant, A. Eryilmaz, and G.E. Dullerud. A large deviation analysis of scheduling in wireless networks. *Preprint*, 2005.

Paul Dupuis, Kevin Leder and Hui Wang
Lefshetz Center for Dynamical Systems
Brown University
Providence, RI 02912, USA
e-mail: dupuis@dam.brown.edu

e-mail: kleder@dam.brown.edu

e-mail: huiwang@dam.brown.edu

Progress in Probability, Vol. 60, 257–269

Exponential Inequalities for Empirical Unbounded Context Trees

Antonio Galves and Florencia Leonardi

Abstract. In this paper we obtain non-uniform exponential upper bounds for the rate of convergence of a version of the algorithm Context, when the underlying tree is not necessarily bounded. The algorithm Context is a well-known tool to estimate the context tree of a Variable Length Markov Chain. As a consequence of the exponential bounds we obtain a strong consistency result. We generalize in this way several previous results in the field.

Mathematics Subject Classification (2000). 62M09, 60G99.

Keywords. Variable memory processes, unbounded context trees, algorithm Context.

1. Introduction

In this paper we present an exponential bound for the rate of convergence of the algorithm Context for a class of unbounded variable memory models, taking values on a finite alphabet A. From this it follows a strong consistency result for the algorithm Context in this setting. Variable memory models were first introduced in the information theory literature by Rissanen [11] as a universal system for data compression. Originally called by Rissanen *finite memory source* or *probabilistic tree*, this class of models recently became popular in the statistics literature under the name of *Variable Length Markov Chains (VLMC)* [1].

The idea behind the notion of variable memory models is that the probabilistic definition of each symbol only depends on a finite part of the past and the length of this relevant portion is a function of the past itself. Following Rissanen we call

This work is part of PRONEX/FAPESP's project *Stochastic behavior, critical phenomena and rhythmic pattern identification in natural languages* (grant number 03/09930-9) and CNPq's projects *Stochastic modeling of speech* (grant number 475177/2004-5) and *Rhythmic patterns, prosodic domains and probabilistic modeling in Portuguese Corpora* (grant number 485999/2007-2). AG is partially supported by a CNPq fellowship (grant 308656/2005-9) and FL is supported by a FAPESP fellowship (grant 06/56980-0).

context the minimal relevant part of each past. The set of all contexts satisfies the suffix property which means that no context is a proper suffix of another context. This property allows to represent the set of all contexts as a rooted labeled tree. With this representation the process is described by the tree of all contexts and a associated family of probability measures on A, indexed by the tree of contexts. Given a context, its associated probability measure gives the probability of the next symbol for any past having this context as a suffix. From now on the pair composed by the context tree and the associated family of probability measures will be called *probabilistic context tree.*

Rissanen not only introduced the notion of variable memory models but he also introduced the algorithm Context to estimate the probabilistic context tree. The way the algorithm Context works can be summarized as follows. Given a sample produced by a chain with variable memory, we start with a maximal tree of candidate contexts for the sample. The branches of this first tree are then pruned until we obtain a minimal tree of contexts well adapted to the sample. We associate to each context an estimated probability transition defined as the proportion of time the context appears in the sample followed by each one of the symbols in the alphabet. From Rissanen [11] to Galves et al. [10], passing by Ron et al. [12] and Bühlmann and Wyner [1], several variants of the algorithm Context have been presented in the literature. In all the variants the decision to prune a branch is taken by considering a *cost* function. A branch is pruned if the cost function assumes a value smaller than a given threshold. The estimated context tree is the smallest tree satisfying this condition. The estimated family of probability transitions is the one associated to the minimal tree of contexts.

In his seminal paper Rissanen proved the weak consistency of the algorithm Context in the case where the contexts have a bounded length, i.e., where the tree of contexts is finite. Bühlmann and Wyner [1] proved the weak consistency of the algorithm also in the finite case without assuming a priori known bound on the maximal length of the memory, but using a bound allowed to grow with the size of the sample. In both papers the cost function is defined using the log likelihood ratio test to compare two candidate trees and the main ingredient of the consistency proofs was the chi-square approximation to the log likelihood ratio test for Markov chains of fixed order. A different way to prove the consistency in the finite case was introduced in [10], using exponential inequalities for the estimated transition probabilities associated to the candidate contexts. As a consequence they obtain an exponential upper bound for the rate of convergence of their variant of the algorithm Context.

The unbounded case as far as we know was first considered by Ferrari and Wyner [8] who also proved a weak consistency result for the algorithm Context in this more general setting. The unbounded case was also considered by Csiszár and Talata [3] who introduced a different approach for the estimation of the probabilistic context tree using the Bayesian Information Criterion (BIC) as well as the Minimum Description Length Principle (MDL). We refer the reader to this last paper for a nice description of other approaches and results in this field, including the context tree maximizing algorithm by Willems et al. [14]. With exception of Weinberger et al. [13], the issue of the rate of convergence of the algorithm estimating the probabilistic context tree was not addressed in the literature until recently. Weinberger et al. proved in the bounded case that the probability that the estimated tree differs from the finite context tree generating the sample is summable as a function of the sample size. Duarte et al. in [6] extends the original weak consistency result by Rissanen [11] to the unbounded case. Assuming weaker hypothesis than [8], they showed that the on-line estimation of the context function decreases as the inverse of the sample size.

In the present paper we generalize the exponential inequality approach presented in [10] to obtain an exponential upper bound for the algorithm Context in the case of unbounded probabilistic context trees. Under suitable conditions, we prove that the truncated estimated context tree converges exponentially fast to the tree generating the sample, truncated at the same level. This improves all results known until now.

The paper is organized as follows. In section 2 we give the definitions and state the main results. Section 3 is devoted to the proof of an exponential bound for conditional probabilities, for unbounded probabilistic context trees. In section 4 we apply this exponential bound to estimate the rate of convergence of our version of the algorithm Context and to prove its consistency.

2. Definitions and results

In what follows A will represent a finite alphabet of size $|A|$. Given two integers $m \leq n$, we will denote by w_m^n the sequence $(w_m, \ldots, w_n)$ of symbols in A. The length of the sequence w_m^n is denoted by $\ell(w_m^n)$ and is defined by $\ell(w_m^n) = n-m+1$. Any sequence w_m^n with $m > n$ represents the empty string and is denoted by λ. The length of the empty string is $\ell(\lambda) = 0$.

Given two finite sequences w and v, we will denote by vw the sequence of length $\ell(v) + \ell(w)$ obtained by concatenating the two strings. In particular, $\lambda w = w\lambda = w$. The concatenation of sequences is also extended to the case in which v denotes a semi-infinite sequence, that is $v = v_{-\infty}^{-1}$.

We say that the sequence s is a *suffix* of the sequence w if there exists a sequence u, with $\ell(u) \geq 1$, such that $w = us$. In this case we write $s \prec w$. When $s \prec w$ or $s = w$ we write $s \preceq w$. Given a sequence w we denote by $\mathrm{suf}(w)$ the largest suffix of w.

In the sequel A^j will denote the set of all sequences of length j over A and A^* represents the set of all finite sequences, that is

$$A^* = \bigcup_{j=1}^{\infty} A^j.$$

Definition 2.1. A countable subset $\mathcal{T}$ of A^* is a *tree* if no sequence $s \in \mathcal{T}$ is a suffix of another sequence $w \in \mathcal{T}$. This property is called the *suffix property.*

We define the *height* of the tree $\mathcal{T}$ as

$$h(\mathcal{T}) = \sup\{\ell(w) : w \in \mathcal{T}\}.$$

In the case $h(\mathcal{T}) < +\infty$ it follows that $\mathcal{T}$ has a finite number of sequences. In this case we say that $\mathcal{T}$ is *bounded* and we will denote by $|\mathcal{T}|$ the number of sequences in $\mathcal{T}$. On the other hand, if $h(\mathcal{T}) = +\infty$ then $\mathcal{T}$ has a countable number of sequences. In this case we say that the tree $\mathcal{T}$ is *unbounded.*

Given a tree $\mathcal{T}$ and an integer K we will denote by $\mathcal{T}|_K$ the tree $\mathcal{T}$ *truncated* to level K, that is

$$\mathcal{T}|_K = \{w \in \mathcal{T} \colon \ell(w) \leq K\} \cup \{w \colon \ell(w) = K \text{ and } w \prec u, \text{ for some } u \in \mathcal{T}\}.$$

We will say that a tree is *irreducible* if no sequence can be replaced by a suffix without violating the suffix property. This notion was introduced in [3] and generalizes the concept of complete tree.

Definition 2.2. A *probabilistic context tree over* A is an ordered pair $(\mathcal{T}, p)$ such that

1. $\mathcal{T}$ is an irreducible tree;
2. $p = \{p(\cdot|w); w \in \mathcal{T}\}$ is a family of transition probabilities over A.

Consider a stationary stochastic chain $(X_t)_{t\in\mathbb{Z}}$ over A. Given a sequence $w \in A^j$ we denote by

$$p(w) \;=\; \mathbb{P}(X_1^j = w)$$

the stationary probability of the cylinder defined by the sequence w. If $p(w) > 0$ we write

$$p(a|w) \;=\; \mathbb{P}(X_0 = a \mid X_{-j}^{-1} = w)\,.$$

Definition 2.3. A sequence $w \in A^j$ is a *context* for the process (X_t) if $p(w) > 0$ and for any semi-infinite sequence $x_{-\infty}^{-1}$ such that w is a suffix of $x_{-\infty}^{-1}$ we have that

$$\mathbb{P}(X_0 = a \mid X_{-\infty}^{-1} = x_{-\infty}^{-1}) \;=\; p(a|w), \quad \text{for all } a \in A,$$

and no suffix of w satisfies this equation.

Definition 2.4. We say that the process (X_t) is *compatible* with the probabilistic context tree $(\mathcal{T}, \bar{p})$ if the following conditions are satisfied

1. $w \in \mathcal{T}$ if and only if w is a context for the process (X_t).
2. For any $w \in \mathcal{T}$ and any $a \in A$, $\bar{p}(a|w) = \mathbb{P}(X_0 = a \mid X_{-\ell(w)}^{-1} = w)$.

Define the sequence $(\alpha_k)_{k\in\mathbb{N}}$ as

$$\alpha_0 := \sum_{a\in A} \inf_{w\in\mathcal{T}} \{\, p(a|w)\, \},$$

$$\alpha_k := \inf_{u\in A^k} \sum_{a\in A} \inf_{w\in\mathcal{T}, w\succ u} \{\, p(a|w)\, \}.$$

From now on we will assume that the probabilistic context tree $(\mathcal{T}, p)$ satisfies the following assumptions.

Assumption 2.5. Non-nullness, that is $\inf_{w\in\mathcal{T}}\{p(a|w)\} > 0$ for any $a \in A$.

Assumption 2.6. Summability of the sequence $(1-\alpha_k), k \geq 0$. In this case denote by

$$\alpha := \sum_{k\in\mathbb{N}}(1-\alpha_k) \ < +\infty.$$

For a probabilistic context tree satisfying Assumptions 2.5 and 2.6, the maximal coupling argument used in [7], or alternatively the perfect simulation scheme presented in [2], imply the uniqueness of the law of the chain compatible with it.

Given an integer $k \geq 1$ we define

$$\mathcal{C}_k = \{u \in \mathcal{T}|_k : p(a|u) \neq p(a|\text{suf}(u)) \text{ for some } a \in A\}$$

and

$$D_k = \min_{u\in\mathcal{C}_k} \max_{a\in A}\{\, |p(a|u) - p(a|\text{suf}(u))|\,\}.$$

We denote by

$$\epsilon_k = \min\{\, p(w)\colon \ell(w) \leq k \text{ and } p(w) > 0\,\}.$$

In what follows we will assume that $x_0, x_1, \ldots, x_{n-1}$ is a sample of the stationary stochastic chain (X_t) compatible with the probabilistic context tree $(\mathcal{T}, p)$.

For any finite string w with $\ell(w) \leq n$, we denote by $N_n(w)$ the number of occurrences of the string in the sample; that is

$$N_n(w) = \sum_{t=0}^{n-\ell(w)} \mathbf{1}\{X_t^{t+\ell(w)-1} = w\}.$$

For any element $a \in A$, the empirical transition probability $\hat{p}_n(a|w)$ is defined by

$$\hat{p}_n(a|w) = \frac{N_n(wa) + 1}{N_n(w\cdot) + |A|}. \tag{2.7}$$

where

$$N_n(w\cdot) = \sum_{b\in A} N_n(wb)\,.$$

This definition of $\hat{p}_n(a|w)$ is convenient because it is asymptotically equivalent to $\frac{N_n(wa)}{N_n(w\cdot)}$ and it avoids an extra definition in the case $N_n(w\cdot) = 0$.

A variant of Rissanen's *Algorithm Context* is defined as follows. First of all, let us define for any finite string $w \in A^*$:

$$\Delta_n(w) = \max_{a\in A} |\hat{p}_n(a|w) - \hat{p}_n(a|\text{suf}(w))|.$$

The $\Delta_n(w)$ operator computes a distance between the empirical transition probabilities associated to the sequence w and the one associated to the sequence $\text{suf}(w)$.

Definition 2.8. Given $\delta > 0$ and $d < n$, the tree estimated with the Algorithm Context is

$$\begin{aligned}\hat{\mathcal{T}}_n^{\delta,d} = \{w \in A_1^d : N_n(aw\cdot) > 0, \Delta_n(a\,\text{suf}(w)) > \delta \text{ for some } a \in A \text{ and}\\ \Delta_n(uw) \leq \delta \text{ for all } u \in A_1^{d-\ell(w)} \text{ with } N_n(uw\cdot) \geq 1\},\end{aligned}$$

where A_1^r denotes the set of all sequences of length at most r. In the case $\ell(w) = d$ we have $A_1^{d-\ell(w)} = \emptyset$.

It is easy to see that $\hat{\mathcal{T}}_n^{\delta,d}$ is an irreducible tree. Moreover, the way we defined $\hat{p}_n(\cdot|\cdot)$ in (2.7) associates a probability distribution to each sequence in $\hat{\mathcal{T}}_n^{\delta,d}$.

The main result in this article is the following

Theorem 2.9. *Let $(\mathcal{T}, p)$ be a probabilistic context tree satisfying Assumptions 2.5 and 2.6 and let (X_t) be a stationary stochastic chain compatible with $(\mathcal{T}, p)$. Then for any integer K, any d satisfying*

$$d > \max_{u \notin \mathcal{T}, \ell(u) \leq K} \min\{k : \exists w \in \mathcal{C}_k, w \succ u\}, \qquad (2.10)$$

any $\delta < D_d$ and any

$$n > \frac{2(|A|+1)}{\min(\delta, D_d - \delta)\epsilon_d} + d$$

we have that

$$\mathbb{P}(\hat{\mathcal{T}}_n^{\delta,d}|_K \neq \mathcal{T}|_K) \leq 4\, e^{\frac{1}{e}}\, |A|^{d+2} \exp\left[-(n-d)\, \frac{[\min(\frac{\delta}{2}, \frac{D_d-\delta}{2}) - \frac{|A|+1}{(n-d)\epsilon_d}]^2 \epsilon_d^2 C}{4|A|^2(d+1)}\right],$$

where

$$C = \frac{\alpha_0}{8e(\alpha + \alpha_0)}.$$

As a consequence we obtain the following strong consistency result.

Corollary 2.11. *Under the conditions of Theorem 2.9 we have*

$$\hat{\mathcal{T}}_n^{\delta,d}|_K = \mathcal{T}|_K,$$

eventually almost surely as $n \to +\infty$.

3. Exponential inequalities for empirical probabilities

The main ingredient in the proof of Theorem 2.9 is the following exponential upper bound

Theorem 3.1. *For any finite sequence w, any symbol $a \in A$ and any $t > 0$ the following inequality holds*

$$\mathbb{P}(\,|N_n(wa) - (n - \ell(w))p(wa)| > t\,) \leq e^{\frac{1}{e}} \exp\left[\frac{-t^2 C}{(n - \ell(w))\ell(wa)}\right],$$

where

$$C = \frac{\alpha_0}{8e(\alpha + \alpha_0)}. \qquad (3.2)$$

As a direct consequence of Theorem 3.1 we obtain the following corollary.

Corollary 3.3. *For any finite sequence w with $p(w) > 0$, any symbol $a \in A$, any $t > 0$ and any $n > \frac{|A|+1}{tp(w)} + \ell(w)$ the following inequality holds*

$$\mathbb{P}\big(|\hat{p}_n(a|w) - p(a|w)| > t\big) \leq 2\,|A|\,e^{\frac{1}{e}} \exp\left[-(n-\ell(w))\frac{[t - \frac{|A|+1}{(n-\ell(w))p(w)}]^2 p(w)^2 C}{4|A|^2\ell(wa)}\right],$$

where C is given by (3.2).

To prove Theorem 3.1 we need a mixture property for processes compatible with a probabilistic context tree $(\mathcal{T}, p)$ satisfying Assumptions 2.5 and 2.6. This is the content of the following lemma.

Lemma 3.4. *Let (X_t) be a stationary stochastic chain compatible with the probabilistic context tree $(\mathcal{T}, p)$ satisfying Assumptions 2.5 and 2.6. Then, there exists a summable sequence $\{\rho_l\}_{l\in\mathbb{N}}$, satisfying*

$$\sum_{l\in\mathbb{N}} \rho_l \leq 1 + \frac{2\alpha}{\alpha_0}, \tag{3.5}$$

such that for any $i \geq 1$, any $k > i$, any $j \geq 1$ and any finite sequence w_1^j, the following inequality holds

$$\sup_{x_1^i \in A^i} |\mathbb{P}(X_k^{k+j-1} = w_1^j \mid X_1^i = x_1^i) - p(w_1^j)| \leq \sum_{l=0}^{j-1} \rho_{k-i-1+l}\,. \tag{3.6}$$

Proof. First note that

$$\inf_{u\in A^\infty} \mathbb{P}(X_k^{k+j-1} = w_1^j \mid X_{-\infty}^i = u_{-\infty}^0 x_1^i) \leq \mathbb{P}(X_k^{k+j-1} = w_1^j \mid X_1^i = x_1^i)$$

$$\leq \sup_{u\in A^\infty} \mathbb{P}(X_k^{k+j-1} = w_1^j \mid X_{-\infty}^i = u_{-\infty}^0 x_1^i).$$

where A^∞ denotes the set of all semi-infinite sequences $u_{-\infty}^0$. The reader can find a proof of the inequalities above in [7, Proposition 3]. Using this fact and the condition of stationarity it is sufficient to prove that for any $k \geq 0$,

$$\sup_{x\in A^\infty} |\mathbb{P}(X_k^{k+j-1} = w_1^j \mid X_{-\infty}^{-1} = x_{-\infty}^{-1}) - p(w_1^j)| \leq \sum_{l=0}^{j-1} \rho_{k+l}.$$

Note that for all pasts $x_{-\infty}^{-1}$ we have

$$\begin{aligned}
&|\mathbb{P}(X_k^{k+j-1} = w_1^j \mid X_{-\infty}^{-1} = x_{-\infty}^{-1}) - p(w_1^j)| \\
&\quad = \Big|\int_{u\in A^\infty} \big[\mathbb{P}(X_k^{k+j-1} = w_1^j \mid X_{-\infty}^{-1} = x_{-\infty}^{-1}) \\
&\qquad\qquad - \mathbb{P}(X_k^{k+j-1} = w_1^j \mid X_{-\infty}^{-1} = u_{-\infty}^{-1})\big]\,dp(u)\Big| \\
&\quad \leq \int_{u\in A^\infty} \big|\mathbb{P}(X_k^{k+j-1} = w_1^j \mid X_{-\infty}^{-1} = x_{-\infty}^{-1}) \\
&\qquad\qquad - \mathbb{P}(X_k^{k+j-1} = w_1^j \mid X_{-\infty}^{-1} = u_{-\infty}^{-1})\big|\,dp(u).
\end{aligned}$$

Therefore, applying the loss of memory property proved in [2, Corollary 4.1] we have that

$$\left|\mathbb{P}(X_k^{k+j-1} = w_1^j \mid X_{-\infty}^{-1} = x_{-\infty}^{-1}) - \mathbb{P}(X_k^{k+j-1} = w_1^j \mid X_{-\infty}^{-1} = u_{-\infty}^{-1})\right| \leq \sum_{l=0}^{j-1} \rho_{k+l},$$

where ρ_m is defined as the probability of return to the origin at time m of the Markov chain on $\mathbb{N}$ starting at time zero at the origin and having transition probabilities

$$p(x,y) = \begin{cases} \alpha_x, & \text{if y = x+1,} \\ 1-\alpha_x, & \text{if y=0,} \\ 0, & \text{otherwise.} \end{cases} \tag{3.7}$$

This concludes the proof of (3.6). To prove (3.5), let(Z_n) be the Markov chain with probability transitions given by (3.7). By definition, we have

$$\begin{aligned}\prod_{l\geq 1}(1-\rho_l) &= \prod_{l\geq 1}\sum_{j=1}^{l} \mathbb{P}(Z_l = j | Z_{l-1} = j-1)\mathbb{P}(Z_{l-1} = j-1) \\ &\geq \prod_{l\geq 1} \alpha_{l-1} \prod_{i=0}^{l-2} \alpha_i \prod_{l\geq 0} \alpha_l^2.\end{aligned}$$

From this, using the inequality $x \leq -\ln(1-x) \leq \frac{x}{1-c}$ which holds for any $x \in (-1, c]$, it follows that

$$\sum_{l\geq 1} \rho_l \leq \sum_{l\geq 0} \log \alpha_l \leq \sum_{l\geq 0} \frac{1-\alpha_l}{\alpha_0}.$$

This concludes the proof of the lemma. □

We are now ready to prove Theorem 3.1.

Proof of Theorem 3.1. Let w be a finite sequence and a any symbol in A. Define the random variables

$$U_j = \mathbf{1}\{X_j^{j+\ell(w)} = wa\} - p(wa),$$

for $j = 0, \ldots, n-\ell(wa)$. Then, using [4, Proposition 4] we have that, for any $p \geq 2$ $\|N_n(wa) - (n-\ell(w))p(wa)\|_p$

$$\begin{aligned}&\leq \Big(2p \sum_{i=0}^{n-\ell(wa)} \sum_{k=i}^{n-\ell(wa)} \|\, \mathbb{E}(U_k \mid U_0, \ldots, U_i)\, \|_\infty\Big)^{\frac{1}{2}} \\ &\leq \Big(2p \sum_{i=0}^{n-\ell(wa)} \sum_{k=i}^{n-\ell(wa)} \sup_{u\in A^{i+\ell(wa)}} \left|\mathbb{P}(X_k^{k+\ell(w)} = wa \mid X_0^{i+\ell(w)} = u) - p(wa)\right|\Big)^{\frac{1}{2}} \\ &\leq \Big(2p\,\ell(wa)(n-\ell(w))\frac{2(\alpha+\alpha_0)}{\alpha_0}\Big)^{\frac{1}{2}}.\end{aligned}$$

Then, as in [5, Proposition 5] we also obtain that, for any $t > 0$,

$$\mathbb{P}(|N_n(wa) - (n - \ell(w))p(wa)| > t) \leq e^{\frac{1}{e}} \exp\Big[\frac{-t^2 C}{(n-\ell(w))\ell(wa)}\Big],$$

where

$$C = \frac{\alpha_0}{8e(\alpha + \alpha_0)}. \qquad \square$$

Proof of Corollary 3.3. First observe that

$$\Big| p(a|w) - \frac{(n-\ell(w))p(wa) + 1}{(n-\ell(w))p(w) + |A|} \Big| \leq \frac{|A|+1}{(n-\ell(w))p(w)}.$$

Then, for all $n \geq (|A|+1)/tp(w) + \ell(w)$ we have that

$$\mathbb{P}\big(|\hat{p}_n(a|w) - p(a|w)| > t \big)$$
$$\leq \mathbb{P}\Big(\Big| \frac{N_n(wa)+1}{N_n(w\cdot) + |A|} - \frac{(n-\ell(w))p(wa)+1}{(n-\ell(w))p(w)+|A|} \Big| > t - \frac{|A|+1}{(n-\ell(w))p(w)} \Big)$$

Denote by $t' = t - (|A|+1)/(n-\ell(w))p(w)$. Then

$$\mathbb{P}\Big(\Big| \frac{N_n(wa)+1}{N_n(w\cdot) + |A|} - \frac{(n-\ell(w))p(wa)+1}{(n-\ell(w))p(w)+|A|} \Big| > t' \Big)$$

$$\leq \mathbb{P}\big(|N_n(wa) - (n-\ell(w))p(wa)| > \frac{t'}{2}[(n-\ell(w))p(w) + |A|] \big)$$

$$+ \sum_{b \in A} \mathbb{P}\big(|N_n(wb) - (n-\ell(w))p(wb)| > \frac{t'}{2|A|}[(n-\ell(w))p(w) + |A|] \big).$$

Now, we can apply Theorem 3.1 to bound above the last sum by

$$2\,|A|\, e^{\frac{1}{e}} \exp\left[-(n-\ell(w)) \frac{[t - \frac{|A|+1}{(n-\ell(w))p(w)}]^2 p(w)^2 C}{4|A|^2 \ell(wa)} \right],$$

where

$$C = \frac{\alpha_0}{8e(\alpha + \alpha_0)}.$$

This finishes the proof of the corollary. $\square$

4. Proof of the main results

Proof of Theorem 2.9. Define

$$O_n^{\delta,d} = \bigcup_{\substack{w \in \mathcal{T} \\ \ell(w) < K}} \ \bigcup_{uw \in \hat{\mathcal{T}}_n^{\delta,d}} \{\Delta_n(uw) > \delta\}, \quad \text{and} \quad U_n^{\delta,d} = \bigcup_{\substack{w \in \hat{\mathcal{T}}_n^{\delta,d} \\ \ell(w) < K}} \ \bigcap_{uw \in \mathcal{T}|_d} \{\Delta_n(uw) \leq \delta\}.$$

Then, if $d < n$ we have that

$$\{\hat{\mathcal{T}}_n^{\delta,d}|_K \neq \mathcal{T}|_K\} = O_n^{\delta,d} \cup U_n^{\delta,d}.$$

The result follows from a succession of lemmas.

Lemma 4.1. *For any* $n > \frac{2(|A|+1)}{\delta\epsilon_d} + d$, *for any* $w \in \mathcal{T}$ *with* $\ell(w) < K$ *and for any* $uw \in \hat{\mathcal{T}}_n^{\delta,d}$ *we have that*

$$\mathbb{P}(\Delta_n(uw) > \delta) \;\leq\; 4\,|A|^2\, e^{\frac{1}{e}} \exp\bigl[-(n-d)\,\frac{[\frac{\delta}{2} - \frac{|A|+1}{(n-d)\epsilon_d}]^2 \epsilon_d^2 C}{4|A|^2(d+1)}\bigr],$$

where C is given by (3.2).

Proof. Recall that

$$\Delta_n(uw) = \max_{a\in A} |\hat{p}_n(a|uw) - \hat{p}_n(a|\mathrm{suf}(uw))|.$$

Note that the fact $w \in \mathcal{T}$ implies that for any finite sequence u with $p(u) > 0$ and any symbol $a \in A$ we have $p(a|w) = p(a|uw)$. Hence,

$$\begin{aligned}\mathbb{P}(\Delta_n(uw) > \delta) \;\leq\; \sum_{a\in A} \bigl[\, &\mathbb{P}\bigl(|\hat{p}_n(a|w) - p(a|w)| > \frac{\delta}{2}\bigr) \\ &+ \mathbb{P}\bigl(|\hat{p}_n(a|uw) - p(a|uw)| > \frac{\delta}{2}\bigr)\bigr].\end{aligned}$$

Using Corollary 3.3 we can bound above the right-hand side of the last inequality by

$$4\,|A|^2\, e^{\frac{1}{e}} \exp\left[-(n-d)\,\frac{[\frac{\delta}{2} - \frac{|A|+1}{(n-d)\epsilon_d}]^2 \epsilon_d^2 C}{4|A|^2(d+1)}\right],$$

where C is given by (3.2). □

Lemma 4.2. *For any* $n > \frac{2(|A|+1)}{(D_d-\delta)\epsilon_d} + d$ *and for any* $w \in \hat{\mathcal{T}}_n^{\delta,d}$ *with* $\ell(w) < K$ *we have that*

$$\mathbb{P}\Bigl(\bigcap_{uw\in\mathcal{T}|_d} \{\Delta_n(uw) \leq \delta\}\Bigr) \leq 4\,|A|\, e^{\frac{1}{e}} \exp\left[-(n-d)\,\frac{[\frac{D_d-\delta}{2} - \frac{|A|+1}{(n-d)\epsilon_d}]^2 \epsilon_d^2 C}{4|A|^2(d+1)}\right],$$

where C is given by (3.2).

Proof. As d satisfies (2.10) there exists $\bar{u}\bar{w} \in \mathcal{T}|_d$ such that $p(a|\bar{u}\bar{w} \neq p(a|\mathrm{suf}\,(\bar{u}\bar{w}))$ for some $a \in A$. Then

$$\mathbb{P}\Bigl(\bigcap_{uw\in\mathcal{T}|_d} \{\Delta_n(uw) \leq \delta\}\Bigr) \;\leq\; \mathbb{P}(\Delta_n(\bar{u}\bar{w}) \leq \delta).$$

Observe that for any $a \in A$,

$$\begin{aligned}|\hat{p}_n(a|\mathrm{suf}(\bar{u}\bar{w})) - \hat{p}_n(a|\bar{u}\bar{w})| \;\geq\; &|p(a|\mathrm{suf}(\bar{u}\bar{w})) - p(a|\bar{u}\bar{w})| \\ &- |\hat{p}_n(a|\mathrm{suf}(\bar{u}\bar{w})) - p(a|\mathrm{suf}(\bar{u}\bar{w}))| - |\hat{p}_n(a|\bar{u}\bar{w}) - p(a|\bar{u}\bar{w})|.\end{aligned}$$

Hence, we have that for any $a \in A$

$$\Delta_n(u\bar{w}) \geq D_d - |\hat{p}_n(a|\mathrm{suf}(u\bar{w})) - p(a|\mathrm{suf}(u\bar{w}))| - |\hat{p}_n(a|u\bar{w}) - p(a|u\bar{w})|.$$

Therefore,

$$\begin{aligned}\mathbb{P}(\Delta_n(u\bar{w}) \leq \delta) \;\leq\; & \mathbb{P}\Big(\bigcap_{a\in A}\{\,|\hat{p}_n(a|\mathrm{suf}(u\bar{w})) - p(a|\mathrm{suf}(u\bar{w}))| \geq \frac{D_d-\delta}{2}\}\Big)\\ & +\mathbb{P}\Big(\bigcap_{a\in A}\{\,|\hat{p}_n(a|u\bar{w}) - p(a|u\bar{w})| \geq \frac{D_d-\delta}{2}\}\Big).\end{aligned}$$

As $\delta < D_d$ and $n > \frac{2(|A|+1)}{(D_d-\delta)\epsilon_d} + d$ we can use Corollary 3.3 to bound above the right-hand side of this inequality by

$$4\,|A|\,e^{\frac{1}{e}}\exp\left[-(n-d)\,\frac{[\frac{D_d-\delta}{2} - \frac{|A|+1}{(n-d)\epsilon_d}]^2\epsilon_d^2 C}{4|A|^2(d+1)}\right],$$

where C is given by (3.2). This concludes the proof of the lemma. □

Now we can finish the proof of Theorem 2.9. We have that

$$\mathbb{P}(\hat{\mathcal{T}}_n^{\delta,d}|_K \neq \mathcal{T}|_K) = \mathbb{P}(O_n^{\delta,d}) + \mathbb{P}(U_n^{\delta,d}).$$

Using the definition of $O_n^{\delta,d}$ and $U_n^{\delta,d}$ we have that

$$\mathbb{P}(\hat{\mathcal{T}}_n^{\delta,d}|_K \neq \mathcal{T}|_K) \leq \sum_{\substack{w\in\mathcal{T}\\ \ell(w)<K}}\ \sum_{uw\in\hat{\mathcal{T}}_n^{\delta,d}} \mathbb{P}(\Delta_n(uw) > \delta) + \sum_{\substack{w\in\hat{\mathcal{T}}_n^{\delta,d}\\ \ell(w)<K}} \mathbb{P}\Big(\bigcap_{uw\in\mathcal{T}|_d} \Delta_n(uw) \leq \delta\Big).$$

Applying Lemma 4.1 and Lemma 4.2 we can bound above the last expression by

$$\mathbb{P}(\hat{\mathcal{T}}_n^{\delta,d}|_K \neq \mathcal{T}|_K) \leq 4\,e^{\frac{1}{e}}\,|A|^{d+2}\exp\left[-(n-d)\,\frac{[\min(\frac{\delta}{2},\frac{D_d-\delta}{2}) - \frac{|A|+1}{(n-d)\epsilon_d}]^2\epsilon_d^2 C}{4|A|^2(d+1)}\right],$$

where C is given by (3.2). We conclude the proof of Theorem 2.9. □

Proof of Corollary 2.11. It follows from Theorem 2.9, using the first Borel-Cantelli Lemma and the fact that the bounds for the error estimation of the context tree are summable in n for a fixed d satisfying (2.10) and $\delta < D_d$. □

5. Final remarks

The present paper presents an upper bound for the rate of convergence of a version of the algorithm Context, for unbounded context trees. This generalizes previous results obtained in [10] for the case of bounded variable memory processes. We obtain an exponential bound for the probability of incorrect estimation of the truncated context tree, when the estimator is given by Definition (2.8). Note that the definition of the context tree estimator depends on the parameter δ, and this parameter appears in the exponent of the upper bound. To assure the consistency of the estimator we need to choose a δ sufficiently small, depending on the transition probabilities of the process. Therefore, our estimator is not universal, in the sense that for any fixed δ it fails to be consistent for any process having $D_d < \delta$. The same happens with the parameter d. In order to choose δ and d not depending on the process, we can allow these parameters to be a function of n, in such a way δ_n goes to zero and d_n goes to $+\infty$ as n diverges. When we do this, we loose the exponential property of the upper bound.

As an anonymous referee has pointed out, Finesso et al. [9] proved that in the simpler case of estimating the order of a Markov chain, it is not possible to obtain pure exponential bounds for the overestimation event with a universal estimator. The above discussion illustrates this fact.

Acknowledgments

We thank Pierre Collet, Imre Csiszár, Nancy Garcia, Aurélien Garivier, Bezza Hafidi, Véronique Maume-Deschamps, Eric Moulines, Jorma Rissanen and Bernard Schmitt for many discussions on the subject. We also thank an anonymous referee that attracted our attention to the interesting paper [9].

References

[1] P. Bühlmann and A. J. Wyner. Variable length Markov chains. *Ann. Statist.*, 27:480–513, 1999.

[2] F. Comets, R. Fernández, and P. Ferrari. Processes with long memory: Regenerative construction and perfect simulation. *Ann. Appl. Probab.*, 12(3):921–943, 2002.

[3] I. Csiszár and Z. Talata. Context tree estimation for not necessarily finite memory processes, via BIC and MDL. *IEEE Trans. Inform. Theory*, 52(3):1007–1016, 2006.

[4] J. Dedecker and P. Doukhan. A new covariance inequality and applications. *Stochastic Process. Appl.*, 106(1):63–80, 2003.

[5] J. Dedecker and C. Prieur. New dependence coefficients. examples and applications to statistics. *Probab. Theory Related Fields*, 132:203–236, 2005.

[6] D. Duarte, A. Galves, and N.L. Garcia. Markov approximation and consistent estimation of unbounded probabilistic suffix trees. *Bull. Braz. Math. Soc.*, 37(4):581–592, 2006.

[7] R. Fernández and A. Galves. Markov approximations of chains of infinite order. *Bull. Braz. Math. Soc.*, 33(3):295–306, 2002.

[8] F. Ferrari and A. Wyner. Estimation of general stationary processes by variable length Markov chains. *Scand. J. Statist.*, 30(3):459–480, 2003.

[9] L. Finesso, C-C. Liu, and P. Narayan. The optimal error exponent for Markov order estimation. *IEEE Trans. Inform. Theory*, 42(5):1488–1497, 1996.

[10] A. Galves, V. Maume-Deschamps, and B. Schmitt. Exponential inequalities for VLMC empirical trees. *ESAIM Prob. Stat.* (*accepted*), 2006.

[11] J. Rissanen. A universal data compression system. *IEEE Trans. Inform. Theory*, 29(5):656–664, 1983.
[12] D. Ron, Y. Singer, and N. Tishby. The power of amnesia: Learning probabilistic automata with variable memory length. *Machine Learning*, 25(2-3):117–149, 1996.
[13] M.J. Weinberger, J. Rissanen, and M. Feder. A universal finite memory source. *IEEE Trans. Inform. Theory*, 41(3):643–652, 1995.
[14] F.M. Willems, Y.M. Shtarkov, and T.J. Tjalkens. The context-tree weighting method: basic properties. *IEEE Trans. Inform. Theory*, IT-44:653–664, 1995.

Antonio Galves and Florencia Leonardi
Instituto de Matemática e Estatística
Universidade de São Paulo
BP 66281, 05315-970
São Paulo, Brasil
e-mail: galves@ime.usp.br

e-mail: leonardi@ime.usp.br

Progress in Probability, Vol. 60, 271–298

Spatial Point Processes and the Projection Method

Nancy L. Garcia and Thomas G. Kurtz

Abstract. The projection method obtains non-trivial point processes from higher-dimensional Poisson point processes by constructing a random subset of the higher-dimensional space and projecting the points of the Poisson process lying in that set onto the lower-dimensional region. This paper presents a review of this method related to spatial point processes as well as some examples of its applications. The results presented here are known for sometime but were not published before. Also, we present a backward construction of general spatial pure-birth processes and spatial birth and death processes based on the projection method that leads to a perfect simulation scheme for some Gibbs distributions in compact regions.

Mathematics Subject Classification (2000). Primary 60G55, 60H99; Secondary 60J25, 60G40, 60G44.

Keywords. Poisson point processes, time change, projection method, perfect simulation.

1. Introduction

A point process is a model of indistinguishable points distributed randomly in some space. The simplest assumption that there are no multiple points and events occurring in disjoint regions of space are independent leads to the well-known Poisson point process. However, it is obvious that not all phenomena can be modelled by a process with independent increments. In $[0,\infty)$, any simple point process N with continuous compensator Λ, that is, $N([0,t])-\Lambda(t)$ is a local martingale, can be obtained as a random time change $N([0,t]) = Y(\Lambda(t))$, where Y is a unit rate Poisson process. (See, for example, Proposition 13.4.III, Daley and Vere-Jones, 1988.) More generally, multivariate counting processes with continuous compensators and without simultaneous jumps can be obtained as multiple random time

This research is partially supported by NSF under grant DMS 05-03983 and CNPq grant 301054/1993-2.

changes of independent unit Poisson processes (Meyer (1971), Aalen and Hoem (1978), Kurtz (1980b), Kurtz (1982)).

The projection method introduced by Kurtz (1989) can be seen as a generalization of the random time change representations. It constructs point processes through projections of underlying Poisson processes using stopping sets. These projections are made carefully in order that the projected process inherits many of the good properties of the Poisson processes. Garcia (1995a and 1995b) used this method to construct birth and death processes with variable birth and death rates and to study large population behavior for epidemic models. However, Kurtz (1989) is an unpublished manuscript and the generality of Garcia (1995a) hides the beauty of the ideas and methods behind technicalities. The goal of this paper is to present the projection method in detail as well as some examples. Another form of stochastic equation for spatial birth processes is considered by Massouliè (1998) and Garcia and Kurtz (2006). The latter paper also considers birth and death processes.

This paper is organized as follows:

Section 3 is based on Kurtz(1989) and provides a proper presentation of the projection method and states the basic theorems regarding martingale properties and moment inequalities for the projected processes. Simple examples, such as inhomogeneous Poisson processes, $M/G/\infty$ queues, and Cox processes, are presented.

Section 4 characterizes some of the processes that can be obtained by the projection method, the result obtained here is more general than the similar one in Garcia(1995a).

Section 5 presents spatial pure birth processes as projections of Poisson processes and derives some consequences, such as ergodicity. Although these results can be seen as particular cases from Garcia (1995a) the proofs in this case are simpler and provide much better insight into the use of the projection method.

Section 6 deals with birth and death processes in the special case when the stationary distribution is a Gibbs distribution, and for the finite case, a backward scheme provides perfect simulation.

2. Basic definitions

In this work, we are going to identify a point process with the counting measure N given by assigning unit mass to each point, that is, $N(A)$ is the number of points in a set A. With this identification in mind, let (S, r) be a complete, separable metric space, and let $\mathcal{N}(S)$ be the collection of σ-finite, Radon counting measures on S. $\mathcal{B}(S)$ will denote the Borel subsets of S. Typically, $S \subset \mathbb{R}^d$.

For counting measures, the Radon condition is simply the requirement that for each compact $K \subset S$, there exists an open set $G \supset K$ such that $N(G) < \infty$. For a general discussion, see Daley and Vere-Jones (1988). We topologize $\mathcal{N}(S)$ with the vague topology.

Definition 2.1. A sequence $\{\xi_k\} \subset \mathcal{N}(S)$ *converges vaguely* to $\xi \in \mathcal{N}(S)$ if and only if for each compact K there exists an open $G \supset K$ such that

$$\lim_{k\to\infty} \int f d\xi_k = \int f d\xi$$

for all $f \in C_b(S)$ with $\text{supp}(f) \subset G$.

Definition 2.2. Let μ be a σ-finite, Radon measure on S. A point process N on S is a *Poisson process* with mean measure μ if the following conditions hold:

(i) For $A_1, A_2, \ldots, A_k \in \mathcal{B}(S)$ disjoint sets, $N(A_1), N(A_2), \ldots, N(A_k)$ are independent random variables.
(ii) For each $A \in \mathcal{B}(S)$ and $k \geq 0$,

$$\mathbb{P}[N(A) = k] = e^{-\mu(A)} \frac{\mu(A)^k}{k!}.$$

Assuming that μ is diffuse, that is, $\mu\{x\} = 0$ for all $x \in S$, the strong independence properties of a Poisson process imply that N conditional on n points of N being sited at $x_1, x_2, \ldots, x_n$ has the properties of $N + \sum_{k=1}^{n} \delta_{x_k}$. Thus, the process "forgets" where it had the n points and behaves as if it were N with the n points adjoined. The notion of conditioning in this case is not straightforward since the the event "having a point at x" has probability zero. Assuming that $\mu(B_\epsilon(x_k)) > 0$ for all $\epsilon > 0$ and $k = 1, \ldots, n$,

$$\lim_{\epsilon\to 0+} \mathbb{E}[F(N)|N(B_\epsilon(x_k)) > 0, k = 1, \ldots, n] = \mathbb{E}\left[F\left(N + \sum_{k=1}^{n} \delta_{x_k}\right)\right]$$

for all $F \in C_b(\mathcal{N}(S))$. As a consequence, we have the following basic identity for Poisson processes.

Proposition 2.3. *Let N be a Poisson process on S with diffuse mean measure μ. Then*

$$\mathbb{E}\left[\int_S f(z, N) N(dz)\right] = \mathbb{E}\left[\int_S f(z, N + \delta_z)\mu(dz)\right]. \tag{2.1}$$

For example, let $c : S \to [0, \infty)$, $\rho : S \times S \to [0, \infty)$ and $\phi : [0, \infty) \to [0, \infty)$ be Borel measurable functions. Then (cf. Garcia (1995a))

$$\mathbb{E}\left[\int_S c(z)\phi\left(\int_S \rho(x, z)N(dx)\right) N(dz)\right]$$
$$= \int_S c(z)\mathbb{E}\left[\phi(\rho(z, z) + \int_S \rho(x, z)N(dx))\right] \mu(dz).$$

Alternatively, we may condition in a formal way in terms of Palm probabilities and Palm distributions (see Karr (1986, Section 1.7) or Daley and Vere-Jones (1988, Chapter 12)).

If we have a sequence of Poisson processes N_n with mean measures $n\mu$, defining the random signed measure W_n by $W_n(A) = n^{-1/2}(N_n(A) - n\,\mu(A))$, for $A_1, \ldots, A_m$ Borel sets with $\mu(A_i) < \infty$, the central limit theorem gives

$$(W_n(A_1), \ldots, W_n(A_m)) \xrightarrow{\mathcal{D}} (W(A_1), \ldots, W(A_m)), \tag{2.2}$$

where W is the mean zero Gaussian process indexed by Borel sets with covariance $\mathbb{E}[W(A)W(B)] = \mu(A \cap B)$.

3. Projection method

The basic idea of the projection method is to obtain point processes from higher-dimensional Poisson processes by constructing a random subset of the higher-dimensional space and projecting the points of the Poisson process lying in that set onto the lower-dimensional subspace. This general approach can be used to construct, for example, Cox processes (the random set is independent of the Poisson process), a large class of Gibbs distributions, and birth and death processes with variable birth and death rates.

This construction gives a natural coupling among point processes and hence a method to compare results and prove limit theorems. The law of large numbers and central limit theorem for Poisson processes can be exploited to obtain corresponding results for the point processes under study. Garcia (1995b) studied large population behavior for epidemic models using the underlying limit theorems for Poisson processes. Ferrari and Garcia (1998) applied the projection method to the study of loss networks.

Even though the basic concepts and ideas described in the remainder of this section were introduced in Kurtz (1989) and were used in Garcia (1995a and 1995b), a number of results appear here for the first time.

3.1. Representation of inhomogeneous Poisson processes as projections of higher-dimensional homogeneous Poisson processes

Let N be a Poisson random measure on $\mathbb{R}^{d+1}$ with Lebesgue mean measure m. Let $C \in \mathcal{B}(\mathbb{R}^{d+1})$ and define N_C, a point process on $\mathbb{R}^d$, by

$$N_C(A) = N(C \cap A \times \mathbb{R}), \quad A \in \mathcal{B}(\mathbb{R}^d). \tag{3.1}$$

Note that N_C is a random measure corresponding to the point process obtained by taking the points of N that lie in C and projecting them onto $\mathbb{R}^d$. Clearly, $N_C(A)$ is Poisson distributed with mean

$$\mu_C(A) = m(C \cap A \times \mathbb{R}).$$

If μ is an absolutely continuous measure (with respect to Lebesgue measure) in $\mathbb{R}^d$ with density f and $C = \{(x, y); x \in \mathbb{R}^d, 0 \le y \le f(x)\}$, then N_C is a Poisson random measure on $\mathbb{R}^d$ with mean measure μ given by

$$\mu_C(A) = m(C \cap A \times \mathbb{R}) = \int_A \int_0^{f(x)} dy\, dx = \int_A f(x)dx = \mu(A).$$

3.2. $M/G/\infty$ queues

A particular application of the projection method for inhomogeneous Poisson processes is the $M/G/\infty$ queue. In fact, Foley (1986) exploits precisely this observation. Consider a process where clients arrive according to a λ-homogeneous Poisson process and are served according to a service distribution ν. For simplicity, assume that there are no clients at time $t = 0$. Let $\mu = \lambda m \times \nu$ and $S = [0, \infty) \times [0, \infty)$, and let N be the Poisson process on S with mean measure μ. Define

$$B(t) = \{(x, y) \in S; x \leq t\} \tag{3.2}$$

and

$$A(t) = \{(x, y) \in S; y \leq t - x\}. \tag{3.3}$$

We can identify the points of N in $B(t)$ with customers that arrive by time t (note that the distribution of $N(B(t))$ is Poisson with parameter λt) and the points in $A(t)$ are identified with customers that complete service by time t. Therefore, the points of N in

$$C(t) = B(t) - A(t) = \{(x, y) : x \leq t, y > t - x\}$$

correspond to the customers in the queue and hence the queue length is

$$Q(t) = N(C(t)). \tag{3.4}$$

Notice that we can construct this process starting at $-T$ instead of 0. Therefore, the system at time 0 in this new construction has the same distribution as the system at time T in the old construction. In fact, defining

$$C_T(t) = \{(x, y) \in [-T, \infty) \times [0, \infty); -T \leq x \leq t, y \geq t - x\}, \tag{3.5}$$

and

$$Q_T(t) = N(C_T(t)), \tag{3.6}$$

we have that

$$Q_T(0) \stackrel{\mathcal{D}}{=} Q(T). \tag{3.7}$$

Letting $T \to \infty$, we have

$$C_T(0) \to \{(x, y); x \leq 0, y \geq -x\}$$

and

$$Q_T(0) \to N(\{(x, y) \in \mathbb{R} \times [0, \infty); y \geq -x \geq 0\}) \tag{3.8}$$

which implies

$$Q(T) \Rightarrow N(\{(x, y) \in \mathbb{R} \times [0, \infty); y \geq -x \geq 0\}) \tag{3.9}$$

in the original $M/G/\infty$ queue. Even though this result is well known by other arguments, this "backward construction" can be used for other non trivial cases. (See Section 6.3.)

3.3. Projections through random sets

The projected process N_C defined by (3.1) was a Poisson process due to the fact that the projection set was deterministic. The construction, however, still makes sense if the projection set C is random. Let N denote a Poisson process on $S = S_1 \times S_2$ with mean measure μ, where S_1 and S_2 are complete separable metric spaces. Let Γ be a random subset of S (in general, not independent of N). Define a point process N_Γ on S_1 by $N_\Gamma(B) = N(\Gamma \cap B \times S_2)$.

For example, if the set Γ is independent of the process N, then N_Γ will be a *Cox process* (or doubly stochastic Poisson process). In fact, conditioned on Γ, in the independent case, N_Γ is a Poisson process with mean measure $\mu(\Gamma \cap \cdot \times S_2)$ and the mean measure, $\mu_\Gamma(B) \equiv \mathbb{E}[N_\Gamma(B)]$, for N_Γ is

$$\mu_\Gamma(B) = \mathbb{E}[\mu(\Gamma \cap B \times S_2)]. \tag{3.10}$$

It is tempting to conjecture that (3.10) holds for other random sets, but in general, that is not true (e.g., let Γ be the support of N). However, there is a class of sets for which the identity does hold, the class of *stopping sets.*

Assume that $(\Omega, \mathcal{F}, \mathbb{P})$ is complete and that $\{\mathcal{F}_A\}$ is a family of complete sub-σ-algebras indexed by Borel subsets, $A \in \mathcal{B}(S)$, having the property that if $A \subset B$ then $\mathcal{F}_A \subset \mathcal{F}_B$. A Poisson point process N on S is *compatible* with $\{\mathcal{F}_A\}$, if for each $A \in \mathcal{B}(S)$, $N(A)$ is $\mathcal{F}_A$-measurable and, for each $C \in \mathcal{B}(S)$ such that $C \cap A = \emptyset$, $N(C)$ is independent of $\mathcal{F}_A$.

For technical reasons, we will restrict attention to Γ with values in the closed sets $\mathcal{C}(S)$, and we will assume that $\{\mathcal{F}_A, A \in \mathcal{C}(S)\}$ is right continuous in the sense that if $\{A_k\} \subset \mathcal{C}(S)$, $A_1 \supset A_2 \supset \cdots$, then $\cap_k \mathcal{F}_{A_k} = \mathcal{F}_{\cap A_k}$. If $\mathcal{F}_A$ is the completion of the σ-algebra $\sigma(N(B) : B \in \mathcal{B}(S), B \subset A)$, then $\{\mathcal{F}_A, A \in \mathcal{C}(S)\}$ is right continuous.

Definition 3.1. A $\mathcal{C}(S)$-valued random variable Γ is a $\{\mathcal{F}_A\}$*-stopping set,* if $\{\Gamma \subset A\} \in \mathcal{F}_A$ for each $A \in \mathcal{C}(S)$.

Γ is *separable* if there exists a countable set $\{H_n\} \subset \mathcal{C}(S)$ such that $\Gamma = \cap\{H_n : H_n \supset \Gamma\}$. Then Γ is *separable with respect to* $\{H_n\}$.

The *information σ-algebra*, $\mathcal{F}_\Gamma$, is given by

$$\mathcal{F}_\Gamma = \{G \in \mathcal{F} : G \cap \{\Gamma \subset A\} \in \mathcal{F}_A \text{ for all } A \in \mathcal{C}(S)\}.$$

Remark 3.2. The definition of stopping set used here is from Garcia (1995a). It differs from the definition used by Ivanoff and Merzbach (1995). A stopping set in the sense used here is a special case of a generalized *stopping time* as used in Kurtz (1980a), and is essentially the same as an *adapted set* as used in Balan (2001).

The significance of separability of Γ is that we can approximate Γ by a decreasing sequence of discrete stopping sets. Without loss of generality, we can always assume $H_1 = S$.

Lemma 3.3. *Let Γ be separable with respect to $\{H_n\}$, and define $\Gamma_n = \cap\{H_k : k \leq n, H_k \supset \Gamma\}$. Then Γ_n is a stopping set with $\Gamma_1 \supset \Gamma_2 \supset \cdots \supset \Gamma$ and $\cap_n \Gamma_n = \Gamma$.*

Proof. Let $\mathcal{S}_n = \{H_{i_1} \cap \cdots \cap H_{i_m} : 1 \leq i_1, \ldots, i_m \leq n\}$. Then

$$\{\Gamma_n \subset A\} = \cup_{C \subset A, C \in \mathcal{S}_n} \{\Gamma \subset C\} \in \mathcal{F}_A,$$

so Γ_n is a stopping set. Separability then implies $\cap_n \Gamma_n = \Gamma$. □

These definitions are clear analogs of the definitions for real-valued stopping times, and stopping sets have many properties in common with stopping times.

Lemma 3.4. *Let $\Gamma, \Gamma_1, \Gamma_2, \ldots$ be $\{\mathcal{F}_A\}$-stopping sets. Then*

(i) *For $A \in \mathcal{C}(S)$, $\{\Gamma \subset A\} \in \mathcal{F}_\Gamma$.*
(ii) *For $n = 1, 2, \ldots$, $\Gamma_1 \cup \cdots \cup \Gamma_n$ is a $\{\mathcal{F}_A\}$-stopping set.*
(iii) *The closure of $\cup_{i=1}^\infty \Gamma_i$ is a stopping set.*
(iv) *If $\Gamma_1 \subset \Gamma_2$, then $\mathcal{F}_{\Gamma_1} \subset \mathcal{F}_{\Gamma_2}$.*
(v) *If the range of Γ is countable, say $\mathcal{R}(\Gamma) = \{C_k\}$, then*

$$\mathcal{F}_\Gamma = \{B \in \mathcal{F} : B = \cup_k \{\Gamma = C_k\} \cap B_k, B_k \in \mathcal{F}_{C_k}\}.$$

(vi) *If Γ is separable and $\{\Gamma_n\}$ are defined as in Lemma 3.3, then for $B \in \mathcal{B}(S)$, $N(\Gamma \cap B)$ is $\cap_n \mathcal{F}_{\Gamma_n}$-measurable.*

Remark 3.5. The intersection of two stopping sets need not be a stopping set.

Proof. $\{\Gamma \subset A\} \cap \{\Gamma \subset B\} = \{\Gamma \subset A \cap B\} \in \mathcal{F}_{A \cap B} \subset \mathcal{F}_B$, for all $B \in \mathcal{C}(S)$, and hence, $\{\Gamma \subset A\} \in \mathcal{F}_\Gamma$.

Since $\{\Gamma_i \subset A\} \in \mathcal{F}_A$, $\{\cup_{i=1}^n \Gamma_i \subset A\} = \cap_{i=1}^n \{\Gamma_i \subset A\} \in \mathcal{F}_A$. Similarly, since A is closed,

$$\{\mathrm{cl} \cup_i \Gamma_i \subset A\} = \{\cup_i \Gamma_i \subset A\} = \cap_i \{\Gamma_i \subset A\} \in \mathcal{F}_A.$$

Suppose $\Gamma_1 \subset \Gamma_2$, and let $G \in \mathcal{F}_{\Gamma_1}$. Then

$$G \cap \{\Gamma_2 \subset A\} = G \cap \{\Gamma_1 \subset A\} \cap \{\Gamma_2 \subset A\} \in \mathcal{F}_A.$$

If $\mathcal{R}(\Gamma) = \{C_k\}$, then

$$\{\Gamma = C_k\} = \{\Gamma \subset C_k\} - \cup_{C_j \subsetneq C_k} \{\Gamma \subset C_j\} \in \mathcal{F}_{C_k}.$$

Similarly, if $B \in \mathcal{F}_\Gamma$, $B_k \equiv B \cap \{\Gamma = C_k\} \in \mathcal{F}_{C_k}$, and hence $B = \cup_k \{\Gamma = C_k\} \cap B_k$ with $B_k \in \mathcal{F}_{C_k}$. Conversely, if $B = \cup_k \{\Gamma = C_k\} \cap B_k$ with $B_k \in \mathcal{F}_{C_k}$,

$$B \cap \{\Gamma \subset A\} = \cup_{C_k \subset A} (B_k \cap \{\Gamma = C_k\}) \in \mathcal{F}_A,$$

so $B \in \mathcal{F}_\Gamma$.

Finally, $\mathcal{R}(\Gamma_n) = \{C_k^n\}$ is countable, so

$$\{N(\Gamma_n \cap B) = l\} \cap \{\Gamma_n \subset A\} = \cup_{C_k^n \subset A} \{N(C_k^n \cap B) = l\} \cap \{\Gamma_n = C_k^n\} \in \mathcal{F}_A$$

and $\{N(\Gamma_n \cap B) = l\} \in \mathcal{F}_{\Gamma_n}$. Since $\Gamma_1 \supset \Gamma_2 \supset \cdots$ and $\cap_n \Gamma_n = \Gamma$, $\lim_{n\to\infty} N(\Gamma_n \cap B) = N(\Gamma \cap B)$ and $N(\Gamma \cap B)$ is $\cap_n \mathcal{F}_{\Gamma_n}$-measurable. □

Lemma 3.6. *If $K \subset S$ is compact, then $\{\Gamma \cap K = \emptyset\} \in \mathcal{F}_\Gamma$. In particular, for each $x \in S$, $\mathbf{1}_\Gamma(x)$ is $\mathcal{F}_\Gamma$-measurable.*

Proof. If $\Gamma \cap K = \emptyset$, then $\inf\{r(x,y) : x \in \Gamma, y \in K\} > 0$. Consequently, setting $G_n = \{y : \inf_{x\in K} r(x,y) < n^{-1}\}$,

$$\{\Gamma \cap K = \emptyset\} = \cup_n \{\Gamma \cap G_n = \emptyset\} = \cup_n \{\Gamma \subset G_n^c\} \in \mathcal{F}_\Gamma.$$

For the second statement, note that $\{\mathbf{1}_\Gamma(x) = 0\} = \{\Gamma \cap \{x\} = \emptyset\} \in \mathcal{F}_\Gamma$. □

We will need to know that, in some sense, limits of stopping sets are stopping sets. If we assume that S is locally compact, then this result is simple to formulate.

Lemma 3.7. *Assume that S is locally compact. Suppose $\{\Gamma_k\}$ are stopping sets, and define*

$$\Gamma = \limsup_{k\to\infty} \Gamma_k \equiv \cap_m \mathrm{cl} \cup_{k\geq m} \Gamma_k.$$

Then Γ is a stopping set.

Proof. Let $G_1 \subset G_2 \subset \cdots$ be open sets with compact closure satisfying $\cup_n G_n = S$, and for $A \in \mathcal{C}(S)$, let $A_n = \{y \in S : \inf_{x\in A} r(x,y) \leq n^{-1}\}$. Noting that $A = \cap_n (A_n \cup G_n^c)$, $\Gamma \subset A$ if and only if for each n, there exists m such that $\cup_{k\geq m} \Gamma_k \subset A_n \cup G_n^c$. Otherwise, for some n, there exist $x_m \in \cup_{k\geq m} \Gamma_k$ such that $x_m \in A_n^c \cap G_n$, and by the compactness of $\mathrm{cl} G_n$, a limit point x of $\{x_m\}$ such that $x \in \Gamma \cap \mathrm{cl}(A_n^c \cap G_n) \subset A^c$. Consequently,

$$\{\Gamma \subset A\} = \cap_n \cup_m \{\cup_{k\geq m} \Gamma_k \subset A_n \cup G_n^c\} \in \cap_n \mathcal{F}_{A_n \cup G_n^c} = \mathcal{F}_A.$$ □

Local compactness also simplifies issues regarding separability of stopping sets. Note that the previous lemma implies that the intersection of a decreasing sequence of stopping sets is a stopping set.

Lemma 3.8. *Let S be locally compact. Then all stopping sets are separable, and if $\{\Gamma_n\}$ is a decreasing sequence of stopping sets with $\Gamma = \cap_n \Gamma_n$, then Γ is a stopping set and $\mathcal{F}_\Gamma = \cap_n \mathcal{F}_{\Gamma_n}$.*

Proof. Let $\{x_i\}$ be dense in S, $\epsilon_j > 0$ with $\lim_{j\to\infty} \epsilon_j = 0$, and $\{G_n\}$ be as in the proof of Lemma 3.7. Let $\{H_k\}$ be some ordering of the countable collection of sets of the form $G_n^c \cup \cup_{l=1}^m \bar{B}_{\epsilon_{j_l}}(x_{i_l})$ with $H_1 = S$. Then for $A \in \mathcal{C}(S)$ and $A_n = \cap \{H_k : k \leq n, A \subset H_k\}$, $A = \cap_n A_n$. If $G \in \cap_n \mathcal{F}_{\Gamma_n}$, then

$$G \cap \{\Gamma \subset A\} = \cap_n G \cap \{\Gamma \subset A_n\} = \cap_n G \cap \{\Gamma_n \subset A_n\} \in \mathcal{F}_A.$$ □

Theorem 3.9. *Let N be a Poisson process in S with mean measure μ and compatible with $\{\mathcal{F}_A\}$. If Γ is a separable $\{\mathcal{F}_A\}$-stopping set and N_Γ is the point process in S_1 obtained by projecting the points of N that lie in Γ onto S_1, then the mean measure for N_Γ satisfies (3.10) (allowing $\infty = \infty$).*

More generally, if $\Gamma^{(1)}$ and $\Gamma^{(2)}$ are stopping sets with $\Gamma^{(1)} \subset \Gamma^{(2)}$, then for each D satisfying $\mu(D) < \infty$,

$$E[N(\Gamma^{(2)} \cap D) - \mu(\Gamma^{(2)} \cap D)|\mathcal{F}_{\Gamma^{(1)}}] = N(\Gamma^{(1)} \cap D) - \mu(\Gamma^{(1)} \cap D) \tag{3.11}$$

and

$$E[N_{\Gamma^{(2)}}(B) - N_{\Gamma^{(1)}}(B)] = E[\mu((\Gamma^{(2)} - \Gamma^{(1)}) \cap B \times S_2], \tag{3.12}$$

again allowing $\infty = \infty$.

Proof. The independence properties of the Poisson process imply that for each $D \in \mathcal{B}(S)$ such that $\mu(D) < \infty$,

$$M^D(A) = N(A \cap D) - \mu(A \cap D) \tag{3.13}$$

is a $\{\mathcal{F}_A\}$-martingale. Let $\{\Gamma_n\}$ be the stopping sets with countable range defined in Lemma 3.3. Then the optional sampling theorem (see Kurtz (1980a)) implies $\mathbb{E}[N(\Gamma_n \cap D) - \mu(\Gamma_n \cap D)] = 0$, and since $\Gamma_1 \supset \Gamma_2 \supset \cdots$ and $\Gamma = \cap_n \Gamma_n$,

$$\mathbb{E}[N(\Gamma \cap D) - \mu(\Gamma \cap D)] = 0.$$

(Note that Γ being a $\{\mathcal{F}_A\}$-stopping set does not imply that $\Gamma \cap D$ is a $\{\mathcal{F}_A\}$-stopping set.) For $B \in \mathcal{B}(S_1)$, assume $\{D_k\} \subset \mathcal{B}(S)$ are disjoint, satisfy $\mu(D_k) < \infty$, and $\cup_k D_k = B \times S_2$. Then, by the monotone convergence theorem,

$$\mathbb{E}[N(\Gamma \cap B \times S_2)] = \sum_{k=1}^{\infty} \mathbb{E}[N(\Gamma \cap D_k)] = \sum_{k=1}^{\infty} \mathbb{E}[\mu(\Gamma \cap D_k)] = \mathbb{E}[\mu(\Gamma \cap B \times S_2)],$$

and the same argument gives (3.11) and (3.12). □

Some martingale properties of this process:

Theorem 3.10. *Let N and Γ be as in Theorem 3.9.*

(a) *If $\mu(D) < \infty$, then L^D defined by*

$$L^D(A) = (N(A \cap D) - \mu(A \cap D))^2 - \mu(A \cap D) \tag{3.14}$$

is an $\{\mathcal{F}_A\}$-martingale, and if $\mu_\Gamma(B) = E[N_\Gamma(B)] < \infty$,

$$\mathbb{E}\big[(N_\Gamma(B) - \mu(\Gamma \cap (B \times S_2)))^2\big] = \mu_\Gamma(B). \tag{3.15}$$

(b) *If $D_1, D_2 \in \mathcal{B}(S)$ are disjoint with $\mu(D_1) + \mu(D_2) < \infty$, then M^{D_1} and M^{D_2} (as defined by Equation (3.13)) are orthogonal martingales in the sense that their product is a martingale. Consequently, if $B_1, B_2 \in \mathcal{B}(S_1)$ are disjoint and $\mathbb{E}[N_\Gamma(B_1)] + \mathbb{E}[N_\Gamma(B_2)] < \infty$, then*

$$\mathbb{E}\big[(N_\Gamma(B_1) - \mu(\Gamma \cap (B_1 \times S_2)))(N_\Gamma(B_2) - \mu(\Gamma \cap (B_2 \times S_2)))\big] = 0.$$

(c) *Let $f : S_1 \times S_2 \to \mathbb{R}_+$, and let $\mathcal{I}_f = \{A \subset S_1 \times S_2; \int_A |e^{f(z)} - 1|\, \mu(dz) < \infty\}$. Then,*

$$M_f(A) = \exp\left\{\int_A f(z)\, N(dz) - \int_A (e^{f(z)} - 1)\, \mu(dz)\right\} \tag{3.16}$$

is an $\{\mathcal{F}_A\}$*-martingale for* $A \in \mathcal{I}_f$*, and therefore, for* $g : S_1 \to \mathbb{R}_+$ *satisfying* $\mathbb{E}[\exp\{\int g(x) N_\Gamma(dx)\}] < \infty$,

$$\mathbb{E}\left[\exp\left\{\int g(x)\, N_\Gamma(dx) - \int_\Gamma (e^{g(x)} - 1)\, \mu(dx \times dy)\right\}\right] = 1. \tag{3.17}$$

Proof. Let $\{D_k\}$ be as in the proof of Theorem 3.9. Then $\{N(\Gamma \cap (B \times S_1) \cap D_k) - \mu(\Gamma \cap (B \times S_1) \cap D_k)\}$ are orthogonal random variables in L^2, with

$$\mathbb{E}[(N(\Gamma \cap (B \times S_1) \cap D_k) - \mu(\Gamma \cap (B \times S_1) \cap D_k))^2] = \mathbb{E}[\mu(\Gamma \cap (B \times S_1) \cap D_k)].$$

Consequently, if $\mathbb{E}[\mu(\Gamma \cap (B \times S_1))] < \infty$, the series converges in L^2, giving (3.15).

Part (b) follows similarly.

The moment generating functional of N is given by

$$\mathbb{E}[e^{\int f\, dN}] = \exp\left\{\int (e^{f(z)} - 1)\, \mu(dz)\right\},$$

which shows that (3.16) is a martingale. Observing that $\{M_f(\cdot \cap D_k)\}$ are orthogonal martingales, and hence that $\mathbb{E}[M_f(\Gamma \cap \cup_{k=1}^{l} D_k)] = 1$, (3.17) follows by the dominated convergence theorem. □

Definition 3.11. Assume that $S = S_1 \times [0, \infty)$. A random function $\phi : S_1 \to [0, \infty)$ is a $\{\mathcal{F}_A\}$*-stopping surface* if the set $\Gamma_\phi = \{(x, y); x \in S_1, 0 \le y \le \phi(x)\}$ is a $\{\mathcal{F}_A\}$-stopping set. (Note that the requirement that Γ_ϕ be closed implies ϕ is upper semicontinuous.)

For simplicity, we will write $\mathcal{F}_\phi$ in place of $\mathcal{F}_{\Gamma_\phi}$. Furthermore, since $\Gamma_\phi \subset A$ if and only if for each $x \in S_1$,

$$\phi(x) \le f(x) = \sup\{z \ge 0 : \{x\} \times [0, z] \subset A\},$$

we only need to consider A of the form $A_f = \{(x, y) : y \le f(x)\}$ for nonnegative, upper semicontinuous f, that is, we only need to verify that $\{\phi \le f\} \in \mathcal{F}_f$ for each nonnegative, upper-semicontinuous f. Again we write $\mathcal{F}_f$ rather than $\mathcal{F}_{A_f}$. Furthermore, since an upper-semicontinuous f is the limit of a decreasing sequence of continuous f_n and $\{\phi \le f\} = \cap_n \{\phi \le f_n\}$ and $\mathcal{F}_f = \cap_n \mathcal{F}_{f_n}$, it is enough to verify $\{\phi \le f\} \in \mathcal{F}_f$ for continuous f.

Lemma 3.12. *Let* ϕ *be a stopping surface. Then*

(i) *For each* $x \in S_1$, $\phi(x)$ *is* $\mathcal{F}_\phi$*- measurable.*
(ii) *For each compact* $K \subset S_1$, $\sup_{x \in K} \phi(x)$ *is* $\mathcal{F}_\phi$*-measurable.*
(iii) *If* $a \ge 0$ *is deterministic and upper semicontinuous, then* $\phi + a$ *is a stopping surface.*

Proof. Since $\{\phi(x) < y\} = \{\Gamma_\phi \cap \{(x, y)\} = \emptyset\} \in \mathcal{F}_\phi$, by Lemma 3.6, $\phi(x)$ is $\mathcal{F}_\phi$-measurable.

Since ϕ assumes its supremum over a compact set, for $y \ge 0$,

$$\{\sup_{x \in K} \phi(x) < y\} = \{\Gamma_\phi \cap K \times \{y\} = \emptyset\} \in \mathcal{F}_\phi.$$

For Part (iii), first assume that a is continuous. Then for f continuous, $\{\phi + a \leq f\} = \{\phi \leq f - a\} \in \mathcal{F}_f$, and for f upper semicontinuous, there exists a decreasing sequence of continuous f_n converging to f, so

$$\{\phi + a \leq f\} = \cap_n \{\phi + a \leq f_n\} \in \mathcal{F}_\phi.$$

But this implies that for every nonnegative, upper-semicontinuous g,

$$\{\phi + a \leq f\} \cap \{\phi \leq g\} \in \mathcal{F}_g,$$

so in particular, $\{\phi + a \leq f\} = \{\phi + a \leq f\} \cap \{\phi \leq f\} \in \mathcal{F}_f$, and hence, $\phi + a$ is a stopping surface.

Finally, for a upper semicontinuous, there is a decreasing sequence of continuous functions a_n converging to a, so $\Gamma_{\phi+a} = \cap_n \Gamma_{\phi+a_n}$, and $\Gamma_{\phi+a}$ is a stopping set by Lemma 3.7. □

For a signed measure γ, let $T(\gamma, B)$ denote the total variation of γ over the set B. Let $\phi_i : S_1 \to [0, \infty)$, $i = 1, 2$, be stopping surfaces.

Corollary 3.13. *Suppose $\mu = \nu \times m$, and write N_{ϕ_i} instead of $N_{\Gamma_{\phi_i}}$. Then,*

$$\mathbb{E}[T(N_{\phi_1} - N_{\phi_2}, B)] = \mathbb{E}\left[\int_B |\phi_1(x) - \phi_2(x)| \nu(dx)\right], \tag{3.18}$$

and consequently,

$$\mathbb{E}\left[\left|\int f(x) N_{\phi_1}(dx) - \int f(x) N_{\phi_2}(dx)\right|\right] \leq \mathbb{E}\left[\int |f(x)||\phi_1(x) - \phi_2(x)| \nu(dx)\right].$$

Proof. Note that

$$T(N_{\phi_1} - N_{\phi_2}, B) = 2N_{\phi_1 \vee \phi_2}(B) - N_{\phi_1}(B) - N_{\phi_2}(B),$$

and since the union of two stopping sets is a stopping set, (3.12) gives (3.18). We then have

$$\begin{aligned} \mathbb{E}\left[\left|\int f(x) N_{\phi_1}(dx) - \int f(x) N_{\phi_2}(dx)\right|\right] &\leq \mathbb{E}\left[\int |f(x)|\, T(N_{\phi_1} - N_{\phi_2}, dx)\right] \qquad (3.19) \\ &\leq \mathbb{E}\left[\int |f(x)||\phi_1(x) - \phi_2(x)|\, \nu(dx)\right]. \end{aligned}$$ □

4. Characterization of point processes as projections of higher-dimensional Poisson processes

It would be interesting to know which point processes can be obtained as projections of higher-dimensional Poisson point processes. Gaver, Jacobs and Latouche (1984) characterized finite birth and death models in randomly changing environments as Markov processes in a higher-dimensional space. Garcia (1995a) generalizes this idea to other counting processes. We consider processes in Euclidean space, although most results have analogs in more general spaces. We assume that all processes are defined on a fixed probability space $(\Omega, \mathcal{F}, \mathbb{P})$.

Let η be a point process on $\mathbb{R}^d \times [0,\infty)$, and let $\mathcal{A}(\mathbb{R}^d) = \{B \in \mathcal{B}(\mathbb{R}^d) : m(B) < \infty\}$. Define $\eta_t(B) = \eta(B \times [0,t])$, $B \in \mathcal{A}(\mathbb{R}^d)$. Assume that $\eta_t(B)$ is a counting process for each $B \in \mathcal{A}(\mathbb{R}^d)$ and if $B_1 \cap B_2 = \emptyset$, that $\eta(B_1)$ and $\eta(B_2)$ have no simultaneous jumps.

Let $\{\mathcal{F}_t\}$ be a filtration, and let $\lambda : \mathbb{R}^d \times [0,\infty) \times \Omega \to [0,\infty)$ be progressive in the sense that for each $t \geq 0$, $\lambda : \mathbb{R}^d \times [0,t] \times \Omega$ is $\mathcal{B}(\mathbb{R}^d) \times \mathcal{B}([0,t]) \times \mathcal{F}_t$-measurable. Assume that λ is locally integrable in the sense that $\int_{B\times[0,t]} \lambda(x,s)ds < \infty$ a.s. for each $B \in \mathcal{A}(\mathbb{R}^d)$ and each $t > 0$. Then η has *$\{\mathcal{F}_t\}$-intensity* λ (with respect to Lebesgue measure) if and only if for each $B \in \mathcal{A}(\mathbb{R}^d)$,

$$\eta_t(B) - \int_{B\times[0,t]} \lambda(x,s)dxds$$

is a $\{\mathcal{F}_t\}$-local martingale.

Theorem 4.1. *Suppose that η has $\{\mathcal{F}_t\}$-intensity λ, λ is upper semicontinuous as a function of x, and there exists $\epsilon > 0$ such that $\lambda(x,t) \geq \epsilon$ for all x, t, almost surely. Then there exists a Poisson random measure N on $\mathbb{R}^d \times [0,\infty)$ such that for*

$$\begin{aligned} \Gamma_t &= \left\{(x,s) : x \in \mathbb{R}^d, 0 \leq s \leq \int_0^t \lambda(x,s)ds\right\} \\ \eta_t(B) &= N(\Gamma_t \cap (B \times [0,\infty)), \end{aligned} \tag{4.1}$$

that is, setting $\phi_t(x) = \int_0^t \lambda(x,s)ds$, $\eta_t = N_{\phi_t}$.

Remark 4.2. The condition that λ is bounded away from zero is necessary to ensure that $\gamma(t,x) = \inf\{s; \int_0^s \lambda(x,u)du \geq t\}$ is defined for all $t > 0$. If this condition does not hold, we can define

$$\eta_t^\epsilon(B) = \eta_t(B) + \xi(B \times [0,\epsilon t]),$$

where ξ is a Poisson process on $\mathbb{R}^d \times [0,\infty)$ with Lebesgue mean measure that is independent of $\vee_t \mathcal{F}_t$. Then it is clear that

$$\eta_t^\epsilon(B) - \int_0^t \int_B (\lambda(x,s) + \epsilon)dx\, ds$$

is a martingale for all $B \in \mathcal{A}(\mathbb{R}^d)$. The theorem then gives the representation

$$\eta_t^\epsilon(B) = N^\epsilon(\Gamma_t^\epsilon \cap B \times [0,\infty))$$

where $\Gamma_t^\epsilon = \{(x,s); x \in \mathbb{R}^d, s \leq \int_0^t (\lambda(x,u) + \epsilon)du\}$. Letting $\epsilon \to 0$, gives the result without the lower bound on λ.

Proof. The proof is similar to that of Theorem 2.6 of Garcia (1995a). □

5. Pure birth processes

The primary approach to building a model using a spatial, pure-birth process with points in a set K (which we will take to be a subset of $\mathbb{R}^d$) is to specify the intensity in a functional form, that is, as a function of the desired process η and, perhaps, additional randomness ξ in a space E. Assume we are given a jointly measurable mapping

$$\tilde{\lambda} : (x, z, u, s) \in K \times D_{\mathcal{N}(K)}[0, \infty) \times E \times [0, \infty) \to \tilde{\lambda}(x, z, u, s) \in [0, \infty).$$

Intuitively, in specifying $\tilde{\lambda}$, we are saying that in the next time interval $(t, t + \Delta t]$, the probability of there being a birth in a small region A is approximately $\int_A \tilde{\lambda}(x, \eta, \xi, t) dx \Delta t$. One can make this precise by requiring the process η to have the property that there is a filtration $\{\mathcal{G}_t\}$ such that

$$\eta_t(B) - \int_0^t \int_B \tilde{\lambda}(x, \eta, \xi, s) dx\, ds$$

is a $\{\mathcal{G}_t\}$-martingale for each $B \in \mathcal{A}(K)$. For this *martingale problem* to make sense, $\tilde{\lambda}$ must depend on η in a nonanticipating way, that is, $\tilde{\lambda}(x, \eta, \xi, s)$ depends on η_r only for $r \leq s$.

The representation (4.1) suggests formulating a stochastic equation by requiring η and a "stopping-surface-valued function" τ to satisfy

$$\begin{aligned} \tau(t, x) &= \int_0^t \tilde{\lambda}(x, \eta, \xi, s) ds \\ \eta_t &= N_{\tau(t)}. \end{aligned} \tag{5.1}$$

If $\tilde{\lambda}$ is appropriately nonanticipating,

$$\int_0^t \int_K \tilde{\lambda}(x, z, u, s) dx ds < \infty \text{ for all } z \in D_{\mathcal{N}(K)}[0, \infty), u \in E,$$

and ξ and N are independent, then the solution of (5.1) and verification of the martingale properties are straightforward. (Just "solve" from one birth to the next.) However, we are interested in situations, say with $K = \mathbb{R}^d$, in which

$$\int_0^t \int_{\mathbb{R}^d} \tilde{\lambda}(x, z, u, s) dx = \infty,$$

that is, there will be infinitely many births in a finite amount of time.

5.1. Existence and uniqueness of time-change equation

To keep the development as simple as possible, we will focus on $\tilde{\lambda}$ such that $\tilde{\lambda}(x, z, u, s) = \lambda(x, z_s)$, $\lambda : \mathbb{R}^d \times \mathcal{N}(\mathbb{R}^d) \to [0, \infty)$, that is, the intensity depends only on the current configuration of points.

In this setting, (5.1) becomes

$$\begin{cases} \dot{\tau}(t,x) = \lambda(x, N_{\tau(t)}) \\ \tau(0,x) = 0 \\ N_{\tau(t)}(B) = N(\Gamma_{\tau(t)} \cap B \times [0,\infty)) \\ \Gamma_{\tau(t)} = \{(x,y); x \in \mathbb{R}^d, 0 \le y \le \tau(t,x)\} \end{cases} \tag{5.2}$$

where we write the system in this form to emphasize the fact that τ is the solution of a random, but autonomous differential equation. For the earlier analysis to work, the solution must also have the property that $\Gamma_{\tau(t)}$ be a stopping set with respect to the filtration $\{\mathcal{F}_A\}$.

We need the following regularity condition.

Condition 5.1. For $\zeta \in \mathcal{N}(\mathbb{R}^d)$, $x \to \lambda(x,\zeta)$ is upper semicontinuous. For $\zeta_1, \zeta_2 \in \mathcal{N}(\mathbb{R}^d)$ and $\{y_i\} \subset \mathbb{R}^d$ satisfying $\zeta_2 = \zeta_1 + \sum_{i=1}^{\infty} \delta_{y_i}$ and $\lambda(x,\zeta_1) < \infty$,

$$\lambda(x,\zeta_2) = \lim_{n\to\infty} \lambda\left(x, \zeta_1 + \sum_{i=1}^{n} \delta_{y_i}\right). \tag{5.3}$$

The following theorem extends conditions of Liggett for models on a lattice (Liggett (1972), Kurtz (1980b)) to the present setting. Garcia (1995a) proves a similar theorem for a general case of birth and death processes. However, this theorem is not a particular case of the general case, since the conditions and techniques used there are not directly applicable for the case in which the death rate equals 0.

Theorem 5.2. *Assume that Condition* 5.1 *holds. Let*

$$a(x,y) = \sup_{\zeta \in \mathcal{N}(\mathbb{R}^d)} |\lambda(x,\zeta+\delta_y) - \lambda(x,\zeta)| \tag{5.4}$$

and $\bar{a}(x) = \sup_{\zeta_1,\zeta_2} |\lambda(x,\zeta_1) - \lambda(x,\zeta_2)|$. *Suppose there exists a positive function c such that* $\sup_x c(x)\bar{a}(x) < \infty$ *and*

$$M = \sup_x \int_{\mathbb{R}^d} \frac{c(x)a(x,y)}{c(y)} dy < \infty.$$

Then, there exists a unique solution of (5.2) *with* $\tau(t,\cdot)$ *a stopping surface for all* $t \ge 0$.

Remark 5.3. For example, suppose $a(x,y) = a(x-y)$ and $\int |y|^p a(y)\,dy < \infty$. Then we can take $c(x) = (1+|x|^p)^{-1}$. Setting $c_1 = \int |y|^p a(y)dy$ and $c_2 = \int a(y)dy$,

$$\begin{aligned} \int (1+|z|^p)a(x-z)dz &= \int a(x-z)dz + \int |z|^p a(x-z)dz \\ &= \int a(y)dy + \int |y-x|^p a(y)dy \\ &\le c_2 + a_p \int |y|^p a(y)dy + a_p|x|^p) \int a(y)dy \\ &\le c_2 + c_1 a_p + c_2 a_p |x|^p \le M(1+|x|^p), \end{aligned}$$

where a_p satisfies $|y-x|^p \leq a_p(|y|^p + |x|^p)$ and $M = \max\{c_2 + c_1 a_p, c_2 a_p\}$. Consequently,

$$\sup_x \int \frac{(1+|z|^p)a(x-z)}{1+|x|^p} < M.$$

Outline of the proof: Let $\mathcal{H}$ be the space of real-valued, measurable processes indexed by $\mathbb{R}^d$, and let $\mathcal{H}_0$ be the subset of $\xi \in \mathcal{H}$ such that

$$\|\xi\| \equiv \sup_{x\in\mathbb{R}^d} c(x)\mathbb{E}[|\xi(x)|] < \infty.$$

Define $\mathcal{H}^+ = \{\xi \in \mathcal{H} : \xi \geq 0\}$, and similarly for $\mathcal{H}_0^+$. Let $F : \phi \in \mathcal{H}_+ \to \tilde{\phi} \in \mathcal{H}_+$, where $\tilde{\phi}(x) = \lambda(x, N_\phi)$. That is,

$$\Gamma_\phi = \{(x,y); x \in \mathbb{R}^d, 0 \leq y \leq \phi(x)\},$$

$$N_\phi(B) = N(\Gamma_\phi \cap B \times [0,\infty)),$$

and

$$F(\phi)(x) = \lambda(x, N_\phi). \tag{5.5}$$

Then the system (5.2) is equivalent to

$$\dot{\tau}(t) = F(\tau(t)). \tag{5.6}$$

We are only interested in solutions for which $\tau(t)$ is a stopping surface. Let τ_1 and τ_2 be two such solutions. We will show that F is Lipschitz, so that we can find estimates of $\|\tau_1(t) - \tau_2(t)\|$ in terms of $\int_0^t \|\tau_1(s) - \tau_2(s)\|ds$ and apply Gronwall's inequality to obtain the uniqueness of the solution. To prove existence, we are going to construct a sequence of stopping surfaces $\tau^{(n)}(t)$ whose limit is a solution of the system.

The proof of the theorem relies on several lemmas.

Lemma 5.4. *Let $\lambda : \mathbb{R}^d \times \mathcal{N}(\mathbb{R}^d) \to \mathbb{R}^+$ satisfy Condition* 5.1, *and define $a(x,y)$ as in* (5.4). *Then*

$$\sup_\zeta \left|\lambda(x, \zeta + \sum_{i=1}^m \delta_{y_i}) - \lambda(x,\zeta)\right| \leq \sum_{i=1}^m a(x, y_i) \tag{5.7}$$

$$\sup_\zeta \left|\lambda(x, \zeta + \sum_{i=1}^\infty \delta_{y_i}) - \lambda(x,\zeta)\right| \leq \sum_{i=1}^\infty a(x, y_i). \tag{5.8}$$

Proof. The definition of $a(x,y)$ and induction give (5.7), and (5.8) then follows by Condition 5.1. ☐

Lemma 5.5. *For γ_1 and γ_2 stopping surfaces,*

$$\mathbb{E}[|F(\gamma_1)(x) - F(\gamma_2)(x)|] \leq \int_{\mathbb{R}^d} a(x,y)\mathbb{E}[|\gamma_1(y) - \gamma_2(y)|]dy\,.$$

Proof. Since N is a Poisson point processes with a σ-finite mean measure, it has only countably many points almost surely. Let, $\{y_1, y_2, \ldots\}$ and $\{z_1, z_2, \ldots\}$ be such that $y_i \neq z_j$, for all i, j, and

$$N_{\gamma_2} = N_{\gamma_1} + \sum_i \delta_{y_i} - \sum_j \delta_{z_j}.$$

Then

$$T(N_{\gamma_1} - N_{\gamma_2}, B) = \sum_i \delta_{y_i}(B) + \sum_j \delta_{z_j}(B)$$

and

$$\begin{aligned}|\lambda(x,N_{\gamma_1}) - \lambda(x,N_{\gamma_2})| &= \Bigg|\lambda(x,N_{\gamma_1}) - \lambda\left(x, N_{\gamma_1} + \sum_i \delta_{y_i}\right) \\ &\quad + \lambda\left(x, N_{\gamma_1} + \sum_i \delta_{y_i}\right) - \lambda\left(x, N_{\gamma_1} + \sum_i \delta_{y_i} - \sum_j \delta_{z_j}\right)\Bigg| \\ \text{(by Lemma 5.4(b))} &\leq \sum_i a(x,y_i) + \sum_j a(x,z_j) = \int_{\mathbb{R}^d} a(x,y) T(N_{\gamma_1} - N_{\gamma_2}, dy).\end{aligned}$$

Therefore,

$$\begin{aligned}\mathbb{E}[|F(\gamma_1)(x) - F(\gamma_2)(x)|] &= \mathbb{E}[|\lambda(x, N_{\gamma_1}) - \lambda(x, N_{\gamma_2})] \\ &\leq \mathbb{E}\left[\int_{\mathbb{R}^d} a(x,y)\, T(N_{\gamma_1} - N_{\gamma_2}, dy)\right] \\ \text{by Corollary 3.13} &\leq \mathbb{E}\left[\int_{\mathbb{R}^d} a(x,y)\, |\gamma_1(y) - \gamma_2(y)|\, dy\right],\end{aligned}$$

and the result follows by Fubini's theorem. □

Lemma 5.6. *The mapping $F : \mathcal{H}_+ \to \mathcal{H}_+$ given by (5.5) is Lipschitz for stopping surfaces in $\mathcal{H}_+$.*

Proof. By Lemma 5.5,

$$\begin{aligned}\|F(\gamma_1) - F(\gamma_2)\| &= \sup_x c(x)\mathbb{E}[|F(\gamma_1)(x) - F(\gamma_2)(x)|] \\ &\leq \sup_x c(x) \int_{\mathbb{R}^d} a(x,y)\, \mathbb{E}[|\gamma_1(y) - \gamma_2(y)|]\, dy \\ &\leq \sup_x \int \frac{c(x)a(x,y)}{c(y)} dy \, \sup_y c(y)\mathbb{E}[|\gamma_1(y) - \gamma_2(y)|] \\ &\leq M\|\gamma_1 - \gamma_2\|.\end{aligned}$$

□

Proof of Theorem 5.2. *Uniqueness.* By Lemma 5.6 we have

$$\begin{aligned}\| \tau_1(t) - \tau_2(t) \| &= \sup_x c(x)\mathbb{E}[|\tau_1(t,x) - \tau_2(t,x)|] \\ &\leq \sup_x c(x)\mathbb{E}[\int_0^t |F(\tau_1(s))(x) - F(\tau_2(s))(x)|] \\ &\leq \int_0^t \|F(\tau_1(s)) - F(\tau_2(s))\| ds \\ &\leq M\int_0^t \|\tau_1(s) - \tau_2(s)\| ds.\end{aligned}$$

Note that $|F(\tau_1(s))(x) - F(\tau_2(s))(x)| \leq \bar{a}(x)$, so $\sup_s \|F(\tau_1(s)) - F(\tau_2(s))\| < \infty$. Consequently, uniqueness follows by Gronwall's inequality.

Existence. Let $\underline{\lambda}(x) = \inf_\zeta \lambda(x,\zeta)$. Then $|\lambda(x,\zeta) - \underline{\lambda}(x)| \leq \bar{a}(x)$. Define

$$\tau^{(n)}(t) = \int_0^t F(s\underline{\lambda})ds, \quad \text{for } 0 \leq t \leq 1/n \tag{5.9}$$

$$\tau^{(n)}(t) = \tau^{(n)}\left(\frac{k}{n}\right) + \int_{\frac{k}{n}}^t F\left(\tau^{(n)}\left(\frac{k}{n}\right) + \left(s - \frac{k}{n}\right)\underline{\lambda}\right) ds, \quad \text{for } \frac{k}{n} < t \leq \frac{k+1}{n}.$$

Set,

$$\gamma^{(n)}(t) = \tau^{(n)}\left(\frac{[nt]}{n}\right) + \left(t - \frac{[nt]}{n}\right)\underline{\lambda}.$$

Then,

$$\tau^{(n)}(t) = \int_0^t F(\gamma^{(n)}(s))ds. \tag{5.10}$$

Note that $\tau^{(n)}(t)$ and $\gamma^{(n)}(t)$ are stopping surfaces (see the proof below). Consequently,

$$\begin{aligned}\| \tau^{(n)}(t) - \gamma^{(n)}(t) \| &= \left\|\tau^{(n)}(t) - \tau^{(n)}([nt]/n) + \tau^{(n)}([nt]/n) - \gamma^{(n)}(t)\right\| \\ &= \left\|\int_{[nt]/n}^t F(\gamma^{(n)}(s))ds - (t - [nt]/n)\underline{\lambda}\right\| \\ &= \left\|\int_{[nt]/n}^t (F(\gamma^{(n)}(s)) - \underline{\lambda})ds\right\| \leq \int_{[nt]/n}^t \| F(\gamma^{(n)}(s)) - \underline{\lambda} \| ds \\ &\leq \int_{[nt]/n}^t \sup_x c(x)\mathbb{E}[|F(\gamma^{(n)}(s))(x) - \underline{\lambda}(x)|]ds \\ &\leq \int_{[nt]/n}^t \sup_x c(x)\bar{a}(x)ds \leq \frac{\sup_x c(x)\bar{a}(x)}{n}.\end{aligned}$$

Also,

$$\begin{aligned}\| \gamma^{(n)}(t) - \gamma^{(m)}(t) \| &\leq \| \tau^{(n)}(t) - \gamma^{(n)}(t) \| + \| \tau^{(m)}(t) - \gamma^{(m)}(t) \| \\ &\quad + \| \int_0^t F(\gamma^{(n)}(s))ds - \int_0^t F(\gamma^{(m)}(s))ds \| \\ &\leq 2\frac{\sup_x c(x)\bar{a}(x)}{n} + M \int_0^t \| \gamma^{(n)}(s) - \gamma^{(m)}(s) \| ds \\ \text{(by Gronwall's inequality)} &\leq 2\frac{\sup_x c(x)\bar{a}(x)}{n} e^{Mt}.\end{aligned}$$

Therefore,

$$\begin{aligned}\| \tau^{(n)}(t) - \tau^{(m)}(t) \| &\leq \| \tau^{(n)}(t) - \gamma^{(n)}(t) \| + \| \gamma^{(n)}(t) - \gamma^{(m)}(t) \| \\ &\quad + \| \gamma^{(m)}(t) - \tau^{(m)}(t) \| \\ &\leq 2\frac{\sup_x c(x)\bar{a}(x)}{n} + 2\frac{\sup_x c(x)\bar{a}(x)}{n} e^{Mt}.\end{aligned}$$

Then, fixing l, $\{\tau^{(n)}(t) - \tau^{(l)}(t)\}$ is a Cauchy sequence in $\mathcal{H}_0$. Completeness of $\mathcal{H}_0$ follows by standard arguments, and so there exists $\tau^*(t) \in \mathcal{H}_0$

$$\tau^*(t) = \lim_{n\to\infty} (\tau^{(n)}(t) - \tau^{(l)}(t)).$$

Then, taking $\tau(t) = \tau^*(t) + \tau^{(l)}(t) = \min_m \sup_{n\geq m} \tau^{(n)}(t)$, $\tau(t)$ is a stopping surface by Lemma 3.7,

$$\lim_{n\to\infty} \|\tau^{(n)}(t) - \tau(t)\| = \lim_{n\to\infty} \|\gamma^{(n)}(t) - \tau(t)\| = 0,$$

and since

$$\left\| \int_0^t F(\gamma^{(n)}(s)ds - \int_0^t F(\tau(s))ds \right\| \leq \int_0^t \|F(\gamma^{(n)}(s)) - F(\tau(s))\|ds \to 0,$$

$$\dot{\tau}(t) = F(\tau(t)).$$

Proof that $\tau^{(n)}(t)$ and $\gamma^{(n)}(t)$ are stopping surfaces. By definition, $\tau^{(n)}(t)$ is a stopping surface if and only if $\{\tau^{(n)}(t) \leq f\} \in \mathcal{F}_f$ for each nonnegative, upper-semicontinuous f.

(i) $0 \leq t \leq 1/n$

$$\tau^{(n)}(t) = \int_0^t F(s\underline{\lambda})ds \qquad \tau^{(n)}(t,x) = \int_0^t \lambda(x, N_{s\underline{\lambda}})ds$$

$$N_{s\underline{\lambda}}(B) = N(\Gamma_{s\underline{\lambda}} \cap B \times [0,\infty)) \qquad \Gamma_{s\underline{\lambda}} = \{(x,u); x \in \mathbb{R}^d, 0 \leq u \leq s\underline{\lambda}(x)\}.$$

By the measurability of λ, the mapping

$$(x,s,\omega) \in \mathbb{R}^d \times [0,t] \times \Omega \to \lambda(x, N_{s\underline{\lambda}})$$

is $\mathcal{B}(\mathbb{R}^d) \times \mathcal{B}([0,t]) \times \mathcal{F}_{t\underline{\lambda}}$-measurable, and hence, $(x,\omega) \to \tau^{(n)}(t,x)$ is $\mathcal{B}(\mathbb{R}^d) \times \mathcal{F}_{t\underline{\lambda}}$-measurable. By the completeness of $\mathcal{F}_{t\underline{\lambda}}$,

$$\{\tau^{(n)}(t) \leq f\}^c = \{\omega : \exists x \ni \tau^{(n)}(t,x) > f(x)\} \in \mathcal{F}_{t\underline{\lambda}}.$$

Note that $\tau^{(n)}(t,x) \geq \underline{\lambda}(x)t$ for all x, by the definition of $\underline{\lambda}$. Consequently, if $f \geq t\underline{\lambda}$, $\{\tau^{(n)}(t) \leq f\} \in \mathcal{F}_{t\underline{\lambda}} \subset \mathcal{F}_f$, and if $f(x) < t\underline{\lambda}(x)$ for some x, $\{\tau^{(n)}(t) \leq f\} = \emptyset \in \mathcal{F}_f$.

(ii) $k/n < t \leq (k+1)/n$.

Proceeding by induction, assume that $\tau^{(n)}(k/n)$ is a stopping surface. Then for $k/n \leq s < (k+1)/n$, by Lemma 3.12, $\gamma^{(n)}(s)$ is a stopping surface. By the definition of $\tau^{(n)}(t,x)$ we have

$$\tau^{(n)}(t,x) = \tau^{(n)}(k/n,x) + \int_{k/n}^{t} \lambda(x, N_{\tau^{(n)}(k/n)+(s-k/n)\underline{\lambda}})ds,$$

and by the definition of $\underline{\lambda}$

$$\tau^{(n)}(t) \geq \gamma^{(n)}(t)$$

Therefore, since $\{\tau^{(n)}(t) \leq f\} \in \mathcal{F}_{\gamma^{(n)}(t)}$

$$\{\tau^{(n)}(t) \leq f\} = \{\tau^{(n)}(t) \leq f\} \cap \{\gamma^{(n)}(t) \leq f\} \in \mathcal{F}_f. \qquad \square$$

An immediate consequence of the existence proof is:

Corollary 5.7. *Let $\tau^{(n)}(t)$ be defined by (5.9) and let*

$$\eta_t^n(B) = N_{\tau^n(t)} = N(\Gamma_{\tau^{(n)}(t)} \cap B \times [0,\infty)).$$

Then $\eta_t^n \to \eta_t$, in probability as $n \to \infty$, uniformly in $t \leq T$.

Another important characteristic of the solution of the time-change problem is that two births cannot occur at the same time.

Theorem 5.8. *Under the conditions of Theorem 5.2, for each $B \in \mathcal{A}(\mathbb{R}^d)$, the process $t \to \eta_t(B)$ is a counting process with intensity $\int_B \lambda(x,\eta_t)dx$. If $B \cap C = \emptyset$, then $\eta_t(B)$ and $\eta_t(C)$ have no simultaneous jumps.*

Proof. Since

$$M_B(D) = N(D \cap B \times [0,\infty)) - \int_{D\cap B\times[0,\infty)} du\,dx$$

and

$$M_C(D) = N(D \cap C \times [0,\infty)) - \int_{D\cap C\times[0,\infty)} du\,dx$$

are orthogonal martingales with respect to $\{\mathcal{F}_A\}$, that is M_B, M_C, and $M_B M_C$ are all martingales, The optional sampling theorem implies $M_B(\Gamma_{\tau(t)})$, $M_C(\Gamma_{\tau(t)})$, and $M_B(\Gamma_{\tau(t)})M_C(\Gamma_{\tau(t)})$ are all martingales with respect to the filtration $\{\mathcal{F}_{\tau(t)}, t \geq 0\}$. Noting that

$$M_B(\tau(t)) = \eta_t(B) - \int_0^t \int_B \tau(x,s)dxds$$

and

$$M_C(\tau(t)) = \eta_t(C) - \int_0^t \int_C \tau(x,s)dxds,$$

the result follows. $\square$

5.2. Equivalence between time change solution and the solution of the martingale problem

Usually, Markov processes are described through their infinitesimal generator. The pure birth process described in the beginning of this section can be characterized as the solution of the martingale problem corresponding to the generator defined by

$$AF(\zeta) = \int_{\mathbb{R}^d} (F(\zeta + \delta_y) - F(\zeta))\lambda(y, \zeta)\, dy, \tag{5.11}$$

for $F \in \mathcal{D}(A) = \{F : \mathcal{N}(\mathbb{R}^d) \to \mathbb{R}; F(\zeta) = \exp\{-\int g\, d\zeta\}, g \geq 0, g \in B_c(\mathbb{R}^d)\}$, where $B_c(\mathbb{R}^d)$ is the set of bounded functions on $\mathbb{R}^d$ with compact support. Our goal in this section is to prove that the two characterizations are equivalent, that is, the solution of the time change problem is the solution of the martingale problem and vice-versa. To avoid issues of integrability and explosion, we assume that for each compact $K \subset \mathbb{R}^d$, $\sup_{x \in K, \zeta \in \mathcal{N}(\mathbb{R}^d)} \lambda(x, \zeta) < \infty$.

Theorem 5.2 gives conditions for existence and uniqueness of random time changes in the strong sense, that is, given the process N. This result will imply the existence and uniqueness of the solution for the martingale problem. However, existence and uniqueness of the solution of the martingale problem implies existence and uniqueness of the time change solution in the weak sense.

Let N be a Poisson point process with Lebesgue mean measure in $\mathbb{R}^d \times [0, \infty)$ defined on a complete probability space $(\Omega, \mathcal{F}, \mathbb{P})$. Let $\lambda : \mathbb{R}^d \times \mathcal{N}(\mathbb{R}^d) \to \mathbb{R}$ be non-negative Borel measurable function. We are interested in solutions of the system:

$$\begin{cases} \tau(0, x) = 0 \\ \dot{\tau}(t, x) = \lambda(x, N_{\tau(t)}) \\ \Gamma_{\tau(t)} = \{(x, y); x \in \mathbb{R}^d, 0 \leq y \leq \tau(t, x)\} \\ N_{\tau(t)}(B) = N(\Gamma_{\tau(t)} \cap B \times [0, \infty)), \end{cases} \tag{5.12}$$

where $\tau(t, \cdot)$ is a stopping surface with respect to the filtration $\{\mathcal{F}_A, A \in \mathcal{C}(\mathbb{R}^d)\}$ as defined in Section 3.3.

Definition 5.9. An $\mathcal{N}(\mathbb{R}^d)$-valued process η is a *weak solution* of (5.12) if there exists a probability space $(\Omega^*, \mathcal{F}^*, \mathbb{P}^*)$ on which are defined processes N^* and τ^* such that N^* is a version of N, (5.12) is satisfied with (N, τ) replaced by (N^*, τ^*), and $N^*_{\tau^*}$ has the same distribution as η.

Theorem 5.10. *Let N be Poisson point process in $\mathbb{R}^d \times [0, \infty)$ with Lebesgue mean measure defined in $(\Omega, \mathcal{F}, \mathbb{P})$. Let $\lambda : \mathbb{R}^d \times \mathcal{N}(\mathbb{R}^d) \to \mathbb{R}$ be a non-negative Borel function, and let A be given by* (5.11).

(a) *If τ and N satisfy* (5.12), *then $\eta(t) = N_{\tau(t)}$ gives a solution of the martingale problem for A.*
(b) *If η is a $\mathcal{N}(\mathbb{R}^d)$-valued process which is a solution of the $D_{\mathcal{N}(\mathbb{R}^d)}[0, \infty)$-martingale problem for A, then η is a weak solution of* (5.12).

Proof. (a) We want to prove that for each $F \in \mathcal{D}(A)$

$$M(t) = F(N_{\tau(t)}) - \int_0^t AF(N_{\tau(s)})ds$$

is a martingale with respect to the filtration $\{\mathcal{G}_t^\tau\} = \{\mathcal{F}_{\Gamma_{\tau(t)}}, t \geq 0\}$. We have as a direct consequence of Theorem 3.9 that for $B \in \mathcal{B}(\mathbb{R}^d)$ with compact closure,

$$M(B \times [0,t]) = M_t(B) = N_{\tau(t)}(B) - \int_B \tau(t,x)\, dx$$

is a martingale. Hence, for $g \in B_c(\mathbb{R}^d)$, that is, $g \geq 0$, bounded and with compact support, we have

$$\begin{aligned}
&\int_{\mathbb{R}^d \times [0,t]} e^{-\int g(y) N_{\tau(s-)}(dy)} (e^{-g(x)} - 1) M(dx \times ds) \\
&\quad = e^{-\int g(y) N_{\tau(t)}(dy)} - 1 \\
&\qquad - \int_0^t \int_{\mathbb{R}^d} \lambda(x, N_{\tau(s)}) (e^{-\int g(y) N_{\tau(s)}(dy) + g(x)} - e^{-\int g(y) N_{\tau(s)}(dy)}) dx \\
&\quad = F(N_{\tau(t)}) - 1 - \int_0^t AF(N_{\tau(s)}) ds
\end{aligned}$$

is a martingale with respect to the filtration $\{\mathcal{G}_t^\tau\}$.

(b) Conversely, if η is a solution of the martingale problem, then there exists a filtration $\{\mathcal{G}_t\}$ such that for $B \in \mathcal{B}(\mathbb{R}^d)$ with compact closure,

$$\eta_t(B) - \int_0^t \int_B \lambda(x, \eta_s) ds$$

is a martingale with respect to $\{\mathcal{G}_t\}$.

Therefore, by the characterization theorem (Theorem 4.1) there exists a Poisson random measure N such that

$$\eta_t(B) = N(\Gamma_t \cap B \times [0, \infty))$$

where $\Gamma_t = \{(x,y); x \in \mathbb{R}^d, y \leq \int_0^t \lambda(x, \eta_s) ds\}$. □

5.3. Stationarity and ergodicity

In the classical theory of stochastic processes, stationarity and ergodicity play important roles in applications. For example, the ergodic theorem asserts the convergence of averages to a limit which is invariant under measure preserving transformations. In the case that the process is ergodic, the ergodic limit is a constant. Important applications arise in establishing consistency of non-parametric estimates of moment densities, and in discussing the frequency of specialized configurations of points (see Example 10.2(a), Daley and Vere-Jones (1988)).

In our case, we are interested in stationarity and invariance under translations (or shifts) in $\mathbb{R}^d$. Our objective is to prove that the process obtained by the time change transformation is stationary and spatially ergodic if λ is translation invariant.

Definition 5.11. λ is *translation invariant*, if for all $x, z \in \mathbb{R}^d, \zeta \in \mathcal{N}(\mathbb{R}^d)$,

$$\lambda(x+z, \zeta) = \lambda(x, S_z\zeta). \tag{5.13}$$

For a general discussion about stationary point processes, see Chapter 10 of Daley and Vere-Jones (1988).

We have a point process $N_{\tau(t)}$ defined on $\mathbb{R}^d$, and we would like to study invariance properties with respect to translations (or shifts) in $\mathbb{R}^d$. For arbitrary $x, z \in \mathbb{R}^d$ and $A \in \mathcal{B}(\mathbb{R}^d)$, write

$$T_x z = x + z \text{ and } T_x A = A + x = \{z + x; z \in A\}.$$

Then, T_x induces a transformation S_x of $\mathcal{N}(\mathbb{R}^d)$ through the equation

$$(S_x\zeta)(A) = \zeta(T_x A), \ \zeta \in \mathcal{N}(\mathbb{R}^d), A \in \mathcal{B}(\mathbb{R}^d). \tag{5.14}$$

Note that if $\zeta = \sum_i \delta_{x_i}$, then $S_x\zeta = \sum_i \delta_{x_i - x}$.

By Lemma 10.1.I (Daley and Vere-Jones (1988)), if $x \in \mathbb{R}^d$, the mapping $S_x : \mathcal{N}(\mathbb{R}^d) \to \mathcal{N}(\mathbb{R}^d)$ defined at (5.14) is continuous and one-to-one. And we have:

(i) $(S_x\delta_z)(\cdot) = \delta_{z-x}$ (where δ is the Dirac measure);
(ii) $\int f(z)(S_x\mu)(dz) = \int f(z)\mu(d(z+x)) = \int f(z-x)\mu(dz)$.

Definition 5.12. A point process ξ with state space $\mathbb{R}^d$ is *stationary* if, for all $u \in \mathbb{R}^d$, the finite-dimensional distributions of the random measures ξ and $S_u\xi$ coincide.

Let $\psi \in \mathcal{N}(\mathbb{R}^d \times [0,\infty))$, and let $\tau^{(n)}(t,\psi)$ be defined by (5.9) with the sample path of the Poisson process N replaced by the counting measure ψ. That is, with reference to (5.5), $F(\phi)(x)$ is replaced by $\lambda(x, \psi_\phi)$, and (5.10) becomes

$$\tau^{(n)}(t,x,\psi) = \int_0^t \lambda(x, \psi_{(\tau^{(n)}(\frac{[ns]}{n},\psi) + (s - \frac{[ns]}{n})\underline{\lambda})})ds. \tag{5.15}$$

For $n = 1, 2, \ldots$, define ξ_n by $\xi_n = \psi_{\tau^{(n)}(t,\psi)}$, that is,

$$\xi_n(B,t,\psi) = \psi(\Gamma_{\tau^{(n)}(t,\psi)} \cap B \times [0,\infty)). \tag{5.16}$$

Lemma 5.13. *If λ is translation invariant, then for $\psi \in \mathcal{N}(\mathbb{R}^d \times [0,\infty))$,*

$$\tau^{(n)}(t, x+z, \psi) = \tau^{(n)}(t, T_z x, \psi) = \tau^{(n)}(t, x, S_{(z,0)}\psi) \tag{5.17}$$

and

$$S_z\xi_n(\cdot, t, \psi) = \xi_n(\cdot, t, S_{(z,0)}\psi). \tag{5.18}$$

Proof. Since $\psi_\phi(B) = \psi\{(x,y) : x \in B, 0 \le y \le \phi(x)\}$,

$$\begin{aligned} S_z(\psi_\phi)(B) &= \psi\{(x,y) : x \in T_z B, 0 \le y \le \phi(x)\} \\ &= \psi\{(x,y) : x - z \in B, 0 \le y \le \phi(T_z(x-z))\} \\ &= S_{(z,0)}\psi\{(x,y) : x \in B, 0 \le y \le \phi(T_z x)\}, \end{aligned}$$

and (5.17) implies (5.18).

To verify (5.17), note that by (5.13),

$$\lambda(x+z,\psi_\phi)=\lambda(T_z x,\psi_\phi)=\lambda(x,S_z(\psi_\phi))=\lambda(x,[S_{(z,0)}\psi]_{\phi\circ T_z})$$

and that by translation invariance of λ, $\underline{\lambda}$ is constant. Then (5.15) gives

$$\begin{aligned}\tau^{(n)}(t,x+z,\psi) &= \int_0^t \lambda(x,S_z(\psi_{(\tau^{(n)}(\frac{[ns]}{n},\psi)+(s-\frac{[ns]}{n})\underline{\lambda})}))ds\\ &= \int_0^t \lambda(x,[S_{(z,0)}\psi]_{(\tau^{(n)}(\frac{[ns]}{n},T_z\cdot,\psi)+(s-\frac{[ns]}{n})\underline{\lambda})}))ds.\end{aligned}$$

Proceeding by induction, (5.17) trivially holds for $t=0$. Assume that it holds for $s\le k/n$. Then for $k/n\le t\le (k+1)/n$, we have

$$\begin{aligned}\tau^{(n)}(t,x+z,\psi) &= \int_0^t \lambda(x,[S_{(z,0)}\psi]_{(\tau^{(n)}(\frac{[ns]}{n},T_z\cdot,\psi)+(s-\frac{[ns]}{n})\underline{\lambda})}))ds\\ &= \int_0^t \lambda(x,[S_{(z,0)}\psi]_{(\tau^{(n)}(\frac{[ns]}{n},S_{(z,0)}\psi)+(s-\frac{[ns]}{n})\underline{\lambda})}))ds\\ &= \tau^{(n)}(t,x,S_{(z,0)}\psi),\end{aligned}$$

and the conclusion follows. □

Theorem 5.14. *Suppose that λ is translation invariant and that the solution of (5.2) is weakly unique. Then for each $t\ge 0$, $N_{\tau(t)}$ is stationary under spatial shifts.*

Remark 5.15. Weak uniqueness is the assertion that all solutions have the same distribution.

Under the assumptions of Theorem 5.2, the conclusion of the theorem follows by Lemma 5.13, the fact that the distribution of N is invariant under shifts, and the convergence of $N_{\tau^{(n)}(t)}$ to $N_{\tau(t)}$.

Proof. By Theorem 5.10, weak existence and uniqueness for the stochastic equation is equivalent to existence and uniqueness of solutions of the martingale problem. If λ is translation invariant and η is a solution of the martingale problem, then $S_z\eta$ is also a solution, so η and $S_z\eta$ must have the same distribution. □

In practice, the useful applications of the ergodic theorem are to those situations where the ergodic limit is a constant.

A stationary process is ergodic if and only if the invariant σ-algebra is trivial. Due to the independence properties of the Poisson process on $\mathbb{R}^d\times[0,\infty)$, it is ergodic for the measure preserving transformations $S_{(x,0)}$. That is, let $(\mathcal{N}(\mathbb{R}^d\times[0,\infty)),\mathcal{B}(\mathcal{N}(\mathbb{R}^d\times[0,\infty))),P)$ be the probability space corresponding to the underlying Poisson process N. Let $\mathcal{I}$ be σ-algebra of *invariant sets* under $S_{(x,0)}$,

$$\mathcal{I}=\{E\in\mathcal{B}(\mathcal{N}(\mathbb{R}^d\times[0,\infty)));P(S_{(x,0)}E\,\triangle\,E)=0,\forall x\in\mathbb{R}^d\}.$$

Then, $\mathcal{I}$ is a trivial σ-algebra, that is, $P(E)=0$ or 1 if $E\in\mathcal{I}$.

Under any conditions that allow us to write $\eta_t=H(t,N)$ for some $H:[0,\infty)\times\mathcal{N}(\mathbb{R}^d\times[0,\infty))\to\mathcal{N}(\mathbb{R}^d)$ so that $S_z\eta_t=H(t,S_{(z,0)}N)$, the spatial ergodicity of η_t follows from the ergodicity of N. Theorem 5.2 give such conditions.

Theorem 5.16. *Let λ be translation invariant, and suppose that the conditions of Theorem 5.2 are satisfied. Then for each $t \geq 0$, $N_{\tau(t)}$ is spatially ergodic.*

Proof. For $\tau^{(n)}(t, x, \psi)$ given by (5.15), $\tau^{(n)}(t)$ in (5.10) is given by $\tau^{(n)}(t, x, N)$, and

$$S_z N_{\tau^{(n)}(t)} = S_{(z,0)} N_{\tau^{(n)}(t, S_{(z,0)}N)}. \tag{5.19}$$

At least along a subsequence $\{n_k\}$, $\tau^{(n_k)}(t)$ converges almost surely to $\tau(t)$. Define $G(t,\psi) = \lim_{k\to\infty} \tau^{(n_k)}(t,\psi)$ if the limit exists, and define $G(t,\psi) \equiv 0$ otherwise. (Note that the collection of ψ for which the limit exists is closed under $S_{(z,0)}$, $z \in \mathbb{R}^d$.) Then $\tau(t) = G(t, N)$ almost surely. Define $H(t,\psi) = \psi_{G(t,\psi)}$, and by (5.19), $S_z H(t,\psi) = H(t, S_{(z,0)}\psi)$. Finally, $N_{\tau(t)} = H(t, N)$, and the theorem follows. □

6. Birth and death processes – constant birth rate, variable death rate

6.1. Gibbs distribution

Consider a spatial point process on $K \subset \mathbb{R}^d$ given by a Gibbs distribution corresponding to a pairwise interaction potential $\rho(x_1, x_2) \geq 0$. The process in which we are interested has a distribution that is absolutely continuous with respect to the spatial Poisson process with mean measure λm on K, with Radon-Nikodym derivative

$$\begin{aligned} L(\zeta) &= C \exp\left\{ -\frac{1}{2} \left[\int\!\!\int \rho(x,y)\zeta(dx)\zeta(dy) - \int \rho(x,x)\zeta(dx) \right] \right\} \\ &= C \exp\left\{ -\sum_{i<j} \rho(x_i, x_j) \right\}, \end{aligned}$$

where $x_1, x_2, \ldots$ are the locations of the point masses in $\zeta \in \mathcal{N}(K)$ and C is a normalizing constant depending only on λ and ρ.

The usual approach to simulating this process is to first identify a spatial birth-death process for which the desired Gibbs distribution is the stationary distribution and then to simulate the birth-death process over a "sufficiently long" time. A significant difficulty with this approach is the need to know what "sufficiently long" is. It is desirable to find a new approach for determining when to terminate the simulation or, if possible, to design a perfect simulation scheme.

There are a variety of birth-death processes which give the same stationary distribution. Consider the process in which points are "born" at a rate λ uniformly over the region, that is, the probability of a birth occurring in a region of area ΔA in a time interval of length Δt is approximately $\lambda \, \Delta A \, \Delta t$. The intensity for the death of a point at x is $\exp\{\sum_i \rho(x, x_i)\}$, where the sum is over all points other than the one at x.

The generator for the process takes the form

$$Af(\zeta) = \int_K (f(\zeta + \delta_y) - f(\zeta))\lambda\, dy + \sum_i (f(\zeta - \delta_{x_i}) - f(\zeta))e^{\sum_{j\neq i} \rho(x_i,x_j)}.$$

To see that the Gibbs distribution is the stationary distribution of this process, let ξ be a Poisson process on K with mean measure λm and apply (2.1) to obtain

$$\begin{aligned}\mathbb{E}&\left[\int_K (f(\xi - \delta_x) - f(\xi))e^{\int \rho(x,y)\xi(dy) - \rho(x,x)}\xi(dx)L(\xi)\right]\\ &= \mathbb{E}\left[\int_K (f(\xi) - f(\xi + \delta_x))e^{\int \rho(x,y)\xi(dy)}L(\xi + \delta_x)\lambda dx\right]\\ &= \mathbb{E}\left[\int_K (f(\xi) - f(\xi + \delta_x))\lambda dx L(\xi)\right]\end{aligned}$$

which implies

$$\int Af(\zeta)L(\zeta)\eta_\lambda(d\zeta) = 0$$

where η_λ is the distribution of the Poisson process with mean measure λm on K. It follows by Echeverria's theorem (Ethier and Kurtz (1986), Theorem 4.9.17) that $L(\zeta)\eta_\lambda(d\zeta)$ is a stationary distribution for the birth-death process. Uniqueness follows by a regeneration argument (see Lotwick and Silverman (1981)).

The following problem was proposed and solved by Kurtz (1989) for the case when K is compact. One motivation for this work was to solve the following problem for infinite regions K, particularly $K = \mathbb{R}^d$. However, we could not obtain the desired result and a slightly different problem was considered. Our hope is that the same technique can be applied to obtain the general result.

6.2. Embedding of birth and death process in Poisson process

Given a Poisson process on $K \times [0,\infty)^2$, we want to construct a family of random sets Γ_t in such way that the birth-death process is obtained by projecting the points of a Poisson process lying in Γ_t onto $\mathbb{R}^d$. We use a Poisson process N on $K\times[0,\infty)\times[0,\infty)$ with mean measure $m^d\times m\times e$ (m^d is Lebesgue measure on $\mathbb{R}^d$, m is Lebesgue measure on $[0,\infty)$, e is the exponential distribution on $[0,\infty)$). For $f : \mathbb{R}^d \times [0,\infty) \to [0,\infty)$ and $t \geq 0$, let $\mathcal{F}_{(f,t)}$ be the completion of the σ-algebra generated by $N(A)$, where either $A \in \mathcal{B}(K \times [0,\infty)^2)$ and $A \subset \{(x,y,s) : s \leq f(x,y), y \leq \lambda t\}$ or $A = A_1 \times [0,\infty)$ with $A_1 \in \mathcal{B}(K \times [0,\lambda t])$.

Let

$$\alpha(x,\zeta) = e^{\int \rho(x,z)\,\zeta(dz)} \tag{6.1}$$

with $\rho(x,x) = 0$, and consider the following system:

$$\begin{cases} \tau(x,y,t) = 0, & \text{if } \lambda t < y\\ \dot\tau(x,y,t) = \alpha(x,N_t), & \text{if } \lambda t \geq y\\ \Gamma_t = \{(x,y,s); x\in K, 0\leq y\leq \lambda t, s > \tau(x,y,t)\} & \\ N_t(B) = N(\Gamma_t \cap B\times\mathbb{R}\times[0,\infty)) & \end{cases} \tag{6.2}$$

where $(\tau(\cdot,t),t)$ is a stopping time with respect to the filtration $\{\mathcal{F}_{(f,t)}\}$ in the sense that $\{\tau(\cdot,t)\leq f, t\leq r\}\in\mathcal{F}_{(f,r)}$.

Interpretation. Each point (x,y,s) of N corresponds to an "individual" who is born at time y/λ and is located at x. The individual dies at time t satisfying $\tau(x,y,t) = s$. Note that in N, conditioned on the (x,y)-coordinates, the s-coordinates are independent and exponentially distributed random variables. Consequently, the probability that a point at x which was born at time y/λ and is still alive at time t, dying in the interval $(t, t+\Delta t)$ is approximately $\alpha(x,N_t)\Delta t$.

6.3. Backwards simulation of Gibbs processes

Fix $T > 0$, and consider the following modification of the system (6.2):

$$\begin{cases} \tau_T(x,y,t) = 0, & \text{if } \lambda(t-T) < y \\ \dot{\tau}_T(x,y,t) = \alpha(x,N_t^T), & \text{if } \lambda(t-T)\geq y \\ \Gamma_t^T = \{(x,y,s); x\in K, -\lambda T\leq y\leq \lambda(t-T), s>\tau_T(x,y,t)\} & \\ N_t^T(B) = N(\Gamma_t^T \cap B\times\mathbb{R}\times[0,\infty)) & \end{cases} \tag{6.3}$$

Note that the system (6.3) is essentially the same as (6.2), except that it is defined using the Poisson process on $K\times[-\lambda T,\infty)\times[0,\infty)$ while the system in (6.2) uses the Poisson process on $K\times[0,\infty)\times[0,\infty)$. That is, the construction is simply shifted to the left by λT. In particular, N_t^T has the same distribution as N_t. A regeneration argument shows that N_T converges in distribution as $T\to\infty$ to the desired Gibbs process. The process N_T^T converges almost surely as $T\to\infty$, since for almost every ω there exists a T_0 such that N_T^T is fixed for $T > T_0$. To see this, let $H_t = \{(x,y,s); x\in K, y\leq -\lambda t, s\geq t-y/\lambda\}$, and let T_0 be the smallest $t>0$ such that $N(H_t)=0$, ($T_0<\infty$ with probability 1). Note that $\Gamma_t^T\subset H_{T-t}$ so that for $T > T_0$, $N_{T-T_0}^T = 0$.

Notice that "backward simulation" is the key idea behind the original CFTP (Coupling from the Past) algorithm proposed by Propp and Wilson (1996) and all related work on perfect simulation. However, the basic CFTP algorithm, sometimes called *vertical CFTP*, is in general not applicable to processes with infinite state space. To deal with this situation, Kendall (1997 and 1998) introduced *dominated CFTP* (also called *horizontal CFTP* and *coupling into and from the past*). This extension also requires the state space to have a partial order, as well as the existence of a monotone coupling among the target process and two reversible *sandwiching processes*, which must be easy to sample. Algorithms of this type are available for attractive point processes and, through a minor modification, also for repulsive point processes (Kendall, 1998). Similarly, Häggström, van Lieshout and Møller (1999) combined ideas from CFTP and the two-component Gibbs sampler to perfectly simulate from processes in infinite spaces which do not have maximal (or minimal) elements.

In our case, we sample directly from a time stationary realization of the process. There is no coalescence criterion, either between coupled realizations or between sandwiching processes. The scheme neither requires nor takes advantage of monotonicity properties. Our construction has the same spirit as the clan of ancestors algorithm proposed by Fernández, Ferrari and Garcia (2002) where the stopping time T_0 at which we know that the invariant measure is achieved is a regeneration time for the process.

The existence and uniqueness of the process for $\mathbb{R}^d$ is obtained by refining the arguments of Section 5 (see Garcia, 1995a), but convergence to the invariant measure is not at all clear. The above backward argument does not work for the infinite case since $T_0 = \infty$ with probability 1. At this point we have no general results for $\mathbb{R}^d$; however, we hope that some of the theory developed here may be useful in giving such results.

References

[1] Aalen, Odd O. and Hoem, Jan M. (1978). Random time changes for multivariate counting processes. *Scand. Actuar. J.*, no. 2, 81–101.

[2] Balan, R.M. (2001). A strong Markov property for set-indexed processes. *Statist. Probab. Lett.* **53** (2001), no. 2, 219–226.

[3] Daley, D.J. and Vere-Jones, D. (1988). *An Introduction to the Theory of Point Processes.* Springer-Verlag, New York, NY.

[4] Ethier, S.N. and Kurtz, T.G. (1986). *Markov Processes: Characterization and Convergence.* John Wiley & Sons.

[5] Fernández, R., Ferrari, P.A. and Garcia, N.L. (2002) Perfect simulation for interacting point processes, loss networks and Ising models. *Stoch. Process. Appl.*, **102**(1), 63–88.

[6] P.A. Ferrari and N. Garcia (1998). One-dimensional loss networks and conditioned $M/G/\infty$ queues. J. Appl. Probab. **35**, no. 4, 963–975.

[7] Foley, Robert D. (1986). Stationary Poisson departure processes from nonstationary queues. *J. Appl. Probab.* **23**, no. 1, 256–260.

[8] Garcia, N.L. (1995a). Birth and Death Processes as Projections of Higher Dimensional Poisson Processes. *Adv. in Applied Probability*, **27**, No. 4, pp. 911–930.

[9] Garcia, N.L. (1995b). Large Population Results for Epidemic Models. *Stochastic Processes and their Applications*, **60**, No. 1, pp. 147–160.

[10] Garcia, N.L. and Kurtz, T.G. (2006) Spatial birth and death processes as solutions of stochastic equations *ALEA*, **1**, pp. 281-303.

[11] Gaver, D.P., Jacobs, P.A. and Latouche, G. (1984). Finite birth-and-death models in randomly changing environments. *Adv. Appl. Prob.*, **16**, pp. 715–731.

[12] Häggström, O. , van Lieshout, M.N.M. and Møller, J. (1999). Characterization results and Markov chain Monte Carlo algorithms including exact simulation for some spatial point processes. *Bernoulli*, **5**(4):641–658.

[13] Ivanoff, B. Gail; Merzbach, Ely. (1995), Stopping and set-indexed local martingales. *Stochastic Process. Appl.* **57** no. 1, 83–98.

[14] Karr, A.F. (1986). *Point Processes and Their Statistical Inference.* Marcel Dekker.

[15] Kendall, W.S. (1997). On some weighted Boolean models. In D. Jeulin, editor, *Proceedings of the International Symposium on Advances in Theory and Applications of Random Sets* (*Fontainebleau,* 1996), pages 105–120. World Sci. Publishing, River Edge, NJ.

[16] Kendall, W.S. (1998). Perfect simulation for the area-interaction point process. In L. Accardi and C.C. Heyde, editors, *Probability Towards* 2000, pages 218–234. Springer.

[17] Kurtz, T.G. (1980a). The optional sampling theorem for martingales indexed by direct sets. *Ann. Probab.*, **8**, pp. 675–681.

[18] Kurtz, T.G. (1980b). Representation of Markov processes as multiparameter time changes. *Ann. of Probability,* **8**, pp. 682–715.

[19] Kurtz, Thomas G. (1982). Representation and approximation of counting processes. *Advances in filtering and optimal stochastic control* (*Cocoyoc,* 1982), 177–191, *Lecture Notes in Control and Inform. Sci.*, 42, Springer, Berlin.

[20] Kurtz, T.G. (1989). Stochastic processes as projections of Poisson random measures. Special invited paper at IMS meeting, Washington, D.C. Unpublished.

[21] Liggett, T.M. (1972). Existence theorems for infinite particle systems. *Trans. Amer. Math. Soc.*, **165**, pp. 471–481.

[22] Lotwick, H.W. and Silverman, B.W. (1981). Convergence of spatial birth-and-death processes. *Math. Proc. Cambridge Phil. Soc.*, **90**, pp. 155–165.

[23] Massoulié, Laurent. (1998). Stability for a general class of interacting point process dynamics and applications. *Stoch. Processes Appl.*, **75**, pp. 1–30.

[24] Meyer, P.A. (1971). Démonstration simplifiée d'un théorème de Knight. Séminaire de Probabilités, V (Univ. Strasbourg, année universitaire 1969–1970), pp. 191–195. *Lecture Notes in Math.*, Vol. 191, Springer, Berlin.

[25] Propp, J.G. and Wilson, D.B. (1996). Exact sampling with coupled Markov chains and applications to statistical mechanics. *Random Structures and Algorithms*, **9**, pp. 223–252.

Nancy L. Garcia
Departamento de Estatística
IMECC – UNICAMP
Caixa Postal 6065
13.081-970 – Campinas, SP
BRAZIL
e-mail: `nancy@ime.unicamp.br`

Thomas G. Kurtz
Departments of Mathematics and Statistics
University of Wisconsin – Madison
480 Lincoln Drive
Madison, WI 53706
USA
e-mail: `kurtz@math.wisc.edu`

Progress in Probability, Vol. 60, 299–304

An Improvement on Vajda's Inequality

Gustavo L. Gilardoni

Abstract. Let D and V be respectively information divergence and variational distance. It is shown that $D \geq \log \frac{2}{2-V} - \frac{2-V}{2} \log \frac{2+V}{2}$, hence improving Vajda's inequality $D \geq \log \frac{2+V}{2-V} - \frac{2V}{2+V}$. The proof is based on a lemma which states that for any f-divergence symmetric in the sense that $D_f(P,Q) = D_f(Q,P)$, one has that $\inf\{D_f(P,Q)\colon V(P,Q) = v\} = \frac{2-v}{2} f(\frac{2+v}{2-v}) - f'(1)v$. This lemma has interest on its own and implies precise lower bounds for several well-known divergences.

Mathematics Subject Classification (2000). 94A17, 26D15.

Keywords. Information inequalities, information or Kullback-Leibler divergence, relative entropy, symmetric f-divergences, variational or L^1 distance.

1. Introduction

All throughout P and Q will be probability measures on a measurable space $(\Omega, \mathcal{A})$ and p and q their densities or Radon-Nikodym derivatives with respect to a common dominating measure μ. The *information divergence* is $D(P,Q) = \int p \log(p/q)\, d\mu$ and the *variational distance* is $V(P,Q) = \int |q-p|\, d\mu$.

There is an extensive literature regarding lower bounds on the information divergence D in terms of variational distance V. Most of it concentrates on bounds which are accurate when V is close to zero. Maybe the best known result is *Pinsker's inequality* $D \geq \frac{1}{2}V^2$, due to Csiszár [1] and Kemperman [2] and so called because it was Pinsker [3] who first showed that $D \geq cV^2$ for sufficiently small c. Shortly after [1] and [2] showed that the constant $c = \frac{1}{2}$ is best possible, efforts began to improve Pinsker's inequality by adding further powers of V to the bound. For instance, Kullback [4, 5] and Vajda [6] showed that $D \geq \frac{1}{2}V^2 + \frac{1}{36}V^4$ and Topsøe [7] that $D \geq \frac{1}{2}V^2 + \frac{1}{36}V^4 + \frac{1}{270}V^6 + \frac{221}{340200}V^8$. Recently, Fedotov, Harremoës and Topsøe [8] obtained a parametrization of the

Research partially supported by CAPES, CNPq and FINATEC grants.

curve $v \mapsto L(v) = \inf\{D(P,Q)\colon V(P,Q) = v\}$ in terms of hyperbolic trigonometric functions and used this to obtain in [9] a bound containing terms up to and including V^{48}. We note that in all these inequalities the coefficients involved are best possible. While these bounds are very useful when one has convergence in the topology induced by D and wants to study convergence in the total variation norm, they say little about the behavior of D when V is away from 0. To see this, we note that $V(P,Q) = 2\sup\{|Q(A) - P(A)| : A \in \mathcal{A}\} = 2\,[Q(B) - P(B)]$, where $B = \{\omega \in \Omega : q(\omega) \geq p(\omega)\}$, and hence $0 \leq V(P,Q) \leq 2$ with equality holding respectively if and only if $P = Q$ or $P \perp Q$. On the other side, the Information Divergence satisfies $0 \leq D(P,Q) \leq +\infty$. $D(P,Q) = 0$ can occur if and only if $P = Q$, while $P \not\ll Q$ implies that $D(P,Q) = +\infty$ (the reciprocal does not hold here though). Therefore, one expects that $D(P,Q) \to \infty$ as $V(P,Q) \to 2$, while any lower bound that is a polynomial in V must be finite as $V(P,Q) \to 2$. Motivated by this fact Vajda showed in [6] that $D \geq B(V) = \log\frac{2+V}{2-V} - \frac{2V}{2+V}$. Since $B(V) = \frac{1}{2}V^2 + O(V^3)$, this lower bound is about as accurate as Pinsker's inequality when V is close to zero. Still, $\lim_{V\to 2} B(V) = +\infty$.

In this note we will show that $D \geq M(V) = \log\frac{2}{2-V} - \frac{2-V}{2}\log\frac{2+V}{2} \geq B(V)$. In due fairness, we must mention that $M(V)$ represents only a slight improvement over $B(V)$. Indeed, $M(V)$ and $B(V)$ are equivalent both as $V \to 0$ and $V \to 2$, while the ratio $M(V)/B(V)$ for $V \in (0,2)$ is concave and attains a maximum $M(V_0)/B(V_0) \approx 1.147$ for $V_0 \approx 1.357$. The fact that improvement over $B(V)$ can be only slight should be no surprise, because in his short but beautiful paper Vajda also showed that $L(v) = \inf\{D(P,Q)\colon V(P,Q) = v\} \leq B_*(v) = B(v) + \frac{2v^3}{(2+v)^2}$. Note that $B_*(V)/B(V)$ attains a maximum for $V_1 \approx 1.228$ with $B_*(V_1)/B(V_1) \approx 1.530$ and that any lower bound for D cannot exceed $B_*(V)$.

Maybe more interesting than the bound itself is the method that we used to derive it, based on approximating D by a symmetric f-divergence and computing exactly the minimum of this f-divergence for fixed V. f-divergences were introduced by Csiszár [10] and Ali and Silvey [11] and besides D and V include many widely used distances and discrepancy measures between probability measures. The f-divergence generated by f is $D_f(P,Q) = \int pf(q/p)\,d\mu$, where $f : (0,\infty) \to \boldsymbol{R}$ is convex and $f(1) = 0$. Jensen's inequality implies that $D_f(P,Q) \geq 0$ with equality holding if and only if $P = Q$, provided that f is strictly convex at $u = 1$. In general, f-divergences are not symmetric, in the sense that $D_f(P,Q)$ does not necessarily equal $D_f(Q,P)$. Indeed, we will show in Lemma 2.2 that D_f is symmetric if and only if there exists a constant b such that $f(u) = uf(1/u) + b(u-1)$. For instance, V is symmetric but D is not.

Using the previous characterization we will show in Lemma 2.3 that for any D_f symmetric one has that

$$\inf\{D_f(P,Q)\colon V(P,Q) = v\} = \frac{2-v}{2}\,f\Big(\frac{2+v}{2-v}\Big) - f'(1)v\,, \tag{1.1}$$

where $f'(1) = 2^{-1}[f'(1-) + f'(1+)]$ is the average between the left and the right derivatives of f at $u = 1$, which exist because of the convexity of f. This result is a special case of a more general lower bound for f-divergences which we have obtained in [12], although for sake of completeness we give here a different, more direct proof (note however that, unlike the proof in [12], here we do not need the hypothesis of differentiability of f). Equation (1.1) gives a lower bound which is best possible for any fixed V and has interest on its own. For instance, it implies that $J \geq V \log \frac{2+V}{2-V}$, $h^2 \geq 2 - \sqrt{4 - V^2}$, and $C \geq \frac{2-V}{2} \log \frac{2-V}{2} + \frac{2+V}{2} \log \frac{2+V}{2}$, where $J(P,Q) = D(P,Q) + D(Q,P)$ is Jeffrey's divergence and $h^2(P,Q) = \frac{1}{2}\int(\sqrt{q} - \sqrt{p})^2\, d\mu = 1 - \int \sqrt{qp}\, d\mu$ and $C(P,Q) = D(P,(P+Q)/2) + D(Q,(P+Q)/2)$ are known respectively as Hellinger and capacitory discrimination. Some of these bounds are known: $h^2 \geq 2 - \sqrt{4 - V^2}$ is equivalent to $V^2 \leq h^2(4 - h^2)$, which dates back at least to LeCam [13] (see also Dragomir, Gluščević and Pearce [14]), while the bound for the capacitory discrimination is equivalent to $C(P,Q) \geq \sum_{n=1}^{\infty} [n(2n-1)2^{2n}]^{-1} V^{2n}$, which was proved by Topsøe [15]. We note however that even for these cases Lemma 2.3 gives a stronger result since it implies that the inequalities are best possible for any fixed V.

The rest of this note contains the statements and proofs of the above mentioned results. Besides showing that $D \geq M(V)$ and the lemmas about symmetric f-divergences, we also state a simple lemma regarding the minimum f-divergence for fixed V when the generating function $f(u)$ is null to the right of $u = 1$.

2. Main results

To avoid unnecessary discussion, we will assume below the usual conventions

$$f(0) = \lim_{u \downarrow 0} f(u), \quad 0 \cdot f(0/0) = 0 \quad \text{and} \quad 0 \cdot f(a/0) = \lim_{\epsilon \downarrow 0} \epsilon f(a/\epsilon) = a \lim_{u \to +\infty} f(u)/u.$$

We also note that an f-divergence does not determine univocally the associated f. Indeed, for any b fixed, D_f and $D_{f-b(u-1)}$ are identical. For instance, $D = D_{-\log u} = D_{u-1-\log u}$.

Theorem 2.1. $D \geq \log \frac{2}{2-V} - \frac{2-V}{2} \log \frac{2+V}{2} \geq \log \frac{2+V}{2-V} - \frac{2V}{2+V}$.

Proof. We prove first the last inequality. Let $h(v) = 2v - \frac{1}{2}(2+v)(4-v)\log(1+v/2)$ $(0 \leq v \leq 2)$. Then $h(0) = h'(0) = h''(0) = 0$ and $h'''(v) = (2+v)^{-2}(5+v) \geq 0$. Therefore $h''(v) \geq h''(0) = 0$, implying that $h'(v) \geq h'(0) = 0$ and hence $h(v) \geq h(0) = 0$. It is easy to see that this implies the desired inequality.

To prove the first inequality, define

$$f_1(u) = \begin{cases} 1 - u + u \log u & u \leq 1 \\ u - 1 - \log u & u > 1 \end{cases} \quad ; \quad f_2(u) = \begin{cases} 2(u-1) - (1+u)\log u & u \leq 1 \\ 0 & u > 1 \end{cases}$$

and check that $f_i(1) = f_i'(1) = 0$ $(i = 1,2)$, $f_1(u) + f_2(u) = u - 1 - \log u$ and, by differentiating two times, that both f_1 and f_2 are convex. Hence we can write that $D = D_{u-1-\log u} = D_{f_1} + D_{f_2}$. Now, since $f_1(u) = u f_1(1/u)$, Lemma 2.3 below implies that $D_{f_1} \geq \frac{2-V}{2} f_1(\frac{2+V}{2-V}) = V - \frac{2-V}{2} \log \frac{2+V}{2-V}$, while Lemma 2.4

gives that $D_{f_2} \geq f_2(\frac{2-V}{2}) = -V - \frac{4-V}{2}\log\frac{2-V}{2}$. This completes the proof, since $V - \frac{2-V}{2}\log\frac{2+V}{2-V} + [-V - \frac{4-V}{2}\log\frac{2-V}{2}] = \log\frac{2}{2-V} - \frac{2-V}{2}\log\frac{2+V}{2}$. □

Before proving (1.1) we will show that if D_f is symmetric one should have that $f(u) = uf(1/u) + b(u-1)$ for some b. Note that in this case we must have that $b = f'(1-) + f'(1+)$. To see this, write

$$\frac{f(u)-f(1)}{u-1} = \frac{uf(1/u)+b(u-1)-f(1)}{u-1} = b - \frac{f(1/u)-f(1)}{1/u-1}$$

and take limits as $u \downarrow 1$.

Lemma 2.2. *$D_f(P,Q) = D_f(Q,P)$ for every P and Q if and only if the generating function f satisfies that $f(u) = uf(1/u) + b(u-1)$ for every $u > 0$ and some constant b.*

Proof. The *if* part follows by direct substitution. To prove the *only if* part, let $\Omega = \{0,1\}$ and for $u > 0$ and $0 < p < \min\{1/u, 1\}$ consider probability measures P and Q with $p(0) = p$, $p(1) = 1-p$, $q(0) = pu$ and $q(1) = 1-pu$. Then $D_f(P,Q) = pf(u) + (1-p)f(\frac{1-pu}{1-p})$ and $D_f(Q,P) = puf(1/u) + (1-pu)f(\frac{1-p}{1-pu})$. Hence, $D_f(P,Q) = D_f(Q,P)$ implies that

$$f(u) - uf(1/u) = \frac{1-p}{p}\left[\frac{1-pu}{1-p}f(\frac{1-p}{1-pu}) - f(\frac{1-pu}{1-p})\right]$$

$$= \left[f(\frac{1-p}{1-pu}) - f(1)\right]\frac{1-pu}{p(u-1)}(u-1) - \left[f(\frac{1-pu}{1-p}) - f(1)\right]\frac{1-p}{p(1-u)}(1-u)\,.$$

Taking limits as $p \downarrow 0$ shows that f has the desired form with $b = f'(1-) + f'(1+)$. □

Lemma 2.3. *Let the f-divergence D_f be such that $D_f(P,Q) = D_f(Q,P)$ for any P and Q. Then (1.1) holds.*

Proof. Let $\tilde{f}(u) = f(u) - f'(1)(u-1)$ so that the symmetry of $D_f = D_{\tilde{f}}$ and the remark before Lemma 2.2 imply now that $\tilde{f}(u) = u\tilde{f}(1/u)$.

We show first that $\inf\{D_f(P,Q)\colon V(P,Q) = v\} \geq \frac{2-v}{2}f(\frac{2+v}{2-v}) - f'(1)v$. Let $B = \{\omega \in \Omega : q(\omega) \geq p(\omega)\}$ and $v = V(P,Q) = 2[Q(B) - P(B)]$. Note that this implies that $\frac{Q(B)}{P(B)} = \frac{2P(B)+v}{2P(B)}$, $\frac{1-Q(B)}{1-P(B)} = \frac{2[1-P(B)]-v}{2[1-P(B)]}$ and $0 \leq P(B) \leq 1 - v/2$. Hence, it follows from the convexity of f and Jensen's inequality that

$$\begin{aligned}
D_f(P,Q) &= \int_B pf(q/p)\,d\mu + \int_{B^c} pf(q/p)\,d\mu \\
&= P(B)\int_B \frac{p}{P(B)}f(q/p)\,d\mu + [1-P(B)]\int_{B^c}\frac{p}{1-P(B)}f(q/p)\,d\mu \\
&\geq P(B)\,f\left(\int_B \frac{q}{p}\frac{p}{P(B)}\,d\mu\right) + [1-P(B)]\,f\left(\int_{B^c}\frac{q}{p}\frac{p}{1-P(B)}\,d\mu\right) \\
&= P(B)\,f\left(\frac{Q(B)}{P(B)}\right) + [1-P(B)]\,f\left(\frac{1-Q(B)}{1-P(B)}\right)
\end{aligned}$$

$$= P(B)\, f\left(\frac{2P(B)+v}{2P(B)}\right) + [1-P(B)]\, f\left(\frac{2[1-P(B)]-v}{2[1-P(B)]}\right)$$
$$= d_v(P(B))\,, \tag{2.1}$$

where for $0 < a < 1 - v/2$ we define $d_v(a) = a\, f(\frac{2a+v}{2a}) + (1-a)\, f(\frac{2(1-a)-v}{2(1-a)}) = a\, \tilde{f}(\frac{2a+v}{2a}) + (1-a)\, \tilde{f}(\frac{2(1-a)-v}{2(1-a)})$. Using now the convexity and definition of $\tilde{f}$ and the fact that $\tilde{f}(u) = u\tilde{f}(1/u)$, we have that

$$\begin{aligned}
d_v(a) &= a\, \tilde{f}\left(\frac{2a+v}{2a}\right) + (1-a)\, \tilde{f}\left(\frac{2(1-a)-v}{2(1-a)}\right)\\
&= a\, \tilde{f}\left(\frac{2a+v}{2a}\right) + (1-a)\, \frac{2(1-a)-v}{2(1-a)}\, \tilde{f}\left(\frac{2(1-a)}{2(1-a)-v}\right)\\
&= \frac{2-v}{2}\left\{\frac{2a}{2-v}\, \tilde{f}\left(\frac{2a+v}{2a}\right) + \frac{2(1-a)-v}{2-v}\, \tilde{f}\left(\frac{2(1-a)}{2(1-a)-v}\right)\right\}\\
&\ge \frac{2-v}{2}\, \tilde{f}\left(\frac{2a}{2-v}\, \frac{2a+v}{2a} + \frac{2(1-a)-v}{2-v}\, \frac{2(1-a)}{2(1-a)-v}\right)\\
&= \frac{2-v}{2}\, \tilde{f}\left(\frac{2a+v}{2-v} + \frac{2(1-a)}{2-v}\right) = \frac{2-v}{2}\, \tilde{f}\left(\frac{2+v}{2-v}\right)\\
&= \frac{2-v}{2}\, f\left(\frac{2+v}{2-v}\right) - f'(1)v\,.
\end{aligned} \tag{2.2}$$

The inequalities (2.1) and (2.2) imply that $D_f(P,Q) \ge \frac{2-v}{2}\, f(\frac{2+v}{2-v}) - f'(1)v$, as was to be shown.

To show that $\inf\{D_f(P,Q)\colon V(P,Q) = v\} \le \frac{2-v}{2}\, f(\frac{2+v}{2-v}) - f'(1)v$, consider a binary space $\Omega = \{0,1\}$ and probability measures P and Q with $p(0) = \frac{2+v}{4}$, $p(1) = \frac{2-v}{4}$, $q(0) = \frac{2-v}{4}$ and $q(1) = \frac{2+v}{4}$, so that $V(P,Q) = v$ and

$$\begin{aligned}
&D_f(P,Q) = D_{\tilde{f}}(Q,P)\\
&= \frac{2+v}{4}\tilde{f}\left(\frac{2-v}{2+v}\right) + \frac{2-v}{4}\tilde{f}\left(\frac{2+v}{2-v}\right) = \frac{2+v}{4}\tilde{f}\left(\frac{2-v}{2+v}\right) + \frac{2-v}{4}\frac{2-v}{2+v}\tilde{f}\left(\frac{2+v}{2-v}\right)\\
&= \frac{2-v}{2}\tilde{f}\left(\frac{2+v}{2-v}\right) = \frac{2-v}{2}f\left(\frac{2+v}{2-v}\right) - f'(1)v.
\end{aligned}$$

□

Lemma 2.4. *Let f be convex with $f(u) = 0$ for $u \ge 1$. Then* $\inf\{D_f(P,Q) : V(P,Q) = v\} = f(\frac{2-v}{2})$.

Proof. To show that $\inf\{D_f(P,Q)\colon V(P,Q) = v\} \le f(\frac{2-v}{2})$, consider as before $B = \{\omega \in \Omega : q(\omega) \ge p(\omega)\}$ and $v = V(P,Q) = 2[Q(B) - P(B)]$. The same argument that led to equation (2.1) implies that $D_f(P,Q) \ge d_v(P(B))$. Now, the convexity of f implies that $\frac{f(y)-f(x)}{y-x}$ is increasing in y and taking $y = \frac{2(1-a)-v}{2(1-a)}$ and $x = 1$ we have that $-\frac{2(1-a)}{v} f(\frac{2(1-a)-v}{2(1-a)})$ is increasing in $(1 - \frac{v}{2(1-a)})$, so that $d_v(a) = (1-a)f(\frac{2(1-a)-v}{2(1-a)})$ is increasing in a for $0 < a < 1 - v/2$ and v fixed. This implies that $d_v(a) \geq d_v(0) = f(\frac{2-v}{2})$ and hence $D_f(P,Q) \geq d_v(P(B)) \geq f(\frac{2-v}{2})$, as was to be shown.

To show that $\inf\{D_f(P,Q)\colon V(P,Q) = v\} \leq f(\frac{2-v}{2})$, consider again $\Omega = \{0,1\}$ and probability measures P and Q with $p(0) = 1$, $p(1) = 0$, $q(0) = \frac{2-v}{2}$ and $q(1) = \frac{v}{2}$, so that $V(P,Q) = v$ and $D_f(P,Q) = f(\frac{2-v}{2})$. □

References

[1] I. Csiszár, "A note on Jensen's inequality," *Studia Sci. Math. Hungar.*, vol. 1, pp. 185–188, 1966.

[2] J.H.B. Kemperman, "On the optimal rate of transmitting information," *Ann. Math. Statist*, vol. 40, pp. 2156–2177, Dec. 1969.

[3] M.S. Pinsker, *Information and Information Stability of Random Variables and Processes.* A. Feinstein, tr. and ed., San Francisco: Holden-Day, 1964.

[4] S. Kullback, "A lower bound for discrimination information in terms of variation," *IEEE Trans. Inf. Theory*, vol. IT-13, pp. 126–127, Jan. 1967.

[5] S. Kullback, "Correction to "a lower bound for discrimination information in terms of variation"," *IEEE Trans. Inf. Theory*, vol. IT-16, p. 652, 1970.

[6] I. Vajda, "Note on discrimination information and variation," *IEEE Trans. Inf. Theory*, vol. 16, no. 6, 1970.

[7] F. Topsøe, "Bounds for entropy and divergence of distributions over a two-element set," *J. Ineq. Pure Appl. Math.*, vol. 2, Article 25, 2001.

[8] A. Fedotov, P. Harremoës, and F. Topsøe, "Refinements of Pinsker's Inequality," *IEEE Trans. Inf. Theory*, vol. 49, pp. 1491–1498, June 2003.

[9] A. Fedotov, P. Harremoës, and F. Topsøe, "Best Pinsker Bound equals Taylor Polynomial of Degree 49," *Computational Technologies*, vol. 8, pp. 3–14, 2003.

[10] I. Csiszár, "Information-type measures of difference of probability distributions and indirect observations," *Studia Sci. Math. Hungar.*, vol. 2, pp. 299–318, 1967.

[11] S.M. Ali and S.D. Silvey, "A general class of coefficients of divergence of one distribution from another," *J. Roy. Statist. Soc. Ser B*, vol. 28, pp. 131–142, 1966.

[12] G.L. Gilardoni, "On the minimum *f*-divergence for given total variation," *C. R. Acad. Sci. Paris, Ser. I*, vol. 343, pp. 763–766, 2006.

[13] L. LeCam, *Asymptotic Methods in Statistical Theory.* New York: Springer-Verlag, 1986.

[14] S.S. Dragomir, V. Gluščević, and C.E.M. Pearce, "Csiszár f-divergence, Ostrowski's inequality and mutual information," *Nonlinear Analysis*, vol. 47, pp. 2375–2386, 2001.

[15] F. Topsøe, "Some inequalities for information divergence and related measures of discrimination," *IEEE Trans. Inf. Theory*, vol. 46, pp. 1602–1609, 2000.

Gustavo L. Gilardoni
Departamento de Estatística, Universidade de Brasília
Brasília, DF 70910–900, Brazil
e-mail: gilardon@unb.br

Progress in Probability, Vol. 60, 305–320

Space-Time Percolation

Geoffrey R. Grimmett

Abstract. The one-dimensional contact model for the spread of disease may be viewed as a directed percolation model on $\mathbb{Z} \times \mathbb{R}$ in which the continuum axis is oriented in the direction of increasing time. Techniques from percolation have enabled a fairly complete analysis of the contact model at and near its critical point. The corresponding process when the time-axis is unoriented is an undirected percolation model to which now standard techniques may be applied. One may construct in similar vein a random-cluster model on $\mathbb{Z} \times \mathbb{R}$, with associated continuum Ising and Potts models. These models are of independent interest, in addition to providing a path-integral representation of the quantum Ising model with transverse field. This representation may be used to obtain a bound on the entanglement of a finite set of spins in the quantum Ising model on $\mathbb{Z}$, where this entanglement is measured via the entropy of the reduced density matrix. The mean-field version of the quantum Ising model gives rise to a random-cluster model on $K_n \times \mathbb{R}$, thereby extending the Erdős–Rényi random graph on the complete graph K_n.

Mathematics Subject Classification (2000). Primary 60K35; Secondary 82B20.

Keywords. Percolation, contact model, random-cluster model, Ising model, Potts model, quantum Ising model, entanglement, quantum random graph, mean field.

1. Introduction

Brazil is justly famous for its beach life and its probability community. In harnessing the first to support the second, a summer school of intellectual distinction and international visibility in probability theory has been created. The high scientific stature of the organizers and of the wider Brazilian community has ensured the attendance of a host of wonderful lecturers during ten years of the Brazilian School of Probability, and the School has attracted an international audience including many young Brazilians who continue to leave their marks within this crossroads subject of mathematics. The warmth and vitality of Brazilian culture have been attractive features of these summer schools, and invitations to participate are greatly valued. This short review concerns two topics of recurring interest at the School, namely percolation and the Ising model (in both its classical and quantum forms), subject to the difference that one axis of the underlying space is allowed to vary continuously.

The percolation process is arguably the most fundamental of models for a disordered medium. Its theory is now well established, and several mathematics books have been written on and near the topic, see [16, 18, 27, 47]. Percolation is at the source of one of the most exciting areas of contemporary probability theory, namely the theory of Schramm–Löwner evolutions (SLE). This theory threatens to explain the relationship between probabilistic models and conformal field theory, and is expected to lead ultimately to rigorous explanations of scaling theory for a host of two dimensional models including percolation, self-avoiding walks, and the Ising/Potts and random-cluster models. See [37, 46, 48, 49] and the references therein.

Percolation theory has contributed via the random-cluster model to the study of Ising/Potts models on a given graph G, see [28]. The methods developed for percolation have led also to solutions of several of the basic questions about the contact model on $G \times \mathbb{R}$, see [1, 9, 10, 39]. It was shown in [2] that the quantum Ising model with transverse field on G may be reformulated in terms of a random-cluster model on $G \times \mathbb{R}$, and it has been shown recently in [30] that random-cluster arguments may be used to study entanglement in the quantum Ising model.

In this short account of percolative processes on $G \times \mathbb{R}$ for a lattice G, greater emphasis is placed on the probability theory than on links to statistical mechanics. We shall recall in Sections 2–3 the problems of percolation on $G \times \mathbb{R}$, and of the contact model on G. This is followed in Section 4 by a description of the continuum random-cluster model on $G \times \mathbb{R}$, and its application to continuum Ising/Potts models. In Section 5 we present a summary of the use of random-cluster techniques to study entanglement in the quantum Ising model on $\mathbb{Z}$. An account is included of a recent result of [30] stating that the entanglement entropy of a line of L spins has order not exceeding $\log L$ in the strong-field regime. The proof relies on a property of random-cluster measures termed 'ratio weak-mixing', studied earlier in [4, 5] for the random-cluster model on a lattice. The corresponding mean-field model is considered in Section 6 under the title 'quantum random graph', and a conjecture is presented for such a model.

2. Continuum percolation

Let $G = (V, E)$ be a finite or countably infinite graph which, for simplicity, we take to be connected with neither loops nor multiple edges. We shall usually take G to be a subgraph of the hypercubic lattice $\mathbb{Z}^d$ for some $d \geq 1$. The models of this paper inhabit the space $G \times \mathbb{R}$, which we refer to as space-time, and we think of $G \times \mathbb{R}$ as being obtained by attaching a 'time-line' $(-\infty, \infty)$ to each vertex $x \in V$.

Let $\lambda, \delta \in (0, \infty)$. The continuum percolation model on $G \times \mathbb{R}$ is constructed via processes of 'cuts' and 'bridges' as follows. For each $x \in V$, we select a Poisson process D_x of points in $\{x\} \times \mathbb{R}$ with intensity δ; the processes $\{D_x : x \in V\}$ are independent, and the points in the D_x are termed 'cuts'. For each $e = \langle x, y \rangle \in E$, we select a Poisson process B_e of points in $\{e\} \times \mathbb{R}$ with intensity λ; the processes $\{B_e : e \in E\}$ are independent of each other and of the D_x. Let $\mathbb{P}_{\lambda,\delta}$ denote the probability measure associated with the family of such Poisson processes indexed by $V \cup E$.

For each $e = \langle x, y \rangle \in E$ and $(e, t) \in B_e$, we think of (e, t) as an edge joining the endpoints (x, t) and (y, t), and we refer to this edge as a 'bridge'. For $(x, s), (y, t) \in V \times \mathbb{R}$, we write $(x, s) \leftrightarrow (y, t)$ if there exists a path π with endpoints (x, s), (y, t) such that: π is a union of cut-free sub-intervals of $G \times \mathbb{R}$ and bridges. For $\Lambda, \Delta \subseteq V \times \mathbb{R}$, we write $\Lambda \leftrightarrow \Delta$ if there exist $a \in \Lambda$ and $b \in \Delta$ such that $a \leftrightarrow b$.

For $(x, s) \in V \times \mathbb{R}$, let $C_{x,s}$ be the set of all points (y, t) such that $(x, s) \leftrightarrow (y, t)$. The clusters $C_{x,s}$ have been studied in [10], where the case $G = \mathbb{Z}^d$ was considered in some detail. Let 0 denote the origin $(0, 0) \in \mathbb{Z}^d \times \mathbb{R}$, and let $C = C_0$ denote the cluster at the origin. Noting that C is a union of line-segments, we write $|C|$ for the Lebesgue measure of C. The *radius* $\mathrm{rad}(C)$ of C is given by

$$\mathrm{rad}(C) = \sup\big\{\|x\| + |t| : (x, t) \in C\big\},$$

where

$$\|x\| = \sup_i |x_i|, \qquad x = (x_1, x_2, \ldots, x_d) \in \mathbb{Z}^d,$$

is the supremum norm on $\mathbb{Z}^d$.

The critical point of the process is defined by

$$\lambda_c(\delta) = \sup\{\lambda : \theta(\lambda, \delta) = 0\},$$

where

$$\theta(\lambda, \delta) = \mathbb{P}_{\lambda,\delta}(|C| = \infty).$$

It is immediate by time-scaling that $\theta(\lambda, \delta) = \theta(\lambda/\delta, 1)$, and we shall use the abbreviations $\lambda_c = \lambda_c(1)$ and $\theta(\lambda) = \theta(\lambda, 1)$.

The following exponential-decay theorem will be useful for the study of the quantum Ising model in Section 5.

Theorem 2.1. [10] *Let $G = \mathbb{Z}^d$ where $d \geq 1$, and consider continuum percolation on $G \times \mathbb{R}$.*

(i) *Let $\lambda, \delta \in (0, \infty)$. There exist γ, ν satisfying $\gamma, \nu > 0$ for $\lambda/\delta < \lambda_c$ such that:*

$$\mathbb{P}_{\lambda,\delta}(|C| \geq k) \leq e^{-\gamma k}, \qquad k > 0, \tag{2.2}$$

$$\mathbb{P}_{\lambda,\delta}(\mathrm{rad}(C) \geq k) \leq e^{-\nu k}, \qquad k > 0. \tag{2.3}$$

(ii) *When $d = 1$, $\lambda_c = 1$ and $\theta(1) = 0$.*

The situation is rather different when the environment is chosen at random. With $G = (V, E)$ as above, suppose that the Poisson process of cuts at a vertex $x \in V$ has some intensity δ_x, and that of bridges parallel to the edge $e = \langle x, y \rangle \in E$ has some intensity λ_e. Suppose further that the δ_x, $x \in V$, are independent, identically distributed random variables, and the λ_e, $e \in E$ also. Write Δ and Λ for independent random variables having the respective distributions, and P for the probability measure governing the environment. [As before, $\mathbb{P}_{\lambda,\delta}$ denotes the measure associated with the percolation model in the given environment. The above use of the letters Δ, Λ to denote random variables is temporary only.]

If there exist $\lambda', \delta' \in (0,\infty)$ such that $\lambda'/\delta' < \lambda_c$ and $P(\Lambda \le \lambda') = P(\Delta \ge \delta') = 1$, then the process is almost surely dominated by a subcritical percolation process, whence there is (almost sure) exponential decay in the sense of Theorem 2.1(i). This may fail in an interesting way if there is no such almost-sure domination, in that one may prove exponential decay in the space-direction but only a weaker decay in the time-direction.

For any probability measure μ and function f, we write $\mu(f)$ for the expectation of f under μ. For $(x,s),(y,t) \in \mathbb{Z}^d \times \mathbb{R}$ and $q \ge 1$, we define

$$d_q(x,s;y,t) = \max\bigl\{\|x-y\|, [\log(1+|s-t|)]^q\bigr\}.$$

Theorem 2.4. [35, 36] *Let $G = \mathbb{Z}^d$ where $d \ge 1$. Suppose that*

$$K = \max\bigl\{P\bigl([\log(1+\Lambda)]^\beta\bigr), P\bigl([\log(1+\Delta^{-1})]^\beta\bigr)\bigr\} < \infty,$$

for some $\beta > 2d^2\bigl(1+\sqrt{1+d^{-1}}+(2d)^{-1}\bigr)$. There exists $Q = Q(d,\beta) > 1$ such that the following holds. For $q \in [1,Q)$ and $m > 0$, there exists $\epsilon = \epsilon(d,\beta,K,m,q) > 0$ and $\eta = \eta(d,\beta,q) > 0$ such that: if

$$E\Bigl(\bigl[\log(1+(\Lambda/\Delta))\bigr]^\beta\Bigr) < \epsilon,$$

there exist identically distributed random variables $D_x \in L^\eta(P)$, $x \in \mathbb{Z}^d$, such that

$$\mathbb{P}_{\lambda,\delta}\bigl((x,s) \leftrightarrow (y,t)\bigr) \le \exp\bigl[-m d_q(x,s;y,t)\bigr] \qquad \text{if } d_q(x,s;y,t) \ge D_x,$$

for $(x,s),(y,t) \in \mathbb{Z}^d \times \mathbb{R}$.

The corresponding theorem of [35] contains no estimate for the tail of the D_x. The above moment property may be derived from the Borel–Cantelli argument used in the proof of [35], which proceeds by a so-called multiscale analysis, see [30], Section 8. Explicit values may be given for the constants Q and η, namely

$$Q = \frac{\beta(\alpha - d + \alpha d)}{\alpha d(\alpha+\beta+1)},$$

where $\alpha = d + \sqrt{d^2+d}$, and one may take any $\eta = \eta(d,\beta,q)$ satisfying

$$0 < \alpha^2\eta < \beta\left(\frac{\alpha - d + \alpha d}{q} - \alpha d\right) - \alpha d(\alpha+1).$$

Complementary accounts of the *survival* of the process in a random environment may be found in [2, 7, 17, 42].

We mention two further types of 'continuum' percolation that arise in applications and have attracted the attention of probabilists. Let Π be a Poisson process of points in $\mathbb{R}^d$ with intensity 1. Two points $x, y \in \Pi$ are joined by an edge, and said to be *adjacent*, if they satisfy a given condition of proximity. One now asks for conditions under which the resulting random graph possesses an unbounded component.

The following conditions of proximity have been studied in the literature.

1. *Lily-pond model.* Fix $r > 0$, and join x and y if and only if $|x - y| \leq r$, where $|\cdot|$ denotes Euclidean distance. There has been extensive study of this process, and of its generalization, the *random connection model*, in which x and y are joined with probability $g(|x-y|)$ for some given non-increasing function $g : (0,\infty) \to [0,1]$. See [27, 41, 43].
2. *Voronoi percolation.* To each $x \in \Pi$ we associate the *tile*

$$T_x = \{z \in \mathbb{R}^d : |z - x| \leq |z - y| \text{ for all } y \in \Pi \setminus \{x\}\}.$$

Two tiles T_x, T_y are declared *adjacent* if their boundaries share a facet of a hyperplane of $\mathbb{R}^d$. We color each tile *red* with probability ρ, different tiles receiving independent colors, and we ask for conditions under which there exists an infinite path of red tiles.

This model has a certain property of conformal invariance when $d = 2, 3$, see [8]. When $d = 2$, there is an obvious property of self-matching, leading to the conjecture that the critical point is given by $\rho_c = \frac{1}{2}$, and this has been proved recently in [15].

3. The contact model

Just as directed percolation on $\mathbb{Z}^d$ arises by allowing only open paths that are 'stiff' in one direction, so the contact model on G is obtained from percolation on $G \times \mathbb{R}$ by requiring that open paths traverse time-lines in the direction of *increasing* time.

As before, we let D_x, $x \in V$, be Poisson processes with intensity δ, and we term points in the D_x 'cuts'. We replace each $e = \langle x, y\rangle \in E$ by two oriented edges $[x,y\rangle$, $[y,x\rangle$, the first oriented from x to y, and the second from y to x. Write $\vec{E}$ for the set of oriented edges thus obtained from E. For each $\vec{e} = [x,y\rangle \in \vec{E}$, we let $B_{\vec{e}}$ be a Poisson process with intensity λ; members of $B_{\vec{e}}$ are termed 'directed bridges' from x to y.

For $(x,s), (y,t) \in V \times \mathbb{R}$, we write $(x,s) \to (y,t)$ if there exists an oriented path π from (x,s) to (y,t) such that: π is a union of cut-free sub-intervals of $V \times \mathbb{R}$ traversed in the direction of increasing time, together with directed bridges in the directions of their orientations. For $\Lambda, \Delta \subseteq V \times \mathbb{R}$, we write $\Lambda \to \Delta$ if there exist $a \in \Lambda$ and $b \in \Delta$ such that $a \to b$.

The directed cluster $\vec{C}$ at the origin is the set

$$\vec{C} = \{(x,s) \in V \times \mathbb{R} : 0 \to (x,s)\},$$

of points reachable from the origin 0 along paths directed away from 0. The percolation probability is given by

$$\vec{\theta}(\lambda, \delta) = \mathbb{P}_{\lambda,\delta}(|\vec{C}| = \infty),$$

and the critical point by

$$\vec{\lambda}_c(\delta) = \sup\{\lambda : \vec{\theta}(\lambda, \delta) = 0\}.$$

As before, we write $\vec{\theta}(\lambda) = \vec{\theta}(\lambda, 1)$ and $\vec{\lambda}_c = \vec{\lambda}_c(1)$.

The conclusion of Theorem 2.1(i) is valid in this new setting, with C replaced by $\vec{C}$, etc, see [9]. The exact value of the critical point is unknown even when $d = 1$, although there are physical reasons to believe in this case that $\vec{\lambda}_c = 1.694\ldots$, the critical value of the so-called reggeon spin model, see [26, 38]. In compensation, it is known that $\vec{\theta}(\vec{\lambda}_c) = 0$ in all dimensions, [9]. The contact model in a random environment may be studied as in Theorem 2.4, see [35, 36].

Further theory of the contact model may be found in [38, 39]. Sakai and van der Hofstad [45] have shown how to apply the lace expansion to the spread-out contact model on $\mathbb{Z}^d$ for $d > 4$, and related results are valid for directed percolation even when the connection function has unbounded domain, see [19, 20].

4. Random-cluster and Ising/Potts models

The percolation model on a graph $G = (V, E)$ may be generalized to obtain the random-cluster model on G, see [28]. Similarly, the continuum percolation model on $G \times \mathbb{R}$ may be extended to a continuum random-cluster model. Let W be a finite subset of V that induces a connected subgraph of G, and let E_W denote the set of edges joining vertices in W. Let $\beta \in (0, \infty)$, and let Λ be the 'box' $\Lambda = W \times [0, \beta]$. Let $\mathbb{P}_{\Lambda,\lambda,\delta}$ denote the probability measure associated with the Poisson processes D_x, $x \in W$, and B_e, $e = \langle x, y\rangle \in E_W$. As sample space we take the set Ω_Λ comprising all finite sets of cuts and bridges in Λ, and we may assume without loss of generality that no cut is the endpoint of any bridge. For $\omega \in \Omega_\Lambda$, we write $B(\omega)$ and $D(\omega)$ for the sets of bridges and cuts, respectively, of ω. The appropriate σ-field $\mathcal{F}_\Lambda$ is that generated by the open sets in the associated Skorohod topology, see [10, 22].

For a given configuration $\omega \in \Omega_\Lambda$, let $k(\omega)$ be the number of its clusters under the connection relation $\leftrightarrow$. Let $q \in (0, \infty)$, and define the 'continuum random-cluster' probability measure $\mathbb{P}_{\Lambda,\lambda,\delta,q}$ by

$$d\mathbb{P}_{\Lambda,\lambda,\delta,q}(\omega) = \frac{1}{Z} q^{k(\omega)} d\mathbb{P}_{\Lambda,\lambda,\delta}(\omega), \qquad \omega \in \Omega_\Lambda, \tag{4.1}$$

for an appropriate normalizing constant, or 'partition function', $Z = Z_\Lambda(\lambda, \delta, q)$. The quantity q is called the *cluster-weighting factor*. The continuum random-cluster model may be studied in very much the same way as the random-cluster model on a lattice, see [28].

The space Ω_Λ is a partially ordered space with order relation given by: $\omega_1 \le \omega_2$ if $B(\omega_1) \subseteq B(\omega_2)$ and $D(\omega_1) \supseteq D(\omega_2)$. A random variable $X : \Omega_\Lambda \to \mathbb{R}$ is called *increasing* if $X(\omega) \le X(\omega')$ whenever $\omega \le \omega'$. An event $A \in \mathcal{F}_\Lambda$ is called *increasing* if its indicator function 1_A is increasing. Given two probability measures μ_1, μ_2 on a measurable pair $(\Omega_\Lambda, \mathcal{F}_\Lambda)$, we write $\mu_1 \le_{\rm st} \mu_2$ if $\mu_1(X) \le \mu_2(X)$ for all bounded increasing continuous random variables $X : \Omega_\Lambda \to \mathbb{R}$.

The measures $\mathbb{P}_{\Lambda,\lambda,\delta,q}$ have certain properties of stochastic ordering as the parameters Λ, λ, δ, q vary. The basic theory will be assumed here, and the reader is referred to [11] for further details. In rough terms, the $\mathbb{P}_{\Lambda,\lambda,\delta,q}$ inherit the properties of stochastic ordering and positive association enjoyed by their counterparts on discrete graphs. Of particular value later will be the stochastic inequality

$$\mathbb{P}_{\Lambda,\lambda,\delta,q} \le_{\rm st} \mathbb{P}_{\Lambda,\lambda,\delta} \qquad \text{when } q \ge 1. \tag{4.2}$$

While it will not be important for what follows, we note that the thermodynamic limit may be taken in much the same manner as for the discrete random-cluster model, whenever $q \ge 1$. Suppose, for example, that W is a finite connected subgraph of the lattice $G = \mathbb{Z}^d$, and assign to the box $\Lambda = W \times [0, \beta]$ a suitable boundary condition. As in [28], if the boundary condition τ is chosen in such a way that the measures $\mathbb{P}^\tau_{\Lambda,\lambda,\delta,q}$ are monotonic as $W \uparrow \mathbb{Z}^d$, then the weak limit $\mathbb{P}^\tau_{\lambda,\delta,q,\beta} = \lim_{W\uparrow\mathbb{Z}^d} \mathbb{P}^\tau_{\Lambda,\lambda,\delta,q}$ exists. One may similarly allow the limit as $\beta \to \infty$ to obtain a measure $\mathbb{P}^\tau_{\lambda,\delta,q} = \lim_{\beta\to\infty} \mathbb{P}^\tau_{\lambda,\delta,q,\beta}$.

Let $G = \mathbb{Z}^d$. Restricting ourselves for convenience to the case of free boundary conditions, we define the percolation probability by

$$\theta(\lambda, \delta, q) = \mathbb{P}_{\lambda,\delta,q}(|C_0| = \infty),$$

and the critical point by

$$\lambda_{\rm c}(\mathbb{Z}^d, q) = \sup\{\lambda : \theta(\lambda, 1, q) = 0\}.$$

Here, $|C|$ denotes the aggregate (one-dimensional) Lebesgue measure of the time intervals comprising C. In the special case $d = 1$, the random-cluster model has a property of self-duality that leads to the following conjecture.

Conjecture 4.3. *The continuum random-cluster model on $\mathbb{Z} \times \mathbb{R}$ with cluster-weighting factor satisfying $q \ge 1$ has critical value $\lambda_{\rm c}(\mathbb{Z}, q) = q$.*

It may be proved by standard means that $\lambda_{\rm c}(\mathbb{Z}, q) \ge q$. See [28], Section 6.2, for the corresponding result on the discrete lattice $\mathbb{Z}^2$.

The *continuum Potts model* on $G\times\mathbb{R}$ is given as follows. Let $q \in \{2, 3, \dots\}$. To each cluster of the random-cluster model with cluster-weighting factor q is assigned a 'spin' from the space $\Sigma = \{1, 2, \dots, q\}$, different clusters receiving independent spins. The outcome is a function $\sigma : V \times \mathbb{R} \to \Sigma$, and this is the spin-vector of a 'continuum q-state Potts model' with parameters λ and δ. When $q = 2$, we refer to the model as a *continuum Ising model.*

It may be seen that the law of the above spin model on $\Lambda = W \times [0, \beta]$ is given by

$$d\mathbb{P}(\sigma) = \frac{1}{Z} e^{\lambda L(\sigma)}\, d\mathbb{P}_{\Lambda,\delta}(D_\sigma),$$

where D_σ is the set of $(x, s) \in W \times [0, \beta]$ such that $\sigma(x, s-) \ne \sigma(x, s+)$, $\mathbb{P}_{\Lambda,\delta}$ is the law of a family of independent Poisson processes on the time-lines $\{x\} \times [0, \beta]$,

$x \in W$, with intensity δ, and

$$L(\sigma) = \sum_{\langle x,y\rangle \in E_W} \int_0^\beta 1_{\{\sigma(x,u)=\sigma(y,u)\}}\, du$$

is the aggregate Lebesgue measure of those subsets of pairs of adjacent time-lines on which the spins are equal. As usual, Z is an appropriate constant.

The continuum Ising model has arisen in the study by Aizenman, Klein, and Newman, [2], of the quantum Ising model with transverse field, as described in the next section.

5. The quantum Ising model

Aizenman, Klein, and Newman reported in [2] a representation of the quantum Ising model in terms of the $q = 2$ continuum random-cluster and Ising models. This was motivated in part by arguments of [17] and by earlier work referred to therein. We summarise this here, and we indicate how it may be used to study the property of entanglement in the quantum Ising model on $\mathbb{Z}$. Related representations may be constructed for a variety of quantum spin systems, see [3].

The quantum Ising model on a finite graph $G = (V, E)$ is defined as follows. To each vertex $x \in V$ is associated a quantum spin-$\frac{1}{2}$ with local Hilbert space $\mathbb{C}^2$. The Hilbert space $\mathcal{H}$ for the system is therefore the tensor product $\mathcal{H} = \bigotimes_{x\in V} \mathbb{C}^2$. As basis for the copy of $\mathbb{C}^2$ labelled by $x \in V$, we take the two eigenstates, denoted as $|+\rangle_x = \begin{pmatrix}1\\0\end{pmatrix}$ and $|-\rangle_x = \begin{pmatrix}0\\1\end{pmatrix}$, of the Pauli matrix

$$\sigma_x^{(3)} = \begin{pmatrix} 1 & 0 \\ 0 & -1 \end{pmatrix}$$

at the site x, with corresponding eigenvalues ± 1. The other two Pauli matrices with respect to this basis are:

$$\sigma_x^{(1)} = \begin{pmatrix} 0 & 1 \\ 1 & 0 \end{pmatrix}, \qquad \sigma_x^{(2)} = \begin{pmatrix} 0 & -i \\ i & 0 \end{pmatrix}. \tag{5.1}$$

In the following, $|\phi\rangle$ denotes a vector and $\langle\phi|$ its adjoint.

Let D be the set of $2^{|V|}$ basis vectors $|\eta\rangle$ for $\mathcal{H}$ of the form $|\eta\rangle = \bigotimes_x |\pm\rangle_x$. There is a natural one-one correspondence between D and the space $\Sigma = \Sigma_V = \prod_{x\in V}\{-1, +1\}$. We may speak of members of Σ as basis vectors, and of $\mathcal{H}$ as the Hilbert space generated by Σ.

The Hamiltonian of the quantum Ising model with transverse field is the operator

$$H = -\tfrac{1}{2}\lambda \sum_{e=\langle x,y\rangle\in E} \sigma_x^{(3)}\sigma_y^{(3)} - \delta \sum_{x\in V} \sigma_x^{(1)}, \tag{5.2}$$

generating the operator $e^{-\beta H}$ where β denotes inverse temperature. Here, $\lambda, \delta \geq 0$ are the spin-coupling and transverse-field intensities, respectively. The Hamiltonian has a unique pure ground state $|\psi_G\rangle$ defined at zero-temperature (that is, in the limit as $\beta \to \infty$) as the eigenvector corresponding to the lowest eigenvalue of H.

Let

$$\rho_G(\beta) = \frac{1}{Z_G(\beta)} e^{-\beta H}, \tag{5.3}$$

where

$$Z_G(\beta) = \mathrm{tr}(e^{-\beta H}) = \sum_{\eta \in \Sigma} \langle \eta | e^{-\beta H} | \eta \rangle.$$

It turns out that the matrix elements of $\rho_G(\beta)$ may be expressed in terms of a type of 'path integral' with respect to the continuum random-cluster model on $G \times [0, \beta]$ with parameters λ, δ and $q = 2$. Let $\Lambda = V \times [0, \beta]$, write Ω_Λ for the configuration space of the latter model, and let $\phi_{G,\beta}$ be the appropriate continuum random-cluster measure on Ω_Λ (with free boundary conditions). For $\omega \in \Omega_\Lambda$, let S_ω denote the space of all functions $s : V \times [0, \beta] \to \{-1, +1\}$ that are constant on the clusters of ω, and let S be the union of the S_ω over $\omega \in \Omega_\Lambda$. Given ω, we may pick an element of S_ω uniformly at random, and we denote this random element as σ. We shall abuse notation by using $\phi_{G,\beta}$ to denote the ensuing probability measure on the coupled space $\Omega_\Lambda \times S$. For $s \in S$ and $W \subseteq V$, we write $s_{W,0}$ (respectively, $s_{W,\beta}$) for the vector $(s(x, 0) : x \in W)$ (respectively, $(s(x, \beta) : x \in W)$). We abbreviate $s_{V,0}$ and $s_{V,\beta}$ to s_0 and s_β, respectively.

The following representation of the matrix elements of $\rho_G(\beta)$ is obtained by the Lie–Trotter expansion of the exponential in (5.3), and it permits the use of random-cluster methods to study the matrix $\rho_G(\beta)$. For example, as pointed out in [2], it implies the existence of the low-temperature limits

$$\langle \eta' | \rho_G | \eta \rangle = \lim_{\beta \to \infty} \langle \eta' | \rho_G(\beta) | \eta \rangle, \qquad \eta, \eta' \in \Sigma.$$

Theorem 5.4. [2] *The elements of the density matrix $\rho_G(\beta)$ are given by*

$$\langle \eta' | \rho_G(\beta) | \eta \rangle = \frac{\phi_{G,\beta}(\sigma_0 = \eta,\ \sigma_\beta = \eta')}{\phi_{G,\beta}(\sigma_0 = \sigma_\beta)}, \qquad \eta, \eta' \in \Sigma. \tag{5.5}$$

This representation may be used to study the degree of entanglement in the quantum Ising model on G. Let $W \subseteq V$, and consider the *reduced density matrix*

$$\rho_G^W(\beta) = \mathrm{tr}_{V \setminus W}(\rho_G(\beta)), \tag{5.6}$$

where the trace is performed over the Hilbert space $\mathcal{H}_{V \setminus W} = \bigotimes_{x \in V \setminus W} \mathbb{C}^2$ of the spins belonging to $V \setminus W$. By an analysis parallel to that leading to Theorem 5.4, we obtain the following.

Theorem 5.7. [30] *The elements of the reduced density matrix $\rho_G^W(\beta)$ are given by*

$$\langle \eta' | \rho_G^W(\beta) | \eta \rangle = \frac{\phi_{G,\beta}(\sigma_{W,0} = \eta,\ \sigma_{W,\beta} = \eta' \mid E)}{\phi_{G,\beta}(\sigma_0 = \sigma_\beta \mid E)}, \qquad \eta, \eta' \in \Sigma_W, \tag{5.8}$$

where E is the event that $\sigma_{V \setminus W,0} = \sigma_{V \setminus W,\beta}$.

Let D_W be the set of $2^{|W|}$ vectors $|\eta\rangle$ of the form $|\eta\rangle = \bigotimes_{x\in W} |\pm\rangle_x$, and write $\mathcal{H}_W$ for the space generated by D_W. Just as before, there is a natural one-one correspondence between D_W and the space $\Sigma_W = \prod_{x\in W}\{-1,+1\}$, and we shall regard $\mathcal{H}_W$ as the Hilbert space generated by Σ_W.

We may write

$$\rho_G = \lim_{\beta\to\infty} \rho_G(\beta) = |\psi_G\rangle\langle\psi_G|$$

for the density matrix corresponding to the ground state of the system, and similarly

$$\rho_G^W = \mathrm{tr}_{V\setminus W}(|\psi_G\rangle\langle\psi_G|) = \lim_{\beta\to\infty} \rho_G^W(\beta). \tag{5.9}$$

There has been extensive study of entanglement in the physics literature, see the references in [30]. The entanglement of the spins in W may be defined as follows.

Definition 5.10. *The* entanglement *of the vertex-set W relative to its complement $V \setminus W$ is the entropy*

$$S_G^W = -\,\mathrm{tr}(\rho_G^W \log_2 \rho_G^W). \tag{5.11}$$

The behavior of S_G^W, for general G and W, is not understood at present. We specialize here to the case of a finite subset of the one-dimensional lattice $\mathbb{Z}$. Let $m, L \geq 0$ and take $V = [-m, m+L]$ and $W = [0, L]$, viewed as subsets of $\mathbb{Z}$. We obtain G from V by adding edges between each pair $x, y \in V$ with $|x-y| = 1$. We write $\rho_m(\beta)$ for $\rho_G(\beta)$, and S_m^L for S_G^W. A key step in the study of S_m^L for large m is a bound on the norm of the difference $\rho_m^L - \rho_n^L$. For a Hermitian matrix A, let

$$\|A\| = \sup_{\|\psi\|=1} \left|\langle\psi|A|\psi\rangle\right|,$$

where the supremum is over all $\psi \in \mathcal{H}_L$ with L^2-norm 1.

Theorem 5.12. [30] *Let $\lambda, \delta \in (0,\infty)$ and write $\theta = \lambda/\delta$. There exist constants C, α, γ depending on θ and satisfying $\gamma > 0$ when $\theta < 1$ such that:*

$$\|\rho_m^L - \rho_n^L\| \leq \min\{2, CL^\alpha e^{-\gamma m}\}, \qquad 2 \leq m \leq n < \infty,\ L \geq 1. \tag{5.13}$$

One would expect that γ may be taken in such a manner that $\gamma > 0$ under the weaker assumption $\lambda/\delta < 2$, but this has not yet been proved (cf. Conjecture 4.3). The constant γ is, apart from a constant factor, the reciprocal of the correlation length of the associated random-cluster model.

Inequality (5.13) is proved in [30] by the following route. Consider the random-cluster model with $q = 2$ on the space-time graph $\Lambda = V \times [0,\beta]$ with 'partial periodic top/bottom boundary conditions'; that is, for each $x \in V \setminus W$, we identify the two vertices $(x,0)$ and (x,β). Let $\phi^{\mathrm{p}}_{m,\beta}$ denote the associated random-cluster measure on Ω_Λ. To each cluster of ω $(\in \Omega_\Lambda)$ we assign a random spin from $\{-1,+1\}$ in the usual manner, and we abuse notation by using $\phi^{\mathrm{p}}_{m,\beta}$ to denote the measure governing both the random-cluster configuration and the spin configuration. Let

$$a_{m,\beta} = \phi^{\mathrm{p}}_{m,\beta}(\sigma_{W,0} = \sigma_{W,\beta}),$$

noting that

$$a_{m,\beta} = \phi_{m,\beta}(\sigma_0 = \sigma_\beta \mid E)$$

as in (5.8).

By Theorem 5.7,

$$\langle\psi|\rho_m^L(\beta) - \rho_n^L(\beta)|\psi\rangle = \frac{\phi_{m,\beta}^{\mathrm{p}}(c(\sigma_{W,0})\overline{c(\sigma_{W,\beta})})}{a_{m,\beta}} - \frac{\phi_{n,\beta}^{\mathrm{p}}(c(\sigma_{W,0})\overline{c(\sigma_{W,\beta})})}{a_{n,\beta}}, \tag{5.14}$$

where $c : \{-1, +1\}^W \to \mathbb{C}$ and

$$\psi = \sum_{\eta\in\Sigma_W} c(\eta)\eta \in \mathcal{H}_W.$$

The random-cluster property of ratio weak-mixing is used in the derivation of (5.13) from (5.14). At the final step of the proof of Theorem 5.12, the random-cluster model is compared with the continuum percolation model of Section 2, and the exponential decay of Theorem 5.12 follows by Theorem 2.1. A logarithmic bound on the entanglement entropy follows for sufficiently small λ/δ.

Theorem 5.15. [30] *Let $\lambda, \delta \in (0, \infty)$ and write $\theta = \lambda/\delta$. There exists $\theta_0 \in (0, \infty)$ such that: for $\theta < \theta_0$, there exists $K = K(\theta) < \infty$ such that*

$$S_m^L \le K \log_2 L, \qquad m \ge 0, \ L \ge 2. \tag{5.16}$$

A stronger result is expected, namely that the entanglement S_m^L is bounded above, uniformly in L, whenever θ is sufficiently small, and perhaps for all $\theta < \theta_{\mathrm{c}}$ where $\theta_{\mathrm{c}} = 2$ is the critical point. See Conjecture 4.3 above, and the references in [30], especially [6]. There is no rigorous picture known of the behavior of S_m^L for large θ, or of the corresponding quantity in dimensions $d \ge 2$, although Theorem 5.12 has a counterpart in this setting. Theorem 5.15 may be extended to the disordered system in which the intensities λ, δ are independent random variables indexed by the vertices and edges of the underlying graph, subject to certain conditions on these variables (cf. Theorem 2.4 and the preceding discussion). See also [24].

6. The mean-field continuum model

The term 'mean-field' is often interpreted in percolation theory as percolation on either a tree (see [27], Chapter 10) or a complete graph. The latter case is known as the Erdős–Rényi random graph $G_{n,p}$, and this is the random graph obtained from the complete graph K_n on n vertices by deleting each edge with probability $1 - p$. The theory of $G_{n,p}$ is well developed and rather refined, see [12, 34], and particular attention has been paid to the emergence of the giant cluster for $p = \lambda/n$ and $\lambda \simeq 1$. A similar theory has been developed for the random-cluster model on K_n with parameters p, q, see [13, 28, 40].

Unless boundary conditions are introduced in the manner of [28, 29, 31], the continuum random-cluster model on a tree may be solved exactly by standard means. We therefore concentrate here on the case of the complete graph K_n on n vertices. Let $\beta > 0$, and attach to each vertex the line $[0, \beta]$ with its endpoints identified; thus, the line forms a circle. We now consider the continuum random-cluster model on $K_n \times [0, \beta]$ with parameters $p = \lambda/n$, $\delta = 1$, and q. [The convention of setting $\delta = 1$ differs from that of [32] but is consistent with that adopted in earlier work on related models.]

Suppose that $q \geq 1$, so that we may use methods based on stochastic comparisons. It is natural to ask for the critical value $\lambda_c = \lambda_c(\beta, q)$ of λ above which the model possesses a giant cluster. This has been answered thus by Ioffe and Levit, [32], in the special case $q = 1$. Let $F(\beta, \lambda)$ be given by

$$F(\beta, \lambda) = \lambda\big[2(1 - e^{-\beta}) - \beta e^{-\beta}\big],$$

and let $\lambda_c = \lambda_c(\beta)$ be chosen so that $F(\beta, \lambda_c) = 1$.

Theorem 6.1. [32] *Let M be the maximal (one-dimensional) Lebesgue measure of the clusters of the process with parameters β, $p = \lambda/n$, $\delta = 1$, $q = 1$. Then, as $n \to \infty$,*

$$\frac{1}{n} M \to \begin{cases} 0 & \text{if } \lambda < \lambda_c, \\ \beta\pi & \text{if } \lambda > \lambda_c, \end{cases}$$

where $\pi = \pi(\beta, \lambda) \in (0, 1)$ when $\lambda > \lambda_c$, and the convergence is in probability.

When $\lambda > \lambda_c$, the density of the giant cluster is π, in that there is probability π that any given point of $K_n \times [0, \beta]$ lies in this giant cluster. The claim of Theorem 6.1 has a straightforward motivation (the proof is more complicated). Let 0 be a vertex of K_n, and let I be the maximal cut-free interval of $0 \times [0, \beta]$ (viewed as a circle) containing the point 0×0. Given I, the mean number of bridges leaving I is $\lambda|I|(n-1)/n \sim \lambda|I|$, where $|I|$ is the Lebesgue measure of I. One may thus approximate to the cluster at 0×0 by a branching process with mean family-size $\lambda E|I|$. It is elementary that $\lambda E|I| = F(\beta, \lambda)$, which is to say that the branching process is subcritical (respectively, supercritical) if $\lambda < \lambda_c$ (respectively, $\lambda > \lambda_c$). The full proof may be found in [32], and a further proof has appeared in [33]. The quantity π is of course the survival probability of the above branching process, and this may be calculated in the standard way on noting that $|I|$ is distributed as $\min\{U + V, \beta\}$ where U, V are independent, exponentially distributed, random variables with mean 1.

What is the analogue of Theorem 6.1 when $q \neq 1$? Indications are presented in [32] of the critical value when $q = 2$, and the problem is posed there of proving this value by calculations of the random-cluster type to be found in [13]. There is a simple argument that yields upper and lower bounds for the critical value for any $q \in [1, \infty)$. We present this next, and also explain our reason for believing the upper bound to be exact when $q \in [1, 2]$.

Consider the continuum random-cluster model on $K_n \times [0,\beta]$ with parameters $p = \lambda/n$, $\delta = 1$, and $q \in (0,\infty)$. Let

$$F_q(\beta,\lambda) = \frac{\lambda}{q^2} \cdot \frac{2e^{\beta q} - 2 + \beta q(q-2)}{e^{\beta q} + q - 1}, \tag{6.2}$$

noting that $F_1 = F$.

Theorem 6.3. *Let M_q be the maximal (one-dimensional) Lebesgue measure of the clusters of the process with parameters β, $p = \lambda/n$, $\delta = 1$, $q \in [1,\infty)$.*

(i) *We have that $\lim_{n\to\infty} n^{-1} M_q = 0$ if $F_q < q^{-1}$, where the convergence is in probability.*

(ii) *There exists $\pi_q = \pi_q(\beta,\lambda)$, satisfying $\pi_q > 0$ whenever $F_q > 1$, such that*

$$\liminf_{n\to\infty} P\left(\frac{1}{n} M_q \geq \beta \pi_q\right) \to 1.$$

The bound π_q may be calculated by a branching-process argument, in the same manner as was $\pi = \pi_1$, above. We conjecture that $n^{-1}M_q \to 0$ in probability if $F_q < 1$ and $q \in [1,2]$. This conjecture is motivated by the evidence of [13] that, in the second-order phase transition occurring when $q \in [1,2]$, the location of the critical point is given by the branching-process approximation described in the sketch proof below. This amounts to the claim that the critical value $\lambda_c(q)$ of the continuum random-cluster model with cluster-weighting factor q satisfies

$$\lambda_c(q) = q^2 \frac{e^{\beta q} + q - 1}{2e^{\beta q} - 2 + \beta q(q-2)}, \qquad q \in [1,2]. \tag{6.4}$$

This is implied by Theorem 6.1 when $q = 1$, and by the claim of [32] when $q = 2$. Note the relatively simple formula when $q = 2$,

$$\lambda_c(2) = \frac{2}{\tanh \beta}, \tag{6.5}$$

which might be termed the critical point of the *quantum random graph*. Dmitry Ioffe has pointed out that the exact calculation (6.5) may be derived from the results of [21, 23]. Results similar to those of Theorem 6.3 may be obtained for $q < 1$ also.

Sketch proof of Theorem 6.3. We begin with part (ii). The idea is to bound the process below by a random graph to which the results of [14, 33] may be applied directly. The bounding process is obtained as follows. First, we place the cuts on each of the time-lines $x \times [0,\beta]$, and we place no bridges. Thus, the cuts on a given time-line are placed in the manner of the continuum random-cluster model on that line. It may be seen that the number D of cuts on any given time-line has mass function

$$P(D = k) = \frac{e^{-\beta}}{Z} \cdot \frac{q^{k \vee 1} \beta^k}{k!}, \qquad k \geq 0,$$

where $a \vee b = \max\{a,b\}$, and Z is the requisite constant,

$$Z = (q-1)e^{-\beta} + e^{\beta(q-1)}.$$

It is an easy calculation that the maximal cut-free interval I containing the point 0×0 satisfies $E|I| = qF_q/\lambda$.

We next place edges between pairs of time-lines according to independent Poisson processes with intensity λ/q. We term the ensuing graph a 'product random-cluster model', and we claim that this model is dominated (stochastically) by the continuum random-cluster model. This may be seen in either of two ways: one may apply suitable comparison inequalities (see [28], Section 3.4) to a discrete approximation of $K_n \times [0, \beta]$ and then pass to the continuum limit, or one may establish it directly for the continuum model. Related material has appeared in [25, 44].

If this 'product' random-cluster model possesses a giant cluster, then so does the original random-cluster model. The former model may be studied either via the general techniques of [14, 33] for inhomogeneous random graphs, or using the usual branching process approximation. We follow the latter route here, but omit the details. In the limit as $n \to \infty$, the mean number of offspring of 0×0 approaches $(\lambda/q)E|I| = F_q$, so that the branching process is supercritical if $F_q > 1$. The claim of part (ii) follows.

For part (i) one proceeds similarly, but with λ/q replaced by λ and the domination reversed. □

Acknowledgements

The author thanks Carol Bezuidenhout for encouraging him to persevere with continuum percolation and the contact model many years ago, and Tobias Osborne and Petra Scudo for elaborating on the relationship between the quantum Ising model and the continuum random-cluster model. He is grateful to Svante Janson for their discussions of random-cluster models on complete graphs. Dima Ioffe has pointed out the link between the quantum random graph and the work of [21, 23], and Akira Sakai and the referee have kindly commented on an earlier version of this paper.

References

[1] M. Aizenman and P. Jung. On the critical behavior at the lower phase transition of the contact process. 2006. arxiv:math.PR/0603227.

[2] M. Aizenman, A. Klein, and C.M. Newman. Percolation methods for disordered quantum Ising models. In R. Kotecký, editor, *Phase Transitions: Mathematics, Physics, Biology, ...*, pages 129–137. World Scientific, Singapore, 1992.

[3] M. Aizenman and B. Nachergaele. Geometric aspects of quantum spin states. *Communications in Mathematical Physics*, 164:17–63, 1994.

[4] K. Alexander. On weak mixing in lattice models. *Probability Theory and Related Fields*, 110:441–471, 1998.

[5] K. Alexander. Mixing properties and exponential decay for lattice systems in finite volumes. *Annals of Probability*, 32:441–487, 2004.

[6] L. Amico, R. Fazio, A. Osterloh, and V. Vedral. Entanglement in many-body systems. arxiv:quant-ph/0703044.

[7] E. Andjel. Survival of multidimensional contact process in random environments. *Boletim da Sociedade Brasileira de Matemática*, 23:109–119, 1992.

[8] I. Benjamini and O. Schramm. Conformal invariance of Voronoi percolation. *Communications in Mathematical Physics*, 197:75–107, 1998.

[9] C.E. Bezuidenhout and G.R. Grimmett. The critical contact process dies out. *Annals of Probability*, 18:1462–1482, 1990.

[10] C.E. Bezuidenhout and G.R. Grimmett. Exponential decay for subcritical contact and percolation processes. *Annals of Probability*, 19:984–1009, 1991.

[11] J. Björnberg. 2007. In preparation.

[12] B. Bollobás. *Random Graphs.* Cambridge University Press, 2nd edition, 2001.

[13] B. Bollobás, G.R. Grimmett, and S. Janson. The random-cluster model on the complete graph. *Probability Theory and Related Fields*, 104:283–317, 1996.

[14] B. Bollobás, S. Janson, and O. Riordan. The phase transition in inhomogeneous random graphs. *Random Structures and Algorithms*, 2007. arxiv:math.PR/0504589.

[15] B. Bollobás and O. Riordan. The critical probability for random Voronoi percolation in the plane is $\frac{1}{2}$. *Probability Theory and Related Fields*, 136:417–468, 2006.

[16] B. Bollobás and O. Riordan. *Percolation.* Cambridge University Press, 2006.

[17] M. Campanino, A. Klein, and J.F. Perez. Localization in the ground state of the Ising model with a random transverse field. *Communications in Mathematical Physics*, 135:499–515, 1991.

[18] R. Cerf. *The Wulff Crystal in Ising and Percolation Models.* Springer, Berlin, 2006.

[19] L.-C. Chen and A. Sakai. Critical behavior and the limit distribution for long-range oriented percolation, I. 2007.

[20] L.-C. Chen and N.-R. Shieh. Critical behavior for an oriented percolation with long-range interactions in dimension $d > 2$. *Taiwanese Journal of Mathematics*, 10:1345–1378, 2006.

[21] T. Dorlas. Probabilistic derivation of a noncommutative version of Varadhan's theorem. 2002.

[22] S.N. Ethier and T.G. Kurtz. *Markov Processes.* Wiley, New York, 1986.

[23] M. Fannes, H. Spohn, and A. Verbeure. Equilibrium states for mean field models. *Journal of Mathematical Physics*, 21:355–358, 1980.

[24] D.S. Fisher. Critical behavior of random transverse-field Ising spin chains. *Physical Review B*, 51:6411–6461, 1995.

[25] H.-O. Georgii and T. Küneth. Stochastic comparison of point random fields. *Journal of Applied Probability*, 34:868–881, 1997.

[26] P. Grassberger and A. de la Torre. Reggeon field theory (Schögl's first model) on a lattice: Monte Carlo calculations of critical behaviour. *Annals of Physics*, 122:373–396, 1979.

[27] G.R. Grimmett. *Percolation.* Springer, Berlin, 1999.

[28] G.R. Grimmett. *The Random-Cluster Model.* Springer, Berlin, 2006.

[29] G.R. Grimmett and S. Janson. Branching processes, and random-cluster measures on trees. *Journal of the European Mathematical Society*, 7:253–281, 2005.

[30] G.R. Grimmett, T.J. Osborne, and P.F. Scudo. Entanglement in the quantum Ising model. *Journal of Satistical Physics*, 131:305–339, 2008.

[31] O. Häggstrom. The random-cluster model on a homogeneous tree. *Probability Theory and Related Fields*, 104:231–253, 1996.

[32] D. Ioffe and A. Levit. Long range order and giant components of quantum random graphs. *Markov Processes and Related Fields*, 13:469–492, 2007.
[33] S. Janson. On a random graph related to quantum theory. *Combinatorics, Probability and Computing*, 16:757–766, 2007.
[34] S. Janson, T. Łuczak, and A. Ruciński. *Random Graphs.* John Wiley, New York, 2000.
[35] A. Klein. Extinction of contact and percolation processes in a random environment. *Annals of Probability*, 22:1227–1251, 1994.
[36] A. Klein. Multiscale analysis in disordered systems: percolation and contact process in random environment. In G.R. Grimmett, editor, *Disorder in Physical Systems*, pages 139–152. Kluwer, Dordrecht, 1994.
[37] G. Lawler. *Conformally Invariant Processes in the Plane.* American Mathematical Society, 2005.
[38] T.M. Liggett. *Interacting Particle Systems.* Springer, Berlin, 1985.
[39] T.M. Liggett. *Stochastic Interacting Systems: Contact, Voter and Exclusion Processes.* Springer, Berlin, 1999.
[40] M. Luczak and T. Łuczak. The phase transition in the cluster-scaled model of a random graph. *Random Structures and Algorithms*, 28:215–246, 2006.
[41] R. Meester and R. Roy. *Continuum Percolation.* Cambridge University Press, 1996.
[42] C.M. Newman and S. Volchan. Persistent survival of one-dimensional contact processes in random environments. *Annals of Probability*, 24:411–421, 1996.
[43] M.D. Penrose. *Random Geometric Graphs.* Oxford University Press, 2003.
[44] C.J. Preston. Spatial birth-and-death processes. *Bulletin of the International Statistical Institute*, 46:371–391, 405–408, 1975.
[45] A. Sakai and R. van der Hofstad. Critical points for spread-out self-avoiding walk, percolation and the contact process above the upper critical dimensions. *Probability Theory and Related Fields*, 132:438–470, 2005.
[46] O. Schramm. Conformally invariant scaling limits: an overview and a collection of open problems. In *Proceedings of the International Congress of Mathematicians, Madrid,* 2006, volume I, pages 513–544. European Mathematical Society, Zürich, 2007.
[47] G. Slade. *The Lace Expansion and its Applications.* Springer, Berlin, 2006.
[48] S. Smirnov. Towards conformal invariance of 2D lattice models. In *Proceedings of the International Congress of Mathematicians, Madrid,* 2006, volume II, pages 1421–1452. European Mathematical Society, Zürich, 2007.
[49] W. Werner. Random planar curves and Schramm–Loewner evolutions. In J. Picard, editor, *Ecole d'Eté de Probabilités de Saint Flour* XXXII–2002, pages 107–195. Springer, Berlin, 2004.

Geoffrey R. Grimmett
Centre for Mathematical Sciences
University of Cambridge
Wilberforce Road
Cambridge CB3 0WB, United Kingdom
e-mail: grg@statslab.cam.ac.uk

Progress in Probability, Vol. 60, 321–329

Computability of Percolation Thresholds

Olle Häggström

Abstract. The critical value for Bernoulli percolation on the $\mathbb{Z}^d$ lattice in any dimension d is shown to be a computable number in the sense of the Church–Turing thesis.

Mathematics Subject Classification (2000). 60K35, 03D10.

Keywords. Percolation, Church–Turing computability, critical value, renormalization.

1. Introduction

In 2004, at the 8th Brazilian School of Probability, I gave a lecture series entitled *Percolation theory: the number of infinite clusters*, based mainly on a draft version of Häggström and Jonasson [6]. This was a highly rewarding experience, not only because of the beautiful location but also because of one of the most stimulating audiences I have ever had. During one of the breaks, and later at an open problems session, Andrei Toom asked whether the critical value p_c for Bernoulli percolation on the $\mathbb{Z}^d$ lattice is computable in the sense of the Church–Turing thesis for all d, and described (probably somewhat tongue-in-cheek) the lack of a known answer to this question as a serious shortcoming of the subject of percolation theory. The purpose of this note is to show how an affirmative answer to Toom's question can be deduced relatively easily from some of the percolation technology that had been developed for other purposes in the 1980's and 1990's. To prove the desired computability result from scratch is a different story, and I suspect it would be quite involved.

The rest of this section is devoted to describing the setup and stating the main result. Then, in Section 2, the algorithm that is used to establish the result is described, and finally in Section 3 the algorithm is shown to halt in finite time with the desired output.

Research supported by the Swedish Research Council and by the Göran Gustafsson Foundation for Research in the Natural Sciences and Medicine.

Percolation theory (see Grimmett [4] for an introduction) deals with connectivity properties of random media, and the most basic setup, known as Bernoulli percolation, is as follows. Let $G = (V, E)$ be a finite or infinite but locally finite graph with vertex set V and edge set E, fix $p \in [0, 1]$, and remove each edge $e \in E$ independently with probability $1 - p$, thus keeping it with probability p, and consider the resulting subgraph of G. It will sometimes be convenient to represent an outcome of this percolation process as an element of $\{0, 1\}^E$, where a 0 denotes the removal of an edge, and a 1 its retention. In the following, we will conform to standard terminology by speaking of retained edges as open, and deleted edges as closed. For $x, y \in V$, we write $x \leftrightarrow y$ for the event that there exists a path of retained edges between x and y.

What we have defined here, and will be concerned with in the following unless otherwise stated, is bond percolation. Alternatively, one may consider site percolation, where it is the vertices rather than the edges that are retained (declared open) or removed (declared closed) independently with retention probability p.

When G is infinite, it is natural to ask for the probability of the existence of at least one infinite connected component of the resulting subgraph. Kolmogorov's 0-1-law implies that this probability $\psi(p)$ is either 0 or 1 for any p, and a simple coupling argument shows that $\psi(p)$ is nondecreasing in p. Combining these two observations yields the existence of a critical value $p_c = p_c(G) \in [0, 1]$ such that

$$\psi(p) = \begin{cases} 0 & \text{for } p < p_c \\ 1 & \text{for } p > p_c\,. \end{cases}$$

Much of percolation theory deals specifically with the case where G is the $\mathbb{Z}^d$ lattice, i.e., the graph with vertex set $\mathbb{Z}^d$ and with edge set $E_{\mathbb{Z}^d}$ consisting of edges connecting vertices at Euclidean distance 1 from each other. The case $d = 1$ is fairly trivial with $p_c = 1$, but already the case $d = 2$ turns out to be extremely intricate, with Kesten's [7] 1980 result that $p_c(\mathbb{Z}^2) = \frac{1}{2}$ standing out as one of the classical achievements in the subject. For higher dimensions $d \geq 3$ no exact expressions for $p_c(\mathbb{Z}^d)$ are known. This makes it natural to ask for upper and lower bounds for p_c as well as properties such as the computability considered here.

Beginning in 1936, a number of formal models of computing – the most well-known ones being *Turing machines* and *λ-calculus* – were introduced that were later shown to yield equivalent notions of computability. The *Church–Turing thesis* states that the set of functions $f : \mathbb{Z} \to \mathbb{Z}$ computable according to one (hence all) of these models exhausts the set of functions that would naturally be regarded as computable. Due to the vagueness of the statement, the thesis cannot be formally proven, but it is held in high esteem among computer scientists, and certainly any function that allows computation by a program written in (pseudo-) Pascal or other standard programming languages can also be computed on a Turing machine. See, e.g., Knuth [8] or Blass and Gurevich [3] for more on this topic. Actually programming a Turing machine is an extremely tedious task, and we will instead follow tradition by reverting to describing our algorithms in a more informal language, yet specifically enough to make it evident that they can be implemented on a computer.

The notion of computability of a function is easily extended to that of a real number x with binary expansion $x = \sum_i x_i 2^{-i}$: we say that x is computable if $f(i) = x_i$ is a computable function (and it is easy to see that this computability property is unchanged if we switch to, e.g., base 3 or any other integer base). We can now state our main result:

Theorem 1.1. *The critical value $p_c = p_c(\mathbb{Z}^d)$ for Bernoulli bond percolation on $\mathbb{Z}^d$ is computable for any d.*

Our choice to state and prove the result only for bond percolation is just a matter of convenience: the result and its proof allow almost verbatim translation to the site percolation setting.

One peculiarity of our proof of Theorem 1.1 is a slight lack of constructiveness: Either p_c is dyadic (i.e., equals $j2^{-i}$ for some integers i and j), or it is not. If it is, then obviously it is also computable, while if it is not, then the algorithm in Section 2 will compute p_c. We are thus unable to point at a single algorithm and with confidence say that *this* algorithm computes p_c. This peculiarity is an artifact of the precise choice of definition of computability of a real number x. If instead (as some authors prefer) we choose the equivalent definition of saying that x is computable if there exists an algorithm which given any i produces an interval of length 2^{-i} that contains x, then a minor variation of the algorithm in Section 2 will suffice to achieve this regardless of whether x is dyadic or not; see Remark 2.3.

The algorithm outlined in Section 2 – or more precisely the variant given in Remark 2.3 – improves a (randomized) scheme for estimating p_c due to Meester and Steif [11]. Their scheme produces a sequence of estimates $\hat{p}_c^{(1)}, \hat{p}_c^{(2)}, \ldots$ that converges (almost surely) to p_c, but at no stage is there any guarantee that $\hat{p}_c^{(i)}$ is within a given distance ε from p_c. In contrast, our algorithm yields a sequence $\hat{p}_c^{(1)}, \hat{p}_c^{(2)}, \ldots$ for which we know that $|\hat{p}_c^{(i)} - p_c| \leq 2^{-i}$ for each i.

How far – i.e., to which lattices and graphs – can Theorem 1.1 be extended? Certainly not to all graphs, because, as observed by van den Berg [2], there exists for any $p \in [0,1]$ a graph G with $p_c(G) = p$. It might be tempting to hope that $p_c(G)$ is computable for every *transitive* graph, but I suspect that even this is false, in view of Leader and Markström's [9] construction of uncountable families of non-isomorphic transitive graphs.

2. The algorithm and some basic properties

Fix the dimension d. For N a multiple of 8, define

$$\Lambda_N = \left\{x = (x_1, \ldots, x_d) \in \mathbb{Z}^d : -\tfrac{5N}{8} \leq x_j \leq \tfrac{5N}{8} \text{ for } j = 1, \ldots, d\right\},$$

and define E_{Λ_N} as the set of edges in the $\mathbb{Z}^d$ lattice whose endpoints are both in Λ_N. For Bernoulli bond percolation on $\mathbb{Z}^d$ consider the two events A_N and B_N defined in terms of the edges in E_{Λ_N} as follows.

- A_N is the event that at least one connected component of the set of open edges in E_{Λ_N} contains two vertices at Euclidean distance more than $N/10$ from each other.
- B_N is the event that the set of open edges in E_{Λ_N} contains a connected component intersecting all the $2d$ sides of the cube Λ_N, but that no other connected component contains two vertices at Euclidean distance more than $N/10$ from each other.

For $p \in [0,1]$, define $\mathbb{P}_p$ as the probability measure on $\{0,1\}^{E_{\mathbb{Z}^d}}$ corresponding to Bernoulli bond percolation on $\mathbb{Z}^d$. Note that for rational p, the probability $\mathbb{P}_p(A_N)$ is easy to compute: just go through all the $2^{d(\frac{5N}{4})^d}$ different configurations $\omega \in \{0,1\}^{E_{\Lambda_N}}$, check for each of them whether A_N happens, and sum

$$p^{n(\omega)}(1-p)^{d(\frac{5N}{4})^d - n(\omega)}$$

over those ω's for which A_N happens; here $n(\omega)$ is the number of 1's in ω. By the same token, we can compute $\mathbb{P}_p(B_N)$.

Our algorithm which, given i, produces the i first binary digits of p_c – or equivalently, gives an interval of the form $[j2^{-i},(j+1)2^{-i})$ containing p_c – is as follows.

(I) Set $N = 8$.
(II) Compute $\mathbb{P}_{j2^{-i}}(A_N)$ and $\mathbb{P}_{j2^{-i}}(B_N)$ for $j = 0, 1, \ldots, 2^i$.
(III) If for some $j \in \{0, \ldots, 2^i - 1\}$ we have

$$\mathbb{P}_{j2^{-i}}(A_N) < (2d-1)^{-3^d}$$

and

$$\mathbb{P}_{(j+1)2^{-i}}(B_N) > 1 - 8^{-9},$$

then let j' be the smallest such j, output the interval $[j'2^{-i},(j'+1)2^{-i})$ and stop. Otherwise increase N by 8 and continue with (II).

We need to show, under the assumption that p_c is non-dyadic, that the algorithm terminates after some finite number of cycles, and that the interval $[j2^{-i},(j+1)2^{-i})$ it outputs satisfies

$$p_c \in [j2^{-i},(j+1)2^{-i}). \tag{2.1}$$

The termination property follows from Proposition 2.1 below applied to $p = j2^{-i}$ and $p^* = (j+1)2^{-i}$ where $j = \max\{j' : j'2^{-i} < p_c\}$, and property (2.1) follows from Proposition 2.2. Hence, once the two propositions are proved, we know that non-dyadicity of p_c implies its computability, and as explained in Section 1 this implies Theorem 1.1. We defer the proofs of the propositions to Section 3.

Proposition 2.1.

(a) *For any* $p < p_c$*, we have* $\lim_{N\to\infty} \mathbb{P}_p(A_N) = 0$.
(b) *For any* $p > p_c$*, we have* $\lim_{N\to\infty} \mathbb{P}_p(B_N) = 1$.

Proposition 2.2.

(a) *For no* $p < p_c$ *and no* $N \in 8, 16, 24, \ldots$ *do we have* $\mathbb{P}_p(B_N) > 1 - 8^{-9}$.

(b) *For no* $p > p_c$ *and no* $N \in 8, 16, 24, \ldots$ *do we have* $\mathbb{P}_p(A_N) < (2d-1)^{-3^d}$.

Remark 2.3. The reason why the above algorithm doesn't necessarily work in case p_c is dyadic is that if $j2^{-i} = p_c$, then we may end up having $\mathbb{P}_{j2^{-i}}(A_N) > (2d-1)^{-3^d}$ and $\mathbb{P}_{j2^{-i}}(B_N) < 1 - 8^{-9}$ for all N, causing the algorithm to keep running without ever terminating. If we are content with an interval of width 2^{-i+1} (which can still be made as small as we wish) containing p_c, then regardless of dyadicity the algorithm will terminate and produce an interval containing p_c if we simply replace step (III) above by

(III′) If for some $j \in \{0, \ldots, 2^i - 1\}$ we have $\mathbb{P}_{j2^{-i}}(A_n) < (2d-1)^{-3^d}$ and $\mathbb{P}_{(j+2)2^{-i}}(B_n) > 1 - 8^{-9}$, then take j' to be the smallest such j, output the interval $[j'2^{-i}, (j'+2)2^{-i})$ and stop. Otherwise increase N by 8 and continue with (II).

Remark 2.4. The proposed algorithm is obviously incredibly slow – so slow that nobody in her right mind would use it in practice to estimate p_c. It might nevertheless be of some theoretical interest to find bounds for its running time. Such bounds can presumably be obtained by inspecting the proofs of the limit theorems from [12] and [1] used in the next section, and extracting convergence rates (though it might require hard work). For the algorithm in Remark 2.3 the bounds can probably be obtained independently of any detailed information about p_c, whereas for the original algorithm p_c would have to enter the bound in one way or another. To see this, suppose for instance that $p_c \in (\frac{3}{8}, \frac{3}{8} + 10^{-1000})$. To get the third binary digit in place we would have to decide whether $\frac{3}{8}$ is sub- or supercritical, and since $\frac{3}{8}$ is so close to p_c that would require a stupendously large N – and the running time is exponential in N.

3. Proofs

The definitions of the events A_N and B_N are tailored to fit into known percolation technology to make the proof of especially Proposition 2.1 as streamlined as possible. In particular, the proof of Proposition 2.1 (a) is based on the famous exponential decay result for subcritical percolation, which was proved by Menshikov [12] and is explained at greater length by Grimmett [4]. Writing D for the radius of the connected component containing the origin 0, i.e.,

$$D = \sup\{\text{dist}(0, x) : x \in \mathbb{Z}^d, x \leftrightarrow 0\}$$

where dist denotes Euclidean distance, the result states that for any $p < p_c$ there exists a $C = C(p) > 0$ such that

$$\mathbb{P}_p(D > n) < e^{-Cn} \tag{3.1}$$

for all n.

Proof of Proposition 2.1 (a). Fix $p < p_c$ and choose $C > 0$ in such a way that (3.1) holds for all n. Write Z_N for the number of vertices $x \in \Lambda_N$ that are connected to at least one vertex at distance at least $N/10$ away. Since Λ_N contains $\left(\frac{5N}{4}+1\right)^d$ vertices, (3.1) implies that the expected value of Z_N satisfies

$$\mathbb{E}_p[Z_N] < \left(\tfrac{5N}{4}+1\right)^d e^{-CN/10}$$

so that

$$\begin{aligned}\lim_{N\to\infty} \mathbb{P}_p(A_N) &\leq \lim_{N\to\infty} \mathbb{P}_p(Z_N \geq 1)\\ &\leq \lim_{N\to\infty} \mathbb{E}_p[Z_N]\\ &\leq \lim_{N\to\infty} \left(\tfrac{5N}{4}+1\right)^d e^{-CN/10} = 0,\end{aligned}$$

as desired. □

The proof of Proposition 2.1 (b) consists in a reference to a result of Antal and Pisztora [1]. The choice of using this particular result is somewhat arbitrary, and could be replaced by any of a number of similar results from the renormalization technology pioneered by Grimmett and Marstrand [5] and discussed at a gentler pace by Grimmett [4].

Proof of Proposition 2.1 (b). This is Antal and Pisztora [1, Prop. 2.1]. □

In order to prove Proposition 2.2, we need the notion of 1-*dependent site percolation.* This is a generalization of ordinary (Bernoulli) site percolation where the independence assumption is weakened as follows. The L_∞-distance between two vertices $(x, y) \in \mathbb{Z}^d$ with coordinates $x = (x_1, \ldots, x_d)$ and $y = (y_1, \ldots, y_d)$ is defined to be $\max_i\{|x_i - y_i|\}$.

Definition 3.1. A $\{0,1\}^{\mathbb{Z}^d}$-valued random object X is said to be 1-dependent if for any finite collection of vertices $x_1, \ldots, x_k \in \mathbb{Z}^d$ such that no two of them are within L_∞-distance 1 from each other we have that $X(x_1), \ldots, X(x_k)$ are independent.

The key to proving Proposition 2.2 is the following lemma about 1-dependent site percolation. By a cluster, we mean a maximal connected component of open vertices.

Lemma 3.2. *Fix $\alpha \in [0,1]$, and let $X \in \{0,1\}^{\mathbb{Z}^d}$ be a translation invariant and 1-dependent site percolation process such that for each $x \in \mathbb{Z}^d$ we have $\mathbb{P}(X(x) = 1) = \alpha$.*

(a) *If $\alpha > 1 - 8^{-9}$, then a.s. X contains an infinite cluster.*
(b) *If $\alpha < (2d-1)^{-3^d}$, then a.s. X contains no infinite cluster.*

One way to prove this result (with the constants $1 - 8^{-9}$ and $(2d-1)^{-3^d}$ inconsequentially replaced by other constants in $(0,1)$) is to invoke the stochastic domination results of Liggett, Schonmann and Stacey [10]. Here we opt, instead, for a simple modification of the classical path and contour counting arguments for proving $p_c \in (0, 1)$ in standard Bernoulli percolation, as outlined, e.g., in the introductory chapter of Grimmett [4]. We begin with part (b) of the lemma.

Proof of Lemma 3.2 (b). For any n, the number of non-selfintersecting paths in the $\mathbb{Z}^d$ lattice from the origin is at most $2d(2d-1)^{n-1}$; this is because the first vertex to go to can be chosen in $2d$ ways, and from then on there are at most $2d-1$ vertices to choose from in each step. In each such path R, we can find a subset S of its vertices that has cardinality at least $n/3^d$ and such that no two $x, y \in S$ are L_∞-neighbors; such a subset can be found by picking vertices in R sequentially and deleting all their L_∞-neighbors. By the definition of 1-dependence, we get that each such R has probability at most $\alpha^{n/3^d}$ of being open, in the sense that all its vertices are open. The expected number of such open paths of length n from the origin is therefore bounded by $2d(2d-1)^{n-1}\alpha^{n/3^d}$, which when $\alpha < (2d-1)^{-3^d}$ tends to 0 as $n \to \infty$. Hence, the origin has probability 0 of being in an infinite cluster, and the same argument applies with the origin replaced by any $x \in \mathbb{Z}^d$, so the existence of an infinite cluster has probability 0. □

Proof of Lemma 3.2 (a). We proceed similarly as in part (b). First note that by restricting to a two-dimensional hyperplane in $\mathbb{Z}^d$, we reduce the problem to only having to consider the case $d = 2$. Define a $*$-path as a sequence of vertices with consecutive vertices at L_∞-distance 1 from each other, and a $*$-circuit as one that ends within L_∞-distance 1 from its starting point. Next, define a *$*$-contour* in $\mathbb{Z}^2$ as a non-selfintersecting $*$-circuit that surrounds the origin, and a *closed $*$-contour* as one whose vertices are all closed. Similarly as in (b), we get that the number of contours of length n is bounded by $n8^n$, and that each of them has probability at most $(1-\alpha)^{n/9}$ of being closed. The expected number of closed contours is thus bounded by

$$\sum_{n=1}^{\infty} n8^n(1-\alpha)^{n/9}$$

which is finite when $\alpha > 1 - 8^{-9}$. For such α the number of closed contours is therefore a.s. finite, whence either the origin itself or some open vertex just outside the "outermost" contour is in an infinite cluster. □

Proposition 2.2 will now be proved using two simple renormalization representations (X and Y below) of Bernoulli bond percolation. Given $N \in \{8, 16, 24, \ldots\}$ and $p \in [0, 1]$, consider a Bernoulli bond percolation process $X \in \{0, 1\}^{E_{\mathbb{Z}^d}}$ with retention parameter p. For $x \in \mathbb{Z}^d$, define the box

$$\Lambda_{N,x} = \{y \in \mathbb{Z}^d : y - Nx \in \Lambda_N\}$$

and define the events $A_{N,x}$ and $B_{N,x}$ analogously to A_N and B_N but pertaining to the edges in $\Lambda_{N,x}$ rather than in Λ_N. Define two site percolation processes $Y, Z \in \{0, 1\}^{\mathbb{Z}^d}$ by setting, for each $x \in \mathbb{Z}^d$,

$$Y(x) = \begin{cases} 1 & \text{if } A_{N,x} \\ 0 & \text{otherwise} \end{cases}$$

and

$$Z(x) = \begin{cases} 1 & \text{if } B_{N,x} \\ 0 & \text{otherwise.} \end{cases}$$

The boxes $\Lambda_{N,x}$ and $\Lambda_{N,y}$ intersect only if $x, y \in \mathbb{Z}^d$ are L_∞-neighbors, whence Y and Z are both 1-dependent percolation processes.

Proof of Proposition 2.2 (a). Assume for contradiction that $p < p_c$ and that $\mathbb{P}_p(B_N) > 1-8^{-9}$. For $x \in \mathbb{Z}^d$ such that $B_{N,x}$ happens, write C_x for the connected component of X whose restriction to $\Lambda_{N,x}$ connects all $2d$ sides of $\Lambda_{N,x}$. Note that if $x, y \in \mathbb{Z}^d$ are L_1-neighbors, then $\{Z(x) = 1\} \cup \{Z(y) = 1\}$ implies that C_x and C_y coincide. By iterating this argument, we get for arbitrary $x, y \in \mathbb{Z}^d$ that if $Z(x) = 1$, $Z(y) = 1$ and x and y are in the same connected component of Z, then C_x and C_y coincide. In particular, if $x \in \mathbb{Z}^d$ satisfies $Z(x) = 1$ and sits in an infinite connected component of Z, then C_x is infinite. But since

$$\begin{aligned} \mathbb{P}(Z(x) = 1) &= \mathbb{P}_p(B_{N,x}) \\ &= \mathbb{P}_p(B_N) > 1 - 8^{-9} \end{aligned}$$

for all x, Lemma 3.2 (a) tells us that Z contains a.s. some infinite cluster. Hence C_x is infinite for some $x \in \mathbb{Z}^d$, so X contains an infinite cluster, contradicting $p < p_c$. □

Proof of Proposition 2.2 (b). Assume for contradiction that $p > p_c$ and that $\mathbb{P}_n(A_N) < (2d-1)^{-3^d}$. If the origin $\mathbf{0}$ is in an infinite cluster of the bond percolation process X, then we must (by the definition of $A_{N,x}$ and Y) have $Y(\mathbf{0}) = 1$, and furthermore that $\mathbf{0}$ belongs to an infinite cluster of Y. But since

$$\begin{aligned} \mathbb{P}(Y(x) = 1) &= \mathbb{P}_p(A_{N,x}) \\ &= \mathbb{P}_p(A_N) < (2d-1)^{-3^d} \end{aligned}$$

for all x, we get by Lemma 3.2 (b) that Y contains a.s. no infinite cluster. Hence $\mathbf{0}$ is a.s. not in an infinite cluster of X, contradicting $p > p_c$. □

References

[1] Antal, P. and Pisztora, A., *On the chemical distance for supercritical Bernoulli percolation*, Ann. Probab. **24** (1996), 1036–1048.

[2] van den Berg, J., *A note on percolation*, J. Phys. A **15** (1982), 605–610.

[3] Blass, A. and Gurevich, Y., *Algorithms: a quest for absolute definitions*, Bull. Europ. Assoc. Theoret. Comp. Sci. **81** (2003), 195–225.

[4] Grimmett, G.R., *Percolation* (2nd ed), Springer, New York, (1999).

[5] Grimmett, G.R. and Marstrand, J.M., *The supercritical phase of percolation is well behaved*, Proc. Royal Soc. A **430** (1990), 439–457.

[6] Häggström, O. and Jonasson, J., *Uniqueness and non-uniqueness in percolation theory*, Probab. Surveys **3** (2006), 289–344.

[7] Kesten, H., *The critical probability of bond percolation on the square lattice equals* $\frac{1}{2}$, Comm. Math. Phys. **74** (1980), 41–59.

[8] Knuth, D.E., *The Art of Computer Programming, Second Edition, Volume* 1 – *Fundamental Algorithms*, Addison-Wesley, New York, (1973).

[9] Leader, I. and Markström, K., *Uncountable families of vertex-transitive finite degree graphs*, Discrete Math. **306** (2006), 678–679.

[10] Liggett, T.M., Schonmann, R.H. and Stacey, A.M., *Domination by product measures*, Ann. Probab. **25** (1997), 71–95.

[11] Meester, R. and Steif, J.E., *Consistent estimation of percolation quantities*, Statist. Neerlandica **52** (1998), 226–238.

[12] Menshikov, M.V., *Coincidence of critical points in percolation problems*, Soviet Math. Dokl. **33** (1986), 856–859.

Olle Häggström
Dept. of Mathematics
Chalmers University of Technology
412 96 Göteborg
Sweden
http://www.math.chalmers.se/~olleh/
e-mail: `olleh@math.chalmers.se`

Progress in Probability, Vol. 60, 331–364

Chip-Firing and Rotor-Routing on Directed Graphs

Alexander E. Holroyd, Lionel Levine, Karola Mészáros, Yuval Peres, James Propp and David B. Wilson

Abstract. We give a rigorous and self-contained survey of the abelian sandpile model and rotor-router model on finite directed graphs, highlighting the connections between them. We present several intriguing open problems.

Mathematics Subject Classification (2000). Primary: 82C20; secondary: 20K01, 05C25.

Keywords. Abelian sandpile model, rotor-router model, chip firing, Eulerian walkers.

1. Introduction

The abelian sandpile and rotor-router models were discovered several times by researchers in different communities operating independently. The abelian sandpile model was invented by Dhar [Dha90] as a test-bed for the concept of self-organized criticality introduced in [BTW87]. Related ideas were explored earlier by Engel [Eng75, Eng76] in the form of a pedagogical tool (the "probabilistic abacus"), by Spencer [Spe87, pp. 32–35], and by Lorenzini [Lor89, Lor91] in connection with arithmetic geometry. The rotor-router model was first introduced by Priezzhev *et al.* [PDDK96] (under the name "Eulerian walkers model") in connection with self-organized criticality. It was rediscovered several times: by Rabani, Sinclair and Wanka [RSW98] as an approach to load-balancing in multiprocessor systems, by Propp [Pro01] as a way to derandomize models such as internal diffusion-limited aggregation (IDLA) [DF91, LBG92], and by Dumitriu, Tetali, and Winkler as part of their analysis of a graph-based game [DTW03]. Articles on the chip-firing game in the mathematical literature include [Big99, Big97, BLS91, BL92]. Those on the

---

A.E. H. was funded in part by an NSERC discovery grant and Microsoft Research.
L. L. was supported by an NSF Graduate Research Fellowship.
J. P. was supported by an NSF research grant.

rotor-router model include [Lev02, LP05, HP08, LP07a, LP07b]. Below we briefly describe the two models, deferring the formal definitions to later sections.

The **abelian sandpile model** on a directed graph G, also called the **chip-firing game**, starts with a collection of chips at each vertex of G. If a vertex v has at least as many chips as outgoing edges, it can **fire**, sending one chip along each outgoing edge to a neighboring vertex. After firing a sequence of vertices in turn, the process stops when each vertex with positive out-degree has fewer chips than out-going edges. The order of firings does not affect the final configuration, a fact we shall discuss in more detail in Section 2.

To define the **rotor-router model** on a directed graph G, for each vertex of G, fix a cyclic ordering of the outgoing edges. To each vertex v we associate a **rotor** $\rho(v)$ chosen from among the outgoing edges from v. A chip performs a walk on G according to the **rotor-router rule**: if the chip is at v, we first increment the rotor $\rho(v)$ to its successor $e = (v, w)$ in the cyclic ordering of outgoing edges from v, and then route the chip along e to w. If the chip ever reaches a **sink**, i.e., a vertex of G with no outgoing edges, the chip will stop there; otherwise, the chip continues walking forever.

A common generalization of the rotor-router and chip-firing models, the **height arrow model**, was proposed in [PDDK96] and studied in [DR04].

We develop the basic theory of the abelian sandpile model in section 2 and define the main algebraic object associated with it, the sandpile group of G [Dha90] (also called the "critical group" by some authors, e.g., [Big99, Wag00]). Furthermore, we establish the basic results about **recurrent** chip configurations, which play an important role in the theory. In Section 3 we define a notion of recurrent configurations for the rotor-router model on directed graphs and give a characterization for them in terms of oriented spanning trees of G. The sandpile group acts naturally on recurrent rotor configurations, and this action is both transitive and free. We deduce appealing proofs of two basic results of algebraic graph theory, namely the Matrix-Tree Theorem [Sta99, 5.6.8] and the enumeration of Eulerian tours in terms of oriented spanning trees [Sta99, Cor. 5.6.7]. We also derive a family of bijections between the recurrent chip configurations of G and the recurrent rotor configurations of G. Such bijections have been constructed before, for example in [BW97]; however, our presentation differs significantly from the previous ones. Section 4 establishes stronger results for both models on Eulerian digraphs and undirected graphs. In Section 5 we present an alternative view of the rotor-router model in terms of "cycle-popping," borrowing an idea from Wilson's work on loop-erased random walk; see [PW98]. We conclude in Section 6 by presenting some open questions.

2. Chip-firing

In a finite directed graph (**digraph**) $G = (V, E)$, a directed edge $e \in E$ points from the vertex tail(e) to the vertex head(e). We allow self-loops (head(e) = tail(e)) as well as multiple edges (head(e) = head(e') and tail(e) = tail(e')) in G. The **out-degree** outdeg(v) of a vertex v (also denoted by d_v) is the number of edges e with tail(e) = v, and the **in-degree** indeg(v) of v is the number of edges e with head(e) = v. A vertex is a **sink** if its out-degree is zero. A **global sink** is a sink s such that from every other vertex there is a directed path leading to s. Note that if there is a global sink, then it is the unique sink.

If G has the same number of edges from v to w as from w to v for all vertices $v \neq w$ then we call G **bidirected**. In particular, a bidirected graph is obtained by replacing each edge of an undirected graph with a pair of directed edges, one in each direction.

Label the vertices of G as $v_1, v_2, \ldots, v_n$. The **adjacency matrix** A of G is the $n \times n$ matrix whose (i, j)-entry is the number of edges from v_i to v_j, which we denote by a_{v_i,v_j} or a_{ij}. The (graph) **Laplacian** of G is the $n \times n$ matrix $\Delta = D - A$, where D is the diagonal matrix whose (i, i)-entry is the out-degree of v_i. That is,

$$\Delta_{ij} = \begin{cases} -a_{ij} & \text{for } i \neq j, \\ d_i - a_{ii} & \text{for } i = j. \end{cases}$$

Note that the entries in each row of Δ sum to zero. If the vertex v_i is a sink, then the ith row of Δ is zero.

A **chip configuration** σ on G, also called a **sandpile** on G, is a vector of non-negative integers indexed by the non-sink vertices of G, where $\sigma(v)$ represents the number of chips at vertex v. A chip configuration σ is **stable** if $\sigma(v) < d_v$ for every non-sink vertex v. We call a vertex v **active** in σ if v is not a sink and $\sigma(v) \geq d_v$. An active vertex v can **fire**, resulting in a new chip configuration σ' obtained by moving one chip along each of the d_v edges emanating from v; that is, $\sigma'(w) = \sigma(w) + a_{vw}$ for all $w \neq v$ and $\sigma'(v) = \sigma(v) - d_v + a_{vv}$. We call the configuration σ' a **successor** of σ.

By performing a sequence of firings, we may eventually arrive at a stable chip configuration, or we might continue firing forever, as the following examples show.

Example 2.1. Consider the complete directed graph on three vertices (without self-loops). Then placing three chips at a vertex gives a configuration that stabilizes in one move, while placing four chips at a vertex gives a configuration that never stabilizes (see Figure 1).

It might appear that the choice of the order in which we fire vertices could affect the long-term behavior of the system; however, this is not the case, as the following lemma shows (and Figure 2 illustrates).

Lemma 2.2 ([Dha90], [DF91]). *Let G be any digraph, let $\sigma_0, \sigma_1, \ldots, \sigma_n$ be a sequence of chip configurations on G, each of which is a successor of the one before, and let $\sigma'_0, \sigma'_1, \ldots, \sigma'_m$ be another such sequence with $\sigma'_0 = \sigma_0$.*

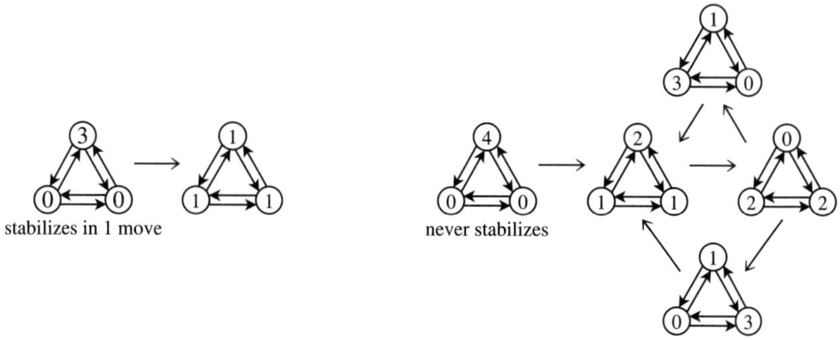

FIGURE 1. Some chip configurations eventually stabilize, while others never stabilize.

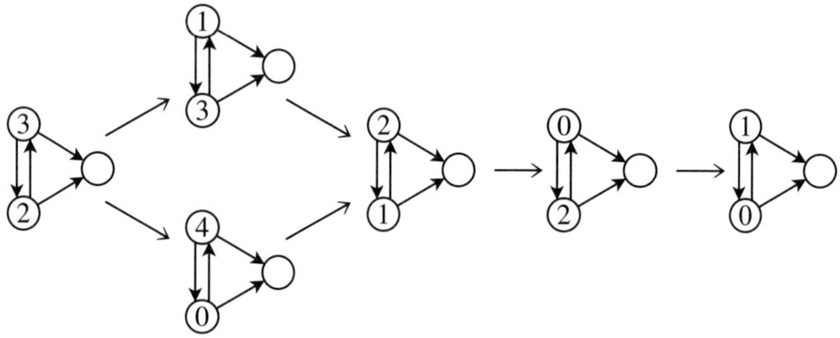

FIGURE 2. Commutation of the chip-firing operations.

1. *If σ_n is stable, then $m \leq n$, and moreover, no vertex fires more times in $\sigma'_0, \ldots, \sigma'_m$ than in $\sigma_0, \ldots, \sigma_n$.*
2. *If σ_n and σ_m are both stable, then $m = n$, $\sigma_n = \sigma'_n$, and each vertex fires the same number of times in both histories.*

Proof. Part 2 is an immediate corollary of part 1, which we now prove. If part 1 fails, then consider a counterexample with $m+n$ minimal. Let v_i be the vertex that fires when σ_{i-1} becomes σ_i, and v'_i be the vertex that fires when σ'_{i-1} becomes σ'_i. Vertex v'_1 must be fired at some stage (say the ith) in the sequence of configurations $\sigma_0, \ldots, \sigma_n$, since σ_n is stable; say $v_i = v'_1$. Then $v_i, v_1, v_2, \ldots, v_{i-1}, v_{i+1}, \ldots, v_n$ is a permissible firing sequence that turns σ_0 into σ_n with the same number of firings at each site as the unpermuted sequence. The firing sequences $v_1, \ldots, v_{i-1}, v_{i+1}, \ldots, v_n$ and $v'_2, v'_3, \ldots, v'_m$ then constitute a smaller counterexample to the lemma (with initial configuration σ'_1), contradicting minimality. □

Definition 2.3. Starting from a configuration σ, Lemma 2.2 shows that there is at most one stable configuration that can be reached by a finite sequence of firings (and that if such a configuration exists, then no infinite sequence of firings is possible). If such a stable configuration exists we denote it σ° and call it the **stabilization** of σ.

Thus far, the presence or absence of sinks was irrelevant for our claims. For the rest of this section, we assume that the digraph G has a global sink s.

Lemma 2.4. *If digraph G has a global sink, then every chip configuration on G stabilizes.*

Proof. Let N be the number of chips in the configuration. Given a vertex v of G, let $v_0, v_1, \ldots, v_{r-1}, v_r$ be a directed path from $v_0 = v$ to $v_r = s$. Every time v_{r-1} fires, it sends a chip to the sink which remains there forever. Thus v_{r-1} can fire at most N times. Every time v_{r-2} fires, it sends a chip to v_{r-1}, and $d_{v_{r-1}}$ such chips will cause v_{r-1} to fire once, so v_{r-2} fires at most $d_{v_{r-1}}N$ times. Iterating backward along the path, we see that v fires at most $d_{v_1} \cdots d_{v_{r-1}}N$ times. Thus each vertex can fire only finitely many times, so by Lemma 2.2 the configuration stabilizes. □

We remark that when G is connected and the sink is the only vertex with in-degree exceeding its out-degree, the bound one gets from the above argument on the total number of firings is far from optimal; see Theorem 4.8 for a better bound.

Define the **chip addition operator** E_v as the map on chip configurations that adds a single chip at vertex v and then lets the system stabilize. In symbols,

$$E_v\sigma = (\sigma + \mathbf{1}_v)^\circ$$

where $\mathbf{1}_v$ is the configuration consisting of a single chip at v.

Lemma 2.5. *On any digraph with a global sink, the chip addition operators commute.*

Proof. Given a chip configuration σ and two vertices v and w, whatever vertices are active in $\sigma + \mathbf{1}_v$ are also active in configuration $\sigma' = \sigma + \mathbf{1}_v + \mathbf{1}_w$. Applying to σ' a sequence of firings that stabilizes $\sigma + \mathbf{1}_v$, we obtain the configuration $E_v\sigma + \mathbf{1}_w$. Stabilizing this latter configuration yields $E_wE_v\sigma$. Thus $E_wE_v\sigma$ is a stabilization of σ'. Interchanging the roles of v and w, the configuration $E_vE_w\sigma$ is also a stabilization of σ'. From Lemma 2.2 we conclude that $E_wE_v\sigma = E_vE_w\sigma$. □

Lemma 2.5 is called the **abelian property**; it justifies Dhar's coinage "abelian sandpile model". From the above proof we also deduce the following.

Corollary 2.6. *Applying a sequence of chip addition operators to σ yields the same result as adding all the associated chips simultaneously and then stabilizing.*

Let G be a digraph on n vertices with global sink s. The **reduced Laplacian** Δ' of G is obtained by deleting from the Laplacian matrix Δ the row and column corresponding to the sink. Note that firing a non-sink vertex v transforms a chip configuration σ into the configuration $\sigma - \Delta'_v$, where Δ'_v is the row of the reduced Laplacian corresponding to v. Since we want to view the configurations before and after firing as equivalent, we are led to consider the group quotient $\mathbb{Z}^{n-1}/H$, where $H = \mathbb{Z}^{n-1}\Delta'$ is the integer row-span of Δ'.

Definition 2.7. Let G be a digraph on n vertices with global sink s. The **sandpile group** of G is the group quotient

$$\mathcal{S}(G) = \mathbb{Z}^{n-1}/\mathbb{Z}^{n-1}\Delta'(G).$$

The connection between the sandpile group and the dynamics of sandpiles on G is made explicit in Corollary 2.16. For the graph in Figure 3, the sandpile group is the cyclic group of order 3. The group structure of $\mathcal{S}(G)$ when G is a tree is investigated in [Lev07].

Lemma 2.8. *The order of $\mathcal{S}(G)$ is the determinant of the reduced Laplacian $\Delta'(G)$.*

Proof. The order of $\mathcal{S}(G)$ equals the index of the lattice $H = \mathbb{Z}^{n-1}\Delta'$ in $\mathbb{Z}^{n-1}$, and, recalling that the volume of a parallelepiped is the determinant of the matrix formed from its edge-vectors, we deduce that this in turn equals the determinant of Δ'. □

Lemma 2.9. *Let G be a digraph with a global sink. Every equivalence class of $\mathbb{Z}^{n-1}$ modulo $\Delta'(G)$ contains at least one stable chip configuration of G.*

Proof. Let δ be the configuration given by $\delta(v) = d_v$ for all v, and let δ° be its stabilization. Then $\delta^\circ(v) < d_v$ for all $v \neq s$, so $\delta - \delta^\circ$ is a positive vector equivalent to the zero configuration. Given any $\alpha \in \mathbb{Z}^{n-1}$, let m denote the minimum of all the coordinates of α together with 0 (so that $m \leq 0$). Then the vector

$$\beta = \alpha + (-m)(\delta - \delta^\circ)$$

is nonnegative and equivalent to α. Hence β° is a stable chip configuration in the same equivalence class as α. □

Example 2.10. An equivalence class may contain more than one stable chip configuration. For example, consider the complete directed graph on three vertices, with one of the vertices made into a sink by deletion of its two outgoing edges (see Figure 3). It is easy to see that there are two stable configurations in the equivalence class of the identity: the configuration in which each of the two non-sink vertices has 0 chips and the configuration in which each of the two vertices has 1 chip. It might seem natural that, if either of these two configurations is to be preferred as a representative of the identity element in the sandpile group, it should be the former. However, this instinct is misleading, as we now explain.

Definition 2.11. A chip configuration σ is **accessible** if from any other chip configuration it is possible to obtain σ by a combination of adding chips and selectively firing active vertices. A chip configuration that is both stable and accessible is called **recurrent**.

FIGURE 3. Two stable chip configurations in the equivalence class of the identity.

Remark 2.12. There are several definitions of "recurrent" that are used in the literature. Lemma 2.17 below shows that these definitions (including the one above) are equivalent for any digraph with a global sink.

We will see shortly (in Lemmas 2.13 and 2.15) that each equivalence class in $\mathbb{Z}^{n-1}/\mathbb{Z}^{n-1}\Delta'$ contains a unique recurrent chip configuration. It is customary to represent each element of the sandpile group by its unique recurrent element. In Example 2.10 the all-1 configuration is accessible, but the all-0 configuration is not. Therefore the all-1 configuration is taken as the canonical representative of the identity. (However, in the context of cluster-firing and superstabilization as described in Definition 4.3 and Lemma 4.6, the all-0 configuration will be the preferred representative.)

Lemma 2.13. *Let G be a digraph with a global sink. Every equivalence class of $\mathbb{Z}^{n-1}$ modulo $\Delta'(G)$ contains at least one recurrent chip configuration of G.*

Proof. Given $\alpha \in \mathbb{Z}^{n-1}$, let m denote the minimum of all the coordinates of α together with 0, so that $m \leq 0$. Write $d_{\max}$ for the maximum out-degree of a vertex in G. Then α is equivalent to the configuration

$$\beta = \alpha + [d_{\max} + (-m)](\delta - \delta^\circ)$$

with δ as in the proof of Lemma 2.9. Since $\delta - \delta^\circ$ has all entries positive, we have $\beta \geq [d_{\max} + (-m)](\delta - \delta^\circ) \geq d_{\max}(\delta - \delta^\circ) \geq d_{\max}$. (Inequalities between two vectors or between a vector and a scalar are interpreted componentwise.) In particular, β is accessible, since any chip configuration can first be stabilized, so that each vertex has fewer than $d_{\max}$ chips, and then supplemented with extra chips to obtain β. Therefore any configuration obtained from β by firing is also accessible. In particular the stabilization β° is thus recurrent and equivalent to α. □

Next we will show that every equivalence class modulo Δ' contains at most one recurrent configuration, making use of the following lemma.

Lemma 2.14. *Let $\epsilon = (2\delta) - (2\delta)^\circ$, where δ is given by $\delta(v) = d_v$ as before. If σ is recurrent, then $(\sigma + \epsilon)^\circ = \sigma$.*

Proof. If σ is recurrent then it is accessible, so it can be reached from δ by adding some (non-negative) configuration ζ and selectively firing. But since σ is also stable this implies that $(\zeta + \delta)^\circ = \sigma$. Consider the configuration

$$\gamma = (\zeta + \delta) + \epsilon = 2\delta + \zeta + \delta - (2\delta)^\circ.$$

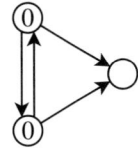

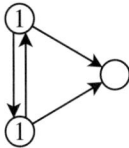

Since $\epsilon \geq 0$, we may start from γ and fire a sequence of vertices that stabilizes $\zeta + \delta$, to obtain the configuration $\sigma + \epsilon$. On the other hand, since $\delta - (2\delta)^\circ \geq 0$ we may start from γ and fire a sequence of vertices that stabilizes 2δ, to obtain the configuration $(2\delta)^\circ + \zeta + \delta - (2\delta)^\circ = \zeta + \delta$, which in turn stabilizes to σ. By Lemma 2.2 it follows that $(\sigma + \epsilon)^\circ = \sigma$. □

Lemma 2.15. *Let G be a digraph with a global sink. Every equivalence class of $\mathbb{Z}^{n-1}$ modulo $\Delta'(G)$ contains at most one recurrent chip configuration of G.*

Proof. Let σ_1 and σ_2 be recurrent and equivalent mod Δ'. Label the non-sink vertices $v_1, \ldots, v_{n-1}$. Then $\sigma_1 = \sigma_2 + \sum_{i \in J} c_i \Delta'_i$, where the c_i are nonzero constants, Δ'_i is the row of the reduced Laplacian Δ' corresponding to v_i, and the index i runs over some subset J of the integers $1, \ldots, n-1$. Write $J = J_- \cup J_+$, where $J_- = \{i : c_i < 0\}$ and $J_+ = \{i : c_i > 0\}$, and let

$$\sigma = \sigma_1 + \sum_{i \in J_-} (-c_i)\Delta'_i = \sigma_2 + \sum_{i \in J_+} c_i \Delta'_i.$$

Let ϵ denote the everywhere-positive chip configuration defined in Lemma 2.14. Take k large enough so that $\sigma' = \sigma + k\epsilon$ satisfies $\sigma'(v_i) \geq |c_i| d_{v_i}$ for all i. Starting from σ', we may fire each vertex v_i for $i \in J_-$ a total of $-c_i$ times, and each of the intermediate configurations is a valid chip configuration because all the entries are nonnegative. The resulting configuration $\sigma_1 + k\epsilon$ then stabilizes to σ_1 by Lemma 2.14. Likewise, starting from σ' we may fire each vertex v_i for $i \in J_+$ a total of c_i times to obtain $\sigma_2 + k\epsilon$, which stabilizes to σ_2. By Lemma 2.2 it follows that $\sigma_1 = \sigma_2$. □

Corollary 2.16. *Let G be a digraph with a global sink. The set of all recurrent chip configurations on G is an abelian group under the operation $(\sigma, \sigma') \mapsto (\sigma + \sigma')^\circ$, and it is isomorphic via the inclusion map to the sandpile group $\mathcal{S}(G)$.*

Proof. Immediate from Lemmas 2.13 and 2.15. □

In view of this isomorphism, we will henceforth use the term "sandpile group" to refer to the group of recurrent configurations.

It is of interest to consider the **identity element** I of the sandpile group, i.e., the unique recurrent configuration equivalent to the all-0 configuration. Here is one method to compute I. Let σ be the configuration $2\delta - 2$. (Arithmetic combinations of vectors and scalars are to interpreted componentwise.) Since $\sigma^\circ \leq \delta - 1$ we have $\sigma - \sigma^\circ \geq \delta - 1$, so $\sigma - \sigma^\circ$ is accessible. Since $\sigma - \sigma^\circ$ is equivalent to 0, the identity element is given by $I = (\sigma - \sigma^\circ)^\circ$.

Figure 4 shows identity elements for the $L \times L$ square grid with "wired boundary," for several values of L. (To be more precise, the graph G is obtained by replacing each edge of the undirected square grid with a pair of directed edges, and adjoining a sink vertex s along with two edges from each of the four corner vertices to s and one edge from each of the other boundary vertices to s.) The identity element of this graph was studied by Le Borgne and Rossin [BR02], but most basic properties of this configuration, such as the existence of the large square in the center, remain unproved.

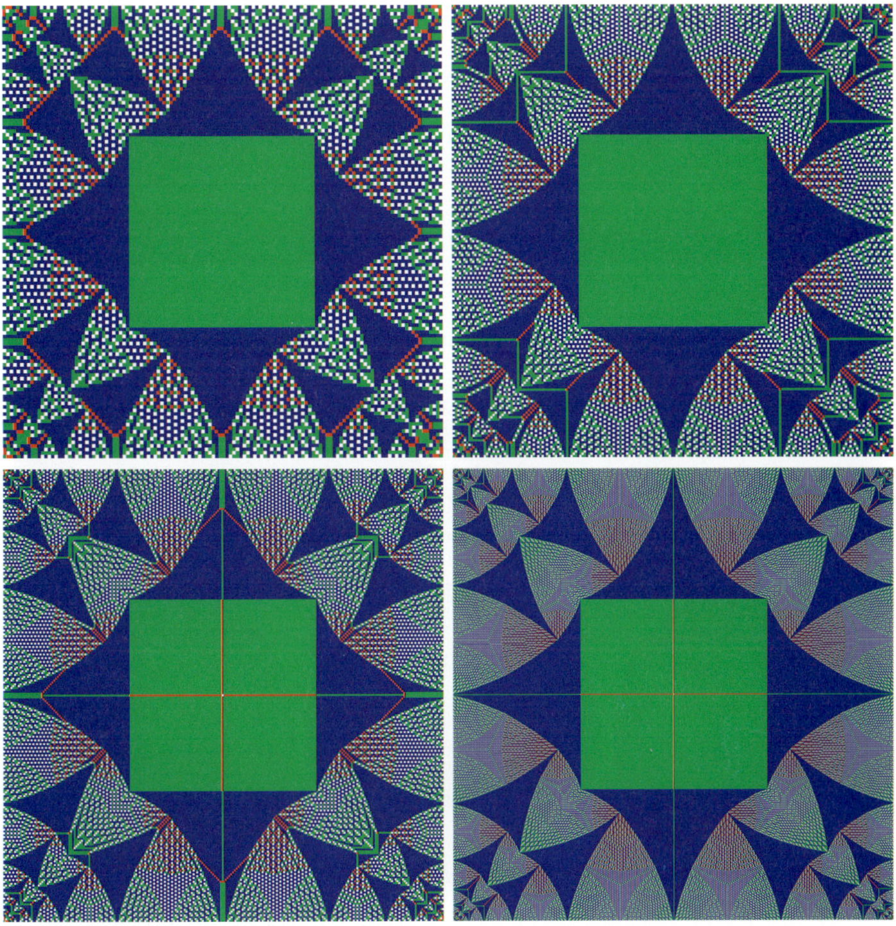

FIGURE 4. The identity element of the sandpile group of the $L \times L$ square grid for different values of L, namely $L = 128$ (upper left), 198 (upper right), 243 (lower left), and 521 (lower right). The color scheme is as follows: orange=0 chips, red=1 chip, green=2 chips, and blue=3 chips.

Figure 5 shows another example, the identity element for the 100×100 directed torus. (That is, for each vertex $(i, j) \in \mathbb{Z}/100\mathbb{Z} \times \mathbb{Z}/100\mathbb{Z}$, there are directed edges from (i, j) to $(i+1 \bmod 100, j)$ and to $(i, j+1 \bmod 100)$, and we make $(0, 0)$ (the lower-left vertex) into a sink by deleting its two outgoing edges.)

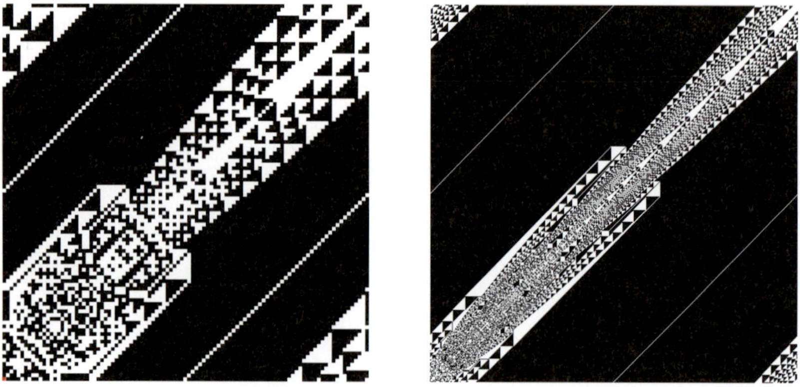

FIGURE 5. The identity element of the sandpile group of the 100×100 directed torus (left) and the 500×500 directed torus (right). The color scheme is as follows: white=0 chips, black=1 chip, and the sink, which is at the lower-left corner, is shown in red.

Figure 6 shows a third example, the identity element for a disk-shaped region of $\mathbb{Z}^2$ with wired boundary. Examples of identity elements for graphs formed from portions of lattices other than the square grid can be found in [LKG90].

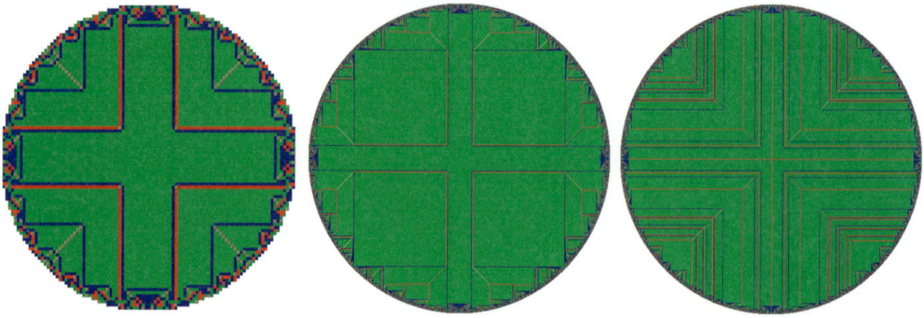

FIGURE 6. The identity element of the sandpile group of disk-shaped regions of diameter 100 (left), 512 (middle), and 521 (right). The color scheme is as follows: orange=0 chips, red=1 chip, green=2 chips, and blue=3 chips.

Also of interest is the **inverse** of a recurrent chip configuration σ (that is, the recurrent chip configuration $\bar{\sigma}$ such that $(\sigma + \bar{\sigma})^\circ = I$). One way to compute the inverse is via $\bar{\sigma} = (\zeta - \zeta^\circ - \sigma)^\circ$, where ζ is any chip configuration satisfying $\zeta \geq 3\delta - 3$. (Here $\zeta - \zeta^\circ - \sigma$ is accessible, since it has at least $d_v - 1$ chips at each vertex v.)

Given two chip configurations σ and ζ, we say that σ is **reachable** from ζ (via excitation-relaxation operations) if there exists a configuration β such that

$\sigma = (\zeta+\beta)^\circ$. Note that this implies that σ is stable. A digraph is **strongly connected** if for any two distinct vertices v, w there are directed paths from v to w and from w to v. We write $G \setminus s$ for the graph obtained from G by deleting the vertex s along with all edges incident to s.

Lemma 2.17. *Let G be a digraph with a global sink s, and let σ be a chip configuration on G. The following are equivalent.*

(1) *σ is recurrent; that is, σ is reachable from any configuration ζ.*
(2) *If ζ is any configuration reachable from σ, then σ is reachable from ζ.*
(3) *σ is reachable from any configuration of the form $E_v\sigma$, where v is a non-sink vertex of G.*
(4) *Each strongly connected component of $G \setminus s$ contains a vertex v such that σ is reachable from $E_v\sigma$.*

Proof. Since trivially (1) $\Rightarrow$ (2) $\Rightarrow$ (3) $\Rightarrow$ (4), it suffices to show (4) $\Rightarrow$ (1).

If (4) holds, there is a chip configuration α such that $(\sigma+\alpha)^\circ = \sigma$ and α is nonzero on at least one vertex of each strongly connected component of G. There exists a positive integer k such that selective firing from $k\alpha$ results in a chip configuration β with at least one chip at each vertex. Moreover, $(\sigma+\beta)^\circ = \sigma$.

Now let ζ be any chip configuration. Since β has at least one chip at each vertex, we have $\zeta \le \sigma + \ell\beta$ for some integer ℓ. Thus we may add chips to ζ and then stabilize to obtain the configuration $(\sigma+\ell\beta)^\circ = \sigma$. Hence σ is recurrent. □

We also note that, for a digraph G with a global sink, the sandpile group is isomorphic to the additive group of harmonic functions modulo 1 on G that vanish on the sink [Sol99]. A function $f : V(G) \to [0,1)$ is **harmonic modulo 1** if $d_v f(v) = \sum_w a_{v,w} f(w) \mod 1$ for all vertices v. For a sandpile configuration σ, the associated harmonic function f is the solution of

$$\sum_w \Delta'_{v,w} f(w) = \sigma(v).$$

For the graph in Figure 3, the three harmonic functions are $(f(v_1), f(v_2)) = (0,0)$, $(f(v_1), f(v_2)) = (1/3, 2/3)$, and $(f(v_1), f(v_2)) = (2/3, 1/3)$.

We conclude this section by pointing out a link between the sandpile group and spanning trees. By Lemma 2.8 the order of the sandpile group of G equals the determinant of the reduced Laplacian Δ' of G. By the **matrix-tree theorem** [Sta99, 5.6.8], this determinant equals the number of **oriented spanning trees** of G rooted at the sink (that is, acyclic subgraphs of G in which every non-sink vertex has out-degree 1). Various bijections have been given for this correspondence; see, for example, [BW97]. In Section 3 we will use the rotor-router model to describe a particularly natural bijection, and deduce the matrix-tree theorem as a corollary.

3. Rotor-routing

Chip-firing is a way of routing chips through a directed graph G in such a fashion that the chips emitted by any vertex v travel in equal numbers along each of the outgoing edges. In order to ensure this equality, however, chips must wait at a vertex v until sufficiently many additional chips have arrived to render v active. Rotor-routing is an alternative approach to distributing chips through G which dispenses with this waiting step. Since we cannot ensure exact equality without waiting, we settle for the condition that the chips emitted by any vertex v travel in *nearly* equal numbers along each of the edges emanating from v. We ensure that this near-equality holds by using a rotor mechanism to decide where each successive chip emitted from a vertex v should be routed.

Given a directed graph G, fix for each vertex v a cyclic ordering of the edges emanating from v. For an edge e with tail v we denote by e^+ the next edge after e in the prescribed cyclic ordering of the edges emanating from v.

Definition 3.1. A **rotor configuration** is a function ρ that assigns to each non-sink vertex v of G an edge $\rho(v)$ emanating from v. If there is a chip at a non-sink vertex v of G, **routing the chip** at v (for one step) consists of updating the rotor configuration so that $\rho(v)$ is replaced with $\rho(v)^+$, and then moving the chip to the head of $\rho(v)^+$. A **single-chip-and-rotor state** is a pair consisting of a vertex w (which represents the location of the chip) and a rotor configuration ρ. The **rotor-router operation** is the map that sends a single-chip-and-rotor state (w, ρ) (where w is not a sink) to the state (w^+, ρ^+) obtained by routing the chip at w for one step. (See Figure 7 for examples of the rotor-router operation.)

As we will see, there is an important link between chip-firing and rotor-routing. A hint at this link comes from a straightforward count of configurations. Recall that a stable chip configuration is a way of assigning some number of chips between 0 and $d_v - 1$ to each non-sink vertex v of G. Thus, the number of stable configurations is exactly $\prod_v d_v$, where the product runs over all non-sink vertices. This is also the number of rotor configurations on G. Other connections become apparent when one explores the appropriate notion of recurrent states for the rotor-router model. We will treat two cases separately: digraphs with no sink, and digraphs with a global sink (Lemma 3.6 applies to both settings).

Definition 3.2. Let G be a **sink-free** digraph, i.e., one in which each vertex has at least one outgoing edge. Starting from the state (w, ρ), if iterating the rotor-router operation eventually leads back to (w, ρ) we say that (w, ρ) is **recurrent**; otherwise, it is **transient**.

Our first goal is to give a combinatorial characterization of the recurrent states, Theorem 3.8. We define a **unicycle** to be a single-chip-and-rotor state (w, ρ) for which the set of edges $\{\rho(v)\}$ contains a unique directed cycle, and w lies on this cycle. (Equivalently, ρ is a connected functional digraph, and w is a vertex on the unique cycle in ρ.) The following lemma shows that the rotor-router operation takes unicycles to unicycles.

Lemma 3.3. *Let G be a sink-free digraph. If (w,ρ) is a unicycle on G, then (w^+,ρ^+) is also a unicycle.*

Proof. Since (w,ρ) is a unicycle, the set of edges $\{\rho(v)\}_{v\neq w} = \{\rho^+(v)\}_{v\neq w}$ contains no directed cycles. The set of edges $\{\rho^+(v)\}$ forms a subgraph of G in which every vertex has out-degree one, so it contains a directed cycle. Since any such cycle must contain the edge $\rho^+(w) = \rho(w)^+$, this cycle is unique, and w^+ lies on it. □

Lemma 3.4. *Let G be a sink-free digraph. The rotor-router operation is a permutation on the set of unicycles of G.*

Proof. Since the set of unicycles is finite, by Lemma 3.3 it is enough to show surjectivity. Given a unicycle $U = (w,\rho)$, let $U^- = (w^-,\rho^-)$ be the state obtained by moving the chip from w to its predecessor w^- in the unique cycle through w, and replacing the rotor at w^- with its predecessor in the cyclic ordering of outgoing edges from w^-. Then the rotor-router operation applied to U^- yields U. It remains to show that U^- is a unicycle; for this it suffices to show that every directed cycle in ρ^- passes through w^-. Suppose that there is a directed cycle of rotors in ρ^- which avoids w^-. Since ρ^- agrees with ρ except at w^-, this same directed occurs within ρ and avoids w^-, a contradiction since w^- is on ρ's unique cycle. □

Corollary 3.5. *Let G be a sink-free digraph. If (w,ρ) is a unicycle on G, then (w,ρ) is recurrent.*

In Lemma 3.7, below, we show that the converse holds when G is strongly connected. We will need the following lemma, which is analogous to Lemma 2.4 for the abelian sandpile. A vertex w is **globally reachable** if for each other vertex v there is a directed path from v to w.

Lemma 3.6. *Let G be a digraph with a globally reachable vertex w. For any starting vertex and rotor configuration, iterating the rotor-router operation a suitable number of times yields a state in which the chip is at w.*

Proof. Since w is globally reachable, either G is sink-free or w is the unique sink. Thus either we can iterate the rotor-router operation indefinitely, or the chip eventually visits w. In the former case, since G is finite, the chip visits some vertex v infinitely often. But if x is a vertex that is visited infinitely often and there is an edge from x to y, then y is also visited infinitely often. Inducting along a path from v to w, we conclude that the chip eventually visits w. □

Lemma 3.7. *Let G be a strongly connected digraph. If (w,ρ) is a recurrent single-chip-and-rotor state on G, then it is a unicycle.*

Proof. Since G is strongly connected, every vertex is globally reachable. Hence by Lemma 3.6, if we start from any initial state and iterate the rotor-router rule sufficiently many times, the chip visits every vertex of G.

Suppose (w,ρ) is a recurrent state. Once every vertex has been visited and we return to the state (w,ρ), suppose the rotors at vertices $v_1,\dots,v_k$ form a directed cycle. If w does not lie on this cycle, then for each i, the last time the chip was at v_i it moved to v_{i+1}, and hence the edge from v_i to v_{i+1} was traversed more recently than the edge from v_{i-1} to v_i. Carrying this argument around the cycle leads to a contradiction. Thus, every directed cycle in the rotor configuration must pass through w. But now if we start from w and follow the rotors, the first vertex we revisit must be w. Hence (w, ρ) is a unicycle. □

Combining Corollary 3.5 and Lemma 3.7, we have proved the following.

Theorem 3.8. *Let G be a strongly connected digraph. Then (w, ρ) is a recurrent single-chip-and-rotor state on G if and only if it is a unicycle.*

Next we consider the case when G is a digraph with a global sink. Note that we cannot apply the rotor-router operation to states in which the chip is at the sink. We call these **absorbing states**. For any starting state, if we iterate the rotor-router operation sufficiently many times, the chip must eventually arrive at the sink by Lemma 3.6.

A **chip-and-rotor state** is a pair $\tau = (\sigma, \rho)$ consisting of a chip configuration σ and rotor configuration ρ on G. A non-sink vertex is **active** in τ if it has at least one chip. If v is active, then **firing** v results in a new chip-and-rotor state given by replacing the rotor $\rho(v)$ with $\rho(v)^+$ and moving a single chip from v to the head of $\rho(v)^+$ (and removing the chip if $\rho(v)^+$ is a sink). We say that τ' is a **successor** of τ if it is obtained from τ by firing an active vertex. We say that τ is **stable** if no vertex can fire, i.e., all chips have moved to a sink and disappeared. The rotor-router operation has the following abelian property analogous to Lemma 2.2.

Lemma 3.9. *Let G be a digraph with a global sink. Let $\tau_0, \tau_1, \dots, \tau_n$ be a sequence of chip-and-rotor states of G, each of which is a successor of the one before. If $\tau_0, \tau'_1, \dots, \tau'_m$ is another such sequence, and τ_n is stable, then $m \le n$. If in addition τ'_m is stable, then $m = n$ and $\tau_n = \tau'_n$, and for each vertex w, the number of times w fires is the same for both histories.*

Proof. Let v_i and v'_i be the vertices that are fired in τ_{i-1} and τ'_{i-1} to obtain τ_i and τ'_i, respectively. We will show that if τ_n is stable and the sequences v and v' agree in the first $i-1$ terms for some $i \le m$, then some permutation of v agrees with v' in the first i terms. Since v'_i is active in $\tau_{i-1} = \tau'_{i-1}$, it must be active in $\tau_i, \tau_{i+1}, \dots,$ until it is fired. Since τ_n is stable, it follows that $v_j = v'_i$ for some $j > i$. Let j be the minimal such index. Starting from τ_0, the vertices $v_1, v_2, \dots, v_{i-1}, v_j, v_i, v_{i+1}, \dots, v_{j-1}, v_{j+1}, \dots, v_n$ can be fired in that order, resulting in the same stable configuration τ_n. Moreover, this sequence agrees with v' in the first i terms.

By induction, it follows that the sequence v' is an initial subsequence of a permutation of v. In particular, $m \le n$. If τ'_m is also stable, by interchanging the roles of τ and τ', we obtain that v' is a permutation of v. □

Given a non-sink vertex v in G, the **chip addition operator** E_v is the map on rotor configurations given by adding a chip at vertex v and iterating the rotor-

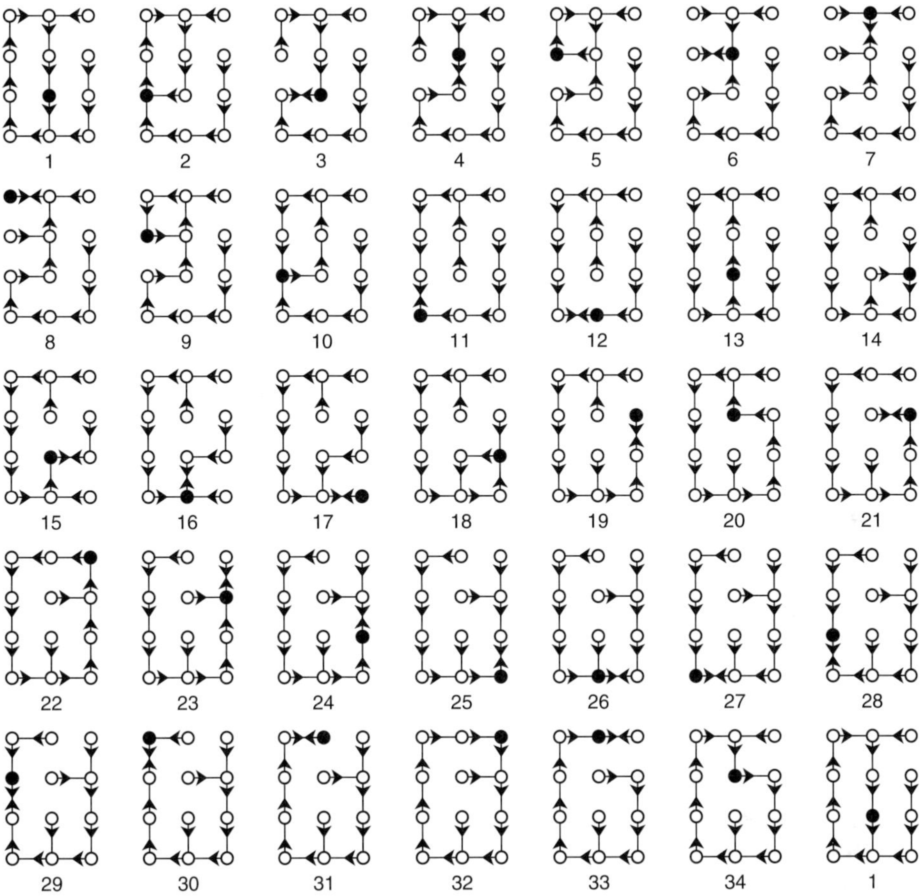

FIGURE 7. The unicycle configurations resulting from the evolution of a particular unicycle on the bidirected 3-by-4 rectangular grid. By Lemma 4.9, the chip traverses each directed edge exactly once before the original unicycle is obtained. Thus the number of distinct unicycle configurations equals the number of directed edges, in this case 34. From Lemma 4.11 it follows that from any given unicycle, after some number of steps, the state will be the same but with the cycle's direction reversed. This occurs, for example, with unicycles 1 and 13.

router operation until the chip moves to the sink. By Lemma 3.9 and the reasoning used in the proof of Lemma 2.5, the operators E_v commute. This is the **abelian property** of the rotor-router model.

If, rather than running the chips until they reach the sink, each chip is run for a fixed number of steps, then the abelian property fails, as the example in Figure 8 illustrates. (The proof of Lemma 3.9 requires that chips be indistinguishable, and it is not possible to run each chip for a fixed number of steps without distinguishing between them.) Despite the failure of commutativity, this way of routing chips has some interesting properties, similar to the bound given in Proposition 3.21; see work of Cooper and Spencer [CS06].

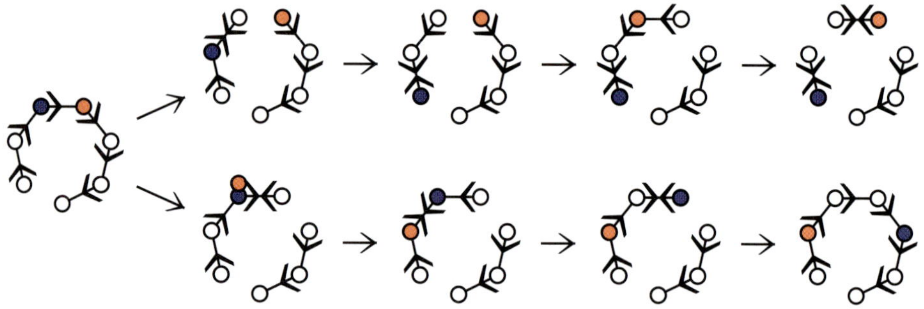

FIGURE 8. Failure of the abelian property for rotor-router walk stopped after two steps.

A rotor configuration ρ on G is **acyclic** if the rotors do not form any directed cycles. If G has a global sink, then ρ is acyclic if and only if the rotors form an oriented spanning tree rooted at the sink.

Lemma 3.10. *Let G be a digraph with a global sink, and let v be a vertex of G. The chip addition operator E_v is a permutation on the set of acyclic rotor configurations on G.*

Proof. We first argue that applying E_v to an acyclic rotor configuration yields an acyclic rotor configuration: this is proved by induction on the number of rotor-routing steps, where the induction hypothesis states that every path leads to either the sink or to the chip.

Since the set of acyclic rotor configurations is finite, it suffices to show surjectivity. Let ρ be an acyclic rotor configuration on G, add an edge e from the sink s to v to form a sink-free digraph G', and assign the rotor $\rho(s) = e$. Then $U = (s, \rho)$ is a unicycle on G'. Starting from U, we can iterate the *inverse* of the rotor-router operation (which by Lemma 3.4 is well defined for unicycles) until the next time we reach a state $U' = (s, \rho')$ with the chip at s. Because U' is a unicycle with chip at s, deleting the edge e leaves an acyclic rotor configuration ρ'. Observe that running the rotor-router operation from U' to U, upon ignoring the edge e, is equivalent to applying E_v to ρ' and obtaining ρ. □

Definition 3.11. We next describe an action of the sandpile group on acyclic rotor configurations. Given a chip configuration σ and a rotor configuration ρ on G, write $\sigma(\rho)$ for the rotor configuration obtained by adding $\sigma(v)$ chips at each vertex v and routing them all to the sink. By Lemma 3.9 the order of these routings is immaterial. Thus we may write $\sigma(\rho)$ as

$$\sigma(\rho) = \left(\prod_{v \in V(G)} E_v^{\sigma(v)} \right) \rho,$$

where the product symbol represents composition of the operators.

It is trivial that $\sigma_2(\sigma_1(\rho)) = (\sigma_1 + \sigma_2)(\rho)$.

Since acyclic rotor configurations on G can be identified with oriented spanning trees rooted at the sink, Lemma 3.10 implies that every chip configuration σ acts as a permutation on the set of oriented spanning trees of G rooted at the sink.

Lemma 3.12. *Let G be a digraph with a global sink, and let ρ be an acyclic rotor configuration on G. If the chip configurations σ_1 and σ_2 are equivalent modulo the reduced Laplacian Δ' of G, then $\sigma_1(\rho) = \sigma_2(\rho)$.*

Proof. If $\sigma(v) \geq d_v$, and we route (for one step) d_v of the chips at v, then the rotor at v makes one full turn and one chip is sent along each outgoing edge from v. By Lemma 3.9, it follows that if σ' is a successor to σ (that is, σ' is obtained from σ by firing some active vertex v), then $\sigma(\rho) = \sigma'(\rho)$ for any rotor configuration ρ. Inducting, we obtain $\sigma(\rho) = \sigma^\circ(\rho)$ for any rotor configuration ρ.

In particular, if I is the recurrent chip configuration that represents the identity element of the sandpile group, we have

$$I(I(\rho)) = (I + I)(\rho) = (I + I)^\circ(\rho) = I(\rho)$$

for any rotor configuration ρ. By Lemma 3.10 the map $\rho \mapsto I(\rho)$ is a permutation on the set of acyclic rotor configurations, so it must be the identity permutation. Now if σ_1, σ_2 are equivalent modulo Δ', then $(\sigma_1 + I)^\circ$ and $(\sigma_2 + I)^\circ$ are recurrent and equivalent modulo Δ', hence equal by Lemma 2.15. Since

$$\sigma_i(\rho) = \sigma_i(I(\rho)) = (\sigma_i + I)(\rho) = (\sigma_i + I)^\circ(\rho), \qquad i = 1, 2,$$

we conclude that $\sigma_1(\rho) = \sigma_2(\rho)$. □

It follows from Lemma 3.12 that the sandpile group of G acts on the set of oriented spanning trees of G rooted at the sink. Our next lemma shows that this action is transitive.

Lemma 3.13. *Let G be a digraph with a global sink. For any two acyclic rotor configurations ρ and ρ' on G, there exists a chip configuration σ on G such that $\sigma(\rho) = \rho'$.*

Proof. For a non-sink vertex v, let $\alpha(v)$ be the number of edges e such that $\rho(v) < e \leq \rho'(v)$ in the cyclic ordering of outgoing edges from v. Starting with rotor configuration ρ, and with $\alpha(v)$ chips at each vertex v, allow each chip to take just one step. The resulting rotor configuration is ρ'; let β be the resulting chip configuration, so that $\alpha(\rho) = \beta(\rho')$, and let γ be the inverse of the corresponding element $(\beta + I)^\circ$ of the sandpile group. By Lemma 3.12 and the fact that $\beta + \gamma$ is equivalent to 0 modulo Δ', we have

$$(\alpha + \gamma)(\rho) = (\beta + \gamma)(\rho') = \rho'. \qquad \square$$

Next we define recurrent rotor configurations on a digraph with a global sink, and show they are in bijection with oriented spanning trees.

Definition 3.14. Let G be a digraph with a global sink. Given rotor configurations ρ and ρ' on G, we say that ρ is **reachable** from ρ' if there is a chip configuration σ such that $\sigma(\rho') = \rho$. We say that ρ is **recurrent** if it is reachable from any other configuration ρ'.

Note that in contrast to Definition 3.2, the location of the chip plays no role in the notion of recurrent states on a digraph with global sink.

Lemma 3.15. *Let G be a digraph with a global sink. A rotor configuration ρ on G is recurrent if and only if it is acyclic.*

Proof. By Lemma 3.10, any configuration reachable from an acyclic configuration must be acyclic, so recurrent implies acyclic. Conversely, if ρ is acyclic and ρ' is any rotor configuration, the configuration $\mathbf{1}(\rho')$ (where $\mathbf{1}$ denotes the configuration with one chip at each vertex) is acyclic, since the rotor at each vertex points along the edge by which a chip last exited. By Lemma 3.13 there is a chip configuration σ such that $\sigma(\mathbf{1}(\rho')) = \rho$, so ρ is reachable from ρ' and hence recurrent. $\square$

Just as for the sandpile model, there are several equivalent definitions of recurrence for the rotor-router model.

Lemma 3.16. *Let G be a digraph with a global sink s, and let ρ be a rotor configuration on G. The following are equivalent.*

(1) *ρ is acyclic.*
(2) *ρ is recurrent; that is, ρ is reachable from any rotor configuration ρ'.*
(3) *If ρ' is reachable from ρ, then ρ is reachable from ρ'.*
(4) *ρ is reachable from any rotor configuration of the form $E_v\rho$, where v is a vertex of G.*
(5) *Each strongly connected component of $G \setminus s$ contains a vertex v such that ρ is reachable from $E_v\rho$.*

Proof. By Lemma 3.15 we have (1) $\Rightarrow$ (2), and trivially (2) $\Rightarrow$ (3) $\Rightarrow$ (4) $\Rightarrow$ (5).

If property (5) holds, let $C_1, \ldots, C_\ell$ be the strongly connected components of $G \setminus s$, and for each i, let $v_i \in C_i$ be such that ρ is reachable from $E_{v_i}\rho$. Choose an integer k large enough so that if we start k chips at any v_i and route them to the sink, every vertex in C_i is visited at least once. Let

$$\rho' = \left(\prod_i E_{v_i}\right)^k \rho.$$

Then in ρ', the rotor at each vertex points along the edge by which a chip last exited, so ρ' is acyclic. Since ρ is reachable from ρ', by Lemma 3.10 it follows that ρ is acyclic. Thus (5) $\Rightarrow$ (1), completing the proof. □

Next we show that the action of the sandpile group on the set of oriented spanning trees of G is free.

Lemma 3.17. *Let G be a digraph with a global sink, and let σ_1 and σ_2 be recurrent chip configurations on G. If there is an acyclic rotor configuration ρ of G such that $\sigma_1(\rho) = \sigma_2(\rho)$, then $\sigma_1 = \sigma_2$.*

Proof. Let $\sigma = \sigma_1 + \overline{\sigma_2}$ (recall that $\overline{\sigma_2}$ is the inverse of σ_2.) By Lemma 3.12, $\sigma(\rho) = \overline{\sigma_2}(\sigma_1(\rho)) = \overline{\sigma_2}(\sigma_2(\rho)) = (\sigma_2 + \overline{\sigma_2})(\rho) = \rho$. Since $\sigma(\rho) = \rho$, after adding σ to ρ, for each vertex v, the rotor at v makes some integer number c_v of full rotations. Each full rotation results in d_v chips leaving v, one along each outgoing edge. Hence $\sigma = \sum_v c_v \Delta'_v$, which is in the row span of the reduced Laplacian, so σ is equivalent to 0 modulo $\Delta'(G)$, and hence $\sigma_1 = \sigma_2$ by Lemma 2.15. □

Corollary 3.18 (Matrix Tree Theorem). *Let G be a digraph and v a vertex of G. The number of oriented spanning trees of G rooted at v is equal to the determinant of the reduced Laplacian $\Delta'(G)$ obtained by deleting from $\Delta(G)$ the row and column corresponding to v.*

Proof. Without loss of generality we may assume the graph is loopless, since loops affect neither the graph Laplacian nor the number of spanning trees.

If v is not globally reachable, then there are no spanning trees rooted at v, and there is a set of vertices S not containing v, such that there are no edges in G from S to S^c. The rows of $\Delta'(G)$ corresponding to vertices in S sum to zero, so $\Delta'(G)$ has determinant zero.

If v is globally reachable, delete all outgoing edges from v to obtain a digraph G' with global sink v. Note that G and G' have the same reduced Laplacian, and the same set of oriented spanning trees rooted at v.

Fix an oriented spanning tree ρ of G'. The mapping $\sigma \mapsto \sigma(\rho)$ from $\mathcal{S}(G')$ to the set of oriented spanning trees of G' is a surjection by Lemma 3.13 and is one-to-one by Lemma 3.17, and by Lemma 2.8, $|\mathcal{S}(G')| = \det \Delta'(G')$. □

Given a digraph G with global sink, define its **rotor-router group** as the subgroup of permutations of oriented spanning trees of G generated by the chip addition operators E_v.

Lemma 3.19. [LL07] *The rotor-router group for a digraph G with a global sink is isomorphic to the sandpile group $\mathcal{S}(G)$.*

Proof. The action of the sandpile group on oriented spanning trees is a homomorphism from the sandpile group $\mathcal{S}(G)$ onto the rotor-router group. For any two distinct sandpile group elements σ_1 and σ_2, for any oriented spanning tree

ρ (there is at least one), by Lemma 3.17 we have $\sigma_1(\rho) \neq \sigma_2(\rho)$, so the associated rotor-router group elements are distinct, i.e., the group homomorphism is an isomorphism. □

Since the number of recurrent chip configurations of G equals the number of oriented spanning trees, it is natural to ask for a bijection. Although there is no truly "natural" bijection, since in general there is no canonical spanning tree to correspond to the identity configuration, we can use the rotor-router model to define a family of bijections. Fix any oriented spanning tree ρ rooted at the sink, and associate it with the identity configuration I. For any other oriented spanning tree ρ', by Lemma 3.13 there exists $\sigma \in \mathcal{S}(G)$ with $\sigma(\rho) = \rho'$; moreover, σ is unique by Lemma 3.17. Associate σ with ρ'. Since this defines a surjective map from recurrent configurations to oriented spanning trees, it must be a bijection.

Remark 3.20. A variant of the rotor-router rule relaxes the cyclic ordering of edges emanating from a vertex, and merely requires one to choose *some* edge emanating from the current location of the chip as the new rotor-setting and move the chip along this edge. This is the **branching operation** introduced by Propp and studied by Athanasiadis [Ath97]. Alternatively, one can put a probability distribution on the edges emanating from each vertex, and stipulate that the new edge is to be chosen at random. This gives the tree-walk introduced by Anantharam and Tsoucas [AT89] in their proof of the Markov chain tree theorem of Leighton and Rivest [LR86].

We conclude this section with the following result of Holroyd and Propp [HP08], which illustrates another area of application of the rotor-router model.

Proposition 3.21. *Let $G = (V, E)$ be a digraph and let $Y \subseteq Z$ be sets of vertices. Assume that from each vertex there is a directed path to Z. Let σ be a chip configuration on G. If the chips perform independent simple random walks on G stopped on first hitting Z, let $H(\sigma, Y)$ be the expected number of chips that stop in Y. If the chips perform rotor-router walks starting at rotor configuration ρ and stopped on first hitting Z, let $H_\rho(\sigma, Y)$ be the number of chips that stop in Y. Then*

$$|H_\rho(\sigma, Y) - H(\sigma, Y)| \leq \sum_{\text{edges } e} |h(\text{head of } e) - h(\text{tail of } e)| \tag{1}$$

where $h(v) := H(1_v, Y)$.

Note that the bound on the right side does not depend on ρ or σ.

Proof. To each edge $e = (u, v)$ we assign a *weight*

$$\text{wt}(e) = \begin{cases} 0 & \text{if } e = \rho(u), \\ h(u) - h(v) + \text{wt}(e^-) & \text{otherwise.} \end{cases}$$

Here e^- is the edge preceding e in the cyclic ordering of edges emanating from u. Since h is a harmonic function on G, the sum of $h(u)-h(v)$ over all edges $e = (u, v)$ emanating from u is zero, so the formula $\text{wt}(e) = h(u) - h(v) + \text{wt}(e^-)$ remains valid even when $e = \rho(u)$. We assign weight $\sum_v \mathrm{wt}(\rho(v))$ to a rotor configuration ρ, and weight $h(v)$ to a chip located at v. By construction, the sum of rotor and chip weights in any configuration is invariant under the operation of rotating the rotor at a chip and then routing the chip. Initially, the sum of all chip weights is $H(\sigma, Y)$. After all chips have stopped, the sum of the chip weights is $H_\rho(\sigma, Y)$. Their difference is thus at most the change in rotor weights, which is bounded above by the sum in (1). □

Similar bounds hold even for some infinite directed graphs in which the right side of (1) is not finite. Thus rotor-routing can give estimates of hitting probabilities with very small error. See [HP08] for more details.

4. Eulerian graphs

A digraph $G = (E, V)$ is **Eulerian** if it is strongly connected, and for each vertex $v \in V$ the in-degree and the out-degree of v are equal. We call G an **Eulerian digraph with sink** if it is obtained from an Eulerian digraph by deleting all the outgoing edges from one vertex; equivalently, G has a sink and every other vertex has out-degree that is at least as large as its in-degree. An **Eulerian tour** of a digraph G is a cycle in G that uses each edge exactly once. Such a tour exists if and only if G is Eulerian. Note that for any connected undirected graph, the corresponding bidirected graph is Eulerian. In this section we show some results that do not hold for general digraphs, but are true for Eulerian ones. We first treat the sandpile model, and then the rotor-router model.

Lemma 4.1 (Burning algorithm [Dha90]**).** *Let G be an Eulerian digraph with sink. A chip configuration σ is recurrent if and only if $(\sigma + \beta)^\circ = \sigma$, where*

$$\beta(v) = \mathrm{outdeg}(v) - \mathrm{indeg}(v) \geq 0.$$

If σ is recurrent, each vertex fires exactly once during the stabilization of $\sigma + \beta$.

Proof. By the "(4) ⇒ (1)" part of Lemma 2.17, if $(\sigma+\beta)^\circ = \sigma$, then σ is recurrent. Conversely, suppose σ is recurrent. Label the non-sink vertices $v_1, \ldots, v_{n-1}$. Since

$$\beta = \sum_{i=1}^{n-1} \Delta'_i, \tag{2}$$

the configurations σ and $(\sigma + \beta)^\circ$ are both recurrent and equivalent modulo Δ'. By Lemma 2.15 it follows that they are equal.

Let c_i be the number of times vertex v_i fires during the stabilization of $\sigma + \beta$. Then

$$\sigma = (\sigma + \beta)^\circ = \sigma + \beta - \sum_{i=1}^{n-1} c_i \Delta'_i.$$

The rows of Δ' are linearly independent, so from (2) we deduce $c_i = 1$ for all i. □

Informally, the burning algorithm can be described as follows: to determine whether σ is recurrent, first "fire the sink" to obtain the configuration $\sigma+\beta$. Then σ is recurrent if and only if every non-sink vertex fires in the stabilization of $\sigma+\beta$. In the non-Eulerian case, there is a generalization of the burning algorithm known as the **script algorithm**, due to Eugene Speer [Spe93].

Let H be an induced subgraph of G not containing the sink. We say that H is **ample** for a chip configuration σ on G if there is a vertex v of H that has at least as many chips as the in-degree of v in H.

Lemma 4.2. *Let G be an Eulerian digraph with sink s. A stable chip configuration σ on G is recurrent if and only if every nonempty induced subgraph of $G \setminus s$ is ample for σ.*

Proof. If σ is recurrent, there is a chip configuration α such that $(\delta+\alpha)^\circ = \sigma$, where $\delta(v) = d_v$. Each vertex of G fires at least once in the process of stabilizing $\delta+\alpha$. Given a nonempty induced subgraph H of G, let v be the vertex of H which first finishes firing. After v finishes firing, it must receive at least as many chips from its neighbors as its in-degree in H, so $\sigma(v)$ is at least the in-degree of v in H. Thus H is ample for σ.

Conversely, suppose that every nonempty induced subgraph of G is ample for σ. Let β be the chip configuration defined in Lemma 4.1. Starting from $\sigma+\beta$, fire as many vertices as possible under the condition that each vertex be allowed to fire only once. Let H be the induced subgraph on the set of vertices that do not fire. Since each vertex v of H is unable to fire even after receiving one chip from each incoming edge whose other endpoint lies outside H, we have

$$\sigma(v) + d_v - \mathrm{indeg}_H(v) \leq d_v - 1.$$

Thus H is not ample and consequently must be empty. So every vertex fires once, after which we obtain the configuration $\sigma+\beta-\sum_{i=1}^{n-1}\Delta'_i = \sigma$. Hence $(\sigma+\beta)^\circ = \sigma$, which implies σ is recurrent by Lemma 2.17. $\square$

Next we define a variant of chip-firing called cluster-firing (see Figure 9), and we use Lemma 4.2 to characterize the stable states for cluster-firing. This gives rise to a notion of "superstable states" which are in some sense dual to the recurrent states.

Definition 4.3. Let G be a digraph with a global sink. Let σ be a chip configuration on G, and let A be a nonempty subset of the non-sink vertices of G. The **cluster-firing** of A yields the configuration

$$\sigma' = \sigma - \sum_{i\in A} \Delta'_i.$$

If σ' is nonnegative we say that the cluster A is **allowed to fire**. We say that σ is **superstable** if no cluster is allowed to fire.

Note that a cluster A may be allowed to fire even if no subset of A is allowed to fire. For example, in the first confiuration in Figure 9, a cluster of two vertices

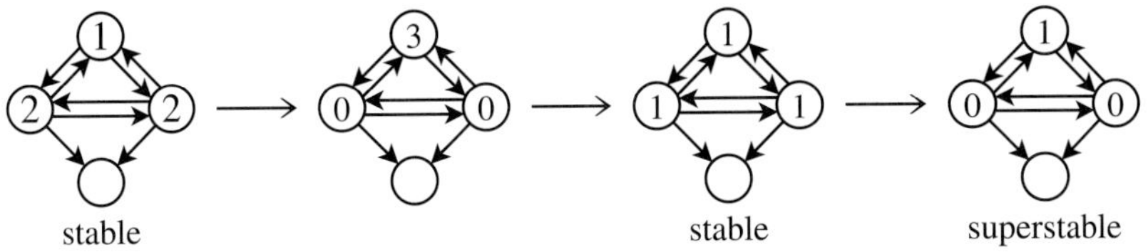

FIGURE 9. A sequence of cluster-firings resulting in a superstable chip configuration. The bottom vertex is the sink. The clusters that fire are first the two neighbors of the sink, next the top vertex, and finally all three non-sink vertices.

is allowed to fire even though the configuration is stable, so no single vertex is allowed to fire.

One could consider an even more general operation, "multicluster-firing," in which different vertices can be fired different numbers of times, so long as at the end of the firings, each vertex has a nonnegative number of chips. However, this further-generalized firing operation does not yield anything new for Eulerian digraphs, since any multicluster-firing can be expressed as a sequence of cluster-firings: Let m denote the maximal number of times that a vertex fires in the multicluster-firing, and C_j denote the set of vertices that fire at least j times in the multicluster-firing. Since the digraph is Eulerian, C_m may be cluster-fired, and so by induction the sets $C_m, C_{m-1}, \ldots, C_1$ can be cluster-fired in that order.

Denote by δ the chip configuration $\delta(v) = d_v$ in which each vertex has as many chips as outgoing edges, and by $\mathbf{1}$ the configuration with a single chip at each vertex.

Theorem 4.4. *Let G be an Eulerian digraph with sink. A chip configuration σ on G is superstable if and only if $\delta - \mathbf{1} - \sigma$ is recurrent.*

Proof. A cluster A is allowed to fire if and only if for each vertex $v \in A$ we have

$$\sigma(v) - d_v + \operatorname{indeg}_A(v) \geq 0.$$

This is equivalent to $d_v - 1 - \sigma(v) < \operatorname{indeg}_A(v)$, i.e., the induced subgraph on A is not ample for $\delta - \mathbf{1} - \sigma$. By Lemma 4.2 the proof is complete. □

By Lemmas 2.13 and 2.15, every equivalence class modulo Δ' contains a unique recurrent configuration, so we obtain the following.

Corollary 4.5. *Let G be an Eulerian digraph with sink. Every equivalence class modulo $\Delta'(G)$ contains a unique superstable configuration.*

As a consequence, we obtain that the cluster-firing model on Eulerian digraphs is abelian; this was proved by Paoletti [Pao07] in the bidirected case.

Corollary 4.6. *Let G be an Eulerian digraph with sink. Let $\sigma_0, \sigma_1, \ldots, \sigma_n$ be a sequence of chip configurations on G, each of which is obtained from the one before*

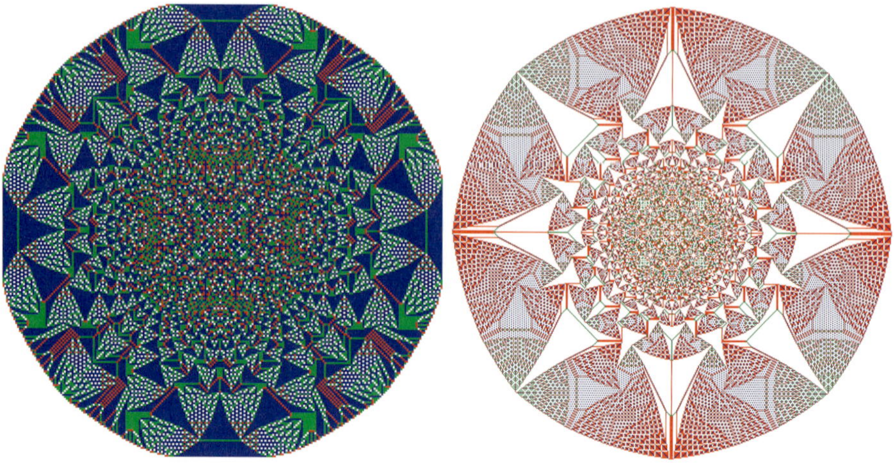

FIGURE 10. Stable sandpile (on left) and superstable sandpile (on right) of 100,000 chips, obtained by placing 100,000 chips at the origin of the integer lattice $\mathbb{Z}^2$ and (super)stabilizing. The color scheme is as follows: white=0 chips, red=1 chip, green=2 chips, and blue=3 chips.

by a cluster-firing, with σ_n superstable. Then any sequence of cluster-firings that starts from σ_0 and ends in a superstable configuration ends in σ_n.

We call the configuration σ_n in Corollary 4.6 the **superstabilization** of σ_0. The following result provides a way to compute the superstabilization.

Proposition 4.7. *Let σ be a chip configuration on an Eulerian digraph with sink. The superstabilization of σ is given by*

$$\sigma^* = \delta - \mathbf{1} - (\delta - \mathbf{1} - \sigma^\circ + I)^\circ$$

where I is the identity element of the sandpile group.

Proof. Since the configuration $\zeta = (\delta - \mathbf{1} - \sigma^\circ + I)^\circ$ is reachable from the identity element, it is recurrent, hence $\sigma^* = \delta - \mathbf{1} - \zeta$ is superstable by Theorem 4.4. Since σ and σ^* are equivalent modulo Δ', it follows from Corollary 4.5 that σ^* is the superstabilization of σ. □

Our final result concerning the sandpile model on Eulerian digraphs is a theorem of Van den Heuvel [vdH01]; see also [Tar88]. We give a shorter and more direct proof than that presented in [vdH01]. By an **bidirected graph with sink** s, we will mean the digraph obtained from an undirected graph by first replacing each edge by a pair of directed edges in opposite directions, and then deleting all outgoing edges from s. The **effective resistance** between two vertices of G is an important quantity in electrical network theory; see, e.g., [DS84]. In particular, the quantity $R_{\max}$ appearing in the proposition below is always bounded above by the diameter of G, but for many graphs it is substantially smaller than the diameter.

Proposition 4.8. *Let G be a bidirected graph with sink, and let σ be a chip configuration on G. The total number of chip moves needed to stabilize σ is at most*

$$2m\,|\sigma|\,R_{\max}$$

where m is the number of edges, $|\sigma|$ is the total number of chips, and $R_{\max}$ is the maximum effective resistance between any vertex of G and the sink.

Note that firing a vertex v consists of d_v chip moves.

Proof. Let $\sigma = \sigma_0, \sigma_1, \dots, \sigma_k = \sigma^\circ$ be a sequence of chip configurations with σ_{i+1} obtained from σ_i by firing a single active vertex x_i. Define the weight of σ_i to be

$$\mathrm{wt}(\sigma_i) = \sum_x \sigma_i(x)\,\mathrm{wt}(x),$$

where

$$\mathrm{wt}(x) = \mathbb{E}_x T_s$$

is the expected time for a simple random walk started at x to hit the sink. By conditioning on the first step X_1 of the walk, we compute

$$\Delta\,\mathrm{wt}(x) = \mathbb{E}_x(\mathbb{E}_{X_1} T_s - T_s) = -1,$$

so firing the vertex x_i decreases the total weight by d_{x_i}. Thus

$$\mathrm{wt}(\sigma_i) - \mathrm{wt}(\sigma_{i+1}) = d_{x_i}. \tag{3}$$

By [CRR[+]97], the function wt is bounded by $2mR_{\max}$. Since the final weight $\mathrm{wt}(\sigma^\circ)$ is nonnegative, summing (3) over i we obtain that the total number of chip moves N needed to stabilize σ is at most

$$N = \sum_{i=0}^{k-1} d_{x_i} = \mathrm{wt}(\sigma) - \mathrm{wt}(\sigma^\circ) \le 2m\,|\sigma|\,R_{\max}. \qquad \square$$

Next we present results about the rotor-router model specific to the Eulerian case. An example of the next lemma is illustrated in Figure 7.

Lemma 4.9. *Let G be an Eulerian digraph with m edges. Let $U = (w, \rho)$ be a unicycle in G. If we iterate the rotor-router operation m times starting from U, the chip traverses an Eulerian tour of G, each rotor makes one full turn, and the state of the system returns to U.*

Proof. Iterate the rotor-router operation starting from U until some rotor makes more than a full turn. Let it be the rotor at vertex v. During this process, v must emit the chip more than d_v times. Hence if $v \neq w$, then v must also receive the chip more than d_v times. Since G is Eulerian, this means that some neighboring vertex u must send the chip to v more than once. However, when the chip goes from u to v for the second time, the rotor at u has executed more than a full turn, contradicting our choice of v. Thus when the rotor at w has made a full turn, the rotors at the other sites have made at most a full turn.

We can now repeat this argument starting from the configuration obtained after the rotor at w has made a full turn. In this way, the future history of the system is divided up into segments, each of length at most m, where the chip is at w at the start of each segment. It follows that over the course of the future history of the system, the chip is at w at least d_w/m of the time.

Since G is strongly connected, we may apply this same argument to every state in the future history of the system, with every vertex of G playing the role of w. As the system evolves, the chip is at v at least d_v/m of the time. Since $\sum_v d_v/m = 1$, the chip is at v exactly d_v/m of the time. Hence, as the rotor at w executes a full turn, the rotors at the other sites also execute a full turn. Since every rotor makes a full turn, every edge is traversed exactly once, so the chip traverses an Eulerian tour. □

We can use Lemma 4.9 to give a bijective proof of a classical result in enumerative combinatorics relating the number of Eulerian tours of an Eulerian graph G to the number of oriented spanning trees of G (see, e.g., [Sta99, Cor. 5.6.7]).

Corollary 4.10. *Let $G = (V, E)$ be an Eulerian digraph. Fix an edge $e \in E$ and let* $\mathrm{tail}(e) = w$. *Let $\mathcal{T}(G, v)$ denote the number of oriented spanning trees in G rooted at w, and let $\epsilon(G, e)$ be the number of Eulerian tours in G starting with the edge e. Then*

$$\epsilon(G, e) = \mathcal{T}(G, w) \prod_{v \in V} (d_v - 1)!.$$

Proof. There are $\prod_{u \in V}(d_v - 1)!$ ways to fix cyclic orderings of the outgoing edges from each vertex. There are $\mathcal{T}(G, w)$ ways to choose a unicycle $U = (w, \rho)$ with the chip at w and the rotor $\rho(w) = e^-$, where e^- is the edge preceding e in the cyclic ordering of outgoing edges from w. Given these data, we obtain from Lemma 4.9 an Eulerian tour of G starting with the edge e, namely the path traversed by the rotor-router walk in m steps.

To show that this correspondence is bijective, given an Eulerian tour starting with the edge e, cyclically order the outgoing edges from each vertex v in the order they appear in the tour. Let $\rho(w) = e^-$ and for $v \neq w$ let $\rho(v)$ be the outgoing edge from v that occurs last in the tour. Then $U = (w, \rho)$ is a unicycle. □

The following result was first announced in [PPS98], in the case of rotor-router walk on a square lattice.

Corollary 4.11. *Let G be a bidirected planar graph with the outgoing edges at each vertex ordered clockwise. Let (w, ρ) be a unicycle on G with the cycle $\mathcal{C}$ oriented clockwise. After the rotor-router operation is iterated some number of times, each rotor internal to $\mathcal{C}$ has performed a full rotation, each rotor external to $\mathcal{C}$ has not moved, and each rotor on $\mathcal{C}$ has performed a partial rotation so that $\mathcal{C}$ is now oriented counter-clockwise.*

Proof. Let G' be the graph obtained from G by deleting all vertices and edges external to $\mathcal{C}$. Note that G', like G, is Eulerian. Let ρ^- be the rotor configuration on G' obtained from ρ by "regressing" each rotor whose vertex lies on $\mathcal{C}$; that is, if v lies on $\mathcal{C}$ and $\rho(v)$ is the edge e, let $\rho(v) = e^-$, the edge immediately preceding

e in the cyclic ordering of the edges emanating from v. Consider the unicycle $U^- = (w, \rho^-)$ on G' and note that it has $\mathcal{C}$ oriented counter-clockwise. Starting from U^- and applying the rotor-router operation $\#\mathcal{C}$ times, the chip will traverse the cycle $\mathcal{C}'$, resulting in the state $U = (w, \rho|_{G'})$. By Lemma 4.9, further iteration of the rotor-router operation on G' returns the system to the state U^-. Since the outgoing edges at each vertex are ordered clockwise, it is straightforward to see that applying the rotor-router rule to U on G' and to (w, ρ) on G results in the same evolution up until the time that state U^- on G' is reached. □

Lemma 4.12. *Let G be an Eulerian digraph, and let G_v be the Eulerian digraph with sink obtained by deleting the outgoing edges from vertex v. Then the abelian sandpile groups $\mathcal{S}(G_v)$ and $\mathcal{S}(G_w)$ corresponding to different choices of sink are isomorphic.*

Proof. Recall that the sandpile group $\mathcal{S}(G_v)$ is isomorphic to $\mathbb{Z}^{n-1}/\mathbb{Z}^{n-1}\Delta'(G)$; we argue that for Eulerian digraphs G it is also isomorphic to $\mathbb{Z}^n/\mathbb{Z}^n\Delta(G)$. Vectors in $\mathbb{Z}^{n-1}$ are isomorphic to vectors in $\mathbb{Z}^n$ whose coordinates sum to 0, and modding out a vector in $\mathbb{Z}^{n-1}$ by a row of the reduced Laplacian Δ' corresponds to modding out the corresponding vector in $\mathbb{Z}^n$ by the corresponding row of the full Laplacian Δ. For Eulerian digraphs G, the last row of the full Laplacian Δ is the negative of the sum of the remaining rows, so modding out by this extra row has no effect. □

We mention one other result that applies to undirected planar graphs, due originally to Berman [Ber86, Prop. 4.1]; see also [CR00].

Theorem 4.13. *If G and G^* are dual undirected planar graphs, then the sandpile groups of G and G^* are isomorphic.* (*By Lemma* 4.12, *the locations of the sink are irrelevant.*)

5. Stacks and cycle-popping

Let G be a digraph with a global sink. In this section we describe a more general way to define rotor-router walk on G, using arbitrary stacks of rotors at each vertex in place of periodic rotor sequences. To each non-sink vertex v of G we assign a bi-infinite **stack** $\rho(v) = (\rho_k(v))_{k\in\mathbb{Z}}$ of outgoing edges from v. To **pop** the stack, we shift it to obtain $(\rho_{k+1}(v))_{k\in\mathbb{Z}}$. To **reverse pop** the stack, we shift it in the other direction to obtain $(\rho_{k-1}(v))_{k\in\mathbb{Z}}$. The rotor-router walk can be defined in terms of stacks as follows: if the chip is at vertex v, pop the stack $\rho(v)$, and then move the chip along the edge $\rho_1(v)$. We recover the ordinary rotor-router model in the case when each stack $\rho(v)$ is a periodic sequence of period d_v in which each outgoing edge from v appears once in each period.

The collection of stacks $\rho = (\rho(v))$, where v ranges over the non-sink vertices of G, is called a **stack configuration** on G. We say that ρ is **infinitive** if for each edge $e = (v, w)$, and each positive integer K, there exist stack elements

$$\rho_k(v) = \rho_{k'}(v) = e$$

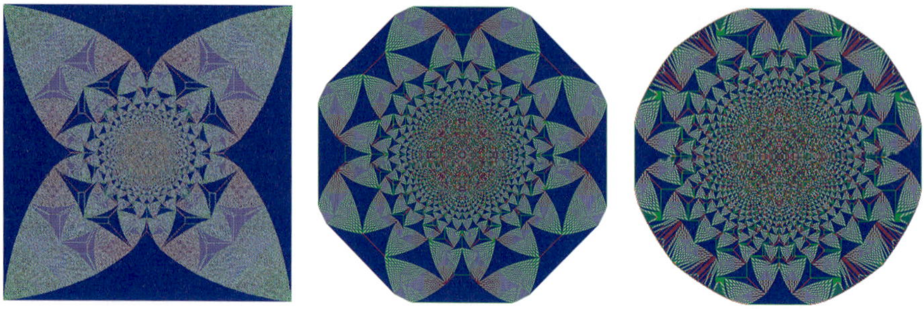

$H = -2$ $H = -1$ $H = 0$

FIGURE 11. Sandpile aggregates of 250,000 chips in $\mathbb{Z}^2$ at hole depths $H = -2$ (left), $H = -1$ (center), and $H = 0$.

with $k \geq K$ and $k' \leq -K$. This condition guarantees that rotor-router walk eventually reaches the sink.

Given a stack configuration ρ, the stack elements $\rho_0(v)$ define a rotor configuration on G. We say that ρ is **acyclic** if ρ_0 contains no directed cycles. If $\mathcal{C} = \{v_1, \ldots, v_m\}$ is a directed cycle in ρ_0, define $\mathcal{C}\rho$ to be the stack configuration obtained by reverse popping each of the stacks $\rho(v_i)$; we call this reverse popping the cycle $\mathcal{C}$. (If $\mathcal{C}$ is not a directed cycle in ρ_0, set $\mathcal{C}\rho = \rho$.)

Theorem 5.1. [PW98] *Let G be a digraph with a global sink, and let ρ^0 be an infinitive stack configuration on G. There exist finitely many cycles $\mathcal{C}_1, \ldots, \mathcal{C}_m$ such that the stack configuration*

$$\rho = \mathcal{C}_m \cdots \mathcal{C}_1 \rho^0$$

is acyclic. Moreover, if $\mathcal{C}'_1, \ldots, \mathcal{C}'_n$ is any sequence of cycles such that the stack configuration $\rho' = \mathcal{C}'_n \cdots \mathcal{C}'_1 \rho^0$ is acyclic, then $\rho' = \rho$.

If v is a non-sink vertex of G, the **chip addition operator** E_v applied to the infinitive stack configuration ρ is the stack configuration ρ' obtained by adding a chip at v and performing rotor-router walk until the chip reaches the sink. The next lemma shows that these operators commute with cycle-popping.

Lemma 5.2. *Let G be a digraph with a global sink, let ρ be an infinitive stack configuration on G, and let $\mathcal{C}$ be a directed cycle in G. Then*

$$E_v(\mathcal{C}\rho) = \mathcal{C}(E_v\rho).$$

Proof. Write $\rho' = E_v\rho$. Let $v = v_0, v_1, \ldots, v_n = s$ be the path taken by a chip performing rotor-router walk from v to the sink starting with stack configuration ρ. If this path is disjoint from $\mathcal{C}$, then the chip performs the same walk starting with stack configuration $\mathcal{C}\rho$, and the cycle $\mathcal{C}$ is present in ρ'_0 if and only if it is present in ρ_0, so the proof is complete.

Otherwise, choose k minimal and ℓ maximal with $v_k, v_\ell \in \mathcal{C}$. The rotor $\rho_0'(v_\ell)$ points to a vertex not in $\mathcal{C}$, so the cycle $\mathcal{C}$ is not present in ρ_0'. Thus we must show $E_v(\mathcal{C}\rho) = \rho'$. With stack configuration $C\rho$, the chip will first travel the path $v_0, \ldots, v_k$, next traverse the cycle $\mathcal{C}$, and finally continue along the remainder of the path $v_k, \ldots, v_n$. Thus the stack at each vertex $w \in \mathcal{C}$ is popped one more time in going from $\mathcal{C}\rho$ to $E_v(\mathcal{C}\rho)$ than in going from ρ to $E_v\rho$; the stack at each vertex $w \notin \mathcal{C}$ is popped the same number of times in both cases. □

The next lemma uses cycle-popping to give a constructive proof of the injectivity of the chip addition operators E_v on acyclic stack configurations. In the case of periodic rotor stacks, we gave a non-constructive proof in Lemma 3.10.

Lemma 5.3. *Let G be a digraph with a global sink. Given an acyclic infinitive stack configuration ρ on G and a non-sink vertex v, there exists an acyclic infinitive stack configuration ρ' such that $E_v\rho' = \rho$.*

Proof. Let ρ^0 be the stack configuration obtained from ρ by reverse popping the stack at each of the vertices on the unique path in ρ_0 from v to the sink. A rotor-router walk started at v with stack configuration ρ^0 will travel directly along this path to the sink, so $E_v\rho^0 = \rho$. If ρ^0 is acyclic, the proof is complete. Otherwise, by Theorem 5.1 there are cycles $\mathcal{C}_1, \ldots, \mathcal{C}_m$ such that $\rho' = \mathcal{C}_m \cdots \mathcal{C}_1\rho^0$ is acyclic. By Lemma 5.2, we have

$$E_v\rho' = \mathcal{C}_m \cdots \mathcal{C}_1(E_v\rho^0) = \mathcal{C}_m \cdots \mathcal{C}_1\rho = \rho$$

where in the last equality we have used that ρ is acyclic. □

Note that the proof shows the following: if ρ, ρ' are acyclic infinitive stack configurations and $E_v\rho' = \rho$, then the unique path in ρ_0 from v to the sink is the **loop-erasure** of the path taken by rotor-router walk started at v with initial configuration ρ'.

6. Conjectures and open problems

In this section we discuss some natural questions about chip-firing and rotor-routing that remain unanswered.

Fey-den Boer and Redig [FR07] consider **aggregation** in the sandpile model on $\mathbb{Z}^d$. In their setup, the underlying graph for the chip-firing game is the infinite undirected d-dimensional cubic lattice $\mathbb{Z}^d$. Start with each site containing $h \leq 2d-2$ chips. Here h may be even be taken negative, corresponding to starting with a "hole" of depth $H = -h$ at each lattice site; that is, each site absorbs the first H chips it receives, and thereafter fires every time it receives an additional four chips. Now add n chips to the origin and stabilize. Denote by $S_{n,H}$ the set of sites in $\mathbb{Z}^d$ which fired in the process of stabilization.

Theorem 6.1 ([FR07]). *Let $\mathcal{C}(r)$ denote the cube of side length $2r+1$ centered at the origin in $\mathbb{Z}^d$. For each n there exists an integer r_n such that $S_{n,2-2d} = \mathcal{C}(r_n)$.*

In two dimensions, Theorem 6.1 states that $S_{n,-2}$ is a square. Simulations indicate that for general $H \geq -2$, the limiting shape of $S_{n,H}$ in $\mathbb{Z}^2$ may be a polygon with $4H + 12$ sides.

Question 6.2. *In $\mathbb{Z}^2$, is the limiting shape of $S_{n,H}$ as $n \to \infty$ a regular $(4H+12)$-gon? Simulations indicate a regular $(4H+12)$-gon with some "rounding" at the corners; it remains unclear if the rounded portions of the boundary become negligible in the limit. Even if the limiting shape is not a polygon, it would still be very interesting to establish the weaker statement that it has the dihedral symmetry D_{4H+12}.*

The square, octagon and dodecagon corresponding to the cases $H = -2, -1, 0$ are illustrated in Figure 11 (page 358). Regarding Question 6.2, we should note that even the existence of a limiting shape for $S_{n,H}$ has not been proved in the case $H > -2$. On the other hand, as $H \to \infty$ the limiting shape is a ball in all dimensions, as shown by Fey and Redig [FR07] and strengthened in [LP07b]. In the theorem below, ω_d denotes the volume of the unit ball in $\mathbb{R}^d$, and B_r denotes the discrete ball

$$B_r = \{x \in \mathbb{Z}^d \mid x_1^2 + \ldots + x_d^2 < r^2\}.$$

Theorem 6.3. [LP07b]. *Fix an integer $H \geq 2 - 2d$. Let $S_{n,H}$ be the set of sites in $\mathbb{Z}^d$ that fire in the process of stabilizing n particles at the origin, if every lattice site begins with a hole of depth H. Write $n = \omega_d r^d$. Then*

$$B_{c_1 r - c_2} \subset S_{n,H}$$

where

$$c_1 = (2d - 1 + H)^{-1/d}$$

and c_2 is a constant depending only on d. Moreover if $H \geq 1 - d$, then for any $\epsilon > 0$ we have

$$S_{n,H} \subset B_{c_1' r + c_2'}$$

where

$$c_1' = (d - \epsilon + H)^{-1/d}$$

and c_2' is independent of n but may depend on d, H and ϵ.

In particular, note that the ratio $c_1/c_1' \uparrow 1$ as $H \uparrow \infty$.

For many classes of graphs, the identity element of the sandpile group has remarkable properties that are not well understood. Let I_n be the identity element of the $n \times n$ grid graph G_n with wired boundary; the states I_n for four different values of n are pictured in Figure 4. Comparing the pictures of I_n for different values of n, one is struck by their extreme similarity. In particular, we conjecture that as $n \to \infty$ the pictures converge in the following sense to a limiting picture on the unit square $[0, 1] \times [0, 1]$.

Conjecture 6.4. *Let* a_n *be a sequence of integers such that* $a_n \uparrow \infty$ *and* $\frac{a_n}{n} \downarrow 0$. *For* $x \in [0,1] \times [0,1]$ *let*

$$f_n(x) = \frac{1}{a_n^2} \sum_{\substack{y \in G_n \\ ||y-nx|| \leq a_n}} I_n(y).$$

There is a sequence a_n *and a function* $f : [0,1] \times [0,1] \to \mathbb{R}_{\geq 0}$ *which is locally constant almost everywhere, such that* $f_n \to f$ *at all continuity points of* f.

Most intriguing is the apparent fractal structure in the conjectural f. Recent progress has been made toward understanding the fractal structure of the identity element of a certain orientation of the square grid; see [CPS07].

By Lemma 4.9, the recurrent orbits of the rotor-router operation on an Eulerian digraph are extremely short: although the number of unicycles is typically exponential in the number of vertices, the orbits are all of size equal to the number of edges. One would expect that such short orbits are not the norm for general digraphs.

Question 6.5. *Does there exist an infinite family of non-Eulerian strongly connected digraphs* G_n*, such that for each* n*, all the unicycles of* G_n *lie in a single orbit of the rotor-router operation?*

Another question stemming from Lemma 4.9 is the following. Fix two edges e_0 and e_1 of a digraph G. Starting from a unicycle on G, record a 0 each time the chip traverses the edge e_0, and record a 1 each time it traverses e_1. If G is Eulerian, then Lemma 4.9 implies that the resulting sequence will simply alternate $0, 1, 0, 1, \ldots$. For a general digraph, the sequence is periodic, since the initial unicycle must recur; what can be said about the period?

Lastly, the articles [PDDK96] and [PPS98] contain several conjectures that are supported by both credible heuristics and computer experiments, but that have not been rigorously proved. For instance, it appears that, with random initial rotor orientations, the set of sites visited by a rotor-router walk of length n in the plane typically has diameter on the order of $n^{1/3}$ [PDDK96] (compare this with the corresponding growth rate for random walk in the plane, which is $n^{1/2}$).

References

[AT89] V. Anantharam and P. Tsoucas. A proof of the Markov chain tree theorem. *Statist. Probab. Lett.*, 8(2):189–192, 1989.

[Ath97] C.A. Athanasiadis. Iterating the branching operation on a directed graph. *J. Graph Theory*, 24(3):257–265, 1997.

[Ber86] K.A. Berman. Bicycles and spanning trees. *SIAM J. Algebraic Discrete Methods*, 7(1):1–12, 1986.

[Big97] N. Biggs. Algebraic potential theory on graphs. *Bull. London Math. Soc.*, 29(6):641–682, 1997.

[Big99] N.L. Biggs. Chip-firing and the critical group of a graph. *J. Algebraic Combin.*, 9(1):25–45, 1999.

[BL92] A. Björner and L. Lovász. Chip-firing games on directed graphs. *J. Algebraic Combin.*, 1(4):305–328, 1992.

[BLS91] A. Björner, L. Lovász, and P.W. Shor. Chip-firing games on graphs. *European J. Combin.*, 12(4):283–291, 1991.

[BR02] Y. Le Borgne and D. Rossin. On the identity of the sandpile group. *Discrete Math.*, 256(3):775–790, 2002.

[BTW87] P. Bak, C. Tang, and K. Wiesenfeld. Self-organized criticality: an explanation of the $1/f$ noise. *Phys. Rev. Lett.*, 59(4):381–384, 1987.

[BW97] N. Biggs and P. Winkler. Chip-firing and the chromatic polynomial. Technical Report LSE-CDAM-97-03, London School of Economics, Center for Discrete and Applicable Mathematics, 1997.

[CPS07] S. Caracciolo, G. Paoletti, and A. Sportiello. Closed-formula identities for the abelian sandpile model, 2007. Preprint.

[CR00] R. Cori and D. Rossin. On the sandpile group of dual graphs. *European J. Combin.*, 21(4):447–459, 2000.

[CRR+97] A.K. Chandra, P. Raghavan, W.L. Ruzzo, R. Smolensky, and P. Tiwari. The electrical resistance of a graph captures its commute and cover times. *Comput. Complexity*, 6(4):312–340, 1996/97.

[CS06] J.N. Cooper and J. Spencer. Simulating a random walk with constant error. *Combin. Probab. Comput.*, 15(6):815–822, 2006. http://www.arxiv.org/abs/math.CO/0402323.

[DF91] P. Diaconis and W. Fulton. A growth model, a game, an algebra, Lagrange inversion, and characteristic classes. *Rend. Sem. Mat. Univ. Politec. Torino*, 49(1):95–119 (1993), 1991.

[Dha90] D. Dhar. Self-organized critical state of sandpile automaton models. *Phys. Rev. Lett.*, 64(14):1613–1616, 1990.

[DR04] A. Dartois and D. Rossin. Height arrow model. *Formal Power Series and Algebraic Combinatorics*, 2004.

[DS84] P. Doyle and J. Snell. *Random Walks and Electric Networks*. Mathematical Association of America, 1984.

[DTW03] I. Dumitriu, P. Tetali, and P. Winkler. On playing golf with two balls. *SIAM J. Discrete Math.*, 16(4):604–615 (electronic), 2003.

[Eng75] A. Engel. The probabilistic abacus. *Ed. Stud. Math.*, 6(1):1–22, 1975.

[Eng76] A. Engel. Why does the probabilistic abacus work? *Ed. Stud. Math.*, 7(1–2):59–69, 1976.

[FR07] A. Fey-den Boer and F. Redig. Limiting shapes for deterministic centrally seeded growth models, 2007. http://arxiv.org/abs/math.PR/0702450.

[HP08] A.E. Holroyd and J. Propp. Rotor-router walks, 2008. Manuscript.

[LBG92] G.F. Lawler, M. Bramson, and D. Griffeath. Internal diffusion limited aggregation. *Ann. Probab.*, 20(4):2117–2140, 1992.

[Lev02] L. Levine. The Rotor-Router Model, 2002. Harvard University senior thesis, http://arxiv.org/abs/math/0409407.

[Lev07] L. Levine. The sandpile group of a tree, 2007.
`http://front.math.ucdavis.edu/0703.5868`.

[LKG90] S.H. Liu, T. Kaplan, and L.J. Gray. Geometry and dynamics of deterministic sand piles. *Phys. Rev. A* (3), 42(6):3207–3212, 1990.

[LL07] I. Landau and L. Levine. The rotor-router model on regular trees, 2007.
`http://arxiv.org/abs/0705.1562`.

[Lor89] D.J. Lorenzini. Arithmetical graphs. *Math. Ann.*, 285(3):481–501, 1989.

[Lor91] D.J. Lorenzini. A finite group attached to the Laplacian of a graph. *Discrete Math.*, 91(3):277–282, 1991.

[LP05] L. Levine and Y. Peres. The rotor-router shape is spherical. *Math. Intelligencer*, 27(3):9–11, 2005.

[LP07a] L. Levine and Y. Peres. Spherical asymptotics for the rotor-router model in $\mathbb{Z}^d$. *Indiana Univ. Math. Journal,* to appear, 2007.
`http://arxiv.org/abs/math/0503251`.

[LP07b] L. Levine and Y. Peres. Strong spherical asymptotics for rotor-router aggregation and the divisible sandpile, 2007.
`http://arxiv.org/abs/0704.0688`.

[LR86] F.T. Leighton and R.L. Rivest. Estimating a probability using finite memory. *IEEE Trans. Inf. Theor.*, 32(6):733–742, 1986.

[Pao07] G. Paoletti. Abelian sandpile models and sampling of trees and forests. Master's thesis, Università degli Studi di Milano, 2007.

[PDDK96] V.B. Priezzhev, D. Dhar, A. Dhar, and S. Krishnamurthy. Eulerian walkers as a model of self-organised criticality. *Phys. Rev. Lett.*, 77:5079–5082, 1996.

[PPS98] A.M. Povolotsky, V.B. Priezzhev, and R.R. Shcherbakov. Dynamics of Eulerian walkers. *Phys. Rev. E*, 58:5449–5454, 1998.

[Pro01] J. Propp, 2001. Correspondence with David Griffeath.

[PW98] J.G. Propp and D.B. Wilson. How to get a perfectly random sample from a generic Markov chain and generate a random spanning tree of a directed graph. *J. Algorithms*, 27(2):170–217, 1998.

[RSW98] Y. Rabani, A. Sinclair, and R. Wanka. Local divergence of markov chains and the analysis of iterative load-balancing schemes. In *IEEE Symp. on Foundations of Computer Science*, pages 694–705, 1998.

[Sol99] R. Solomyak. Essential spanning forests and electrical networks on groups. *J. Theoret. Probab.*, 12(2):523–548, 1999.

[Spe87] J. Spencer. *Ten Lectures on the Probabilistic Method*, volume 52 of *CBMS-NSF Regional Conference Series in Applied Mathematics.* Society for Industrial and Applied Mathematics (SIAM), Philadelphia, PA, 1987.

[Spe93] E.R. Speer. Asymmetric abelian sandpile models. *J. Statist. Phys.*, 71(1–2): 61–74, 1993.

[Sta99] R.P. Stanley. *Enumerative Combinatorics. Vol.* 2, volume 62 of *Cambridge Studies in Advanced Mathematics.* Cambridge University Press, 1999.

[Tar88] G. Tardos. Polynomial bound for a chip firing game on graphs. *SIAM J. Discrete Math.*, 1(3):397–398, 1988.

[vdH01] J. van den Heuvel. Algorithmic aspects of a chip-firing game. *Combin. Probab. Comput.*, 10(6):505–529, 2001.

[Wag00] D. Wagner. The critical group of a directed graph, 2000. http://arxiv.org/abs/math/0010241.

Alexander E. Holroyd
Department of Mathematics
University of British Columbia
e-mail: holroyd@math.ubc.ca

Lionel Levine
Department of Mathematics
University of California, Berkeley
e-mail: levine@math.berkeley.edu

Karola Mészáros
Department of Mathematics
Massachusetts Institute of Technology
e-mail: karola@math.mit.edu

Yuval Peres
Theory Group
Microsoft Research
e-mail: peres@microsoft.com

James Propp
Department of Mathematical Sciences
University of Massachusetts at Lowell
e-mail: propp@cs.uml.edu

David B. Wilson
Theory Group
Microsoft Research
e-mail: dbwilson@microsoft.com

Progress in Probability, Vol. 60, 365–381

Ising Model Fog Drip: The First Two Droplets

Dmitry Ioffe and Senya Shlosman

Abstract. We present here a simple model describing coexistence of solid and vapour phases. The two phases are separated by an interface. We show that when the concentration of supersaturated vapour reaches the dew-point, the droplet of solid is created spontaneously on the interface, adding to it a monolayer of a "visible" size.

Mathematics Subject Classification (2000). Primary: 82B20, 82B24 Secondary: 60G60 .

Keywords. Ising model, Phase segregation, Condensation.

1. Introduction: Condensation phenomenon in the Ising model

The phenomenon of droplet condensation in the framework of the Ising model was first described in the papers [DS1], [DS2]. It deals with the following situation. Suppose we are looking at the Ising spins $\sigma_t = \pm 1$ at low temperature β^{-1}, occupying a d-dimensional box T_N^d of the linear size $2N$ with periodic boundary conditions. If we impose the canonical ensemble restriction, fixing the total mean magnetization,

$$\frac{M_N}{\left|T_N^d\right|} \triangleq \frac{1}{\left|T_N^d\right|} \sum \sigma_t,$$

to be equal to the spontaneous magnetization, $m^*(\beta) > 0$, then the typical configuration that we see will look as a configuration of the $(+)$-phase. That means that the spins are taking mainly the values $+1$, while the values -1 are seen rarely, and the droplets of minuses in the box T_N^d are at most of the size of $K(d) \ln N$. We want now to put more -1 particles into the box T_N^d, and we want to see how the above droplet picture would evolve. That means, we want to look at the model with a different canonical constraint:

$$M_N = m^*(\beta) \left|T_N^d\right| - b_N,$$

$b_N > 0$. It turns out that if b_N-s are small, nothing is changed in the above picture; namely, if

$$\frac{b_N}{\left|T_N^d\right|^{\frac{d}{d+1}}} \to 0 \text{ as } N \to \infty,$$

then in the corresponding canonical ensemble all the droplets are still microscopic, not exceeding $K(d)\ln N$ in linear size. On the other hand, once

$$\liminf_{N\to\infty} \frac{b_N}{\left|T_N^d\right|^{\frac{d}{d+1}}} \geq a,$$

for some $a = a(\beta)$ large enough, the situation becomes very different: among many $(-)$-droplets there is one of the linear size of the order of $(b_N)^{1/d} \geq a^{1/d} N^{\frac{d}{d+1}}$, while all the rest of the droplets are still at most logarithmic. Therefore the scale $b_N \sim \left|T_N^d\right|^{\frac{d}{d+1}}$ can be called the *condensation threshold, or dew-point.*

The behavior of the system *at the threshold scale,* i.e., for

$$b_N = c\left|T_N^d\right|^{\frac{d}{d+1}}(1 + o_N(1)),$$

is considered in the $d = 2$ case in [BCK]. It is shown there the existence of the critical value c_{cr} such that for $b_N = cN^{4/3}$ with $c < c_{cr}$ there is no large droplets, while for $c > c_{cr}$ there is exactly one (with probability going to 1 as $N \to \infty$). Furthermore, it turns out that the transition is of the first order (in the size of the large droplet, which does not vanish on the critical scale as $c \searrow c_{cr}$).

Sharp description of the transition *inside the threshold* is considered in [HIK]. The buildup of the probability to see the critical droplet takes place within a window $b_N = c_{cr}N^{4/3} + \delta_{cr}N^{2/3}\log N + \rho N^{2/3}$ (as ρ goes from $-\infty$ to ∞).

The above condensation picture suffers from one (largely esthetic) defect: both below and immediately above the condensation threshold the droplets are "too small to be visible", i.e., they are of the size sublinear with respect to the system size. This defect was to some degree bypassed in [BSS]. It is argued there on heuristic level, that in the low-temperature 3D Ising model in the regime when b_N is already of the volume order, i.e., $b_N \sim \nu N^3$, the sequence of condensations happens, with "visible" results. In such regime one expects to find in the box T_N^3 a droplet Γ of $(-)$-phase, of linear size of the order of N, having the approximate shape of the Wulff crystal, which crystal at low temperatures has 6 flat facets. One expects furthermore that the surface Γ itself has 6 flat facets, at least for some values of b_N. However, when one further increases the "supersaturation parameter" b_N, by an increment of the order of N^2, one expects to observe the condensation of extra $(-)$-particles on one of the flat facets of Γ (randomly chosen), forming a monolayer $\mathfrak{m}$ of thickness of one lattice spacing, and of linear size to be cN, with $c \geq c_{\text{crit}} = c_{\text{crit}}(\beta)$, with $c_{\text{crit}}N$ being smaller than the size of the facet. As b_N increases further, the monolayer $\mathfrak{m}$ grows, until all the facet is covered by it. So one expects to see here the condensation of the supersaturated gas of $(-)$-particles into a monolayer of linear size $\sim c_{\text{crit}}N$, which is "visible". (Indeed, such monolayers were observed in the experiments of condensation of the Pb.) The rigorous results obtained in [BSS] are much more modest: the model studied there is the Solid-on-Solid model, and even in such simplified setting the evidence of appearance of the monolayer $\mathfrak{m}$ of linear size is indirect.

The purpose of the present paper is to consider another 3D lattice model, where one can completely control the picture and prove the above behavior to happen. Namely, we consider a system of ideal particles in the phase transition regime, and we put these phases – the vapour phase and the solid phase – into coexistence by applying the canonical constraint, i.e., by fixing the total number of particles. We study the interface Γ, separating them, and we show that when we increase the total number of particles, the surface Γ changes in the way described above. More precisely, we show that for some values of concentration the surface Γ is essentially flat, but when the concentration increases up to the dew-point, a monolayer $\mathfrak{m}$ of a size at least $c_{\text{crit}}N$ appears on Γ, with N being the linear size of our system.

2. Informal description of the main result

In this section we describe our results informally. We will use the language of the Ising model, though below we treat rigorously a simpler model of the interface between two ideal particles phases. Ising model language makes the description easier; moreover, we believe that our picture holds for the Ising spins as well.

Suppose we are looking at the Ising spins $\sigma_t = \pm 1$ at low temperature β^{-1} in a 3D box B_N of the linear sizes $RN \times RN \times 2N$. The parameter R should be chosen sufficiently large in order to be compatible with the geometry of monolayer creation as described below. We impose $(+)$-boundary conditions in the upper half-space ($z > 0$), and $(-)$-boundary conditions in the lower half-space ($z < 0$). These $(\pm)$-boundary conditions force an interface Γ between the $(+)$ and the $(-)$ phases in V_N, and the main result of the paper [D1] is a claim that the interface Γ is rigid. It means that at any location, with probability going to 1 as the temperature $\beta^{-1} \to 0$, the interface Γ coincides with the plane $z = 0$. If we impose the canonical ensemble restriction, fixing the total mean magnetization M_N to be zero, then the properties of Γ stay the same.

We will now put more -1 particles into V_N; that is, we fix M_N to be

$$M_N = -b_N = -\delta N^2,$$

and we will describe the evolution of the surface Γ as the parameter $\delta > 0$ grows. The macroscopic image of this evolution is depicted on Figure 1.

0. $\mathbf{0} \leq \delta < \delta^{\mathbf{1}}$. Nothing is changed in the above picture – namely, the interface Γ stays rigid. It is essentially flat at $z = 0$; the local fluctuations of Γ are rare and do not exceed $K \ln N$ in linear size.

I. $\delta^{\mathbf{1}} < \delta < \delta^{\mathbf{2}}$. The monolayer $\mathfrak{m}_1$ appears on Γ. This is a random outgrowth on Γ, of height one. Inside $\mathfrak{m}_1$ the height of Γ is typically $z = 1$, while outside it we have typically $z = 0$.

For δ close to δ^1 the shape of $\mathfrak{m}_1$ is *the Wulff shape,* given by the Wulff construction, with the surface tension function $\tilde{\tau}^{2D}(n)$, $n \in \mathbb{S}^1$, given by

$$\tilde{\tau}(n) = \frac{d}{dn}\tau^{3D}(m)\Big|_{m=(0,0,1)}. \tag{2.1}$$

Here $\tau^{3D}(m)$, $m \in \mathbb{S}^2$ is the surface tension function of the 3D Ising model, the derivatives in (2.1) are taken at the point $(0,0,1) \in \mathbb{S}^2$ along all the tangents $n \in \mathbb{S}^1$ to the sphere $\mathbb{S}^2$. The "radius" of $\mathfrak{m}_1$ is of the order of N, i.e., it equals to $r_1(\delta)N$, and as $\delta \searrow \delta^1$ we have $r_1(\delta) \searrow r_{cr} > 0$. In particular, we never see a monolayer $\mathfrak{m}$ of radius smaller than $r_{cr}N$.

As we explain below r_{cr} should scale like $R^{2/3}$. In particular, it is possible to choose R in such a fashion that $R > 2r_{cr}$ or, in other words, for values of R sufficiently large the critical droplet fits into B_N.

As δ increases, the monolayer $\mathfrak{m}_1$ grows in size, and at a certain moment $\delta = \delta^{\mathbf{1.5}}$ it touches the faces of the box B_N After that moment the shape of $\mathfrak{m}_1$ is different from the Wulff shape. Namely, it is *the Wulff plaquette* (see [SchS]), made from four segments on the four sides of the $RN \times RN$ square, connected together by the four quarters of the Wulff shape of radius $\tilde{r}_1(\delta)N$. We have evidently $\tilde{r}_1(\delta^{\mathbf{1.5}}) = R/2$. As $\delta \nearrow \delta^{\mathbf{2}}$, the radius $\tilde{r}_1(\delta)$ decreases to some value $\tilde{r}_1(\delta^{\mathbf{2}})N$, with $\tilde{r}^1(\delta^{\mathbf{2}}) > 0$.

II. $\delta^{\mathbf{2}} < \delta < \delta^{\mathbf{2.5}}$. The second monolayer $\mathfrak{m}_2$ is formed on the top of $\mathfrak{m}_1$. Asymptotically it is of Wulff shape with the radius $r_2(\delta)N$, with $r_2(\delta) \searrow r_2^+(\delta^{\mathbf{2}})$ as $\delta \searrow \delta^2$, with $r_2^+(\delta^2) > 0$. The first monolayer $\mathfrak{m}_1$ has a shape of Wulff plaquette with radius $\tilde{r}_1(\delta)$, which satisfies

$$\tilde{r}_1(\delta) = r_2(\delta).$$

A somewhat curious relation is:

$$r_2^+(\delta^{\mathbf{2}}) \text{ is strictly bigger than } \tilde{r}_1(\delta^{\mathbf{2}}).$$

In other words, the Wulff-plaquette-shaped monolayer $\mathfrak{m}_1$ undergoes a jump in its size and shape as the supersaturation parameter δ crosses the value δ^2. In fact, the monolayer $\mathfrak{m}_1$ shrinks in size: the radius $\tilde{r}_1(\delta)$ increases as δ grows past $\delta^{\mathbf{2}}$.

II.5 $\delta^{\mathbf{2.5}} < \delta < \delta^{\mathbf{3}}$. At the value $\delta = \delta^{\mathbf{2.5}}$ the growing monolayer $\mathfrak{m}_2$ meets the shrinking monolayer $\mathfrak{m}_1$, i.e., $r_2(\delta^{\mathbf{2.5}}) = \tilde{r}_1(\delta^{\mathbf{2.5}}) = R/2$. Past the value $\delta^{\mathbf{2.5}}$ the two monolayers $\mathfrak{m}_2 \subset \mathfrak{m}_1$ are in fact asymptotically equal, both having the shape of the Wulff plaquette with the same radius $\tilde{r}_1(\delta) = \tilde{r}_2(\delta)$, decreasing to the value $\tilde{r}_1(\delta^{\mathbf{3}}) = \tilde{r}_2(\delta^{\mathbf{3}})$ as δ increases up to $\delta^{\mathbf{3}}$.

III. $\delta^{\mathbf{3}} < \delta < \delta^{\mathbf{4}}$. The third monolayer $\mathfrak{m}_3$ is formed, of the asymptotic radius $r_3(\delta)N$, with $r_3(\delta) \searrow r_3^+(\delta^{\mathbf{3}})$ as $\delta \searrow \delta^{\mathbf{3}}$, with $r_3^+(\delta^{\mathbf{3}}) > 0$. The radii of two bottom Wulff plaquettes $\tilde{r}_1(\delta) = \tilde{r}_2(\delta) = r_3(\delta)$ decrease to the value $r_3^+(\delta^{\mathbf{3}})$ as δ decreases down to $\delta^{\mathbf{3}}$, with $r_3^+(\delta^{\mathbf{3}}) > \tilde{r}_i(\delta^{\mathbf{3}})$, so the two Wulff plaquettes $\mathfrak{m}_1, \mathfrak{m}_2$ shrink, jumping to a smaller area, as δ passes the threshold value $\delta^{\mathbf{3}}$.

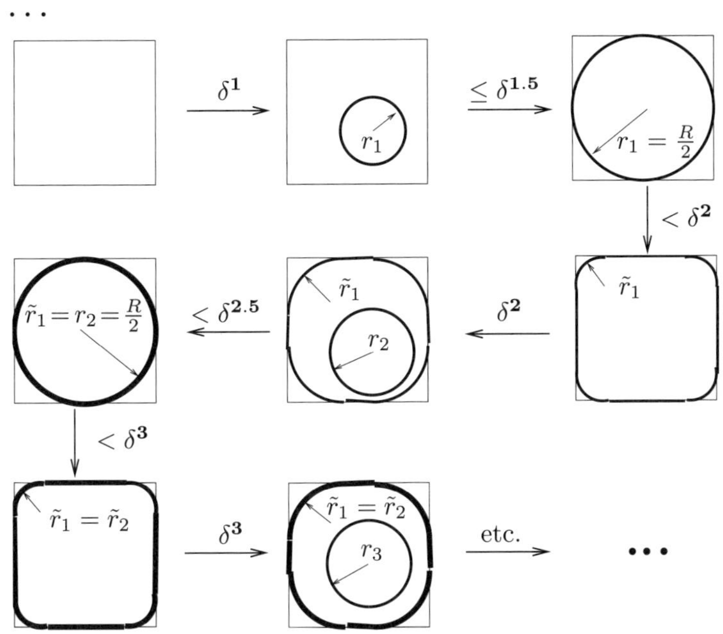

FIGURE 1. Creation and evolution of macroscopic monolayers on Γ as δ grows

A complete investigation of the restricted Wulff variational problem (see (7.5) below) and, accordingly, a rigorous treatment of the interface repulsion phenomenon which shows up on the microscopic level in all the regimes from **II.5** on is relegated to a forthcoming paper [IS]. For the rest of the paper we shall focus on the regimes **0, I** and **II** in the context of a simplified model which we proceed to introduce.

3. Our model

We consider the following lattice model of two-phase coexistence. The 3D box

$$B_N = \Lambda_N \times \{-N - 1/2, -N + 1/2, \dots, N - 1/2, N + 1/2\}$$

is filled with two kinds of particles: v-particles (vapour phase) and s-particles (solid phase). Here Λ_N is a two-dimensional $RN \times RN$ box;

$$\Lambda_N = \{0, 1, \dots, RN - 1\}^2 ,$$

and R is a constant, which we shall set later on to be big enough, in order to make our picture reacher. We have $|B_N| = 2R^2N^3$. Vapour v-particles are occupying the upper part of B_N, while solid s-particles – the lower part. Some sites of the box B_N can be empty. In our model the two phases are separated by an interface Γ, which is supposed to be an SOS-type surface; it is uniquely defined by a function

$$h_\Gamma : \Lambda_N^\circ \to \{-N, -N+1, \dots, N\},$$

where Λ_N° is the interior of Λ_N. We assume that the interface Γ is pinned at zero height on the boundary $\partial\Lambda_N$, that is $h_\Gamma \equiv 0$ on $\partial\Lambda_N$.

Such a surface Γ splits B_N into two parts; let us denote by $V_N(\Gamma)$ and $S_N(\Gamma)$ the upper and the lower halves. The set of configurations of our model consists thus from a surface Γ plus a choice of two subsets, $\sigma_v \subset V_N(\Gamma)$ and $\sigma_s \subset S_N(\Gamma)$; we have a vapour particle at a point $x \in B_N$ iff $x \in \sigma_v$, and similarly for solid particles.

The partition function $Z_N(\beta)$ of our model is now given by

$$Z_N(\beta) = \sum_{(\Gamma,\sigma_v,\sigma_s)} \exp\{-\beta|\Gamma| - (a|\sigma_v| + b|V_N(\Gamma)\setminus\sigma_v| + c|\sigma_s| + d|S_N(\Gamma)\setminus\sigma_s|)\}. \quad (3.1)$$

Here $|\Gamma|$ is the surface area of Γ, $|\sigma_v|$ is the number of vapour particles,..., while a, b, c, d are four chemical potentials. We want the two phases to be in the equilibrium, so we suppose that

$$e^{-a} + e^{-b} = e^{-c} + e^{-d} \equiv e^{-f},$$

where the last equality is our definition of the free energy f. Accordingly, let us define microscopic occupation probabilities in vapour and solid states as

$$p^v = e^{f-a} \quad \text{and} \quad p^s = e^{f-c}.$$

To mimic the fact that the density of the solid state has to be higher, we impose the relation $p^v < p^s$.

We will study our model under the condition that the total number of particles is fixed, and in the leading order of N it is $2\rho R^2 N^3$, with ρ between the values p^v and p^s. Of course, flat interface at level zero should correspond to the choice

$$\rho_0 = \frac{p^s + p^v}{2}.$$

More generally, given $\rho = \rho_0 + \Delta$, one expects to find Γ to be located approximately at the height ℓN above zero level, where ℓ satisfies

$$\frac{\ell}{2}(p^s - p^v) = \Delta.$$

The above reasoning suggests that in our model the formation of macroscopic monolayers over flat interface should happen in canonical ensemble with total number of particles being fixed at

$$2\rho_0 R^2 N^3 + \delta N^2 \triangleq a_0 N^3 + \delta N^2 \quad (3.2)$$

with varying δ.

We will denote by $\mathbb{P}$ the ("grand canonical") probability distribution on triples $\{\Gamma, \sigma_v, \sigma_s\}$ corresponding to the above partition function. Our main interest in this paper is the study of the conditional distribution of the random surface Γ, under condition that the total *number of particles*

$$\Sigma \triangleq |\sigma_v| + |\sigma_s| \triangleq \Sigma_v + \Sigma_s,$$

is fixed, i.e., the distribution $\mathbb{P}\left(\Gamma \,\middle|\, \Sigma = a_0 N^3 + \delta N^2\right)$.

To study this conditional distribution we rely on Bayes' rule,

$$\mathbb{P}\left(\Gamma \,\middle|\, \Sigma = a_0 N^3 + \delta N^2\right) = \frac{\mathbb{P}\left(\Sigma = a_0 N^3 + \delta N^2 \,\middle|\, \Gamma\right) \mathbb{P}(\Gamma)}{\sum_{\Gamma'} \mathbb{P}\left(\Sigma = a_0 N^3 + \delta N^2 \,\middle|\, \Gamma'\right) \mathbb{P}(\Gamma')}.$$

The control over the conditional probabilities $\mathbb{P}\left(\bullet \,\middle|\, \Gamma\right)$ comes from volume order local limit theorems for independent Bernoulli variables, whereas a priori probabilities $\mathbb{P}(\Gamma)$ are derived from representation of Γ in terms of a gas of non-interacting contours.

In the sequel $c_1, c_2, \ldots$ are positive constants which appear in various inequalities and whose values are fixed in such a way that the corresponding bounds hold true.

4. Volume order limit theorems

The study of probabilities $\Pr\left(\Sigma = a_0 N^3 + \delta N^2 \,\middle|\, \Gamma\right)$ is easy, since we are dealing with independent variables. Indeed, let $B_N = S_N \cup V_N$ be the decomposition of B_N induced by Γ. Then, the $\mathbb{P}\left(\bullet \,\middle|\, \Gamma\right)$-conditional distribution of the overall number of particles is

$$\Sigma = \sum_{i \in S_N} \xi_i^s + \sum_{j \in V_N} \xi_j^v,$$

with iid Bernoulli(p^s) random variables ξ_i^s, and iid Bernoulli(p^v) random variables ξ_j^v.

Let $\alpha(\Gamma)$ be the signed volume under the interface Γ,

$$\alpha(\Gamma) = \int\int h_\Gamma(x, y) dx dy, \tag{4.1}$$

where we set h_Γ to be equal to $h_\Gamma(i)$ in the unit box $i + [1/2, 1/2]^2$. Clearly, $|S_N| = R^2 N^3 + \alpha(\Gamma)$ and $|V_N| = R^2 N^3 - \alpha(\Gamma)$. Accordingly,

$$\mathbb{E}\left(\Sigma \,\middle|\, \Gamma\right) = a_0 N^3 + \alpha(\Gamma) p^{sv},$$

where $p^{sv} \triangleq p^s - p^v$. Introducing the variances $D^s = p^s(1-p^s)$, $D^v = p^v(1-p^v)$ and $D = D^s + D^v$, we infer from the Local Limit Theorem (LLT) behavior: For every K fixed there exist two positive constants c_1 and c_2, such that

$$c_1 \leq \frac{\mathbb{P}\left(\Sigma = a_0 N^3 + \delta N^2 \,\middle|\, \Gamma\right)}{\frac{1}{\sqrt{\pi D|B_N|}} \exp\left\{-\frac{(\alpha(\Gamma)p^{sv} - \delta N^2)^2}{D|B_N|}\right\}} \leq c_2, \tag{4.2}$$

uniformly in N, $|\delta| \leq K$ and Γ, provided $|\alpha(\Gamma)| \leq KN^2$.

5. Surface weights

We now want to describe the a priori probability distribution $\mathbb{P}(\Gamma)$. It is convenient and natural to express it via the weights $\{w(\Gamma)\}$, so that

$$\mathbb{P}(\Gamma) \triangleq \Pr(\Gamma) = \frac{w(\Gamma)}{\sum_\Gamma w(\Gamma)}, \tag{5.1}$$

where we shall use an additional symbol Pr in order to stress that the corresponding probabilities are computed in the contour model we are going to introduce now.

For our purposes it is necessary to introduce a contour parameterization of the set of all surfaces Γ. Contours will live on the bonds of the dual (two dimensional) box $\Lambda_N^* = \{1/2, 3/2, \dots, RN - 3/2\}^2$, and they are defined as follows: Given an interface Γ and, accordingly, the height function h_Γ which, by definition, is identically zero outside Λ_N°, define the following semi-infinite subset $\widehat{\Gamma}$ of $\mathbb{R}^3$,

$$\widehat{\Gamma} = \bigcup_{\substack{(x,y,k) \\ k < h_\Gamma(x,y)}} \left((x,y,k) + \widehat{C}\right),$$

where $\widehat{C} = [-1/2, 1/2]^3$ is the unit cube. The above union is over all $(x,y) \in \mathbb{Z}^2$ and $k \in 1/2 + \mathbb{Z}$.

Consider now the level sets of Γ, i.e., the sets

$$H_k = H_k\left(\widehat{\Gamma}\right) = \left\{(x,y) \in \mathbb{R}^2 : (x,y,k) \in \widehat{\Gamma}\right\}, \; k = -N, -N+1, \dots, N.$$

We define *contours* as the connected components of sets ∂H_k. The length $|\gamma|$ of a contour is defined in an obvious way. Since, by construction all contours are closed polygons composed of the nearest neighbour bonds of Λ_N^*, the notions of interior $\mathrm{int}(\gamma)$ and exterior $\mathrm{ext}(\gamma)$ of a contour γ are well defined. A contour γ is called a $\oplus$-contour ($\ominus$-contour), if the values of the function h_Γ at the immediate exterior of γ are smaller (bigger) than those at the immediate interior of γ.

Alternatively, let us orient the bonds of each contours $\gamma \subseteq \partial H_k$ in such a way that when we traverse γ the set H_k remains to the right. Then $\oplus$-contours are those which are clockwise oriented with respect to their interior, whereas $\ominus$-contours are counter-clockwise oriented with respect to their interior.

Let us say that two oriented contours γ and γ' are compatible, $\gamma \sim \gamma'$, if

1. Either $\mathrm{int}(\gamma) \cap \mathrm{int}(\gamma') = \emptyset$ or $\mathrm{int}(\gamma) \subseteq \mathrm{int}(\gamma')$ or $\mathrm{int}(\gamma') \subseteq \mathrm{int}(\gamma)$.
2. Whenever γ and γ' share a bond b, b has the same orientation in both γ and γ'.

A family $\Gamma = \{\gamma_i\}$ of oriented contours is called consistent, if contours of Γ are pair-wise compatible. It is clear that the interfaces Γ are in one-to-one correspondence with consistent families of oriented contours. The height function h_Γ could be reconstructed from a consistent family $\Gamma = \{\gamma\}$ in the following way: For every contour γ the sign of γ, which we denote as $\mathrm{sign}(\gamma)$, could be read from it orientation. Then,

$$h_\gamma(x,y) = \mathrm{sign}(\gamma)\chi_{\mathrm{int}(\gamma)}(x,y) \quad \text{and} \quad h_\Gamma = \sum_{\gamma\in\Gamma} h_\gamma,$$

where χ_A is the indicator function of A.

We are finally ready to specify the weights $w(\Gamma)$ which appear in (5.1): Let $\Gamma = \{\gamma\}$ be a consistent family of oriented (signed) contours, Then,

$$w(\Gamma) = \exp\left\{-\beta \sum_{\gamma\in\Gamma} |\gamma|\right\}. \tag{5.2}$$

By definition the weight of the flat interface $w(\Gamma_0) = 1$.

6. Estimates in the contour ensemble

In order to make the contour model (5.1) , (5.2) tractable one should, evidently, make certain assumptions on the largeness of β, e.g., e^β should be certainly larger than the connective constant of self-avoiding random walks on $\mathbb{Z}^2$ [MS]. In fact, it would be possible to push for optimal results in terms of the range of β along the lines of recent developments in the Ornstein-Zernike theory [I, CIV1, CIV2]. However, in order to facilitate the exposition and in order to focus on the phenomenon of monolayer creation per se, we shall just conveniently assume that β is so large that one or another form of cluster expansion goes through, see eg. [D2]. Due to the ($\pm$-contour) symmetry of the model the corresponding techniques would be quite similar to those developed in the context of the 2D low temperature Ising model in [DKS]. Consequently, instead of stating conditions on β explicitly we shall just assume that $\beta > \beta_0$, where β_0 is so large that all the claims formulated below are true.

In the sequel we shall employ the following notation: $\mathcal{C}$ for clusters of non-compatible contours and $\Phi_\beta(\mathcal{C})$ for the corresponding cluster weights which shows up in the cluster expansion representation of partition functions.

Peierls estimate on appearance of γ. Given a contour γ and a consistent family of contours Γ, let us say that $\gamma \overset{k}{\in} \Gamma$, if γ appears in Γ exactly k times. Then,

$$\Pr\left(\gamma \overset{k}{\in} \Gamma\right) \le e^{-k\beta|\gamma|}. \tag{6.1}$$

Indeed, every Γ satisfying $\gamma \overset{k}{\in} \Gamma$ can be decomposed as $\Gamma = \Gamma' \cup \gamma \cup \cdots \cup \gamma$. Therefore,

$$\Pr\left(\gamma \overset{k}{\in} \Gamma\right) \le \frac{\sum_{\Gamma'} w(\Gamma') e^{-k\beta|\gamma|}}{\sum_{\Gamma'} w(\Gamma')},$$

where the sums are over all consistent families which are compatible with γ, but do not contain it.

Fluctuations of $\alpha(\Gamma)$ and absence of intermediate contours. The following rough a priori statement is a consequence of (6.1): There exist positive ν such that for every $b_0 > 0$ fixed,

$$\Pr\left(|\alpha(\Gamma)| > bN^2\right) \le c_3 e^{-\nu N\sqrt{b}}, \tag{6.2}$$

uniformly in $b \ge b_0$ and in N large enough.

In view of (4.2) (computed with respect to the flat interface Γ_0 with $\alpha(\Gamma_0) = 0$) the bound (6.2) implies that the canonical distribution $\mathbb{P}\left(\bullet \,\middle|\, \Sigma = a_0 N^3 + \delta N^2\right)$ is concentrated on Γ with

$$\alpha(\Gamma) \le N^2 \max\left\{\frac{\delta^4}{\nu^2 D^2 R^4}, b_0\right\}. \tag{6.3}$$

Now let the interface Γ be given by a consistent collection of contours, and assume that $\gamma \sim \Gamma$. Of course $\alpha(\Gamma \cup \gamma) = \alpha(\Gamma) + \alpha(\gamma)$. Let us assume that the surface Γ satisfies the estimate (6.3). Then

$$\begin{aligned}
&\mathbb{P}\left(\Gamma \cup \gamma \,\middle|\, \Sigma = a_0 N^3 + \delta N^2\right) \\
&\le \frac{\mathbb{P}\left(\Sigma = a_0 N^3 + \delta N^2 \,\middle|\, \Gamma \cup \gamma\right)}{\mathbb{P}\left(\Sigma = a_0 N^3 + \delta N^2 \,\middle|\, \Gamma\right)} \cdot \frac{\Pr(\Gamma \cup \gamma)}{\Pr(\Gamma)} \\
&\le c_4 \exp\left\{c_5 \frac{|\alpha(\gamma)|}{N} - \beta|\gamma|\right\} \le c_4 \exp\left\{c_6 \frac{|\gamma|^2}{N} - \beta|\gamma|\right\}
\end{aligned}$$

where we have successively relied on Bayes' rule, (4.2) and on the isoperimetric inequality.

It follows that for every K there exists $\epsilon = \epsilon(\beta) > 0$ such that intermediate contours γ with

$$\frac{1}{\epsilon} \log N < |\gamma| < \epsilon N \tag{6.4}$$

are, uniformly in $|\delta| < K$, improbable under the conditional distribution

$$\mathbb{P}\left(\bullet \,\middle|\, \Sigma = a_0 N^3 + \delta N^2\right).$$

In the sequel we shall frequently *ignore* intermediate contours, as if they do not contribute at all to the distribution (5.1). To avoid confusion, we shall use $\widehat{\Pr}$ for the restricted contour ensemble, which is defined exactly as in (5.1), except that the intermediate contours γ satisfying (6.4) are suppressed.

7. The surface tension and the Wulff shape

Since we are anticipating formation of a monolayer droplet on the interface, we are going to need the surface tension function in order to study such a droplet and to determine its shape. It is defined in the following way: Let λ be an oriented site self avoiding path on the dual lattice $\mathbb{Z}^2_*$. An oriented contour γ is said to be compatible with λ; $\gamma \sim \lambda$, if $\lambda \cap \text{int}(\gamma) = \emptyset$ and if whenever λ and γ share a bond b, the orientation of b is the same in both λ and γ. Accordingly, if $\mathcal{C}$ is a cluster of (incompatible) contours, then $\mathcal{C} \sim \lambda$ if $\gamma \sim \lambda$ for every $\gamma \in \mathcal{C}$.

In the sequel $0^* = (1/2, 1/2)$ denotes the origin of $\mathbb{Z}^2_*$. Let $x \in \mathbb{Z}^2_*$. Set,

$$T_\beta(x) = \sum_{\lambda:0^* \to x} exp\left\{ -\beta|\lambda| - \sum_{\mathcal{C} \not\sim \lambda} \Phi_\beta(\mathcal{C}) \right\},$$

where the sum is over all oriented self-avoiding paths from 0^* to x.

Let $n \in \mathbb{S}^1$ be a unit vector, and $n^\perp \in \mathbb{S}^1$ is orthogonal to it. The surface tension τ_β in direction n is defined as

$$\tau_\beta(n) = -\lim_{L \to \infty} \frac{1}{L} \log T_\beta(\lfloor Ln^\perp \rfloor).$$

Consider the Wulff variational problem, which is a question of finding the minimum $w_\beta(S)$ of the functional,

$$w_\beta(S) \equiv \min_{\{\lambda:\text{Area}(\lambda)=S\}} \mathcal{W}(\lambda).$$

Here

$$\mathcal{W}(\lambda) = \int_\lambda \tau_\beta(n_s)\, ds,$$

n_s being the unit normal to λ at the point $\lambda(s)$, and the minimum is taken over all closed self-avoiding loops λ, enclosing the area S. Of course, $w_\beta(S) = \sqrt{S} w_\beta(1)$. Let us denote by W_β the Wulff shape, which is the minimizing loop with area $S = 1$.

As in [DKS] it could be shown that if β is sufficiently large, then τ_β is well defined and strictly positive. Furthermore, the boundary of the optimal loop W_β is locally analytic and has uniformly positive and bounded curvature.

One can now apply to the present setting the machinery and the results of [DKS], [DS1], [DS2], [SchS] and [ISch]. They allow us to study the probabilities of the events

$$\Pr(A_b) \equiv \Pr\{\Gamma : \alpha(\Gamma) = b\}, \tag{7.1}$$

where we consider here the probability distribution (5.1).

As it follows from local limit results in the restricted phase [DKS] without intermediate contours (6.4), for all values of b, the probability $\widehat{\Pr}(A_b)$ is bounded above by

$$\widehat{\Pr}(A_b) \le c_7 \exp\left\{ -c_8 \frac{b^2}{N^2} \wedge N \right\}. \tag{7.2}$$

In particular, for the values of $b \ll N^{3/2}$ the main contribution to $\widehat{\Pr}(A_b)$ comes from small contours; $|\gamma| < \epsilon^{-1}\log N$. In other words, for such values of b, conditional distribution $\widehat{\Pr}\left(\cdot \,\middle|\, A_b\right)$ is concentrated on the interfaces Γ which are essentially flat: all contours γ of a typical surface Γ are less than $\epsilon^{-1}\log N$ in length, while their density goes to zero as $\beta \to \infty$.

On the other hand, for values of $b \gg N^{3/2}$ long contours contribute, and the probabilities $\Pr(A_b)$ satisfy

$$\log \Pr(A_b) = -\sqrt{b}w_\beta(1)(1 + o_N(1)), \tag{7.3}$$

provided, of course, that the scaled Wulff shape $\sqrt{b/N^2}\, W_\beta$ fits into the square $[0, R]^2$. Under these two restrictions on b the analysis of [DKS] implies that the conditional distribution $\Pr\left(\cdot \,\middle|\, A_b\right)$ is concentrated on the interfaces Γ which are "occupying two consecutive levels". Namely, the set $\{\gamma_i\}$ of contours, comprising Γ, contains exactly one large contour, γ_0, of diameter $\sim \sqrt{b}$, while the rest of them have their lengths not exceeding $\epsilon^{-1}\ln N$. The contour γ_0 is of $\oplus$-type, so for the majority of points inside γ_0 the value of the height function h_Γ is 1, while outside γ_0 it is mainly zero. Finally, the contour γ_0 has

- *Asymptotic shape*: The contour γ_0 is of size $\sim \sqrt{b}$, and it follows very close the curve $\sqrt{b}W_\beta$. Namely, the latter can be shifted in such a way that the Hausdorff distance
$$\rho_H\left(\gamma_0, \sqrt{b}W_\beta\right) \le \sqrt[3]{b}. \tag{7.4}$$

Of course, all the claims above should be understood to hold only on the set of typical configurations, i.e., on the sets of (conditional) probabilities going to 1 as $N \to \infty$.

In the present paper we also we need to consider such values of $b \sim 2R^2N^2$, when the scaled Wulff shape $\sqrt{b/N^2}\, W_\beta$ does not fit into the square $[0, R]^2$. This situation was partially treated in the paper [SchS], and the technique of that paper provides us with the following information about the typical behavior of Γ under the distribution $\widehat{\Pr}\left(\cdot \,\middle|\, A_b\right)$ for the remaining values of b.

Namely, instead of the Wulff variational problem we have to consider the following *restricted* Wulff variational problem, which is a problem of finding the minimum

$$w_\beta^{rst}(S) \equiv \min_{\{k;\lambda_1,\dots,\lambda_k\}} \mathcal{W}_S^{rst}(k;\lambda_1,\dots,\lambda_k) \equiv \mathcal{W}(\lambda_1) + \dots + \mathcal{W}(\lambda_k), \tag{7.5}$$

where

- the curves $\lambda_1, \dots, \lambda_k$ are closed piecewise smooth loops *inside* the unit square Q_1;
- the loops λ_i are nested: $\mathrm{Int}(\lambda_k) \subseteq \mathrm{Int}(\lambda_{k-1}) \subseteq \dots \subseteq \mathrm{Int}(\lambda_1)$;
- $\mathrm{Area}(\lambda_k) + \mathrm{Area}(\lambda_{k-1}) + \dots + \mathrm{Area}(\lambda_1) = S.$

The parameter k is not fixed; we have to minimize over k as well. For the area parameter S small enough, the minimum in (7.5) is attained at $k = 1$, while λ_1 is the scaled Wulff shape, $\sqrt{S}W_\beta$. In other words, in this regime $w_\beta^{rst}(S) = w_\beta(S)$. Let S_1 be the maximal value, for which the inclusion $\sqrt{S}W_\beta \subset Q_1$ is possible. In the range $S_1 < S < 1$ the solution to (7.5) is given by $k = 1$, while the loop λ_1 is the corresponding Wulff plaquette, described above. In the range $1 < S < 2S_1$ the solution has the value $k = 2$, the curve λ_1 is the Wulff plaquette, while the curve $\lambda_2 \subset \lambda_1$ is the Wulff shape; they are uniquely defined by the two conditions:

1. $\text{Area}(\lambda_2) + \text{Area}(\lambda_1) = S$,
2. the curved parts of λ_1 are translations of the corresponding quarters of λ_2.

In the range $2S_1 < S < 2$ we have $k = 2$, while the loops $\lambda_2 = \lambda_1$ are identical Wulff plaquettes.

The relation (7.3) is generalized to

$$\log \Pr(A_b) = -RNw_\beta^{rst}\left(\frac{b}{R^2N^2}\right)(1 + o_N(1)). \tag{7.6}$$

The function $w_\beta^{rst}(S)$ is evidently increasing in S. For S small it behaves as $c'\sqrt{S}$. In the vicinity of the point $S = 1$ it behaves as $c''\sqrt{S-1}$ for $S > 1$, and as $c'''\sqrt{1-S}$ for $S < 1$. Otherwise it is a smooth function of S, $0 \leq S < 2$. The two singularities we just pointed out, are responsible for the interesting geometric behavior of our model, which has been described informally in Section 2, and will be explicitly formulated in the next Section. Namely, each one is responsible for the appearance of the corresponding droplet.

Accordingly, once the Wulff shape $\sqrt{b/N^2}\, W_\beta$ does not fit into the square $[0, R]^2$, while $b \leq c(\beta) R^2N^2$ (where the constant $c(\beta) \to 1$ as $\beta \to \infty$) the conditional distribution $\Pr\left(\cdot \,\middle|\, A_b\right)$ is concentrated on the interfaces Γ which again are occupying two consecutive levels. The set $\{\gamma_i\}$ of contours, comprising Γ, contains one large contour, γ_0, this time of diameter $\sim R$, which in some places is going very close to the boundary of our box. The rest of contours have their lengths not exceeding $\epsilon^{-1} \ln N$. The contour γ_0 is of $\oplus$-type, and for the majority of points inside γ_0 the value of the height function h_Γ is 1, while outside γ_0 it is mainly zero. Finally, the contour γ_0 has asymptotic shape of the Wulff plaquette, in the same sense as in (7.4).

In the remaining range $R^2N^2 \leq b \leq 2R^2N^2$ the set $\{\gamma_i\}$ of contours, comprising Γ, contains exactly two large contours, γ_0 and γ_1, with $\gamma_1 \subset \text{Int}(\gamma_0)$, both of the $\oplus$-type. The interface Γ is, naturally, occupying three consecutive levels: it is (typically) at the height 2 inside γ_1, at height 1 between γ_0 and γ_1, and at height 0 outside γ_0. Note that for b close to R^2N^2 the contour γ_1 is free to move inside γ_0, so its location is random (as is also the case in the regime of the unique large contour, when the scaled Wulff shape $\sqrt{b/N^2}\, W_\beta$ fits into the square $[0, R]^2$). The contour γ_0, on the other hand, is (nearly) touching all four sides of the boundary of our box, so it is relatively less free to fluctuate.

In the complementary regime, when b is close to $2R^2N^2$, the two contours γ_0 and γ_1 have the same size in the leading order (which is linear in N), while the

Hausdorff distance between them is only $\sim N^{1/2}$; it is created as a result of the entropic repulsion between them. In particular, in the limit as $N \to \infty$, and under the $\frac{1}{N}$ scaling, the two contours coincide, going in asymptotic shape to the same Wulff plaquette. The study of this case needs the technique, additional to that contained in [DKS], [DS1], [DS2], [SchS] and [ISch], since the case of two repelling large contours was not considered there. The case of the values b above $2R^2N^2$ is even more involved, since there we have to deal with several large mutually repelling contours. We will return to it in a separate publication, see [IS].

8. Main result

We are ready now to describe the monolayers creation in our model: Let us fix $p^s > p^v$ (and hence p^{sv} and D), and let β be sufficiently large. Let us also fix R large enough, so that the rescaled Wulff shape of area

$$\sqrt[3]{\frac{D^2 w_\beta^2(1)}{p^{sv}}}\, R^{4/3}$$

fits into the $R \times R$ square.

Theorem 8.1. *Let Γ be a typical interface drawn from the conditional distribution $\mathbb{P}\left(\bullet \,\middle|\, \Sigma = a_0 N^3 + \delta N^2\right)$. Define*

$$\delta^{\mathbf{1}} = \frac{3}{2}\sqrt[3]{D^2 w_\beta^2 p^{sv}}\, R^{4/3}. \tag{8.1}$$

- *For values of δ satisfying $0 < \delta < \delta^{\mathbf{1}}$, the interface Γ is essentially flat: all contours of Γ have lengths bounded above by $\epsilon^{-1}\log N$.*
- *There exists $\delta^{\mathbf{2}} > \delta^{\mathbf{1}}$, such that for $\delta^{\mathbf{1}} < \delta < \delta^{\mathbf{2}}$ the interface Γ has one monolayer. Precisely, Γ contains exactly one large contour γ_0 of approximately Wulff shape (or Wulff plaquette shape), such that*

$$\alpha(\gamma_0) > \frac{2\delta}{3p^{sv}} N^2. \tag{8.2}$$

  The rest of contours of Γ are small; their lengths are bounded above by $\epsilon^{-1}\log N$.
- *Similarly, there exists a value δ_R, such that for $\delta^{\mathbf{2}} < \delta < \delta_R$ the interface Γ has two monolayers, and contains exactly two large contours, γ_0 and $\gamma_1 \subset \mathrm{Int}(\gamma_0)$. The bigger one, γ_0, has the shape of the Wulff plaquette, while the smaller one has the Wulff shape. Again, $\alpha(\gamma_1) > \frac{2\delta}{3p^{sv}} N^2$.*

9. Proof of the main result

Let us fix δ and consider the surface distribution $\mathbb{P}\left(\bullet \,\middle|\, \Sigma = a_0 + \delta N^2\right)$. Since we can ignore intermediate contours (6.4) and since we already know how the typical surfaces looks like in the constraint ensembles $\widehat{\Pr}\left(\bullet \,\middle|\, A_b\right)$, it would be enough to study conditional probabilities $\mathbb{P}\left(A_b \,\middle|\, \Sigma = a_0 + \delta N^2 \right)$. Namely, for every δ we need to know the range of the typical values of the "volume" observable b. To do this we will compare the probabilities $\mathbb{P}\left(A_b \,,\, \Sigma = a_0 + \delta N^2 \right) \sim \mathbb{P}\left(\Sigma = a_0 + \delta N^2 \middle| A_b \right) \widehat{\Pr}\left(A_b\right)$ for various values of b, in order to find the dominant one.

There are three regimes to be worked out: Fix $\eta \in (0, 1/2)$ and c_9 small enough.

Case 1. $b \le N^{1+\eta}$. By (4.2) and (7.2),

$$\begin{aligned} & c_{10} exp\left\{ -\frac{\delta^2}{2DR^2} N - O\left(\frac{b^2}{N^2}\right) \right\} \\ & \le N^{3/2} \mathbb{P}\left(\Sigma = a_0 + \delta N^2 \,\middle|\, A_b \right) \widehat{\Pr}\left(A_b\right) \\ & \le c_{11} exp\left\{ -\frac{\delta^2}{2DR^2} N \right\}. \end{aligned} \tag{9.1}$$

Case 2. $N^{1+\eta} < b \le c_9 N^2$. By (4.2) and (7.2),

$$\mathbb{P}\left(\Sigma = a_0 + \delta N^2 \,\middle|\, A_b \right) \widehat{\Pr}\left(A_b\right) \le c_{12} exp\left\{ -\frac{\delta^2}{2DR^2} N + \frac{\delta p^{sv} b}{NR^2 D} - c_8 \frac{b^2}{N^2} \wedge N \right\}. \tag{9.2}$$

Obviously, once c_9 is chosen to be sufficiently small, the right-hand side of (9.2) is negligible with respect to the lower bound on left-hand side of (9.1) (computed at $b \ll N^{1+\eta}$).

Case 3. $b = \rho N^2$ with $\rho > c_9$. By (7.6) and, once again, by volume order local limit result (4.2),

$$\begin{aligned} & exp\left\{ -\frac{(\delta - p^{sv}\rho)^2}{DR^2} N - RN w_\beta^{rst}\left(\frac{\rho}{R^2}\right) - o(N) \right\} \\ & \le \mathbb{P}\left(\Sigma = a_0 + \delta N^2 \,\middle|\, A_b \right) \Pr\left(A_b\right) \\ & \le exp\left\{ -\frac{(\delta - p^{sv}\rho)^2}{2DR^2} N - RN w_\beta^{rst}\left(\frac{\rho}{R^2}\right) + o(N) \right\}. \end{aligned} \tag{9.3}$$

Therefore, in order to figure out the dominant contribution between (9.1) and (9.3), we have to find the global minimum of the function

$$\frac{(\delta - p^{sv}\rho)^2}{2DR^2} + R w_\beta^{rst}\left(\frac{\rho}{R^2}\right) \tag{9.4}$$

on the interval $\rho \in [0, 2R^2]$. Similar minimization problem was investigated in [BCK]. It needs just an elementary calculus: For small values of ρ our function reduces to $\frac{(\delta - p^{sv}\rho)^2}{2DR^2} + w_\beta(1)\sqrt{\rho}$. After the following change of variables:

$$\lambda = \frac{p^{sv}\rho}{\delta} \quad \text{and} \quad \kappa = \kappa(\delta) = \frac{\delta^{3/2}}{2DR^2 w_\beta(1)\sqrt{p^{sv}}},$$

we have to look for global minimizers of

$$\phi_\kappa(\lambda) \triangleq \kappa(1-\lambda)^2 + \sqrt{\lambda}.$$

Set

$$\kappa_c = \kappa\left(\delta^{\mathbf{1}}\right) = \frac{1}{2}\left(\frac{3}{2}\right)^{3/2}. \tag{9.5}$$

One easily sees that

- If $\kappa < \kappa_c$, then the global minimizer is 0.
- If $\kappa = \kappa_c$ then there are exactly two global minimizers; 0 and $\lambda_c = 2/3$.
- If $\kappa > \kappa_c$, then the global minimizer λ_m is the maximal solution of

  $$4\kappa\sqrt{\lambda}\,(1-\lambda)\,,$$

  which, in particular, satisfies $\lambda_m > 2/3$.

A similar analysis applies in the vicinity of the singularity of the function $w_\beta^{rst}\left(\frac{\rho}{R^2}\right)$ at $\frac{\rho}{R^2} \sim 1$. Since the function $w_\beta^{rst}(S)$ is monotone, and has the derivative equal to $+\infty$ at $S = 1$, the point of the global minimum of (9.4), which is a monotone function of δ, never belongs to some neighborhood of the point $\frac{\rho}{R^2} = 1$. Therefore at some $\delta = \delta^{\mathbf{2}}$ it jumps from some value $\rho_- < R^2$ to $\rho_+ > R^2$.

The proof of Theorem 1 is, thereby, completed.

10. Conclusions

In this paper we have described a model of the interface between the vapour and liquid phases, evolving as the total number of particles increases. We have shown that the evolution of the interface goes via the spontaneous formation on it of one monolayer of the size of the system. We believe that the same result can be proven for the 3D Ising model with the same boundary conditions, i.e., periodic in two horizontal directions and $\pm$ in the vertical one. It will be very interesting to establish the phenomenon of the monolayer formation in the 3D Ising model with (+)-boundary conditions, when the monolayer attaches itself to a facet of the Wulff-like (random) crystal. This problem, however, seems to be quite difficult, since one needs to control the rounded part of the crystal. This rounded part is probably behaving as a massless Gaussian random surface (compare with [K]), and this alone indicates enough the complexity of the problem.

References

[BCK] Biskup, M., Chayes, L. and Kotecky, R.: Critical Region for Droplet Formation In the Two-Dimensional Ising Model, Comm. Math. Phys., v. 242, pp. 137–183, 2003.

[BSS] Bodineau, T., Schonmann, R. and Shlosman, S.: 3D Crystal: How Flat its Flat Facets Are? Comm. Math. Phys., v. 255, Number 3, pp. 747–766, 2005.

[CIV1] M. Campanino, D. Ioffe and Y. Velenik: *Ornstein-Zernike theory for finite range Ising models above T_c*, Probab. Theory Related Fields 125 (2003), no. 3, 305–349.

[CIV2] M. Campanino, D. Ioffe and Y. Velenik: *Fluctuation theory of connectivities in sub-critical random cluster models*, to appear in Annals of Probability (2008).

[D1] R.L. Dobrushin. *Gibbs states describing a coexistence of phases for the three-dimensional Ising model*, Teor. Ver. i ee Primeneija, 17, 582–600, (1972).

[D2] R. Dobrushin, P. Groeneboom and M. Ledoux: *Lectures on probability theory and statistics* Lectures from the 24th Saint-Flour Summer School held July 7–23, 1994. Edited by P. Bernard. Lecture Notes in Mathematics, 1648. Springer-Verlag, Berlin, (1996).

[DKS] R.L. Dobrushin, R. Kotecky and S.B. Shlosman: *Wulff construction: a global shape from local interaction,* AMS translations series, Providence (Rhode Island), 1992.

[DS1] R.L. Dobrushin and S. Shlosman: *Large and moderate deviations in the Ising model*, In: "Probability contributions to statistical mechanics", R.L. Dobrushin ed., "Advances in Soviet Mathematics", v. 18, pp. 91–220, AMS, Providence, RI, 1994

[DS2] R.L. Dobrushin and S. Shlosman: *Droplet condensation in the Ising model: moderate deviations point of view*, Proceedings of the NATO Advanced Study Institute: "Probability theory of spatial disorder and phase transition", G. Grimmett ed., Kluwer Academic Publishers, vol. 20, pp. 17–34, 1994

[HIK] O. Hryniv, D. Ioffe and R. Kotecky, *in preparation* (2007).

[K] R. Kenyon: *Dominos and the Gaussian free field*, Ann. Prob. 29, no. 3 (2001), 1128–1137.

[I] D. Ioffe: *Ornstein-Zernike behaviour and analyticity of shapes for self-avoiding walks on Z^d*, Markov Process. Related Fields 4 (1998), no. 3, 323–350.

[ISch] D. Ioffe and R.H. Schonmann: *Dobrushin-Kotecký-Shlosman theorem up to the critical temperature*, Comm. Math. Phys. 199 (1998), no. 1, 117–167.

[IS] D. Ioffe and S. Shlosman: *Ising model fog drip, II: the puddle.* In preparation.

[MS] N. Madras and G. Slade: *The self-avoiding walk*: Probability and its Applications. Birkhäuser Boston, Inc., Boston, MA, (1993).

[SchS] R.H. Schonmann and S. Shlosman: *Constrained variational problem with applications to the Ising model*: J. Statist. Phys. 83 (1996), no. 5-6, 867–905.

Dmitry Ioffe
Faculty of Industrial Engineering and Management,
Technion
Haifa 32000, Israel
e-mail: `ieioffe@ie.technion.ac.il`

Senya Shlosman
Centre de Physique Théorique, CNRS,
Luminy Case 907,
F-13288 Marseille, Cedex 9, France
e-mail: `shlosman@cpt.univ-mrs.fr`

Progress in Probability, Vol. 60, 383–427

Convergence to Fractional Brownian Motion and to the Telecom Process: the Integral Representation Approach

Ingemar Kaj and Murad S. Taqqu

Abstract. It has become common practice to use heavy-tailed distributions in order to describe the variations in time and space of network traffic workloads. The asymptotic behavior of these workloads is complex; different limit processes emerge depending on the specifics of the work arrival structure and the nature of the asymptotic scaling. We focus on two variants of the infinite source Poisson model and provide a coherent and unified presentation of the scaling theory by using integral representations. This allows us to understand physically why the various limit processes arise.

Mathematics Subject Classification (2000). Primary 60F05; Secondary 60K10, 60G18, 60G52.

Keywords. Computer traffic, self-similar processes, fractional Brownian motion, stable processes, Poisson point processes, heavy tails, limit theorems.

1. Introduction

Our understanding of the random variation in packet networks computer traffic has improved considerably in the last decade. Mathematical models were developed, which capture patterns observed in traffic data such as self-similarity. An essential element of these models is the use of heavy-tailed distributions at the microscopic scale. Because the mathematics can be involved, it is often difficult to understand physically why heavy-tailed distributions yield the different stochastic processes that appear at the macroscopic scale. We shall use integral representations in order to clarify this mechanism. We aim to give a *coherent and unified presentation* of a large spectrum of approximation results, so that the features and the dependence structure of the limiting processes are convincingly "explained" by the underlying model assumptions including heavy tails. This approach will also allow us to solve some open problems.

A number of different models have been suggested to capture the essential characteristics of packet traffic on high-speed links. A popular view of network traffic is an aggregate of packet streams, each generated by a source that is either in an active on-state transmitting data or an inactive off-state. In reality separate flows of packets interact because of the influence of transport protocols or other mechanisms, but in modeling work it is a standard approach to assume statistical independence between flows. This leads naturally to considering the cumulative workload as the result of adding independent on-off processes that are integrated over time. The superposition of independent renewal-reward processes have a similar interpretation, where the sources are not necessarily switching between on and off but rather change transmission rates randomly at random times. A third category of models is based on Poisson arrivals of independent sessions, where the sessions are typically long-lived and carry workload continuously or in discrete packets. Such models of Poisson shot noise type, called infinite source Poisson processes, have been specifically proposed for modeling noncongested Internet backbone links at the flow level, Barakat *et al.* 2003.

The preceding models have heavy-tailed versions, obtained by assuming that the on/off periods, the interrenewal times, or the session durations are given by heavy-tailed distributions and one can define stationary versions of these traffic models. Through detailed studies, the asympotic behavior of the workload fluctuations around its mean has been investigated and a pattern has emerged with certain generic characteristics. Taqqu (2002) and Willinger *et al.* (2003) provide summaries including details on the relevant networking concepts and observed characteristics of measured traffic. Stegeman (2002), Pipiras *et al.* (2004) and Mikosch *et al.* (2002) give a variety of results while investigating the range of possible asymptotic growth conditions. Briefly, whenever the number of multiplexing flows grows at a fast rate relative to time, fractional Brownian motion appears as a canonical limit process. If the rewards, i.e., the transmission rates, have heavy tails, then a more general stable process with dependent increments, called the Telecom process, appears instead of fractional Brownian motion, see Levy and Taqqu (2000) and Pipiras and Taqqu (2000). Whenever the degree of aggregation is slow compared to time, the natural limit process is a stable Lévy process with independent increments. In an intermediate scaling regime another type of Telecom process appears, which is neither Gaussian nor stable, Gaigalas and Kaj (2003).

Some further papers dealing with fractional Brownian limit processes under fast growth are Rosenkrantz and Horowitz (2002) and Çağlar (2004). Results on approximation by the stable Lévy motion under slow growth conditions are derived in Jedidi *et al.* (2004), and the intermediate scaling regime is further investigated in Kaj and Martin-Löf (2005). The many results in the literature use a variety of mathematical techniques, often complicated and specialized for the particular model studied, offering limited intuition as to the origin of the limit processes and their physical explanation in terms of first principles of the underlying models.

The purpose of this paper is to consider a physical model which shows clearly why these various limiting scaling processes arise. For this purpose we use integral representations and focus on two variants of the infinite source Poisson model. Because integral representations are interpretable physically, they shed light on the structure of the resulting limit processes. By using this approach, we can derive all the above asymptotics in a unified manner. We are also able to provide the solution to an open problem: finding the intermediate process when the rewards have infinite variance. Some of the approximation techniques we use have been unified in Pipiras and Taqqu (2006). In addition, our approach for the case of fixed rewards has been successfully extended in a spatial setting of Poisson germ-grain models and recast in a more abstract formulation involving random fields in Kaj *et al.* (2007), further developed in Biermé *et al.* (2006, 2007).

The paper is organized as follows. In Section 1 we develop the models and derive some basic properties. We state the main results in Section 2 and prove them in Section 3. In Section 4, the convergence in finite-dimensional distributions of the continuous flow model is extended to weak convergence in function space.

1.1. The infinite source Poisson model

Infinite source Poisson models are arrival processes with $M/G/\infty$ input obtained by integrating the standard $M/G/\infty$ queueing system size. The resulting class of Poisson shot noise processes are widely used traffic models which describe the amount of workload accumulating over time. Such models have been suggested as realistic workload processes for Internet traffic, where is is natural to assume that while web sessions are initiated according to a Poisson process, duration lengths and transmission rates could vary considerably. More exactly, the aggregated traffic consists of sessions with starting points distributed according to a Poisson process on the real time line. Each session lasts a random length of time and involves workload arriving at a random transmission rate. There are two slightly different sets of assumptions that are natural to make regarding the precise traffic pattern during a session. The first is that the workload arrives continuously at a randomly chosen transmission rate, which is fixed throughout the session and independent of the session length. The second type of model assumes that the workload arrives in discrete entities, packets, according to a Poisson process throughout the session, and such that the size of each packet is chosen independently from a given packet size distribution. The duration and the continuous or discrete rate of traffic in one session is independent of the traffic in any other session, although in general the sessions overlap. One novelty in this work is that we point out how these two types of models differ in their asymptotic behavior and that we explain the origin of the qualitative differences.

We are going to introduce the workload models using directly an integral representation with respect to Poisson measures, as in Kurtz (1996) and Çağlar (2004), rather than working with a more traditional Poisson shot noise representation, as in Kaj (2005). This approach is designed to help in understanding the scaling limit behavior of the models, and leads to useful representations of the limit processes. In formalizing the traffic pattern, the starting points of sessions will be called arrival times and the session lengths their durations. The traffic rate will be described in terms of a reward distribution, either continuous flow rewards or compound Poisson rewards. With each session in the continuous flow rate model we associate an arrival time S, a duration U and a reward R. A session in the case of compound Poisson packet arrivals is characterized by an arrival time S, a duration U, and a compound Poisson process $\Xi(t)$ constructed from copies of the reward R.

The basic notation and assumptions are as follows:

Arrivals: Workload sessions start according to a Poisson process on the real line with intensity $\lambda > 0$. The arrival times are denoted $\dots, S_j, S_{j+1}, \dots$.

Durations: The session length distribution is represented by the random variable $U > 0$ with distribution function $F_U(u) = P(U \leq u)$ and expected value

$$\nu = E(U) < \infty.$$

We have either

$$E(U^2) < \infty$$

or

$$P(U > u) \sim L_U(u)u^{-\gamma}/\gamma$$

as $u \to \infty$, where $1 < \gamma < 2$. We extend the parameter range to $1 < \gamma \leq 2$, by letting $\gamma = 2$ represent the case $E(U^2) < \infty$.

Rewards: (1) Continuous flow rewards. The transmission rate valid during a session is given by a random variable $R > 0$ with $F_R(r) = P(R \leq r)$ and

$$E(R) < \infty.$$

We suppose either

$$E(R^2) < \infty$$

or

$$P(R > r) \sim L_R(r)r^{-\delta}/\delta$$

as $r \to \infty$, where $1 < \delta < 2$. Again the parameter range extends to $1 < \delta \leq 2$ by letting $\delta = 2$ be the case $E(R^2) < \infty$. Observe that the aggregated workload in a session is the product UR.

(2) Compound Poisson rewards. The packet stream in a session is a compound Poisson process

$$\Xi(t) = \sum_{i=1}^{M(t)} R_i,$$

where the packet sizes (R_i) are independent and identically distributed with distribution $F_R(dr)$ having the same properties as above for continuous flow rewards, and $\{M(t), t \geq 0\}$ is a standard Poisson process of intensity one. In this case, the aggregated workload in a session is $\sum_{i=1}^{M(U)} R_i$.

Remark 1. We use γ, δ as basic parameters for renewals and rewards for a number of reasons: (1) there will be no confusion with other works that used α, β. (2) It maintains the order used in other works: $\gamma \leftrightarrow \alpha$, $\delta \leftrightarrow \beta$. (3) Since the limit processes can be γ-stable or δ-stable, it is preferable to use indices such as γ and δ which do not have the intrinsic meaning that α and β have in relation to stable distributions. We suggest in fact that, in the future, γ and δ be used instead of α and β.

Remark 2. To simplify the presentation of our work and the statements of our results we will set $L_U = L_R = 1$. In the proofs section, however, we deal with the the modifications that one has to do when L_U and L_R are general slowly varying functions.

We are now prepared to define the infinite Poisson source workload process using integrals with respect to a Poisson measure. The aim is to define an infinite source Poisson process, W^*_λ, such that for $t \geq 0$,

$$W^*_\lambda(t) = \text{the aggregated workload in the time interval } [0,t].$$

1.1.1. The continuous flow reward model. Let $N(ds, du, dr)$ denote a Poisson point measure on $R \times R_+ \times R_+$ with intensity measure

$$n(ds, du, dr) = \lambda ds\, F_U(du)\, F_R(dr). \tag{1}$$

We use S, U, R as generic notation for the random quantities and s, u, r for a particular session outcome so that a Poisson event in (s, u, r) represents a session arriving at time s of duration u and with reward size r. With the choice of (1), we obtain a fluid model for network traffic where sessions begin successively on the (physical) time line labeled s at Poisson rate λ. A session is active during the time interval $[s, s+u]$ and transmits traffic at rate r throughout the session, where (u, r) is an outcome of independent random variables (U, R). For example,

$$\int_{-\infty}^{t}\int_0^\infty\int_0^\infty 1_{\{s<t<s+u\}}\, N(ds, du, dr) = \text{the number of active sessions at time } t.$$

To express similarly W^*_λ in terms of the point measure N, we fix $t > 0$ and partition the total traffic streams into traffic originating from sessions that began in the infinite past, $s \leq 0$, and traffic from sessions starting at a time s with $0 < s < t$. In the former case, sessions do not count if $s + u \leq 0$, the contribution to $W^*_\lambda(t)$ is $(u - |s|)r = (s+u)r$ if $0 < s + u \leq t$, and it is tr if $s + u > t$. In the latter case, the amount of traffic workload that counts for $W^*_\lambda(t)$ is ur if $u < t - s$ and $(t-s)r$ otherwise. Hence

$$\begin{aligned} W^*_\lambda(t) &= \int_{-\infty}^{0}\int_0^\infty\int_0^\infty (t \wedge (s+u)_+)r\, N(ds, du, dr) \\ &\quad + \int_0^t\int_0^\infty\int_0^\infty ((t-s)\wedge u)r\, N(ds, du, dr). \end{aligned} \tag{2}$$

Recall (Campbell Theorem, Kingman (1993), Section 3.2) that an integral of the form $I(f) = \int_S f(x)\, N(dx)$, where N is a Poisson random measure on a space S, exists with probability 1 if and only if $\int_S \min(|f(x)|, 1)\, n(dx) < \infty$ where $n(dx) = EN(dx)$. Moreover, if $\int_S |f(x)|\, n(dx) < \infty$ then the expected value of the integral equals $EI(f) = \int_S f(x)\, n(dx)$. Thus,

$$\begin{aligned}
&EW_\lambda^*(t)\\
&= E(R)\Big(\int_{-\infty}^0 \int_0^\infty t \wedge (s+u)_+ \,\lambda ds\, F_U(du) + \int_0^t \int_0^\infty (t-s) \wedge u\, \lambda ds\, F_U(du)\Big)\\
&= \lambda E(R)\Big(\int_0^t \int_s^\infty P(U > u)\, duds + \int_0^t \int_0^s P(U > u)\, duds\Big)\\
&= \lambda \nu E(R) t, \qquad (3)
\end{aligned}$$

by performing in each of the two terms an integration by parts in the variable u. For example,

$$\begin{aligned}
\int_0^\infty (t-s) \wedge u\, F_U(du) &= \int_0^{t-s} u\, F_U(du) + (t-s) P(U > t-s)\\
&= \int_0^{t-s} P(U > u)\, du.
\end{aligned}$$

The two integral terms in (2) may be combined into a single integral, by writing

$$W_\lambda^*(t) = \int_{-\infty}^\infty \int_0^\infty \int_0^\infty ((t-s)_+ \wedge u - (-s)_+ \wedge u) r\, N(ds, du, dr). \qquad (4)$$

The kernel

$$K_t(s,u) = (t-s)_+ \wedge u - (-s)_+ \wedge u \qquad (5)$$

is such that

$$0 \le K_t(s,u) = \begin{cases} 0 & \text{if } s+u \le 0 \text{ or } s \ge t\\ s+u & \text{if } s \le 0 \le s+u \le t\\ t & \text{if } s \le 0,\ t \le s+u\\ u & \text{if } 0 \le s,\ s+u \le t\\ t-s & \text{if } 0 \le s \le t \le s+u. \end{cases}$$

Hence $K_t(s,u)$ is a function of the starting time s and the duration u of a session that measures the length of the time interval contained in $[0,t]$ during which the session is active. Figure 1 indicates the shape of $K_t(s,u)$ defined on the (s,u)-plane when we have fixed a value of t. Write

$$\widetilde{N}(ds, du, dr) = N(ds, du, dr) - n(ds, du, dr) \qquad (6)$$

for the compensated Poisson measure with intensity measure $n(ds, du, dr)$. By (4) and (3),

$$W_\lambda^*(t) = \int_{-\infty}^\infty \int_0^\infty \int_0^\infty K_t(s,u) r\, N(ds, du, dr) \qquad (7)$$

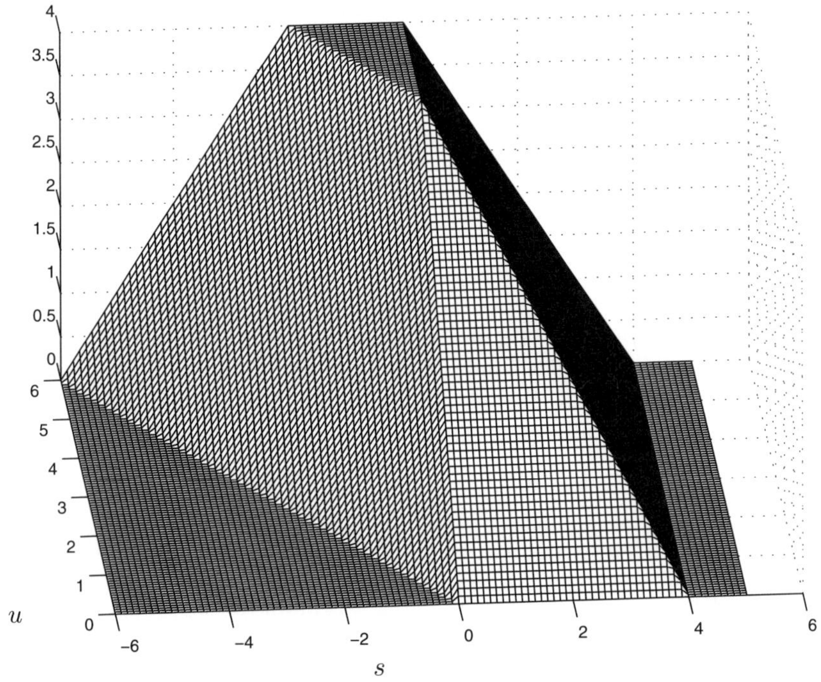

FIGURE 1. The kernel function $K_t(s,u)$, $t=4$, $-6 \le s \le 6$, $0 \le u \le 6$

with

$$\int_{-\infty}^{\infty}\int_0^{\infty}\int_0^{\infty} K_t(s,u)r\, n(ds,du,dr) < \infty,$$

and

$$W_\lambda^*(t) = \lambda\nu E(R)t + \int_{-\infty}^{\infty}\int_0^{\infty}\int_0^{\infty} K_t(s,u)r\,\widetilde{N}(ds,du,dr), \tag{8}$$

which represents the workload in the form of a linear drift and random Poisson fluctuations.

Note that the case of fixed unit rewards, $R \equiv 1$, is contained as a special case of the above by setting F_R equal to the Dirac measure

$$F_R(dr) = \delta_1(dr),$$

which then gives

$$W_\lambda^*(t) = \int_{-\infty}^{\infty}\int_0^{\infty} K_t(s,u)\,N(ds,du)$$

with

$$\int_{-\infty}^{\infty}\int_0^{\infty} K_t(s,u)\,\lambda ds\,F_U(du) = \lambda\nu t < \infty.$$

Here $N(ds, du)$ is the marginal of the Poisson measure $N(ds, du, dr)$ restricted to its first two coordinates.

1.1.2. The compound Poisson arrival workload model. This model results from nesting two Poisson measures as follows. During each session, we allow packets to be generated at discrete Poisson time points. More precisely, consider the compound Poisson process

$$\Xi(t) = \sum_{i=1}^{M(t)} R_i, \quad t \geq 0, \tag{9}$$

where $(R_i)_{i\geq 1}$ is an i.i.d. sequence from the distribution F_R and $M(t)$ is a unit rate Poisson process on R_+. The paths of Ξ are elements in the space D of right-continuous functions with left limits, $t \mapsto \xi(t)$, $t \geq 0$, and we let μ denote the distribution of Ξ defined on D. Let $N_\sharp(ds, du, d\xi)$ be a Poisson measure on $R \times R_+ \times D$ with intensity measure

$$n_\sharp(ds, du, d\xi) = \lambda ds\, F_U(du)\, \mu(d\xi). \tag{10}$$

A Poisson event of $N_\sharp$ at (s, u, ξ) represents a session that starts at s, has duration u, and generates packets according to ξ. The length of time in $[0, t]$ during which the session is active is given by $K_t(s, u)$ defined in (5), and the resulting workload is therefore given by $\xi(K_t(s, u))$. Thus, the accumulated workload $W^*_\lambda(t)$ under compound Poisson packet generation is

$$W^*_{\sharp,\lambda}(t) = \int_{-\infty}^{\infty}\int_0^{\infty}\int_D \xi(K_t(s,u))\, N_\sharp(ds, du, d\xi). \tag{11}$$

Since

$$E\Xi(t) = EM(t)E(R) = tE(R), \tag{12}$$

the expected value of $W^*_{\sharp,\lambda}(t)$ equals

$$\begin{aligned} EW^*_{\sharp,\lambda}(t) &= \int_{-\infty}^{\infty}\int_0^{\infty}\int_D \xi(K_t(s,u))\, \lambda ds\, F_U(du)\, \mu(d\xi) \\ &= \int_{-\infty}^{\infty}\int_0^{\infty} K_t(s,u)\, \lambda ds\, F_U(du)\, E(R) \\ &= \lambda\nu E(R)t, \end{aligned}$$

just as in the continuous flow model. By analogy with (8) we have the representation

$$W^*_{\sharp,\lambda}(t) = \lambda\nu E(R)t + \int_{-\infty}^{\infty}\int_0^{\infty}\int_D \xi(K_t(s,u))\, \widetilde{N}_\sharp(ds, du, d\xi) \tag{13}$$

in terms of the compensated Poisson measure

$$\widetilde{N}_\sharp(ds, du, d\xi) = N_\sharp(ds, du, d\xi) - n_\sharp(ds, du, d\xi). \tag{14}$$

Kurtz (1996) introduced general workload input models of this form, Çağlar (2004) considers the above model with a specific choice of duration distribution F_U. Poissonized integral representations are discussed in Cohen and Taqqu (2003) and Wolpert and Taqqu (2004).

1.2. Preliminary observations

We now represent the continuous flow model as an integral of an instantaneous arrival rate process, show that the workload models have stationary increments, and provide alternative representations which do not involve the presence of an infinite stretch of past arrivals.

1.2.1. Instantaneous arrival rate for continuous flow workload. The integration kernel K_t in (5) has several useful alternative representations. The relation

$$(t-s)_+ \wedge u - (-s)_+ \wedge u = \int_{-s}^{t-s} 1_{\{0<y<u\}}\, dy$$

yields

$$K_t(s,u) = \int_0^t 1_{\{s<y<s+u\}}\, dy \tag{15}$$

and the geometric interpretation

$$K_t(s,u) = |(0,t) \cap (s, s+u)|.$$

The resulting bounds

$$0 \le K_t(s,u) \le t \wedge u \tag{16}$$

are used repeatedly in the proofs below. A further equivalent representation of the kernel function $K_t(s,u)$ is given by

$$K_t(s,u) = \int_0^u 1_{\{0<y+s<t\}}\, dy. \tag{17}$$

As a consequence of relation (15) applied to (7), one can represent the accumulated workload of the continuous flow model as

$$W_\lambda^*(t) = \int_0^t W_\lambda(y)\, dy, \quad W_\lambda(y) = \int_{-\infty}^{\infty}\int_0^{\infty}\int_0^{\infty} 1_{\{s<y<s+u\}}\, r\, N(ds,du,dr). \tag{18}$$

Here the integrand $W_\lambda(y)$, $-\infty < y < \infty$, is itself a well-defined random instantaneous workload arrival rate process and $W_\lambda^*(t)$ is the corresponding cumulative workload. The expressions (18) provide a physical interpretation of W_λ and W_λ^*. The instantaneous rate $W_\lambda(y)$ is the Poisson aggregation of rewards of all sessions that are active at time y, and the cumulative workload W_λ^* builds up accordingly during the time integration over $[0,t]$.

1.2.2. Stationarity of the increments of the workloads.

Lemma 1. *In the continuous flow workload model, the instantaneous arrival rate process* $\{W_\lambda(y), -\infty < y < \infty\}$ *is stationary and the cumulative workload process* $\{W_\lambda^*(t), t \ge 0\}$ *has stationary increments.*

Proof. Because of the time-homogeneity of $n(ds, du, dr)$ in the variable s the shifted process

$$W_\lambda(y+\tau) = \int_{-\infty}^{\infty}\int_0^{\infty}\int_0^{\infty} 1_{\{s<y+\tau<s+u\}}\, r\, N(ds, du, dr)$$

has the same finite-dimensional distributions as

$$\int_{-\infty}^{\infty}\int_0^{\infty}\int_0^{\infty} 1_{\{s<y<s+u\}}\, r\, N(ds, du, dr) = W_\lambda(y). \qquad \square$$

Remark 3. Consider a link of maximal traffic capacity $C > 0$. The process

$$C_\lambda(t) = \int_0^t (W_\lambda(y) - C)_+\, dy, \quad t \geq 0,$$

represents the cumulative workload loss up to time t on the congested link where any traffic of instantaneous rate in excess of C is lost.

Remark 4. In the case $R \equiv 1$, the stationary process W_λ measures the system size of the standard M/G/∞ service model running on the real line with service distribution $G = F_U$. For each fixed y, $W_\lambda(y)$ is Poisson distributed with expected value $\lambda\nu$ because for $R \equiv 1$,

$$W_\lambda(y) = \int\int 1_{\{s<t<s+u\}}\, N(ds, du) = N(A), \quad A = \{(s,u) : s < t < s+u\},$$

with

$$EN(A) = \int\int_A \lambda ds\, F_U(du) = \lambda \int_{-\infty}^t P(U > t-s)\, ds = \lambda \int_0^\infty P(U > s)\, ds = \lambda\nu.$$

For the discrete packet generation workload model we apply a similar but slightly different argument.

Lemma 2. *The compound Poisson arrival workload model $\{W^*_{\sharp,\lambda}, t \geq 0\}$ has stationary increments.*

Proof. By (17),

$$K_{t+\tau}(s,u) - K_t(s,u) = \int_0^u 1_{\{t<y+s<t+\tau\}}\, dy = K_\tau(s-t, u),$$

and hence by (9),

$$\Xi(K_{t+\tau}(s,u)) - \Xi(K_t(s,u)) \stackrel{d}{=} \Xi(K_\tau(s-t,u)).$$

Since $n_\sharp(ds, du, d\xi)$ is time-homogeneous in the variable s, it now follows from (11) that

$$\begin{aligned}
W^*_{\sharp,\lambda}(t+\tau) - W^*_{\sharp,\lambda}(t) &\stackrel{d}{=} \int_{-\infty}^{\infty}\int_0^{\infty}\int_D \xi(K_\tau(s-t,u))\, N_\sharp(ds, du, d\xi) \\
&\stackrel{d}{=} \int_{-\infty}^{\infty}\int_0^{\infty}\int_D \xi(K_\tau(s,u))\, N_\sharp(ds, du, d\xi) \\
&\stackrel{d}{=} W^*_{\sharp,\lambda}(\tau) \qquad \square
\end{aligned}$$

Here and elsewhere, the notation $\stackrel{d}{=}$ denotes equality in the sense of the finite-dimensional distributions.

At this point, having established the property of stationary increments for $W_\lambda^*(t)$ and $W_{\sharp,\lambda}^*(t)$, we comment on the special case $E(R^2) < \infty$ when the reward distribution F_R has a finite second moment. Then, for $s < t$,

$$\mathrm{Cov}(W_\lambda^*(s), W_\lambda^*(t)) = \frac{1}{2}\left(\mathrm{Var}(W_\lambda^*(s)) + \mathrm{Var}(W_\lambda^*(t)) - \mathrm{Var}(W_\lambda^*(t-s))\right)$$

where

$$\mathrm{Var}(W_\lambda^*(t)) = E(R^2)\int_{-\infty}^{\infty}\int_0^\infty K_t(s,u)^2\,ds\,F_U(du).$$

Also,

$$\mathrm{Var}(W_{\sharp,\lambda}^*(t)) = \lambda\nu E(R^2)\,t + (ER)^2\int_{-\infty}^{\infty} K_t(s,u)^2\,\lambda ds\,F_U(du).$$

The crucial property of regular variation which determines the large time behavior of these processes in the finite variance case ($\delta = 2$) is the asymptotic power law

$$\mathrm{Var}(W_\lambda^*(t)) \sim \mathrm{Var}(W_{\sharp,\lambda}^*(t)) \sim \mathrm{const}\, t^{2H}, \quad t \to \infty,$$

where we apply the convention of using a Hurst index H, which in our case is related to the tail parameter γ as

$$H = \frac{3-\gamma}{2} \in (1/2, 1).$$

Our limit results will show that the parameter H appears as a self-similarity index in those cases where the limit process is Gaussian. However, our results cover several other cases as well and hence we will keep γ and δ as basic parameters. A line of research of current interest is that of estimation of such key parameters based on observations of the process. For example, Fay, Roueff and Soulier (2007), study a wavelet-based estimator of the Hurst index for the continuous rate flow model based on the infinite source Poisson process and of the corresponding instantaneous arrival rate process described above. They show consistency of the estimator and study the rate of convergence. Some of the results allow for specific dependencies between durations and rewards. Simulation technique for these processes is a related and relevant direction of research. Here, we restrict to mentioning the references Bardet *et al.* (2003a, 2003b) which survey estimation and simulation techniques for long-range dependent random processes.

1.2.3. Representations based on an equilibrium distribution. The workload processes $W_\lambda^*(t)$ and $W_{\sharp,\lambda}^*$, as defined in (2) and (11), involve sessions arriving at any time s in the infinite past. We now provide an alternative representation of the workload, such that for each t the underlying random mechanism generating $W_\lambda^*(t)$ or $W_{\sharp,\lambda}^*(t)$ consists of sessions with arrival times restricted to the time interval $[0,t]$. To do this, recall the two terms leading to (2). One term

$$\int_0^t\int_0^\infty\int_0^\infty ((t-s)\wedge u)r\,N(ds,du,dr)$$

represents a nonstationary workload process only governed by session arrivals in $(0,t]$. We focus here on the other term, which represents arrivals in the past, is a Poisson integral with expected value

$$\int_{-\infty}^0\int_0^\infty\int_0^\infty (t\wedge(s+u)_+)r\,n(ds,du,dr) = \lambda E(R)\int_0^t\int_u^\infty P(U>v)\,dvdu.$$

To express this as an integral of sessions starting at $s=0$ and with respect to a different Poisson measure, we introduce the notation $\widetilde{U}$ for the equilibrium distribution associated to U having distribution function $F_{\widetilde{U}}(u)=P(\widetilde{U}\le u)$ such that

$$1-F_{\widetilde{U}}(u)=\frac{1}{\nu}\int_u^\infty P(U>v)\,dv\sim\frac{1}{\nu\gamma(\gamma-1)u^{\gamma-1}},\quad u\to\infty. \tag{19}$$

Let $M(dv,du,dr)$ be a Poisson measure on $[0,1]\times R_+\times R_+$ with intensity measure

$$m(dv,du,dr)=\lambda\nu dv\,F_{\widetilde{U}}(du)\,F_R(dr)$$

and independent of $N(ds,du,dr)$. The measure $M(dv,du,dr)$ produces a Poisson distributed number of independent sessions each with duration taken from $\widetilde{U}$ and reward R. One has

$$\int_{-\infty}^0\int_0^\infty\int_0^\infty (t\wedge(s+u)_+)r\,N(ds,du,dr)\overset{d}{=}\int_0^1\int_0^\infty\int_0^\infty (t\wedge u)r\,M(dv,du,dr) \tag{20}$$

To see this, use the fact that the characteristic function of a Poisson integral satisfies

$$\ln E\exp\Big\{i\theta\int f(x)N(dx)\Big\}=\int(e^{i\,f(x)}-1)\,n(dx)$$

and observe that

$$\begin{aligned}
&\log E\exp\Big\{\sum_{j=1}^n\theta_j\int_{-\infty}^0\int_0^\infty\int_0^\infty (t_j\wedge(s+u)_+)r\,N(ds,du,dr)\Big\}\\
&=\int_0^\infty\int_0^\infty\int_0^\infty (e^{i\sum_{j=1}^n\theta_j(t_j\wedge(u-s)_+)r}-1)\lambda ds\,F_U(du)\,F_R(dr)\\
&=\lambda\int_0^\infty F_U(du)\int_0^u ds\int_0^\infty F_R(dr)(e^{i\sum_{j=1}^n\theta_j(t_j\wedge s)r}-1)\\
&=\lambda\int_0^\infty ds\int_s^\infty F_U(du)\int_0^\infty F_R(dr)(e^{i\sum_{j=1}^n\theta_j(t_j\wedge s)r}-1)\\
&=\lambda\nu\int_0^\infty F_{\widetilde{U}}(ds)\int_0^\infty F_R(dr)(e^{i\sum_{j=1}^n\theta_j(t_j\wedge s)r}-1)\\
&=\log E\exp\Big\{\sum_{j=1}^n\theta_j\int_0^1\int_0^\infty\int_0^\infty (t_j\wedge s)r\,M(dv,ds,dr)\Big\},
\end{aligned}$$

where M has intensity measure $m(dv, ds, dr)$. Therefore we can express $W_\lambda^*(t)$ as

$$W_\lambda^*(t) \stackrel{d}{=} \int_0^1 \int_0^\infty \int_0^\infty (t \wedge u) r \, M(dv, du, dr) + \int_0^t \int_0^\infty \int_0^\infty ((t-s) \wedge u) r \, N(ds, du, dr).$$

The expected number of sessions contributing to the first term in this alternative representation is $\lambda\nu$ and we have the following interpretation. A random number of sessions, Poisson distributed with mean $\lambda\nu$, arrive at time $s = 0$. They last independently over time durations $\widetilde{U}$ and transmit independently at rate R, hence a Poisson event at (v, u, r) contributes the workload $(t \wedge u)r$ to $W_\lambda^*(t)$. The number $v \in [0, 1]$ assigned to each session is an auxiliary part of the construction for generating the correct number of initial sessions at time $s = 0$, and has no physical meaning in itself.

With $\widetilde{M}(dv, du, dr) = M(dv, du, dr) - m(dv, du, dr)$, this can also be expressed as

$$W_\lambda^*(t) - \lambda\nu E(R)t \stackrel{d}{=} \int_0^1 \int_0^\infty \int_0^\infty (t \wedge u) r \, \widetilde{M}(dv, du, dr) + \int_0^\infty \int_0^\infty \int_0^\infty ((t-s)_+ \wedge u) r \, \widetilde{N}(ds, du, dr). \quad (21)$$

Similarly, the compound Poisson arrival workload process (11) has the representation

$$W_{\sharp,\lambda}^*(t) \stackrel{d}{=} \lambda\nu E(R)t + \int_0^1 \int_0^\infty \int_D \xi(t \wedge u) \, \widetilde{M}_\sharp(dv, du, d\xi) + \int_0^t \int_0^\infty \int_D \xi((t-s) \wedge u) \, \widetilde{N}_\sharp(ds, du, d\xi), \quad (22)$$

where $\widetilde{M}_\sharp(dv, du, d\xi) = M_\sharp(dv, du, d\xi) - \lambda\nu dv \, F_{\widetilde{U}}(du) \, \mu(d\xi)$.

2. Scaling behavior of the workload process

We are interested in the various limit processes that arise when the speed of time increases in proportion to the intensity of traffic sessions. Heuristically, these approximation results describe the random variation in traffic patterns that correspond to larger and larger volumes of Internet traffic being transmitted over networks of higher and higher capacity.

The traffic fluctuations in an infinite source Poisson system are expressed by the workload process centered around its average value, $W_\lambda^*(t) - \lambda\nu E(R)t$. To balance the increasing session intensity λ, we will speed up time by a factor a and simultaneously normalize the size using a factor b. It follows from (8) and (18) that the scaled continuous flow workload process has the form

$$\frac{1}{b}(W_\lambda^*(at) - \lambda\nu E(R)at) = \frac{1}{b}\int_0^{at} (W_\lambda(y) - \lambda\nu E(R))\,dy \tag{23}$$
$$= \frac{1}{b}\int_{-\infty}^{\infty}\int_0^{\infty}\int_0^{\infty} K_{at}(s,u)r\,\widetilde{N}(ds,du,dr), \quad t \geq 0.$$

Similarly, the scaled compound Poisson workload process is given by

$$\frac{1}{b}(W_{\sharp,\lambda}^*(at) - \lambda at\nu E(R)) = \frac{1}{b}\int_{-\infty}^{\infty}\int_0^{\infty}\int_D \xi(K_{at}(s,u))\widetilde{N}_\sharp(ds,du,d\xi). \tag{24}$$

We are going to study both as λ tends to infinity with a and b appropriately chosen functions of λ, which also tend to infinity. Observe that there are several ways to change variables in the integrals. We will use

$$\frac{1}{b}(W_\lambda^*(at) - \lambda at\nu E(R))$$
$$\stackrel{d}{=} \frac{1}{b}\int_{-\infty}^{\infty}\int_0^{\infty}\int_0^{\infty} K_{at}(as,u)r\,(N(ads,du,dr) - \lambda ads\,F_U(du)\,F_R(dr)) \tag{25}$$
$$\stackrel{d}{=} \frac{a}{b}\int_{-\infty}^{\infty}\int_0^{\infty}\int_0^{\infty} K_t(s,u)r\,(N(ads,adu,dr) - \lambda ads\,F_U(adu)\,F_R(dr)) \tag{26}$$
$$\stackrel{d}{=} \int_{-\infty}^{\infty}\int_0^{\infty}\int_0^{\infty} K_t(s,u)r\Big(N(ads,adu,\frac{b}{a}dr) - \lambda ads\,F_U(adu)\,F_R(\frac{b}{a}dr)\Big), \tag{27}$$

and other variations. We used here the scaling property

$$K_{at}(as,au) = aK_t(s,u). \tag{28}$$

Thus, turning to the compound Poisson arrival model (24) we obtain, e.g.,

$$\frac{1}{b}(W_{\sharp,\lambda}^*(at) - \lambda at\nu E(R)) \stackrel{d}{=} \frac{1}{b}\int_{-\infty}^{\infty}\int_0^{\infty}\int_D \xi(aK_t(s,u))\widetilde{N}_\sharp(ads,adu,d\xi)$$

instead of (26). An interesting feature of our approximation results is that the choice of either continuous flow rate or compound Poisson packet generation during sessions *does affect* the limit process. In fact, we will see that for the compound Poisson model there is an additional averaging effect that takes place during sessions, which changes the asymptotic behavior relative to that of the continuous flow model. This means that the influence of heavy-tailed distributions acting over long time scales alone does not dictate limit results. Rather, the local workload structure over short time scales has an impact on the asymptotics.

Remark 5. To simplify the notation the following useful convention will be used in the sequel: the presence of the term $N(dx) - n(dx)$ will imply, in particular, that $N(dx)$ is a Poisson random measure with intensity measure $n(dx)$.

2.1. Gaussian and stable random measures and processes

We will be using Gaussian and α-stable random measures $M(dx)$ with control measure $m(dx)$ defined for $x \in R^d$. The measure M has the following properties. If $A_1, \ldots, A_n$ are disjoint Borel sets in R^d, then $M(A_1), \ldots, M(A_n)$ are independent random variables. If $\alpha = 2$ (Gaussian case), then for any Borel set A in R^d, the random variable $M(A)$ is normal with mean 0 and variance $m(A)$. If $\alpha < 2$, then

$$\sigma_\alpha M(A) \stackrel{d}{=} \int_A \int_0^\infty r \left(N(dv, dr) - m(dv)\, r^{-(1+\alpha)} dr \right) \tag{29}$$

where

$$\sigma_\alpha = \left(\frac{2\Gamma(2-\alpha)}{\alpha(\alpha-1)} (-\cos \pi\alpha/2) \right)^{1/\alpha}, \tag{30}$$

and thus $M(A)$ has an α-stable distribution which is totally skewed to the right (this is because $r > 0$).

The characteristic function of $M(A)$ is given by

$$\ln E(e^{i\theta M(A)}) = -m(A)|\theta|^\alpha k_\alpha(\theta), \tag{31}$$

where

$$k_\alpha(\theta) = 1 - i(\operatorname{sign}\theta) \tan \pi\alpha/2 \tag{32}$$

(For more details, see Samorodnitsky and Taqqu (1994), pages 156, 119 and 5.) We will write M_2 to denote a Gaussian random measure and M_α to denote an α-stable random measure with $\alpha < 2$. The index α will be either γ or δ.

We will also consider a Lévy-stable process $\Lambda_\alpha(t)$ with index $1 < \alpha < 2$ totally skewed to the right (here again α will be either γ or δ). This is a process with independent increments which can be represented as

$$\Lambda_\alpha(t) = \sigma_\alpha \int_0^t M_\alpha(ds) \stackrel{d}{=} \int_0^t \int_0^\infty r(N(ds, dr) - ds\, r^{-(1+\alpha)} dr), \tag{33}$$

where σ_α is given by (30) and $M(ds)$ is an α-stable random measure with control measure ds and $N(ds, dr)$ is a Poisson random measure with intensity measure $ds\, r^{-1-\alpha} dr$ (see Samorodnitsky and Taqqu, Theorem 3.12.2).

We will also use (standard) fractional Brownian motion $B_H(t)$, which is a Gaussian, mean 0 process, with stationary increments and covariance

$$EB_H(t_1)B_H(t_2) = \frac{1}{2} \left\{ |t_1|^{2H} + |t_2|^{2H} - |t_1 - t_2|^{2H} \right\},$$

where $0 < H < 1$. Fractional Brownian motion is H-self-similar, that is, for any $a > 0$, the processes $B_H(at)$, $t \geq 0$ and $a^H B_H(t)$, $t \geq 0$ have identical finite-dimensional distributions. Fractional Brownian motion reduces to Brownian motion when $H = \frac{1}{2}$.

2.2. Results on fast, intermediate, and slow connection rates

When we let the session intensity λ increase to infinity and simultaneously scale time, letting a tend to infinity, and scale size, letting b tend to infinity, it is possible to obtain several different limit processes in (23) and (24). A crucial feature of these limiting schemes is the relative speed at which λ and a increase. Namely, in most cases it is the asymptotic behavior of the ratio

$$\lambda/a^{\gamma-1}$$

which determines the proper normalizing sequence b and the limit process. More precisely, we will see that this is the case for the continuous flow model in the situation $1<\gamma<\delta\leq 2$, when the durations length has a heavier tail than that of the rewards, and for the compound Poisson model for any set of parameters $1<\gamma,\delta\leq 2$ except $\gamma=\delta=2$. To understand why the ratio $\lambda/a^{\gamma-1}$ enters in the picture, consider the representations (21) and (22) of the workload using the equilibrium session lengths $\widetilde{U}$. At time zero, or at any fixed time point, there is a Poisson number of independent sessions of mean $\lambda\nu$. The remaining length of each session has the distribution $\widetilde{U}$. Hence, letting M be a Poisson random measure with mean $\lambda\nu$, we have

$$\#(\lambda,a) = \text{number of initial sessions still active at time } a \stackrel{d}{=} \sum_{i=1}^{M} 1_{\{\widetilde{U}_i>a\}}.$$

For a given choice of sequences λ and a, $\#(\lambda,a)$ measures the degree to which very long sessions are present and contribute to the total workload. The expected value of the random variable $\#(\lambda,a)$ is

$$E(\#(\lambda,a)) = \lambda\nu P(\widetilde{U}>a) \sim \frac{1}{\gamma(\gamma-1)}\frac{\lambda}{a^{\gamma-1}}, \tag{34}$$

in view of (19). This makes it natural to distinguish three limit regimes based on whether $E(\#(\lambda,a))$ tends to a finite and positive constant, tends to infinity, or vanishes to zero as λ and a goes to infinity. We will introduce a parameter c to quantify the relative speed in the scaling of time and size, and refer to the three cases as:

$$\begin{aligned}
&\text{intermediate connection rate:} && \lambda/a^{\gamma-1}\to c^{\gamma-1}, \quad 0<c<\infty,\\
&\text{fast connection rate:} && \lambda/a^{\gamma-1}\to\infty,\\
&\text{slow connection rate:} && \lambda/a^{\gamma-1}\to 0.
\end{aligned}$$

2.2.1. Intermediate connection rate (ICR). We consider the asymptotics

$$E(\#(\lambda,a)) \sim \text{const}, \quad \lambda,a\to\infty,$$

in which case the number of very long sessions stays bounded. In this situation two kinds of summation schemes influence the workload. First, the aggregation of traffic corresponding to a large value of λ consists of many overlapping sessions, all active at the same fixed time. Secondly, for large a the accumulated traffic in the interval $[0,at]$ involves many sessions that were active during some period in the past. To clarify this structure using heuristic arguments before stating the precise results, let us consider the case $E(R^2) < \infty$, take $c = 1$, and recall the representation (18) of $W_\lambda^*(t)$. We have

$$\frac{1}{a}(W_\lambda^*(at) - \lambda\nu E(R)at) = \frac{1}{a}\int_0^{at} (W_\lambda(y) - \lambda\nu E(R))\,dy$$
$$\sim \frac{1}{a^{(3-\gamma)/2}}\int_0^{at} \frac{W_\lambda(y) - \lambda\nu E(R)}{\sqrt{\lambda}}\,dy,$$

since $\lambda a^{3-\gamma} \sim a^2$. For each y, $W_\lambda(y)$ has a compound Poisson distribution with finite variance and hence for large λ the distribution of the integrand $(W_\lambda(y) - \lambda\nu E(R))/\sqrt{\lambda}$ is approximately Gaussian. The subsequent integration over y affects the covariance structure but preserves the Gaussian distribution. On the other hand, the following argument indicates that we should expect a stable distribution in the limit. Suppose for convenience that λ is an integer and decompose $W_\lambda = \sum_{i=1}^{\lambda} W_1^i$ as a sum of i.i.d. components W_1^i, $1 \le i \le \lambda$. Then

$$\frac{1}{a}(W_\lambda^*(at) - \lambda\nu E(R)at) = \frac{1}{a}\int_0^{at} \sum_{i=1}^{\lambda}(W_1^i(y) - \nu E(R))\,dy$$
$$\sim \frac{1}{\lambda^{1/\gamma}}\sum_{i=1}^{\lambda}\frac{1}{a^{1/\gamma}}\Big(\int_0^{at} W_1^i(y)\,dy - \nu E(R)at\Big),$$

where we use $\lambda a \sim a^\gamma$. The integral process $\int_0^t W_1^i(y)\,dy$, $t \ge 0$, that appears in the last expression is increasing with expected value νt, but since the integrand $W_1^i(y)$ typically stays constant for intervals of length U and the distribution of U has infinite variance, there is no Gaussian central limit law for the corresponding centered process. Instead, we note that $\int_0^t W_1^i(y)\,dy$, after centering and scaling by a as above, should behave as a renewal process having interrenewal times with the heavy-tailed distribution F_U of index γ. For such processes it is known that the limit distribution as $a \to \infty$ is stable with stable index γ. The additional summation over i preserves the stable distribution. For a more detailed discussion in a similar case (of inverse Lévy processes), see Kaj and Martin-Löf (2004).

Turning now to the statement of our first result, it turns out that the limit processes under ICR scaling are neither Gaussian nor stable. In fact new limit processes arise. A further interesting consequence is that the limits are different for the continuous flow rate model and for the compound Poisson model.

Theorem 1. *Consider a pair of parameters* $1 < \gamma < 2$ *and* $1 < \delta \le 2$, *fix an arbitrary constant* c, $0 < c < \infty$, *and assume*

$$\lambda \to \infty, \quad a \to \infty, \quad \frac{\lambda}{a^{\gamma-1}} \to c^{\gamma-1}.$$

Take $b = a$ *as size factor.*

(i) *If $1 < \gamma < \delta \leq 2$, the continuous flow rate model, scaled and normalized as in (23), has the limit*

$$\frac{1}{a}(W_\lambda^*(at) - \lambda\nu E(R)at) \Rightarrow c\, Y_{\gamma,R}(t/c),$$

where

$$\begin{aligned} Y_{\gamma,R}(t) &= \int_{-\infty}^{\infty}\int_0^{\infty}\int_0^{\infty} K_t(s,u) r \left(N(ds,du,dr) - ds\, u^{-(1+\gamma)} du\, F_R(dr)\right) \\ &= \int_{-\infty}^{\infty}\int_0^{\infty} K_t(s,u)\Big(\int_0^{\infty} r\, N(ds,du,dr) - E(R)\, ds\, u^{-(1+\gamma)} du\Big). \end{aligned} \quad (35)$$

In the special case of fixed rewards, $R \equiv 1$, the limit process is

$$Y_\gamma(t) = \int_{-\infty}^{\infty}\int_0^{\infty} K_t(s,u)\left(N(ds,du) - ds\, u^{-(1+\gamma)} du\right). \quad (36)$$

(ii) *The compound Poisson workload model in (24) has the limit process*

$$\frac{1}{a}(W_{\sharp,\lambda}^*(at) - \lambda\nu E(R)at) \Rightarrow E(R)\, c\, Y_\gamma(t/c),$$

where Y_γ is defined in (36).

Convention. The convergence is in the sense of the finite-dimensional distributions in this theorem and in the following one. Weak convergence in function space will be established in Section 4.

Remark 6. The limit process $Y_{\gamma,R}$ is not self-similar, because N does not have the scaling properties that a Gaussian or a stable process has. However, if we assume that the reward distribution $F_R(dr)$ has finite variance then

$$\begin{aligned} \mathrm{Var}(Y_{\gamma,R}(t)) &= E(R^2)\int_{-\infty}^{\infty}\int_0^{\infty} K_t(s,u)^2\, ds\, u^{-(1+\gamma)} du \\ &= E(R^2)\,\sigma^2\, t^{2H}, \qquad H = \frac{3-\gamma}{2}, \end{aligned}$$

where σ^2 is given in (38). Thus, in this case $Y_{\gamma,R}$ is second order self-similar with Hurst index H.

Benassi *et al.* (1997) introduced local asymptotic self-similarity as another means of generalizing the class of self-similar processes. It is shown in Gaigalas and Kaj (2003) and with a proof more adapted to the present setting in Gaigalas (2006), that the process Y_γ is locally asymptotically self-similar with index H and with fractional Brownian motion as tangent process, in the sense that

$$\left\{\frac{Y_\gamma(t+\lambda u) - Y_\gamma(t)}{\lambda^H},\ u \in \mathbf{R}\right\} \Rightarrow \{B_H(u),\ u \in \mathbf{R}\}, \quad \text{as} \quad \lambda \to 0.$$

Benassi *et al.* (2002) defined a stochastic process $X(t)$ to be asymptotically self-similar at infinity with index H if there exists a process $R(t)$ such that

$$\lambda^{-H} X(\lambda t) \to R(t), \quad \text{as } \lambda \to \infty.$$

The intermediate limit process $Y_\gamma(t)$ is asymptotically self-similar at infinity with index $H = 1/\gamma$ and asymptotic process $R(t)$ given by an γ-stable Lévy process, totally skewed to the the right, see Gaigalas (2006).

Remark 7. The difference between the representation of $W_\lambda^*(t) - EW_\lambda^*(t)$ in (8) and that of $Y_{\gamma,R}(t)$ in (35) is that the control measure $F_U(du)$ is now replaced by $u^{-(1+\gamma)}$ which is not a probability measure anymore.

Remark 8. The process $Y_{\gamma,R}$ will be called the *Intermediate Telecom process.* We are thus able to identify the limit process in the case of general reward distributions, which has been an open problem in Pipiras *et al.* (2004). The special case of fixed rewards $R \equiv 1$, has been solved earlier. It can be obtained by combining results in Gaigalas (2006), Kaj (2005) and Gaigalas and Kaj (2003).

Remark 9. The limit for the compound Poisson workload model is a scaled version of Y_γ defined in (36) and, as noted in the theorem, Y_γ is $Y_{\gamma,R}$ in (35) in the special case of fixed rewards $R \equiv 1$.

2.2.2. Fast connection rate (FCR). In this case, a large number of very long sessions contribute in the asymptotic limit of aggregating the traffic workload. Essentially, we will have a summation scheme for processes as in the ordinary central limit theorem, but with strong dependencies building up over time. For the continuous flow model the limit is Gaussian in the case of finite variance rewards and the limit is stable if the reward distribution does not possess finite variance. For the compound Poisson packet generation model, the limit is Gaussian whether the rewards have finite variance or not.

Theorem 2. *Let* $1 < \gamma < 2$, $1 < \delta \leq 2$, *and assume*

$$\lambda \to \infty, \qquad a \to \infty, \qquad \frac{\lambda}{a^{\gamma-1}} \to \infty. \tag{37}$$

Set

$$b = \lambda^{1/\delta}\, a^{(\delta+1-\gamma)/\delta} \quad \textit{so that} \quad b/a = (\lambda/a^{\gamma-1})^{1/\delta} \to \infty.$$

(i) *In the case of finite variance rewards,*

$$1 < \gamma < \delta = 2,$$

so $b = \lambda^{1/2} a^{(3-\gamma)/2}$, *then the limit process for* (23) *is the fractional Brownian motion*

$$\sqrt{E(R^2)}\, \sigma\, B_H(t)$$

with index

$$H = \frac{3-\gamma}{2} \in (1/2, 1),$$

where

$$\sigma^2 = \int_{-\infty}^{\infty} \int_0^{\infty} K_1(s,u)^2\, ds\, u^{-(\gamma+1)} du = \frac{2}{\gamma(\gamma-1)(2-\gamma)(3-\gamma)}. \tag{38}$$

Alternatively, the limit process can be represented as

$$E(R^2)^{1/2} \int_{-\infty}^{\infty} \int_0^{\infty} K_t(s,u)\, M_2(ds,du), \tag{39}$$

where $K_t(s,u)$ *is the kernel defined in* (5) *and* $M_2(ds,du)$ *is a Gaussian random measure with control measure*

$$ds\, u^{-(1+\gamma)} du.$$

(ii) *If the reward distribution has infinite variance with a lighter tail than that of the durations,*

$$1 < \gamma < \delta < 2,$$

then the limit of (23) *is the Telecom process*

$$\begin{aligned} Z_{\gamma,\delta}(t) & \\ &= \int_{-\infty}^{\infty} \int_0^{\infty} \int_0^{\infty} K_t(s,u) r \left(N(ds,du,dr) - ds\, u^{-(1+\gamma)} du\, r^{-(1+\delta)} dr\right) \qquad (40) \\ &= \sigma_\delta \int_{-\infty}^{\infty} \int_0^{\infty} K_t(s,u)\, M_\delta(ds,du), \qquad (41) \end{aligned}$$

where the random measure $M_\delta(ds,du)$ *is* δ*-stable and has the control measure*

$$ds\, u^{-(\gamma+1)} du.$$

The process $Z_{\gamma,\delta}(t)$ *is a* δ*-stable process, which is* H*-self-similar with*

$$H = \frac{\delta + 1 - \gamma}{\delta} \in (1/\delta, 1).$$

The factor σ_δ *is given in* (30) (*with* $\alpha = \delta$).

(iii) *If we replace* W^*_λ *by* $W^*_{\sharp,\lambda}$, *then for arbitrary parameters*

$$1 < \gamma < 2,\ 1 < \delta \leq 2,$$

the limit process of (24) *is the fractional Brownian motion*

$$\big(E(R)\,\sigma\big)\, B_H(t), \quad t \geq 0.$$

Remark 10. The symmetric δ-stable version of the Telecom process appeared in Pipiras and Taqqu (2002). The Telecom process reduces to $CB_H(t)$ when $\delta = 2$. The easiest way to see this is to note that the random measure M_δ is Gaussian when $\delta = 2$ and hence the process $Z_{\gamma,2}$ is Gaussian, has stationary increments and is H-self-similar with $H = (3-\gamma)/2$.

Remark 11. The kernel $K_t(s,u)$ appears both in the representations (40) of the Telecom process and in the representation (35) of the intermediate Telecom process. In (40), the control measure involving r in the stable density $r^{-(1+\delta)}dr$ and thus the Telecom process is a δ-stable process. For the intermediate Telecom process (35), however, the part of the control measure involving r is $F_R(dr)$ which has finite variance in the case $\delta = 2$ and while it has infinite variance in the case $\delta < 2$, the process is not necessarily stable.

2.2.3. Slow connection rate (SCR). The remaining case SCR leads to stable Lévy processes in the asymptotic limit. The interpretation of the scaling condition $E(\#(\lambda, a)) \to 0$, $\lambda, a \to \infty$, in (34) is that there are essentially no sessions that survive the scaling whose remaining durations are so long that they could cause long-range dependence in the limit process. Rather, the additive terms that contribute to the cumulative workload are asymptotically independent and belong to a stable domain of attraction.

The multiplicative constants appearing in the limit depend on the local traffic structure during sessions. Again the limit process for the compound Poisson model depends only on the expected reward $E(R)$, which shows that this is a general property valid for all choices of scaling.

Theorem 3. *Consider the scaling regime*

$$a \to \infty, \qquad \frac{\lambda}{a^{\gamma-1}} \to 0$$

or, if λ is bounded away from zero, just

$$\frac{\lambda}{a^{\gamma-1}} \to 0,$$

and take

$$b = (\lambda a)^{1/\gamma} \quad \text{so that} \quad a/b = (a^{\gamma-1}/\lambda)^{1/\gamma} \to \infty.$$

(i) *If*

$$1 < \gamma < \delta \leq 2,$$

or, more generally, if

$$E(R^\gamma) < \infty,\ 1 < \gamma < 2$$

(including $\gamma = \delta$ with slowly varying functions such that $E(R^\gamma)$ is finite), then the limit for the continuous flow rate model (23) *is*

$$[E(R^\gamma)]^{1/\gamma}\, \Lambda_\gamma(t),$$

where Λ_γ is a Lévy-stable process with stable index γ. The limit process can be represented as

$$\begin{aligned} &E(R^\gamma)^{1/\gamma}\, \Lambda_\gamma(t) \\ &\quad = \int_{-\infty}^{\infty}\int_0^{\infty}\int_0^{\infty} 1_{\{0<s<t\}} ur\Big(N(ds, du, dr) - ds\, u^{-(1+\gamma)} du\, F_R(dr)\Big) \\ &\quad \overset{d}{=} \sigma_\gamma \int_{-\infty}^{\infty}\int_0^{\infty} 1_{\{0<s<t\}} r M_\gamma(ds, dr) \end{aligned}$$

where σ_γ is defined in (30) *(with $\alpha = \gamma$) and $M_\gamma(ds, dr)$ is γ-stable with control measure $ds\, F_R(dr)$, as defined in* (33).

(ii) *For any choice of parameters*

$$1 < \gamma < 2,\, 1 < \delta \leq 2,$$

the compound Poisson workload model (24) *has the limit*

$$E(R)\,\Lambda_\gamma(t).$$

2.3. Remaining choices for the parameters γ and δ

For the continuous model we supposed earlier that $1 < \gamma < \delta \leq 2$. We will now consider the remaining cases

$$\gamma = \delta = 2, \quad \textit{and} \quad 1 < \delta < \gamma \leq 2$$

The first of these, $\gamma = \delta = 2$, remains also for the compound Poisson model. The second, $1 < \gamma < \delta \leq 2$ will be applied to the continuous flow model, together with $1 < \gamma = \delta < 2$, given that proper moments exist. The generic choice of normalization is $b = (\lambda a)^{1/\delta}$ in each of the remaining cases. As $\lambda \to \infty$ and $a \to \infty$, and with this b, the convergence results hold regardless of the limit behavior of $\lambda/a^{\gamma-1}$. Hence the distinctions FCR, ICR, SCR are now irrelevant.

Theorem 4. *Set*

$$b = (\lambda a)^{1/\delta} \tag{42}$$

and assume

$$\lambda \to \infty,\ a \to \infty \quad \textit{or} \quad a \to \infty,\ b \to \infty \quad \textit{in any arbitrary way.}$$

(i) *Assume*

$$\gamma = \delta = 2.$$

Here $E(U^2) < \infty$, $E(R^2) < \infty$. *The continuous flow model in* (23) *has the limit*

$$\sqrt{E(U^2)E(R^2)}\,B(t) \quad t \geq 0,$$

and the compound Poisson model in (24) *has the limit*

$$\sqrt{E(U^2)}E(R)\,B(t) \quad t \geq 0,$$

where $B(t)$, $t \geq 0$, *denotes standard Brownian motion.*

(ii) *Assume*

$$1 < \delta < 2,$$

and that either γ satisfies

$$\delta < \gamma \leq 2$$

or, more generally, that U satisfies

$$E(U^\delta) < \infty$$

(*thus including $\gamma = \delta$ with slowly varying functions making U^δ have finite mean*). *The limit process for the continuous flow model is*

$$[E(U^\delta)]^{1/\delta}\,\Lambda_\delta(t), \quad t \geq 0,$$

where $\Lambda_\delta(t)$ is a Lévy stable process with index δ.

Remark 12. For the case of fixed rewards, $R = 1$, higher-dimensional versions of Theorems 1–3 have been obtained in Kaj *et al.* (2007). Spatial versions of the continuous flow reward model are obtained by replacing the collection of sessions on the real line by a family of sets $\{x_j + u_j C\}_j$ on $\mathbf{R}^d$, where C is a fixed bounded set of volume $|C| = 1$ and vanishing boundary $|\partial C| = 0$. The location and size of the sets are given by a Poisson measure $N(dx, du)$ on $\mathbf{R}^d \times \mathbf{R}^+$ with intensity $\lambda dx\, F(du)$ such that the size distribution $F(du)$ is heavy-tailed at infinity. The analog of the workload functional W^*_λ is taken to be a stochastic integral

$$X(\mu) = \int_{\mathbf{R}^d} \int_{\mathbf{R}^+} \mu(x + uC)\, N(dx, du), \quad \mu \in \mathcal{M},$$

where $\mathcal{M}$ is a suitable subset of signed measures on $\mathbf{R}^d$. Here, $X(\mu)$ represents the configuration of mass on R^d of a Poisson germ-grain model with germs uniformly located in space and heavy-tailed grain size. The choice $d = 1$, $C = [0, 1]$ and $\mu_t(dy) = 1_{\{0<y<t\}}\, dy$ yields

$$\mu_t(s + uC) = \int_{[s,s+u]} 1_{\{0<y<t\}}\, dy = K_t(s, u)$$

in view of (15), which shows for this example $X(\mu_t) = W^*_\lambda(t)$.

By choosing properly the spatial scale, or equivalently, the size of the grains, in relation to the Poisson intensity and taking a limit the fluctuations of $X(\mu)$ again exhibit three different scaling regimes. The limiting operations are carried out with the use of generalized random fields based on a careful choice of the space of measures $\mathcal{M}$. The results in Kaj *et al.* (2007) generalize the Gaussian, stable and intermediate limits obtained here to a spatial setting and are in complete analogy to those of Theorems 1, 2 and 3, for the case of fixed rewards.

Biermé et al. (2006, 2007), extend the Gaussian and the intermediate scaling limit results further for an analogous model where the intensity of the size of grains has a specified power law behavior close to zero. It turns out that for such models one can obtain in the scaling limit, for example, the family of fractional Brownian fields $\{B_H(x),\, x \in \mathbf{R}^d\}$ with Hurst index H, $0 < H < 1/2$. Here, $B_H(x)$, $x \in \mathbf{R}^d$, are zero mean Gaussian random variables such that

$$\mathrm{Cov}(B_H(x), B_H(y)) = \frac{1}{2}\left(|x|^{2H} + |y|^{2H} - |x - y|^{2H}\right).$$

3. Proof of the theorems

The proofs in our setting provide an intuitive feeling for why the various limits appear. We will focus on the characteristic functions of the scaled and normalized workload process. By performing the appropriate limit operation for each choice of limiting scheme and deriving the limiting characteristic functions, we are able to identify the limit processes. We begin by stating characteristic functions for the processes $W^*_\lambda(t)$ and $W^*_{\sharp,\lambda}(t)$ centered at their expected values. We will then consider each case separately. This includes the intermediate, fast, and slow connection rates when the tails of the durations are heavier than the tails of the rewards. Further cases arise when the reward tails are heavier. We will have to consider separately the continuous flow model and the compound Poisson model.

3.1. Characteristic functions

The formulas given in the next two lemmas, which will be used repeatedly in the sequel, are consequences of a general property of Poisson integrals $\int f(x)(N(dx)-n(dx))$, namely that

$$\ln E\exp\Big\{i\int f(x)(N(dx)-n(dx))\Big\} = \int (e^{i\,f(x)}-1-if(x))\,n(dx),$$

which is well defined if

$$\int (f(x)^2\wedge|f(x)|)\,n(dx)<\infty,$$

and in particular, if either $\int f^2(x)\,n(dx)<\infty$ or $\int |f(x)|\,n(dx)<\infty$.

Lemma 3. *The characteristic function for the finite-dimensional distributions of the centered continuous flow workload process $W^*_\lambda(t)-\lambda\nu E(R)t$, $t\geq 0$, is given for arbitrary $n\geq 1$, $0\leq t_1\leq\cdots\leq t_n$, and real $\theta_1,\ldots,\theta_n$, by the relation*

$$\ln E\,\exp\Big\{i\sum_{j=1}^n\theta_j(W^*_\lambda(t_j)-\lambda\nu E(R)t_j)\Big\} = \int_{-\infty}^\infty\int_0^\infty\int_0^\infty h(s,u,r)\,n(ds,du,dr),$$

where

$$h(s,u,r)=\exp\Big\{i\sum_{j=1}^n\theta_jK_{t_j}(s,u)r\Big\}-1-i\sum_{j=1}^n\theta_jK_{t_j}(s,u)r \tag{43}$$

and $n(ds,du,dr)=\lambda ds\,F_U(du)\,F_R(dr)$ is the intensity measure defined in (1).

Lemma 4. *The characteristic function for the finite-dimensional distributions of the centered compound Poisson workload process $W^*_{\sharp,\lambda}(t)-\lambda\nu E(R)t$, $t\geq 0$, is given for arbitrary $n\geq 1$, $0\leq t_1\leq\cdots\leq t_n$, and real $\theta_1,\ldots,\theta_n$, by*

$$\ln E\,\exp\Big\{i\sum_{j=1}^n\theta_j(W^*_{\sharp,\lambda}(t_j)-\lambda\nu E(R)t_j)\Big\} = \int_{-\infty}^\infty\int_0^\infty g(s,u)\,n(ds,du),$$

where

$$n(ds,du)=\lambda ds\,F_U(du)$$

and

$$\begin{aligned} g(s,u) &= E\Big(\exp\Big\{i\sum_{i=1}^{n}\theta_j\Xi(K_{t_j}(s,u))\Big\} - 1 - i\sum_{i=1}^{n}\theta_j\Xi(K_{t_j}(s,u))\Big) \\ &= \exp\Big\{\int_0^\infty\int_0^\infty\Big(\exp\Big\{i\sum_{j=1}^{n}\theta_j 1_{\{w\le K_{t_j}(s,u)\}}r\Big\} - 1\Big)dw\,F_R(dr)\Big\} \\ &\qquad\qquad -1 - i\sum_{j=1}^{n}\theta_j K_{t_j}(s,u)E(R). \end{aligned}$$

Observe that the expressions for the logarithmic characteristic functions stated in Lemmas 3 and 4 above are well defined, because the inequality

$$\big|e^{iu} - 1 - iu\big| \le 2|u|, \quad u \in R, \tag{44}$$

and the relation

$$\int_{-\infty}^{\infty} K_t(s,u)\,ds = ut, \tag{45}$$

which is readily derived from (15), imply

$$\begin{aligned} &\int_{-\infty}^{\infty}\int_0^\infty\int_0^\infty |h(s,u,r)|\,n(ds,du,dr) \\ &\qquad \le 2\sum_{i=1}^{n}|\theta_j|\int_{-\infty}^{\infty}\int_0^\infty\int_0^\infty K_t(s,u)r\,n(ds,du,dr) = 2\sum_{i=1}^{n}|\theta_j|\,\lambda\nu E(R)t \end{aligned}$$

and

$$\begin{aligned} &\int_{-\infty}^{\infty}\int_0^\infty |g(s,u)|\,n(ds,du) \\ &\qquad \le 2\sum_{i=1}^{n}|\theta_j|E(R)\int_{-\infty}^{\infty}\int_0^\infty K_t(s,u)\,n(ds,du) = 2\sum_{i=1}^{n}|\theta_j|\,\lambda\nu E(R)t. \end{aligned}$$

More refined estimates will be needed to carry out the various scaling limit operations.

3.2. Proof of Theorem 1 (ICR)

We can lump together the finite and infinite variance cases but we will need to distinguish between the continuous flow model and the compound Poisson model.

3.2.1. The continuous flow model. Applying (26) with $b = a$ and Lemma 3, we have

$$\begin{aligned} &\ln E\,\exp\Big\{i\sum_{j=1}^{n}\theta_j(W_\lambda^*(at_j) - \lambda\nu E(R)at_j)/a\Big\} \\ &\qquad = \int_{-\infty}^{\infty}\int_0^\infty\int_0^\infty h(s,u,r)\,EN(ads,adu,dr), \end{aligned} \tag{46}$$

where h is defined in (43). Under the ICR assumption $\lambda, a \to \infty$ with $\lambda/a^{\gamma-1} \to c^{\gamma-1}$, the scaled intensity measure has the asymptotic form

$$EN(ads, adu, dr) = \lambda ads\, F_U(adu)\, F_R(dr) \sim c^{\gamma-1}\, dsu^{-(1+\gamma)} du\, F_R(dr).$$

The logarithmic characteristic function of the process $Y_{\gamma,R}$ defined by the Poisson integral expression (35) is given by

$$\ln E \exp\Big\{i \sum_{j=1}^{n} \theta_j\, Y_{\gamma,R}(t_j)\Big\} = \int_{-\infty}^{\infty}\int_0^{\infty}\int_0^{\infty} h(s,u,r)\, dsu^{-(1+\gamma)} du\, F_R(dr),$$

in complete analogy to the result of Lemma 3. Thus,

$$\begin{aligned} &\ln E \exp\Big\{i \sum_{j=1}^{n} \theta_j\, cY_{\gamma,R}(t_j/c)\Big\} \\ &= \int_{-\infty}^{\infty}\int_0^{\infty}\int_0^{\infty} h(cs,cu,r)\, dsu^{-(1+\gamma)} du\, F_R(dr) \qquad (47) \\ &= c^{\gamma-1}\int_{-\infty}^{\infty}\int_0^{\infty}\int_0^{\infty} h(s,u,r)\, dsu^{-(1+\gamma)} du\, F_R(dr), \qquad (48) \end{aligned}$$

where (47) follows from (28) expressed as $cK_{t/c}(s,u) = K_t(cu,cs)$. Hence to prove Theorem 1 i), it is enough to verify that (46) converges to (48) under the ICR scaling.

Integration by parts in the variable u shows that the right-hand side of (46) equals

$$\int_{-\infty}^{\infty}\int_0^{\infty}\int_0^{\infty} \frac{\partial}{\partial u} h(s,u,r)\, ds\, \lambda a P(U > au) du\, F_R(dr), \qquad (49)$$

where U, which has the distribution $F_U(du)$, satisfies by assumption

$$\lambda a P(U > au) \to c^{\gamma-1} u^{-\gamma}/\gamma.$$

If we are allowed to take this limit inside the integral in (49), then another integration by parts will revert the resulting integral into the form (48) and hence conclude the proof. To justify the last steps it remains to demonstrate that the integrand in (49) is appropriately dominated. The proofs of the required estimates simplify somewhat if we first make the change of variable $s \to s + u$. Hence we agree to consider instead of (49) the integral

$$\int_{-\infty}^{\infty}\int_0^{\infty}\int_0^{\infty} \frac{\partial}{\partial u} h(s-u,u,r)\, ds\, \lambda a P(U > au) du\, F_R(dr). \qquad (50)$$

(Note that the function in the integrand is the derivative of h with respect to its second argument u, evaluated in the point $(s-u,u,r)$.) We will use the Potter bounds, see Bingham *et al.* (1987). Since the function $P(U > u)$ is regularly varying at $u \to \infty$ with tail behavior $u^{-\gamma}$, the Potter bound yields for any $\epsilon > 0$ a number $a_0 > 0$ such that

$$\frac{P(U > au)}{P(U > a)} \le 2\, u^{-\gamma} \max(u^{-\epsilon}, u^{\epsilon})$$

for $a \geq a_0$ and $au \geq a_0$. Moreover, since $\lambda a P(U > a) \to c^{\gamma-1}/\gamma$, using possibly a larger a_0 we have

$$\lambda a P(U > au) \leq 2(c^{\gamma-1}/\gamma + \epsilon)u^{-\gamma} \max(u^{-\epsilon}, u^{\epsilon}), \quad a \geq a_0, \ au \geq a_0. \tag{51}$$

Since $\frac{\partial}{\partial u} K_t(s,u) = 1_{\{0<s+u<t\}}$ by (17),

$$\frac{\partial}{\partial u} h(s-u,u,r) = i\Big(\exp\Big\{i\sum_{j=1}^{n} \theta_j K_{t_j}(s-u,u)r\Big\} - 1\Big) \sum_{k=1}^{n} \theta_k 1_{\{0<s<t_k\}} r.$$

For any $0 \leq \kappa \leq 1$, we have $|e^{ix} - 1| \leq 2^{1-\kappa}|x|^{\kappa}$ and $(\sum_{i=1}^{n} |x_i|)^{\kappa} \leq \sum_{i=1}^{n} |x_i|^{\kappa}$. Since (16) implies $0 \leq K_t(s,u) \leq u$,

$$\Big|\exp\Big\{i\sum_{j=1}^{n} \theta_j K_{t_j}(s,u)r\Big\} - 1\Big| \leq \Big(2^{1-\kappa}\sum_{j=1}^{n} |\theta_j|^{\kappa} u^{\kappa} r^{\kappa}\Big) \wedge 2 \tag{52}$$

and so

$$\Big|\frac{\partial}{\partial u} h(s-u,u,r)\Big| \leq 2 \min\Big(\sum_{j=1}^{n} |\theta_j|^{\kappa} u^{\kappa} r^{\kappa}, 1\Big) \sum_{k=1}^{n} |\theta_k| 1_{\{0<s<t_k\}} \, r. \tag{53}$$

We may assume that $t_1 > 0$ and for convenience that $a \geq a_0$ is so large that $a_0/a \leq t_1$. Relations (51) and (53) now imply that the integrand in (49) is bounded on $\{u \geq a_0/a\}$,

$$\Big|\frac{\partial}{\partial u} h(s,u,r)\Big| \lambda a P(U > au) \, 1_{\{u \geq a_0/a\}} \leq B_1(s,u,r),$$

where

$$B_1(s,u,r) = C_{\epsilon,\kappa} u^{-\gamma} \max(u^{-\epsilon}, u^{\epsilon}) \min\Big(\sum_{j=1}^{n} |\theta_j|^{\kappa} u^{\kappa} r^{\kappa}, 1\Big) \sum_{k=1}^{n} |\theta_k| 1_{\{0<s<t_k\}} \, r$$

and

$$C_{\epsilon,\kappa} = 4(c^{\gamma-1}/\gamma + \epsilon).$$

Now

$$\begin{aligned} &\int_{-\infty}^{\infty}\int_{0}^{\infty}\int_{0}^{\infty} B_1(s,u,r)\, ds du F_R(dr) \\ &\quad \leq C_{\epsilon,\kappa} \sum_{k=1}^{n} |\theta_k| t_k \Big(E(R^{1+\kappa}) \int_{0}^{t_1} \sum_{j=1}^{n} |\theta_j|^{\kappa} u^{\kappa-\gamma} \max(u^{-\epsilon}, u^{\epsilon})\, du \\ &\qquad + E(R) \int_{t_1}^{\infty} u^{-\gamma} \max(u^{-\epsilon}, u^{\epsilon})\, du\Big) \end{aligned}$$

Since $1 < \gamma < \delta \leq 2$ we may choose ϵ and κ such that

$$1 + \epsilon < \gamma, \quad \gamma + \epsilon < 1 + \kappa < \delta.$$

Then $E(R^{1+\kappa}) < \infty$ and the du-integrals are finite.

Next we must find a dominating function which applies to $0 < u \le a_0/a \le 1$. By (53),

$$\left|\frac{\partial}{\partial u}h(s-u,u,r)\right| \le 2\sum_{j=1}^n |\theta_j|^\kappa \sum_{k=1}^n |\theta_k| 1_{\{0<s<t_k\}}\, u^\kappa r^{1+\kappa}.$$

Fix $\epsilon > 0$. Using

$$u^\kappa \le (a_0/a)^{\gamma-1+\epsilon} u^{1+\kappa-\gamma-\epsilon}, \quad u \le a_0/a,$$

it follows from Markov's inequality,

$$\lambda a P(U > au) \le \lambda u^{-1} E(U),$$

that

$$u^\kappa \lambda a P(U > au) \le a_0^{\gamma-1+\epsilon} \nu\, u^{\kappa-\gamma-\epsilon} \lambda / a^{\gamma-1+\epsilon}.$$

Recall that the general scaling assumption for ICR is $\lambda L_U(a)/a^{\gamma-1} \to c^{\gamma-1}$, where L_U is a slowly varying function related to the asymptotic form of the duration U. By a general property of slowly varying functions, $a^{-\epsilon} \le L(a)$ for a sufficiently large. Hence we end up with

$$\left|\frac{\partial}{\partial u}h(s-u,u,r)\right| \lambda a P(U > au)\, 1_{\{u \le a_0/a\}} \le B_2(s,u,r),$$

where

$$B_2(s,u,r) = C_{\epsilon,\kappa} \sum_{k=1}^n |\theta_k| 1_{\{0<s<t_k\}}\, u^{\kappa-\gamma-\epsilon} r^{1+\kappa} 1_{\{0<u\le 1\}}$$

and

$$C_{\epsilon,\kappa} = 2 a_0^{\gamma-1+\epsilon}\, \nu (c^{\gamma-1} + \epsilon) \sum_{j=1}^n |\theta_j|^\kappa.$$

The bound B_2 is integrable with respect to $ds\, du\, F_R(dr)$ for $\gamma + \epsilon < 1 + \kappa < \delta$. Since ϵ is arbitrary, this allows us to apply the dominated convergence theorem and complete the proof of part i) of Theorem 1.

3.2.2. The compound Poisson workload model. We turn to part ii) of Theorem 1. By Lemma 4,

$$\ln E \exp\Big\{ i \sum_{j=1}^n \theta_j (W^*_{\sharp,\lambda}(at_j) - \lambda \nu E(R) a t_j)/a \Big\} = \int_{-\infty}^{\infty}\int_0^\infty g_a(s,u)\, n(ads, adu),$$

where

$$\begin{aligned} g_a(s,u) &= \exp\Big\{ \int_0^\infty \int_0^\infty a\Big(\exp\Big\{ i \sum_{j=1}^n \theta_j 1_{\{w \le K_{t_j}(s,u)\}} r/a \Big\} - 1\Big) dw F_R(dr) \Big\} \\ &\qquad -1 - i \sum_{j=1}^n \theta_j K_{t_j}(s,u) E(R) \end{aligned}$$

after making the change of variables $w \to aw$, $s \to as$, $u \to au$ and using (28). Since

$$g_a(s,u) \sim \exp\Big\{i\sum_{j=1}^{n}\theta_j K_{t_j}(s,u)\,E(R)\Big\} - 1 - i\sum_{j=1}^{n}\theta_j K_{t_j}(s,u)E(R), \quad a \to \infty,$$

we can complete the proof in much the same way as in the previous part i), noticing that this case is in fact simpler in the sense that only the expected reward $E(R)$ and not the full distribution $F_R(dr)$ enters the limiting characteristic function. Since, as $a \to \infty$,

$$n(ads, adu) = \lambda ads\, F_U(adu) \sim c^{\gamma-1}\, ds\, u^{-(\gamma+1)}du,$$

the result in this case is

$$\begin{aligned}
&\int_{-\infty}^{\infty}\int_0^{\infty} g_a(s,u)\, n(ads, adu)\\
&\to \quad c^{\gamma-1}\int_{-\infty}^{\infty}\int_0^{\infty}\Big(\exp\Big\{i\sum_{j=1}^{n}\theta_j K_{t_j}(s,u)\,E(R)\Big\}\\
&\qquad\qquad -1 - i\sum_{j=1}^{n}\theta_j K_{t_j}(s,u)E(R)\Big)\, ds\, u^{-(\gamma+1)}du\\
&= \quad \int_{-\infty}^{\infty}\int_0^{\infty}\Big(\exp\Big\{i\sum_{j=1}^{n}\theta_j c K_{t_j/c}(s,u)\,E(R)\Big\}\\
&\qquad\qquad -1 - i\sum_{j=1}^{n}\theta_j c K_{t_j/c}(s,u)E(R)\Big)\, ds\, u^{-(\gamma+1)}du\\
&= \ln E\,\exp\Big\{i\sum_{j=1}^{n}\theta_j\, cY_\gamma(t_j/c)\,E(R)\Big\},
\end{aligned}$$

where we used (28) and where the process Y_γ is defined in (36).

3.3. Proof of Theorem 2 (FCR)

In the asymptotic regime of fast connection rate and for the continuous flow model it is necessary to study the cases $\delta = 2$ and $\delta < 2$ separately.

3.3.1. The continuous flow model, finite variance rewards. We start with the case $\delta = 2$ of finite second moment rewards and we use representation (26) of the workload process, to avoid scaling in the variable r. Observe first that

$$\begin{aligned}
EN(ads, adu, dr) &= \lambda ads\, F_U(adu)\, F_R(dr)\\
&\sim \lambda a a^{-\gamma} ds\, u^{-\gamma-1} du\, F_R(dr)\\
&= \left(\frac{b}{a}\right)^2 ds\, u^{-\gamma-1} du\, F_R(dr).
\end{aligned}$$

Hence, setting $\zeta = b/a$, by (26) and Lemma 3,

$$\begin{aligned}\ln E \exp\Big\{i \sum_{j=1}^{n} \theta_j (W_\lambda^*(at_j) - \lambda\nu E(R) at_j)/b\Big\} \\ = \int_{-\infty}^{\infty}\int_0^{\infty}\int_0^{\infty} h(s,u,\zeta^{-1}r)\, EN(ads, adu, dr) \\ \sim \int_{-\infty}^{\infty}\int_0^{\infty}\int_0^{\infty} h(s,u,\zeta^{-1}r)\, \zeta^2\, ds\, u^{-(\gamma+1)} du\, F_R(dr).\end{aligned}$$

To justify taking the limit inside of the integral a similar argument applies as in the proof of Theorem 1. The task is to dominate the integrand in

$$\int_{-\infty}^{\infty}\int_0^{\infty}\int_0^{\infty} \frac{\partial}{\partial u} h(s-u,u,\zeta^{-1}r)\, \lambda a P(U > au)\, ds\, du\, F_R(dr), \tag{54}$$

just as we did earlier for ICR in (50). Because of the finite variance condition $E(R^2) < \infty$, this case is simpler and we can use (53) with $\kappa = 1$. Potter's theorem and the Markov inequality apply again to obtain bounds for the tail probability $P(U > au)$. The resulting estimates together justify using the dominated convergence theorem. Hence the Taylor expansion

$$h(s,u,\zeta^{-1}r) = -\zeta^{-2}\frac{1}{2}\Big(\sum_{j=1}^{n} \theta_j K_{t_j}(s,u) r\Big)^2 + o(\zeta^{-2}), \quad \zeta \to \infty,$$

shows that

$$\begin{aligned}\ln E \exp\Big\{i \sum_{j=1}^{n} \theta_j (W_\lambda^*(at_j) - \lambda\nu E(R) at_j)/b\Big\} \\ \to -\frac{1}{2}\int_{-\infty}^{\infty}\int_0^{\infty}\int_0^{\infty} \Big(\sum_{j=1}^{n} \theta_j K_{t_j}(s,u) r\Big)^2 ds\, u^{-(\gamma+1)} du\, F_R(dr) \\ = -\frac{1}{2} E(R^2) \sum_{i=1}^{n}\sum_{j=1}^{n} \theta_i\theta_j \int_{-\infty}^{\infty}\int_0^{\infty} K_{t_i}(s,u) K_{t_j}(s,u)\, ds\, u^{-\gamma-1} du.\end{aligned}$$

One has

$$\int_{-\infty}^{\infty}\int_0^{\infty} K_{t_i}(s,u) K_{t_j}(s,u)\, ds\, u^{-\gamma-1} du = \frac{\sigma^2}{2}(t_i^{2H} + t_j^{2H} - |t_i - t_j|^{2H}),$$

where $H = (3-\gamma)/2$ and σ is given by (38), and therefore the limit process is the fractional Brownian motion

$$E(R^2)^{1/2}\, \sigma\, B_H(t).$$

An alternative way to see that the limit is fractional Brownian motion is to observe that the process (39) is Gaussian, H-self-similar and has stationary increments.

3.3.2. Continuous flow model, infinite variance rewards ($\delta < 2$). In the case $1 < \gamma < \delta < 2$ of infinite variance rewards, we have $F_R(dr) \sim r^{-\delta-1}\,dr$ as $r \to \infty$. Lemma 3 and the scaling representation (27) yield

$$\ln E\,\exp\Big\{i\sum_{j=1}^{n}\theta_j(W_\lambda^*(at_j) - \lambda\nu E(R)at_j)/b\Big\}$$
$$= \int_{-\infty}^{\infty}\int_0^{\infty}\int_0^{\infty} h(s,u,r)\,EN(ads, adu, (b/a)dr)$$

where h is defined in (43). Because of the choice of the normalization factor b,

$$\begin{aligned} EN(ads, adu, (b/a)dr) &= \lambda ads\,F_U(adu)\,F_R((b/a)dr) \\ &\sim ds\,u^{-\gamma-1}du\,r^{-\delta-1}dr, \quad \frac{b}{a} \to \infty. \end{aligned}$$

We need to verify that the limiting log-characteristic function is given by

$$\int_{-\infty}^{\infty}\int_0^{\infty}\int_0^{\infty} h(s,u,r)\,ds\,u^{-\gamma-1}du\,r^{-\delta-1}dr,$$

which is the logarithm of the characteristic function of the Telecom process as defined in (40). In view of (29) this also yields the representation (41). The corresponding δ-stable form of the characteristic function is obtained by integrating over r (Samorodnitsky and Taqqu (1994), Exercise 3.24):

$$\int_{-\infty}^{\infty}\int_0^{\infty}\int_0^{\infty} h(s,u,r)\,ds\,u^{-\gamma-1}du\,r^{-\delta-1}dr$$
$$= -\frac{1}{2}(\sigma_\delta)^\delta \int_{-\infty}^{\infty}\int_0^{\infty}\Big|\sum_{j=1}^{n}\theta_j K_{t_j}(s,u)\Big|^\delta k_\delta\Big(\sum_{j=1}^{n}\theta_j K_{t_j}(s,u)\Big)\,ds\,u^{-\gamma-1}du,$$

where σ_δ is given by (30) and $k_\delta(\theta)$ by (32), with $\alpha = \delta$.

To establish the limit result we fix $\epsilon > 0$ and split the integral in three parts,

$$\int_{-\infty}^{\infty}\int_0^{\infty}\int_0^{\infty} h(s,u,r)(\lambda ads\,F_U(adu)\,F_R((b/a)dr)$$
$$-ds\,u^{-\gamma-1}du\,r^{-\delta-1}dr) = I_\epsilon^1 + I_\epsilon^2 + I_\epsilon^3,$$

corresponding to the three domains of integration $A_\epsilon^1 = \{u > \epsilon, r > \epsilon\}$, $A_\epsilon^2 = \{u < \epsilon < r\}$ and $A_\epsilon^3 = \{r < \epsilon\}$, not involving the integration over s.

Writing

$$\begin{aligned} \mu_\lambda(du, dr) &= \lambda a\,uF_U(adu)\,rF_R((b/a)dr), \\ \mu(du, dr) &= u^{-\gamma}du\,r^{-\delta}dr, \end{aligned}$$

and

$$H(u,r) = \frac{1}{ur}\int_{-\infty}^{\infty} h(s,u,r)\,ds,$$

we have

$$I_\epsilon^1 = \int_\epsilon^\infty \int_\epsilon^\infty H(u,r)(\mu_\lambda(du,dr) - \mu(du,dr)).$$

Here, $|H(u,r)| \le 2\sum_{j=1}^n \theta_j t_j < \infty$ in view of (44) and (45). It follows similarly that $H(u,r)$ is jointly continuous in A_ϵ^1. Since

$$\int\int_{A_\epsilon^1} \mu_\lambda(du,dr) < \infty, \quad \int\int_{A_\epsilon^1} \mu(du,dr) < \infty,$$

and the measure μ_λ converges weakly to μ, we obtain $I_\epsilon^1 \to 0$ by weak convergence.

We now consider I_ϵ^2. Using the more general inequality $|e^{ix} - 1 - ix| \le c_\kappa |x|^{1+\kappa}$ where c_κ is a constant and $\kappa \in [0,1]$, we obtain

$$\Big|\int_{-\infty}^\infty \int_0^\epsilon \int_\epsilon^\infty h(s,u,r)\, ds\, u^{-\gamma-1} du\, r^{-\delta-1} dr\Big|$$
$$\le c_\kappa 2^\kappa \sum_{j=1}^n |\theta_j|^{1+\kappa} \int_\epsilon^\infty r^{\kappa-\delta}\, dr \int_{-\infty}^\infty \int_0^\epsilon K_{t_j}(s,u)^{1+\kappa} ds\, u^{-\gamma-1} du. \quad (55)$$

Using (16) in the form $0 \le K_t(s,u) \le u$ together with (45) we may continue with

$$\int_{-\infty}^\infty K_{t_j}(s,u)^{1+\kappa}\, ds \le u^\kappa \int_{-\infty}^\infty K_{t_j}(s,u)\, ds = u^{1+\kappa} t_j,$$

and then compute the remaining integrals on the right-hand side of (55). Under the assumption $\gamma < 1+\kappa < \delta$, this yields a constant c'_κ such that

$$\Big|\int_{-\infty}^\infty \int_0^\epsilon \int_\epsilon^\infty h(s,u,r)\, ds\, u^{-\gamma-1} du\, r^{-\delta-1} dr\Big| \le c'_\kappa \epsilon^{2(1+\kappa)-\gamma-\delta}.$$

Similarly,

$$\Big|\int_{-\infty}^\infty \int_0^\epsilon \int_\epsilon^\infty h(s,u,r) \lambda a ds\, F_U(adu)\, F_R((b/a)dr)\Big|$$
$$\le d_\kappa \lambda a \int_\epsilon^\infty r^{1+\kappa}\, F_R((b/a)dr) \int_0^\epsilon u^{1+\kappa}\, F_U(adu)$$

for a suitable constant d_κ. By the properties of regularly varying functions, we have

$$(b/a)^\delta \int_\epsilon^\infty r^{1+\kappa}\, F_R((b/a)dr) \to \int_\epsilon^\infty r^{\kappa-\delta}\, dr, \quad b/a \to \infty$$

if $1+\kappa < \delta$, and

$$a^\gamma \int_0^\epsilon u^{1+\kappa}\, F_U(adu) \to \int_0^\epsilon u^{\kappa-\gamma}\, du, \quad a \to \infty$$

if $\gamma < 1+\kappa$. Hence we can find d'_κ such that

$$\Big|\int_{-\infty}^\infty \int_0^\epsilon \int_\epsilon^\infty h(s,u,r) \lambda a ds\, F_U(adu)\, F_R((b/a)dr)\Big|$$
$$\le d'_\kappa \epsilon^{2(1+\kappa)-\gamma-\delta}, \quad \gamma < 1+\kappa < \delta.$$

By taking in addition κ such that $(\gamma+\delta)/2 < 1+\kappa < \delta$ this shows

$$I_\epsilon^2 \leq (c'_\kappa + d'_\kappa)\, \epsilon^{2(1+\kappa)-\gamma-\delta} \to 0 \quad \epsilon \to 0.$$

Finally,

$$\Big| \int_{-\infty}^{\infty} \int_0^{\infty} \int_0^{\epsilon} h(s,u,r)\, ds\, u^{-\gamma-1} du\, r^{-\delta-1} dr \Big|$$
$$\leq c_2 \sum_{i,j=1}^{n} \theta_i\theta_j \int_{-\infty}^{\infty} \int_0^{\infty} K_{t_i}(s,u) K_{t_j}(s,u)\, ds\, u^{-\gamma-1} du \int_0^{\epsilon} r^{1-\delta}\, dr.$$

Since the $dsdu$-integral is the finite covariance function $\mathrm{Cov}(B_H(t_i), B_H(t_j))$ of fractional Brownian motion with $H = (3-\gamma)/2$, the right-hand side takes the form $\mathrm{const}\, \epsilon^{2-\delta} \to 0$, $\epsilon \to 0$. Similarly, for λ and a sufficiently large, we obtain

$$\Big| \int_{-\infty}^{\infty} \int_0^{\infty} \int_0^{\epsilon} h(s,u,r) \lambda a ds\, F_U(adu)\, F_R((b/a)dr) \Big| \leq \mathrm{const}\, \epsilon^{2-\delta}.$$

Thus $I_\epsilon^3 \to 0$ as $\epsilon \to 0$, which concludes the proof of the desired convergence of characteristic functions for this case.

3.3.3. The compound Poisson model. For the compound Poisson model the limit process is the same for all parameters in the range $1 < \gamma < 2$, $1 < \delta \leq 2$. By Lemma 4 and (28),

$$\ln E \exp\Big\{ i\frac{1}{b} \sum_{j=1}^{n} \theta_j (W^*_{\sharp,\lambda}(at_j) - \lambda\nu E(R) a t_j) \Big\} = \int_{-\infty}^{\infty} \int_0^{\infty} g_{a,b}(s,u)\, \lambda a ds\, F_U(adu),$$

where

$$g_{a,b}(s,u) = \exp\Big\{ \int_0^{\infty} \int_0^{\infty} \Big(\exp\Big\{ i \sum_{j=1}^{n} \theta_j 1_{\{w \leq aK_{t_j}(s,u)\}} r/b \Big\} - 1 \Big) dw\, F_R(dr) \Big\}$$
$$-1 - i \sum_{j=1}^{n} \theta_j a K_{t_j}(s,u) E(R)/b$$
$$\sim \exp\Big\{ \int_0^{\infty} \int_0^{\infty} i \sum_{j=1}^{n} \theta_j 1_{\{w \leq K_{t_j}(s,u)\}}\, r(a/b)\, dw\, F_R(dr) \Big\}$$
$$-1 - i \sum_{j=1}^{n} \theta_j a K_{t_j}(s,u) E(R)/b$$
$$= \exp\Big\{ i \sum_{j=1}^{n} \theta_j K_{t_j}(s,u) E(R)(a/b) \Big\} - 1 - i \sum_{j=1}^{n} \theta_j K_{t_j}(s,u) E(R)(a/b).$$

Hence, by Taylor expansion as $a/b \to 0$, the log-characteristic function converges to

$$-\frac{1}{2}\int_{-\infty}^{\infty}\int_{0}^{\infty}\Big(\sum_{j=1}^{n}\theta_j K_{t_j}(s,u)E(R)\Big)^2\,ds\,u^{-\gamma-1}du$$

$$=-\frac{1}{2}E(R)^2\sum_{i=1}^{n}\sum_{j=1}^{n}\theta_i\theta_j\int_{-\infty}^{\infty}\int_{0}^{\infty}K_{t_i}(s,u)K_{t_j}(s,u)\,ds\,u^{-\gamma-1}du.$$

The limit is therefore the fractional Brownian motion $E(R)\,\sigma\,B_H(t)$, $t\geq 0$.

3.4. Proof of Theorem 3 (SCR)

The proofs in the regime of slow connection rate are similar to the previous ones. To see which limit to expect we shall scale directly the integral representations instead of the characteristic functions.

3.4.1. The continuous flow model. The relevant scaling for any choice of parameters $1<\gamma<\delta\leq 2$, is

$$\frac{1}{b}\int_{-\infty}^{\infty}\int_{0}^{\infty}\int_{0}^{\infty}K_{at}(s,u)r\,\widetilde{N}(ds,du,dr)$$

$$\int_{-\infty}^{\infty}\int_{0}^{\infty}\int_{0}^{\infty}K_{(a/b)t}(s/b,u/b)r\,\widetilde{N}(ds,du,dr)$$

$$=\int_{-\infty}^{\infty}\int_{0}^{\infty}\int_{0}^{\infty}K_{(a/b)t}((a/b)s,u)r\,\widetilde{N}(a\,ds,b\,du,dr),$$

where $\widetilde{N}$ is defined by (6). Here, the compensator n scales as

$$\begin{aligned} n(a\,ds,b\,du,dr) &= \lambda a ds\,F_U(b\,du)\,F_R(dr)\\ &\sim ds\,u^{-1-\gamma}du\,F_R(dr),\end{aligned}$$

since $b=(\lambda a)^{1/\gamma}\to\infty$. Moreover, if we write $z=a/b$ then $z\to\infty$ and, using (17),

$$\begin{aligned} K_{zt}(zs,u) &= \int_0^u 1_{\{0<y+zs<zt\}}\,dy\\ &\to \int_0^u 1_{\{0<s<t\}}\,dy = u1_{\{0<s<t\}},\qquad z\to\infty. \end{aligned}\tag{56}$$

This suggests that the limit process is given by

$$\int_{-\infty}^{\infty}\int_{0}^{\infty}\int_{0}^{\infty}u1_{\{0<s<t\}}\,r\,\big(N(ds,du,dr)-ds\,u^{-1-\gamma}\,du\,F_R(dr)\big)$$

$$\stackrel{d}{=}\sigma_\gamma\int_0^t\int_0^\infty r\,M_\gamma(ds,dr)\stackrel{d}{=}E(R^\gamma)^{1/\gamma}\sigma_\gamma\int_0^t M_\gamma(ds)\tag{57}$$

$$\stackrel{d}{=}E(R^\gamma)^{1/\gamma}\,\Lambda_\gamma(t),\tag{58}$$

where σ_γ is defined in (30) and $M_\gamma(ds, dr)$ and $M_\gamma(ds)$ are γ-stable random measures with control measures $ds\, F_R(dr)$ and ds respectively and where $\Lambda_\gamma(t)$ is a Lévy-stable process with index γ. The limit process is well defined for any distribution F_R with $E(R^\gamma) < \infty$, in particular if we keep our assumption on R being regularly varying with a tail of index δ, such that $\gamma < \delta \leq 2$.

In order to establish the convergence, we begin as in (50), with the representation

$$\ln E\, \exp\Big\{i \sum_{j=1}^n \theta_j (W^*_\lambda(t_j) - \lambda\nu E(R)t_j)\Big\}$$
$$= \int_{-\infty}^{\infty}\int_0^\infty\int_0^\infty \frac{\partial}{\partial u} h(s-u,u,r)\, \lambda ds\, P(U>u) du\, F(dr).$$

Applying the scaling parameters a and b it follows that

$$\ln E\, \exp\Big\{i \sum_{j=1}^n \theta_j (W^*_\lambda(at_j) - \lambda\nu E(R)at_j)/b\Big\}$$
$$= \int_{-\infty}^{\infty}\int_0^\infty\int_0^\infty i\Big(\exp\Big\{i\sum_{j=1}^n \theta_j K_{at_j}(s-u,u)r/b\Big\} - 1\Big)$$
$$\times \frac{1}{b}\sum_{k=1}^n \theta_k 1_{\{0<s<at_k\}} r\, \lambda ds\, P(U>u) du\, F(dr)$$
$$= \int_{-\infty}^{\infty}\int_0^\infty\int_0^\infty i\Big(\exp\Big\{i\sum_{j=1}^n \theta_j K_{zt_j}(zs-u,u)r\Big\} - 1\Big)$$
$$\times \sum_{k=1}^n \theta_k 1_{\{0<s<t_k\}} r\, ds\, b^\gamma P(U>bu) du\, F(dr),$$

where $z = a/b \to \infty$ and we have used the normalization $b^\gamma = \lambda a$ valid under SCR. Now

$$K_{zt}(zs-u,u) \to u 1_{\{0<s<t\}}, \quad z \to \infty$$

and

$$b^\gamma P(U > bu) \to \frac{1}{\gamma u^\gamma}, \quad b \to \infty.$$

This shows that the above integrand with respect to $ds\, du\, F_R(dr)$,

$$f_\lambda(s,u,r) = i\Big(\exp\Big\{i\sum_{j=1}^n \theta_j K_{zt_j}(zs-u,u)r\Big\} - 1\Big)\sum_{k=1}^n \theta_k 1_{\{0<s<t_k\}} r\, b^\gamma P(U>bu),$$

has the pointwise limit

$$f(s,u,r) = i\Big(\exp\Big\{i\sum_{j=1}^n \theta_j 1_{\{0<s<t_j\}}\, ur\Big\} - 1\Big)\sum_{k=1}^n \theta_k 1_{\{0<s<t_k\}} r\, \gamma^{-1} u^{-\gamma}$$

as λ and hence z and b tend to infinity. Since the logarithmic characteristic function of the limit process in Theorem 3 i) is given by

$$\int_{-\infty}^{\infty}\int_0^{\infty}\int_0^{\infty} f(s,u,r)\,ds\,du\,F_R(dr),$$

by Lemma 3, it remains to show that $|f_\lambda(s,u,r)|$ is dominated by an integrable function.

Since $b^\gamma P(U>b)\to 1/\gamma$, $b\to\infty$, it follows from the Potter bound as in (51) that for any $\epsilon>0$ there is a number b_0, such that

$$b^\gamma P(U>ub)\le 2(1/\gamma+\epsilon)\,u^{-\gamma}\max(u^{-\epsilon},u^{\epsilon}) \qquad b\ge b_0,\quad ub\ge b_0.$$

There is no restriction to assume $t_1>0$ and that λ and thus b are so large that $b_0/b\le t_1$. The task of estimating $|f_\lambda(s,u,r)|$ will be split accordingly in the three cases $0<u<b_0/b$, $b_0/b\le u<t_1$ and $t_1\le u<\infty$, where Potter's bound is applicable in the two latter but not in the first interval.

As in (52), for any $0<\kappa<1$,

$$\Big|\exp\Big\{i\sum_{j=1}^n \theta_j K_{zt_j}(zs-u,u)r\Big\}-1\Big|\le 2\min\Big(\sum_{j=1}^n|\theta_j|^\kappa u^\kappa r^\kappa,1\Big).$$

This shows

$$|f_\lambda(s,u,r)|\,1_{\{b_0/b\le u\}}\le f_1(s,u,r)$$

where

$$f_1(s,u,r)=4(1/\gamma+\epsilon)\sum_{k=1}^n|\theta_k|1_{\{0<s<t_k\}}\,r\min\Big(\sum_{j=1}^n|\theta_j|^\kappa u^\kappa r^\kappa,1\Big)u^{-\gamma}\max(u^{-\epsilon},u^{\epsilon}).$$

This upper bound is integrable, since

$$\begin{aligned}&\int_{-\infty}^{\infty}\int_0^{\infty}\int_0^{\infty} f_1(s,u,r)\,ds\,du\,F_R(dr)\\ &\quad\le 4(1/\gamma+\epsilon)\sum_{k=1}^n|\theta_k|t_k\Big(ER^{1+\kappa}\int_0^{t_1}u^{-\gamma+\kappa}\max(u^{-\epsilon},u^{\epsilon})\,du\\ &\qquad\qquad +E(R)\int_{t_1}^{\infty}u^{-\gamma}\max(u^{-\epsilon},u^{\epsilon})\,du\Big)<\infty,\end{aligned}$$

if we choose $\gamma+\epsilon<1+\kappa<\delta$ and $1+\epsilon<\gamma$.

It remains to find a dominating function for small u, that is $0<u<b_0/b$. Again by (52), for $0<\kappa<1$,

$$|f_\lambda(s,u,r)|\le\sum_{j=1}^n|\theta_j|^\kappa\sum_{k=1}^n\theta_k 1_{\{0<s<t_k\}}r^{1+\kappa}\,u^\kappa\,b^\gamma P(U>bu).$$

Moreover, by Markov's inequality

$$\begin{aligned} u^{\kappa} b^{\gamma} P(U > bu) &\le u^{1+\kappa-\gamma}(b_0/b)^{\gamma-1} b^{\gamma} \frac{1}{bu} E(U) \\ &= \nu b_0^{\gamma-1} u^{\kappa-\gamma}. \end{aligned}$$

With the choice $\gamma < 1+\kappa < \delta$ we obtain the integrable upper bound

$$|f_\lambda(s,u,r)| 1_{\{0<u<b_0/b\}} \le \nu b_0^{1-\kappa} \sum_{j=1}^{n} |\theta_j|\kappa \sum_{k=1}^{n} \theta_k 1_{\{0<s<t_k\}} r^{1+\kappa} u^{\kappa-\gamma} 1_{\{0<u\le t_1\}}.$$

3.4.2. The compound Poisson workload model. For the compound Poisson model (11) one has

$$\begin{aligned} &\frac{1}{b}\int_{-\infty}^{\infty}\int_0^{\infty}\int_D \xi(K_{at}(s,u))\, \widetilde{N}_\sharp(ds,du,d\xi) \\ &= \int_{-\infty}^{\infty}\int_0^{\infty}\int_D \frac{1}{b}\xi(bK_{(a/b)t}((a/b)s,u))\, \widetilde{N}_\sharp(a\,ds, b\,du, d\xi), \end{aligned}$$

where $\widetilde{N}_\sharp$ is defined in (14). Its compensator $n_\sharp$ in (10) is like the compensator n in (1) but with $F_R(dr)$ replaced by $\mu(d\xi)$. Hence as $a, b \to \infty$, we have as in (56)

$$n_\sharp(a\,ds, b\,du, d\xi) \sim ds\, u^{-1-\gamma} du\, \mu(d\xi)$$

and, again observing that $z = a/b \to \infty$,

$$\begin{aligned} &\int_D \frac{1}{b}\xi(bK_{(a/b)t}((a/b)s,u))\,\mu(d\xi) \\ &= E(R)\, K_{zt}(zs,u) \\ &\sim E(R)\, u\, 1_{\{0<s<t\}}, \end{aligned}$$

by (12) and (56). The limit process is therefore

$$\begin{aligned} E(R)\int_{-\infty}^{\infty}\int_0^{\infty} u 1_{\{0<s<t\}}\,(N(ds,du) - ds\, u^{-(1+\gamma)}\,du) &\stackrel{d}{=} E(R)\,\sigma_\gamma \int_0^t M_\gamma(ds) \\ &\stackrel{d}{=} E(R)\,\Lambda_\gamma(t), \end{aligned}$$

the formal verification of which rests again on studying the scaled characteristic function, this time using Lemma 4. The processes M_γ and Λ_γ are as in (57) and (58).

3.5. Proof of Theorem 4

3.5.1. Finite variance durations and rewards, $\gamma = \delta = 2$. Here, $E(U^2) < \infty$ and $E(R^2) < \infty$. To avoid scaling U and R, we use representation (25), i.e.,

$$\frac{1}{b}(W_\lambda^*(at) - \lambda at\nu E(R)) = \frac{1}{b}\int_{-\infty}^{\infty}\int_0^{\infty}\int_0^{\infty} K_{at}(as,u) r \widetilde{N}(a\,ds, du, dr)$$

where

$$\begin{aligned} EN(ads, du, dr) &= \lambda ads\, F_U(du)\, F_R(dr) \\ &= b^2\, ds\, F_U(du)\, F_R(dr). \end{aligned}$$

By (17),

$$K_{at}(as, u) = \int_0^u 1_{\{0<y+as<at\}}\, dy \to u 1_{\{0<s<t\}} \quad \text{as} \quad a \to \infty.$$

Hence by Lemma 3, as $b \to \infty$,

$$\begin{aligned} &\ln E \exp\Big\{ i \sum_{j=1}^n \theta_j (W_\lambda^*(at_j) - \lambda\nu E(R) at_j)/b \Big\} \\ &= \int_{-\infty}^\infty \int_0^\infty \int_0^\infty h(as, u, r/b)\, EN(ads, du, dr) \\ &\sim -\frac{1}{2} \int_{-\infty}^\infty \int_0^\infty \int_0^\infty \Big(b^{-1} \sum_{j=1}^n \theta_j K_{at_j}(as, u) r \Big)^2 b^2\, ds\, F_U(du)\, F_R(dr) \\ &\sim -\frac{1}{2} \int_{-\infty}^\infty \int_0^\infty F_U(du) \Big(\sum_{j=1}^n \theta_j 1_{\{0<s<t_j\}} u \Big)^2 ds \int_0^\infty r^2 F_R(dr) \\ &= -\frac{1}{2} \int_{-\infty}^\infty \Big(\sum_{j=1}^n \theta_j 1_{\{0<s<t_j\}} \Big)^2 ds\, E(U^2)\, E(R^2), \end{aligned}$$

so the limit is

$$E(U^2)^{1/2}\, E(R^2)^{1/2}\, B(t)$$

where $B(t)$ is Brownian motion.

When we consider instead $W_{\sharp,\lambda}$ and apply Lemma 4, then the resulting expression is slightly different:

$$\begin{aligned} &\ln E \exp\Big\{ i \sum_{j=1}^n \theta_j (W_{\sharp,\lambda}^*(at_j) - \lambda\nu E(R) at_j)/b \Big\} \\ &\sim -\frac{1}{2} (ER)^2 \int_{-\infty}^\infty \int_0^\infty \Big(b^{-1} \sum_{j=1}^n \theta_j K_{at_j}(as, u) \Big)^2 b^2\, ds\, F_U(du) \\ &\sim -\frac{1}{2} \int_{-\infty}^\infty \Big(\sum_{j=1}^n \theta_j 1_{\{0<s<t_j\}} \Big)^2 ds\, E(U^2)\, (ER)^2, \end{aligned}$$

which corresponds to the limit process $E(U^2)^{1/2}\, E(R)\, B(t)$.

3.5.2. Continuous model, rewards have heavier tails than those of durations, $1 < \delta < \gamma \le 2$. Take $1 < \delta < 2$, and assume either $\delta < \gamma \le 2$ or that we have an arbitrary distribution F_U with $E(U^\delta) < \infty$. Recall that

$$b = (\lambda a)^{1/\delta}.$$

Using (17) and (42),

$$\begin{aligned}
&\frac{1}{b}\int_{-\infty}^{\infty}\int_{0}^{\infty}\int_{0}^{\infty} K_{at}(s,u)r\,\widetilde{N}(ds,du,dr)\\
&=\int_{-\infty}^{\infty}\int_{0}^{\infty}\int_{0}^{\infty}(r/b)\int_{0}^{u}1_{\{0<y+s<at\}}\,dy\,\widetilde{N}(ds,du,dr)\\
&=\int_{-\infty}^{\infty}\int_{0}^{\infty}\int_{0}^{\infty}\int_{0}^{u}1_{\{0<y/a+s<t\}}\,dy\,r\,\widetilde{N}(a\,ds,du,b\,dr)\\
&\sim\int_{-\infty}^{\infty}\int_{0}^{\infty}\int_{0}^{\infty}1_{\{0<s<t\}}\,ur(N(ds,du,dr)-ds\,F_U(du)\,r^{-1-\delta}\,dr)\\
&\stackrel{d}{=}\sigma_\delta\int_{-\infty}^{\infty}\int_{0}^{\infty}1_{\{0<s<t\}}\,uM_\delta(ds,du)\\
&\stackrel{d}{=}E(U^\delta)^{1/\delta}\,\Lambda_\delta(t),
\end{aligned}$$

where $M_\delta(ds,du)$ is δ-stable with control measure $m(ds,du) = ds\,F_U(du)$ and $\Lambda_\delta(t)$ is a Lévy stable process with index γ.

4. Weak convergence

This section is devoted to extending our previous results on convergence of the finite-dimensional distributions to weak convergence in function space.

Theorem 5. *For the continuous flow model, which has continuous trajectories, the convergence holds in the sense of weak convergence of stochastic processes in the space of continuous functions.*

4.1. Proof of tightness for the continuous flow model

To prove weak convergence in the continuous case, we are going to establish the following tightness criterion. For some $\alpha > 0$ (in our case $1 < \alpha \le 2$) and $\beta > 1$,

$$E\Big|\frac{1}{b}(W_\lambda^*(at_1)-\lambda\nu E(R)at_1)-\frac{1}{b}(W_\lambda^*(at_2)-\lambda\nu E(R)at_2)\Big|^\alpha \le \text{const}\,|t_2-t_1|^\beta,$$

uniformly in λ, a, b. Clearly, because of stationarity of the increments, it suffices to show for any fixed $t > 0$ the uniform bound

$$E\Big|\frac{1}{b}(W_\lambda^*(at)-\lambda\nu E(R)at)\Big|^\alpha \le \text{const}\,t^\beta. \tag{59}$$

Lemma 5. *For the continuous flow model* (2) *and for any* $1 < \alpha \le 2$, *we have the estimate*

$$E|W_\lambda^*(t)-\lambda\nu E(R)t|^\alpha \le 2E(R^\alpha)\int_{-\infty}^{\infty}\int_{0}^{\infty} K_t(s,u)^\alpha\,\lambda ds\,F_U(du).$$

Proof. Suppose first that $E(R^2) < \infty$. Then we can take $\alpha = 2$. It is readily checked that in this case we have the equality

$$E(W_\lambda^*(t) - \lambda\nu E(R)t))^2 = E(R^2) \int_{-\infty}^{\infty} \int_0^{\infty} K_t(s,u)^2 \, \lambda ds \, F_U(du).$$

For $1 < \alpha < 2$ we will use the estimate

$$E|X|^\alpha \le A(\alpha) \int_0^\infty (1 - |\Phi_X(\theta)|^2)\theta^{-\alpha-1} \, d\theta, \tag{60}$$

where

$$A(\alpha) = \Big(\int_0^\infty (1 - \cos(x))x^{-\alpha-1} \, dx \Big)^{-1} < \infty \tag{61}$$

and $\Phi_X(\theta) = E(e^{i\theta X})$ is the characteristic function of the random variable X. This technique goes back to von Bahr and Esseen (1965), and is used in Gaigalas (2004) in a similar context as here. With $X = W_\lambda^*(t) - \lambda\nu E(R)t$ we have

$$\Phi_X(\theta) = \exp\Big\{ \int_{-\infty}^{\infty} \int_0^{\infty} \int_0^{\infty} (e^{i\theta K_t(s,u)r} - 1 - i\theta K_t(s,u)r) \, n(ds,du,dr) \Big\}$$

and

$$\begin{aligned} 1 - |\Phi_X(\theta)|^2 &= 1 - \exp\Big\{ -2 \int_{-\infty}^{\infty} \int_0^{\infty} \int_0^{\infty} (1 - \cos(\theta K_t(s,u)r)) \, n(ds,du,dr) \Big\} \quad (62) \\ &\le 2 \int_{-\infty}^{\infty} \int_0^{\infty} \int_0^{\infty} (1 - \cos(\theta K_t(s,u)r)) \, n(ds,du,dr). \end{aligned}$$

Since this last relation implies

$$\begin{aligned} &\int_0^\infty (1 - |\Phi_X(\theta)|^2)\theta^{-\alpha-1} \, d\theta \\ &\quad \le 2 \int_{-\infty}^{\infty} \int_0^{\infty} \int_0^{\infty} E(1 - \cos(\theta K_t(s,u)R))\theta^{-\alpha-1} \, d\theta \, \lambda ds \, F_U(du) \\ &\quad = 2E(R^\alpha) \int_{-\infty}^{\infty} \int_0^{\infty} K_t(s,u)^\alpha \, \lambda ds \, F_U(du)/A(\alpha), \end{aligned}$$

we obtain the estimate stated in the lemma by using (60). □

We are now prepared to prove tightness under the scaling of intermediate connection rates. Because of the assumption $\gamma < \delta$ we can apply Lemma 5 with α such that $\gamma < \alpha < \delta$. If $\delta = 2$ we may even take $\alpha = 2$. In all cases $E(R^\alpha) < \infty$ and, using (28) and an integration by parts,

$$\begin{aligned} E\Big| \frac{W_\lambda^*(at) - \lambda\nu E(R)at}{a} \Big|^\alpha &\le 2E(R^\alpha) \int_{-\infty}^{\infty} \int_0^{\infty} K_t(s,u)^\alpha \, \lambda a ds \, F_U(adu) \\ &= 2E(R^\alpha) \int_{-\infty}^{\infty} \int_0^{\infty} \alpha K_t(s,u)^{\alpha-1} 1_{\{0<s+u<t\}} \, \lambda a ds \, P(U > au) \, du. \end{aligned}$$

By using (16) and applying the Potter bound and the fact that $\lambda/a^{\gamma-1} \to c^{\gamma-1} \in (0,\infty)$, it follows that the last double integral is bounded by

$$\text{const} \int_{-\infty}^{\infty}\int_0^{\infty} \alpha(t\wedge u)^{\alpha-1}\, 1_{\{0<s+u<t\}} \max(u^{-\gamma-\epsilon}, u^{-\gamma+\epsilon})\, dsdu < \infty,$$

where the integral is finite since we can take $\epsilon > 0$ such that $\alpha-\epsilon<\gamma<\alpha$. Hence the dominated convergence theorem applies, and we have

$$E\Big|\frac{W_\lambda^*(at) - \lambda\nu E(R)at}{a}\Big|^\alpha \le 3\, c^{\gamma-1}\, E(R^\alpha) \int_{-\infty}^{\infty}\int_0^{\infty} K_t(s,u)^\alpha\, ds\, u^{-\gamma-1} du,$$

say, for sufficiently large λ and a. Using once again (16) and (45),

$$\begin{aligned}\int_{-\infty}^{\infty}\int_0^{\infty} K_t(s,u)^\alpha\, ds\, u^{-\gamma-1} du \;&\le\; \int_0^{\infty} (u\wedge t)^{\alpha-1} u^{-\gamma} t\, du \\ &= \frac{\alpha-1}{(\alpha-\gamma)(\gamma-1)}\, t^{1+\alpha-\gamma}. \qquad (63)\end{aligned}$$

Thus we have found α and $\beta = 1+\alpha-\gamma > 1$, such that (59) holds uniformly in λ and a. This completes the proof of weak convergence for the intermediate Telecom process in Theorem 1 i).

The proof of tightness for the case of fast connection rate scaling and finite variance rewards, that is Theorem 2 i) where the fractional Brownian motion arises in the limit, is very similar to that of the preceding case. When we apply (28) and use the parameters $\gamma < \alpha = \delta = 2$ and $b^2 = \lambda a^{3-\gamma}$ under the scaling FCR, then Lemma 5 yields the estimate

$$\begin{aligned}E\Big[\Big(\frac{W_\lambda^*(at) - \lambda\nu E(R)at}{b}\Big)^2\Big] \;&\le\; 2E(R^2)\frac{1}{a^{3-\gamma}} \int_{-\infty}^{\infty}\int_0^{\infty} K_{at}(s,u)^2\, ds\, F_U(du) \\ &= 2E(R^2) \int_{-\infty}^{\infty}\int_0^{\infty} K_t(s,u)^2\, ds\, a^\gamma F_U(adu).\end{aligned}$$

The same arguments as above lead to the uniform bound const $t^{3-\gamma}$, which verifies the tightness criterion (59) with $\alpha = 2$ and $\beta = 3-\gamma > 1$.

The final case for the continuous flow model is tight convergence to the Telecom process in Theorem 2 ii). In this case we will need the following version of the previous Lemma 5. This is simply the inequality (60), expressed in terms of (62).

Lemma 6. *For the continuous flow model* (2) *and for any* $1 < \alpha < 2$, *we have the estimate*

$$\begin{aligned}&E|W_\lambda^*(t) - \lambda\nu E(R)t|^\alpha \le A(\alpha) \\ &\times \int_0^\infty \Big(1 - \exp\Big\{-2\int_{-\infty}^{\infty}\int_0^{\infty}\int_0^{\infty} (1-\cos(\theta K_t(s,u)r))\, n(ds,du,dr)\Big\}\Big)\theta^{-\alpha-1}\, d\theta,\end{aligned}$$

with $A(\alpha)$ *defined in* (61).

For any $\gamma < \alpha < \delta$, it follows from the lemma that

$$E\Big|\frac{W_\lambda^*(at) - \lambda\nu E(R)at}{b}\Big|^\alpha \le A(\alpha)\int_0^\infty (1 - |\Phi_{\lambda,a,b}(\theta)|^2)\,\theta^{-\alpha-1}\,d\theta,$$

where

$$\begin{aligned}|\Phi_{\lambda,a,b}(\theta)|^2 &= \exp\Big\{-2\int_{-\infty}^\infty\int_0^\infty\int_0^\infty (1-\cos(\theta K_t(s,u)r))\,n(ads,adu,(b/a)dr)\Big\}\\ &\sim \exp\Big\{-2\int_{-\infty}^\infty\int_0^\infty\int_0^\infty (1-\cos(\theta K_t(s,u)r))\,ds\,u^{-(1+\gamma)}du\,r^{-(1+\delta)}dr\Big\}\\ &= \exp\{-2\theta^\delta J_t(\gamma,\delta)\},\end{aligned}$$

where

$$J_t(\gamma,\delta) = A(\delta)^{-1}\int_{-\infty}^\infty\int_0^\infty K_t(s,u)^\delta\,ds\,u^{-(1+\gamma)}du.$$

By the method based on Potter bounds, used repeatedly above, it follows that we can find a constant $C_{\alpha,\gamma,\delta}$ (changing each time it occurs below), such that the inequality

$$E\Big|\frac{W_\lambda^*(at) - \lambda\nu E(R)at}{b}\Big|^\alpha \le C_{\alpha,\gamma,\delta}\int_0^\infty (1 - \exp\{-2J_t(\gamma,\delta)\theta^\delta\})\,\theta^{-\alpha-1}\,d\theta$$

holds uniformly in λ, a and b. Since, for $\alpha < \delta$, the integral $\int_0^\infty (1-e^{-2\theta^\delta})\theta^{-1-\alpha}\,d\theta$ is finite, and since it was shown in (63) that $J_t(\gamma,\delta) \le \text{const}\,t^{1+\delta-\gamma}$, this yields the final estimate

$$E\Big|\frac{W_\lambda^*(at) - \lambda\nu E(R)at}{b}\Big|^\alpha \le C_{\alpha,\gamma,\delta}\,J_t(\gamma,\delta)^{\alpha/\delta} \le C_{\alpha,\gamma,\delta}\,t^{(1+\delta-\gamma)\alpha/\delta}.$$

Now, $1 < \gamma < \alpha < \delta$ implies that $(1+\delta-\gamma)\alpha/\delta > 1$, and hence the growth criterion (59) is fulfilled. This ends the proof of weak convergence of the scaled infinite Poisson process towards the Telecom process in Theorem 2 ii).

Acknowledgements

This research was done in part while Murad S. Taqqu visited Uppsala University within the grant Stochastic methods in telecommunications funded by the Swedish Foundation for Strategic Research. He was also supported by NSF grants DMS-0102410 and DMS 0505747 at Boston University. The authors would like to thank Raimundas Gaigalas for interesting discussions.

References

[1] C. Barakat, P. Thiran, G. Iannaccone, C. Diot and P. Owezarski (2003). Modeling Internet backbone traffic at the flow level. IEEE Transactions on Signal Processing, 51:8, Aug. 2003, 2111–2124.

[2] J.-M. Bardet, G. Lang, G. Oppenheim, A. Philippe, S. Stoev and M.S. Taqqu (2003). Semi-parametric estimation of the long-range dependence parameter: A survey. In: Theory and Applications of Long-Range Dependence, Eds. P. Doukhan, G. Oppenheim, M.S. Taqqu, Birkhäuser, Basel, 2003.

[3] J.-M. Bardet, G. Lang, G. Oppenheim, A. Philippe, S. Stoev and M.S. Taqqu (2003). Generators of long-range dependence processes: A survey. In: Theory and Applications of Long-Range Dependence, Eds. P. Doukhan, G. Oppenheim, M.S. Taqqu, Birkhäuser, Basel, 2003.

[4] A. Benassi, S. Jaffard and D. Roux (1997). Elliptic Gaussian random processes. Rev. Mat. Iberoamericana, 13, 19–90.

[5] A. Benassi, S. Cohen and J. Istas (2002). Identification and properties of real harmonizable fractional Lévy motions. Bernoulli 8, 97–115.

[6] H. Biermé, A. Estrade and I. Kaj (2006). About scaling behavior of random balls models. 6th Int. Conf. on Stereology, Spatial Statistics and Stochastic Geometry, Prague, June 2006.

[7] H. Biermé, A. Estrade and I. Kaj (2007). Self-similar random fields and rescaled random balls models. Preprint.

[8] P. Billingsley (1968). Convergence of probability measures. John Wiley and Sons, New York, 1968.

[9] N.H. Bingham, C.M. Goldie and J.L. Teugels (1987). Regular variation. Cambridge University Press, Cambridge 1987.

[10] M. Çağlar (2004). A long-range dependent workload model for packet data traffic. Mathematics of Operations Research 29:1, 92–105

[11] S. Cohen and M.S. Taqqu (2003). Small and large scale behavior of the Poissonized Telecom process. Methodology and Computing in Applied Probability 6, 363–379.

[12] G. Fay, F. Roueff and Ph. Soulier (2007). Estimation of the memory parameter of the infinite-source Poisson process. Bernoulli 13, 473–491.

[13] R. Gaigalas (2006). A Poisson bridge between fractional Brownian motion and stable Lévy motion. Stochastic Process. Appl. 116, 447–462.

[14] R. Gaigalas and I. Kaj (2003). Convergence of scaled renewal processes and a packet arrival model. Bernoulli 9, 671–703.

[15] W. Jedidi, J. Almhana, V. Choulakian and R. McGorman (2004). A general Poisson shot noise traffic model, and functional convergence to stable processes. Preprint, University of Moncton, NB, Canada.

[16] I. Kaj, A. Martin-Löf (2004). Scaling limit results for the sum of many inverse Lévy subordinators. U.U.D.M. 2004:13, Department of Mathematics, Uppsala University,

[17] I. Kaj (2005). Limiting fractal random processes in heavy-tailed systems. In: Fractals in Engineering, New Trends in Theory and Applications, pp. 199–218. Eds. J. Levy-Lehel, E. Lutton, Springer-Verlag London 2005.

[18] I. Kaj, L Leskelä, I. Norros and W. Schmidt (2007). Scaling limits for random fields with long-range dependence. Ann. Probab. 35:2, 528–550.

[19] J.F.C. Kingman (1993). Poisson Processes. Oxford University Press, Oxford, U.K.

[20] T.G. Kurtz (1996). Limit theorems for workload input models. F.P. Kelly, S. Zachary, I. Ziedins, eds. Stochastic Networks, theory and applications. Clarendon Press, Oxford, U.K.

[21] J.B. Levy and M.S. Taqqu (2000). Renewal reward processes with heavy-tailed interrenewal times and heavy-tailed rewards. Bernoulli 6, 23–44.

[22] K. Maulik, S. Resnick and H. Rootzén (2003). A network traffic model with random transmission rate. Adv. Appl. Probab. 39, 671–699.

[23] Th. Mikosch, S. Resnick, H. Rootzén and A. Stegeman (2002). Is network traffic approximated by stable Lévy motion or fractional Brownian motion? Ann. Appl. Probab. 12, 23–68.

[24] V. Pipiras and M.S. Taqqu (2000). The limit of a renewal-reward process with heavy-tailed rewards is not a linear fractional stable motion. Bernoulli 6, 607–614.

[25] V. Pipiras and M.S. Taqqu (2002). The structure of self-similar stable mixed moving averages. The Annals of Probability 30, 898–932.

[26] V. Pipiras and M.S. Taqqu (2008). Small and Large Scale Asymptotics of some Lévy Stochastic Integrals. Methodol. Comput. Appl. Probab. 10, 299–314.

[27] V. Pipiras, M.S. Taqqu and J.B. Levy (2004). Slow, fast and arbitrary growth conditions for renewal reward processes when the renewals and the rewards are heavy-tailed. Bernoulli 10, 121–163.

[28] W.A. Rosenkrantz and J. Horowitz (2002). The infinite source model for internet traffic: statistical analysis and limit theorems. Methods and applications of analysis, 9:3, 445–462.

[29] G. Samorodnitsky and M.S. Taqqu (1994). Stable Non-Gaussian Processes: Stochastic Models with Infinite Variance. Chapman and Hall, New York 1994.

[30] A. Stegeman (2002). The nature of the beast: analyzing and modeling computer network traffic. Thesis, University of Groningen.

[31] M.S. Taqqu (2002). The modeling of Ethernet data and of signals that are heavy-tailed with infinite variance. Scand. J. Stat. 29, 273–295.

[32] B. von Bahr and C.G. Esseen (1965). Inequalities for the rth absolute moment of a sum of random variables, $1 \leq r \leq 2$. Ann. Math. Statist. 36, 299–303.

[33] W. Willinger, V. Paxson, R.H. Riedi and M.S. Taqqu (2003). Long-range dependence and data network traffic. In: Theory and Applications of Long-Range Dependence, Eds. P. Doukhan, G. Oppenheim, M.S. Taqqu, Birkhäuser, Basel, 2003.

[34] R. Wolpert and M.S. Taqqu (2005). Fractional Ornstein-Uhlenbeck Lévy processes and the Telecom process: upstairs and downstairs. Signal Processing 85:8, 1523–1545.

Ingemar Kaj
Department of Mathematics
Uppsala University
Box 480
SE 751 06 Uppsala, Sweden
e-mail: ikaj@math.uu.se

Murad S. Taqqu
Department of Mathematics
Boston University
Boston, MA 02215-2411, USA
e-mail: murad@math.bu.edu

Progress in Probability, Vol. 60, 429–461

Positive Recurrence of a One-Dimensional Variant of Diffusion Limited Aggregation

Harry Kesten and Vladas Sidoravicius

Abstract. The present paper studies a variant of the DLA model: At time 0 choose i.i.d. random variables $N(x), x \in \mathbb{Z}^+ = \{1, 2, \dots\}$. Each $N(x)$ has a Poisson (μ) distribution. $N(x)$ will be the number of particles at x at time 0. In the model we also put a mark at an integer position. Its position at time t is denoted by $\mathcal{M}(t)$. We take $\mathcal{M}(0) = 0$. We regard the particles as frozen (i.e., they stay in place) till a certain random time, which generally differs for different particles. At such a random time a particle is "thawed", that is, it starts to move according to a continuous time simple random walk. At all times there will be i thawed particles in the system. Assume that at time 0 we have i thawed particles in the system at some positions in $\mathbb{Z}^+$, and that the $N(x), x \in \mathbb{Z}^+$, are i.i.d. as described above. The i thawed particles perform independent, continuous time, simple random walks until the first time τ_1 at which one of them jumps from 1 to 0. At this time we move the mark from 0 to 1 (i.e., $\mathcal{M}(t) = 0$ for $t < \tau_1$, but $\mathcal{M}(\tau_1) = 1$). Also at time τ_1 all particles at 1 are "absorbed by the mark". This includes frozen as well as thawed particles which are at 1 at time τ_1. If r thawed particles are removed at time τ_1, then we thaw another r particles. We take for these the r particles nearest to the mark at time τ_1, with some rule for breaking ties. The particles thawed at time τ_1 start simple random walks at that time.

At any time t there will be i thawed particles strictly to the right of the mark. If $\mathcal{M}(t) = p$, then the mark stays at p till the first time $\tau \geq t$ at which one of the i thawed particles jumps from $p+1$ to $\mathcal{M}(t) = p$. We then move the mark to position $p+1$ and absorb all particles at $p+1$. If r thawed particles are absorbed, then we also thaw r new particles. Again we thaw the particles nearest to the mark. We continue this process forever.

We prove that almost surely $\lim_{t\to\infty} t^{-1}\mathcal{M}(t)$ exists and is strictly positive, provided μ is large enough.

Mathematics Subject Classification (2000). Primary 60K35; secondary 60J15, 82C41.

Keywords. Diffusion limited aggregation, positive recurrence, Lyapounov function, growth model.

1. Introduction

In [9] we studied a one-dimensional version of Diffusion Limited Aggregation (DLA). This model also arose as a limiting case of a model for the spread of an infection (see [8]), but since this connection will not be needed here, we shall not discuss this further. The DLA model is as follows. The aggregate at time t is represented by an interval $[0, \mathcal{M}(t)] \cap \mathbb{Z}$. At time 0, $\mathcal{M}(0) = 0$ and there are $N(x)$ particles at x, for each strictly positive integer x. The $N(x), x \geq 1$ are i.i.d. Poisson variables with mean μ. All particles perform independent symmetric simple, continuous time, random walks. When at some time $t-$ the right edge of the aggregate $\mathcal{M}(t-) = k$ and one of the particles is at $k+1$ and tries to jump to k, then the aggregate moves one step to the right, i.e., $\mathcal{M}$ increases by 1 so that $\mathcal{M}(t) = k+1$. At the same time all particles which were at $k+1$ at time $t-$ are removed from the system (or equivalently, are absorbed by the aggregate). The principal question in [9] was to find the asymptotic behavior of $\mathcal{M}(t)$. We showed that if $\mu < 1$, then $\mathcal{M}(t)$ is of order $\sqrt{t}$. We conjectured that there exists some critical $\mu_c \in [1, \infty)$, such that for $\mu > \mu_c$, $\mathcal{M}(t)$ grows linearly in t.

We showed that for each value of μ, $t^{-1}\mathcal{M}(t)$ is almost surely bounded in $t \geq 1$. However we could not (and still cannot) show that $\liminf t^{-1}\mathcal{M}(t) > 0$ for large μ.

The present paper studies a variant of the DLA model for which we prove that almost surely $\lim_{t\to\infty} t^{-1}\mathcal{M}(t)$ exists and is strictly positive, provided (the analogue of) μ is large enough. We now describe this variant.

At time 0 choose i.i.d. random variables $N(x), x \in \mathbb{Z}^+ = \{1, 2, \dots\}$. Each $N(x)$ has a Poisson (μ) distribution. $N(x)$ will be the number of particles at x at time 0. In the model we also put a mark at an integer position. Its position at time t is denoted by $\mathcal{M}(t)$. We take $\mathcal{M}(0) = 0$. We regard the particles as frozen (i.e., they stay in place) till a certain random time, which generally differs for different particles. At such a random time a particle is "thawed", that is, it starts to move according to a continuous time simple random walk.

At all times there will be i thawed particles in the system. Later we shall specify the initial state. For the time being assume that at time 0 we have i thawed particles in the system at some positions in $\mathbb{Z}^+$, and that the $N(x), x \in \mathbb{Z}^+$, are i.i.d. as described above. The i thawed particles perform independent, continuous time, simple random walks until the first time τ_1 at which one of them jumps from 1 to 0. At this time we move the mark from 0 to 1 (i.e., $\mathcal{M}(t) = 0$ for $t < \tau_1$, but $\mathcal{M}(\tau_1) = 1$). Also at time τ_1 all particles at 1 are "absorbed by the mark". This includes frozen as well as thawed particles which are at 1 at time τ_1. Absorbed particles will not play any further role, so we may as well remove them from the system; we will sometimes call them removed particles. If r thawed particles are removed at time τ_1, then we thaw another r particles. We take for these the r particles nearest to the mark at time τ_1, with some rule for breaking ties. Note that all particles left in the system are strictly to the right of the mark at τ_1 (i.e., in $\{2, 3, \dots\}$). This will be the case at all times. The particles thawed at time τ_1

start simple random walks at that time. All random walks of the thawed particles are independent.

At any time t there will be i thawed particles strictly to the right of the mark. If $\mathcal{M}(t) = p$, then the mark stays at p till the first time $\tau \geq t$ at which one of the i thawed particles jumps from $p+1$ to $\mathcal{M}(t) = p$. We then move the mark to position $p+1$ and absorb all particles at $p+1$. If r thawed particles are absorbed, then we also thaw r new particles. Again we thaw the particles nearest to the mark. We continue this process forever.

A somewhat more formal representation of the system is as follows. Let $X^j(t), 1 \leq j \leq i$, be the positions of the i thawed particles in the system at time t. Denote their relative positions with respect to the mark by $\widehat{X}^j(t) := X^j(t) - \mathcal{M}(t)$. We define $\tau_0 = 0$,

$$\tau_1 = \min\{t \geq 0 : \text{ some thawed particle jumps from 1 to 0}\}$$

and

$$\tau_{k+1} = \min\{t > \tau_k : \text{ some thawed particle jumps from } \mathcal{M}(\tau_k) + 1 \text{ to } \mathcal{M}(\tau_k)\}. \tag{1.1}$$

By construction

$$\mathcal{M}(t) = k \text{ for } \tau_k \leq t < \tau_{k+1}. \tag{1.2}$$

Next we describe which particles are thawed at time τ_k. Suppose there are r_k thawed particles at position k at time τ_k-. These are then absorbed at τ_k and we must find r_k frozen particles from positions $k+1, k+2, \dots$ to thaw. If there are at least r_k particles at $(k+1, \tau_k-)$ (in space-time notation), then we simply select r_k of these by some arbitrary rule and are done with the selection. If there are only $r_{k,1} < r_k$ particles at $(k+1, \tau_k-)$, then all of these are thawed and we now examine how many particles there are at $(k+2, \tau_k-)$. If there are at least $r_k - r_{k,1}$ such particles, then we thaw $r_k - r_{k,1}$ of them, chosen by some arbitrary rule. If there are only $r_{k,2} < r_k - r_{k,1}$ such particles, then we thaw all of these and then search for these successively among the particles at $(k+j, \tau_k-), j = 2, 3, \dots$. When we have found r_k particles to thaw, there will be a "hole" in front of the mark, i.e. an interval $k+1, k+2, \dots, k+H-1$ at which all particles have been thawed, and with some or all particles at $k+H$ also thawed, but with no particles at $k+H+1$ having been thawed. In particular, there was no need to examine how many frozen particles there are at any of the sites $k+H+j, j \geq 1$. Thus, at the end of this procedure the numbers of frozen particles at these sites are still i.i.d. Poisson (μ) random variables. Note that $H = 1$ is possible. In this case the interval $k+1, \dots, k+H-1$ is empty.

We shall write $H(t)$ for the size of the hole at time t. This too is constant on $[\tau_k, \tau_{k+1})$ and we shall write H_k for this constant value. By definition

$$\begin{aligned}&\text{at time } \tau_k \text{ all particles in } [k+1, k+H_k-1 \text{ and some}\\ &\text{particles at } k+H_k \text{ are thawed, but none of the frozen}\\ &\text{particles in } [k+H_k+1, \infty) \text{ were needed so far.}\end{aligned} \tag{1.3}$$

We shall call the replacement of certain particles by newly thawed ones "adjustments" of the positions. These adjustments are defined as follows. Let

$$\widehat{J}_k = \text{the index of the coordinate which jumps from } k \text{ to } k-1 = \mathcal{M}(\tau_k-) \text{ at time } \tau_k.$$

Let there be r_k thawed particles at k at time τ_k- and let these be the particles with indices $1 \le j_1 \le j_2 \le \cdots \le j_{r_k}$ (one of these equals $\widehat{J}_k$; these are the particles which are one unit to the right of the mark at time τ_k- and which will be absorbed at time τ_k). Also let the r_k replacement particles which are thawed at time τ_k be located at $k + x_1 \le k + x_2 \le \cdots \le k + x_{r_k}$. Then set

$$A_k^j = \begin{cases} \widehat{X}^j(\tau_k) - \widehat{X}^j(\tau_k-) = x_\ell - 1 & \text{if } j = j_\ell \text{ but } j \neq \widehat{J}_k \\ \widehat{X}^{\widehat{J}_k}(\tau_k) - \widehat{X}^{\widehat{J}_k}(\tau_k-) + 1 = x_{\widehat{J}_k} & \text{if } j = \widehat{J}_k \\ \widehat{X}^j(\tau_k) - \widehat{X}^j(\tau_k-) = -1 & \text{if } j \notin \{j_1, \dots, j_{r_k}\}. \end{cases} \tag{1.4}$$

With this choice for the adjustments A_k^j we can represent the $\widehat{X}^j$ as

$$\widehat{X}^j(t) = X^j(t) - \mathcal{M}(t) = X^j(0) + S^j(t) + \sum_{k:0<\tau_k\le t} A_k^j, \tag{1.5}$$

where the $\{S^j\}$ are independent simple random walks. Roughly speaking, we want to prove here that the process $\{\widehat{X}^j(t), 1 \le j \le i\}$ is positive recurrent or ergodic.

We shall be particularly interested in the subsequence of $\{\tau_k\}$ at which $H_k = 1$. Accordingly we define $\sigma_0 = 0$, and

$$\sigma_{\ell+1} = \min\{\tau_k > \sigma_\ell : H_k = 1\}. \tag{1.6}$$

We need a number of definitions.

$$W_k := \text{ number of particles at } k + H_k \text{ which are still frozen at time } \tau_k$$

(this is after the replacement of particles at time τ_k has been carried out) and

$$k(n) := \text{ the index for which } \sigma_n = \tau_{k(n)}. \tag{1.7}$$

$$Z_k := (\widehat{X}^1(\tau_k), \dots, \widehat{X}^i(\tau_k), H_k, W_k).$$

Z_k is a $(i+2)$-vector of non-negative integers which describes the state of the system at τ_k. In fact, the first $i+1$ coordinates of Z_k are strictly positive. Finally, if $\tau_k = \sigma_n$ (i.e., $k = k(n)$), then we define

$$V_n := \big(\widehat{X}^1(\tau_{k+1}-), \dots, \widehat{X}^i(\tau_{k+1}-)\big).$$

These describe the positions of the particles just before the τ which follows a σ, i.e., which follows a τ at which $\{H = 1\}$ occurred. The further evolution in time from V_n to V_{n+1} consists first of the adjustment of the $\widehat{X}^j$ from their values at $\tau_{k(n)+1}-$ to their values at $\tau_{k(n)+1}$; next the change due to the motion of the random walks S^j during $(\tau_{k(n)+1}, \tau_{k(n)+2}]$; then another adjustment and random walk motions etc. till we reach $\tau_{k(n+1)+1}-$. We further need the following σ-fields which are analogues of the $\mathcal{H}(t)$ and $\mathcal{K}_k$ in Section 2:

$$\begin{aligned}\mathcal{S}_k = \sigma\text{-field generated by all } \tau_i \text{ with } i \le k \text{ as well as for } i \le k, \text{ the collections}\\ \text{of particles thawed at } \tau_i, \text{ and the paths through time } \tau_k \text{ of all these}\\ \text{particles; in addition, for } i \le k \text{ also } H_i, W_i \text{ and the } N(x) \text{ for } x \le i + H_i;\end{aligned}$$

Thus, $\mathcal{S}_k$ "contains all information which has been observed during $[0, \tau_k]$." In particular, the Z_i with $i \le k$ are $\mathcal{S}_k$-measurable. We shall also use the slightly larger σ-fields

$$\mathcal{T}_k := \sigma\text{-field generated by } \mathcal{S}_k \text{ and the paths } S^j(\tau_k + s) - S^j(\tau_k),$$
$$1 \le j \le i, 0 \le s \le \tau_{k+1} - \tau_k.$$

These σ-fields describe in addition to the information in $\mathcal{S}_k$ also the motions of the particles through time τ_{k+1}, but not the adjustments at time τ_{k+1}. In particular, the number of particles to be thawed at time τ_{k+1} is measurable with respect to $\mathcal{T}_k$. However, we get no information on the $N(x)$ with $x > k + H_k$ from this. The $N(x)$ with $x > k + H(k)$ are still independent of $\mathcal{T}_k$. For later use we note that

$$\mathcal{T}_{k-1} \subset \mathcal{S}_k \subset \mathcal{T}_k. \tag{1.8}$$

Lemma 1. *$\{Z_k\}_{k\ge0}$ and $\{V_\ell\}_{\ell\ge0}$ are Markov chains with stationary transition probabilities and countable state space (and discrete time).*

Proof. The position of the mark at time τ_k is k, so that the sequence $\mathcal{M}(\tau_k)$ is nonrandom. Given Z_k, one can find the future evolution by adding only the following information: (i) the paths in the future of the i particles which are thawed at time τ_k; (ii) the $N(x)$ for $x \in [k + H_k + 1, \infty)$, and (iii) the future paths of these still frozen particles in $[k + H_k + 1, \infty)$. Note that Z_k tells us how many frozen particles there still are at $k+H_k$ at time τ_k. But given all $Z_0, Z_1, \dots, Z_k$ the future paths of the particles at $(X^j(\tau_k), \tau_k)$ are independent simple random walk paths starting at these locations, independent of everything else. Moreover, as we noted before, the numbers of frozen particles at positions in $[k+H_k+1, \infty)$ have not yet been examined at time τ_k, and therefore are still i.i.d. Poisson (μ) variables, also independent of $Z_1, \dots, Z_k$. These numbers together with H_k, W_k and the future paths of the particles determine all H_m, W_m for $m > k$. Thus the conditional distribution of the Z_m for $m > k$, given $Z_0, \dots, Z_k$ depends on Z_k only. Moreover, one can see that the transition probabilities from a state at time τ_k to a state at time τ_ℓ depend on $\ell - k$ only. We omit the details. This proves the statement of the lemma for the Z-process.

Essentially the same proof as for the Z-process shows that $\{V_n, H_{k(n)}, W_{k(n)}\}$ is a Markov chain. However, $H_{k(n)} = 1$, by definition of the σ's. Furthermore, we claim that $W_{k(n)}$ has no influence on the V_ℓ with $\ell > n$. Indeed, since $H_{k(n)} = 1$, $W_{k(n)}$ tells us how many frozen particles there are at position $k(n) + H_{k(n)} = k(n) + 1$ at time $\tau_{k(n)}$. At $\tau_{k(n)+1}$, which is the first τ after $\tau_{k(n)}$, the mark moves to $k(n)+1$ and absorbs all particles at that site, irrespective of $W_{k(n)}$. This proves our claim and the lemma. □

There is a counterpart to the preceding lemma in continuous time. For this we define

$$H(t) = H_k \text{ if } \tau_k \le t < \tau_{k+1}, \quad W(t) = W_k \text{ if } \tau_k \le t < \tau_{k+1},$$

and

$$Z(t) = (\widehat{X}^1(t), \dots, \widehat{X}^i(t), H(t), W(t)).$$

Then $\{Z(t)\}_{t\ge 0}$ is a continuous time Markov process with stationary transition probabilities. Even though this process is the principal subject of investigation here, we shall not prove this Markov property. A proof goes along the lines of the proof of the last lemma. Our aim in this paper is to prove the following theorem and corollary.

Theorem 1. *For sufficiently large i and for $\mu \ge 2i$, the process $\{Z(t)\}_{t\ge 0}$ is positive recurrent.*

Corollary. *For sufficiently large i and $\mu \ge 2i$*

$$\lim_{t\to\infty} \frac{\mathcal{M}(t)}{t} \text{ exists and is strictly positive.}$$

2. Proof

As state space for the process $\{Z(t)\}$ we can take the collection of integer $(k+2)$-tuples $\{(z_1, \dots, z_i, h, w) : z_j \ge 1, 1 \le j \le i, h \ge 1, w \ge 0\}$. It is not hard to see that the process can go from any state to any other state in this collection by a suitable choice of the $N(x)$ and the random walk paths. We merely point out that we can choose random walk paths which go from some $(z_1', \dots z_i')$ to a given $(z_1'', \dots, z_i'')$, without any path jumping from 1 to 0 in between. Thus, for any $N \ge 1$, the process can go from $(z_1', \dots z_i', h', w')$ to $(1, N, \dots, N, h', w')$ and similarly from $(1, N, \dots, N, h'', w'')$ to $(z_1'', \dots, z_i'', h'', w'')$. Thus we only need to prove that the process can go from any state $(1, N, \dots, N, h', w')$ to any $(1, N, \dots, N, h'', w'')$, for some large N. If the process is in $(1, N, \dots, N, h', w')$ there is a positive probability that next the particle at 1 jumps to 0. This will require one replacement. By repeated replacements of one particle at a time one can indeed go from $(1, N, \dots, N, h', w')$ to $(1, N, \dots, N, h'', w'')$. We leave the tedious details to the reader.

The result of the above considerations is that the $\{Z(t)\}$-process is irreducible. The theorem and corollary therefore do not depend on the initial condition, but for the purpose of the proof we choose the initial condition as follows. At time 0 we start in a state with i thawed particles at some positions $y_1, \dots, y_i$ such that

$$\widehat{L}_0 := \sum_{j=1}^{i} y_i \le \alpha := D_{12} i^{24\cdot 15+1} \tag{2.1}$$

and $H_0 = 1$ and some $W_0 \geq 0$. That is, we put W_0 frozen particles at position 1, or we have no frozen particles at 1 and we regard the situation as one in which all particles at 1 have been thawed already, but no particles in $[2, \infty)$ have been thawed yet. So in this state the $N(x), x \geq 2$, are still i.i.d. Poisson (μ) random variables. We take $\mathcal{M}(0) = 0, k(0) = 0$ and regard this initial condition as the positions of the thawed particles in the system at time $\tau_0 = 0$, just after the adjustments at τ_0 have been made. Thus the particles now start to move till the next time τ_1 when a particle wants to jump onto the mark. Since we took $H_0 = 1$ we have also $\sigma_0 = \tau_0$ and $k(0) = 0$. Since we regard the system as just having thawed a particle and with $H_0 = 1$, one of the thawed particles must sit at position one unit to the right of the mark, that is we must have $y_j = 1$ for at least one $1 \leq j \leq i$.

Lemma 2. *Let $\eta = (\mu - i)/2 > 0$ and let*

$$C_1 = \frac{\eta^2}{4\mu}. \tag{2.2}$$

Then for each $k \geq 0$ and

$$m \geq \frac{\mu(H_k + 1)}{\eta}, \tag{2.3}$$

it holds

$$P\{H_\ell > 1 \text{ for all } k+1 \leq \ell < k+m | \mathcal{T}_k\} \leq e^{-C_1 m}. \tag{2.4}$$

Proof. Let

$$q = \min\big\{\ell \geq k + H_k + 1 : \sum_{x=k+H_k+1}^{\ell} N(x) > i(\ell - k - 1)\big\}; \tag{2.5}$$

$q = \infty$ if the set in the right-hand side here is empty. Also, let p be the smallest index exceeding k for which τ_p equals some σ_n (if no such σ_n exists, then we take $p = \infty$). We claim that

$$p \leq q. \tag{2.6}$$

To prove this (deterministic) claim we clearly may assume that $q < \infty$. Assume now that $q < \infty$ and that (2.6) fails for some sample point. By the definition of p this means

$$H_\ell > 1 \text{ for each } \ell \in \{k+1, k+2, \dots, q\}. \tag{2.7}$$

In particular, none of the frozen particles in $\{k+2, k+3, \dots, q+1\}$ which are still frozen at time τ_k are absorbed while they are still frozen. Indeed, a particle at $\ell \in \{k+2, k+3, \dots, q+1\}$ can be absorbed while frozen only at the time τ_ℓ when the mark first reaches ℓ. In particular, such a particle must still have been frozen at $\tau_{\ell-1}$ when the mark came to $\ell - 1$. But if at that time there are still frozen particles left at ℓ, then there are frozen particles one unit to the right of the mark at time $\tau_{\ell-1}$, and therefore $H_{\ell-1} = 1$. This contradicts (2.7). Thus, all particles in $\{k+2, k+3, \dots, q\}$ which are still frozen at time τ_k are thawed at some time no later than τ_{q-1}, which is the last time particles at q can be thawed (since the mark reaches q at τ_q). However, by definition of H_k, none of the frozen particles originally in $[k+H_k+1,\infty)$ are thawed at time τ_k or even at $\tau_{k+1}-$. Therefore, the total number of particles thawed during the time interval $[\tau_{k+1},\tau_{q-1}]$ is at least

$$\text{(the number of frozen particles originally in } [k+H_k+1,q]) = \sum_{x=k+H_k+1}^{q} N(x).$$

On the other hand, at each τ_ℓ at most i particles have to be replaced, and at most i particles are thawed. Therefore there are at least $i^{-1}\sum_{x=k+H_k+1}^{q} N(x)$ times in $[\tau_{k+1},\tau_{q-1}]$ at which some particles are thawed. That is, the mark moves at least $i^{-1}\sum_{x=k+H_k+1}^{q} N(x)$ times in $[\tau_{k+1},\tau_{q-1}]$. But this says

$$q-k-1 \geq \frac{1}{i}\sum_{x=k+H_k+1}^{q} N(x),$$

which contradicts our choice of q. Thus (2.7) must fail, or equivalently, (2.6) must hold.

The result (2.6) immediately implies that the left-hand side of (2.4) is at most

$$\begin{aligned} P\{\tau_q \geq \tau_p \geq \tau_{k+m}|\mathcal{T}_k\} &\leq P\{q \geq k+m|\mathcal{T}_k\} \\ &\leq P\{\sum_{x=k+H_k+1}^{k+m-1} N(x) \leq i(m-2)\big|\mathcal{T}_k\}. \end{aligned} \tag{2.8}$$

However, conditional on $\mathcal{T}_k$, the $N(x)$ with $x > k+H_k$ are i.i.d. Poisson (μ) random variables, and hence $\sum_{x=k+H_k+1}^{k+m-1} N(x)$ is a Poisson $\big((m-H_k-1)\mu\big)$ variable. But under (2.3) $(m-H_k-1)\mu - i(m-2) \geq \eta m$. It follows that for $\rho = (m-H_k-1)\mu$ and $\theta \geq 0$

$$\begin{aligned} &P\{H_\ell > 1 \text{ for all } k+1 \leq \ell < k+m|\mathcal{T}_k\} \\ &\leq P\{(\text{a Poisson } \big((m-H_k-1)\mu\big) \text{ variable}) - (m-H_k-1)\mu \\ &\qquad\qquad \leq i(m-2)-(m-H_k-1)\mu \leq -\eta m\} \\ &\leq \exp[-\theta\eta m + \theta\rho + \rho(e^{-\theta}-1)]. \end{aligned} \tag{2.9}$$

For $\rho < \eta m$ the second member here is even zero, since a Poisson variable has to be non-negative. For $\rho \geq \eta m$ take $\theta = \eta m/(2\rho) \leq 1/2$ and use that $e^{-\theta}-1 \leq -\theta+\theta^2$ for $0 \leq \theta \leq 1/2$ to obtain the inequality (2.4) from (2.9). □

The following definition will be useful in the sequel:

$$K_k(1) = 1, K_k(m) := I[H_\ell > 1 \text{ for all } k+1 \leq \ell < k+m], m > 1.$$

$K_k(m)$ is the indicator function of the event that the first σ after τ_k comes at or after τ_{k+m}. On the event

$$\{\sigma_n = \tau_k\}, \tag{2.10}$$

$k(n) = k$ and $\{K_k(m) = 1\} = \{k(n+1) \geq k+m\}$.

Lemma 3. *With μ, η and C_1 as in Lemma 2, it holds for $m \geq 2\mu/\eta$ on the event* (2.10)*, that*

$$P\{k(n+1) - k \geq m \big| \mathcal{T}_k\} = P\{K_k(m) = 1 \big| \mathcal{T}_k\} \leq e^{-C_1 m}. \tag{2.11}$$

Moreover, if

$$\mu \geq 2i \geq 6, \tag{2.12}$$

and still on (2.10)*, for any integer $m \geq 1$ and $q \geq 0$,*

$$E\{H_{k+m}^q K_k(m) \big| \mathcal{T}_k\} \leq D_2 e^{-D_1 m/2}. \tag{2.13}$$

Finally,

$$P\{H_{k+m} \geq r, K_k(m) = 1 | \mathcal{T}_k\} \leq D_3 e^{-D_1 r} \text{ for } m \geq 1, r \geq 0 \tag{2.14}$$

for some universal constants D_1, D_3, and with D_2 depending on q only.

Proof. The inequality (2.11) is just a special case of (2.4), since $k(n) = k$ on the event (2.10).

For (2.13) we assumed that $\mu \geq 2i$. In this case, $\eta = (\mu - i)/2 \geq \mu/4$, and consequently condition (2.3) is implied by $m \geq 8$, since $H_k = 1$ on (2.10). Moreover, in this situation, the C_1 in (2.2) is at least $\mu/64 \geq i/32 \geq 3/32$. So we will take $D_1 \leq 3/32$. (2.11) then gives

$$P\{K_k(m) = 1 \big| \mathcal{T}_k\} \leq e^{-D_1(m-8)} \text{ for } m \geq 1. \tag{2.15}$$

(Note that this holds automatically for $m \leq 8$.)

The proof of (2.13) starts with an application of Schwarz' inequality:

$$E\{H_{k+m}^q K_k(m) \big| \mathcal{T}_k\} \leq \Big[E\{H_{k+m}^{2q} K_k(m) \big| \mathcal{T}_k\} P\{K_k(m) = 1 \big| \mathcal{T}_k\}\Big]^{1/2}.$$

By (2.15), $\big[P\{K_k(m) = 1 \big| \mathcal{T}_k\}\big]^{1/2} \leq \exp[-D_1(m/2 - 4)]$, so that it suffices for (2.13) to prove

$$E\{H_{k+m}^{2q} | \mathcal{T}_k\} \leq D_2^2 \exp[-8D_1]. \tag{2.16}$$

To this end, observe that $H_{\ell+1} \geq H_\ell - 1$ for any ℓ, by virtue of (1.3). Thus, if $H_{k+m} \geq r$ for some $r > 1$, and $K_k(m) = 1$, then $H_\ell > 1$ for $k + 1 \leq \ell < k + m$ and H_ℓ stays greater than 1 for $k + m \leq \ell < k + m + r - 2$. Therefore, by (2.11), for $m + r - 1 \geq 8$, $r \geq 2$

$$\begin{aligned}
&P\{H_{k+m} \geq r, K_k(m) = 1 | \mathcal{T}_k\} \\
&\leq P\{H_\ell > 1 \text{ for all } k + 1 \leq \ell \leq k + m + r - 2 | \mathcal{T}_k\} \\
&\leq P\{\sigma_{n+1} \geq k + m + r - 1 \big| \mathcal{T}_k\} \\
&\leq \exp[-D_1(m + r - 1)].
\end{aligned}$$

Clearly this implies

$$P\{H_{k+m} \geq r, K_k(m) = 1 | \mathcal{T}_k\} \leq \exp[D_1(8 - r)] \text{ for } m \geq 1, r \geq 0, \tag{2.17}$$

and hence (2.14). (There is nothing to prove for $r < 9$.) It also follows that for all $m \geq 1$,

$$E\{H_{k+m}^{2q} K_k(m) | \mathcal{T}_k\} \leq \sum_{r=0}^{\infty} r^{2q} \exp[D_1(8-r)],$$

which is the desired (2.16) for a suitable choice of D_2. □

Remark 1. We shall apply (2.13) only for integer $q \leq 100$. We may therefore choose one D_2 for which (2.13) holds in all applications of (2.13) in this paper.

It is immediate from (2.11) that the sequence of k's for which $\{H_k = 1\}$ occurs has a.s. a strictly positive lower density, and in particular all $k(n)$ are almost surely finite. In fact the difference between two successive k's for which the event $H_k = 1$ occurs has an exponentially bounded tail, uniformly in the history of the process before the first of the two occurrences. In our notation this means that the sequence of k's for which τ_k equals some σ has a positive lower density. The next major step is to show that even the subsequence of k's for which τ_k equals some σ and for which $\widetilde{X}^j(\tau_k-) < \alpha$ has strictly positive lower density (for a suitable large constant α).

Finally we introduce the analogues of $\widetilde{Q}_k^{(r)}$ and $\widetilde{L}_k$:

$$\widehat{Q}_k^{(r)} := \sum_{j=1}^{i} \big[\widehat{X}^j(\tau_k-) - I[j = \widehat{J}_k]\big]^r = \sum_{j=1}^{i} \big[X^j(\tau_k-) - \mathcal{M}(\tau_k-) - I[j = \widehat{J}_k]\big]^r,$$

and

$$\widehat{L}_k := \sum_{j=1}^{i} \big[[\widehat{X}^j(\tau_k-) - I[j = \widehat{J}_k]\big] = \widehat{Q}_k^{(1)}.$$

We remind the reader that $\mathcal{T}_k$ is defined a few lines before (1.8). Note that for fixed $m \geq 1$, $k(n) + m$ is a stopping time with respect to the increasing sequence of σ-fields $\{\mathcal{T}_{k-1}, k = 1, 2, \dots\}$, so that the σ-fields $\mathcal{T}_{k(n)+m-1}$ are well defined.

Lemma 4. *For $m \geq 1, 2 \leq r \leq 100, n \geq 0, i \geq i_1 := \max_{0 \leq q,r \leq 100} i(r,q,1), \mu \geq 2i$, and for a suitable constant D_4 it holds*

$$\begin{aligned} &E\{|\widehat{Q}_{k(n)+m+1}^{(r)} - \widehat{Q}_{k(n)+m}^{(r)}| | \mathcal{T}_{k(n)+m-1}\} \\ &\leq D_4 \widehat{Q}_{k(n)+m}^{(r-1)} \left(\frac{\widehat{L}_{k(n)+m}}{i}\right)^{1/2} E\{H_{k(n)+m}^r | \mathcal{T}_{k(n)+m-1}\}. \end{aligned} \tag{2.18}$$

In addition for $\varepsilon > 0, i \geq i_2(\varepsilon) := \max_{0 \leq q,r \leq 100} i(r,q,\varepsilon), \mu \geq 2i, m \geq 1, 2 \leq r \leq 100$ and a suitable constant D_5

$$\begin{aligned} E\{\widehat{Q}_{k(n)+m+1}^{(r)} | \mathcal{T}_{k(n)+m-1}\} - \widehat{Q}_{k(n)+m}^{(r)} &\leq D_5 r \varepsilon \widehat{Q}_{k(n)+m}^{(r-1)} E\{H_{k(n)+m}^r | \mathcal{T}_{k(n)+m-1}\} \\ &- r 2^{1-r} \widehat{Q}_{k(n)+m}^{(r-1)} + 3ir E\{H_{k(n)+m}^r | \mathcal{T}_{k(n)+m-1}\}. \end{aligned} \tag{2.19}$$

Remark 2. We have not yet proven – and it is not proven in this lemma either – that $E\{|\widehat{Q}^{(r)}_{k(n)+m+1} - \widehat{Q}^{(r)}_{k(n)+m}|\}$ or $E\{\widehat{Q}^{(r)}_k\}$ are finite. This will only be shown in Lemma 6. Nevertheless the present lemma makes sense because if $\mathcal{B}$ is a σ-field with a sub σ-field $\mathcal{F}$, then for a *non-negative* $\mathcal{B}$-measurable function Q, $E\{Q|\mathcal{F}\}$ is well defined and unique (up to changes on null sets) (see [5], Section 7B).

Proof. We shall repeatedly use the observation that

$$A^j_{k+m} \begin{cases} \leq H_{k+m} & \text{if } \widehat{X}^j(\tau_{k+m}-) = 1 \\ = -1 & \text{if } \widehat{X}^j(\tau_{k+m}-) \geq 2, \end{cases} \tag{2.20}$$

which follows immediately from (1.4). In particular

$$|A^j_k| \leq H_k. \tag{2.21}$$

We shall also need the analogue of Lemma 2 of [9] for the present model. Only some simple changes are necessary to obtain a version valid for the model with frozen particles, namely replacement of k by $k+m$, of $Y^j(\tau_{k+m})$ by H_{k+m} and of X^j by $\widehat{X}^j$. This shows that for $2r \in \{2,3,\dots\}$ and for $i \geq i(r,p,\varepsilon)$,

$$\begin{aligned} &E\{\delta^r_{k+m}[\widehat{X}^j(\tau_{k+m})]^p|\mathcal{T}_{k+m-1}\} \\ &\leq \varepsilon 2^{2r+p-3}\Big(\frac{\widehat{L}_{k+m}}{i}\Big)^{2r-1}\Big[[\widehat{X}^j(\tau_{k+m}-)]^p E\{H_{k+m}|\mathcal{T}_{k+m-1}\} + E\{H^{p+1}_{k+m}|\mathcal{T}_{k+m-1}\}\Big] \\ &\quad + \varepsilon 2^{2r+p-3}\Big[[\widehat{X}^j(\tau_{k+m}-)]^p E\{H^{2r}_{k+m}|\mathcal{T}_{k+m-1}\} + E\{H^{2r+p}_{k+m}|\mathcal{T}_{k+m-1}\}\Big] \\ &\leq D_6\varepsilon\Big(\frac{\widehat{L}_{k+m}}{i}\Big)^{2r-1}[\widehat{X}^j(\tau_{k+m}-)]^p E\{H^{2r+p}_{k+m}|\mathcal{T}_{k+m-1}\}. \end{aligned} \tag{2.22}$$

In the last inequality we used that $H_{k+m} \geq 1$, so that H^q_{k+m} is non-decreasing in q and also that $\widehat{L}_q \geq i-1$; $D_6 = D_6(r,p)$ and $i(r,p,\varepsilon)$ are suitable constants depending only on the indicated arguments.

To prove (2.18) we first use some of the arguments in Lemma 3 in [9]. This, together with (2.20) results in the following (deterministic) inequality:

$$\begin{aligned} |\widehat{Q}^{(r)}_{k(n)+m+1} - \widehat{Q}^{(r)}_{k(n)+m}| &\leq \sum_{j=1}^i \Big|\big[\widehat{X}^j(\tau_{k(n)+m}-) - I[j = \widehat{J}_{k(n)+m}] + A^j_{k(n)+m}\big]^r \\ &\quad - \big[\widehat{X}^j(\tau_{k(n)+m}-) - I[j = \widehat{J}_{k(n)+m}]\big]^r\Big| \\ &\quad + \sum_{j=1}^i\sum_{u=1}^r \binom{r}{u}[\widehat{X}^j(\tau_{k(n)+m})]^{r-u}|S^j(\tau_{k(n)+m+1}) - S^j(\tau_{k(n)+m})|^u \\ &\leq iH^r_{k(n)+m} + r\widehat{Q}^{(r-1)}_{k(n)+m} \\ &\quad + \sum_{j=1}^i\sum_{u=1}^r \binom{r}{u}[\widehat{X}^j(\tau_{k(n)+m})]^{r-u}|S^j(\tau_{k(n)+m+1}) - S^j(\tau_{k(n)+m})|^u. \end{aligned} \tag{2.23}$$

As we mentioned, $k(n)+m$ is a stopping time with respect to the increasing sequence of σ-fields $\{\mathcal{T}_{k-1}, k=1,2,\dots\}$. Simple general properties of conditional expectations therefore show that for fixed n, m, and any positive function Q,

$$E\{Q|\mathcal{T}_{k(n)+m-1}\} = \sum_{k=0}^{\infty} E\{Q|\mathcal{T}_{k+m-1}\}I[k(n)=k]. \tag{2.24}$$

In particular, if we fix n and k, then on the event $\{k(n)=k\}$

$$\begin{aligned} E\{|\widehat{Q}^{(r)}_{k(n)+m+1} &- \widehat{Q}^{(r)}_{k(n)+m}||\mathcal{T}_{k(n)+m-1})\} \\ &\le iE\{H^r_k|\mathcal{T}_{k+m-1})\} + r\widehat{Q}^{(r-1)}_{k+m} \\ &+ \sum_{j=1}^{i}\sum_{u=1}^{r}\binom{r}{u} E\{[\widehat{X}^j(\tau_{k+m})]^{r-u}|S^j(\tau_{k+m+1}) - S^j(\tau_{k+m})|^u\big|\mathcal{T}_{k+m-1}\}. \end{aligned}$$

Following the proof of Lemma 2 in [9] (see also (2.22)) we have for $u \ge 2$ and some constant $D_7 = D_7(r,u)$

$$\begin{aligned} &E\{[\widehat{X}^j(\tau_{k+m})]^{r-u}|S^j(\tau_{k+m+1}) - S^j(\tau_{k+m})|^u\big|\mathcal{T}_{k+m-1}\} \\ &= E\Big\{[\widehat{X}^j(\tau_{k+m})]^{r-u}E\{|S^j(\tau_{k+m+1}) - S^j(\tau_{k+m})|^u\big|\mathcal{S}_{k+m}\}\Big|\mathcal{T}_{k+m-1}\Big\} \\ &\le D_7E\Big\{[\widehat{X}^j(\tau_{k+m}-)]^{r-u}H^r_{k+m}\Big(\frac{\widehat{L}_{k+m}}{i}\Big)^{u-1}\Big|\mathcal{T}_{k+m-1}\Big\} \\ &= D_7\big[\widehat{X}^j(\tau_{k+m}-)\big]^{r-u}\Big(\frac{\widehat{L}_{k+m}}{i}\Big)^{u-1}E\{H^r_{k+m}|\mathcal{T}_{k+m-1})\}. \end{aligned} \tag{2.25}$$

We sum this over j from 1 to i and use that $\sum_{j=1}^{i}[\widehat{X}^j(\tau_q-)]^{r-u} = \widehat{Q}^{(r-u)}_q + 1 \le 2\widehat{Q}^{(r-u)}_q$ and $(\widehat{L}_{k+m}/i)^{u-1}\widehat{Q}^{(r-u)}_{k+m} \le Q^{(r-1)}_{k+m}$ (see (4.59) in [9]). For $u \ge 2$ we then get a contribution $D_8\widehat{Q}^{(r-1)}_{k(n)+m}E\{H^r_{k(n)+m}|\mathcal{T}_{k(n)+m-1}\}$ to (2.18), where D_8 depends on r only.

For $u=1$ we get

$$\begin{aligned} &E\Big\{[\widehat{X}^j(\tau_{k+m})]^{r-1}E\{|S^j(\tau_{k+1}) - S^j(\tau_k)||\mathcal{S}_{k+m}\}\Big|\mathcal{T}_{k+m-1})\Big\} \\ &\le E\Big\{[\widehat{X}^j(\tau_{k+m})]^{r-1}\big[E\{|S^j(\tau_{k+1}) - S^j(\tau_k)|^2|\mathcal{S}_{k+m}\}\big]^{1/2}\Big|\mathcal{T}_{k+m-1}\Big\} \\ &\le D_8[\widehat{X}^j(\tau_{k+m}-)]^{r-1}\Big(\frac{\widehat{L}_{k+m}}{i}\Big)^{1/2}E\{H^r_{k+m}|\mathcal{T}_{k+m-1}\}. \end{aligned}$$

The inequality (2.18) on the event $\{k(n)=k\}$ now follows by summing this contribution also over j from 1 to i and by adding it to the previous contributions.

For (2.19) we again follow the proofs of Lemmas 2 and 3 in [9]. Exactly as in (4.54) and (4.56) of [9] this leads on the event $\{k(n)=k\}$ to the following replacement of (2.23) and (2.25):

$$
\begin{aligned}
E\{\widehat{Q}^{(r)}_{k(n)+m+1}|\mathcal{T}_{k(n)+m-1}\} &\le D_7\varepsilon \sum_{u=2}^{r}\Big(\frac{\widehat{L}_{k+m}}{i}\Big)^{u-1}\widehat{Q}^{(r-u)}_{k+m}E\{H^r_{k+m}|\mathcal{T}_{k+m-1}\}\\
&\quad+\sum_{j=1}^{i}E\Big\{\big[\widehat{X}^j(\tau_{k+m}-)-I[j=\widehat{J}_{k+m}]+A^j_{k+m}\big]^r\\
&\qquad-\big[\widehat{X}^j(\tau_{k+m}-)-I[j=\widehat{J}_{k+m}]\big]^r\big|\mathcal{T}_{k+m-1}\Big\}+\widehat{Q}^{(r)}_{k+m}\\
&\le D_7\varepsilon\sum_{u=2}^{r}\Big(\frac{\widehat{L}_{k+m}}{i}\Big)^{u-1}\widehat{Q}^{(r-u)}_{k+m}E\{H^r_{k+m}|\mathcal{T}_{k+m-1}\}\\
&\quad-r2^{1-r}\widehat{Q}^{(r-1)}_{k+m}+i\big[r2^{1-r}+rE\{H^r_{k+m}\big|\mathcal{T}_{k+m-1}\}\big]+\widehat{Q}^{(r)}_{k+m}\\
&\le D_7r\varepsilon\widehat{Q}^{(r-1)}_{k+m}E\{H^r_{k+m}|\mathcal{T}_{k+m-1}\}-r2^{1-r}\widehat{Q}^{(r-1)}_{k+m}\\
&\quad+3irE\{H^r_{k+m}\big|\mathcal{T}_{k+m-1}\}+\widehat{Q}^{(r)}_{k+m}.
\end{aligned}
\tag{2.26}
$$

Note that there is no contribution for $u = 1$, because without the absolute value signs $E\{[S^j(\tau_{k+m+1})-S^j(\tau_{k+m})]|\mathcal{T}_{k+m-1}\}=0$. The inequality (2.26) holds on $\{k(n)=k\}$ for any fixed k. (2.19) now follows by multiplying both sides with $I[k(n)=k]$ and summing over k (see (2.24)). □

Now note that $[\widehat{L}_u]^{1/2}\le\sum_{j=1}^{i}[X^j(\tau_u-)-I[j=\widehat{J}_u]^{1/2}=Q_u^{(1/2)}$ and hence

$$[\widehat{L}_u]^{1/2}Q_u^{(r-1)}\le iQ_u^{(r-1/2)} \text{ (see (4.65) in [9]).}$$

Therefore, if we multiply both sides of (2.18) by $K_{k(n)}(m)$ and take expectation we get

$$
\begin{aligned}
&E\{|\widehat{Q}^{(r)}_{k(n)+m+1}-\widehat{Q}^{(r)}_{k(n)+m}|K_{k(n)}(m)\}\\
&\le D_4E\Big\{\widehat{Q}^{(r-1)}_{k(n)+m}\Big(\frac{\widehat{L}_{k(n)+m}}{i}\Big)^{1/2}H^r_{k(n)+m}K_{k(n)}(m)\Big\}\\
&\le D_4 i^{1/2}E\Big\{\widehat{Q}^{(r-1/2)}_{k(n)+m}H^r_{k(n)+m}K_{k(n)}(m)\Big\}\\
&\le D_4 i^{1/2}\Big[E\{\widehat{Q}^{(r)}_{k(n)+m}K_{k(n)}(m)\}\Big]^{1-1/(2r)}\Big[E\{H^{2r^2}_{k(n)+m}K_{k(n)}(m)\}\Big]^{1/(2r)}\\
&\text{(Hölder's inequality)}\\
&\le D_9 i^{1/2}\Big[E\{\widehat{Q}^{(r)}_{k(n)+m}K_{k(n)}(m)\}\Big]^{1-1/(2r)},\ m\ge1,
\end{aligned}
$$

where in the last inequality we used that

$$E\{H^{2r^2}_{k(n)+m}K_{k(n)}(m)\}=E\{E\{H^{2r^2}_{k(n)+m}K_{k(n)}(m)|\mathcal{T}_{k(n)}\}\}\le D_{10}$$

(by (2.24) and (2.13)). Now take $n = 0$ and $k(0) = 0$. If the particles at time τ_0 are initially at the (non-random) positions $x^1, \dots, x^i$, then, for $r \geq 1$,

$$E\{\widehat{Q}_1^{(r)}\} \leq \sum_{j=1}^{i} E\{[x^j + S^j(\tau_1)]^r\}, \tag{2.27}$$

and this is finite for any choice of the x^j, provided $E\delta_1^{r/2} < \infty$, i.e., for $r < i$. We shall choose $i(r, q, \varepsilon) > 100$ so that this condition is automatically fulfilled if $i \geq i_2(\varepsilon), r \leq 100$. It then follows by induction on m that

$$E\{\widehat{Q}_{m+1}^{(r)} K_0(m)\} = E\{\widehat{Q}_{m+1}^{(r)} I[m \leq k(1)]\} < \infty \text{ for all } m \geq 0, 1 \leq r \leq 100. \tag{2.28}$$

From now on we fix $\varepsilon > 0$ such that

$$D_5 D_6 \varepsilon \leq 2^{-50} \tag{2.29}$$

and we always assume that

$$i \geq i_2(\varepsilon) \text{ and } \mu \geq 2i. \tag{2.30}$$

In addition we take

$$s = 15, \theta = 16 \text{ and } r \in \{2, 3, \dots, 24\}. \tag{2.31}$$

However, many steps remain valid for more general values of s, θ and r. Next we introduce an important quantity which has no direct analogue in Section 2. We define

$$R_n^{(r)} := \widehat{Q}_{k(n+1)+1}^{(r)} - \widehat{Q}_{k(n)+1}^{(r)} = \sum_{m=1}^{\infty} [\widehat{Q}_{k(n)+m+1}^{(r)} - \widehat{Q}_{k(n)+m}^{(r)}] K_{k(n)}(m). \tag{2.32}$$

Note that $\widehat{Q}_{k(n)+1}^{(r)}$ is a sum of powers of $\widehat{X}^j(\tau_{k(n)+1}-) - I[j = \widehat{J}_{k(n)+1}]$. These are essentially positions evaluated at time $\tau_{k(n)+1}-$, i.e., at the first τ after σ_n, but before the adjustments at that time have been made. They still involve the $S^j(\tau_{k(n)+1})$, but not the $A_{k(n)+1}^j$. In order to use Lemma 4 we need to show that

$$E\{R_n^{(r)} | \mathcal{T}_{k(n)}\} \leq 0. \tag{2.33}$$

In fact we shall need a stronger version of this inequality. This will be given in Lemma 7.

Lemma 5. *For each $q \geq 0$ there exists some constant $D_{12} = D_{12}(q) \geq 1$ (independent of i, n) such that for $2 \leq r \leq 24, i \geq i_2(\varepsilon)$ and $\mu \geq 2i$, on the event*

$$\{\widehat{L}_{k(n)+1} \geq D_{12} i^{24 \cdot 15+2}\}, \tag{2.34}$$

it holds

$$\sum_{m=1}^{\infty} E\{\widehat{Q}_{k(n)+m}^{(r)} H_{k(n)+m}^q K_{k(n)}(m) | \mathcal{T}_{k(n)}\} \leq D_6 \widehat{Q}_{k(n)+1}^{(r)} \text{ a.s.} \tag{2.35}$$

In addition, there exists a constant D_7, such that even without the requirement (2.34)

$$\sum_{m=1}^{\infty} E\{\widehat{Q}^{(r)}_{k(n)+m} H^q_{k(n)+m} K_{k(n)}(m) | \mathcal{T}_{k(n)}\} \le \exp[D_7 i^{361}] \widehat{Q}^{(r)}_{k(n)+1} \quad a.s. \tag{2.36}$$

Proof. We shall give the proof of (2.35) in full. After that we list the few changes which have to be made to obtain (2.36) without using (2.34). We break the proof up into four steps.

Step 1. Reduction to a large deviation estimate. We define

$$\mathcal{I}(p,m) = \mathcal{I}(p,m,n) = I[e^p \widehat{Q}^{(r)}_{k(n)+1} < \widehat{Q}^{(r)}_{k(n)+m} \le e^{p+1} \widehat{Q}^{(r)}_{k(n)+1}],$$

$$\mathcal{J}_p = I[\widehat{Q}^{(r)}_{k(n)+m} > e^p \widehat{Q}^{(r)}_{k(n)+1} \text{ for some } 1 \le m \le k(n+1) - k(n)],$$

$$m_0 = m_0(p,n) = \inf\{m \ge 1 : \widehat{Q}^{(r)}_{k(n)+m} > e^p \widehat{Q}^{(r)}_{k(n)+1}\}$$

(this is ∞ if no such m exist), and

$$\mathcal{W}(p,n) = \big(\text{number of } m \in [1, k(n+1) - k(n)] \text{ with } \widehat{Q}^{(r)}_{k(n)+m} > e^p \widehat{Q}^{(r)}_{k(n)+1}\big).$$

We fix a k and for the time being we work on the event (2.10). On this even the conditional probability with respect to $\mathcal{T}_{k(n)}$ in (2.35) is equal to the conditional probability with respect to $\mathcal{T}_k$ (see (2.24)). We remind the reader that $K_{k(n)}(m) = 1$ if and only if $k(n+1) \ge k+m$, or equivalently, $\sigma_{n+1} \ge \tau_{k+m}$. Since for any $u \ge 0$ and $Q > 0$ either $u \le Q$ or $u \in (e^p Q, e^{p+1} Q]$ for a unique integer $p \ge 0$, the left-hand side of (2.35) is (on (2.10)) at most

$$\begin{aligned}
&\widehat{Q}^{(r)}_{k+1} \sum_{m=1}^{\infty} E\{H^q_{k(n)+m} K_k(m) | \mathcal{T}_k)\} \\
&\qquad + \sum_{m=1}^{\infty} \sum_{p=0}^{\infty} E\{\widehat{Q}^{(r)}_{k+m} H^q_{k(n)+m} K_k(m) \mathcal{I}(p,m) | \mathcal{T}_k\} \\
&\le \widehat{Q}^{(r)}_{k+1} \sum_{m=1}^{\infty} E\{H^q_{k(n)+m} K_k(m) | \mathcal{T}_k\} \\
&\qquad + \widehat{Q}^{(r)}_{k+1} \sum_{m=1}^{\infty} \sum_{p=0}^{\infty} e^{p+1} E\{H^q_{k(n)+m} K_k(m) \mathcal{I}(p,m) | \mathcal{T}_k\}.
\end{aligned} \tag{2.37}$$

By virtue of Lemma 3 there exists some constant $D_{10} = D_{10}(q)$ (independent of n, k and i, as long as (2.12) is satisfied) such that the first sum in the right-hand side here is at most

$$\widehat{Q}^{(r)}_{k+1} \sum_{m=1}^{\infty} E\{H^q_{k+m} K_k(m) | \mathcal{T}_k\} \le D_{10} \widehat{Q}^{(r)}_{k+1}. \tag{2.38}$$

Now, the second sum in the right-hand side of (2.37) is bounded by

$$\widehat{Q}^{(r)}_{k+1}\sum_{p=0}^{\infty} e^{p+1}\Big[E\{\sum_{m=1}^{\infty} H^{2q}_{k+m}K_k(m)\big|\mathcal{T}_k\}E\{\mathcal{W}(p,n)\big|\mathcal{T}_k)\}\Big]^{1/2}$$

$$\le \widehat{Q}^{(r)}_{k+1}\sum_{p=0}^{\infty} e^{p+1}\Big[D_{10}E\{\mathcal{W}(p,n)\big|\mathcal{T}_k\}\Big]^{1/2} \text{ (again by (2.13))}.$$

In addition,

$$\mathcal{W}(p,n) = \mathcal{J}_p \times \big(\text{number of } m \in [1, k(n+1)-k(n)] \text{ with } \widehat{Q}^{(r)}_{k(n)+m} > e^p\widehat{Q}^{(r)}_{k(n)}\big)$$
$$\le \mathcal{J}_p[k(n+1)-k].$$

Hence, by (2.11), there exists a constant D_{13} (independent of p, n and i, as long as $\mu \ge 2i$), such that

$$\begin{aligned}\{\mathcal{W}(p,n)|\mathcal{T}_k\} &\le \{\mathcal{J}_p[k(n+1)-k]\big|\mathcal{T}_k\}\\ &\le \big[P\{\mathcal{J}_p = 1|\mathcal{T}_k\}E\{[k(n+1)-k]^2|\mathcal{T}_k\}\big]^{1/2}\\ &\le D_{13}\big[P\{\mathcal{J}_p = 1|\mathcal{T}_k\}\big]^{1/2}\\ &= D_{13}\big[P\{m_0 \le k(n+1)-k|\mathcal{T}_k\}\big]^{1/2} \end{aligned} \tag{2.39}$$

By combining (2.37)–(2.39) we see that the left-hand side of (2.35) is at most

$$\widehat{Q}^{(r)}_{k+1}\Big\{D_{10} + [D_{10}D_{13}]^{1/2}\sum_{p=0}^{\infty} e^{p+1}\big[P\{m_0 \le k(n+1)-k|\mathcal{T}_k\}\big]^{1/4}\Big\}. \tag{2.40}$$

It follows that it suffices for (2.35) to show that there exists some p_0, independent of n and i, such that for all n, for $p \ge p_0$ and uniformly on the event (2.10),

$$\begin{aligned}&P\{m_0(p,n) \le k(n+1)-k|\mathcal{T}_k\}\\ &= P\Big\{\frac{\widehat{Q}^{(r)}_{k+q}}{\widehat{Q}^{(r)}_{k+1}} > e^p \text{ for some } 1 \le q \le k(n+1)-k\big|\mathcal{T}_k\Big\} \le e^{-6p}.\end{aligned} \tag{2.41}$$

We now fix $r \in \{2, \dots, 24\}$ and introduce the random variables

$$\Gamma_u = \log\Big[\frac{\widehat{Q}^{(r)}_{k+u}}{\widehat{Q}^{(r)}_{k+u-1}}\Big].$$

In terms of these random variables

$$\{m_0 \le k(n+1)-k\} = \Big\{\sum_{u=2}^{q}\Gamma_u > p \text{ for some } 2 \le q \le k(n+1)-k\Big\}, \tag{2.42}$$

and (2.41) can then be given the equivalent form

$$P\Big\{\sum_{u=2}^{q}\Gamma_u > p \text{ for some } 2 \le q \le k(n+1)-k\big|\mathcal{T}_k\Big\} \le e^{-6p}. \tag{2.43}$$

This form resembles a large deviation estimate and we shall indeed prove it by large deviation methods.

Step 2. Some truncations. First we do some truncations in the left-hand side of (2.42). Without loss of generality we take $D_1 \le C_1$ and $D_3 \ge 1$ in Lemma 3. We further fix a constant D_{14} such that $D_1 D_{14} \ge 7$. We define the event

$$\mathcal{A}_1(n,p) = \{k(n+1) - k(n) \le D_{14}p \text{ and } H_{k(n)+u} \le D_{14}p \\ \text{for all } 0 \le u \le k(n+1) - k(n)\}.$$

Note that $H_{k(n)} = 1$, by definition of $k(n)$. Therefore, $H_{k(n)} > D_{14}p$ is impossible, provided we restrict ourselves to $p > 1/D_{14}$, as we may. Then, on the event (2.10),

$$\begin{aligned} P\{[\mathcal{A}_1(n,p)]^c|\mathcal{T}_k\} \le P\{k(n+1) - k(n) > D_{14}p|\mathcal{T}_k\} \\ + \sum_{1\le u\le D_{14}p} P\{H_{k(n)+u} > D_{14}p, K_{k(n)}(u) = 1|\mathcal{T}_k\} \\ \le e^{-C_1 D_{14}p} + D_3[1 + D_{14}p]e^{-D_1 D_{14}p} \\ \le D_3[2 + D_{14}p]\exp[-D_1 D_{14}p] \end{aligned} \tag{2.44}$$

by virtue of Lemma 3 and the fact that $C_1 \ge D_1$. Since we took $D_1 D_{14} \ge 7$, we have $P\{[\mathcal{A}_1(n,p)]^c|\mathcal{T}_k\} \le D_{15}\exp[-7p]$ and

$$\begin{aligned} P\{m_0 \le & k(n+1) - k|\mathcal{T}_k\} \le D_{15}e^{-7p} \\ & + P\{\mathcal{A}_1(n,p) \text{ and } \sum_{u=2}^{q} \Gamma_u > p \text{ for some } 2 \le q \le k(n+1) - k\big|\mathcal{T}_k\} \end{aligned} \tag{2.45}$$

on the event ((2.10).

Next we need some estimates for the Γ_u. These are based on the definition of $\widehat{Q}^{(r)}_{k+u}$ and (1.5). To begin we define m_1 by

$$m_1 - 1 = \sup\{q \in [1, m_0] : \frac{\widehat{Q}^{(r)}_{k+q}}{\widehat{Q}^{(r)}_{k+1}} \le p^{6r^2}\}.$$

In particular,

$$\frac{\widehat{Q}^{(r)}_{k+u}}{\widehat{Q}^{(r)}_{k+1}} > p^{6r^2} \text{ for } m_1 \le u \le m_0. \tag{2.46}$$

Moreover, if p is so large that $p^{6r^2} < e^p$, then $2 \le m_1 \le m_0$ on the event $\{m_0 \le k(n+1) - k\}$. ($m_1 - 1 = m_0$ is impossible, because $\widehat{Q}^{(r)}_{k(n)+m_0} > e^p\widehat{Q}^{(r)}_{k(n)+1}$). If in addition $\mathcal{A}_1$ occurs, then $k < k + m_1 \le k + m_0 \le k(n+1) \le k + D_{14}p$ and $H_{k+m_1-1} \le D_{14}p$.

The following event will be useful:

$$\mathcal{A}_2(u,p,r) := \{m_1 \le D_{14_p}, H_{k+u} \le D_{14}p, \widehat{Q}^{(r)}_{k+u} \le p^{6r^2}\widehat{Q}^{(r)}_{k+1}\}.$$

(We suppress the dependence on $k = k(n)$ in the notation for $\mathcal{A}_2$; the same comment applies to $\mathcal{A}_3$–$\mathcal{A}_5$ below). Note that on the event (2.10)

$$\{m_0 \le k(n+1) - k\} \cap \mathcal{A}_1(n,p) \cap [\mathcal{A}_2(m_1 - 1, p, r)]^c = \emptyset, \tag{2.47}$$

provided $6p^{r^2} < e^p$. Indeed, if $m_0 \le k(n+1) - k$ and $\mathcal{A}_1(n,p)$ occurs, then, as we just saw, $2 \le m_1 \le m_0 \le k(n+1) - k \le D_{14}p$ and $H_{k+m_1-1} \le D_{14p}$. Also, by definition of m_1, $\widehat{Q}^{(r)}_{k+m_1-1} \le p^{6r^2}\widehat{Q}^{(r)}_{k+1}$. Thus $\mathcal{A}_2(m_1 - 1, p, r)$ occurs automatically on $\{m_0 \le k(n+1) - k\} \cap \mathcal{A}_1(n,p)$. (2.47) allows us to work only on the event $\mathcal{A}_2(m_1 - 1, p, r)$.

We now derive a crude bound for $\widehat{Q}^{(r)}_{k+m_1}$ when $r \ge 1$.

$$\begin{aligned}
\widehat{Q}^{(r)}_{k+u} &= \sum_{j=1}^{i} \big[\widehat{X}^j(\tau_{k+u-1}-) - I[j = \widehat{J}_{k+u-1}] + A^j_{k+u-1} + S^j(\tau_{k+u}) - S^j(\tau_{k+u-1})\big]^r \\
&\le 3^r \sum_{j=1}^{i} \Big[[\widehat{X}^j(\tau_{k+u-1}-) - I[j = \widehat{J}_{k+u-1}]^r + \big|A^j_{k+u-1}\big|^r \\
&\qquad\qquad + \big|S^j(\tau_{k+u}) - S^j(\tau_{k+u-1})\big|^r\Big] \\
&= 3^r \widehat{Q}^{(r)}_{k+u-1} + 3^r \sum_{j=1}^{i} \big|A^j_{k+u-1}\big|^r + 3^r \sum_{j=1}^{i} \big|S^j(\tau_{k+u}) - S^j(\tau_{k+u-1})\big|^r.
\end{aligned} \tag{2.48}$$

As we saw in (2.21), $|A^j_q| \le H_q$. Therefore, on $\{H_{k+m_1-1} \le D_{14}p\} \cap \mathcal{A}_1(n,p)$ and with $u = m_1$, the second term in the far right-hand side of (2.48) is at most $3^r i H^r_{k+m_1-1} \le 3^r i (D_{14}p)^r \le 2(3D_{14}p)^r \widehat{Q}^{(r)}_{k+1}$ (because $\widehat{Q}^{(r)}_q \ge i - 1$ for all q).

To bound the third term in the right-hand side of (2.48) we shall apply Markov's inequality. To this end we assume that $k(n) = k$ and we first derive a bound on $E\{\Delta^j_{k+m-1} | \mathcal{S}_{k+m-1}\}$, where

$$\Delta^j_{k+m-1} := S^j(\tau_{k+m}) - S^j(\tau_{k+m-1}). \tag{2.49}$$

To find such a bound we denote the successive times in $(0, \infty)$ at which any of the S^j has a jump by $\lambda_1 < \lambda_2 < \cdots$ and take $\lambda_0 = 0$. Then the $\lambda_{u+1} - \lambda_u, u \ge 0$, are i.i.d exponential variables with mean $1/i$. Also, $\tau_{k+m-1} = \lambda_v$ and $\tau_{k+m} = \lambda_w$ for some $v < w$. Indeed, the τ_u are the successive times at which some $\widehat{X}^\ell$ jumps to 0, and in particular, at each τ_u some S^ℓ has a jump. τ_{k+m} equals the first $\lambda_w > \tau_{k+m-1}$ at which some $\widehat{X}^\ell$ jumps to 0. From the fact that the S^ℓ are independent simple random walks, it follows that at each λ_u, almost surely exactly one of the S^ℓ takes a step of ± 1. Moreover, each of the $S^\ell, 1 \le \ell \le i$, has probability $1/i$ of jumping at time λ_u, and which S^ℓ jumps, as well as the size of the jump of this S^ℓ are independent of all λ_u. Thus, if we define

$$T^\ell_0 = 0, T^\ell_u = S^\ell(\lambda_u),\ u \ge 1, 1 \le \ell \le i,$$

then $(T_u^1, \dots, T_u^i)$ is a discrete time i-dimensional simple random walk on $\mathbb{Z}^i$. Furthermore,

$$\Delta_{k+m-1}^j = T_w^j - T_v^j$$

and on the event $\{\tau_{k+m-1} = \lambda_v, \widehat{X}^\ell(\lambda_v) = x^\ell, 1 \le \ell \le i\}$, we have

$$P\{\Delta_{k+m-1}^j = z | \mathcal{S}_{k+m-1}\} = P\{T_q^j = z | T_0^\ell = x^\ell, 1 \le \ell \le i\},$$

where

$$q := \min\{u \ge 1 : T_u^\ell = 0 \text{ for some } 1 \le \ell \le i\}.$$

In particular, for each fixed $1 \le j \le i$,

$$E\{|\Delta_{k+m-1}^j|^s | \mathcal{S}_{k+m-1}\} = E\{|T_q^j|^s | T_0^\ell = x^\ell, 1 \le \ell \le i\}.$$

Of course if $(\widetilde{S}^1, \dots, \widetilde{S}^i)$ is a discrete time simple random walk on $\mathbb{Z}^i$ starting at $\mathbf{0}$ and if we define

$$q(\mathbf{x}) = q(x^1, \dots, x^i) = \min\{u \ge 1 : \widetilde{S}_u^\ell = -x^\ell \text{ for some } 1 \le \ell \le i\},$$

then also

$$E\{|\Delta_{k+m-1}^j|^s | \mathcal{S}_{k+m-1}\} = E\{|\widetilde{S}_{q(\mathbf{x})}^j|^s | \widetilde{S}_0^\ell = 0, 1 \le \ell \le i\}.$$

We now apply the Burkholder-Davis inequality to the martingale $\widetilde{S}_{u \wedge q(\mathbf{z})}^j$ (see [4], Theorem A.2.2 or [6], Theorem 2.10). Taking into account that for fixed ℓ, $|\widetilde{S}_{u+1}^\ell - \widetilde{S}_u^\ell|^2 \le 1$, we see that there exists for $s \ge 1$ some constant $\widetilde{D}_s$ which depends on s only, such that

$$\begin{aligned} E\{|\Delta_{k+m-1}^j|^s | \mathcal{S}_{k+m-1}\} &= E\Big\{\Big| \sum_{u=0}^{q(\mathbf{x})-1} [\widetilde{S}_{u+1}^j - \widetilde{S}_u^j]\Big|^s \Big| \widetilde{S}_0^\ell = 0, 1 \le \ell \le i\Big\} \\ &\le \widetilde{D}_s E\{[q(\mathbf{x})]^{s/2}\}. \end{aligned} \tag{2.50}$$

We could continue with computations for the discrete time random walk $(\widetilde{S}_u^1, \dots, \widetilde{S}_u^i)$, but since we anyway need bounds for the moments of δ_u later, it is more economical to replace the right-hand side here by a bound in terms of the moments of some δ_u. This can be done easily, by observing that the conditional distribution, given $\mathcal{S}_{k+m-1}$, of

$$\delta_{k+m-1} = \tau_{k+m} - \tau_{k+m-1} = \sum_{u=v}^{w-1} [\lambda_{u+1} - \lambda_u]$$

is on the event

$$\{X^\ell(\tau_{k+m-1}) = x^\ell, 1 \le \ell \le i\}, \tag{2.51}$$

the same as the distribution of

$$\sum_{u=0}^{q(\mathbf{x})-1} \widetilde{\lambda}_u$$

for a sequence $\{\widetilde{\lambda}_u\}$ of i.i.d. exponential variables with mean $1/i$. The variables $\widetilde{\lambda}_u$ are also independent of $q(\mathbf{x})$. Therefore, if we use $\widetilde{E}$ to denote expectation with respect to the $\widetilde{\lambda}_u$, it is the case for $s \geq 2$ that

$$E\{\delta_{k+m-1}^{s/2}|\mathcal{S}_{k+m-1}\} = E\Big\{\Big|\sum_{u=0}^{q(\mathbf{x})-1} \widetilde{\lambda}_u\Big|^{s/2}\Big\} = E\Big\{\widetilde{E}\Big|\sum_{u=0}^{q(\mathbf{x})-1} \widetilde{\lambda}_u\Big|^{s/2}\Big\}$$
$$\geq E\Big\{\Big|\sum_{u=0}^{q(\mathbf{x})-1} \widetilde{E}\widetilde{\lambda}_u\Big|^{s/2}\Big\} = i^{-s/2}E\big[q(\mathbf{x})\big]^{s/2}.$$

Combined with (2.50) this gives for $s \geq 2$

$$E\big\{|\Delta_{k+m-1}^j|^s\big|\mathcal{S}_{k+m-1}\big\} \leq \widetilde{D}_s i^{s/2} E\{\delta_{k+m-1}^{s/2}|\mathcal{S}_{k+m-1}\}. \tag{2.52}$$

We replace s by $2rs$ and remind the reader that we showed in (4.20) of [9], for $2rs \in \{2, 3, \dots\}$ and integer $u \geq 0$, that there exists an $i_0(2rs)$ such that for all $i \geq i_0$ and $u \geq 0$

$$E\{\delta_u^{rs}|\mathcal{S}_u\} \leq \ell^1(\tau_u)\Big(\frac{L_u}{i}\Big)^{2rs-1}. \tag{2.53}$$

In the present model of the frozen particles, $\ell^1(\tau_u) \leq H_u$, because at time τ_u at least one particle is thawed at $\mathcal{M}(\tau_u) + H_u$. Also, by (2.21),

$$L_u = \widehat{L}_u + \sum_{j=1}^{i} A_u^j \leq \widehat{L}_u + iH_u. \tag{2.54}$$

Now define

$$\mathcal{A}_3(u, p, r) = \{H_{k+u} \leq D_{14}p,\ \widehat{Q}_{k+u}^{(r)} \leq p^{6r^2}\widehat{Q}_{k+1}^{(r)}\}.$$

Then, for $u \geq 0, r \geq 1, 2rs \in \{2, 3, \dots\}$, it holds on $\mathcal{A}_3(u, p, r)$

$$E\{\delta_{k+u}^{rs}|\mathcal{S}_{k+u}\} \leq D_{14}p\Big(\frac{\widehat{L}_{k+u} + iD_{14}p}{i}\Big)^{2rs-1} \leq D_{15}p\Big[\Big(\frac{\widehat{L}_{k+u}}{i}\Big)^{2rs-1} + (D_{14}p)^{2rs-1}\Big]$$
$$\leq D_{15}p\Big[\frac{\widehat{Q}_{k+u}^{(r)}}{i}\Big]^{2s-1/r} + D_{15}p(D_{14}p)^{2rs-1}(\text{see (4.59) in [9]})$$
$$\leq D_{15}p\Big[\frac{p^{6r^2}\widehat{Q}_{k+1}^{(r)}}{i}\Big]^{2s-1/r} + D_{15}p(D_{14}p)^{2rs-1}.$$

Now

$$\widehat{Q}_{k+1}^{(r)} \geq \widehat{L}_{k+1} \geq i - 1 \geq i/2 \tag{2.55}$$

(compare (4.10) in [9]) and (again by (4.59) in [9])

$$\frac{\widehat{Q}_{k+1}^{(r)}}{i} \geq \Big(\frac{\widehat{L}_{k+1}}{i}\Big)^r. \tag{2.56}$$

Therefore, for $D_{14}p \le p^{6r}/2$,

$$E\{\delta^{rs}_{k+u}|\mathcal{S}_{k+u}\} \le 2D_{15}p\Big[\frac{p^{6r^2}\widehat{Q}^{(r)}_{k+1}}{i}\Big]^{2s-1/r} \le 2D_{15}p\Big[\frac{p^{6r^2}\widehat{Q}^{(r)}_{k+1}}{i}\Big]^{2s}\frac{i}{\widehat{L}_{k+1}}. \quad (2.57)$$

Finally, if we assume that (2.34) holds, then for $r \ge 1, 2rs \in \{2,3,\dots\}$ and $D_{14}p \le p^{6r}/2$,

$$\begin{aligned} &E\{|\Delta^j_{k+u}|^{2rs}I[\mathcal{A}_2(u,p,r)]\big|\mathcal{S}_{k+1}\} \le E\{|\Delta^j_{k+u}|^{2rs}I[\mathcal{A}_3(u,p,r)]\big|\mathcal{S}_{k+1}\} \\ &= E\Big\{I[\mathcal{A}_3(u,p,r)]E\{|\Delta^j_{k+u}|^{2rs}|\mathcal{S}_{k+u}\}\Big|\mathcal{S}_{k+1}\Big\} \\ &\le \widetilde{D}_{2rs}E\Big\{I[\mathcal{A}_3(u,p,r)]E\{i^{rs}\delta^{rs}_{k+u}|\mathcal{S}_{k+u}\}\Big|\mathcal{S}_{k+1}\Big\} \\ &\le 2D_{15}\widetilde{D}_{2rs}pi^{1+(r-2)s}\frac{\big[p^{6r^2}\widehat{Q}^{(r)}_{k+1}\big]^{2s}}{\widehat{L}_{k+1}} \le D_{16}i^{-2s-1}p^2\big[p^{6r^2}\widehat{Q}^{(r)}_{k+1}\big]^{2s} \\ &\quad \text{(by (2.34) and (2.31))}. \end{aligned} \quad (2.58)$$

We are now ready to apply Markov's inequality to estimate the third term on the right of (2.48). For the remainder of the proof of (2.35) we shall condition on $\mathcal{S}_{k+1}$ instead of $\mathcal{T}_k$. In view of (1.8) a bound of the form $E\{X|\mathcal{S}_{k+1}\} \le \gamma$ for a non-negative random variable X, valid for almost all sample points, automatically implies that $E\{X|\mathcal{T}_k\} \le \gamma$ almost surely.

Define $\mathcal{A}_4 = \mathcal{A}_4(p,r)$ as the event

$$\mathcal{A}_4 = \Big\{\sum_{j=1}^{i}\big|S^j\big(\tau_{k+m_1}\big) - S^j\big(\tau_{k+m_1-1}\big)\big|^r \le e^{p/4}\widehat{Q}^{(r)}_{k+1}\Big\}.$$

Then, on the event (2.10) intersected with (2.34), and for $p \ge p_2$ for a suitable p_2 depending on r only (recall that $s = 15$),

$$\begin{aligned} &P\{m_0 \le k(n+1) - k, \mathcal{A}_1 \cap \mathcal{A}_2(m_1 - 1, p, r) \cap \mathcal{A}_4^c|\mathcal{S}_{k+1}\} \\ &\le \sum_{2\le m\le D_{14}p} P\{\mathcal{A}_2(m-1,p,r) \text{ and for some } j \le i \\ &\qquad\qquad \big|S^j\big(\tau_{k+m}\big) - S^j\big(\tau_{k+m-1}\big)\big|^r > \frac{1}{i}e^{p/4}\widehat{Q}^{(r)}_{k+1}|\mathcal{S}_{k+1}\} \\ &\le \sum_{j=1}^{i}\sum_{2\le m\le D_{14}p}\Big[\frac{1}{i}e^{p/4}\widehat{Q}^{(r)}_{k+1}\Big]^{-2s}E\{|\Delta^j_{k+m-1}|^{2rs}I[\mathcal{A}_2(m-1,p,r)]\big|\mathcal{S}_{k+1}\} \\ &\le iD_{14}\Big[\frac{1}{i}e^{p/4}\widehat{Q}^{(r)}_{k+1}\Big]^{-2s}D_{16}i^{-2s-1}p^{12r^2s+3}[\widehat{Q}^{(r)}_{k+1}]^{2s} \\ &= D_{14}D_{16}p^{12r^2s+3}e^{-ps/2} \le e^{-7p}. \end{aligned} \quad (2.59)$$

Combining (2.47), (2.44), (2.45), (2.59) and (2.42) we have (for large p)

$$\begin{aligned}
&P\{m_0 \le k(n+1) - k|\mathcal{T}_k\}\\
&\le P\{[\mathcal{A}_1(n,p)]^c|\mathcal{T}_k\} + E\Big\{P\{m_0 \le k(n+1)-k, \mathcal{A}_1(n,p)|\mathcal{S}_{k+1}\}\Big|\mathcal{T}_k\Big\}\\
&= P\{[\mathcal{A}_1(n,p)]^c|\mathcal{T}_k\}\\
&\qquad + E\Big\{P\{m_0 \le k(n+1)-k, \mathcal{A}_1(n,p)\cap\mathcal{A}_2(m_1-1,p,r)|\mathcal{S}_{k+1}\}\Big|\mathcal{T}_k\Big\}\\
&\le P\{[\mathcal{A}_1(n,p)]^c|\mathcal{T}_k\}\\
&\qquad + E\Big\{P\{m_0 \le k(n+1)-k, \mathcal{A}_1\cap\mathcal{A}_2(m_1-1,p,r)\cap[\mathcal{A}_4]^c|\mathcal{S}_{k+1}\}\Big|\mathcal{T}_k\Big\}\\
&\qquad + E\Big\{P\{m_0 \le k(n+1)-k, \mathcal{A}_1\cap\mathcal{A}_2(m_1-1,p,r)\cap\mathcal{A}_4|\mathcal{S}_{k+1}\}\Big|\mathcal{T}_k\Big\}\\
&\le (D_{15}+1)e^{-7p} + E\Big\{P\{\sum_{u=2}^{q}\Gamma_u > p \text{ for some } 2\le q\le k(n+1)-k,\\
&\qquad\qquad \mathcal{A}_1\cap\mathcal{A}_2(m_1-1,p,r)\cap\mathcal{A}_4\big|\mathcal{S}_{k+1}\}\Big|\mathcal{T}_k\Big\}. \qquad (2.60)
\end{aligned}$$

Step 3. Reformulation of problem after the truncations. The reason for doing the estimates following (2.48) is that it follows from (2.48), the definition of m_1 and the lines following (2.48) that on $\{m_0 \le k(n+1)-k\}\cap\mathcal{A}_1\cap\mathcal{A}_4$

$$\widehat{Q}^{(r)}_{k+m_1} \le \big[3^r p^{6r^2} + 2(3D_{14}p)^r + 3^r e^{p/4}\big]\widehat{Q}^{(r)}_{k+1} \le 2\cdot 3^r e^{p/4}\widehat{Q}^{(r)}_{k+1} \qquad (2.61)$$

(provided $3^r p^{6r^2} + 2(3D_{14}p)^r \le 3^r e^{p/4}$). If (2.61) holds and $\widehat{Q}^{(r)}_{k+m_0}/\widehat{Q}^{(r)}_{k+1} \ge e^p$, then $\widehat{Q}^{(r)}_{k+m_0}/\widehat{Q}^{(r)}_{k+m_1} \ge [2\cdot 3^r]^{-1}e^{3p/4}$. It follows from this that the probability in the extreme right of (2.60) is at most

$$\begin{aligned}
&P\{\mathcal{A}_1\cap\mathcal{A}_2(m_1-1,p,r)\cap\mathcal{A}_4 \text{ and}\\
&\qquad \sum_{u=m_1+1}^{q}\Gamma_u > 3p/4 - \log(2\cdot 3^r) \text{ for some } q\in(m_1, D_{14}p]\big|\mathcal{S}_{k+1}\}\\
&\le \sum_{2\le m\le D_{14}p}\ \sum_{m<q\le D_{14}p} P\{\sum_{u=m+1}^{q}\Gamma_u > 3p/4-\log(2\cdot 3^r), \text{ and } H_{k+v-1}\le D_{14}p\\
&\qquad \text{as well as } \widehat{Q}^{(r)}_{k+v-1} \ge p^{6r^2}\widehat{Q}^{(r)}_{k+1} \text{ for } m < v \le |\mathcal{S}_{k+1}\} \text{ (cf. (2.46))}. \qquad (2.62)
\end{aligned}$$

We now work on an upper bound for the summand in the right-hand side of (2.62) corresponding to some given m, q. We define

$$Y^j_{k+u-1} := \widehat{X}^j(\tau_{k+u-1}-) - I[j = \widehat{J}_{k+u-1}].$$

and the events

$$\mathcal{A}_5 = \mathcal{A}_5(u,p,r) := \{H_{k+u-1} \le D_{14}p \text{ as well as } \widehat{Q}^{(r)}_{k+u-1} \ge p^{6r^2}\widehat{Q}^{(r)}_{k+1}\}.$$

We remind the reader that Δ^j is defined in (2.49). We first show that the influence of the A^j_q on the Γ_u is small, as far as proving (2.43) is concerned. We use that

$Y^j_{k+u-1} \geq 0$, and for $r \geq 1$

$$\begin{aligned}\widehat{Q}^{(r)}_{k+u} &= \sum_{j=1}^{i} [Y^j_{k+u-1}]^r + \sum_{j=1}^{i}\sum_{s=1}^{r} \binom{r}{s} \big[A^j_{k+u-1} + \Delta^j_{k+u-1}\big]^s [Y^j_{k+u-1}]^{r-s} \\ &\leq \widehat{Q}^{(r)}_{k+u-1} + \sum_{j=1}^{i}\sum_{s=1}^{r} \binom{r}{s} 2^s |A^j_{k+u-1}|^s [Y^j_{k+u-1}]^{r-s} \\ &\qquad + \sum_{j=1}^{i}\sum_{s=1}^{r} \binom{r}{s} 2^s |\Delta^j_{k+u-1}|^s [Y^j_{k+u-1}]^{r-s}. \end{aligned} \tag{2.63}$$

Note that

$$\begin{aligned}\sum_{s=1}^{r} \binom{r}{s} 2^s |A^j_{k+u-1}|^s [Y^j_{k+u-1}]^{r-s} &= [Y^j_{k+u-1} + 2|A^j_{k+u-1}|]^r - [Y^j_{k+u-1}]^r \\ &\leq r2|A^j_{k+u-1}|[Y^j_{k+u-1} + 2|A^j_{k+u-1}|]^{r-1}. \end{aligned} \tag{2.64}$$

If $\widehat{X}^j(\tau_{k+u-1}-) \geq 2$, then $Y^j_{k+u-1} \geq 1$ and $A^j_{k+u-1} = -1$ and

$$[Y^j_{k+u-1} + 2|A^j_{k+u-1}|]^{r-1} \leq 3^{r-1}[Y^j_{k+u-1}]^{r-1} \leq 3^{r-1} H^{r-1}_{k+u-1} [Y^j_{k+u-1}]^{r-1}.$$

If $\widehat{X}^j(\tau_{k+u-1}-) = 1$, then $|A^j_{k+u-1}| \leq H_{k+u-1}$ (see (2.21)). In this case we have two subcases: $Y^j_{k+u-1} = 0$ if $j = \widehat{J}_{k+u-1}$, and $Y^j_{k+u-1} = 1$ otherwise. In both subcases

$$[Y^j_{k+u-1} + 2|A^j_{k+u-1}|]^{r-1} \leq [2H_{k+u-1} + 1]^{r-1}\big\{[Y^j_{k+u-1}]^{r-1} + I[j = \widehat{J}^{k+u-1}]\big\}.$$

Thus there exists a constant $D_{17} = D_{17}(r)$, so that in all cases

$$\sum_{s=1}^{r} \binom{r}{s} 2^s |A^j_{k+u-1}|^s [Y^j_{k+u-1}]^{r-s} \leq D_{17} H^r_{k+u-1}\big\{[Y^j_{k+u-1}]^{r-1} + I[j = \widehat{J}_{k+u-1}]\big\},$$

and

$$\begin{aligned}&\sum_{j=1}^{i}\sum_{s=1}^{r} \binom{r}{s} 2^s |A^j_{k+u-1}|^s [Y^j_{k+u-1}]^{r-s} \\ &\qquad \leq D_{17} H^r_{k+u-1}\big[\widehat{Q}^{(r-1)}_{k+u-1} + 1\big] \leq 2D_{17} H^r_{k+u-1} \widehat{Q}^{(r-1)}_{k+u-1}.\end{aligned}$$

Going back to (2.63) we see that

$$\begin{aligned}\Gamma_u = \log \frac{\widehat{Q}^{(r)}_{k+u}}{\widehat{Q}^{(r)}_{k+u-1}} &\leq \log \left[1 + \frac{2D_{17} H^r_{k+u-1} \widehat{Q}^{(r-1)}_{k+u-1}}{\widehat{Q}^{(r)}_{k+u-1}}\right] \\ &\quad + \log \left[1 + \frac{\sum_{j=1}^{i}\sum_{s=1}^{r} \binom{r}{s} |2\Delta^j_{k+u-1}|^s [Y^j_{k+u-1}]^{r-s}}{\widehat{Q}^{(r)}_{k+u-1}}\right].\end{aligned}$$

On $\mathcal{A}_5(u,p,r)$ we also have the following string of inequalities:

$$\widehat{L}_{k+u-1} \geq \big[\widehat{Q}^{(r)}_{k+u-1}\big]^{1/r} \geq \big[p^{6r^2}\widehat{Q}^{(r)}_{k+1}\big]^{1/r} = p^{6r}\big[\widehat{Q}^{(r)}_{k+1}\big]^{1/r} \geq p^{6r} i^{(1-r)/r}\widehat{L}_{k+1} \tag{2.65}$$

(use Hölder for the last inequality). Consequently, still on $\mathcal{A}_5(u,p,r)$,

$$\frac{\widehat{Q}^{(r-1)}_{k+u-1}}{\widehat{Q}^{(r)}_{k+u-1}} \leq \frac{i}{\widehat{L}_{k+u-1}} \text{ (see (4.59) in [9])} \leq \frac{i^{2-1/r}}{p^{6r}\widehat{L}_{k+1}}. \tag{2.66}$$

If also (2.34) holds (with $D_{12} \geq 1$), then the last member of these inequalities is at most p^{-5r} and (still on $\mathcal{A}_5(u,p,r)$)

$$\begin{aligned}\Gamma_u &\leq \frac{2D_{17}H^r_{k+u-1}}{p^{5r}} + \log\Big[1 + \frac{\sum_{j=1}^i\sum_{s=1}^r\binom{r}{s}|2\Delta^j_{k+u-1}|^s[Y^j_{k+u-1}]^{r-s}}{\widehat{Q}^{(r)}_{k+u-1}}\Big]\\ &\leq \frac{2D_{17}D^r_{14}}{p^{4r}} + \log\Big[1 + \frac{\sum_{j=1}^i\sum_{s=1}^r\binom{r}{s}|2\Delta^j_{k+u-1}|^s[Y^j_{k+u-1}]^{r-s}}{\widehat{Q}^{(r)}_{k+u-1}}\Big].\end{aligned} \tag{2.67}$$

This finishes our estimate of the influence of the A^j_q on the Γ_u.

We shall set

$$\zeta_u = \sum_{j=1}^i\sum_{s=1}^r\binom{r}{s}|2\Delta^j_{k+u-1}|^s[Y^j_{k+u-1}]^{r-s}.$$

Note that ζ_u is $\mathcal{S}_{k+u}$-measurable. We then see that the (m,q)-summand in (2.62) is (for $p/4 > \log(2\cdot 3^r) + 2D_{17}D^{r+1}_{14}p^{-4r+1}$) at most

$$P\Big\{\mathcal{A}_5(v,p,r) \text{ for } m < v \leq q \text{ and } \sum_{u=m+1}^q \log\Big[1+\frac{\zeta_u}{\widehat{Q}^{(r)}_{k+u-1}}\Big] > \frac{p}{2}\Big|\mathcal{S}_{k+1}\Big\}. \tag{2.68}$$

Step 4. A large deviation estimate. Guided by large deviation estimates we finally use the following bounds, for any integer $\theta \geq 1$, on the event $\mathcal{A}_5(u,p,r)$, which is $\mathcal{S}_{k+u-1}$-measurable.

$$\begin{aligned}&E\Big\{\exp\Big[\theta\log\Big[1+\frac{\zeta_u}{\widehat{Q}^{(r)}_{k+u-1}}\Big]\Big]\Big|\mathcal{S}_{k+u-1}\Big\}\\ &= E\Big\{\Big[1+\frac{\zeta_u}{\widehat{Q}^{(r)}_{k+u-1}}\Big]^\theta\Big|\mathcal{S}_{k+u-1}\Big\} = \sum_{v=0}^\theta\binom{\theta}{v}\frac{E\{\zeta_u^v|\mathcal{S}_{k+u-1}\}}{[\widehat{Q}^{(r)}_{k+u-1}]^v}.\end{aligned} \tag{2.69}$$

Furthermore,

$$E\{\zeta_u^v|\mathcal{S}_{k+u-1}\} = \sum_{\mathbf{j}}\sum_{\mathbf{s}} E\Big\{2^v\prod_{w=1}^v\Big[\binom{r}{s_w}|\Delta^{j_w}_{k+u-1}|^{s_w}[Y^{j_w}_{k+u-1}]^{r-s_w}\Big]\Big|\mathcal{S}_{k+u-1}\Big\}, \tag{2.70}$$

where the sums over $\mathbf{j}$ and $\mathbf{s}$ are over all $j_1, \dots, j_v$ from 1 to i, and over $s_1, \dots, s_v$ from 1 to r, respectively. Now the Y^j_{k+u-1} are $\mathcal{S}_{k+u-1}$-measurable. So, in the above expectation we only need to estimate

$$E\Big\{\prod_{w=1}^{v} |\Delta^{j_w}_{k+u-1}|^{s_w} \Big| \mathcal{S}_{k+u-1}\Big\} \le \prod_{w=1}^{v} \Big[E\{|\Delta^{j_w}_{k+u-1}|^{\sigma} \big| \mathcal{S}_{k+u-1}\}\Big]^{s_w/\sigma}, \tag{2.71}$$

where $\sigma := \sum_{w=1}^{v} s_w$ and we applied Hölder's inequality. By (2.52) it holds for $\sigma \ge 2$

$$E\{|\Delta^{j_w}_{k+u-1}|^{\sigma} \big| \mathcal{S}_{k+u-1}\} \le \widetilde{D}_\sigma i^{\sigma/2} E\{\delta^{\sigma/2}_{k+u-1} | \mathcal{S}_{k+u-1}\} \tag{2.72}$$

for some constant $\widetilde{D}_\sigma$ which depends on σ only. In our sums $\sigma \le vr \le \theta r$, so that we can choose one $\widetilde{D}_\sigma$ which works for all estimates here. Further, as in (2.53) and (2.54), if $\sigma \in \{2, 3, \dots\}$, then

$$E\{\delta^{\sigma/2}_{k+u-1} | \mathcal{S}_{k+u-1}\} \le H_{k+u-1}\Big(\frac{\widehat{L}_{k+u-1} + iH_{k+u-1}}{i}\Big)^{\sigma-1}.$$

On $\mathcal{A}_5(u,p,r)$ intersected with (2.34), we have from (2.66) that for $p \ge D_{14}$,

$$\widehat{L}_{k+u-1} \ge p^{6r} i^{(1-r)/r} \widehat{L}_{k+1} \ge D_{14} p i \ge i H_{k+u-1}. \tag{2.73}$$

The right-hand side of (2.71) is therefore on the intersection of $\mathcal{A}_5(u,p,r)$ and (2.34) at most

$$\begin{aligned} & D_{18} i^{\sigma/2} H_{k+u-1}\Big(\frac{\widehat{L}_{k+u-1}}{i}\Big)^{\sigma-1} \\ & \le D_{18} i^{\sigma/2} \frac{D_{14} p i}{\widehat{L}_{k+u-1}} \prod_{w=1}^{v} \Big(\frac{\widehat{L}_{k+u-1}}{i}\Big)^{s_w} \\ & \le \frac{D_{18} D_{14} p i^{2-1/r+\sigma/2}}{p^{6r}\widehat{L}_{k+1}} \prod_{w=1}^{v} \Big(\frac{\widehat{L}_{k+u-1}}{i}\Big)^{s_w} \\ & \le \frac{D_{18} D_{14} i^{2-1/r+\theta r/2}}{p^5 D_{12} i^{24 \cdot 15+1}} \prod_{w=1}^{v} \Big(\frac{\widehat{L}_{k+u-1}}{i}\Big)^{s_w} \\ & \le \frac{D_{19}}{p^5 i} \prod_{w=1}^{v} \Big(\frac{\widehat{L}_{k+u-1}}{i}\Big)^{s_w} \quad \text{(recall the choices in (2.31)).} \end{aligned} \tag{2.74}$$

Substitution of this estimate into (2.70) yields

$$\begin{aligned} & E\{\zeta^v_u | \mathcal{S}_{k+u-1}\} \\ & \le \sum_{\mathbf{s}} \frac{D_{19}}{p^5 i} 2^v \prod_{w=1}^{v} \Big[\binom{r}{s_w}\Big(\frac{\widehat{L}_{k+u-1}}{i}\Big)^{s_w} \sum_{j_w=1}^{i} \big[Y^{j_w}_{k+u-1}\big]^{r-s_w}\Big] \\ & = \sum_{\mathbf{s}} \frac{D_{19}}{p^5 i} 2^v \prod_{w=1}^{v} \Big[\binom{r}{s_w}\Big(\frac{\widehat{L}_{k+u-1}}{i}\Big)^{s_w} \widehat{Q}^{(r-s_w)}_{k+u-1}\Big] \end{aligned}$$

$$\leq \sum_{\mathbf{s}} \frac{D_{20}}{p^5 i} \prod_{w=1}^{v} \widehat{Q}^{(r)}_{k+u-1} \text{ (see (4.59) in [9])} = \frac{D_{21}}{p^5 i} \big[\widehat{Q}^{(r)}_{k+u-1}\big]^v, \tag{2.75}$$

provided $\sigma \geq 2$ and the intersection of $\mathcal{A}_5(u,p,r)$ and (2.34) occurs. Terms with $\sigma < 2$ occur in (2.70) only if $v = 0$, for which the left-hand side of (2.70) equals 1, and if $v = 1$. In the latter case, we have to replace (2.72) and (2.75) by

$$E\big\{|\Delta^{j_w}_{k+u-1}|\big|\mathcal{S}_{k+u-1}\big\} \leq \big[E\big\{|\Delta^{j_w}_{k+u-1}|^2\big|\mathcal{S}_{k+u-1}\big\}\big]^{1/2}$$
$$\leq \widetilde{D}_2\big[iE\{\delta_{k+u-1}|\mathcal{S}_{k+u-1}\}\big]^{1/2} \leq \widetilde{D}_2\big[H_{k+u-1}\widehat{L}_{k+u-1}\big]^{1/2},$$

and

$$E\{\zeta_u\big|\mathcal{S}_{k+u-1}\} \leq \frac{D_{21}}{p^5 i}\widehat{Q}^{(r)}_{k+u-1} + D_{20} i (D_{14}p)^{1/2}\Big(\frac{1}{\widehat{L}_{k+u-1}}\Big)^{1/2}\widehat{Q}^{(r)}_{k+u-1}$$
$$\leq \frac{D_{21}}{p^5 i}\widehat{Q}^{(r)}_{k+u-1} + \frac{D_{21}}{p^2 i^{1/2}}\widehat{Q}^{(r)}_{k+u-1}. \tag{2.76}$$

To simplify our formula we write

$$\Psi_u(\theta) := I[\mathcal{A}_5(u,p,r)] \exp\Big[\theta \log\big[1 + \frac{\zeta_u}{\widehat{Q}^{(r)}_{k+u-1}}\big]\Big].$$

Substitution of the preceding estimates in the right-hand side of (2.69) then yields

$$E\Big\{\Psi_u(\theta)\Big|\mathcal{S}_{k+u-1}\Big\} \leq 1 + \frac{D_{21}}{p^2 i^{1/2}} + \sum_{v=1}^{\theta}\binom{\theta}{v}\frac{D_{21}}{p^5 i} \leq \exp\big[\frac{D_{22}}{p^2 i^{1/2}}\big]. \tag{2.77}$$

Finally we use this estimate successively for $u = q, u = q-1, \dots, u = m+1$. Since ζ_u and ψ_u are $\mathcal{S}_{k+u}$-measurable, this gives, on the event (2.10),

$$E\Big\{\prod_{u=m+1}^{q} \Psi_u(\theta)\Big|\mathcal{S}_{k+1}\Big\}$$
$$= E\Big\{\Big[\prod_{u=m+1}^{q-1} \Psi_u(\theta)\Big] E\{\Psi_q(\theta)|\mathcal{S}_{k+q-1}\}\Big|\mathcal{S}_{k+1}\Big\}$$
$$\leq \exp\big[\frac{D_{22}}{p^2 i^{1/2}}\big] E\Big\{\prod_{u=m+1}^{q-1} \Psi_u(\theta)\Big|\mathcal{S}_{k+1}\Big\}$$
$$< \cdots < \exp\big[\frac{(q-m)D_{22}}{p^2 i^{1/2}}\big].$$

Thus, for $q - m \leq D_{14}p$ the probability (2.68) is at most

$$e^{-\theta p/2} E\Big\{\prod_{u=m+1}^{q} \Psi_u(\theta)\Big|\mathcal{S}_{k+1}\Big\} \leq \exp\big[-\theta p/2 + \frac{(q-m)D_{22}}{p^2 i^{1/2}}\big]$$
$$\leq \exp\big[-\theta p/2 + \frac{D_{22}D_{14}}{p i^{1/2}}\big].$$

By taking $\theta = 16$ and p so large that

$$(D_{14}p)^2 \exp[-8p + D_{22}D_{14}/p] \leq \exp[-7p],$$

we see that the right-hand side of (2.62) (and hence the left-hand side also) is at most $\exp[-7p]$. In view of (2.60), (2.41) and (1.8) this proves (2.35) under the assumption (2.34).

We now drop the assumption (2.34). Then no changes are necessary until we come to (2.41). We will not be able to prove that this holds for all $p \geq p_0$ for some p_0 which is independent of i. However, (2.40) is still an upper bound for the left-hand side of (2.36) (which equals the left-hand side of (2.35)). Therefore, it is enough for (2.36) to show that (2.41) holds for all n and all $p \geq D_{23}i^{361}$ on (2.10), with D_{23} independent of n, i, p. All the estimates through (2.57) remain valid; assumption (2.34) is not used in their proof. However, to obtain (2.58) we required $\widehat{L}_{k+1} \geq i^{rs+2}p^{-1}$. This last inequality will hold anyway for $p \geq 2i^{rs+1}$ (by (2.55)). Thus also (2.58)–(2.60) hold under this last restriction, even without (2.34). Similarly we required

$$\frac{i^{2-1/r}}{p^{6r}\widehat{L}_{k+1}} \leq \frac{1}{p^{5r}}.$$

to obtain (2.67). Since $\widehat{L}_{k+1} \geq i-1 \geq i/2$ holds in any case, this last inequality and (2.67) will hold for $p \geq 2i$. Next, the proof of (2.74) relies on (2.73). (2.34) was used to prove the second inequality of (2.73) and again for the one but last inequality in (2.74). But the second inequality of (2.73) is obviously valid for $p \geq 2D_{14}i$. Similarly we can ignore the one but last inequality in (2.74) and immediately obtain the last inequality of (2.74) for all $p \geq i^{3+\theta r/2}$ (recall $r \geq 2$ and $\sigma \leq \theta r$). Lastly, for the last inequality in (2.76) we require

$$i\Big(\frac{p}{\widehat{L}_{k+u-1}}\Big)^{1/2} \leq \frac{1}{p^2 i^{1/2}}.$$

In view of (2.65) this will hold if $p \geq 2i^3$. The remaining part of the proof does not make use of assumption (2.34). All these conditions on p are indeed satisfied if $p \geq D_{23}i^{361}$ for a suitable constant D_{23}. Hence (2.36) holds in general. □

Lemma 6. *For all* $n \geq 0$, $2 \leq r \leq 100$,

$$E\Big\{\sum_{m=1}^{\infty} |\widehat{Q}^{(r)}_{k(n)+m+1} - \widehat{Q}^{(r)}_{k(n)+m}|K_{k(n)}(m)\Big|\mathcal{T}_{k(n)}\Big\} < \infty \text{ a.s.} \tag{2.78}$$

Moreover,

$$E\{|R^{(r)}_n|\} < \infty \text{ (see (2.32))} \tag{2.79}$$

and

$$E\{\widehat{Q}^{(r)}_{k(n)+1}\} < \infty \tag{2.80}$$

for all $n \geq 0, r \geq 0$.

Proof. We note that for integers $k \geq 0, m \geq 1$ and real $x \geq 0$ we have

$$\begin{aligned}&\{K_{k(n)}(m) = 1\} \cap \{\tau_{k(n)+m} \leq x\} \cap \{k(n) = k\}\\&= \{K_{k(m)} = 1, \tau_{k+m} \leq x, k(n) = k\}\\&= \{k(n) = k, \text{ but } H_{k+\ell} > 1 \text{ for } 1 \leq \ell \leq m-1, \tau_{k+m} \leq x\} \in \mathcal{T}_{k(n)+m-1}.\end{aligned}$$

By taking the union over k we see from this that $K_{k(n)}(m)$ is $\mathcal{T}_{k(n)+m-1}$-measurable. We may therefore multiply both sides of (2.18) by $K_{k(n)}(m)$, take conditional expectation with respect to $\mathcal{T}_{k(n)}$ and sum over $m \geq 1$. This yields

$$\begin{aligned}&\sum_{m=1}^{\infty} E\{|\widehat{Q}^{(r)}_{k(n)+m+1} - \widehat{Q}^{(r)}_{k(n)+m}|K_{k(n)}(m)|\mathcal{T}_{k(n)}\}\\&\qquad \leq D_4 \sum_{m=1}^{\infty} E\Big\{\widehat{Q}^{(r-1)}_{k(n)+m}\Big(\frac{\widehat{L}_{k(n)+m}}{i}\Big)^{1/2} K_{k(n)}(m) H^r_{k(n)+m}|\mathcal{T}_{k(n)}\Big\} \text{ a.s.} \qquad (2.81)\end{aligned}$$

for $2 \leq r \leq 100$. But $\widehat{L}_q \geq i-1 \geq i/2$, so that $[\widehat{L}_q/i]^{1/2} \leq \sqrt{2}\widehat{L}_q/i$. Moreover, $\widehat{Q}^{(r-1)}_q \widehat{L}_q/i \leq \widehat{Q}^{(r)}_q$, by (4.59) in [9]. We can therefore continue (2.81) to obtain for $2 \leq r \leq 100$ that

$$\begin{aligned}&\sum_{m=1}^{\infty} E\{|\widehat{Q}^{(r)}_{k(n)+m+1} - \widehat{Q}^{(r)}_{k(n)+m}|K_{k(n)}(m)|\mathcal{T}_{k(n)}\}\\&\leq D_4\sqrt{2}\sum_{m=1}^{\infty} E\{\widehat{Q}^r_{k(n)+m} H^r_{k(n)+m} K_{k(n)}(m)|\mathcal{T}_{k(n)}\}\\&\leq D_4\sqrt{2}\exp[D_7 i^{361}]\widehat{Q}^{(r)}_{k(n)+1} \text{ a.s. (by (2.36)).}\end{aligned}$$

This proves (2.78).

Next we show by induction on n that $E\{R^{(r)}_n\}$ is finite for all $2 \leq r \leq 100$. To start the induction we note that the starting positions $\widehat{X}^j(0), 1 \leq j \leq i$, are assumed to be non-random, and if $i \geq i_0(2r)$, then

$$E\{\widehat{Q}^{(r)}_{k(0)+1}\} = E\{\widehat{Q}^{(r)}_1\} \leq 2^r \sum_{j=0}^{i} [\widehat{X}^j(0)]^r + 2^r E\{\sum_{j=1}^{i} |\Delta^j_0|^r\} \qquad (2.82)$$

is finite by the calculations following (2.49) (see also (2.27) and following lines).

Now assume it has already been proven that $E\{|R^{(r)}_q|\} < \infty$ for all $2 \leq r \leq 100$ and $q \leq n-1$. By the definition (2.32) of R_n we then have

$$\begin{aligned}&E\{|R^{(r)}_n|\} \leq E\Big\{E\Big\{\sum_{m=1}^{\infty} |\widehat{Q}^{(r)}_{k(n)+m+1} - \widehat{Q}^{(r)}_{k(n)+m}|K_{k(n)}(m)\Big|\mathcal{T}_{k(n)}\Big\}\Big\}\\&\leq E\Big\{D_4\sqrt{2}\exp[D_7 i^{361}]\widehat{Q}^{(r)}_{k(n)+1}\Big\} = E\Big\{D_4\sqrt{2}\exp[D_7 i^{361}]\big[\widehat{Q}^{(r)}_{k(1)+1} + \sum_{q=1}^{n-1} R^{(r)}_q\big]\Big\}.\end{aligned}$$

The right-hand side here is finite by the induction hypothesis, so that (2.79) follows for $2 \le r \le 100$. It then follows for $0 \le r < 2$ as well, by Jensen's inequality. The statement about the finiteness of the expectations of the $\widehat{Q}^{(r)}_{k(n)+1}$ also follows because $\widehat{Q}^{(r)}_{k(n)+1} = \widehat{Q}^{(r)}_{k(1)+1} + \sum_{q=1}^{n-1} R^{(r)}_q$. □

Lemma 7. *Choose* ε *in Lemma* 4 *so that*

$$D_5 D_6 \varepsilon \le 2^{-100}. \tag{2.83}$$

For $i \ge i_2(\varepsilon), \mu \ge 2i, 3 \le r \le 100$ *and all* $n, m \ge 0$ *it holds on the event* (2.34)

$$E\{R^{(r)}_n | \mathcal{T}_{k(n)}\} \le -1. \tag{2.84}$$

Proof. We multiply (2.19) by $K_{k(n)}(m)$ and take the conditional expectation with respect to $\mathcal{T}_{k(n)}$ on both sides. We then sum over $m \ge 1$. This gives

$$E\{R^{(r)}_n | \mathcal{T}_{k(n)}\} = E\Big\{ \sum_{m=1}^{\infty} [\widehat{Q}^{(r)}_{k(n)+m+1} - \widehat{Q}^{(r)}_{k(n)+m}] K_{k(n)}(m) | \mathcal{T}_{k(n)} \Big\}$$

(interchanging the expectation and the sum is justified by (2.78))

$$\le D_5 r \varepsilon \sum_{m=1}^{\infty} E\Big\{ \widehat{Q}^{(r-1)}_{k(n)+m} E\{H^r_{k(n)+m} K_{k(n)}(m) | \mathcal{T}_{k(n)+m-1})\} \Big| \mathcal{T}_{k(n)} \Big\}$$

$$- r 2^{1-r} \sum_{m=1}^{\infty} E\{\widehat{Q}^{(r-1)}_{k(n)+m} K_{k(n)}(m) | \mathcal{T}_{k(n)}\}$$

$$+ 3ir \sum_{m=1}^{\infty} E\Big\{ E\{H^r_{k(n)+m} K_{k(n)}(m) | \mathcal{T}_{k(n)+m-1}\} | \mathcal{T}_{k(n)} \Big\}. \tag{2.85}$$

But $\widehat{Q}^{(r-1)}_{k(n)+m} K_{k(n)}(m)$ is $\mathcal{T}_{k(n)+m-1}$-measurable, so that by Lemma 5,

$$\sum_{m=1}^{\infty} E\Big\{ \widehat{Q}^{(r-1)}_{k(n)+m} E\{H^r_{k(n)+m} K_{k(n)}(m) | \mathcal{T}_{k(n)+m-1}\} | \mathcal{T}_{k(n)} \Big\}$$

$$= \sum_{m=1}^{\infty} E\Big\{ E\{\widehat{Q}^{(r-1)}_{k(n)+m} H^r_{k(n)+m} K_{k(n)}(m) | \mathcal{T}_{k(n)+m-1}\} | \mathcal{T}_{k(n)} \Big\}$$

$$= \sum_{m=1}^{\infty} E\Big\{ \widehat{Q}^{(r-1)}_{k(n)+m} H^r_{k(n)+m} K_{k(n)}(m) | \mathcal{T}_{k(n)} \Big\}$$

$$\le D_6 \widehat{Q}^{(r-1)}_{k(n)+1}.$$

Since always $K_{k(n)}(1) = 1$, we further have

$$\sum_{m=1}^{\infty} E\{\widehat{Q}^{(r-1)}_{k(n)+m} K_{k(n)}(m) | \mathcal{T}_{k(n)}\}$$

$$\ge E\{\widehat{Q}^{(r-1)}_{k(n)+1} K_{k(n)}(1) | \mathcal{T}_{k(n)}\} = E\{\widehat{Q}^{(r-1)}_{k(n)+1} | \mathcal{T}_{k(n)}\}.$$

These relations show that

$$E\{R_n^{(r)}|\mathcal{T}_{k(n)}\} \le [D_5 r\varepsilon D_6 - r2^{1-r}]\widehat{Q}^{(r-1)}_{k(n)+1} + 3ir\sum_{m=1}^{\infty} E\{H^r_{k(n)+m}K_{k(n)}(m)\big|\mathcal{T}_{k(n)}\}.$$

By (2.13) there exists some constant D_{10} such that for $\mu \ge 2i$

$$\sum_{m=1}^{\infty} E\{H^r_{k(n)+m}K_{k(n)}(m)\big|\mathcal{T}_{k(n)}\} \le D_{10}.$$

Since $\widehat{Q}^{(r-1)}_{k(n)+1} \ge \widehat{L}_{k(n)+1} \ge 2^{r+2}D_{10}ir$ on the event (2.34), for i large enough, it follows that

$$\begin{aligned}E\{R_n^{(r)}|\mathcal{T}_{k(n)}\} &\le [D_5 r\varepsilon D_6 - r2^{1-r} + 2^{-r-1}]\widehat{Q}^{(r-1)}_{k(n)+1}\\ &\le -r2^{-1-r}\widehat{Q}^{(r-1)}_{k(n)+1} \text{ (see (2.83))}\\ &\le -r2^{-1-r}\widehat{L}_{k(n)+1} \le -1 \text{ (on (2.34))}.\end{aligned}$$

□

The remaining part of this paper imitates the proof of Theorem 2 in Section 2 of [9]. We remind the reader of our choice of the initial state just after the statement of Theorem 1. We define

$$\widehat{\nu} = \inf\{n \ge 1 : \widehat{L}_{k(n)+1} < D_{12}i^{361}\}.$$

$\widehat{\nu}$ is a stopping time with respect to the filtration $\{\mathcal{T}_{k(n)}\}$.

Lemma 8. *If $i \ge i_2(\varepsilon), \mu \ge 2i$, and $3 \le r \le 100$, then*

$$E\{\widehat{Q}^{(r)}_{k(n)+1}I[\widehat{\nu} > n-1]\} \le E\{\widehat{Q}^{(r)}_{k(1)+1}\} < \infty \textit{ for all } n \ge 1. \tag{2.86}$$

Proof. By Lemma 6, the expectations in (2.86) make sense. Now note that the event (2.34) occurs for all $1 \le n < \widehat{\nu}$, by definition of $\widehat{\nu}$, and note that $I[\widehat{\nu} > n]$ is $\mathcal{T}_{k(n)}$-measurable. Lemma 15 therefore shows that for $n \ge 1$

$$\begin{aligned}E\{\widehat{Q}^{(r)}_{k(n+1)+1}|\mathcal{T}_{k(n)}\}I[\widehat{\nu} > n]\} &= \widehat{Q}^{(r)}_{k(n)+1}I[\widehat{\nu} > n] + E\{R_n^{(r)}|\mathcal{T}_{k(n)}\}I[\widehat{\nu} > n]\\ &\le \widehat{Q}^{(r)}_{k(n)+1}I[\widehat{\nu} > n] - I[\widehat{\nu} > n] \le \widehat{Q}^{(r)}_{k(n)+1}I[\widehat{\nu} > n-1].\end{aligned}$$

By taking expectations we obtain that $E\{\widehat{Q}^{(r)}_{k(n)+1}I[\widehat{\nu} > n-1]\}$ is non-increasing in $n \ge 1$, as desired. □

Lemma 9. *For $i \ge i_2(\varepsilon), \mu \ge 2i$, there exists a constant C_1 such that*

$$P\{\widehat{\nu} \ge n\} \le \frac{C_1}{n^{25}}, \quad n \ge 1. \tag{2.87}$$

Proof. We define $M_0 = 0$ and for $q \geq 1$

$$M_q := \sum_{n=0}^{q-1} \Big[R_n^{(3)} - E\{R_n^{(3)} | \mathcal{T}_{k(n)}\}\Big] I[\widehat{\nu} > n]$$
$$= \sum_{n=0}^{q-1} \Big[\big[\widehat{Q}^{(3)}_{k(n+1)+1} - \widehat{Q}^{(3)}_{k(n)+1}\big] - E\Big\{\big[\widehat{Q}^{(3)}_{k(n+1)+1} - \widehat{Q}^{(3)}_{k(n)+1}\big]\Big|\mathcal{T}_{k(n)}\Big\}\Big] I[\widehat{\nu} > n]$$
$$= \widehat{Q}^{(3)}_{k(q)\wedge k(\widehat{\nu})+1} - \widehat{Q}^{(3)}_{k(0)+1} - \sum_{n=0}^{(q-1)\wedge(\widehat{\nu}-1)} E\{R_n^{(3)} | \mathcal{T}_{k(n)}\}, \quad q \geq 1. \tag{2.88}$$

Note that $\{M_q\}$ is a well-defined $\{\mathcal{T}_q\}$-martingale by virtue of Lemma 6 (see (2.32) for $R_n^{(3)}$). Now

$$\sum_{n=0}^{(q-1)\wedge(\widehat{\nu}-1)} E\{R_n^{(3)} | \mathcal{T}_{k(n)}\} \leq E\{R_0^{(3)} | \mathcal{T}_0\} - [(q-1)\wedge(\widehat{\nu}-1)] + 1$$
$$\leq E\{\widehat{Q}^{(3)}_{k(1)+1} | \mathcal{T}_0\} - [(q-1)\wedge(\widehat{\nu}-1)] + 1,$$

by virtue of Lemma 7 and the fact that (2.34) occurs for all $1 \leq n < \widehat{\nu}$. Moreover, $\widehat{Q}^{(3)}_{k(q)+1} \geq 0$. Therefore,

$$P\{\widehat{\nu} \geq q\} \leq P\{M_q + \widehat{Q}^{(3)}_{k(0)+1} + E\{\widehat{Q}^{(3)}_{k(1)} | \mathcal{T}_0\} \geq q-2\}$$
$$\leq D_{21}\frac{E\{|M_q|^r\}}{q^r} + D_{21}\frac{E\{|\widehat{Q}^{(3)}_{k(0)+1}|^r\}}{q^r} + D_{21}\frac{E\{|\widehat{Q}^{(3)}_{k(1)+1}|^r\}}{q^r}. \tag{2.89}$$

By (2.80), (2.82) and the lines following it the second and third term in the right-hand side of (2.89) are $O(q^{-r})$ and we only have to prove that

$$E\{|M_q|^r\} \leq C_2 q^{r/2} \tag{2.90}$$

for some $C_2 = C_2(i, \widehat{X}^1(0), \dots, \widehat{X}^i(0)) < \infty$. This is shown as at the end of Section 2 of [9]. By Burkholder's inequality,

$$E\{|M_q|^r\}$$
$$\leq E\Big\{\Big[\sum_{n=0}^{q-1}\big[R_n^{(3)} - E\{R_n^{(3)} | \mathcal{T}_{k(n)}\}\big]^2 I[\widehat{\nu} > n]\Big]^{r/2}\Big\}$$
$$\leq q^{r/2-1}\sum_{n=0}^{q-1} E\big\{\big|R_n^{(3)} - E\{R_n^{(3)} | \mathcal{T}_{k(n)}\}\big|^r I[\widehat{\nu} > n]\big\} \text{ (by Hölder)}$$
$$\leq C_3 q^{r/2-1}\sum_{n=0}^{q-1} E\big\{\big|R_n^{(3)}\big|^r I[\widehat{\nu} > n]\big\} \text{ (if } r \geq 1)$$
$$= C_3 q^{r/2-1}\sum_{n=0}^{q-1} E\big\{\big|\widehat{Q}^{(3)}_{k(n+1)+1} - \widehat{Q}^{(3)}_{k(n)+1}\big|^r I[\widehat{\nu} > n]\big\}$$

$$\leq C_4 q^{r/2-1} \sum_{n=0}^{q} E\{|\widehat{Q}^{(3)}_{k(n)+1}|^r I[\widehat{\nu} > n-1]\}$$
$$\leq C_4 q^{r/2-1} \sum_{n=0}^{q} i^{r-1} E\{\widehat{Q}^{(3r)}_{k(n)+1} I[\widehat{\nu} > n-1]\} \text{ (by Hölder)}$$
$$\leq C_5 q^{r/2} \text{ (by (2.86)).} \qquad \square$$

Finally we come to the *proof of Theorem* 1.

Proof. It suffices to prove

$$E\{\tau_{k(\widehat{\nu})+1}\} < \infty. \tag{2.91}$$

But

$$E\{\tau_{k(\widehat{\nu})+1}\} = E\Big\{\sum_{n=0}^{\infty}[\tau_{k(n+1)+1} - \tau_{k(n)+1}] I[\widehat{\nu} > n]\Big\} + E\tau_1$$
$$= \sum_{n=0}^{\infty} E\Big\{E\{\sum_{m=1}^{\infty} \delta_{k(n)+m} K_{k(n)}(m) \big| \mathcal{T}_{k(n)}\} I[\widehat{\nu} > n]\Big\} + E\tau_1. \tag{2.92}$$

Now

$$E\{\delta_{k(n)+m} K_{k(n)}(m) \big| \mathcal{T}_{k(n)}\}$$
$$= E\Big\{E\{\delta_{k(n)+m} \big| \mathcal{T}_{k(n)+m-1}\} K_{k(n)}(m) \Big| \mathcal{T}_{k(n)}\Big\}$$
$$\leq E\Big\{H^2_{k(n)+m}\Big(\frac{\widehat{L}_{k(n)+m}}{i}\Big) K_{k(n)}(m) \Big| \mathcal{T}_{k(n)}\Big\} \text{ (see (2.22))}$$
$$\leq \Big[E\{H^3_{k(n)+m} K_{k(n)}(m) \big| \mathcal{T}_{k(n)}\}\Big]^{2/3} \Big[E\{\Big(\frac{\widehat{L}_{k(n)+m}}{i}\Big)^3 K_{k(n)}(m) \big| \mathcal{T}_{k(n)}\}\Big]^{1/3}$$
$$\leq D_2 e^{-D_1 m/3} \Big[\frac{1}{i} E\{\widehat{Q}^{(3)}_{k(n)+m} K_{k(n)}(m) \big| \mathcal{T}_{k(n)}\}\Big]^{1/3} \text{ (by (2.13) and Jensen)}$$
$$\leq C_6 e^{-D_1 m/3} \big[\widehat{Q}^{(3)}_{k(n)+1}\big]^{1/3} \text{ (by (2.36)).}$$

Substitution of this estimate into (2.92) gives

$$E\{\tau_{k(\widehat{\nu})+1}\}$$
$$\leq C_7 \sum_{n=0}^{\infty} E\Big\{\big[\widehat{Q}^{(3)}_{k(n)+1}\big]^{1/3} I[\widehat{\nu} > n]\Big\}$$
$$\leq C_7 \sum_{n=0}^{\infty} \Big[E\{\widehat{Q}^{(3)}_{k(n)+1} I[\widehat{\nu} > n-1]\}\Big]^{1/3} \Big[P\{\widehat{\nu} > n\}\Big]^{2/3}$$
$$\leq C_8 \sum_{n=0}^{\infty} \big[P\{\widehat{\nu} > n\}\big]^{2/3} \text{ (by (2.86))} < \infty \text{ (by (2.87)).} \qquad \square$$

References

[1] Chow, Y.S. and Teicher, H. (1986), Probability Theory, second ed., Springer-Verlag.

[2] Fayolle, G., Malyshev, V.A. and Menshikov, M.V. (1995), Topics in the Constructive Theory of Countable Markov Chains, Cambridge University Press.

[3] Freedman, D. (1973), Another note on the Borel-Cantelli lemma and the strong law, with the Poisson approximation as a by-product, vol. 1, Ann. Probab., 910–925.

[4] Gut, A. (1988), Stopped Random Walks, Springer-Verlag.

[5] Hunt, G.A. (1966), Martingales et Processus de Markov, Monographies de la Société Mathématique de France, vol. 1, Dunod.

[6] Hall, P. and Heyde, C.C. (1980), Martingale Limit Theory and its Application, Academic Press.

[7] Kesten, H. and Sidoravicius, V. (2003), Branching random walk with catalysts, vol. 8, Elec. J. Probab.

[8] Kesten, H. and Sidoravicius, V. (2005), The spread of a rumor or infection in a moving population (DLA) and positive recurrence of Markov chains, vol. 33 Ann. Probab., 2402–2462.

[9] Kesten, H. and Sidoravicius, V. (2008), A problem in one-dimensional Diffusion Limited Aggregation (DLA) and positive recurrence of Markov chains, to appear in Ann. Probab.

[10] Zygmund, A. (1959), Trigonometric Series, vol I, Cambridge University Press, second ed.

Harry Kesten
Department of Mathematics
Malott Hall
Cornell University
Ithaca NY 14853, USA
e-mail: kesten@math.cornell.edu

Vladas Sidoravicius
IMPA
Estr. Dona Castorina 110
Jardim Botânico
22460-320, Rio de Janeiro, RJ, Brasil
e-mail: vladas@impa.br

Progress in Probability, Vol. 60, 463–480

Gibbsian Description of Mean-Field Models

Arnaud Le Ny

Abstract. We introduce a new framework to describe mean-field models in the spirit of the DLR description of probability measures on infinite product probability spaces used for lattice spin systems. The approach, originally introduced by C. Kuelske in 2003, is inspired by the generalized Gibbsian formalism recently developed in the context of the Dobrushin program of restoration of Gibbsianness, and enables the recovery of many of its features in the mean-field context. It is based on a careful study of the continuity properties of the limiting conditional probabilities of the finite-volume mean-field measures as a function of empirical averages, when the limiting procedure is properly done to avoid trivialities. This contribution is an extended version of a poster presented at the Xth Brazilian school of probability and has mainly a review character.

Mathematics Subject Classification (2000). Primary 60K35; Secondary 60G09, 82B23.

Keywords. Gibbs vs. non-Gibbs, mean-field models, generalized Gibbs measures, Curie-Weiss random field Ising model, renormalization, Glauber dynamics, exchangeability.

1. Introduction

This review-type paper can also be seen as an advertisement for an as yet unachieved theory that needs to be developed. We want to motivate a recent approach to mean-field models based on the DLR construction of measures on lattice spin-systems, originally due to C. Kuelske. Using a particular way of performing the infinite-volume limit of the conditional probabilities in the mean-field set-up, considered as functions of empirical averages, we use the characterization of Gibbsianness in terms of continuity of its conditional probabilities ('quasilocality') to propose a new 'DLR-like' description of mean-field models, already studied in [13, 17, 19]. Before discussing this notion, we justify our choice by a description of the standard mathematical Gibbs formalism, emphasizing the importance of the topological property of quasilocality in lattice spin-systems, as described in

[6, 9, 11]. Its importance is also highlighted by the Dobrushin program of restoration of Gibbsianness whose results indicate its crucial role to get a proper notion of equilibrium states, that Gibbs measures are intended to represent. It has indeed been shown during the last decades [6] that this quasilocality property can be lost under very natural scaling transformations of Gibbs measures, the *renormalization group (RG) transformations*, a fact which contradicts a lot of physics folklore. In this program, it has been established that any relevant weakening of the Gibbs property, to make it stable under scaling RG transformations, should incorporate nice topological properties involving some almost-sure quasilocality [20]. This importance of continuity properties has motivated our new notion of Gibbsianness for mean-field models, based on a careful limiting procedure in the investigation of the continuity properties of the conditional probabilities of the mean-field measures. We next describe the features of the Gibbs vs. non-Gibbs picture that this mean-field approach shares with the DLR one on lattice systems, and eventually describe how far from a specification is our new notion. This theory is still under construction, and in particular the four first sections have a review character, and we hope one could go further in this direction.

In Section 2, we describe the standard DLR description of Gibbs and quasilocal measures and provide results on the 2d-Ising model needed thereafter. In Section 3, we describe a few occurrences of non-Gibbsianness on lattice systems and emphasize the importance of quasilocality through the recent developments of generalized Gibbs measures. Motivated by these developments, we introduce our notion of Gibbsianness for mean-field models in Section 4 and illustrate it with the mean-field counter-part of the non-Gibbsian examples of Section 3. We conclude this paper by an extension of this theory in the case of the Curie-Weiss Ising model in Section 5, where we also discuss some limitations of this approach. We eventually propose to couple this approach to a DLR description of exchangeable measures proposed by H.O. Georgii among others, briefly described in Section 5 too.

2. DLR description of probability measures

2.1. Gibbs measures for lattice systems

Gibbs measures have been introduced in statistical mechanics to model phase transitions and more generally to describe equilibrium states of systems consisting of a large number of interacting components [6, 9, 11]. In order to get a modelization of the latter in a probabilistic framework, one considers a formalism at infinite volume, restricted here to the case of the lattice $S = \mathbb{Z}^d$. The description of phase transitions will be made possible by prescribing the conditional probabilities for given boundary conditions in the outside of any finite set. The set of all finite subsets of S is denoted by $\mathcal{S}$ and in our set-up a random variable (or a *spin*) σ_i is attached to each site i of the lattice, taking values in a finite probability space $(E, \mathcal{E}, \rho_0)$ equipped with the discrete topology.

The family of random variables $\sigma = (\sigma_i)_{i \in \mathbb{Z}}$ belongs then to the *configuration space* $(\Omega, \mathcal{F}, \rho) = (E^S, \mathcal{E}^{\otimes S}, \rho_0^{\otimes S})$, which is an infinite product probability space equipped with the product topology. Its elements, the *configurations*, will be denoted by Greek letters σ, ω, η, etc. We denote the configuration space at finite volume $\Lambda \in \mathcal{S}$ by $(\Omega_\Lambda, \mathcal{F}_\Lambda, \rho_\Lambda) = (E^\Lambda, \mathcal{E}^{\otimes \Lambda}, \rho_0^{\otimes \Lambda})$ and the projection of a configuration $\sigma \in \Omega$ to Ω_Λ by σ_Λ. Similarly, we described by $\sigma_\Lambda \omega_{\Lambda^c}$ a configuration which agrees with σ in Λ and with ω outside. The microscopic quantities will be described by (limits of) *local* functions, i.e., functions $f : \Omega \longrightarrow \mathbb{R}$ that are $\mathcal{F}_\Lambda$-measurable for some $\Lambda \in \mathcal{S}$. We also define in a standard way the σ-algebras $\mathcal{F}_{\Lambda^c}$ and macroscopic quantities will be somehow related to the famous *σ-algebra at infinity* or *tail σ-algebra* $\mathcal{F}_\infty := \cap_{\Lambda \in \mathcal{S}} \mathcal{F}_{\Lambda^c}$.

In order to model interactions of our system, one introduces *potentials*. In the usual Gibbs formalism [6, 11], they have to satisfy a summability condition in order to define *Hamiltonians* at infinite volume with prescribed boundary conditions.

Definition 2.1 (U.A.C. potentials). A *Uniformly Absolutely Convergent (U.A.C.) potential* is a family $\Phi = (\Phi_A)_{A \in \mathcal{S}}$ of local functions $\Phi_A : \Omega \longrightarrow \mathbb{R}$, such that each Φ_A is $\mathcal{F}_A$-measurable, and satisfying

$$\sum_{A \in \mathcal{S}, A \ni 0} \sup_{\omega \in \Omega} \mid \Phi_A(\omega) \mid < +\infty. \tag{2.1}$$

For such a potential, it is possible to define the following

Definition 2.2 (Hamiltonian at volume $\Lambda \in \mathcal{S}$ and boundary condition $\omega \in \Omega$).

$$\forall \sigma \in \Omega, \; H_\Lambda^\Phi(\sigma \mid \omega) = \sum_{A \in \mathcal{S}, A \cap \Lambda \neq \emptyset} \Phi_A(\sigma_\Lambda \omega_{\Lambda^c}).$$

A candidate to represent the conditional probabilities of the equilibrium states at some temperature $\beta^{-1} > 0$ is then derived in accordance with the fundamental laws of thermodynamics, and inspired by the finite-volume Boltzmann-Gibbs weights (see [11]), one introduces:

Definition 2.3 (Gibbs distribution at vol. $\Lambda \in \mathcal{S}$, boundary condition ω).

$$\forall \sigma \in \Omega, \; \gamma_\Lambda^{\beta\Phi}(\sigma \mid \omega) := \frac{1}{Z_\Lambda^{\beta\Phi}(\omega)} \cdot e^{-\beta H_\Lambda(\sigma \mid \omega)}. \tag{2.2}$$

It is normalized by the usual partition function $Z_\Lambda^{\beta\Phi}(\omega)$ in order to get a probability measure $\gamma_\Lambda^{\beta\Phi}(\cdot \mid \omega)$ defined on $(\Omega, \mathcal{F})$ for fixed boundary condition $\omega \in \Omega$:

$$\forall B \in \mathcal{F}, \gamma_\Lambda^{\beta\Phi}(B \mid \omega) = \int_B \gamma_\Lambda^{\beta\Phi}(\sigma \mid \omega) \rho_\Lambda(d\sigma_\Lambda). \tag{2.3}$$

The family $(\gamma_\Lambda^{\beta\Phi})_{\Lambda \in \mathcal{S}}$ is called a *Gibbs specification* because it specifies (or prescribes) the conditional probabilities of the expected equilibrium state in the following DLR framework[1]:

[1]DLR comes from R. Dobrushin, O. Lanford and D. Ruelle who introduced it in the late sixties.

Definition 2.4 (Gibbs measure for (β, Φ)). A probability measure μ on $(\Omega, \mathcal{F})$ is said to be a *Gibbs measure* at temperature $\beta^{-1} > 0$ for a UAC potential Φ iff for all $\Lambda \in \mathcal{S}$ a version of its conditional probability w.r.t. $\mathcal{F}_{\Lambda^c}$ is given by (2.3), i.e.,

$$\forall B \in \mathcal{F},\ \mu[B \mid \mathcal{F}_{\Lambda^c}](\omega) = \gamma_\Lambda^{\beta\Phi}(B \mid \omega) \quad \mu\text{−a.e.}(\omega).$$

More generically, a measure is said to be Gibbs if there exists a UAC potential such that the previous definition holds. We shall see that equivalent characterizations involving continuity properties of conditional probabilities instead of convergence properties of the potential hold (Theorem 2.11). The main example of Gibbs measures and phase transitions is provided by the Ising model of ferromagnetism, described later on in this note.

2.2. Specification and DLR consistency

The former construction of Gibbs measures on an infinite product probability space can be generalized to avoid the use of potentials and Hamiltonians, defining measures by the prescription of their conditional probabilities w.r.t. the outside of any finite set.

Definition 2.5 (Specification). A specification on $(\Omega, \mathcal{F})$ is a family $\gamma = (\gamma_\Lambda)_{\Lambda \in \mathcal{S}}$ of probability kernels on $(\Omega, \mathcal{F})$ satisfying:

1. Properness: $\forall B \in \mathcal{F}_{\Lambda^c},\ \forall \omega \in \Omega,\ \gamma_\Lambda(B \mid \omega) = \mathbf{1}_B(\omega)$.
2. DLR consistency: If $\Lambda \subset \Delta \in \mathcal{S}$, then $\gamma_\Delta \gamma_\Lambda = \gamma_\Delta$.

We recall that the product of the kernels is given for all $B \in \mathcal{F}$ and $\omega \in \Omega$ by

$$\gamma_\Delta \gamma_\Lambda(B \mid \omega) = \int_\Omega \gamma_\Lambda(B \mid \sigma)\gamma_\Delta(d\sigma \mid \omega). \tag{2.4}$$

Properties 1 and 2 allow specifications to be natural candidates to represent conditional probabilities of measures w.r.t. the outside of any finite set, with the notable exception that there are defined *everywhere* and not only almost everywhere. In particular, consistency has to do with the so-called *tower property* of double conditionings[2]. Our main example of specification is of course provided by Gibbs specifications $\gamma^{\beta\Phi}$ given by (2.2) and (2.3) for a convergent potential. Consistency condition 2. holds thanks to the particular sum involved in the exponential weight, but one could prove that it is not so particular (Theorem 2.11 or [11]).

Definition 2.6 (Measures specified by a specification). A probability measure μ on $(\Omega, \mathcal{F})$ is *specified by* a specification γ (or *consistent with* γ) if the latter is a realization of its finite-volume conditional probabilities, i.e., if

$$\forall B \in \mathcal{F}, \forall \Lambda \in \mathcal{S},\ \mu[B \mid \mathcal{F}_{\Lambda^c}](\cdot) = \gamma_\Lambda(B \mid \cdot) \quad \mu\text{−a.s.} \tag{2.5}$$

The set of all the measures specified by a specification γ is denoted by $\mathcal{G}(\gamma)$. Equivalently, μ is consistent with γ if it satisfies the *DLR equation:*

$$\mu \;=\; \mu\gamma_\Lambda, \quad \forall \Lambda \in \mathcal{S} \tag{2.6}$$

---

[2]For a more precise discussion of these statements and on conditional probabilities, see [9].

where the kernels γ_Λ transform μ into the probability measures $\mu\gamma_\Lambda$ on $(\Omega, \mathcal{F})$, defined, e.g., by its action on bounded measurable functions f via

$$(\mu\gamma_\Lambda)f(\omega) = \int_\Omega \gamma_\Lambda f(\omega)\mu(d\omega).$$

In the former, for any bounded measurable functions f and for all $\Lambda \in \mathcal{S}$, $\gamma_\Lambda f$ is the function of the boundary condition ω defined by

$$\gamma_\Lambda f(\omega) = \int_\Omega f(\sigma)\gamma_\Lambda(d\sigma \mid \omega).$$

The existence of a measure in $\mathcal{G}(\gamma)$ is not granted for any specification, but there exists a good topological framework insuring it, the *quasilocality*. On the other hand, and on the contrary to the marginal-based Kolmogorov construction of measures on infinite product probability spaces, one can get non-uniqueness of the measure specified by the specification, interpreted as the manifestation of *phase transition*, see, e.g., the Ising model next section.

2.3. Quasilocality and Gibbs representation theorem

Quasilocality manages to represent microscopic quantities that are in some sense localized and asymptotically insensitive to phenomena occurring arbitrarily far.

Definition 2.7 (Quasilocal functions). A function $f : \Omega \longrightarrow \mathbb{R}$ is said to be *quasilocal* if it is a uniform limit of local functions, or equivalently if and only if

$$\lim_{\Lambda \uparrow \mathcal{S}} \sup_{\sigma_\Lambda = \omega_\Lambda} \mid f(\sigma) - f(\omega) \mid = 0.$$

In our finite-spin setting, quasilocality is equivalent to uniform continuity w.r.t. the product topology on Ω [11]. Indeed, for this topology, a typical neighborhood of a configuration $\omega \in \Omega$ is of the form, for $\Lambda \in \mathcal{S}$,

$$\mathcal{N}_\Lambda(\omega) = \{\sigma \in \Omega : \sigma_\Lambda = \omega_\Lambda, \sigma_{\Lambda^c} \text{ arbitrary}\}.$$

Note that these open sets have a non-zero a priori measure. This will be crucial in the arising of non-Gibbsian measures after natural renormalization transformations. Next we extend this concept and give the

Definition 2.8 (Quasilocal specification and measure). A specification is said to be *quasilocal* if and only if for each $\Lambda \in \mathcal{S}$ and every local function f, $\gamma_\Lambda f$ is quasilocal. Similarly, a probability measure μ is quasilocal if and only if $\mu \in \mathcal{G}(\gamma)$ with γ quasilocal.

This property induces an important consequence on the conditional expectations of local functions: using the DLR relation (2.5), one gets that for any local function f, any $\Lambda \in \mathcal{S}$ and any configuration $\omega \in \Omega$, *there always exists a version of the conditional expectations* $\mu[f \mid \mathcal{F}_{\Lambda^c}]$ which is *continuous* at ω. The failure of this property, called the presence of an *essential discontinuity*, will be the main tool to detect non-Gibbsianness in the next section.

Due to its relationship with (uniform) continuity w.r.t. the product topology, this quasilocality property can be seen as an extension of the (two-sided) Markov property[3] : conditioning w.r.t. the outside of any finite sets might depend on spins arbitrarily far away, but this dependence is getting asymptotically negligible. This nice topological framework enables to get a few properties on the set of measures specified by a quasilocal specification γ. We gather now a few results from [11]:

Theorem 2.9. [11] *Let γ be a quasilocal specification. Then:*

1. $\mathcal{G}(\gamma) \neq \emptyset$ *is a Choquet simplex, i.e., a convex set whose elements admit a unique convex combination of its extreme elements.*
2. $\mu \in \mathcal{G}(\gamma)$ *is extreme if and only if μ is trivial on the tail σ-field $\mathcal{F}_\infty$.*
3. *If μ is an extreme element of $\mathcal{G}(\gamma)$, then for all quasilocal function[4] f,*

$$\lim_{\Lambda \uparrow \mathcal{S}} \gamma_\Lambda(f \mid \omega) = \mu[f] \quad \mu\text{−a.e.}(\omega)$$

The proofs can be found in Chapter 7 of [11]. Item 1 requires non-trivial measurability and convexity concepts. Item 2 is very interesting in view of our physical motivations: the events of the tail σ-fields being those who does not depend on what happens on any finite (microscopic) part of the lattice, one considers that their probabilities should not fluctuate for a reasonable physical phase. Thus item 2 tells us that the true physical phases of the system are the extreme elements of $\mathcal{G}(\gamma)$, and item 3 provides a way to select them by conditioning with *typical* boundary conditions. For a wider discussions of these concepts, see [9, 11]; we use them now to describe the main example of phase transition, in the sense that $\mathcal{G}(\gamma)$ is not a singleton.

Ising model on $\mathbb{Z}^2$: Spins take value in $E = \{-1, +1\}$ and the a priori measure of the configuration space is the product of the Bernoulli measures $\rho_0 = \frac{1}{2}\delta_{-1} + \frac{1}{2}\delta_{+1}$. The interaction is a *nearest-neighbor* one, where for any $\sigma \in \Omega,\ i, j \in S$,

$$\Phi_{\{i\}}(\sigma) = -h\,\sigma_i, \quad \Phi_{\{i,j\}}(\sigma) = -\sigma_i\sigma_j \text{ if } \mid \mathrm{i} - \mathrm{j} \mid = 1 \tag{2.7}$$

and $\Phi_A(\sigma) = 0$ for other $A \in \mathcal{S}$. The real h is called external magnetic field. The following theorem is the combination of many famous results, the most complete being those of [1, 12] on the structure of the simplex. We denote by $+$ (resp. $-$) the configuration whose value is $+1$ (resp -1) at any site of S.

Theorem 2.10. *Let $\gamma^{\beta\Phi}$ be a Gibbs specification with an Ising potential (2.7) on $\Omega = \{-1, +1\}^{\mathbb{Z}^2}$ and zero external field $h = 0$. Then there exists a critical inverse temperature $0 < \beta_c < +\infty$ such that*

---

[3]This is precisely the reason why W. Sullivan uses the terminology *almost Markovian* for quasilocality in [24].

[4]We denote by $\mu[f]$ the μ-expectation of any bounded measurable function f.

- $\mathcal{G}(\gamma^{\beta\Phi}) = \{\mu_\beta\}$ *for all* $\beta < \beta_c$.
- $\mathcal{G}(\gamma^{\beta\Phi}) = [\mu_\beta^-, \mu_\beta^+]$ *for all* $\beta > \beta_c$ *where* $\mu_\beta^- \neq \mu_\beta^+$ *can be selected via '−' or '+' boundary conditions: for all* f *local,*

$$\lim_{\Lambda \uparrow S} \gamma_\Lambda[f \mid -] = \mu_\beta^-[f] \text{ and } \lim_{\Lambda \uparrow S} \gamma_\Lambda[\mathrm{f} \mid +] = \mu_\beta^+[\mathrm{f}].$$

In particular, they have opposite magnetizations $\mu_\beta^+[\sigma_0] = -\mu_\beta^-[\sigma_0] > 0$.

It is straightforward to see that the nearest-neighbor dependence immediately leads to quasilocality. In fact Gibbsian specifications are the archetype of quasilocal ones. The summability condition (2.1) on the potential insures that every Gibbs measure is quasiocal, and the converse requires only the additional property of non-nullness: A specification is said to be *uniformly non-null* if, for each $\Lambda \in \mathcal{S}$ and $A \in \mathcal{F}$, there exists constants $0 < a_\Lambda \leq b_\Lambda < \infty$ such that $a_\Lambda \cdot \rho(A) \leq \gamma_\Lambda(A|\cdot) \leq b_\Lambda \cdot \rho(A)$.

Theorem 2.11 (Gibbs representation theorem [9, 11]**).** *Let* γ *be a specification. Then the followings are equivalent:*

1. *There exists a UAC potential* Φ *such that* γ *is the Gibbsian specification for* Φ *at some inverse temperature* $\beta > 0$.
2. γ *is quasilocal and uniformly non-null.*

The construction of a UAC potential from the quasilocal specification is based on a careful use of the Moebius inclusion-exclusion principle (see, e.g., [9, 14, 24]).

Thus, an equivalent characterization of Gibbsianness exists in terms of continuity properties of conditional properties. One of our goal in this note and in the extension of this theory to mean-field is to emphasize on the importance of this characterization to get a proper Gibbs formalism, see Section 3.4 on generalized Gibbs measures.

Hence, a sufficient condition for non-Gibbsianness is the existence of a point of *essential discontinuity*, i.e., a configuration at which *every* realization of some conditional probability of μ with respect to the outside of finite-sets is discontinuous. It appeared during the last decades that such a configuration could arise after some natural transformations of Gibbs measures.

3. Non-Gibbsianness in lattice spin systems

While on the positive side the former DLR description of Gibbs measures provides a good framework to model phase transitions and equilibrium states for lattice systems[5], it appeared a few decades ago that the Gibbs property was not necessarily conserved under natural scaling transformations, a fact which was at the

[5] The fact that equilibrium properties of Gibbs measures are satisfactory is more easily seen in the so-called variational approach, where a variational principle translates the second law of thermodynamics in terms of zero relative entropy of measures. It is equivalent to the DLR approach up to the restriction to translation-invariant potentials, specifications and measures, see [6, 11].

origin of the so-called *renormalization group pathologies.* This has been widely described in the seminal paper [6] and we describe here a few basic examples of these pathologies, due to a violation of quasilocality for the transformed measure. In all our examples, the initial measure is a Gibbs measure for the Ising model on $\mathbb{Z}^2$ but similar phenomena also occur for other models and other dimensions.

3.1. Decimation of the $2d$-Ising model [6]

It corresponds to the projection of the 2d-Ising model on the sublattice of even sites. More formally, define the decimation transformation to be

$$T : \Omega \longrightarrow \Omega; \omega \longmapsto \tilde{\omega}; \ \ \tilde{\omega}_i = \omega_{2i}, \quad \forall i \in \mathbb{Z}^2.$$

Theorem 3.1. [6] *Let* $\beta > \frac{1+\sqrt{5}}{2} \cdot \beta_c$ *and denote by* $\nu_\beta^+ = T\mu_\beta^+$ *the decimation of the '+' phase of the* $2d$*-Ising model. Then* ν_β^+ *is not quasilocal, hence non-Gibbs.*

Sketch of the proof. To prove non-quasilocality of the renormalized measure, one exhibits a so-called *bad configuration*, i.e., a configuration where the conditional expectation, w.r.t. the outside of a finite set, of a local function, is essentially discontinuous, or equivalently is discontinuous in arbitrary neighborhoods [6]. The role of a bad configuration is played here by the *alternating configuration* $\tilde{\omega}_{\rm alt}$ defined for all $i = (i_1, i_2) \in \mathbb{Z}^2$ by $\tilde{\omega}_{{\rm alt}_i} = (-1)^{i_1+i_2}$. Computing the magnetization under the decimated measure, conditioned on the boundary condition $\tilde{\omega}_{\rm alt}$ outside the origin, gives different limits when one approaches this configuration with all + (resp. all −) arbitrarily far away, as soon as phase transition is possible in the decorated lattice, a version of $\mathbb{Z}^2$ where even sites have been removed. The global neutrality of this bad configuration leaves the door open to such a phase transition, and is crucial in its badness.

3.2. Stochastic evolution of Gibbs measures [5]

For the sake of simplicity, we describe the results for infinite temperature Glauber dynamics of low temperature phases of the Ising model at dimension $d \geq 2$, but the results extend to high temperature [5]. Starting from the +-phase μ_β^+ of the Ising model at low enough temperature $\beta^{-1} > 0$, we apply a stochastic spin-flip dynamics at rate 1, independently over the sites. The time evolved measure is then described by

$$\mu_{\beta,t}(\eta) := \sum_{\sigma \in \Omega} \mu_\beta^+(\sigma) \prod_{i \in \mathbb{Z}^d} \frac{e^{\eta_i \sigma_i h_t}}{2 \cosh h_t}, \quad \text{with } \mathrm{h_t} = \frac{1}{2} \log \frac{1 + \mathrm{e}^{-2t}}{1 - \mathrm{e}^{-2t}}. \tag{3.1}$$

The product kernel in (3.1) is a special case of a Glauber dynamics for infinite temperature [5, 21]. It is known that the time-evolved measure $\mu_{\beta,t}$ tends to a spin-flip invariant product measure on $\{-1, +1\}^{\mathbb{Z}^d}$, with $t \uparrow \infty$, which is trivially

[6]It is automatically of non-zero ν_β^+-measure in our topological setting (see the notion of *strong quasilocality in* [9]) and thus modifying conditional probabilities on a negligible set cannot make it continuous.

quasilocal and Gibbs. Nevertheless, the Gibbs property is lost during this evolution and recovered only at equilibrium:

Theorem 3.2. [5] *Assume that the initial temperature β^{-1} is small compared to the critical temperature of the nearest-neighbor Ising model for $d \geq 2$. Then there exists $t_0(\beta) \leq t_1(\beta)$ such that:*

1. *$\mu_{\beta,t}$ is a Gibbs measure for all $0 \leq t < t_0(\beta)$.*
2. *$\mu_{\beta,t}$ is* **not** *a Gibbs measure for all $t > t_1(\beta)$.*

Similar questions about such evolutions have been recently studied in the physics literature, see [23]. Non-Gibbsianness is here related to the possibility of phase transition in some constrained model. There remains a large interval of time where the validity of the Gibbs property of the time-evolved measure remains unknown for this lattice model. This has motivated the study of similar phenomena for mean-field models in [19], where the sharpness of the Gibbs/non-Gibbs transition has been proved, see Section 4.2. This study has required the introduction of the new notion of Gibbsianness for mean-field models described in this paper, and the relationship between lattice and mean-field results encourages us to investigate it further on.

3.3. Joint measure of short range disordered systems [15, 18]

This example has led to much progress in the Dobrushin program of restoration of Gibbsianness, and has reinforced our philosophy of focusing on continuity properties of conditional probabilities rather than on convergence properties of the potential, as we shall see next section. Non-Gibbsianness has here also been very useful to explain pathologies in the so-called *Morita approach to disordered systems*, see the proceedings volume [7]. The *Random Field Ising Model* (RFIM) is an Ising model where the magnetic field h is replaced by (say i.i.d. ± 1) random variable η_i of common law $\mathbb{P}$ at each site of the lattice. For a given $\eta = (\eta_i)_{i \in S}$, whose law is also denoted by $\mathbb{P}$, the corresponding ('quenched') Gibbs measures depend on this disorder and are denoted by $\mu[\eta]$. The Morita approach considers the joint measure 'configuration-disorder', formally defined by $K(d\eta, d\sigma) = \mu[\eta](d\sigma)\mathbb{P}(d\eta)$, to be Gibbs for a potential of the joint variables, but it has been proved in [15] that this measure can be non-Gibbs for $d \geq 2$ and for a small disorder. The mechanism, although more complicated, is similar to the previous examples, and the arising of non-quasilocality is made possible when a ferromagnetic ordering is itself possible in the quenched system, and thus the conditions on d and on the disorder are those required for such a phase transition in [3].

Before using this example to emphasize how important are continuity properties of conditional probabilities in the Gibbs formalism, we describe recent extensions of the Gibbs property within the so-called *Dobrushin program of restoration of Gibbsianness.*

3.4. Generalized Gibbs measures

In 1996, in view of the RG pathologies described by van Enter *et al.* [6], R.L. Dobrushin launched a program of restauration of Gibbsianness consisting in two parts [2]:

1. To give an alternative (weaker) definition of Gibbsianness that would be stable under scaling transformations.
2. To restore the thermodynamics properties of these measures in order to get a proper definition of equilibrium states.

The first part of this program yields two different restoration notions, one based on convergence properties of the potential and one based on continuity properties of conditional probabilities [22]:

- **Weakly Gibbsian measures**: These are measures μ that are consistent with a Gibbs specification with a potential which is only μ-almost surely convergent (w.r.t. the measure itself).
- **Almost Gibbsian measures**: These are measures μ whose finite-volume conditional probabilities are continuous functions of the boundary condition, except on a set of μ-measure zero. It is also called *almost quasilocality.*

An almost-sure version of the Kozlov-Sullivan [14, 24, 22] use of the inclusion-exclusion principle in Theorem 2.11 proves that *almost Gibbs implies weak Gibbs.*

All our examples of the previous section have been proved to be weakly Gibbs ([22, 16]) and it is also the case of most of the renormalized measures. More interestingly, whereas the decimated measure has been proved to be almost Gibbsian in [10], the contrary has been proved [16] for the joint measure of the RFIM, which has even a set of bad configurations with full measure. The peculiarity of this example appeared to be a good advertisement for the importance of quasilocality in the characterization of equilibrium states for lattice spin systems, and to discriminate the weak Gibbs restoration from almost Gibbs one, due to the consequences it has on the thermodynamic properties of the corresponding measures, as we shall see now in the second part of the Dobrushin program.

This second part aims at the restoration of the thermodynamics properties of Gibbs measures, mostly in terms of a variational principle that allows to identify them as the description of the states who minimize the free energy of the system, thus equilibrium states in virtue of the second law of thermodynamics. In the usual Gibbs formalism, this principle can be expressed in a very closed form in terms of zero relative entropy of two measures [11]: Two (translation-invariant) measures are Gibbs for the same potential if and only if they have zero relative entropy, and in such case they are equilibrium states for the same interacting system in the sense that they minimize its free energy. This result has been extended in [20] to (translation-invariant) quasilocal measures, where a relevant version of it also clearly identifies almost Gibbsian measures as good candidates to represent equilibrium states. It is moreover established there that the joint measures of the RFIM, one of the physically relevant known examples of a weakly Gibbsian measure that is not almost Gibbs, also provide an example of two (non almost

Gibbs) measures described by a different system of conditional probabilities that are equilibrium states for each other. In this situation, we have two candidates to be equilibrium, corresponding to different interactions, that saturate the variational principle of each other, which can be easily seen to be physically irrelevant.

The fact that this happens for a weakly and non almost Gibbsian measure clearly indicates that one has to insist on some continuity properties of the conditional probabilities in order to restore the Gibbs property in the framework of the Dobrushin program. Together with Theorem 2.11, this also emphasizes the relevance of the description of Gibbs measures in the quasilocal framework, i.e., in terms of topological properties of conditional probabilities rather than in terms of potentials. These observations have motivated our new 'DLR-like' approach to mean-field models, that we are now ready to describe.

4. 'DLR-like' description of mean-field models

We present now the recent description of mean-field models used a very few times [13, 17, 19] and still under investigation. We shall focus on the main mean-field model, the Curie-Weiss Ising model, and briefly indicate how this could be extended to general mean-field models.

4.1. Standard description of the CW Ising model

We focus thus on the mean-field approximation of the standard Ising model, the *Curie-Weiss (CW) Ising model.* It is the corresponding model on the complete graph: For any finite volume $V_N = \{1, \ldots, N\}$, consider the sequence of finite volume configuration spaces $(\Omega_N = \{-1,+1\}^{V_N})_{N \in \mathbb{N}^*}$ and define by Kolmogorov consistency the infinite product probability space $(\Omega, \mathcal{F}, \rho)$, where ρ is the a priori normalized product measure. For any $\sigma \in \Omega$, $N \in \mathbb{N}^*$, we denote $\sigma_{[1,N]}$ the usual projection on the corresponding configuration space $(\Omega_N, \mathcal{F}_N, \rho_N)$.

The CW-Ising model is the version of the Ising model where the lattice $\mathbb{Z}^d$ is replaced by the complete graph, forgetting thus the geometry and considering equal interaction between any pair of sites, each site feeling a field with a strength proportional to the mean magnetization outside the considered site[7]. This approximation is known to furnish a paradigm of a phase transition, but it is also known that not all of its properties can be converted into lattice results (cf. van der Waals theory, equivalence of ensembles). In this setting, one first introduces the *Curie-Weiss measures* on the finite-volume configuration space $(\Omega_N, \mathcal{F}_N, \rho_N)$ and at temperature $\beta^{-1} > 0$ by

$$\mu_N^\beta[d\sigma_{[1,N]}] = \frac{1}{Z_{[1,N]}} \cdot e^{\beta \sum_{i=1}^n \sigma_i \left(\frac{1}{N} \sum_{j \neq i} \sigma_j\right)} \rho_N[d\sigma]. \tag{4.1}$$

[7]Thus, at each site, an external field corresponding to the mean magnetization is acting, justifying the denomination *mean-field.* By mean-field model we mean more generally a model where the Hamiltonian is permutation-invariant and volume-dependent. For lattice spin systems, it is a function of the empirical magnetization.

The interaction becomes thus volume-dependent and this non-locality prevents a description of this model and of Gibbs measures in the proper DLR framework describe above. To emphasize this, one rewrites the CW measures in terms of the empirical magnetization at finite volume V_N, defined for all $\sigma \in \Omega$ by $m_N(\sigma) = \frac{1}{N}\sum_{i \in V_N} \sigma_i$, such that on $(\Omega, \mathcal{F}, \rho)$,

$$\mu_N^\beta[d\sigma] = \frac{1}{Z_{[1,N]}} \cdot e^{\beta N\, m_N^2(\sigma)} \rho(d\sigma) \tag{4.2}$$

where the interaction between spins depends only on the empirical magnetization. Thus the CW measure at finite volume V_N is invariant under all the permutations in this volume, and this will lead to exchangeability in the infinite-volume formalism. Standard large deviation arguments [4] show indeed that the sequence of CW measures $(\mu_N^\beta)_{N \in \mathbb{N}}$ weakly converges to an infinite volume probability measure

$$\mu_\beta = \frac{1}{2}\mu_\beta^+ + \frac{1}{2}\mu_\beta^- \tag{4.3}$$

with μ_β^+, μ_β^- are explicitly known to be the product measures $\mu_\beta^\pm = (\mu_0^\pm)^{\otimes \mathbb{N}^*}$ of the single-site marginal given for all $i \in \mathbb{N}^*$ and for all $\sigma \in \Omega$ by

$$\mu_0^\pm(\sigma_i) = \frac{e^{\pm \beta \cdot m_\beta^* \cdot \sigma_i}}{2\cosh(\beta \cdot m_\beta^*)} \tag{4.4}$$

where m_β^* is the solution of the famous *mean-field equation* $m = \tanh(\beta \cdot m)$, or equivalently minimizer of the rate-function (or free energy) of the system [4]. One says that this model shows a phase transition at temperature $\beta_c^{-1} = 1$ in the sense that at high temperature $m_\beta^* = -m_\beta^* = 0$, and μ_β is the uniform product measure ρ, whereas at low temperature, $m_\beta^* \neq -m_\beta^* > 0$ and μ_β is a particular convex combination of a so-called '+' -phase and '−'-phase. The latter has to be compared with the famous theorem of de Finetti [11]: the finite-volume CW measures being invariant by permutations in V_N, the weak limit μ will be *exchangeable*, i.e., invariant over all permutations, and thus a convex combination of product measures. We shall come back to exchangeability next section.

If one considers now the conditional probabilities of μ_β, one easily checks that the infinite-volume measures obtained with this formalism can be trivially non-Gibbs in the sense given on the lattice, because convex combinations mentioned above are known to be non-quasilocal, with all spin configurations being points of discontinuities [6, 8]. To avoid these trivialities and try nevertheless to extract good continuity properties of some conditional probabilities, C. Kuelske has proposed in [17] to do the conditioning at finite-volume first, and to perform the infinite-volume afterwards.

Our aim here is to describe this procedure and provide hints about the relevance of this notion, that will be discussed afterwards.

4.2. (Non-) Gibbsianness in mean-field

We consider a *mean-field model* to be a sequence $(\mu_N^\beta)_{N\in\mathbb{N}^*}$ of probability measures on $(\Omega_N, \mathcal{F}_N, \rho_N)$ where the density of μ_N^β is a smooth function of the empirical magnetization, like, e.g., in (4.2). For the sake of simplicity, our vocabulary will concern the CW-Ising model but the generalization is straightforward.

Although the macroscopic character of the Hamiltonian prevents at a first sight to describe this model in the classical DLR construction, exchangeability leads to nice properties of the conditional probabilities at finite volume, that leave the door open to description of that type. Indeed, it is straightforward to check that the conditional probabilities of the (*finite-volume*) *CW measures* are, when conditioned w.r.t. the outside of finite sets, functions of the empirical magnetization outside this latter finite set only. This led to the following definition in [13, 17, 19], where prescribed (real) values of the magnetization, to which many spin configurations correspond, are called *points*:

Definition 4.1 (MF-Gibbs measures). We call α_0 a **good point** for the MF-model $(\mu_N^\beta)_N$ if and only if

1. The limit
$$\gamma_1^\beta(\sigma \mid \alpha) := \lim_{N\uparrow\infty} \mu_N^\beta(\sigma_1 \mid \sigma_{[2,N]}) \quad \text{whenever} \quad \lim_{\mathrm{N}\uparrow\infty} \frac{1}{\mathrm{N}} \sum_{\mathrm{i}=2}^{\mathrm{N}} \sigma_{\mathrm{i}} = \alpha \tag{4.5}$$
exists for all α in a neighborhood of α_0.
2. The function $\alpha \longmapsto \gamma_1^\beta(\sigma \mid \alpha)$ is continuous at $\alpha = \alpha_0$.

We call the mean-field model $(\mu_N)_N^\beta$ **Gibbs** iff every point is good.

Let us see that the kernels γ_1^β are good candidates to represent an equivalent notion to specifications for MF-models. Before studying its relevance in the last section, we describe how this definition extends many features of Gibbs vs. non-Gibbs from lattice spin-systems to corresponding mean-field models.

Curie-Weiss Ising model: An easy computation on the conditional probabilities leads for all α to the kernel

$$\gamma_1^\beta(\sigma_1 \mid \alpha) = \frac{e^{\beta\cdot\alpha\cdot\sigma_1}}{2\cosh\beta\alpha}.$$

It is trivially continuous at every α and hopefully one recovers a relevant Gibbs property for this basic model.

Decimation of the Curie-Weiss Ising model [17]: Starting from the CW-Ising model at volume V_N, one considers the decimated measure to be the marginal distribution $\mu_N^\beta(\sigma_{[1,M]})$ of the first M spins and is interested in the behavior of its conditional probability when $M = M_N$ grows like a multiple of N, as N goes to infinity. Denote still by m_β^* the largest solution of the mean-field equation $m = \tanh(m)$ and by $m^{\mathrm{CW}}(\beta, h)$ that of the inhomogeneous mean-field equation $m = \tanh(\beta(m + h))$

for $h \in \mathbb{R}$. Then one gets the following Gibbs vs non-Gibbs picture in mean-field, which is very reminiscent to what happens on lattices (Theorem 3.1):

Theorem 4.2. [17] *Assume that* $\lim_{N\to\infty} \frac{M_N}{N} = 1 - p$ *with* $0 \leq p \leq 1$.

If $\lim_{M\uparrow\infty} \sum_{i=2}^{M} \sigma_i = \alpha \neq 0$, *then the limit* (4.5) *of* $\mu_N^\beta(\sigma_1 \mid \sigma_{[2,M_N]})$ *exists and is*

$$\gamma_1^\beta(\sigma \mid \alpha) = \frac{e^{\beta \cdot h_{\beta,p}(\alpha)\cdot \sigma_1)}}{2\cosh(\beta \cdot h_{\beta,p}(\alpha))}$$

with

$$\begin{aligned} h_{\beta,p}(\alpha) &= \alpha && \text{if } p = 0 \\ &= p \cdot m^{\mathrm{CW}}(p\beta, \frac{1-p}{p}\alpha) + (1-p)\cdot\alpha && \text{if } 0 < p < 1 \\ &= m^{\mathrm{CW}}(\beta,\alpha)\,\mathrm{sign}(\alpha) && \text{if } p = 1. \end{aligned}$$

In particular, it is continuous in $\alpha \neq 0$ but discontinuous in $\alpha = 0$ for $p\beta > 1$. The decimation of the Curie-Weiss Ising model is thus non-Gibbs in our MF sense at low enough temperature, in complete agreement to the behavior on lattice systems described in Section 3, whereas it would share exactly the same property as the original model in a naive approach[8].

This similarity has also been established for joint measures of disordered systems in [15] or for other transformations in [13].

Spin-flip dynamics of the Curie-Weiss Ising model [20]: In the same spirit, we have used in [20] this new notion to try to complete the picture of the evolution of the Gibbs property during infinite temperature spin-flip dynamics of low temperature Curie-Weiss Ising models. We have recovered the results already known on the lattice (Theorem 3.2), we have managed to prove that the transition times was sharp, but we also have discovered a new phenomenon, namely that during intermediate times, the time-evolved measure is non-Gibbs, in our mean-field sense, due to the existence of non-neutral bad points. It is an open question whether this type of badness could arise for configurations of short-range models on lattices or if it is a peculiarity of the mean-field framework. In the former, computations can often be carried out more fully and allows here a full analytic description of the time-evolution picture of the Gibbs property. The fact that the features of the lattice model are recovered upgrades the relevance of our new Gibbs notion, discussed next section. Also, this motivates the development of this notion in order to open new perspectives in the original lattice formalism. The proof of the former time-evolution picture suggests for example that phase transition in some constrained models is not necessary to get essential discontinuity, as suspected, but can also be related to the possibility of metastability, see [19].

[8]The decimated measure being a marginal of the original CW-measure, the infinite-volume weak limit will be the same convex combination of the same marginal product measures.

5. Discussion

We have thus a kernel γ_1, candidate to play the role of the one-site elements of specifications in the usual DLR framework, that shares with its lattice correspondent similar topological properties. Nevertheless, much more is needed to get fully satisfied with such a DLR-Gibbs description and we shall indeed see that further developments are needed before getting a complete theory.

Consider the case of the Curie-Weiss Ising model again. Specifications are families of kernels indexed by the finite subsets, so let us first extend our definition and give

Definition 5.1 (MF-kernels). We call *MF-kernel* the family of kernels

$$\gamma^\beta = (\gamma_M^\beta)_{M \in \mathbb{N}^*}$$

defined for all $\sigma \in \Omega$ and $\alpha \in [-1,+1]$ by

$$\gamma_M(\sigma \mid \alpha) := \lim_{N \to \infty} \mu_N^\beta[\sigma_{[1,M]} \mid \sigma_{[M+1,N]}] \text{ provided } \lim_{\mathrm{N}\uparrow\infty} \frac{1}{\mathrm{N}} \sum_{\mathrm{i}=\mathrm{M}+1}^{\mathrm{N}} \sigma_{\mathrm{i}} = \alpha \quad (5.1)$$

whenever this limit exists.

In our set-up, a MF-kernel is then a family of kernels from $([-1,+1],\mathcal{B}[-1,+1])$ to $(\Omega, \mathcal{F})$, where $(\mathcal{B}[-1,+1])$ is the standard Borel σ-algebra of $[-1,+1]$. Before meeting troubles due to the different nature of the latter spaces, let us describe a nice property of factorization of the MF-kernel on correlation functions in the CW-Ising model, whose proof is a straightforward computation.

Theorem 5.2 (Factorization). *Consider the CW-Ising model. Then the limits* (5.1) *in Definition* 5.1 *exist for all* $M \in \mathbb{N}^*$ *and the well-defined MF-kernels* $\gamma^\beta = (\gamma_M^\beta)_{M \in \mathbb{N}^*}$ *have the following factorization property:* $\forall M \in \mathbb{N}^*$, $\forall k = 1, \dots, M$, $\forall \sigma \in \Omega$, $\forall \alpha \in [-1,+1]$,

$$\gamma_M\left(\prod_{i=1}^k \sigma_i \mid \alpha\right) = \prod_{i=1}^k \gamma_1(\sigma_i \mid \alpha) = \prod_{i=1}^k \frac{e^{\beta \cdot \alpha \cdot \sigma_i}}{2\cosh \beta\alpha}. \quad (5.2)$$

We have then a family of kernels indexed by the finite subsets representing some infinite-volume limit of the conditional probabilities of the MF-measure. The extra factorization property of the kernels has to be compared with the product form of the infinite-volume limit obtained by weak limit.

Nevertheless, we encounter here the negative side of our new notion. To get properly speaking a specification of the conditional probabilities of the measure, one would like to get a consistency relation to characterize the infinite-volume measures in terms of invariance through the kernels. Here, due to the different nature of spaces involved in the kernel, DLR expressions like (2.4) and (2.6) are ill defined: The product of the kernel is impossible and the action of a kernel γ_M^β on a measure μ on $(\Omega, \mathcal{F})$ yields a probability measure on $([-1,+1],\mathcal{B}[-1,+1])$.

Thus our approach has to be modified to get a satisfactory notion of mean-field specification. To get more insights on these notions, let us describe briefly an existing DLR construction of exchangeable measures, described in [11] (see Chapter 7 and references therein). It is an adaptation of the notion of specification (Definition 2.5) where the tail σ-algebra $\mathcal{F}_\infty$ is replaced by the σ-algebras of *symmetric events*

$$\mathcal{I} = \cap_{N \in \mathbb{N}^*} \mathcal{I}_N$$

where for all $N \in \mathbb{N}^\star$, $\mathcal{I}_N$ is the σ-algebra generated by the permutation-invariant transformations of Ω_N. Then H.O. Georgii [11] introduces a family of proper kernels $\gamma = (\gamma_N)_N$ from $\mathcal{I}_N$ to $\mathcal{F}$, instead[9] of $\mathcal{F}_{\Lambda^c}$ to $\mathcal{F}$, defined for all $A \in \mathcal{I}_N$ and $\omega \in \Omega$ by

$$\gamma_N(A \mid \omega) = \frac{1}{N!} \sum_{\tau \in I_N} \mathbf{1}_A(\tau\omega)$$

where the sum runs over the set I_N of permutations τ of Ω_N. Up to the change of sub-σ-algebras already mentioned, this family of kernels is indeed a specification [11] and one easily identifies the set $\mathcal{G}(\gamma)$ to be this of all exchangeable measures. It is also a convex set and an adaptation of Theorem 2.9 identifies its extreme elements as the measures that are trivial on $\mathcal{I}$, which are the product measures. By this construction, one recovers thus de Finetti's theorem: The exchangeable measures are convex combinations of product measures.

6. Conclusions

We are thus in the following situation: One has, on one hand, a family of MF-kernels whose topological and factorization properties are consistent with the lattice model they approximate, but for which consistency and characterization of equilibrium measures have not been recovered, and, on the other hand, permutation-invariant proper specifications that define too many measures (all the exchangeable ones) and whose topological properties has not been linked to the lattice results. The next step is thus to find an appropriate way to combine the two approaches to get a satisfactory notion of DLR measures in mean-field, that could be compared to the DLR measures on lattices to try to import new properties occurring in mean-fields, via, e.g., some adapted Kac limits for specifications.

Acknowledgment

The author thanks C.-E. Pfister and C. Kuelske for stimulating discussions on the subject and o Instituto Nacional de Matematica Pura e Aplicada (IMPA, Rio de Janeiro) for hospitality during the redaction of this contribution.

[9]We have defined the kernels of a specification on $\mathcal{F}$ but it can be defined directly on $\mathcal{F}_{\Lambda^c}$ [6].

References

[1] M. Aizenman. *Translation invariance and instability of phase coexistence in the two-dimensional Ising system.* Comm. Math. Phys. **73** (1980), no 1:83–94.

[2] R.L. Dobrushin and S.B. Shlosman. *'Non-Gibbsian' states and their description.* Comm. Math. Phys. **200** (1999), no 1:125–179.

[3] J. Bricmont and A. Kupiainen. *Phase transition in the 3d Random Field Ising Model.* Comm. Math. Phys. **142** (1988), 539–572.

[4] R.S. Ellis. *Entropy, large deviations, and statistical mechanics.* Fundamental Principles of Mathematical Sciences, 271, Springer-Verlag, New-York, 1985.

[5] A.C.D. van Enter, R. Fernández, F. den Hollander, and F. Redig. *Possible loss and recovery of Gibbsianness during the stochastic evolution of Gibbs measures.* Comm. Math. Phys. **226** (2002):101–130.

[6] A.C.D. van Enter, R. Fernández and A.D. Sokal. *Regularity properties of position-space renormalization group transformations: scope and limitations of Gibbsian theory.* J. Stat. Phys. **72** (1993), 879–1167.

[7] A.C.D. van Enter, A. Le Ny and F. Redig (eds). *Proceedings of the workshop 'Gibbs vs. non-Gibbs in statistical mechanics and related fields' (Eurandom* 2003). Mark. Proc. Relat. Fields **10** (2004), no 3.

[8] A.C.D. van Enter and J. Lorinczi. *Robustness of the non-Gibbsian property: some examples.* J. Phys. A: Math. Gen. **29** (1996) 2465–2473.

[9] R. Fernández. *Gibbsianness and non-Gibbsianness in lattice random fields* In Mathematical Statistical Physics (Les Houches LXXXIII, 2005,) A. Bovier, F. Dunlop, F. den Hollander, A. van Enter, J. Dalibard (Eds.), Elsevier, 2006.

[10] R. Fernández, A. Le Ny and F. Redig. *Variational principle and almost quasilocality for renormalized measures.* J. Stat. Phys. **111** (2003), 465–477.

[11] H.O. Georgii. *Gibbs measures and Phase transition*, de Gruyter Studies in Mathematics, vol 9, 1988.

[12] Y. Higuchi. *On the absence of non-translation invariant Gibbs states for the two-dimensional Ising model.* Random fields, Vol I, II (Esztergom, 1979), Colloq. Math. Soc. Janos Bolyai **27**:517–534, 1981.

[13] O. Häggström and C. Kuelske. *Gibbs property of the fuzzy Potts model on trees and in mean-field.* Mark. Proc. Relat. Fields **10** (2004), no 3:477–506.

[14] O. Kozlov. *Gibbs description of a system of random variables.* Problems Inform. Transmission. **10** (1974), 258–265.

[15] C. Kuelske. *(Non-) Gibbsianness and Phase transition in Random Lattice Spin Models.* Mark. Proc. Rel. Field. **5** (1999), 357–383.

[16] C. Kuelske. *Weakly Gibbsian representation for joint measures of quenched lattice spin models.* Probab. Th. Relat. Fields **119** (2001), 1–30.

[17] C. Kuelske. *Analogues of Non-Gibbsianness in Joint-measures of Disordered Mean-Field Models.* J. Stat. Phys. **112** (2003), no 5/6:1079–1108.

[18] C. Kuelske and A.C.D. van Entcr. *Two connections between random systems and non-Gibbsian measures.* J. Stat. Phys. **126** (2007): 1007–1024.

[19] C. Kuelske and A. Le Ny. *Spin-flip dynamics of the Curie-Weiss Model: loss of Gibbsianness with possibly broken symmetry.* Comm. Math. Phys. **271** (2007), 431–454.

[20] C. Kuelske, A. Le Ny and F. Redig. *Relative entropy and variational properties of generalized Gibbs measures.* Ann. Probab. **32** (2004), no 2:1691–1726.

[21] A. Le Ny and F. Redig. *Short times conservation of Gibbsianness under local stochastic evolutions.* J. Stat. Phys. **109** (2002), nos 5/6:1073–1090.

[22] C. Maes, F. Redig and A. Van Moffaert. *Almost Gibbsian versus Weakly Gibbsian.* Stoc. Proc. Appl. **79** (1999), no 1:1–15.

[23] A. Petri and M. De Oliveira. *Temperature of non-equilibrium lattice systems.* Intern. J. Mod. Phys. C **17** (2006), no 12:1703–1715.

[24] W.G. Sullivan. *Potentials for almost Markovian random fields.* Comm. Math. Phys. **33** (1976), 61–74.

Arnaud Le Ny
Laboratoire de mathématiques
Bâtiment 425
Université de Paris-Sud XI
F-91405 Orsay cedex, France
e-mail: `arnaud.leny@math.u-psud.fr`

Progress in Probability, Vol. 60, 481–495

What is the Difference Between a Square and a Triangle?

Vlada Limic and Pierre Tarrès

Dedicated to Persi Diaconis

Abstract. We offer a reader-friendly introduction to the attracting edge problem (also known as the "triangle conjecture") and its most general current solution of Limic and Tarrès (2007). Little original research is reported; rather this article "zooms in" to describe the essential characteristics of two different techniques/approaches verifying the almost sure existence of the attracting edge for the strongly edge reinforced random walk (SERRW) on a square. Both arguments extend straightforwardly to the SERRW on even cycles. Finally, we show that the case where the underlying graph is a triangle cannot be studied by a simple modification of either of the two techniques.

Mathematics Subject Classification (2000). Primary 60G50; Secondary 60J10, 60K35.

Keywords. Reinforced walk, supermartingale, time-line construction, attracting edge.

1. Introduction

We briefly describe the general setting introduced, for example, in [4]. Let $\mathcal{G}$ be a connected graph with set of vertices $V = V(\mathcal{G})$, and set of (unoriented) edges $E = E(\mathcal{G})$. The only assumption on the graph is that each vertex has at most $D(\mathcal{G})$ adjacent vertices (edges), for some $D(\mathcal{G}) < \infty$, so that $\mathcal{G}$ is of bounded degree.

Call two vertices v, v' *adjacent* ($v \sim v'$ in symbols) if there exists an edge, denoted by $\{v, v'\} = \{v', v\}$, connecting them.

Let $\mathbb{N} = \{0, 1, \ldots\}$, and let $W : \mathbb{N} \longrightarrow (0, \infty)$ be the *reinforcement weight* function. Assume we are given initial *edge weights* $X_0^e \in \mathbb{N}$ for all $e \in E$, such that $\sup_e X_0^e < \infty$. Let I_n be a V-valued random variable, recording the position of

The research of V.L. was supported in part by an Alfred P. Sloan Research Fellowship.
The research of P.T. was supported in part by a Leverhulme Prize.

the particle at time $n \in \mathbb{N}$. Set $I_0 := v_0$ for some $v_0 \in \mathcal{G}$. Let $(\mathcal{F}_n, n \geq 0)$ be the filtration generated by I.

The edge reinforced random walk (ERRW) on $\mathcal{G}$ evolves according to a random dynamics with the following properties:

(i) if currently at vertex $v \in \mathcal{G}$, in the next step the particle jumps to a nearest neighbor of v,

(ii) the probability of a jump from v to v' at time n is *"W-proportional"* to the number of previous traversals of the edge connecting v and v', that is,

$$\mathbb{P}(I_{n+1} = v' | \mathcal{F}_n) 1_{\{I_n = v\}} = \frac{W(X_n^{\{v,v'\}})}{\sum_{w \sim v} W(X_n^{\{v,w\}})} 1_{\{I_n = v \sim v'\}},$$

where X_n^e, $e \in E(\mathcal{G})$ equals

$$X_n^e = X_0^e + \sum_{k=0}^{n-1} 1_{\{\{I_k, I_{k+1}\} = e\}}.$$

We recommend a recent survey by Pemantle [7] as an excellent overview of processes with reinforcement: results, techniques, open problems and applications.

Let (H) be the following condition on W:

$$\sum_{k \in \mathbb{N}} \frac{1}{W(k)} < \infty. \tag{H}$$

We call any edge reinforced random walk corresponding to W that satisfies (H) a *strongly* edge reinforced random walk. Denote by

$$A := \{\exists\, n : \{I_k, I_{k+1}\} = \{I_{k+1}, I_{k+2}\}, k \geq n\}$$

the event that eventually the particle traverses a single (random) edge of the graph. On A we call that edge the *attracting edge.* It is easy to see that (H) is the necessary and sufficient condition for

$$\mathbb{P}(\{I_n, I_{n+1}\} = \{I_0, I_1\} \text{ for all } n) > 0.$$

This implies that (H) is necessary and sufficient for $\mathbb{P}(A) > 0$. The necessity can be seen by splitting A into a countable union of events, where each corresponds to getting attracted to a specific edge after a particular time with a specific configuration of weights on the neighbouring edges. Since A is a tail event, it seems natural to wonder whether

$$\mathbb{P}(A) = 1 \tag{A}$$

holds. The authors studied this problem in [4], and concluded that, under additional technical assumptions, (H) implies (A). In particular,

Theorem 1 ([4], Corollary 3). *If $\mathcal{G}$ has bounded degree and if W is non-decreasing, then* (H) *implies* (A).

We denote by $\mathcal{G}_l$ the *cycle* of length l, with vertices $\{0, 1, \ldots, l-1\}$ and edges

$$e_i = \{i, i+1\},\ i = 0, \ldots, l-1,$$

where $l \geq 3$, and where the addition is done modulo l.

Let us now concentrate on the case where the underlying graph $\mathcal{G}$ is the square $\mathcal{G}_4$. The next two sections demonstrate two different techniques of proving the following claim.

Theorem 2. *If $\mathcal{G} = \mathcal{G}_4$, then* (H) *implies* (A).

In fact we will concentrate on a somewhat simpler claim whose proof can be "recycled" (as we henceforth discuss) in order to arrive to the full statement of Theorem 2.

Proposition 3. *If $\mathcal{G} = \mathcal{G}_4$, then* (H) *implies*

$$\mathbb{P}(\textit{all four edges are traversed infinitely often}) = 0.$$

In Section 4 we discuss the reasons why these techniques which are well suited for $\mathcal{G} = \mathcal{G}_4$, or any graph of bounded degree without an odd cycle, cf. [9] or [3], do not extend to the setting where $\mathcal{G} = \mathcal{G}_3$ is a triangle. In fact, the following "triangle conjecture" is still open in its full generality (cf. Theorem 2 and Theorem 1) where W is a general (irregular) weight function satisfying (H):

Open Problem 4. *If $\mathcal{G} = \mathcal{G}_3$, then* (H) *implies* (A).

2. A continuous time-lines technique

This technique adapts a construction due to Rubin, and was invented by Davis [1] and Sellke [9]. It played a key role in his proof of the attracting edge property on $\mathbb{Z}^d$, and was also used by Limic [3] in order to simplify the attracting edge problem on graphs of bounded degree to the same problem on odd cycles. Denote for simplicity

$$e_i := \{i, i+1\},\ i = 0, \ldots, 3,$$

where addition is done modulo 4. For each $i = 0, \ldots, 3$ and $k \geq 1$ let E_k^i be an exponential random variable with mean $1/W(k)$, such that $\{E_k^i, i = 0, \ldots, 3, k \geq 1\}$ is a family of independent random variables. Denote by

$$T_n^i := \sum_{k=X_0^{e_i}}^{X_0^{e_i}+n} E_k^i,\ n \geq 0,\ T_\infty^i := \sum_{k=X_0^{e_i}}^{\infty} E_k^i,\ i = 0, \ldots, 3.$$

Note that the random variables T_n^i, T_∞^i, $i = 0, \ldots, 3$, $n \in \mathbb{N}$, are continuous, independent and finite almost surely (the last property is due to assumption (H)). In Figure 1, the T_n^i are shown as dots, and the "limits" T_∞^i, $i = 0, \ldots, 3$ are indicated.

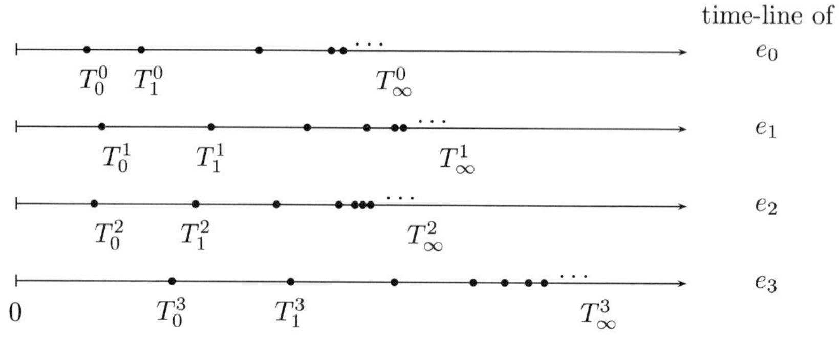

Figure 1

Here is how one can construct a realization of the edge reinforced random walk on $\mathcal{G}_4$ from the above data, or (informally) from the figure. Given the current position of the walk, simultaneously erase (at rate 1) the two time-lines of the incident edges in the chronological direction until encountering the next dot belonging to either of the time-lines. At this point, the walk steps into a new vertex by traversing the edge that corresponds to the time-line containing the dot. The procedure continues inductively.

We next explain why this construction indeed leads to a realization of the edge reinforced random walk, by considering carefully the first three steps,. Assume for concreteness that the initial position is vertex 0 incident to the edges e_0 and e_3. The time-lines of e_0 and e_3 are erased until the minimum of $T_0^0 = E^0_{X_0^{e_0}}$ and $T_0^3 = E^3_{X_0^{e_3}}$. In the figure this minimum happens to be T_0^0. Thus the particle moves from 0 to 1 (traversing edge e_0) in the first step. Due to the properties of exponentials, the probability of this move is exactly $W(X_0^{e_0})/(W(X_0^{e_0})+W(X_0^{e_3}))$. The two incident edges to the current position I_1 are now e_0 and e_1. Continue by simultaneous erasing (the previously non-erased parts of) time-lines corresponding to e_0 and e_1 until the next dot. In the figure, the dot again appears on the line of e_0. Hence the particle traverses the edge e_0 in the second step and therefore jumps back to vertex 0. Note that again the probability of this transition matches the one of the edge reinforced random walk. Continue by the simultaneous erasure of time-lines corresponding to e_0 and e_3. Based on the figure, the particle makes the third step across the edge e_3, since the (residual) length of the interval on the time-line of e_3 until T_0^3 is smaller than $T_2^0 - T_1^0 = E^1_{X_0^{e_0}+2}$. The memoryless property of the exponential distribution insures that the (residual) length of the interval until T_0^3 is again distributed as exponential (rate $W(X_0^{e_3})$) random variable, independent of all other data. Hence, the transition probability again matches that of the ERRW.

Note that the above construction can be done with any number $l \geq 3$ of time-lines (corresponding to the length l of the underlying circular graph), and we make use of this generalization in Section 4.

As a by-product of the above construction, a continuous-time version of the edge reinforced random walk emerges, where the particle makes the jumps exactly at times when the dots are encountered. More precisely, if we denote by $\widetilde{I}(s)$ the position of the particle at time s and if $\tau_0 = 0$ and $0 < \tau_1 < \tau_2 < \cdots$ are the successive jump times of $\widetilde{I}$, then the (discrete time) ERRW constructed above and the continuous-time version are coupled so that

$$I_k \equiv \widetilde{I}(\tau_k), \ k \geq 0.$$

It is worth noting that this continuous-time version is analogous to the Harris construction of a continuous-time Markov chain from the discrete one, yet it is different since the parameters of the exponential clocks vary. In particular, under assumption (H), the total time of evolution of the continuous-time random walk is finite.

Consider the total time of evolution for the continuous time walk,

$$T := \lim_{k\to\infty} \tau_k.$$

Note that at any time $s \geq 0$ the particle is incident to one of the edges e_0 and e_2, and equally it is incident to one of the edges e_1 and e_3, hence

$$\begin{aligned} T &= \sum_{i=0,\, i \text{ even}}^{3} \text{total time spent on boundary vertices of } e_i \\ &= \sum_{i=0,\, i \text{ odd}}^{3} \text{total time spent on boundary vertices of } e_i. \end{aligned}$$

Note that

$$\begin{aligned} &\{\text{all four edges are traversed infinitely often}\} && (1) \\ &= \{\text{the time-line of } e_i \text{ is erased up to time } T^i_\infty \text{ for each } i = 0, \ldots, 3\} \\ &\subset \{T = T^0_\infty + T^2_\infty = T^1_\infty + T^3_\infty\}. && (2) \end{aligned}$$

However, due to the independence and continuity of $T^0_\infty + T^2_\infty$ and $T^1_\infty + T^3_\infty$, the identity (2) happens with probability 0. We conclude that (1) happens with probability 0, and therefore that Proposition 3 holds.

In order to obtain the proof of Theorem 2 now note that there are essentially three possibilities remaining for the asymptotic evolution: the edge reinforced random walk visits infinitely often either one, or two adjacent, or three edges. In the latter two cases, there is at least one vertex j such that both edges e_{j-1} and e_j are traversed infinitely often. Moreover, after finitely many steps, every excursion from j starts and ends with the same edge. Now one can measure the time spent at site j from two perspectives: that of waiting to traverse edge e_{j-1}, and that of waiting to traverse edge e_j. The reader will quickly construct a variation to the above argument (alternatively, consult [9] or [3]) determining that a branching vertex exists with probability 0.

Note that this continuous time-lines technique still works on even cycles $\mathcal{G}_{2k}$. Indeed, given the continuous-time realization of the edge reinforced random walk constructed above, we observe that on the event that all edges are visited infinitely often,

$$T = \sum_{i=0, i \text{ even}}^{2k-1} T_\infty^i = \sum_{i=0, i \text{ odd}}^{2k-1} T_\infty^i, \tag{3}$$

where $T := \lim_{k\to\infty} \tau_k$ is the total time of evolution for the walk. As before, (3) is a consequence of the fact that T equals the total time spent on both the boundary of even and the boundary of odd edges. Now, due to independence and continuity of $\sum_{i=0,i \text{ even}}^{2k-1} T_\infty^i$ and $\sum_{i=0,i \text{ odd}}^{2k-1} T_\infty^i$, the identity (3) happens with probability 0 so that, almost surely, at least one of the edges in the cycle is visited only finitely often, and we conclude (A) as in the case of the square.

3. A martingale technique

Let, for all $n \in \mathbb{N}$,

$$W^*(n) := \sum_{k=0}^{n-1} \frac{1}{W(k)},$$

with the convention that $W^*(0) := 0$.

Assume the setting of Proposition 3. For all $n \in \mathbb{N}$, $i = 0, \ldots, 3$, define the processes

$$Y_n^\pm(i) := \sum_{k=0}^{n-1} \frac{1_{\{I_k = i, I_{k+1} = i\pm 1\}}}{W\big(X_k^{\{i,i\pm 1\}}\big)} \tag{4}$$

$$\kappa_n^i := Y_n^+(i) - Y_n^-(i) \tag{5}$$

Clearly, $\kappa_\cdot^i$ is measurable with respect to the filtration $(\mathcal{F}_n,\ n \geq 0)$. Moreover, it is easy to check that $(\kappa_n^i,\ n \geq 0)$ is a martingale : on $\{I_n = i\}$, $E(\kappa_{n+1}^i - \kappa_n^i | \mathcal{F}_n)$ is equal to

$$\frac{1}{W(X_n^{e_i})} \frac{W(X_n^{e_i})}{W(X_n^{e_i}) + W(X_n^{e_{i-1}})} - \frac{1}{W(X_n^{e_{i-1}})} \frac{W(X_n^{e_{i-1}})}{W(X_n^{e_i}) + W(X_n^{e_{i-1}})} = 0.$$

Therefore the process

$$\kappa_n := \kappa_n^0 - \kappa_n^1 + \kappa_n^2 - \kappa_n^3 + \sum_{i=0}^{3} (-1)^i W^*(X_0^{e_i}), \tag{6}$$

is also a martingale. Due to assumption (H), each of the four processes $\kappa_\cdot^i$ is a difference of bounded non-decreasing processes, and therefore has an almost sure limit as $n \to \infty$. Hence denote by κ_∞ the finite limit $\lim_{n\to\infty} \kappa_n$.

Now

$$\kappa_n = \sum_{i=0}^{3} (-1)^i W^*(X_n^{e_i}). \tag{7}$$

This implies that

$$\{\text{all four edges are traversed infinitely often}\} \subset \{\kappa_\infty = 0\},$$

so that it suffices to show

$$\mathbb{P}(\mathcal{A}_\infty) = 0, \tag{8}$$

where

$$\mathcal{A}_\infty := \{\text{all four edges are traversed infinitely often}\} \cap \{\kappa_\infty = 0\}.$$

In order to prove (8), we now analyze carefully the variance of the increments of the martingale $(\kappa_n)_{n\in\mathbb{N}}$ (decreasing to 0, due to (H)), which will enable us to prove the nonconvergence of this martingale to 0 a.s. on the event that all edges are visited infinitely often. This technique adapts an argument proving almost sure nonconvergence towards unstable points of stochastic approximation algorithms, introduced by Pemantle [6] and generalized by Tarrès [12, 13].

Fix large n, and note that

$$\begin{aligned}\mathbb{E}((\kappa_{n+1})^2 - (\kappa_n)^2|\mathcal{F}_n) &= \mathbb{E}((\kappa_{n+1} - \kappa_n)^2|\mathcal{F}_n)\\ &= \mathbb{E}\left(\sum_{i=0}^{3} \frac{1_{\{I_n=i,I_{n+1}=i+1\}}}{(W(X_n^{e_i}))^2} + \frac{1_{\{I_n=i,I_{n+1}=i-1\}}}{(W(X_n^{e_{i-1}}))^2}\middle| \mathcal{F}_n\right).\end{aligned} \tag{9}$$

From now on abbreviate

$$\alpha_n := \sum_{j=X_n^*}^{\infty} \frac{1}{(W(j))^2},$$

where $X_n^* = \min_{i=0,\dots,3} X_n^{e_i}$. For $\varepsilon > 0$, define the stopping time

$$S := \inf\{k \geqslant n : |\kappa_k| > \varepsilon\sqrt{\alpha_n}\}. \tag{10}$$

Since

$$(\kappa_S)^2 - (\kappa_n)^2 = \sum_{k=n}^{\infty}((\kappa_{k+1})^2 - (\kappa_k)^2)1_{\{S>k\}},$$

by nested conditioning we obtain

$$\mathbb{E}((\kappa_S)^2 - (\kappa_n)^2|\mathcal{F}_n) = \mathbb{E}(\sum_{k=n}^{\infty} \mathbb{E}[(\kappa_{k+1})^2 - (\kappa_k)^2|\mathcal{F}_k]1_{\{S>k\}}|\mathcal{F}_n),$$

so that, due to (9), we obtain

$$\begin{aligned}\mathbb{E}((\kappa_S)^2 - (\kappa_n)^2|\mathcal{F}_n) &= \mathbb{E}\left[\sum_{k=n}^{S-1}\sum_{i=0}^{3} \frac{1_{\{I_k=i,I_{k+1}=i+1\}}}{(W(X_k^{e_i}))^2} + \frac{1_{\{I_k=i,I_{k+1}=i-1\}}}{(W(X_k^{e_{i-1}}))^2}|\mathcal{F}_n\right]\\ &= \sum_{i=0}^{3}\mathbb{E}\left[\sum_{k=X_n^{e_i}}^{X_S^{e_i}-1} \frac{1}{W(k)^2}|\mathcal{F}_n\right] \geq \alpha_n\mathbb{P}(\mathcal{A}_\infty \cap \{S=\infty\}|\mathcal{F}_n).\end{aligned} \tag{11}$$

However, $\kappa_S = 0$ on $\{S = \infty\} \cap \mathcal{A}_\infty$, also $|\kappa_S| = |\kappa_\infty| \leqslant \varepsilon\sqrt{\alpha_n}$ on $\{S = \infty\}$ and, on $\{S < \infty\}$, $|\kappa_S| \leq (1+\varepsilon)\sqrt{\alpha_n}$ since the over(under)shoot of κ at time S is bounded by a term of the type $1/W(l)$ for some random $l \geq X_n^*$, so in particular it is bounded by $\sqrt{\alpha_n}$. Hence

$$\mathbb{E}((\kappa_S)^2|\mathcal{F}_n) \leq \mathbb{E}((\kappa_S)^2 1_{\{S<\infty\}\cup\mathcal{A}_\infty^c}|\mathcal{F}_n) \leq (1+\varepsilon)^2\alpha_n\mathbb{P}(\{S<\infty\}\cup\mathcal{A}_\infty^c|\mathcal{F}_n). \quad (12)$$

Letting $p := \mathbb{P}(\mathcal{A}_\infty \cap \{S=\infty\}|\mathcal{F}_n)$, we conclude by combining inequalities (11) and (12) that $p \leq (1+\varepsilon)^2(1-p)$, or equivalently

$$\mathbb{P}(\mathcal{A}_\infty \cap \{S=\infty\}|\mathcal{F}_n) = p \leq (1+\varepsilon)^2/(1+(1+\varepsilon)^2) < 1, \quad (13)$$

almost surely.

It will be convenient to let $\varepsilon = 5$. Then note that the shifted process $(\kappa_{S+k}, k \geq 0)$ is again a martingale with respect to the filtration $\tilde{\mathcal{F}}_k := \mathcal{F}_{S+k}$. Moreover, due to (9), we have that

$$\mathbb{E}((\kappa_\infty - \kappa_S)^2|\mathcal{F}_S) \leq 4\alpha_S \leq 4\alpha_n,$$

so that by the Markov inequality, a.s. on $\{S<\infty\}$,

$$\mathbb{P}(\mathcal{A}_\infty|\mathcal{F}_S) \leq \mathbb{P}(|\kappa_\infty - \kappa_S| > 5\sqrt{\alpha_n}|\mathcal{F}_S) \leq \frac{4\alpha_n}{25\alpha_n} = \frac{4}{25},$$

thus

$$\mathbb{P}(\mathcal{A}_\infty^c|\mathcal{F}_n) \geq \mathbb{E}[\mathbb{P}(\mathcal{A}_\infty^c|\mathcal{F}_S)1_{\{S<\infty\}}|\mathcal{F}_n] \geq \frac{21}{25}\mathbb{P}(S<\infty|\mathcal{F}_n).$$

Note that (13) now implies

$$\mathbb{P}(\mathcal{A}_\infty^c|\mathcal{F}_n)\left(1+\frac{25}{21}\right) \geq \mathbb{P}(\mathcal{A}_\infty^c|\mathcal{F}_n) + \mathbb{P}(S<\infty|\mathcal{F}_n) \geq 1-(1+\varepsilon)^2/(1+(1+\varepsilon)^2),$$

so finally

$$\mathbb{P}(\mathcal{A}_\infty^c|\mathcal{F}_n) \geq c,$$

almost surely for some constant $c > 0$. By the Lévy 0-1 law, we conclude that Proposition 3 holds.

In order to prove Theorem 2 we can proceed as in the previous section to show that no branching point is possible. In particular, we consider $i, j \in \{0, \ldots, 3\}$ such that $j \neq i, i-1$, and events of the form

$$\{e_i \text{ and } e_{i-1} \text{ both traversed i.o.}\} \cap \{e_j \text{ not visited after time } n\},$$

for some finite n, and then use an appropriate modification of $(\kappa_k^i, k \geq n)$ that would have to converge to a particular limit on the above event, and show in turn that this convergence occurs with probability 0.

Note that again this martingale technique extends in the more general setting of even cycles $\mathcal{G}_{2k}$. Indeed, let $Y_n^{\pm}(i)$ and κ_n^i be defined as in (4) and (5) and, let

$$\kappa_n := \sum_{i=0}^{2k-1}(-1)^i\kappa_n^i + \sum_{i=0}^{2k-1}(-1)^i W^*(X_0^{e_i}).$$

As in equation (7),

$$\kappa_n = \sum_{i=0}^{2k-1} (-1)^i W^*(X_n^{e_i}),$$

so that

$$\{\text{all edges are traversed infinitely often}\} \subset \{\kappa_\infty = 0\}.$$

The study of the variances of the martingale increments explained in Section 3 yields similarly that $\mathbb{P}(\{\kappa_\infty = 0\}) = 0$. Hence, almost surely, at least one of the edges in the cycle is visited only finitely often and, as before, an adaptation of this argument implies (A).

4. Comparing square and triangle

In order to additionally motivate our interest in the evolution of edge reinforced random walk on cycles, we recall that the continuous time-line technique can be adapted in order to prove that, on any graph of bounded degree, almost surely, the strongly edge reinforced random walk either satisfies (A) or it eventually keeps traversing infinitely often all the edges of a (random) odd sub-cycle. The argument was given by Limic in [3], Section 2, using graph-based techniques. The martingale method could be used in a similar way to prove the above fact. In view of this, note that solving the attracting edge problem on odd cycles is necessary and sufficient for obtaining the solution on general bounded degree graphs.

The aim of this section is to explain why the continuous time-line and martingale techniques do not extend easily to the setting where $\mathcal{G}$ is an odd cycle (e.g., a triangle).

4.1. Odd versus even in the time-line technique

The argument in the setting of even cycles relied on the existence of the non-trivial linear identity (3) involving independent continuous random variables. We are going to argue next that no such non-trivial linear relation (and in fact no non-linear smooth relation either) can hold with positive probability in the odd case.

Fix $l \geq 3$ and consider the set

$$\mathcal{X} := \{(x_k^i)_{k\in\mathbb{N}, i\in\{0,\ldots,l-1\}} : \forall i \in \{0,\ldots,l-1\}, \forall k \geq 0, x_k^i > 0, \sum_m x_m^i < \infty\}.$$

Given $\mathbf{x} = (x_k^i)_{k\in\mathbb{N}, i\in\{0,\ldots,l-1\}} \in \mathcal{X}$, define

$$t_n^i \equiv t_n^i(\mathbf{x}) := \sum_{k=0}^{n} x_k^i, n \geq 0,\ t_\infty^i \equiv t_\infty^i(\mathbf{x}) := \sum_{k=0}^{\infty} x_k^i,\ i = 0,\ldots,l-1.$$

After placing dots at points $t_0^i < t_1^i < \ldots$ on the time-line of e_i, $i = 0,\ldots,l-1$, and fixing the starting position ι_0 one can perform, as in Section 2, the time-line construction of the (now deterministic) walk, driven by $\mathbf{x}$, evolving in continuous time. If at any point the erasing procedure encounters more than one dot (on two or more different time-lines) simultaneously, choose to step over the edge corresponding to one of these time-lines in some prescribed way, for example, to the one having the smallest index. Denote by $s^i := s^i(\mathbf{x}, \iota_0)$ the total time this deterministic walk spends visiting vertex i. Similarly, denote by

$$t^{e_i} = t^{e_i}(\mathbf{x}, \iota_0) := s^i(\mathbf{x}, \iota_0) + s^{i+1}(\mathbf{x}, \iota_0) \tag{14}$$

the total time that this deterministic walk spends waiting on the boundary vertices $i, i+1$ of e_i. Of course, $t^{e_i}(\mathbf{x}, \iota_0) \leq t^i_\infty(\mathbf{x})$, where the equality holds if and only if e_i is traversed infinitely often. In the case of even l the identity

$$\sum_{j=0, j \text{ even}}^{l-1} t^{e_j}(\mathbf{x}, \iota_0) = \sum_{j=0, j \text{ odd}}^{l-1} t^{e_j}(\mathbf{x}, \iota_0), \ \mathbf{x} \in \mathcal{X},$$

lead to (3) and was the key for showing that (A) occurs. Let

$$y \equiv y(\mathbf{x}, \iota_0) := \begin{pmatrix} s^0 \\ \vdots \\ s^{l-1} \end{pmatrix}, \ z \equiv z(\mathbf{x}, \iota_0) := \begin{pmatrix} t^{e_0} \\ \vdots \\ t^{e_{l-1}} \end{pmatrix},$$

and

$$M^{(l)} := (\chi_{\{i,i+1\}}(j))_{0 \leqslant i,j \leqslant l-1},$$

where χ_B denotes a characteristic function of a set B, and the addition is done modulo l, for instance

$$M^{(5)} = \begin{bmatrix} 1 & 1 & 0 & 0 & 0 \\ 0 & 1 & 1 & 0 & 0 \\ 0 & 0 & 1 & 1 & 0 \\ 0 & 0 & 0 & 1 & 1 \\ 1 & 0 & 0 & 0 & 1 \end{bmatrix}.$$

Then (14) states that

$$z(\mathbf{x}, \iota_0) = M^{(l)} y(\mathbf{x}, \iota_0), \ \mathbf{x} \in \mathcal{X}.$$

Note that the determinant $\det(M^{(l)}) = 1 - (-1)^l$ can easily be computed explicitly since $M^{(l)}$ is a circular matrix. Hence $M^{(l)}$ is a regular matrix if and only if l is odd. Therefore, for odd l and fixed ι_0, a nontrivial identity

$$\beta \cdot z(\mathbf{x}, \iota_0) = c, \ \mathbf{x} \in \mathcal{X}, \tag{15}$$

for some $\beta \in \mathbb{R}^l \setminus \{0\}, \ c \in \mathbb{R}$, holds if and only if,

$$\beta' \cdot \mathbf{y}(\mathbf{x}, \iota_0) = c, \ \mathbf{x} \in \mathcal{X}, \tag{16}$$

where $\beta' = (M^{(l)})^\tau \beta \in \mathbb{R}^l$ is again $\neq 0$.

Now (16) cannot hold identically on $\mathcal{X}$, and we are about to show a somewhat stronger statement. Let $\mathbf{x} \in \mathcal{X}$ and fix some $r \in (0, \infty)$. Then for $j \in \{0, \dots, l-1\}$, let $\eta_r^j(\mathbf{x}) \equiv \tilde{\mathbf{x}}^{(j)} := (x_k^{i,(j)})_{k\in\mathbb{N}, i=0,\dots,l-1} \in \mathcal{X}$ be defined as follows: if $k \geq 0$, $i \in \{0, \dots, l-1\}$,

$$\tilde{x}_k^{i,(\iota_0)} := x_k^i + r\chi_{\{(\iota_0,0),(\iota_0-1,0)\}}((i,k)),$$

while for $j \neq \iota_0$, if the walk driven by $\mathbf{x}$ visits site j for the first time by traversing e_{j-1}, let

$$\tilde{x}_k^{i,(j)} := x_k^i + r\chi_{\{(j,0),(j-1,1)\}}((i,k)),$$

otherwise let

$$\tilde{x}_k^{i,(j)} := x_k^i + r\chi_{\{(j-1,0),(j,1)\}}((i,k)). \tag{17}$$

Note that (17) comprises also the case where the walk driven by $\mathbf{x}$ never visits site j.

Now we will modify the edge reinforced random walk by delaying the first jump out of a particular site j by some positive amount r, without changing anything else in the behaviour. Informally, the law of the modified version will be absolutely continuous with respect to the law of the original, and this will lead to a contradiction.

More precisely, for each fixed j, consider the two (deterministic time-continuous) walks: the original one that is driven by $\mathbf{x}$, and the new one that is driven by the transformed family $\eta_r^j(\mathbf{x})$. It is easy to check that either neither of the walks visits site j, or they both do. In the latter case, if we denote respectively by $a(\mathbf{x})$ and $a(\eta_r^j(\mathbf{x}))$ the amount of time they spend at site j before leaving, then $a(\eta_r^j(\mathbf{x})) = a(\mathbf{x}) + r$. Everything else in the evolution of the two walks is the same. In particular, if the walk driven by $\mathbf{x}$ ever visits j, then

$$s^j(\eta_r^j(\mathbf{x}), \iota_0) = s^j(\mathbf{x}, \iota_0) + r, \text{ and } s^i(\eta_r^j(\mathbf{x}), \iota_0) = s^i(\mathbf{x}, \iota_0),\ i \neq j. \tag{18}$$

Now one can simply see that if the walk driven by $\mathbf{x}$ visits all sites at least once, then any identity (16) breaks in any open neighborhood of $\mathbf{x}$ due to points $\eta_r^j(\mathbf{x})$ contained in it for sufficiently small positive r.

Recall the setting of Section 2, and to simplify the notation, assume $X_0^{e_i} = 0$, $i = 0, \dots, l-1$, and specify the initial position $\widetilde{I}_0 = v_0 \in \{0, \dots, l-1\}$. The point of the above discussion is that the random walk $\widetilde{I}$ is then the walk driven by a $\mathcal{X}$-valued random family $\mathbf{E} = (E_k^i)_{k\in\mathbb{N}, i=0,\dots,l-1}$, where the random variables E_k^i, $k \geq 0$ are independent exponentials, as specified in Section 2. If

$$A_{\text{all}} := \{\widetilde{I} \text{ visits all vertices at least once}\},$$

then $A_{\text{all}} = \{\mathbf{E} \in \mathbf{A}_{\text{all}}\}$, where $\mathbf{A}_{\text{all}}$ contains all $\mathbf{x} \in \mathcal{X}$ such that the deterministic walk driven by $\mathbf{x}$ visits all vertices at least once. It is natural to ask whether one or more (up to countably many, note that this would still be useful) identities of the form (15) hold on $\mathbf{A}_{\text{all}}$, with positive probability. For l even, we know that the answer is affirmative. For l odd, this is equivalent to asking whether one or more (up to countably many) identities of the form (16) hold on $\mathbf{A}_{\rm all}$, with positive probability. Assume

$$\mathbb{P}(A_{\rm all} \cap \{\beta' \cdot y(\mathbf{E}, v_0) = c\}) = p(\beta', c) > 0, \tag{19}$$

for β', c as in (16). Take j such that $\beta'_j \neq 0$ (at least one such j exists). We will assume that $j \neq v_0$, the argument is somewhat similar and simpler otherwise. Denote by $\mathbf{A}_1^{j,-} \subset \mathcal{X}$ the set of all $\mathbf{x}$ such that the walk driven by $\mathbf{x}$ visits j for the first time by traversing e_{j-1}, and let $A_1^{j,-} = \{\mathbf{E} \in \mathbf{A}_1^{j,-}\}$. In view of (19), without loss of generality, we may assume

$$\mathbb{P}(A_{\rm all} \cap A_1^{j,-} \cap \{\beta' \cdot y(\mathbf{E}, v_0) = c\}) \geq p(\beta', c)/2, \tag{20}$$

As a consequence of the earlier discussion in the deterministic setting we have $A_1^{j,-} = \{\mathbf{E} \in \mathbf{A}_1^{j,-}\} \subset \{\eta_r^j(\mathbf{E}) \in \mathbf{A}_1^{j,-}\}$, and

$$A_{\rm all} \cap \{\beta' \cdot y(\mathbf{E}, v_0) = c\} \subset \{\eta_r^j(\mathbf{E}) \in \mathbf{A}_{\rm all}\} \cap \{\beta' \cdot y(\eta_r^j(\mathbf{E}), v_0) = c + r\beta'_j\},$$

almost surely. Therefore,

$$\mathbb{P}(\{\eta_r^j(\mathbf{E}) \in \mathbf{A}_{\rm all} \cap \mathbf{A}_1^{j,-}\} \cap \{\beta' \cdot y(\eta_r^j(\mathbf{E}), v_0) = c + r\beta'_j\}) \geq p(\beta', c)/2. \tag{21}$$

However, E_k^i are continuous and independent random variables, each taking values in any interval $(a, b) \subset (0, \infty)$, $a < b \leq \infty$ with positive probability. Moreover, since E_0^j and E_1^{j-1} are exponential (rate $W(0)$ and $W(1)$, respectively), one can easily verify that for any cylinder set $\mathbf{B} \subset \mathcal{X}$,

$$\begin{aligned} \mathbb{P}(\eta_r^j(\mathbf{E}) \in \mathbf{B} \cap \mathbf{A}_1^{j,-}) &= \\ \mathbb{P}(\{E_i^k + r\chi_{\{(j,1),(j-1,0)\}}((i,k)), \;\; &k \geq 0, j = 0, \ldots, l-1\} \in \mathbf{B} \cap \mathbf{A}_1^{j,-}) \\ &\leq e^{r(W(0)+W(1))}\mathbb{P}(\mathbf{E} \in \mathbf{B} \cap \mathbf{A}_1^{j,-}). \end{aligned} \tag{22}$$

Now (22) and (21) imply

$$\mathbb{P}(\{\mathbf{E} \in \mathbf{A}_{\rm all}\} \cap \{\beta' \cdot y(\mathbf{E}, v_0) = c + r\beta'_j\}) \geq e^{-r(W(0)+W(1))}p(\beta', c)/2, \tag{23}$$

and this, together with (19), leads to a contradiction, since adding (23) over all rational $r \in (0, 1)$ would imply $\mathbb{P}(A_{\rm all}) = \mathbb{P}(\mathbf{E} \in \mathbf{A}_{\rm all}) = \infty$.

In the absence of a convenient linear identity (15), the reader might be tempted to look for non-linear ones. Yet, the last argument can be extended to a more generalized setting where (16) is replaced by

$$\mathbf{y}(\mathbf{x}, \iota_0) \in M, \;\; \mathbf{x} \in \mathcal{X}, \tag{24}$$

for some $l - 1$-dimensional differentiable manifold $M \subset \mathbb{R}^l$. In particular, this includes the case where $F(\mathbf{y}(\mathbf{x}, \iota_0)) = 0$, $\mathbf{x} \in \mathcal{X}$, for some smooth function F with non-trivial gradient (see, for example, [10] Theorem 5-1). Indeed, assume that, in analogy to (19),

$$\mathbb{P}(A_{\rm all} \cap \{\mathbf{y}(\mathbf{E}, v_0) \in M\}) > 0. \tag{25}$$

Then, since $\mathbf{y}(\mathbf{E}, \upsilon_0)$ is a finite random vector, due to the definition of differential manifolds (cf. [10] p. 109), there exists a point $x \in M$, two bounded open sets $U \ni x, V \subset \mathbb{R}^l$, and a diffeomorphism $h : U \to V$ such that

$$\mathbb{P}(A_{\text{all}} \cap \{\mathbf{y}(\mathbf{E}, \upsilon_0) \in M \cap U\}) =: p(M, U) > 0, \tag{26}$$

and

$$h(U \cap M) = V \cap (\mathbb{R}^{l-1} \times \{0\}) = \{\mathbf{v} \in V : v_l = 0\}.$$

Denote by $\mathbf{e}_j$ the jth coordinate vector in $\mathbb{R}^l$. Then (26) can be written as

$$\mathbb{P}(A_{\text{all}},\, \mathbf{y}(\mathbf{E}, \upsilon_0) \in U,\, h(\mathbf{y}(\mathbf{E}, \upsilon_0)) \cdot \mathbf{e}_l = 0) = p(M, U).$$

As a consequence of the Taylor decomposition, for all $j \in \{0, \ldots, l-1\}$, for any $\mathbf{u} \in U$ and for all small r,

$$h(\mathbf{u} + r\mathbf{e}_j) \cdot \mathbf{e}_l = h(\mathbf{u}) \cdot \mathbf{e}_l + r\, \mathrm{D}h(\mathbf{u})\, \mathbf{e}_j \cdot \mathbf{e}_l + \mathrm{err}(\mathbf{u}, j, r), \tag{27}$$

where for each $\mathbf{u} \in U$, $\mathrm{D}h(\mathbf{u})$ is the differential operator of h at $\mathbf{u}$, and where the error term $\mathrm{err}(\mathbf{u}, j, r) = o(r)$ as $r \to 0$. Since h is a diffeomorphism, given any $\mathbf{u} \in U$, there exists a $j \equiv j(\mathbf{u}) \in \{0, \ldots, l-1\}$ such that $\mathrm{D}h(\mathbf{u})\, \mathbf{e}_{j+1} \cdot \mathbf{e}_l \neq 0$. Therefore (26) implies that, for some $j \in \{0, \ldots, l-1\}$,

$$\mathbb{P}(A_{\text{all}},\, \mathbf{y}(\mathbf{E}, \upsilon_0) \in M \cap U,\, \mathrm{D}h(\mathbf{y}(\mathbf{E}, \upsilon_0))\, \mathbf{e}_{j+1} \cdot \mathbf{e}_l > 0) \geq \frac{p(M, U)}{2\,l}, \tag{28}$$

or

$$\mathbb{P}(A_{\text{all}},\, \mathbf{y}(\mathbf{E}, \upsilon_0) \in M \cap U,\, \mathrm{D}h(\mathbf{E}(\mathbf{x}, \iota_0))\, \mathbf{e}_{j+1} \cdot \mathbf{e}_l < 0) \geq \frac{p(M, U)}{2\,l}.$$

Without loss of generality, suppose (28) and choose $c, d \in (0, \infty)$, $c < d$, and $\delta = \delta(c) > 0$ such that

$$\begin{aligned} \mathbb{P}(A_{\text{all}},\, \mathbf{y}(\mathbf{E}, \upsilon_0) \;\in\; & M \cap U,\, \mathrm{D}h(\mathbf{y}(\mathbf{E}, \upsilon_0))\, \mathbf{e}_{j+1} \cdot \mathbf{e}_l \in (c, d), \\ & \sup_{r \in (0,\delta)} |\mathrm{err}(\mathbf{y}(\mathbf{E}, \upsilon_0), j+1, r)|/r \leq c/2) \geq \frac{p(M, U)}{4\,l}. \end{aligned} \tag{29}$$

Consider the modified processes $\eta_r^j(\mathbf{E})$, $r > 0$, corresponding to this j, and note that $\mathbf{y}(\eta_r^j(\mathbf{E}), \upsilon_0) = \mathbf{y}(\mathbf{E}, \upsilon_0) + r\, \mathbf{e}_{j+1}$. Now, due to (27) and (29), one can choose a decreasing sequence $(r_m)_{m=1}^{\infty}$ of positive numbers converging to 0, so that the intervals defined by $J(r_m) := (c\, r_m/2, d\, r_m + c\, r_m/2)$, for each $m \geq 1$, are mutually disjoint (i.e., $J(r_m) \cap J(r_{m'}) = \emptyset$ for $m < m'$) and such that

$$\mathbb{P}(\eta_{r_m}^j(\mathbf{E}) \in \mathbf{A}_{\text{all}},\, h(\mathbf{y}(\eta_{r_m}^j(\mathbf{E}), \upsilon_0)) \cdot \mathbf{e}_{j+1} \in J(r_m)) \geq \frac{p(M, U)}{4\,l},$$

hence

$$\mathbb{P}(\mathbf{E} \in \mathbf{A}_{\text{all}},\, h(\mathbf{y}(\mathbf{E}, \upsilon_0)) \cdot \mathbf{e}_{j+1} \in J(r_m)) \geq e^{-r_m(W(0)+W(1))} \frac{p(M, U)}{4\,l}.$$

As in the linear case, one arrives to a contradiction.

4.2. Odd versus even in the martingale technique

The reason why the martingale technique fails on odd cycles is similar: there is no non-trivial martingale that can be expressed as a linear combination of the different $W^*(X_n^{e_i})$, $i = 0, \ldots, l-1$, as in identity (7). Indeed, let us fix a time $n \in \mathbb{N}$ and let, for all $i \in \mathbb{Z}/l\mathbb{Z}$,

$$y_i := \mathbb{E}(Y_{n+1}^+(i) - Y_n^+(i)|\mathcal{F}_n) = \mathbb{E}(Y_{n+1}^-(i) - Y_n^-(i)|\mathcal{F}_n),$$
$$z_i := \mathbb{E}(W^*(X_{n+1}^{e_i}) - W^*(X_n^{e_i})|\mathcal{F}_n),$$

$$Y_n := \begin{pmatrix} y_0 \\ \vdots \\ y_{l-1} \end{pmatrix}, \quad Z_n := \begin{pmatrix} z_0 \\ \vdots \\ z_{l-1} \end{pmatrix}.$$

Then, for all $0 \leqslant i \leqslant l-1$,

$$z_i = y_i + y_{i+1},$$

since $W^*(X_{n+1}^{e_i}) = Y_n^+(i) + Y_n^-(i+1)$. This implies again that, almost surely,

$$Z_n = M^{(l)} Y_n, \; n \geq 1.$$

Suppose there is a fixed vector $\beta \in \mathbb{R}^l$ such that the dot product βY_n equals 0, almost surely, for all n. Since, at each time step $n \in \mathbb{N}$, Y_n has (almost surely) only one non-zero coordinate, namely, $y_i > 0$ for $i = I_n$ and $y_j = 0$ for $j \neq I_n$, and since the walk visits each and every vertex at least once with positive probability, we see that β is necessarily the null-vector. As before, if l is odd, $M^{(l)}$ is a regular matrix, and therefore no martingale can be expressed as a non-trivial deterministic linear combination of the different $W^*(X_n^{e_i})$, $i = 0, \ldots, l-1$.

However, we show in [4] that, for all $i = 0, \ldots, l-1$, if t_n^i is the nth return time to the vertex i, the process $W^*(X_{t_n^i}^{e_i}) - W^*(X_{t_n^i}^{e_{i-1}})$ approximates a martingale. The accuracy of this approximation depends on the regularity of the weight function W, hence our argument requires technical assumptions on W. In particular, the main theorem in [4] implies (A) for strongly edge reinforced random walks, where W is nondecreasing.

Even though the time-lines technique is simpler in general, one cannot adapt it similarly, since it uses the independence of random variables and is therefore unstable with respect to small perturbation.

Acknowledgment

P.T. would like to thank Christophe Sabot for an interesting discussion. We are very grateful to the referee for useful comments.

References

[1] B. Davis. Reinforced random walk. *Probab. Theo. Rel. Fields*, 84:203–229, 1990.

[2] D. Coppersmith and P. Diaconis. Random walks with reinforcement. *Unpublished manuscript*, 1986.

[3] V. Limic. Attracting edge property for a class of reinforced random walks. *Ann. Probab.*, 31:1615–1654, 2003.

[4] V. Limic and P. Tarrès. Attracting edge and stronglt edge reinforced walks. *Ann. Probab.*, 35:1783–1806, 2007.

[5] F. Merkl and S. Rolles. Edge-reinforced random walk on a ladder. *Ann. Probab.*, 33:2051–2093, 2005.

[6] R. Pemantle. Nonconvergence to unstable points in urn models and stochastic approximation. *Ann. Probab.*, 18:698–712, 1990.

[7] R. Pemantle. A survey of random processes with reinforcement. *Probab. Surv.*, 4:1–79, 2007.

[8] S. Rolles. On the recurrence of edge-reinforced random walk on Z×G. *Probab. Theo. Rel. Fields*, 135:216–264, 2006.

[9] T. Sellke. Reinforced random walks on the d-dimensional integer lattice. Technical report 94-26, Purdue University, 1994.

[10] M. Spivak. *Calculus on Manifolds*. W.A. Benjamin, Inc., 1965.

[11] P. Tarrès. VRRW on $\mathbb{Z}$ eventually gets stuck at a set of five points. *Ann. Probab.*, 32(3B):2650–2701, 2004.

[12] P. Tarrès. Pièges répulsifs. *Comptes Rendus de l'Académie des Sciences*, Sér. I Math, 330:125–130, 2000.

[13] P. Tarrès. Pièges des algorithmes stochastiques et marches aléatoires renforcées par sommets. Thèse de l'ENS Cachan, France.
Available on http://www.maths.ox.ac.uk/~tarres/.

Vlada Limic
CNRS and Université de Provence
Technopôle de Château-Gombert
39, rue F. Joliot-Curie
F-13453 Marseille cedex 13, France
e-mail: `vlada@cmi.univ-mrs.fr`

Pierre Tarrès
Mathematical Institute
University of Oxford
23–29 St Giles
Oxford OX1 3LB, United Kingdom
e-mail: `tarres@maths.ox.ac.uk`

Progress in Probability, Vol. 60, 497–525

Long-Range Dependence in Mean and Volatility: Models, Estimation and Forecasting

Sílvia R.C. Lopes

Abstract. In this paper we consider the estimation and forecasting future values of some stochastic processes exhibiting *long-range dependence*, both in mean and in volatility. We summarize basic definitions, properties and some results considering ARFIMA and SARFIMA processes, which exhibit *long memory in mean*. We proceed in the same manner considering FIGARCH and Fractionally Integrated Stochastic Volatility (FISV) processes where one can find *long memory in volatility*.

Estimation methods in parametric and semiparametric classes are presented for estimating the fractional parameter based on the classical *Ordinary Least Squares* and two robust methodologies (the *Least Trimmed Squares* and the *MM-estimation*).

An application of the SARFIMA methodology, based on the Nile River monthly flows data, is presented.

Mathematics Subject Classification (2000). Primary 60G10, 62G05, 62G35, 62M10, 62M15; Secondary 62M20.

Keywords. Long-range dependence, long memory, ARFIMA models, SARFIMA models, FIGARCH models, stochastic volatility models, semiparametric and parametric estimation, robust estimation, forecasting.

1. Introduction

Models for *long-range dependence*, or *long memory*, *in mean* were first introduced by Mandelbrot and Wallis [32], Mandelbrot and Taqqu [33], Granger and Joyeux [17] and Hosking [18], following the seminal work of Hurst [21].

We refer the reader to a recent collection of papers in Doukhan et al. [12] which reviews long-range dependence from many different angles, both theoretically and in the applied sense. We also refer the book by Palma [38] for general results on long-range dependence.

Persistence or *long-range dependence* property has been observed in time series in different areas of the science such as meteorology, astronomy, hydrology, and economics, as reported in Beran [4]. The *long-range dependence* can be characterized by two different but equivalent (see Bary [3]) forms given below, where $0.0 < d < 0.5$ is a constant:

- in time domain, the autocorrelation function $\rho_X(\cdot)$ decays hyperbolically to zero, that is, $\rho_X(k) \simeq k^{2d-1}$, when $k \to \infty$;
- in frequency domain, the spectral density function $f_X(\cdot)$ is unbounded when the frequency is near zero, that is, $f_X(w) \simeq w^{-2d}$, when $w \to 0$.

One of the models that exhibits the long-range dependence is the Autoregressive Fractionally Integrated Moving Average (ARFIMA) process. While in the ARFIMA process, the autocorrelation function shows a hyperbolic decay rate, in an ARMA process this function decays in an exponential rate.

However, in an ARFIMA process one can not capture the periodicity frequently present in some real data sets, even though still the long memory feature occurs in these data. The so-called Seasonal Autoregressive Fractionally Integrated Moving Average (SARFIMA) processes are a natural extension of the ARFIMA process. This model takes into account the seasonality inherent to such data.

The ARFIMA framework was also naturally extended towards *volatility models*. The Fractionally Integrated Generalized Autoregressive Conditionally Heteroskedastic (FIGARCH) models were introduced by Baillie, Bollerslev and Mikkelsen [2] and Bollerslev and Mikkelsen [7], motivated by the fact that autocorrelation function of the squared, log-squared, or the absolute value series of an asset return decays slowly, even when the return series has no serial correlation. In order to model long memory in the second moment, Breidt et al. [9] introduced the Fractionally Integrated Stochastic Volatility (FISV) model.

In this paper we will analyze *long-range dependence in mean* and *volatility*. We shall consider estimation and forecasting for different models.

To describe a method that generates SARFIMA processes, it is convenient to have a closed formula for the Durbin-Levinson Algorithm. This algorithm is given by recurrence relations allowing, for the partial autocorrelation function of the process, to go from lag k to lag $(k+1)$. This algorithm relates autocorrelation and partial autocorrelation functions of a process and Brietzke et al. [10] give its closed formula for SARFIMA processes.

Models for heteroskedastic time series with long memory are of great interest in econometrics and finance, where empirical facts about asset returns have motivated the several extensions of GARCH type models (for a review, see Lopes and Mendes [26]). Many empirical papers have detected the presence of long memory in the volatility of risky assets, market indexes, exchange rates. As the number of models available increases, it becomes of interest a simple, fast, and accurate estimation procedure for the fractional parameter d, independent of the specification of a parametric model. The regression based semiparametric (semiparametric in the sense that a full parametric model is not specified for the spectral density function of the process) estimators seem to be the natural candidates. However, the performance of the semiparametric estimators is greatly affected by their asymptotic statistical properties, besides depending on their definition and estimation method and is also heavily dependent on the number of frequencies $m = g(n)$ used for the regression. Lopes and Mendes [26] considers several long memory models in mean and in volatility presenting some light on the heavy dependency of the frequency number m for the semi-parametric estimation procedures.

The regression method was introduced in the pioneer work of Geweke and Porter-Hudak [16], giving rise to several other proposals. Hurvich and Ray [22] introduced a cosine-bell function as a spectral window, to reduce bias in the periodogram function. They found that data tapering and the elimination of the first periodogram ordinate in the regression equation, could increase the estimator accuracy. However, smaller bias was obtained at the cost of a larger variance. Velasco [48] also considered smoothed versions of the periodogram function while in Velasco [49] the consistency and asymptotic normality of the regression estimators was proved for any d, considering non-stationary and non-invertible processes. Reisen et al. [41] carried out an extensive simulation study comparing both the semiparametric and parametric approaches in ARFIMA processes. Monte Carlo methods were also considered by Lopes et al. [28] for non-stationary ARFIMA processes.

However, not always the ultimate interest is just the estimation of the *fractional* or *seasonal fractional parameter*, respectively, denoted by d and D. Frequently, one can also be interested in forecasting values of the processes. Reisen and Lopes [42] present some simulations and applications of forecasting ARFIMA processes while, more recently, Bisognin and Lopes [5] give an account of the estimation and forecasting issues for SARFIMA processes.

The paper is organized as follows. In Section 2 we define ARFIMA, SARFIMA, FIGARCH and FISV processes presenting their definitions together with some properties and results. Section 3 summarizes a closed formula for the Durbin-Levinson's algorithm relating the partial autocorrelation and the autocorrelation functions for the SARFIMA$(0, D, 0)$ processes. In Section 4 we present one parametric and five semi-parametric estimation procedures and their respective robust versions. In Section 5 some forecasting theory for the models presented here is summarized. In Section 6 we present an application and Section 7 contains a summary of the paper.

2. Long memory models

In this section we define both the long memory models in mean and in volatility. We present below the basic definitions, some properties and results. In this section we consider ARFIMA, SARFIMA, FIGARCH and FISV processes.

2.1. ARFIMA(p,d,q) processes

In this sub-section we define the ARFIMA process, which exhibits the *long memory in mean* characteristic.

Definition 2.1. Let $\{\varepsilon_t\}_{t\in\mathbb{Z}}$ be a white noise process with zero mean and variance $\sigma_\varepsilon^2 > 0$, $\mathcal{B}$ the *backward-shift operator*, that is, $\mathcal{B}^k(X_t) = X_{t-k}$, $\phi(\cdot)$ and $\theta(\cdot)$ polynomials of degrees p and q, respectively, given by

$$\phi(z) = \sum_{j=0}^{p}(-\phi_j)z^j \quad \text{and} \quad \theta(z) = \sum_{k=0}^{q}(-\theta_k)z^k,$$

where ϕ_j, $1 \le j \le p$, and θ_k, $1 \le k \le q$, are real constants with $\phi_0 = -1 = \theta_0$. Let $\{X_t\}_{t\in\mathbb{Z}}$ be a linear process given by

$$\phi(\mathcal{B})(1-\mathcal{B})^d(X_t-\mu) = \theta(\mathcal{B})\varepsilon_t,\ t \in \mathbb{Z}, \tag{2.1}$$

where μ is the *mean* of the process, $d \in (-0.5, 0.5)$ and $\nabla^d \equiv (1-\mathcal{B})^d$ is the *difference operator*, defined as the binomial expansion

$$(1-\mathcal{B})^d = \sum_{k=0}^{\infty}\binom{d}{k}(-\mathcal{B})^k = 1 - d\mathcal{B} - \frac{d}{2!}(1-d)\mathcal{B}^2 - \frac{d}{3!}(1-d)(2-d)\mathcal{B}^3 - \cdots, \tag{2.2}$$

for all $d \in \mathbb{R}$, with

$$\binom{d}{k} = \frac{\Gamma(1+d)}{\Gamma(1+k)\Gamma(1+d-k)},$$

and $\Gamma(\cdot)$ the Gamma function (see Brockwell and Davis [11]). Then, the process $\{X_t\}_{t\in\mathbb{Z}}$ is called a *general fractional differenced* ARFIMA(p,d,q) *process*, where d is the *degree* or *parameter of fractional differencing.*

From the expression (2.1), the process

$$U_t = (1-\mathcal{B})^d(X_t-\mu),\ t \in \mathbb{Z},$$

given by

$$\phi(\mathcal{B})U_t = \theta(\mathcal{B})\varepsilon_t,\ t \in \mathbb{Z},$$

is an *autoregressive moving average* ARMA(p,q) *process.*

If $d \in (-0.5, 0.5)$ then the process $\{X_t\}_{t\in\mathbb{Z}}$ is stationary and invertible (see Theorem 2.4) and its *spectral density function* is given by

$$f_X(w) = f_U(w)\left[2\,\sin(\frac{w}{2})\right]^{-2d}, \text{ for } 0 < w \le \pi, \tag{2.3}$$

where $f_U(\cdot)$ is the spectral density function of the ARMA(p,q) process. One observes that $f_X(w) \simeq w^{-2d}$, when $w \to 0$.

The use of the spectral density function can be seen as a practical device for determining the precise rate of decay of the autocorrelation function.

When $p = 0 = q$, in the expression (2.1), one obtain the so-called *pure* ARFIMA$(0,d,0)$ *process.*

It is commonly used the following terminology: the ARFIMA(p, d, q) process exhibits the characteristic of *long memory* when $d \in (0.0, 0.5)$, of *intermediate memory* when $d \in (-0.5, 0.0)$ and of *short memory* when $d = 0$.

If $d \geq 0.5$ the ARFIMA process is non-stationary although for $d \in [0.5, 1.0)$ it is *level-reverting* in the sense that there is no long-run impact of an innovation on the value of the process and in this case the classical estimation procedures presented in Section 4 of this work still hold (see Lopes et al. [28]; Olbermann et al. [36] and Velasco [49]). The level-reversion property no longer holds when $d \geq 1$. If $d \leq -0.5$ the ARFIMA process is non-invertible.

For more details on the properties of the ARFIMA(p, d, q) processes see, for instance, Hosking [18] and Lopes et al. [28]. We refer the reader to Sena Jr et al. [46] for an extensive Monte Carlo simulation study to evaluate the performance of some parametric and semi-parametric estimators for long and short-memory parameters of the ARFIMA(p, d, q) model with conditional heteroskedastic innovation errors (in fact, an ARFIMA-GARCH model).

2.2. SARFIMA$(p, d, q) \times (P, D, Q)_s$ processes

In many practical situations a time series can exhibit a periodic pattern. This is a common feature in fields such as meteorology, economics, hydrology and astronomy. Sometimes, even in these fields, the periodicity can depend on time, that is, the autocorrelation structure of the data varies from season to season. In our analysis, we consider the seasonality period constant over seasons. However, the periodic pattern of such kind of time series can not be described by an ARFIMA(p, d, q) process.

We shall consider the *autoregressive fractionally integrated moving average with seasonality* processes, denoted hereafter by SARFIMA$(p, d, q) \times (P, D, Q)_s$, which are an extension of the ARFIMA(p, d, q) models, proposed by Granger and Joyeux [17] and Hosking [18]. The SARFIMA processes exhibit *long-range dependence in mean* besides the *seasonality of period s*.

We shall give some definitions and some properties for the SARFIMA(p, d, q) $\times (P, D, Q)_s$ processes. We recall, however, that these properties also hold for the ARFIMA(p, d, q) process when one considers $P = Q = 0$, $D = 0$ and $s = 1$. These properties are still true for the pure version ARFIMA$(0, d, 0)$ process, when $p = 0 = q$.

Definition 2.2. Let $\{X_t\}_{t \in \mathbb{Z}}$ be a stochastic process given by the expression

$$\phi(\mathcal{B})\Phi(\mathcal{B}^s)\nabla^d\nabla_s^D(X_t - \mu) = \theta(\mathcal{B})\Theta(\mathcal{B}^s)\varepsilon_t, \quad \text{for} \quad t \in \mathbb{Z}, \tag{2.4}$$

where μ is the *mean* of the process, $\{\varepsilon_t\}_{t \in \mathbb{Z}}$ is a white noise process, $s \in \mathbb{N}$ is the seasonal period, $D, d \in (-0.5, 0.5)$, $\mathcal{B}$ is the *backward-shift operator*, that is, $\mathcal{B}^k(X_t) = X_{t-k}$, and $\mathcal{B}^{sk}(X_t) = X_{t-sk}$, ∇^d and ∇_s^D are, respectively, the *difference* and the *seasonal difference operators*, where ∇_s^D is given by

$$\nabla_s^D \equiv (1 - \mathcal{B}^s)^D = \sum_{k \geq 0} \binom{D}{k} (-\mathcal{B}^s)^k = 1 - D\mathcal{B}^s - \frac{D(1-D)}{2!}\mathcal{B}^{2s} - \cdots \tag{2.5}$$

and ∇^d is given by the expression (2.2). The polynomials $\phi(\cdot)$, $\theta(\cdot)$, $\Phi(\cdot)$, and $\Theta(\cdot)$ have degrees p, q, P, and Q, respectively, and are defined by

$$\begin{aligned}\phi(z) &= \sum_{j=0}^{p}(-\phi_j)\,z^j, \quad \theta(z) = \sum_{k=0}^{q}(-\theta_k)\,z^k,\\ \Phi(z) &= \sum_{l=0}^{P}(-\Phi_l)\,z^l, \quad \Theta(z) = \sum_{m=0}^{Q}(-\Theta_m)\,z^m,\end{aligned}$$

where ϕ_j, $1 \le j \le p$, θ_k, $1 \le k \le q$, Φ_l, $1 \le l \le P$, and Θ_m, $1 \le m \le Q$ are constants and $\phi_0 = \theta_0 = -1 = \Phi_0 = \Theta_0$. Then, $\{X_t\}_{t\in\mathbb{Z}}$ is a *seasonal fractionally integrated ARIMA*$(p,d,q)\times(P,D,Q)_s$ process with period s, denoted by SARFIMA$(p,d,q)\times(P,D,Q)_s$, where d and D are, respectively, *the degree of fractional differencing and of seasonal fractional differencing* parameters.

Remark 2.3. (a) When $P = Q = 0$, $D = 0$ and $s = 1$ the SARFIMA$(p,d,q)\times(P,D,Q)_s$ process is just the ARFIMA(p,d,q) process (see Beran [4]). In this situation it is already known the behavior of the parameter estimators and also the forecasting properties for these models (see Lopes et al. [28]; Reisen et al. [41]; and Reisen and Lopes [42]).

(b) A particular case of the SARFIMA$(p,d,q)\times(P,D,Q)_s$ process is when $p = q = P = Q = 0$. This process is called the *seasonal fractionally integrated ARIMA model with period* s, denoted by SARFIMA$(0,D,0)_s$, which will be the main goal of our expository paper. It is given by

$$\nabla_s^D(X_t-\mu) \equiv (1-\mathcal{B}^s)^D(X_t-\mu) = \varepsilon_t, \quad t\in\mathbb{Z}. \tag{2.6}$$

In what follows we shall describe some of the properties of the SARFIMA-$(0,D,0)_s$ process. We recall that these properties also hold for the ARFIMA$(0,d,0)$ process, when $D = d$ and $s = 1$ (see, for instance, Hosking [18] and [19]).

Without loss of generality, we shall consider $\mu = 0$ in expressions (2.1), (2.4) and in their pure versions. For the extensions of these properties to the complete SARFIMA process, given by Definition 2.2, we refer the reader to Bisognin [6].

Theorem 2.4. *Let $\{X_t\}_{t\in\mathbb{Z}}$ be the SARFIMA$(0,D,0)_s$ process given by the expression* (2.6)*, with zero mean and $s\in\mathbb{N}$ as the seasonal period. Then,*

(i) *when $D > -0.5$, $\{X_t\}_{t\in\mathbb{Z}}$ is an invertible process with infinite autoregressive representation given by*

$$\Pi(\mathcal{B}^s)X_t = \sum_{k\ge 0}\pi_k\mathcal{B}^{sk}(X_t) = \sum_{k\ge 0}\pi_k X_{t-sk} = \varepsilon_t,$$

where

$$\pi_k = \frac{-D(1-D)\cdots(k-D-1)}{k!} = \frac{(k-D-1)!}{k!(-D-1)!} = \frac{\Gamma(k-D)}{\Gamma(k+1)\Gamma(-D)}. \tag{2.7}$$

When $k\to\infty$, $\pi_k \sim \frac{1}{\Gamma(-D)}k^{-D-1}$.

(ii) *when* $D < 0.5$, $\{X_t\}_{t\in\mathbb{Z}}$ *is a stationary process with an infinite moving average representation given by*

$$X_t = \Psi(\mathcal{B}^s)\varepsilon_t = \sum_{k\geq 0} \psi_k \mathcal{B}^{sk}(\varepsilon_t) = \sum_{k\geq 0} \psi_k \varepsilon_{t-sk},$$

where

$$\psi_k = \frac{D(1+D)\cdots(k+D-1)}{k!} = \frac{(k+D-1)!}{k!(D-1)!} = \frac{\Gamma(k+D)}{\Gamma(k+1)\Gamma(D)}. \tag{2.8}$$

When $k \to \infty$, $\psi_k \sim \frac{1}{\Gamma(D)} k^{D-1}$.

In the following, we assume that $D \in (-0.5, 0.5)$.

(iii) *The process* $\{X_t\}_{t\in\mathbb{Z}}$ *has spectral density function given by*

$$f_X(w) = \frac{\sigma_\varepsilon^2}{2\pi}\left[2\sin\left(\frac{sw}{2}\right)\right]^{-2D}, \quad 0 < w \leq \pi. \tag{2.9}$$

At the seasonal frequencies, for $\nu = 0, 1, \ldots, \lceil s/2 \rceil$, *where* $\lceil x \rceil$ *means the integer part of* x, *it behaves as*

$$f_X\left(\frac{2\pi\nu}{s} + w\right) \sim f_\varepsilon\left(\frac{2\pi\nu}{s}\right)(sw)^{-2D}, \quad \textit{when} \quad w \to 0\cdot$$

In the following, let A *be the set* $\{1, 2, \ldots, s-1\}$, *and* $\mathbb{Z}_{\geq}$ *be the set* $\{k \in \mathbb{Z} | k \geq 0\}$.

(iv) *The process* $\{X_t\}_{t\in\mathbb{Z}}$ *has autocovariance and autocorrelation functions of order* k, $k \in \mathbb{Z}_{\geq}$, *given, respectively, by*

$$\gamma_X(sk+\xi) = \begin{cases} \frac{(-1)^k\Gamma(1-2D)}{\Gamma(1+k-D)\Gamma(1-k-D)}\sigma_\varepsilon^2 = \gamma_X(k), & \textit{if } \xi = 0 \\ 0, & \textit{if } \xi \in A, \end{cases} \tag{2.10}$$

and

$$\rho_X(sk+\xi) = \begin{cases} \frac{\Gamma(k+D)\Gamma(1-D)}{\Gamma(1+k-D)\Gamma(D)} = \rho_X(k), & \textit{if } \xi = 0 \\ 0, & \textit{if } \xi \in A. \end{cases} \tag{2.11}$$

When $k \to \infty$, $\rho_X(sk) \sim \frac{\Gamma(1-D)}{\Gamma(D)} k^{2D-1}$.

(v) *The process* $\{X_t\}_{t\in\mathbb{Z}}$ *has partial autocorrelation function given by*

$$\phi_X(sk+\xi, sl+\eta) = \begin{cases} -\binom{k}{l}\frac{\Gamma(l-D)\Gamma(k-l+1-D)}{\Gamma(-D)\Gamma(1+k-D)} = \phi_X(k,l), & \textit{if } \eta = 0 \\ 0, & \textit{if } \eta \in A, \end{cases} \tag{2.12}$$

for any $k, l \in \mathbb{Z}_{\geq}$, *and* $\xi \in A \cup \{0\}$.
From expression (2.12), *when* $k = l$, *the partial autocorrelation function of order* k *is given by*

$$\phi_X(sk, sk) = \frac{D}{k-D} = \phi_X(k,k), \quad \textit{for all} \quad k \in \mathbb{Z}_{\geq}. \tag{2.13}$$

Proof. For a proof see Brietzke et al. [10]. □

Remark 2.5. (a) The spectral density function of the SARFIMA$(0, D, 0)_s$ process in the seasonal frequencies is unbounded when $0.0 < D < 0.5$, and it has zeros when D is negative.

(b) Among seasonal frequencies the SARFIMA process has similar behavior to the ARFIMA process.

(c) The SARFIMA$(p, d, q) \times (P, D, Q)_s$ process is stationary when $d + D$ and D are less than 0.5 and the polynomials $\phi(\mathcal{B}) \cdot \Phi(\mathcal{B}^s) = 0$ and $\theta(\mathcal{B}) \cdot \Theta(\mathcal{B}^s) = 0$ have no roots in common and the roots of $\phi(\mathcal{B}) \cdot \Phi(\mathcal{B}^s) = 0$ are outside of the unit circle. When we consider all the above assumptions and also $d + D, D > 0$, then the process has *seasonal long memory*.

(d) If $\{X_t\}_{t\in\mathbb{Z}}$ is a stationary stochastic SARFIMA$(p, d, q)\times(P, D, Q)_s$ process (see expression (2.4)), with $d, D \in (-0.5, 0.5)$, its spectral density function is given by

$$f_X(w) = \frac{\sigma_\varepsilon^2}{2\pi} \frac{|\theta(e^{-iw})|^2}{|\phi(e^{-iw})|^2} \frac{|\Theta(e^{-isw})|^2}{|\Phi(e^{-isw})|^2} \left[2\sin\left(\frac{w}{2}\right)\right]^{-2d} \left[2\sin\left(\frac{sw}{2}\right)\right]^{-2D},$$

for all $0 < w \leq \pi$, where σ_ε^2 is the variance of the white noise $\{\varepsilon_t\}_{t\in\mathbb{Z}}$ process.

The following theorem shows that the stochastic process $\{X_t\}_{t\in\mathbb{Z}}$, given by expression (2.6), with seasonality $s \in \mathbb{N}$ and $D < 0.5$, is ergodic.

Theorem 2.6. *Let $\{X_t\}_{t\in\mathbb{Z}}$ be a SARFIMA$(0, D, 0)_s$ process given by expression (2.6), with zero mean, seasonal period $s \in \mathbb{N}$ and $D < 0.5$. Then, $\{X_t\}_{t\in\mathbb{Z}}$ is an ergodic process.*

Proof. Let $\{X_t\}_{t\in\mathbb{Z}}$ be a SARFIMA$(0, D, 0)_s$ process, given by expression (2.6), with zero mean and seasonal period $s \in \mathbb{N}$. Let $D < 0.5$ be the seasonal fractional differencing parameter. From item (i) in Theorem 2.4, the process $\{X_t\}_{t\in\mathbb{Z}}$ has an infinite moving average representation given by

$$X_t = \Psi(\mathcal{B}^s)\varepsilon_t = \sum_{k\geq 0} \psi_k \mathcal{B}^{sk}(\varepsilon_t) = \sum_{k\geq 0} \psi_k \varepsilon_{t-sk},$$

where the coefficients $\{\psi_k\}_{k\geq 0}$ are given by the expression (2.8) and $\{\varepsilon_t\}_{t\in\mathbb{Z}}$ is a white noise process. From item (ii) in Theorem 2.4, for $D < 0.5$, this is a stationary process. So, one has

$$\sigma_\varepsilon^2 \sum_{k\geq 0} \psi_k^2 = \gamma_X(0) = \mathbb{E}({X_t}^2) < \infty. \tag{2.14}$$

In fact,

$$\mathbb{E}({X_t}^2) = \mathbb{E}\left[\Big(\sum_{k\geq 0}\psi_k\varepsilon_{k-t}\Big)\Big(\sum_{j\geq 0}\psi_j\varepsilon_{j-t}\Big)\right] = \mathbb{E}\left[\sum_{k\geq 0}\psi_k^2\varepsilon_{k-t}^2 + \sum_{k,j\geq 0, k\neq j}\psi_k\psi_j\varepsilon_{k-t}\varepsilon_{j-t}\right]$$
$$= \sum_{k\geq 0}\psi_k^2\mathbb{E}(\varepsilon_{k-t}^2) = \sigma_\varepsilon^2\sum_{k\geq 0}\psi_k^2.$$

Hence, from expression (2.14), $\sum_{k\geq 0}\psi_k^2 < \infty$. Lemma 3.1 in Olbermann [37] proves the ergodicity for moving average processes of finite and infinite order. This lemma requires the coefficients of an infinite moving average representation to be squared absolutely summable. Therefore, one concludes that the process $\{X_t\}_{t\in\mathbb{Z}}$, given by (2.6), is ergodic. □

For general definition and properties of ergodicity in stochastic processes see Durret [13].

For SARFIMA$(0, D, 0)_s$ processes, the next theorem shows that the conditional expectation and conditional variance depend only on the past values distant from multiples of the seasonality s. This theorem is very important when one needs to generate the mentioned processes.

Theorem 2.7. *Let $\{X_t\}_{t\in\mathbb{Z}}$ be the SARFIMA$(0, D, 0)_s$ process given by the expression (2.6), with zero mean, $s \in \mathbb{N}$ as the seasonal period and $D \in (-0.5, 0.5)$. The conditional expectation and the conditional variance of X_t, given X_l, for all $l < t$, denoted respectively by $m_t \equiv \mathbb{E}(X_t|X_l, l < t)$ and $v_t \equiv \mathrm{Var}(X_t|X_l, l < t)$, are given by*

$$\begin{cases} m_\zeta = 0, \quad \text{for} \quad \zeta = 1,\ldots,s-1, \\ m_{sk} = \displaystyle\sum_{j=1}^{k} \phi_X(sk, sj)\, X_{sk-sj}, \quad \text{for} \quad k \in \mathbb{N}, \\ m_{sk+\zeta} = \displaystyle\sum_{j=1}^{k} \phi_X(sk+\zeta, sj)\, X_{sk+\zeta-sj}, \end{cases} \tag{2.15}$$

and

$$\begin{cases} v_\zeta = \sigma_\varepsilon^2, \quad \text{for} \quad \zeta = 1,\ldots,s-1, \\ v_{sk} = \sigma_\varepsilon^2 \displaystyle\prod_{j=1}^{k} (1 - \phi_X^2(sj, sj)), \quad \text{for} \quad k \in \mathbb{N}, \\ v_{sk+\zeta} = v_{sk}, \end{cases} \tag{2.16}$$

where $t = \zeta$ determines the mean and the variance for lags smaller than s, $t = sk$ for multiple lags of s, and $t = sk+\zeta$ for not multiple lags of s, $\phi_X(\cdot,\cdot)$ is the partial autocorrelation function of the process $\{X_t\}_{t\in\mathbb{Z}}$ given by item (v) *in Theorem* 2.4*, and σ_ε^2 is the variance of the white noise process.*

Proof. For a proof see Bisognin and Lopes [5]. □

2.3. FIGARCH(p, d, q) processes

In this section we shall consider one natural extension of the ARFIMA framework towards volatility models. Models for time series with *long-range dependence in volatility* are of great interest in econometrics and finance. Lopes and Mendes [26] review some extensions of the GARCH class of processes and study the performance of 300 regression type estimators for several long memory models, including

FIGARCH processes. The authors show that the performance of the semiparametric estimators are affected by their asymptotic statistical properties besides by their strong dependency on the number of frequencies used for the regression.

Denote by $\mathcal{F}_t$ the σ-field of events generated by $\{X_l; l \leq t\}$ and assume that $\mathbb{E}(X_t|\mathcal{F}_{t-1}) = 0$ *a.s.* Following Engle [14] and Bollerslev [8] we specify a GARCH(p,q) model by

$$X_t = \sigma_t Z_t, \tag{2.17}$$

where Z_t is an independent identically distributed random variable with zero mean and unit variance such that $X_t|\mathcal{F}_{t-1}$ is an independent random variable with zero mean and variance $\sigma_t^2 \equiv \mathrm{Var}(X_t|\mathcal{F}_{t-1})$ defined by

$$\sigma_t^2 = \omega + \alpha(\mathcal{B})X_t^2 + \beta(\mathcal{B})\sigma_t^2, \tag{2.18}$$

where $\omega > 0$ is a real constant, $\alpha(\mathcal{B}) = \sum_{j=1}^{p} \alpha_j \mathcal{B}^j$ and $\beta(\mathcal{B}) = \sum_{k=1}^{q} \beta_k \mathcal{B}^k$. For a FIGARCH process (see Baillie et al. [2] and Bollerslev and Mikkelsen [7]) the σ_t, in expression (2.17), is defined as

$$\begin{aligned}
\sigma_t^2 &= \omega\,(1-\beta(\mathcal{B}))^{-1} + \{1-(1-\beta(\mathcal{B}))^{-1}[1-\alpha(\mathcal{B})-\beta(\mathcal{B})](1-\mathcal{B})^d\}X_t^2 \\
&= \omega\,(1-\beta(\mathcal{B}))^{-1} + \{1-(1-\beta(\mathcal{B}))^{-1}\phi(\mathcal{B})(1-\mathcal{B})^d\}X_t^2 \\
&= \omega\,(1-\beta(\mathcal{B}))^{-1} + \lambda(\mathcal{B})X_t^2,
\end{aligned} \tag{2.19}$$

where

$$\lambda(\mathcal{B}) = \sum_{k=0}^{\infty} \lambda_k \mathcal{B}^k = 1-(1-\beta(\mathcal{B}))^{-1}\phi(\mathcal{B})(1-\mathcal{B})^d, \tag{2.20}$$

$\phi(\mathcal{B}) = 1-\alpha(\mathcal{B})-\beta(\mathcal{B}))$ and the binomial series expansion in $\mathcal{B}$, denoted by

$$(1-\mathcal{B})^d \equiv 1-\delta_d(\mathcal{B}) = 1-\sum_{k=1}^{\infty} \delta_{d,k}\,\mathcal{B}^k, \tag{2.21}$$

is given by (2.2), with $d \in [0,1]$.

The coefficients $\delta_{d,k} = d\frac{\Gamma(k-d)}{\Gamma(k+1)\Gamma(1-d)}$, in expression (2.21), are such that

$$\delta_{d,k} = \delta_{d,k-1}\left(\frac{k-1-d}{k}\right),$$

for all $k \geq 1$, where $\delta_{d,0} \equiv 1$.

The following proposition totally characterizes any FIGARCH(p,d,q) process and also gives a recurrent formula for the coefficients $\{\lambda_k\}_{k\geq 0}$ given in expression (2.20).

Proposition 2.8. *Let $\{X_t\}_{t\in\mathbb{Z}}$ be any FIGARCH(p,d,q) process, for $d \in [0,1]$, defined by expressions* (2.17) *and* (2.19). *Then, the coefficients $\{\lambda_k\}_{k\geq 0}$, in expression*

(2.20), *are given by*

$$\lambda_0 = 0$$

$$\lambda_k = \sum_{j=1}^{p} \beta_j \lambda_{k-j} + \alpha_k + \delta_{d,k} - \sum_{m=1}^{\max\{p,q\}} \gamma_m \delta_{d,k-m}, \quad \textit{if} \quad 1 \le k \le p$$

$$\lambda_k = \sum_{l=1}^{q} \beta_l \lambda_{k-l} + \delta_{d,k} - \sum_{m=1}^{\max\{p,q\}} \gamma_m \delta_{d,k-m}, \quad \textit{if} \quad k > p,$$

where

$$\gamma_m = \begin{cases} \alpha_m, & \textit{if} \quad p > q, \\ \alpha_m + \beta_m, & \textit{if} \quad p = q, \\ \beta_m, & \textit{if} \quad p < q, \end{cases}$$

with α_j, $1 \le j \le p$, *and* β_l, $1 \le l \le q$, *are given in expression* (2.18) *and* $\delta_{d,k}$, *for* $k \ge 0$, *given in expression* (2.21).

Proof. For a proof see Lopes and Mendes [26]. □

2.4. FISV(p,d,q) processes

In this section we shall consider another natural extension of the ARFIMA framework towards volatility models. We consider the Fractionally Integrated Stochastic Volatility (FISV) model, introduced by Breidt et al. [9]. Lopes and Mendes [26] also consider this model when analyzing the *long-range dependence in volatility*.

Let $\{Y_t\}_{t\in\mathbb{Z}}$ be the stochastic process such that

$$Y_t = \sigma_\epsilon \, g(X_t)\epsilon_t, \tag{2.22}$$

where X_t is a long memory in mean process, $g(\cdot)$ is a continuous function and $\{\epsilon_t\}_{t\in\mathbb{Z}}$ is a white noise process with zero mean and unit variance. Since $\mathrm{Var}(Y_t|X_t) = \sigma_\epsilon^2 \, g(X_t)^2$, for certain functions $g(\cdot)$ the process defined by (2.22) may be described as a long memory stochastic volatility process (see Robinson and Zaffaroni [43]). This large class of volatility models include the long memory nonlinear moving average models of Robinson and Zaffaroni [43] and the FISV process introduced by Breidt et al. [9].

In a FISV(p,d,q) process $\{Y_t\}_{t\in\mathbb{Z}}$, the function $g(\cdot)$ in (2.22) is given by

$$g(X_t) = \exp\left(\frac{X_t}{2}\right), \tag{2.23}$$

where $\{X_t\}_{t\in\mathbb{Z}}$ is an ARFIMA(p,d,q) process given by (2.1), and ϵ_t and ε_t are independent and identically distributed standard normal, and mutually independent. One observes that $\mathrm{Var}(Y_t|X_t) = \sigma_\epsilon^2 \exp(X_t)$. In particular, squaring both sides of equation (2.22), with the function $g(\cdot)$ given by expression (2.23), and taking logarithms,

$$\ln(Y_t^2) = \mu_\xi + X_t + \xi_t, \tag{2.24}$$

where $\mu_\xi = \ln(\sigma_\epsilon^2) + \mathbb{E}[\ln(\epsilon_t^2)]$, and $\xi_t = \ln(\epsilon_t^2) - \mathbb{E}[\ln(\epsilon_t^2)]$. Hence, $\ln(Y_t^2)$ is the sum of a Gaussian ARFIMA process and independent non-Gaussian noise with zero mean. Consequently, the autocovariance function of the process $\ln(Y_t^2)$, when $d \in (-0.5, 0.5)$, is such that

$$\gamma_{\ln(Y_t^2)}(k) \sim k^{2d-1}, \quad \text{when } k \to \infty, \tag{2.25}$$

while its spectral density function has the property that

$$f_{\ln(Y_t^2)}(w) \sim w^{-2d}, \quad \text{when } w \to 0. \tag{2.26}$$

For $d \in (0.0, 0.5)$, the spectral density function in expression (2.26) is unbounded, when $w \to 0$. This point is of great importance for the application of the traditional regression estimation procedures, based on the periodogram function, given in Section 4. Lopes and Mendes [26] also present the performance of the estimators of all parameters in FISV models when the white noise process $\{\epsilon_t\}_{t\in\mathbb{Z}}$ has standard normal distribution or t-Student distribution with 4 degrees of freedom.

3. Durbin-Levinson algorithm

The partial autocorrelation function of a process $\{X_t\}_{t\in\mathbb{Z}}$ with zero mean and autocovariance function $\gamma_X(\cdot)$, such that $\gamma_X(k) \to 0$, as $k \to 0$, is defined below. For more details, see Brockwell and Davis [11]. This definition also holds for any ARFIMA or SARFIMA processes and item (v) of Theorem 2.4, in Section 2, presents the partial autocorrelation function of SARFIMA$(0, D, 0)$ processes.

Definition 3.1. Let $\{X_t\}_{t\in\mathbb{Z}}$ be a stochastic process with zero mean and autocovariance function $\gamma_X(\cdot)$ such that $\gamma_X(k) \to 0$, as $k \to 0$. The *partial autocorrelation function*, denoted by $\phi_X(k, j)$, $k \in \mathbb{Z}_{\geqslant}$ and $j = 1, \ldots, k$, are the coefficients in the equation

$$\mathcal{P}_{\overline{sp}(X_1,X_2,\ldots,X_k)}(X_{k+1}) = \sum_{j=1}^{k} \phi_X(k,j) X_{k+1-j},$$

where $\mathcal{P}_{\overline{sp}(X_1,X_2,\ldots,X_k)}(X_{k+1})$ is the orthogonal projection of X_{k+1} in the closed span $\overline{sp}(X_1, X_2, \ldots, X_k)$ generated by the previous observations. Then, from the equations

$$\langle X_{k+1} - \mathcal{P}_{\overline{sp}(X_1,X_2,\ldots,X_k)}(X_{k+1}), X_j \rangle = 0, \;\; j = 1, \ldots, k,$$

where $\langle \cdot, \cdot \rangle$ defines the internal product on the Hilbert space $\mathcal{L}^2(\Omega, \mathcal{A}, \mathbb{P})$ given by $\langle X, Y \rangle = \mathbb{E}(XY)$, we obtain

$$\begin{bmatrix} 1 & \rho_X(1) & \rho_X(2) & \cdots & \rho_X(k-1) \\ \rho_X(1) & 1 & \rho_X(1) & \cdots & \rho_X(k-2) \\ \vdots & \vdots & \vdots & \vdots & \vdots \\ \rho_X(k-1) & \rho_X(k-2) & \rho_X(k-3) & \cdots & 1 \end{bmatrix} \begin{bmatrix} \phi_X(k,1) \\ \phi_X(k,2) \\ \vdots \\ \phi_X(k,k) \end{bmatrix} = \begin{bmatrix} \rho_X(1) \\ \rho_X(2) \\ \vdots \\ \rho_X(k) \end{bmatrix}, \tag{3.1}$$

with $\rho_X(\cdot)$ the autocorrelation function of the process $\{X_t\}_{t\in\mathbb{Z}}$. The coefficients $\phi_X(k, j)$, $k \in \mathbb{Z}_{\geqslant}$, $j = 1, \cdots, k$, are uniquely determined by (3.1).

The definition of partial autocorrelation function plays an important role in the Durbin-Levinson algorithm (see expressions (3.2) and (3.3) below) and its expression for seasonal fractionally integrated processes is given in item (v) of Theorem 2.4.

Brietzke et al. [10] give a closed formula for the Durbin-Levinson Algorithm for the partial autocorrelation function, defined by the expression (2.12), for seasonal fractionally integrated processes. This is a crucial algorithm and a summary of its description is given as follows.

Let $\{X_t\}_{t\in\mathbb{Z}}$ be a SARFIMA$(0, D, 0)_s$ process, given in expression (2.6), with mean μ equal to zero. We want to show that its partial autocorrelation function $\phi_X(\cdot,\cdot)$, given in item (v) of Theorem 2.4, satisfies the following systems

$$\phi_X(sl, sl) = \frac{\rho_X(sl) - \sum_{j=1}^{sl-1} \phi_X(sl-1, j)\rho_X(sl-j)}{1 - \sum_{j=1}^{sl-1} \phi_X(sl-1, j)\rho_X(j)} \tag{3.2}$$

and

$$\phi_X(k+1, sl) = \phi_X(k, sl) - \phi_X(k+1, k+1)\phi_X(k, k+1-sl), \tag{3.3}$$

for any $l \in \mathbb{Z}_{\geqslant}$ such that $sl < k+1$, where $k+1$ may or may not be a multiple of s, where $\rho_X(\cdot)$ is given in item (iv) of Theorem 2.4.

Recurrence relations (3.2) and (3.3) are known as the Durbin-Levinson algorithm and they explain how to go from lag k to lag $(k+1)$ for the partial autocorrelation function $\phi_X(\cdot,\cdot)$. Brietzke et al. [10] prove the recurrence relation (3.2)–(3.3) for any $D \in (-0.5, 0.5)$, with $D \neq 0$.

Lemma 3.2. *Let $\{X_t\}_{t\in\mathbb{Z}}$ be a process given by (2.6). For any $k, l \in \mathbb{Z}_{\geqslant}$, the partial autocorrelation function of $\{X_t\}_{t\in\mathbb{Z}}$, denoted by $\phi_X(\cdot,\cdot)$, satisfies the system given in (3.3), whenever $l < k+1$.*

Proof. For a proof see Brietzke et al. [10]. □

Lemma 3.3. *Let $\{X_t\}_{t\in\mathbb{Z}}$ be a process given by (2.4), where $D \in (-0.5, 0.5)$ with $D \neq 0$. Then, the quotient in expression (3.2) is given by*

$$\frac{\rho_X(l) - \sum_{j=1}^{l-1} \phi_X(l-1, j)\rho_X(l-j)}{1 - \sum_{j=1}^{l-1} \phi_X(l-1, j)\rho_X(j)} = \frac{\displaystyle\sum_{j=0}^{l-1} \binom{l-1}{j} \frac{\Gamma(j-D)\Gamma(l-j-D)\Gamma(l-j+D)}{\Gamma(l-j-D+1)}}{\displaystyle\sum_{j=0}^{l-1} \binom{l-1}{j} \frac{\Gamma(j-D)\Gamma(l-j-D)\Gamma(j+D)}{\Gamma(j-D+1)}}. \tag{3.4}$$

Proof. For a proof see Brietzke et al. [10]. □

We still need to show that (3.4) is equal to $\phi_X(l, l)$. This follows from Theorem 3.7 below. One can show that the numerator of the left-hand side of expression

(3.4) times $(l - D)$ (or its denominator times D) is equal to $\phi_X(l, l)$, that is,

$$(l - D)\sum_{j=0}^{l-1}\binom{l-1}{j}\frac{\Gamma(j-D)\Gamma(l-j-D)\Gamma(l-j+D)}{\Gamma(l-j-D+1)}$$
$$= D\sum_{j=0}^{l-1}\binom{l-1}{j}\frac{\Gamma(j-D)\Gamma(l-j-D)\Gamma(j+D)}{\Gamma(j-D+1)}. \tag{3.5}$$

Moreover, one can also show that

$$(l - D)\sum_{j=0}^{l-1}\binom{l-1}{j}\frac{\Gamma(j-D)\Gamma(l-j-D)\Gamma(l-j+D)}{\Gamma(l-j-D+1)}$$
$$= D\Gamma(-D)\Gamma(D-l+1)(l-1)!\,2^{l-1}\cdot\prod_{i=0}^{l-2}\left(D-\frac{i+1}{2}\right). \tag{3.6}$$

The equalities (3.5) and (3.6) follow, respectively, from Corollaries 3.10 and 3.8, below. The system (3.2) follows immediately from expressions (3.5) and (2.13). The Durbin-Levinson algorithm is a consequence of Theorem 2.4 and equality (3.6) above. We shall first define the *hypergeometric function*.

Definition 3.4. If a_i, b_i and x are complex numbers, with $b_i \notin \mathbb{Z}_{\leqslant}$, we define the *hypergeometric function* by

$${}_3F_2\big(a_1, a_2, a_3; b_1, b_2; x\big) = \sum_{n=0}^{\infty}\frac{\big(a_1\big)_n\big(a_2\big)_n\big(a_3\big)_n}{\big(b_1\big)_n\big(b_2\big)_n}\frac{x^n}{n!},$$

where $(a)_n$ stands for the *Pochhammer symbol*

$$(a)_n = \frac{\Gamma(a+n)}{\Gamma(a)} = \begin{cases} a(a+1)\cdots(a+n-1), & \text{if } n \geq 1 \\ 1, & \text{if } n = 0. \end{cases}$$

This series is absolutely convergent for all $x \in \mathbb{C}$ such that $|x| < 1$, and also for $|x| = 1$, provided $\Re\big(b_1 + b_2\big) > \Re\big(a_1 + a_2 + a_3\big)$, where $\Re(z)$ means the real part of $z \in \mathbb{C}$. Furthermore, it is said to be *balanced* if $b_1 + b_2 = 1 + a_1 + a_2 + a_3$. Note that in case some a_i is a nonpositive integer the above sum is finite and it suffices to let n range from 0 to $-a_i$.

The following identity for a *terminating balanced hypergeometric sum* is very useful. For the identity's proof we refer the reader to Andrews et al. [1], Thm. 2.2.6, page 69.

Theorem 3.5 (Identity of Pfaff–Saalschütz). *Let $k \in \mathbb{Z}_{\geqslant}$, and a, b, and c be complex numbers such that $c, 1 + a + b - c - k \notin \mathbb{Z}_{\leqslant}$. Then,*

$${}_3F_2\big(-k, a, b; c, 1+a+b-c-k; 1\big) = \frac{\big(c-a\big)_k\big(c-b\big)_k}{\big(c\big)_k\big(c-a-b\big)_k}. \tag{3.7}$$

Remark 3.6. If $\left(c_n\right)_{n\geq 0}$ is a sequence of complex numbers satisfying

$$\frac{c_{n+1}}{c_n} = \frac{(a_1+n)(a_2+n)(a_3+n)x}{(n+1)(b_1+n)(b_2+n)} \text{ for all } n,$$

straightforward computations show that

$$\sum_{n=0}^{\infty} c_n = c_0 \cdot {}_3F_2\big(a_1, a_2, a_3; b_1, b_2; x\big)\,. \tag{3.8}$$

The identity (3.7) and the above remark are fundamental for the proof of the following theorem.

Theorem 3.7. *Let x and z be complex numbers, with $x \notin \mathbb{Z}$ and $z \notin \mathbb{Z}_{\geqslant}$. Then,*

$$\sum_{j=0}^{l-1} \binom{l-1}{j} \frac{\Gamma(j-x)\,\Gamma(l-j+x)}{z-j} = \frac{\Gamma(-x)\,\Gamma(1+x)\,\Gamma(1-z)}{z\,\Gamma(l-z)} \cdot (l-1)! \prod_{i=1}^{l-1} (x-z+i)\,. \tag{3.9}$$

For $z \in \{l, l+1, \ldots\}$ the right-hand side of expression (3.9) *has a removable singularity and by analytic continuation the result is still true.*

Proof. For a proof see Brietzke et al. [10]. □

Corollary 3.8. *If $l \in \mathbb{N} - \{1\}$ and D is a noninteger complex number, then*

$$(l-D) \sum_{j=0}^{l-1} \binom{l-1}{j} \frac{\Gamma(j-D)\,\Gamma(l+D-j)}{l-D-j}$$
$$= D\,\Gamma(-D)\,\Gamma(D-l+1)\,(l-1)!\,2^{l-1} \cdot \prod_{i=0}^{l-2} \left(D - \frac{i+1}{2}\right). \tag{3.10}$$

Corollary 3.9. *If $l \in \mathbb{N} - \{1\}$ and D is a noninteger complex number, then*

$$D \sum_{k=0}^{l-1} \binom{l-1}{k} \frac{\Gamma(l-1-k+D)\,\Gamma(k-D+1)}{l-1-k-D}$$
$$= D\,\Gamma(-D)\,\Gamma(D-l+1)\,(l-1)!\,2^{l-1} \cdot \prod_{i=0}^{l-2} \left(D - \frac{i+1}{2}\right). \tag{3.11}$$

Corollary 3.10. *If $l \in \mathbb{N}$ and D is a noninteger complex number, then*

$$(l-D) \sum_{j=0}^{l-1} \binom{l-1}{j} \frac{\Gamma(j-D)\,\Gamma(l+D-j)}{l-D-j}$$
$$= D \sum_{k=0}^{l-1} \binom{l-1}{k} \frac{\Gamma(l-1-k+D)\,\Gamma(k-D+1)}{l-1-k-D}. \tag{3.12}$$

4. Classical and robust estimation procedures

In the literature of the stochastic long memory processes, there exist several estimation procedures for the fractional parameter d. We now summarize some of these estimation procedures both for long memory models in mean and in volatility: here we present one parametric and four semi-parametric methods. For these methods we also consider their robust versions. For a non-parametric method based on wavelet theory applied to the fractional parameter estimation, in ARFIMA processes, we refer the reader to Lopes and Pinheiro [25]. We also refer the reader to Olbermann et al. [35] for another work where a non-parametric method based on wavelet theory is used to estimate the hyperbolic rate decay parameter for the autocorrelation function of Manneville-Pomeau processes.

The methodology in this section will be presented based on ARFIMA(p, d, q) processes which are the simplest among all the others analyzed in Section 2.

We recall that when $\{Y_t\}_{t\in\mathbb{Z}}$ follows a FISV process with $d \in (-0.5, 0.5)$, $\ln(Y_t^2)$ is the sum of a zero mean Gaussian ARFIMA process and an independent non-Gaussian innovation process. Also, the FIGARCH(p, d, q) process, with $d \in [0, 1]$, has been defined in expression (8) of Baillie et al. [2] as an ARFIMA process on the squared data with a more complicated error structure. Thus, the regression based methods described below also apply to the other processes considered in Section 2.3 and 2.4.

In this section we summarize six methods for the estimation of the fractional differencing parameter:

- The semi-parametric regression method based on the periodogram function proposed by Geweke and Porter-Hudak [16]. This estimator is denoted hereafter by GPH;
- The semi-parametric regression method based on the smoothed periodogram when one considers the Bartlett lag window. This estimator is denoted hereafter by BA;
- The semi-parametric regression method based on GPH with trimming l and bandwidth $g(n)$ proposed by Robinson [44]. This estimator is denoted here by R;
- The semi-parametric method based on the partial sum process proposed by Mandelbrot and Taqqu [33], based on Hurst [21] estimator. This estimator is largely known as the R/S statistics;
- The cosine-bell tapered data method, denoted in the sequel by $GPHT$, considers the cosine-bell function as a transformation of the data and follows similarly to the GPH method. It was proposed by Hurvich and Ray [22];
- The parametric approximated maximum likelihood method, proposed by Fox and Taqqu [15], based on the approximation given by Whittle [50], is denoted hereafter by W.

Let $\{X_t\}_{t\in\mathbb{Z}}$ be a ARFIMA(p,d,q) process with $d \in (-0.5, 0.5)$, given by (2.1). Its spectral density function is given by

$$f_X(w) = f_U(w)\left[2\,\sin\left(\frac{w}{2}\right)\right]^{-2d}, \text{ for } 0 < w \leq \pi, \tag{4.1}$$

where $f_U(\cdot)$ is the spectral density function of the ARMA process.

Consider the set of harmonic frequencies $w_j = \frac{2\pi j}{n}$, $j = 0, 1, \ldots, \lceil n/2 \rceil$, where n is the sample size and $\lceil x \rceil$ means the integer part of x. By taking the logarithm of the spectral density function $f_X(\cdot)$ given by (4.1), and adding $\ln(f_U(0))$, and $\ln(I(w_j))$ to both sides of this expression we obtain

$$\ln(I(w_j)) = \ln(f_U(0)) - d\ln\left[2\,\sin\left(\frac{w_j}{2}\right)\right]^2 + \ln\left\{\frac{f_U(w_j)}{f_U(0)}\right\} + \ln\left\{\frac{I(w_j)}{f_X(w_j)}\right\}, \tag{4.2}$$

where $I(\cdot)$ is the periodogram function given by

$$I(w) = \frac{1}{2\pi}\left(\hat{\gamma}_X(0) + 2\sum_{l=1}^{n-1}\hat{\gamma}_X(l)\cos(l\,w)\right), \tag{4.3}$$

with $\hat{\gamma}_X(h) = \frac{1}{n}\sum_{k=1}^{n-h}(x_k - \bar{x})(x_{k+h} - \bar{x})$, for $h \in \{0, 1, \ldots, n-1\}$, is the sample autocovariance function and $\bar{x} = \frac{1}{n}\sum_{k=1}^{n} x_k$ is the sample mean of the process $\{X_t\}_{t\in\mathbb{Z}}$ in (2.1).

When considering only the frequencies close to zero, the term $\ln\{\frac{f_U(w_j)}{f_U(0)}\}$ may be discarded. Then, we may rewrite (4.2) in the context of a simple linear regression model

$$y_j = a - d\,x_j + e_j,\ j = 1, \ldots, m, \tag{4.4}$$

where $m = g(n) = n^{\alpha}$, for $0 < \alpha < 1$, $(a, -d)$ are the regression coefficients, $a = \ln(f_U(0))$, $y_j = \ln(I(w_j))$, $x_j = \ln\{2\sin(w_j/2)\}^2$ and the errors $e_j = \ln\{\frac{I(w_j)}{f_X(w_j)}\}$ are uncorrelated random variables centered at zero with constant variance.

A semi-parametric regression estimator may be obtained by minimizing some loss function of the residuals $r_j = y_j - \hat{a} + \hat{d}\,x_j$. We will consider three different loss functions. They give rise to the classical *Ordinary Least Squares* method (OLS), and two high breakdown point robust methods, the *Least Trimmed Squares* method (LTS), and the MM-*estimation* method.

The OLS estimators are the values $(\hat{a}, -\hat{d})$ which minimize the loss function

$$L_1(m) = \sum_{j=1}^{m}(r_j)^2, \tag{4.5}$$

where $r_j = y_j - \hat{a} + \hat{d}\,x_j$ is the residual related to the regression (4.4).

Whenever the errors e_i follow a normal distribution, the OLS estimates have the minimum variance among all unbiased estimates. In fact, it is well known (see

Huber [20]) that regression outliers, leverage points, and gross errors are responsible for considerable bias and inefficiency (even in the Gaussian environment) in the OLS estimates.

Robust alternatives to OLS may be obtained by minimizing a robust version of the dispersion of the residuals. The *Least Trimmed Squares* (LTS) estimates of Rousseeuw [45] minimize the loss function

$$L_2(m) = \sum_{j=1}^{m^*} (r^2)_{j:m} \,, \tag{4.6}$$

where $(r^2)_{j:m}$ are the squared and then ordered residuals, that is, $(r^2)_{1:m} \leq \cdots \leq (r^2)_{m^*:m}$, and m^* is the number of points used in the optimization procedure. The constant m^* is responsible both for the breakdown point value and the efficiency. When m^* is approximately $m/2$ the breakdown point is approximately 50%. The LTS estimates have been previously used by Taqqu et al. [47] for the estimation of the long-range parameter in ARFIMA models and by Lopes and Mendes [26] for the estimation of long-range parameter both in mean and in volatility models.

The *MM-estimates* (see Yohai [51]) may present simultaneously high breakdown point and high efficiency. They are defined as the solution $(\hat{a}, -\hat{d})$ which minimizes the loss function

$$L_3(m) = \sum_{j=1}^{m} \rho_2 \left(\frac{r_j}{\kappa}\right)^2 \,, \tag{4.7}$$

subject to the constraint

$$\frac{1}{m}\sum_{j=1}^{m} \rho_1 (\frac{r_j}{\kappa}) \leq b \,, \tag{4.8}$$

where ρ_2 and ρ_1 are symmetric, bounded, nondecreasing functions on $[0, \infty)$ with $\rho_j(0) = 0$ and $\lim_{u \to \infty} \rho_j(u) = 1$, for $j = 1, 2$, κ is a scale parameter, and b is a tuning constant. The breakdown point of the MM-estimator only depends on ρ_1 and it is given by $\min(b, 1 - b)$.

4.1. Classical and robust GPH estimators

The first estimation method based on the periodogram function was introduced in the pioneer work of Geweke and Porter-Hudak [16].

Let $\{X_t\}_{t \in \mathbb{Z}}$ be a ARFIMA(p, d, q) process with $d \in (-0.5, 0.5)$, given by the expression (2.1). From the linear regression given by (4.4) the classical *GPH-LS* estimator of d is then given by

$$GPH - LS = -\frac{\sum_{j=1}^{g(n)} (x_j - \bar{x})(y_j - \bar{y})}{\sum_{j=1}^{g(n)} (x_j - \bar{x})^2} \,, \tag{4.9}$$

where the trimming value $g(n)$ is usually $g(n) = n^{\alpha}$, for $0 < \alpha < 1$, y_j is based on (4.3) and x_j is as previously defined in expression (4.4). Lopes et al. [28] considered α in the interval $[0.55, 0.65]$, and Porter-Hudak [40] considered $\alpha \in \{0.62, 0.75\}$ for the case of seasonal fractionally integrated time series data. Lopes and Mendes [26] consider $\alpha \in \{0.50, 0.52, \ldots, 0.84, 0.86\}$. The version not tunned by α, that is, based on the $\lceil \frac{n}{2} \rceil$ data points, equivalent to set $\alpha = 0.8997$, was also considered in Lopes and Mendes [26].

To obtain the robust versions of the *GPH* estimator we just apply the *LTS* and the *MM* methodologies to the regression model (4.4) with $m = n^{\alpha}$, based on (4.3). This gives rise to the *GPH-LTS* and the *GPH-MM* estimators.

4.2. Classical and robust *BA* estimators

The periodogram function is not a consistent estimator for the spectral density function (see Brockwell and Davis [11]). Lopes and Lopes [29] analyzes the convergence in distribution sense for the periodogram function based on a time series of a stationary process. This process is obtained from the iterations of a continuous transformation invariant for an ergodic probability. In this later work, the authors only assume a certain rate of convergence to zero for the autocovariance function of the stochastic process, that is, it is assumed that there exist $C > 0$ and $\xi > 2$ such that $|\gamma_X(k)| \leq C|k|^{-\xi}$, for all $k \in \mathbb{Z}$, where $\gamma_X(\cdot)$ is the autocovariance function of the process. This result can be applied to a time series obtained from the iteration of a certain class of deterministic transformations (or its natural extension; see, for instance, Lopes and Lopes [30]) whose initial point is distributed according to an ergodic probability.

Returning to the general setting, by considering the Bartlett lag window, a consistent estimator for the spectral density function may be obtained. This smoothed version of the periodogram function is defined by

$$I_{\text{smooth}}(w) = \frac{1}{2\pi} \sum_{j=-\nu}^{\nu} \kappa\left(\frac{j}{\nu}\right) \hat{\gamma}_X(j) \cos(j\, w), \tag{4.10}$$

where $\kappa(\cdot)$ is the Bartlett lag window given by

$$\kappa(x) = \begin{cases} 1 - |x|, & \text{if } |x| \leq 1 \\ 0, & \text{otherwise}, \end{cases} \tag{4.11}$$

with ν being the truncation point of the weighted function.

The classical and robust versions are obtained by applying the *OLS*, the *LTS* and the *MM* methodologies to the regression model (4.4) based on (4.10) and (4.11), producing the *BA-LS*, the *BA-LTS*, and the *BA-MM* estimators.

4.3. Classical and robust *R* estimators

The regression estimator *R*, proposed by Robinson [44] is obtained by applying the Ordinary Least Squares method in (4.4) based on (4.3), but considering only the frequencies $j \in \{l, l+1, \ldots, g(n)\}$, where $l > 1$ is a trimming value that tends to infinity more slowly than $g(n)$.

It is interesting to compare the R and the LTS concepts. The R concept trims the extreme x_j values associated with the frequencies close to zero, which we know are the important ones. On the other hand, the LTS concept trims the extreme ordered residuals which may or may be not associated to small frequencies, but certainly are associated to leverage points. In other words, the LTS procedure identifies which data points associated with small frequencies are outliers and, if they exist, excludes them from the calculations. The R-LTS and R-MM versions are obtained by applying the robust methodologies, as previously.

4.4. Classical and robust R/S estimators

Various methods for estimating the self-similarity parameter H or the intensity of long-range dependence in a time series are available, some of which are described in detail in Beran [4]. In a pioneer work by Mandelbrot and van Ness [31] the authors describe the self-similarity parameter H through the fractional Brownian motion processes. The R/S statistics, the so-called *rescaled adjusted range*, was firstly considered by Mandelbrot and Taqqu [33], based on Hurst [21] estimator. The self-similarity parameter H is related to the fractional parameter d by the equation $d = H + \frac{1}{2}$. Lo [24] proposes the use of the R/S statistics with a different normalization that makes the estimator more robust to some form of short-range dependence. It is based on the range of the partial sum process $S_k = \sum_{j=1}^{k}(X_j - \bar{X}_n)$ and it is defined by

$$R/S(q) = \frac{\max_{1\leq k\leq n} S_k - \min_{1\leq k\leq n} S_k}{\hat{\sigma}(q)}, \tag{4.12}$$

where $\bar{X}_n = \frac{1}{n}\sum_{j=1}^{n} X_j$, $\hat{\sigma}^2(q) = \hat{\gamma}_S(0) + 2\sum_{j=1}^{q}\omega_j(q)\hat{\gamma}_S(j)$, the sample autocovariances $\gamma_S(h) = \frac{1}{n}\sum_{l=1}^{n-h}(S_l - \bar{S}_n)(S_{l+h} - \bar{S}_n)$, for $0 \leq h < n$, account for the possible short-range dependence up to the qth order and the weights $\omega_j(q) = 1 - \frac{1}{q+1}$ correspond to the Bartlett window.

4.5. Classical and robust $GPHT$ estimators

The $GPHT$ method (see Hurvich and Ray [22] and Velasco [49]) uses a modified periodogram function given by

$$I(w_j) = \frac{1}{\sum_{t=0}^{n-1} g(t)^2}\left|\sum_{t=0}^{n-1} g(t) X_t e^{-i\,w_j\,t}\right|^2, \tag{4.13}$$

where the tapered data is obtained from the cosine-bell function $g(\cdot)$ defined by

$$g(t) = \frac{1}{2}\left[1 - \cos\left(\frac{2\pi(t+0.5)}{n}\right)\right]. \tag{4.14}$$

We obtain the classical $GPHT$-LS and the robust versions $GPHT$-LTS and $GPHT$-MM by applying the classical and the robust methodologies on model (4.4) based on (4.13) and (4.14), and setting $m = n^{\alpha}$.

4.6. Classical W estimator

The W estimator was proposed by Whittle [50]. He considered the function

$$Q(\boldsymbol{\eta}) = \int_{-\pi}^{\pi} \frac{I(w)}{f_X(w;\boldsymbol{\eta})} dw,$$

where $\boldsymbol{\eta}$ denotes the vector of unknown parameters, and $f_X(\cdot\,;\boldsymbol{\eta})$ is the spectral density function of $\{X_t\}_{t\in\mathbb{Z}}$, given by (4.1) and $I(\cdot)$ is the periodogram function given by (4.3).

The W estimator is the value of $\boldsymbol{\eta}$ which minimizes the function $Q(\cdot)$. Here $\boldsymbol{\eta}$ is the vector $(\phi_1, \ldots, \phi_p, d, \sigma_\varepsilon, \theta_1, \ldots, \theta_q)$. The estimation procedure is carried out by finding the value $\hat{\boldsymbol{\eta}}$ which minimizes

$$\mathcal{B}_n(\boldsymbol{\eta}) = \sum_{j=1}^{\lceil \frac{n-1}{2} \rceil} \frac{I(w_j)}{f_X(w_j;\boldsymbol{\eta})}. \tag{4.15}$$

More details of this estimator can be found in Fox and Taqqu [15]. Differently from the previous four estimators, the W estimator is in the parametric class.

We point out that the estimation procedures considered for the ARFIMA processes in this section can be easily extended to the stochastic volatility models given in Sections 2.3 and 2.4.

5. Forecasting in long memory processes

Let $\{X_t\}_{t\in\mathbb{Z}}$ be a SARFIMA$(0, D, 0)_s$ process with $D \in (-0.5, 0.5)$, given by the expression (2.6). Suppose one wants to forecast the value X_{t+h} for h-step-ahead. The *minimum mean squared error forecasting value* is given by

$$\widehat{X}_t(h) \equiv \mathbb{E}\,(X_{t+h}\,|\mathcal{F}_t), \tag{5.1}$$

where $\mathcal{F}_t$ is the σ-field of events generated by $\{X_\ell; \ell \leqslant t\}$. This minimizes the *mean squared error of forecasting* $\mathbb{E}(X_{t+h} - \widehat{X}_t(h))$. In this case, the *forecasting error* is given by

$$e_t(h) = X_{t+h} - \widehat{X}_t(h). \tag{5.2}$$

To calculate the forecasting values one uses the following facts:

(a) $\mathbb{E}(X_{t+h}|\mathcal{F}_t) = \begin{cases} X_{t+h}, & \text{if } \ h \leqslant 0, \\ \widehat{X}_t(h), & \text{if } \ h > 0; \end{cases}$

(b) $\mathbb{E}(\varepsilon_{t+h}|\mathcal{F}_t) = \begin{cases} \varepsilon_{t+h}, & \text{if } \ h \leqslant 0, \\ 0, & \text{if } \ h > 0. \end{cases}$

Therefore, to calculate the forecasting values, one

(a) substitutes the past expectations ($h \leqslant 0$) for known values, X_{t+h} and ε_{t+h};

(b) substitutes the future expectations ($h > 0$) for forecasting values $\widehat{X}_t(h)$ and 0.

The following theorem presents some results for forecasting a future value of a SARFIMA$(0, D, 0)_s$ process, given by the expression (2.6).

Theorem 5.1. *Let $\{X_t\}_{t\in\mathbb{Z}}$ be a SARFIMA$(0, D, 0)_s$ process, with zero mean and seasonality $s \in \mathbb{N}$, given in expression (2.6). Consider $D > -0.5$. Then, for all $h \in \mathbb{N}$:*

(i) *The minimum mean squared error forecasting value is given by*

$$\widehat{X}_n(h) = -\sum_{k\geq 0} \pi_k \, \widehat{X}_n(h - sk), \tag{5.3}$$

where π_k is given in expression (2.7).

(ii) *The forecasting error is given by $e_n(h) = \sum_{k=0}^{\lceil \frac{h}{s} \rceil - 1} \psi_k \, \varepsilon_{n+h-sk}$, where ψ_k is given by expression (2.8).*

(iii) *The theoretical and sample variances of the forecast error are given, respectively, by*

$$\mathrm{Var}(e_n(h)) = \sigma_\varepsilon^2 \sum_{k=0}^{\lceil \frac{h}{s} \rceil - 1} \psi_k^2, \quad \text{and} \quad \widehat{Var}(e_n(h)) = \widehat{\sigma}_\varepsilon^2 \sum_{k=0}^{\lceil \frac{h}{s} \rceil - 1} \widehat{\psi}_k^2,$$

where $\widehat{\psi}_k$ is given by expression (2.8) when D is replaced by one of its estimated values, through some of the estimation procedures proposed in Section 4.

(iv) *The* bias *and the percentage bias to estimate the theoretical variance of the forecasting error are given by*

$$\mathrm{bias}(h) = \widehat{Var}(e_n(h)) - \mathrm{Var}(e_n(h)) \qquad \text{and}$$

$$\mathrm{perbias}(h) = \frac{|\widehat{Var}(e_n(h)) - \mathrm{Var}(e_n(h))|}{\mathrm{Var}(e_n(h))} \times 100\ \%.$$

(v) *The mean squared error of forecasting is given by $msef_n = \frac{1}{h}\sum_{k=1}^{h} (e_n(k))^2$.*

(vi) *Moreover, if the process $\{\varepsilon_t\}_{t\in\mathbb{Z}}$ is such that $\varepsilon_t \sim \mathcal{N}(0, \sigma_\varepsilon^2)$, for any $t \in \mathbb{Z}$, then an $100(1-\gamma)\%$ confidence interval for X_{n+h} is given by*

$$\widehat{X}_n(h) - z_{\frac{\gamma}{2}} \widehat{\sigma}_\varepsilon \left[\sum_{k=0}^{\lceil \frac{h}{s} \rceil - 1} \widehat{\psi}_k^2 \right]^{\frac{1}{2}} \leqslant X_{n+h} \leqslant \widehat{X}_n(h) + z_{\frac{\gamma}{2}} \widehat{\sigma}_\varepsilon \left[\sum_{k=0}^{\lceil \frac{h}{s} \rceil - 1} \widehat{\psi}_k^2 \right]^{\frac{1}{2}},$$

where $z_{\frac{\gamma}{2}}$ is the value such that $\mathbb{P}(Z \geqslant z_{\frac{\gamma}{2}}) = \frac{\gamma}{2}$, with $Z \sim \mathcal{N}(0,1)$, and $\widehat{\psi}_k$ is given by the above item (iii).

Proof. For a proof see Bisognin and Lopes [5]. □

We point out that a similar result to Theorem 5.1 can be stated for ARFIMA processes.

6. An application

In this section we analyze an observed time series data, and also a simulated seasonal fractionally integrated ARMA time series. Our goal is to give an application of the SARFIMA methodology, analyzing these two time series in order to detect whether seasonal long memory is present in these data.

In Section 6.1 we analyze an observed time series as an application to the SARFIMA$(p, d, q) \times (P, D, Q)_s$ process. As it is not easy to find observed examples modeled by the pure SARFIMA$(0, D, 0)_s$ process, we simulate a time series and analyzed it in Section 6.2.

6.1. Nile River monthly flows data

We consider the time series reporting the Nile River monthly flows at Aswan, kindly provided by A. Montanari (for the graphic of the data we refer the reader to Montanari et al. [34]). This time series consists of 1,466 observations, from August of 1872 to September of 1994, and it is approximately a Gaussian time series.

Figures 6.1 (a) and (b) present, respectively, the sample autocorrelation function and the periodogram function for the Nile River flows at Aswan. From these figures one can see long memory features for this time series, since its sample autocorrelation has a slowly hyperbolic decay, and its periodogram function exhibits periodic pattern caused by an annual cycle. Figure 6.1 (b) shows the peaks on the Fourier frequencies w_j, where $j = \left[\frac{n}{s}\right] i = \left[\frac{1{,}466}{12}\right] i = 122i$, for $i = 0, 1, \ldots, 6$. These features are also reported in Montanari et al. [34].

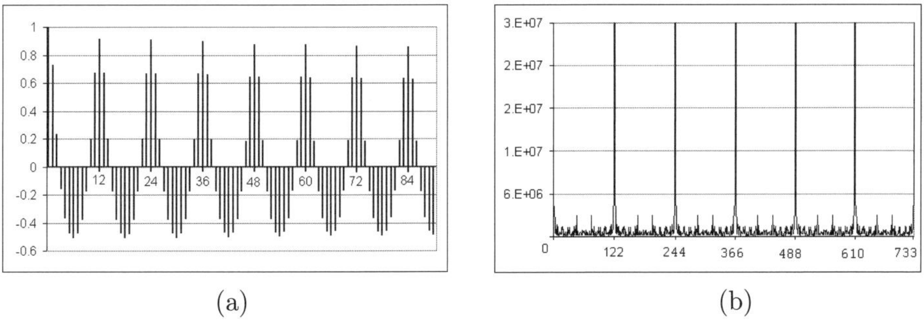

FIGURE 6.1. Nile River Monthly Flows Data at Aswan: (a) sample autocorrelation function; (b) periodogram function.

The model that best fits the original data is a SARFIMA$(p, d, q) \times (P, D, Q)_s$ with $p = q = P = Q = 1$, $d = 0$, $D = W$ and $s = 12$ (for a complete analysis, we refer the reader to Bisognin and Lopes [5]). The long memory parameter is estimated by the approximated maximum likelihood method proposed by Fox and Taqqu [15] (see Section 4), with $W = 0.1980$.

Table 6.1: Estimated Values of D for: (a) Nile River Monthly Flows Data; (b) Simulated Time Series Data.

SARFIMA$(0, D, 0)_s$ with $s = 12$ and $\alpha = 0.55$						
Estimator	GPH	BA	R	R/S	$GPHT$	W
(a) Nile River Monthly Flows Data	0.2399	0.3126	0.2381	0.1638	0.4196	0.1980
(b) Simulated Time Series Data	0.4219	0.4216	0.4398	0.3893	0.4185	0.3834

Table 6.1 (a) gives the estimation results for this time series with seasonality $s = 12$, since Figures 6.1 (a) and (b) exhibits this periodic pattern. All the semi-parametric estimation procedures select the number of regressors $m = g(n)$, in the expression (4.4), from the first seasonal frequency, no matter what value one uses for s.

Table 6.2 gives the estimators and its standard deviation (denoted here by Std. Dev.) values for the parameters in the SARFIMA$(p, d, q) \times (P, D, Q)_s$ model, that best fitted the Nile River monthly flows data at Aswan.

The residual analysis was also performed for the fitted model and it indicates that the errors are approximately Gaussian white noise.

Table 6.2: Fitted Model for the Nile River Flows Data.

SARFIMA$(p, d, q) \times (P, D, Q)_s$ with $p = q = P = Q = 1$, $d = 0$, $D = W$ and $s = 12$					
	ϕ_1	Φ_1	D	θ_1	Θ_1
Estimator	0.6147	0.9944	0.1980	-0.2238	0.9207
Std. Dev.	0.0291	0.0295	0.0011	0.0357	0.0145

6.2. Simulated time series

Here we consider a complete estimation, and also the forecasting analysis, for a simulated seasonal fractionally integrated time series as in expression (2.6), when $n = 1,466$, $D = 0.4$, and $s = 12$.

Figures 6.2 (a), and (b) show the sample autocorrelation, and the periodogram functions of this simulated time series: there exist long memory characteristics in this time series. By analyzing the periodogram function we also observe a periodic pattern with seasonality $s = 12$.

Table 6.1 (b) gives the estimators of the parameter D for a SARFIMA$(0, D, 0)_s$ with $s = 12$, that best fits the simulated time series.

The best estimator for the simulated time series is $W = 0.3834 \simeq 0.4$. In the semi-parametric estimator class the total number of regressors $m = g(n)$, in the expression (4.4), was selected from the first seasonal frequency.

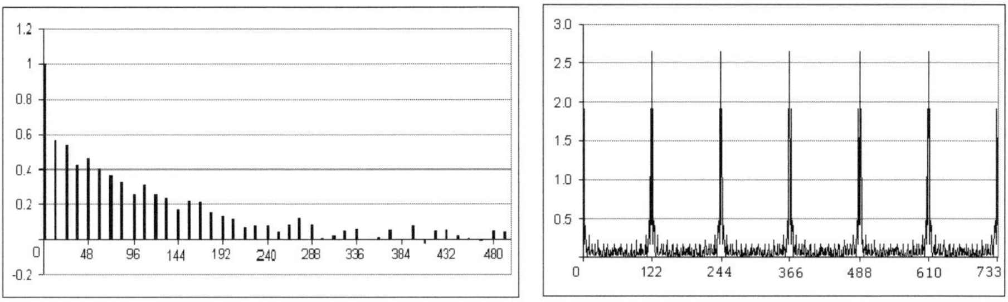

(a) (b)

FIGURE 6.2. Simulated Time Series Data: (a) sample autocorrelation function; (b) periodogram function.

Figure 6.3 shows the confidence interval at 95% confidence level for the 5-step ahead forecasting values based on all estimation procedures considered in Section 4 for the simulated time series data.

We refer the reader to Lopes and Nunes [27] and Pinheiro and Lopes [39] for a long-range dependence studies in DNA sequences. For a self-similar analysis in the Ethernet traffic we refer the work by Leland et al. [23]. We also mention Beran [4], Doukhan et al. [12] and Palma [38] for a series of examples and applications on *long-range dependence*.

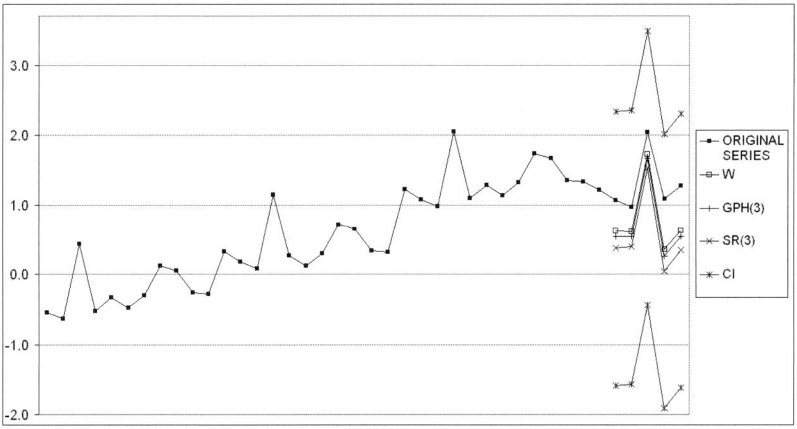

FIGURE 6.3. Confidence interval at 95% confidence level for the 5-step ahead forecasting in the simulated time series data.

7. Conclusions

In this paper we addressed the issue of modeling the *long-range dependence* through a series of different stochastic processes.

We considered models with *long memory in mean* (ARFIMA and SARFIMA processes) and *in volatility* (FIGARCH and FISV processes), with innovations following either a Gaussian or a non-Gaussian distribution.

In studying SARFIMA$(0, D, 0)_s$ processes we emphasize Theorems 2.1–2.3 and 5.1. Theorem 2.2 presents the ergodicity property while Theorem 2.3 presents the conditional expectation, and conditional variance for these processes. Theorem 2.3 is very important for generating any SARFIMA$(0, D, 0)_s$ process or its complete version SARFIMA$(p, d, q) \times (P, D, Q)_s$ process. Theorem 5.1 gives some properties for forecasting the value X_{n+h}, when $h \geqslant 1$, in SARFIMA$(0, D, 0)_s$ processes.

Based on the Pfaff–Saalschütz's Identity and some properties of the hypergeometric functions, we derived a compact and closed formula for the Durbin-Levinson algorithm in order to obtain the partial autocorrelation functions of order k for SARFIMA$(0, D, 0)_s$ processes.

In Section 4 we presented one parametric and four semiparametric methods for estimating the fractional differencing parameter. The classical *Ordinary Least Squares* (*OLS*) method and two robust methodologies (that is, the *Least Trimmed Squares* (*LTS*) and the *MM-estimation*) were presented for each estimation method in the semiparametric class.

We recall that in a FISV process the logarithm transformation of its squared value is the sum of a zero mean Gaussian ARFIMA process and an independent non-Gaussian innovation process. Also, the FIGARCH(p, d, q) process is an ARFIMA process on the squared data with a more complicated error structure. In view of this, all the estimation methods proposed in Section 4, or in any of the author's papers mentioned in the references, can also be applied to FIGARCH and FISV processes besides the ARFIMA process.

As an illustration of the SARFIMA methodology we presented an application in hydrology.

Acknowledgements

S.R.C. Lopes was partially supported by CNPq-Brazil, by *Millennium Institute in Probability*, by Edital Universal *Modelos com Dependência de Longo Alcance: Análise Probabilística e Inferência* (CNPq-No. 476781/2004-3) and also by *Fundação de Amparo à Pesquisa no Estado do Rio Grande do Sul* (FAPERGS Foundation).

The author would like to thank an anonymous referee and both editors for their valuable comments and suggestions that improved the final version of the manuscript.

References

[1] G.E. Andrews, R. Askey and R. Roy, *Special Functions, Encyclopedia of Mathematics and its Applications.* Cambridge University Press, 1999.

[2] R.T. Baillie, T. Bollerslev and H.O. Mikkelsen, *Fractionally Integrated Generalized Autoregressive Conditional Heteroskedasticity.* J. of Econometrics **74** (1996), 3–30.

[3] N.K. Bary, *A Treatise on Trigonometric series.* Pergamon Press, 1964.

[4] J. Beran, *Statistics for Long-Memory Process.* Chapman and Hall, 1994.

[5] C. Bisognin and S.R.C. Lopes, *Estimating and Forecasting the Long Memory Parameter in the Presence of Periodicity.* J. of Forecasting **26** (2007), 405–427.

[6] C. Bisognin, *Estimação e Previsão em Processos SARFIMA*$(p,d,q) \times (P,D,Q)_s$ *na Presença de Outliers.* Ph.D. Thesis in the Mathematics Graduate Program. Federal University of Rio Grande do Sul, Porto Alegre, 2007. URL Address: www.mat.ufrgs.br/~slopes.

[7] T. Bollerslev and H.O. Mikkelsen, *Modeling and pricing long memory in stock market volatility.* J. of Econometrics **73** (1996), 151–184.

[8] T. Bollerslev, *Generalized Autoregressive Conditional Heteroskedasticity.* J. of Econometrics **31** (1986), 307–327.

[9] F.J. Breidt, N. Crato and P. de Lima, *The detection and estimation of long memory in stochastic volatility.* J. of Econometrics **83** (1998), 325–348.

[10] E.H.M. Brietzke, S.R.C. Lopes and C. Bisognin, *A Closed Formula for the Durbin-Levinson's Algorithm in Seasonal Fractionally Integrated Processes.* Mathematical and Computer Modelling **42** (2005), 1191–1206.

[11] P.J. Brockwell and R.A. Davis, *Time Series: Theory and Methods.* Springer-Verlag, 1991.

[12] P. Doukhan, G. Oppenheim and M.S. Taqqu (eds.) *Theory and Applications of Long-Range Dependence.* 1st Edition, Birkhäuser, 2003.

[13] R. Durret, *Probability: Theory and Examples.* Duxbury Press, 2004.

[14] R.F. Engle, *Autoregressive conditional heteroskedasticity with estimates of the variance of U.K. inflation.* Econometrica **50** (1982), 987–1008.

[15] R. Fox and M.S. Taqqu, *Large-sample properties of parameter estimates for strongly dependent stationary Gaussian time series.* Annals of Statistics **14** (1986), 517–532.

[16] J. Geweke and S. Porter-Hudak, *The Estimation and Application of Long Memory Time Series Model.* J. of Time Series Analysis **4** (1983), 221–238.

[17] C.W.J. Granger and R. Joyeux, *An Introduction to Long Memory Time Series Models and Fractional Differencing.* J. of Time Series Analysis **1** (1980), 15–29.

[18] J. Hosking, *Fractional Differencing.* Biometrika **68** (1981), 165–167.

[19] J. Hosking, *Modelling Persistence in Hydrological Time Series using Fractional Differencing.* Water Resources Research **68** (1984), 1898–1908.

[20] P.J. Huber, *Robust Statistics.* John Wiley, 1981.

[21] H.R. Hurst, *Long-term storage in reservoirs.* Trans. Am. Soc. Civil Eng. **116** (1951), 770–799.

[22] C.M. Hurvich and B.K. Ray, *Estimation of the memory parameter for nonstationary or noninvertible fractionally integrated processes.* J. of Time Series Analysis **16** (1995), 017–041.

[23] W.E. Leland, M.S. Taqqu, W. Willinger and D.V. Wilson, *On the self-similar nature of Ethernet traffic (Extended version).* IEEE/ACM Transactions on Networking **2** (1994), 01–15.

[24] A.W. Lo, *Long-term memory in stock market prices.* Econometrica **59** (1991), 1279–1313.

[25] S.R.C. Lopes and A. Pinheiro, *Wavelets for Estimating the Fractional Parameter in Non-stationary ARFIMA Process.* In revision, 2007.

[26] S.R.C. Lopes and B.V.M. Mendes, *Bandwidth Selection in Classical and Robust Estimation of Long Memory.* International Journal of Statistics and Systems **1** (2006), 177–200.

[27] S.R.C. Lopes and M.A. Nunes, *Long Memory Analysis in DNA Sequences.* Physica A: Statistical Mechanics and its Applications **361** (2006), 569–588.

[28] S.R.C. Lopes, B.P. Olbermann and V.A. Reisen, *A Comparison of Estimation Methods in Non-stationary Arfima Processes.* J. of Statistical Computation and Simulation **74** (2004), 339–347.

[29] A.O. Lopes and S.R.C. Lopes, *Convergence in Distribution of the Periodogram for Chaotic Processes.* Stochastics and Dynamics **2** (2002), 609–624.

[30] A.O. Lopes and S.R.C. Lopes, *Parametric Estimation and Spectral Analysis of Piecewise Linear Maps of the Interval.* Advances in Applied Probability **30** (1998), 757–776.

[31] B.B. Mandelbrot and J. van Ness, *Fractional Brownian Motion, Fractional Noises and Applications.* S.I.A.M. Review **10** (1968), 422–437.

[32] B.B. Mandelbrot and J.R. Wallis, *Computer Experiments with Fractional Gaussian Noises. Part* 1, *Averages and Variances.* Water Resources Research **5** (1969), 228–267.

[33] B.B. Mandelbrot and M.S. Taqqu, *Robust R/S analysis of long-run serial correlation.* In Proceedings of the 42nd Session of the International Statistical Institute, Manila. *Bulletin of the International Statistical Institute* **48** (1979), Book 2, 69–104.

[34] A. Montanari, R. Rosso and M.S. Taqqu, *A seasonal fractional ARIMA Model applied to the Nile River monthly flows at Aswan.* Water Resources Research **36** (2000), 1249–1259.

[35] B.P. Olbermann, S.R.C. Lopes and A.O. Lopes, *Parameter Estimation in Manneville-Pomeau Processes.* Submitted, 2007.

[36] B.P. Olbermann, S.R.C. Lopes and V.A. Reisen, *Invariance of the First Difference in ARFIMA Models.* Computational Statistics **21** (2006), 445–461.

[37] B.P. Olbermann, *Estimação em Classes de Processos Estocásticos com Decaimento Hiperbólico da Função de Autocorrelação.* Ph.D. Thesis in the Mathematics Graduate Program. Federal University of Rio Grande do Sul, Porto Alegre, 2002. URL Address: www.mat.ufrgs.br/∼slopes.

[38] W. Palma, *Long-memory Time Series: Theory and Methods.* 1st Edition, John Wiley, 2007.

[39] A. Pinheiro and S.R.C. Lopes, *The Use of Wavelets for Studying Long-Range Dependence in DNA Sequences.* Manuscript in preparation, 2007.

[40] S. Porter-Hudak, *An Application of the Seasonal Fractionally Differenced Model to the Monetary Aggregates.* J. of American Statistical Association **85** (1990), 338–344.

[41] V.A. Reisen, B. Abraham and S.R.C. Lopes, *Estimation of Parameters in ARFIMA Processes: A Simulation Study.* Communications in Statistics: Simulation and Computation **30** (2001), 787–803.

[42] V.A. Reisen and S.R.C. Lopes, *Some Simulations and Applications of Forecasting Long Memory Time Series Models.* J. of Statistical and Planning Inference **80** (1999), 269–287.

[43] P.M. Robinson and P. Zaffaroni, *Nonlinear Time Series with Long Memory: a Model for Stochastic Volatility.* J. of Statistical Planning and Inference **68** (1998), 359–371.

[44] P.M. Robinson, *Log-periodogram regression of time series with long range dependence.* The Annals of Statistics **23** (1995), 1048–1072.

[45] P.J. Rousseeuw, *Least Median of Squares Regression.* J. of the American Statistical Association **79** (1984), 871–880.

[46] M.R. Sena Jr, V.A. Reisen and S.R.C. Lopes, *Correlated Errors in the Parameters Estimation of ARFIMA Model: A Simulated Study.* Communications in Statistics: Simulation and Computation **35** (2006), 789–802.

[47] M.S. Taqqu, V. Teverovsky and W. Willinger, *Estimators for long range dependence: an empirical study.* Fractals **3** (1995), 785–798.

[48] C. Velasco, *Gaussian Semiparametric Estimation of Non-stationary Time Series.* J. of Time Series Analysis **20** (1999a), 87–127.

[49] C. Velasco, *Non-stationary log-periodogram regression.* J. of Econometrics **91** (1999b), 325–371.

[50] P. Whittle, *Estimation and information in stationary time series.* Arkiv för Mathematik **2** (1953), 423–434.

[51] V.J. Yohai, *High Breakdown point and high efficiency robust estimates for regression.* Annals of Statistics **15** (1987), 642–656.

Sílvia R.C. Lopes
Universidade Federal do Rio Grande do Sul
Instituto de Matemática
Av. Bento Gonçalves, 9500
91509-900 Porto Alegre - RS
Brazil
e-mail: `silvia.lopes@ufrgs.br`

Progress in Probability, Vol. 60, 527–542

Percolation in the Sherrington-Kirkpatrick Spin Glass

J. Machta, C.M. Newman and D.L. Stein

Abstract. We present extended versions and give detailed proofs of results about percolation (using various sets of two-replica bond occupation variables) in Sherrington-Kirkpatrick spin glasses (with zero external field) that were first given in an earlier paper by the same authors. We also explain how ultrametricity is manifested by the densities of large percolating clusters. Our main theorems concern the connection between these densities and the usual spin overlap distribution. Their corollaries are that the ordered spin glass phase is characterized by a unique percolating cluster of maximal density (normally coexisting with a second cluster of nonzero but lower density). The proofs involve comparison inequalities between SK multireplica bond occupation variables and the independent variables of standard Erdős-Rényi random graphs.

Mathematics Subject Classification (2000). Primary 82B44, 60K35; Secondary 05C80, 82B43, 60K37.

Keywords. Spin glass, percolation, Sherrington-Kirkpatrick model, Fortuin-Kasteleyn, random graphs.

1. Introduction

In Ising ferromagnets (with no external field), it is well known that the ordered (broken symmetry) phase manifests itself within the associated Fortuin-Kasteleyn (FK) random cluster representation [1] by the occurrence of a single positive density percolating cluster (see [2]). In a recent paper [3], we investigated the nature of spin glass ordering within the FK and other graphical representations and concluded that the percolation signature of the spin glass phase is the presence of a single *two-replica* percolating network of *maximal* density, which typically coexists with a second percolating network of lower density. The evidence presented in that paper for this conclusion was two-fold: suggestive numerical results in the case of the three-dimensional Edwards-Anderson (EA) spin glass [4] and rigorous results for the Sherrington-Kirkpatrick (SK) spin glass [5].

In this paper, we expand on those results for the SK model in several ways. First, we give much more detailed proofs, both for two-replica FK (TRFK) percolation and for the different percolation of "blue" bonds in the two-replica graphical representation studied earlier by Chayes, Machta and Redner [6, 7] (CMR). Second, we go beyond the $\pm J$ SK model (as treated in [3]) to handle quite general choices of the underlying distribution ρ for the individual coupling variables, including the usual Gaussian case. Third, we organize the results (see in particular Theorems 1 and 2) in such a way as to separate out (see Theorems 4 and 5) those properties of the overlap distribution for the supercritical SK model that are needed to prove related properties about percolation structure. Such a separation is called for because many properties of the overlap distribution that are believed to be valid based on the Parisi ansatz (see [8]) for the SK model have not yet been rigorously proved.

Another way we expand on the results of the previous paper is to present (in Section 3) an analysis of the percolation signature of ultrametricity in the SK model, which is expected to occur [9], but has not yet been proved rigorously. That is, we describe (see Theorems 6, 7 and 8) how the percolation cluster structure of multiple networks of differing densities in the context of three replicas would exhibit ultrametricity. We note that as a spinoff of Theorem 8, we have (at least in the SK model – numerical investigations for the EA model have not yet been done) a third graphical percolation signature of the spin glass transition beyond the two analyzed in our earlier work – namely one involving uniqueness of the maximum density percolating network (one out of *four* clusters) in a three replica mixed CMR-FK representation.

In addition to these extensions of our earlier results and as we note in a remark at the end of this introductory section, the technical machinery we develop in this paper can be used to obtain other results for the SK model, such as an analysis of large cluster densities at or near the critical point. Before getting to that, we first give an outline of the other sections of the paper.

In Section 2 we describe the SK models and the CMR and TRFK percolation occupation variables we will be dealing with throughout. We then present our main results, starting with Theorems 1 and 2 which relate, in the limit $N \to \infty$, the densities of the largest percolation clusters to the overlap distribution for the CMR and TRFK representations respectively. Then, after stating a known basic result (Theorem 3) about the vanishing of the overlap in the subcritical (and critical) SK model, we present Theorems 4 and 5, which give respectively the CMR and TRFK percolation signatures of the SK phase transition, under various assumptions about the SK overlap distribution. Results relating percolation structure and ultrametricity in the SK model are presented in Section 3. Then in Section 4, we present all the proofs. That section begins with three lemmas that are the technical heart of the paper and explain why one can compare, via two-sided stochastic domination inequalities, SK percolation occupation variables to the independent variables of Erdős-Rényi random graphs [10]. A key feature of these comparison results is that they are done only *after conditioning on the values of the spin variables in all the replicas being considered*. This feature helps explain why the size of the overlap is crucial – because it determines the sizes of the various "sectors" of vertices (e.g., those where the spins in two replicas agree or those where they disagree) within which the comparisons can be made.

Remark 1.1. *The first part of Theorem 5 says that for $\beta \leq 1$, the size of the largest TRFK doubly occupied cluster is $o(N)$ or equivalently that its density $D_1^{\mathrm{TRFK}} = o(1)$ as $N \to \infty$. But the proof (see the TRFK part of Lemma 4.2) combined with known results about $G(N, p_N)$, the Erdős-Rényi random graph with N vertices and independent edge occupation probability p_N, implies quite a bit more – that D_1^{TRFK} is $O(\log N/N)$ for $\beta < 1$ and $O(N^{2/3}/N)$ for $\beta = 1$. Even more is implied in the critical case. E.g., in the critical scaling window, where $\beta = \beta_N = 1 + \lambda/N^{1/3}$, the largest clusters, of size proportional to $N^{2/3}$, behave exactly like those occurring in a pair of independent random graphs (see, e.g., [11, 12, 13, 14]) – i.e., as $N \to \infty$, the limiting distribution of $(N^{1/3}D_1^{\mathrm{TRFK}}, N^{1/3}D_2^{\mathrm{TRFK}}, \ldots)$ is the same as that obtained by taking two independent copies of $G(N/2, 2(\beta_N)^2/N)$, combining the sizes of the largest clusters in the two copies, then rank ordering them and dividing by $N^{2/3}$. One can also show that for $\beta > 1$, the size of the third largest TRFK cluster behaves like that of the second largest cluster in a single copy of the supercritical Erdős-Rényi random graph – i.e., $O(\log N)$ [10]. But that derivation requires a strengthened version of Lemma 4.2 and further arguments, which will not be presented in this paper.*

2. Main results

Before stating the main results, we specify the random variables we will be dealing with. For specificity, we choose a specific probabilistic coupling so that even though we deal with two different graphical representations, and a range of inverse temperatures β, we define all our random variables for the system of positive integer size N on a single probability space. The corresponding probability measure will be denoted $\mathbf{P}_N$ (with $\mathbf{P}$ denoting probability more generically).

For each N, we have three types of random variables: real-valued couplings $\{J_{ij}\}$ (with $1 \leq i < j \leq N$), $\{\sigma_i\}_{1 \leq i \leq N}$ and $\{\tau_i\}_{1 \leq i \leq N}$ for each of two replicas, and a variety of percolation $\{0,1\}$-valued bond occupation variables which we will define below. These random variables and their joint distributions depend on both N and β (although we have suppressed that in our notation), but to define them, we rely on other sets of real-valued random variables not depending on N or β: $\{K_{ij}\}_{1 \leq i < j < \infty}$ and $\{U_{ij}^{\ell}\}_{1 \leq i < j < \infty}$ for each replica indexed by $\ell = 1$ or 2. (In later sections, we will consider more than two replicas.) Each of these sets is an i.i.d. family and the different sets are mutually independent. The U_{ij}^{ℓ}'s are independent mean one exponentials and will be used to define the bond occupation variables (conditionally on the couplings and spins). The K_{ij}'s, which determine the J_{ij}'s for given N (and β) by $J_{ij} = K_{ij}/\sqrt{N}$, have as their common distribution a probability measure ρ on the real line about which we make the following assumptions: ρ is even ($d\rho(x) = d\rho(-x)$) with no atom at the origin ($\rho(\{0\}) = 0$), variance one ($\int_{-\infty}^{+\infty} x^2 d\rho(x) = 1$) and a finite moment generating function ($\int_{-\infty}^{+\infty} e^{tx} d\rho(x) < \infty$ for all real t). The two most common choices are the Gaussian (where ρ is a mean zero, variance one normal distribution) and the $\pm J$ (where ρ is $(\delta_1 + \delta_{-1})/2$) spin glasses.

For a given N and β, we have already defined the couplings J_{ij}. The conditional distribution, given the couplings, of the spin variables σ, τ for the two replicas, is that of an independent sample from the Gibbs distribution; i.e.,

$$\text{const} \times \exp\left[\beta \sum_{1 \le i < j \le N} J_{ij}(\sigma_i \sigma_j + \tau_i \tau_j)\right] . \tag{1}$$

It remains to define the percolation bond occupation variables of interest, given the couplings and the spins. Two of these are the FK (random cluster) variables – one set for each replica; we will denote these $\mathbf{n}_{ij}^{\ell}$ for $\ell = 1$ (corresponding to the first (σ) replica) and $\ell = 2$ (corresponding to the second (τ) replica). These may be constructed as follows. For a given $i < j$, if the bond $\{i, j\}$ is unsatisfied in the first replica – i.e., if $J_{ij}\sigma_i\sigma_j < 0$, then set $\mathbf{n}_{ij}^1 = 0$; if the bond is satisfied, then set $\mathbf{n}_{ij}^1 = 1$ if $U_{ij}^1 \le 2\beta|J_{ij}|$ (i.e., with probability $1 - \exp(-2\beta|J_{ij}|)$) and otherwise set it to zero. Define $\mathbf{n}_{ij}^2$ similarly using the second (τ) replica. We will be particularly interested in the percolation properties of the variables $\mathbf{n}_{ij} = \mathbf{n}_{ij}^1 \mathbf{n}_{ij}^2$ that describe *doubly FK-occupied* bonds. We will use the acronym TRFK (for Two Replica FK) to denote various quantities built out of these variables.

There is another two-replica graphical representation, introduced by Chayes, Machta and Redner [6, 7] (which we will denote by CMR) that we will also consider. This representation in general is described in terms of three types of bonds which may be thought of as those that are colored blue or red or else are uncolored. One way of defining the blue bonds, whose occupation variables we will denote by $\mathbf{b}_{ij}$, is that $\mathbf{b}_{ij} = 1$ if $\{i, j\}$ is satisfied in *both* replicas *and* also either $\mathbf{n}_{ij}^1 = 1$ or $\mathbf{n}_{ij}^2 = 1$ or both (which occurs with probability $1 - \exp(-4\beta|J_{ij}|)$); otherwise $\mathbf{b}_{ij} = 0$. We will be interested in the percolation properties of the blue bonds. Although we will not be using them in this paper, we note that a bond $\{i, j\}$ is colored red if and only if $\sigma_i\sigma_j\tau_i\tau_j = -1$ (or equivalently $\{i, j\}$ is satisfied in exactly one of the two replicas) *and* also that satisfied bond is FK-occupied (which occurs with probability $1 - \exp(-2\beta|J_{ij}|)$).

A key role in the theory of spin glasses, and this will also be the case for their percolation properties, is played by the Parisi (spin) overlap. For a given N and β, this overlap is the random variable,

$$Q = Q(N, \beta) = N^{-1} \sum_{1 \le i \le N} \sigma_i \tau_i . \tag{2}$$

Closely related to the overlap are the densities (i.e., the fractions of sites out of N) $D_a = D_a(N,\beta)$ and $D_d = D_d(N,\beta)$ of the collections of sites where the spins of the two replicas respectively agree and disagree with each other. Since $Q = D_a - D_d$ and $D_a + D_d = 1$, one can express $D_{\max} = \max\{D_a, D_d\}$ and $D_{\min} = \min\{D_a, D_d\}$ as $D_{\max} = [1 + |Q|]/2$ and $D_{\min} = [1 - |Q|]/2$.

It should be clear from our definitions of the various bond occupation variables that if one of i, j is in the collection of agree sites and the other is in the collection of disagree sites, then the bond $\{i, j\}$ is satisfied in exactly one of the two replicas and so $\{i, j\}$ can neither be a TRFK occupied bond nor a CMR blue bond. So percolation (i.e., occurrence of giant clusters containing order N of the sites) can only occur separately within the agree or within the disagree collections of sites. Our results concern when this happens and its connection with the spin glass phase transition in the SK model via the overlap random variable Q.

We will first state two general theorems relating the occurrence of giant clusters to the behavior of Q and then state a number of corollaries. The corollaries depend for their applicability on results about the nature of Q in the SK model, some of which have been and some of which have not yet been derived rigorously. The first theorem concerns CMR percolation. We denote the density of the kth largest CMR blue cluster by $D_k^{\mathrm{CMR}}(N,\beta)$.

Theorem 1. *In the CMR representation, for any $0 < \beta < \infty$, the following three sequences of random variables tend to zero in probability, i.e., the $\mathbf{P}_N$-probability that the absolute value of the random variable is greater than ε tends to zero as $N \to \infty$ for any $\varepsilon > 0$.*

$$D_1^{\mathrm{CMR}}(N,\beta) - \left[\frac{1 + |Q(N,\beta)|}{2}\right] \to 0\,. \tag{3}$$

$$D_2^{\mathrm{CMR}}(N,\beta) - \left[\frac{1 - |Q(N,\beta)|}{2}\right] \to 0\,. \tag{4}$$

$$D_3^{\mathrm{CMR}}(N,\beta) \to 0\,. \tag{5}$$

To state the next theorem, we define $\theta(c)$ for $c \in [0,\infty)$ to be the order parameter for mean-field percolation – i.e., the asymptotic density (fraction of sites out of N as $N \to \infty$) of the largest cluster in $G(N, c/N)$, the Erdős-Rényi random graph with occupation probability c/N independently for each edge in the complete graph of N sites [10]. It is a standard fact [10] that $\theta(c)$ is zero for $0 \le c \le 1$ and for $c > 1$ is the strictly positive solution of

$$\theta = 1 - e^{-c\theta}\,. \tag{6}$$

Theorem 2. *In the TRFK representation, for any $0 < \beta < \infty$, the following limits (in probability) are valid as $N \to \infty$ for D_k^{TRFK}, the density of the kth largest TRFK doubly occupied cluster.*

$$D_1^{\mathrm{TRFK}}(N,\beta) - \theta(2\beta^2\frac{1 + |Q(N,\beta)|}{2})[\frac{1 + |Q(N,\beta)|}{2}] \to 0\,. \tag{7}$$

$$D_2^{\mathrm{TRFK}}(N,\beta) \;-\; \theta(2\beta^2\frac{1-|Q(N,\beta)|}{2})[\frac{1-|Q(N,\beta)|}{2}] \;\to\; 0\,. \tag{8}$$

$$D_3^{\mathrm{TRFK}}(N,\beta) \;\to\; 0\,. \tag{9}$$

We now denote by $P(N,\beta)$ the probability distribution of the overlap $Q(N,\beta)$. This is the Parisi overlap distribution, averaged over the disorder variables $\mathcal{K} = \{K_{ij}\}$ (with $1 \le i < j < \infty$). The unaveraged overlap distribution requires conditioning on $\mathcal{K}$. So, for example, we have, for $q \in [-1,1]$,

$$P(N,\beta)([-1,q]) \;=\; \mathrm{Av}[\mathbf{P}_N(Q(N,\beta) \le q|\,\mathcal{K})], \tag{10}$$

where Av denotes the average over the disorder distribution of $\mathcal{K}$.

The quantity $E(Q(N,\beta)^2|\,\mathcal{K})$ is closely related to $\sum_{1\le i<j\le N}[E(\sigma_1\sigma_j|\,\mathcal{K})]^2$ (see, e.g., [15], Lemma 2.2.10) which in turn is closely related to the derivative of the finite volume free energy (see, e.g., [16], Prop. 4.1). It then follows that $E(Q(N,\beta)^2) \to 0$ as $N \to \infty$ first for $\beta < 1$ ([16], Prop. 2.1) and then (using results of [17, 18, 19] and of [20, 21] – see also [22]) also for $\beta = 1$. This implies the following theorem, one of the basic facts in the mathematical theory of the SK model.

Theorem 3. *For $\beta \le 1$, $Q(N,\beta) \to 0$ (in probability) as $N \to \infty$, or equivalently $P(N,\beta) \to \delta_0$.*

The situation regarding rigorous results about the nonvanishing of $Q(N,\beta)$ as $N \to \infty$ for $\beta > 1$ is less clean. For example, using results from the references cited just before Theorem 3, it follows that $E(Q(N,\beta)^2)$ has a limit as $N \to \infty$ for all β, related by a simple identity to the derivative at that β of the infinite-volume free energy, $\mathcal{P}(\beta)$, given by the Parisi variational formula, $\mathcal{P}(\beta) = \inf_m \mathcal{P}(m,\beta)$, where the inf is over distribution functions $m(q)$ with $q \in [0,1]$ – see [20]. Furthermore, since for $\beta > 1$, $\mathcal{P}(\beta)$ is strictly below the "annealed" free energy [23] (which equals $\mathcal{P}(\delta_0,\beta)$, where δ_0 is the distribution function for the unit point mass at $q = 0$), it follows (see [22]) from Lipschitz continuity of $\mathcal{P}(m,\beta)$ in m [18, 21] that

$$\lim_{N\to\infty} E(Q(N,\beta)^2) \;>\; 0 \;\text{ for all }\; \beta > 1\,. \tag{11}$$

However, it seems that it is not yet proved in general that $Q(N,\beta)$ has a unique limit (in distribution) nor very much about the precise nature of any limit. In order to explain the corollaries of our main theorems without getting bogged down in these unresolved questions about the SK model, we will list various properties which are expected to be valid for (at least some values of) $\beta > 1$ and then use those as assumptions in our corollaries. Some related comments are given in Remark 2.2 below.

Possible behaviors of the supercritical overlap. For $\beta > 1$, $P(N,\beta)$ converges as $N \to \infty$ to some P_β with the following properties.

- Property P1: $P_\beta(\{0\}) = 0$.
- Property P2: $P_\beta(\{-1,+1\}) = 0$.
- Property P3: $P_\beta([-1,-1+(1/\beta^2)] \cup [1-(1/\beta^2),1]) = 0$.

If one defines $q_{EA}(\beta)$, the Edwards-Anderson order parameter (for the SK model), to be the supremum of the support of P_β, and one assumes that P_β has point masses at $\pm q_{EA}(\beta)$, then Properties P2 and P3 reduce respectively to $q_{EA}(\beta) < 1$ and $q_{EA}(\beta) < 1 - (1/\beta^2)$. Weaker versions of the three properties that do not require existence of a limit for $P(N,\beta)$ as $N \to \infty$ are as follows.

- Property P1′:

$$\lim_{\varepsilon \downarrow 0} \limsup_{N\to\infty} \mathbf{P}_N(|Q(N,\beta)| < \varepsilon) = 0. \tag{12}$$

- Property P2′:

$$\lim_{\varepsilon \downarrow 0} \limsup_{N\to\infty} \mathbf{P}_N(|Q(N,\beta)| > 1-\varepsilon) = 0. \tag{13}$$

- Property P3′

$$\lim_{\varepsilon \downarrow 0} \limsup_{N\to\infty} \mathbf{P}_N(|Q(N,\beta)| > 1-(1/\beta^2)-\varepsilon) = 0. \tag{14}$$

We will state the next two theorems in a somewhat informal manner and then provide a more precise meaning in Remark 2.1 below.

Theorem 4. (Corollary to Theorem 1) *In the CMR representation, for any* $0 < \beta \leq 1$, *there are exactly two giant blue clusters, each of (asymptotic) density* $1/2$. *For* $1 < \beta < \infty$, *there are either one or two giant blue clusters, whose densities add to* 1; *there is a unique one of (maximum) density in* $(1/2, 1]$ *providing Property* P1 *(or* P1′*) is valid and there is another one of smaller density in* $(0, 1/2)$ *providing Property* P2 *(or* P2′*) is valid.*

Theorem 5. (Corollary to Theorem 2) *In the TRFK representation, there are no giant doubly occupied clusters for* $\beta \leq 1$. *For* $1 < \beta < \infty$, *there are either one or two giant doubly occupied clusters with a unique one of maximum density providing Property* P1 *(or* P1′*) is valid and another one of smaller (but nonzero) density providing Property* P3 *(or* P3′*) is valid.*

Remark 2.1. *Theorem* 4 *states that* $(D_1^{\mathrm{CMR}}(N,\beta), D_2^{\mathrm{CMR}}(N,\beta), D_3^{\mathrm{CMR}}(N,\beta))$ *converges (in probability or equivalently in distribution), for* $0 < \beta \leq 1$, *to* $(1/2, 1/2, 0)$ *while Theorem* 5 *states that the corresponding triple of largest TRFK cluster densities converges to* $(0,0,0)$. *A precise statement of the results for* $\beta \in (1,\infty)$ *is a bit messier because it has not been proved that there is a single limit in distribution of these cluster densities, although since the densities are all bounded (in* $[0,1]$*) random variables, there is compactness with limits along subsequences of* N*'s. For example, in the CMR case, assuming Properties* P1′ *and* P2′, *the precise statement is that any limit in distribution of the triplet of densities is supported on* $\{(1/2+a, 1/2-a, 0) : a \in (0,1/2)\}$. *Precise statements for the other cases treated in the two theorems are analogous.*

Remark 2.2. *Although Property* P1′ *does not seem to have yet been rigorously proved (for any* $\beta > 1$*), a weaker property does follow from* (11). *Namely, that for all* $\beta > 1$, *the limit in* (12) *is strictly less than one. Weakened versions of* *portions of Theorems* 4 *and* 5 *for* $\beta > 1$ *follow – e.g., any limit in distribution of the triplet of densities in Remark* 2.1 *must assign strictly positive probability to* $\{(1/2 + a, 1/2 - a, 0) : a \in (0, 1/2]\}$.

3. Ultrametricity and percolation

In this section, in order to discuss ultrametricity, which is expected to occur in the supercritical SK model [9], we consider three replicas, whose spin variables are denoted $\{\sigma_i^\ell\}$ for $\ell = 1, 2, 3$. We denote by $\mathbf{n}_{ij}^\ell$ the FK occupation variables for replica ℓ and by $\mathbf{b}_{ij}^{\ell m}$ the CMR blue bond occupation variables for the pair of replicas ℓ, m. Thus $\mathbf{b}_{ij}^{12}$ corresponds in our previous notation to $\mathbf{b}_{ij}$. We also denote by $Q^{\ell m} = Q^{\ell m}(N, \beta)$ the overlap defined in (2), but with σ, τ replaced by σ^ℓ, σ^m.

Let us denote by $P^3(N, \beta)$ the distribution of the triple of overlaps (Q^{12}, Q^{13}, Q^{23}). Ultrametricity concerns the nature of the limits as $N \to \infty$ of $P^3(N, \beta)$, as follows, where we define

$$\mathbf{R}^3_{\text{ultra}} = \{(x, y, z) : |x| = |y| \le |z| \,\text{or}\, |x| = |z| \le |y| \,\text{or}\, |y| = |z| \le |x|\} . \tag{15}$$

Possible ultrametric behaviors of the supercritical overlap. $P^3(N, \beta)$ converges, for $\beta > 1$, to some P^3_β as $N \to \infty$ with

- Property P4: $P^3_\beta(\mathbf{R}^3_{\text{ultra}}) = 1$.

We will generally replace this property by a weakened version, P4′, in which it is not assumed that there is a single limit P^3_β as $N \to \infty$ but rather the same property is assumed for every subsequence limit. There is another property that simplifies various of our statements about how ultrametricity is manifested in the sizes of various percolation clusters. This property, which, like ultrametricity, is expected to be valid in the supercritical SK model (see [24], where this property is discussed and also numerically tested in the three-dimensional EA model) is the following.

- Property P5: $P^3_\beta(\{(x, y, z) : xyz \ge 0\}) = 1$.

Again we will use a weaker version P5′ in which it is not assumed that there is a single limit P^3_β as $N \to \infty$.

One formulation of ultrametricity using percolation clusters is the next theorem, an immediate corollary of Theorem 1, in which we denote by $D_j^{\ell m} = D_j^{\ell m}(N, \beta)$ the density of sites in $\mathcal{C}_j^{\ell m}$, the jth largest cluster formed by the bonds $\{i, j\}$ with $\mathbf{b}_{ij}^{\ell m} = 1$ (i.e., the jth largest CMR blue cluster for the pair $\{\ell, m\}$ of replicas). Note that D_j^{12} coincides in our previous notation with D_j^{CMR}.

Theorem 6. (Corollary to Theorem 1) *For* $1 < \beta < \infty$, *assuming Property* P4′, *any subsequence limit in distribution as* $N \to \infty$ *of the triple* $(D_1^{12} - D_2^{12}, D_1^{13} - D_2^{13}, D_1^{23} - D_2^{23})$ *is supported on* $\mathbf{R}^3_{\text{ultra}}$.

In our next two theorems, instead of looking at *differences* of densities, we express ultrametricity directly in terms of densities themselves. This is perhaps more interesting because rather than having three density differences, there will be *four* densities. We begin in Theorem 7 with a fully CMR point of view with four natural non-empty *intersections* of CMR blue clusters. Then Theorem 8 mixes CMR and FK occupation variables to yield (four) other natural clusters.

There are a number of ways in which the four sets of sites in our fully CMR perspective can be defined, which turn out to be equivalent (for large N). One definition is as follows. For α, α' each taken to be either the letter a (for agree) or the letter d (for disagree), define $\Lambda_{\alpha\alpha'}(N,\beta)$ to be the set of sites $i \in \{1,\ldots,N\}$ where σ_i^1 agrees (for $\alpha = a$) or disagrees (for $\alpha = d$) with σ_i^2 and σ_i^1 agrees (for $\alpha' = a$) or disagrees (for $\alpha' = d$) with σ_i^3; also denote by $D_{\alpha\alpha'}(N,\beta)$ the density of sites (i.e., the fraction of N) in $\Lambda_{\alpha\alpha'}(N,\beta)$. Then denote by $\mathcal{C}_{\alpha\alpha'}^{\ell m}$ the largest cluster (thought of as the collection of its sites) formed within $\Lambda_{\alpha\alpha'}$ by the $\mathbf{b}^{\ell m} = 1$ blue bonds. Finally, define $\mathcal{C}_{\alpha\alpha'} = \mathcal{C}_{\alpha\alpha'}^{12} \cap \mathcal{C}_{\alpha\alpha'}^{13}$, $D_{\alpha\alpha'}^{\mathrm{CMR}}$ to be the density of sites (fraction of N) in $\mathcal{C}_{\alpha,\alpha'}$ and $\hat{D}^{\mathrm{CMR}}(N,\beta)$ to be the vector of four densities $(D_{aa}^{\mathrm{CMR}}, D_{ad}^{\mathrm{CMR}}, D_{da}^{\mathrm{CMR}}, D_{dd}^{\mathrm{CMR}})$. To state the next theorem, let $\mathbf{R}^{4,+}$ denote $\{(x_1,x_2,x_3,x_4) : \text{each}\, x_i \geq 0\}$ and define

$$\mathbf{R}_{\mathrm{ultra}}^{4,+} = \{(x_1,x_2,x_3,x_4) \in \mathbf{R}^{4,+} : x_{(1)} > x_{(2)} \geq x_{(3)} = x_{(4)}\}, \tag{16}$$

where $x_{(1)}, x_{(2)}, x_{(3)}, x_{(4)}$ are the rank ordered values of x_1, x_2, x_3, x_4.

Theorem 7. *For* $0 < \beta \leq 1$, $\hat{D}^{\mathrm{CMR}}(N,\beta) \to (1/4, 1/4, 1/4, 1/4)$ *(in probability) as* $N \to \infty$. *For* $1 < \beta < \infty$ *and assuming Properties* P1′, P2′, P4′, P5′, *any limit in distribution of* $\hat{D}^{\mathrm{CMR}}(N,\beta)$ *is supported on* $\mathbf{R}_{\mathrm{ultra}}^{4,+}$.

Remark 3.1. *The equilateral triangle case where* $|Q^{12}| = |Q^{13}| = |Q^{23}| = q_u$ *corresponds to* $x_{(1)} = (1+3q_u)/4$ *and* $x_{(2)} = x_{(3)} = x_{(4)} = (1-q_u)/4$. *The alternative isosceles triangle case with, say,* $|Q^{12}| = q_u > |Q^{13}| = |Q^{23}| = q_\ell$ *corresponds to* $x_{(1)} = (1+q_u+2q_\ell)/4$, $x_{(2)} = (1+q_u-2q_\ell)/4$ *and* $x_{(3)} = x_{(4)} = (1-q_u)/4$.

In the next theorem, we consider $\mathcal{C}_{\alpha\alpha'}^*$ defined as the largest cluster in $\Lambda_{\alpha\alpha'}(N,\beta)$ formed by bonds $\{i,j\}$ with $\mathbf{b}_{ij}^{12}\mathbf{n}_{ij}^3 = 1$ – i.e., bonds that are simultaneously CMR blue for the first two replicas and FK-occupied for the third replica. Let $\hat{D}^*(N,\beta)$ denote the corresponding vector of four densities. As in the previous theorem, we note that there are alternative, but equivalent for large N, definitions of the clusters and densities (e.g., as the four largest clusters formed by bonds $\{i,j\}$ with $\mathbf{b}_{ij}^{12}\mathbf{n}_{ij}^3 = 1$ in all of $\{1,\ldots,N\}$ without a priori restriction to $\Lambda_{\alpha\alpha'}(N,\beta)$).

Theorem 8. *For* $0 < \beta \leq 1$, $\hat{D}^*(N,\beta) \to (0,0,0,0)$ *(in probability) as* $N \to \infty$. *For* $1 < \beta < \infty$ *and assuming Properties* P1′, P3′, P4′, P5′, *any limit in distribution of* $\hat{D}^*(N,\beta)$ *is supported on* $\mathbf{R}_{\mathrm{ultra}}^{4,+}$.

Remark 3.2. *Here the limiting densities are of the form* $x_{(j)} = \theta(4\beta^2 d_{(j)})\, d_{(j)}$ *where* $d_{(1)} = (1+q_u+2q_\ell)/4$, $d_{(2)} = (1+q_u-2q_\ell)/4$ *and* $d_{(3)} = d_{(4)} = (1-q_u)/4$, *with* $q_\ell = q_u$ *for the equilateral triangle case.*

4. Proofs

Before giving the proofs of our main results, we present several key lemmas which are the technical heart of our proofs. We will use the notation $\gg$ and $\ll$ to denote stochastic domination (in the FKG sense) for either families of random variables or their distributions. E.g., for the m-tuples $X = (X_1, \dots, X_m)$ and $Y = (Y_1, \dots, Y_m)$ we write $X \ll Y$ or $Y \gg X$ to mean that $\mathbf{E}(h(X_1, \dots, X_m)) \leq \mathbf{E}(h(Y_1, \dots, Y_m))$ for every coordinatewise increasing function h (for which the two expectations exist).

Our key lemmas concern stochastic domination inequalities in the k-replica setting comparing conditional distributions of the couplings $\{K_{ij}\}$ or related bond occupation variables, when the spins $\sigma^1, \dots, \sigma^k$ are fixed, to product measures. These allow us to approximate percolation variables in the SK model by the *independent* variables of Erdős-Rényi random graphs when $N \to \infty$. Given probability measures ν_{ij} on $\mathbf{R}$ for $1 \leq i < j \leq N$, we denote by $\mathrm{Prod}_N(\{\nu_{ij}\})$ the corresponding product measure on $\mathbf{R}^{N(N-1)/2}$. We also will denote by $\rho[\gamma]$ the probability measure defined by $d\rho[\gamma](x) = e^{\gamma x} d\rho(x) / \int_{-\infty}^{+\infty} e^{\gamma x'} d\rho(x')$.

Lemma 4.1. *Fix* N, β, k *and let* $\tilde{\mu}^k_{N,\beta}$ *denote the conditional distribution, given the spins* $\sigma^1 = (\sigma^1_i : 1 \leq i \leq N), \dots, \sigma^k$, *of* $\{K'_{ij} \equiv \varepsilon_{i,j} K_{ij}\}_{1 \leq i < j \leq N}$ *(where the* $\varepsilon_{i,j}$*'s are any given* ± 1 *values). Then*

$$\mathrm{Prod}_N(\{\rho[\gamma^{k,-}_{ij}]\}) \ll \tilde{\mu}^k_{N,\beta} \ll \mathrm{Prod}_N(\{\rho[\gamma^{k,+}_{ij}]\}), \tag{17}$$

where

$$\gamma^{k,\pm}_{ij} = \frac{\beta}{\sqrt{N}} [\varepsilon_{ij}(\sigma^1_i \sigma^1_j + \cdots + \sigma^k_i \sigma^k_j) \pm k]. \tag{18}$$

Proof. Define the partition function

$$Z_{N,\beta} = Z_{N,\beta}(\{K'_{ij}\}) = \sum_\sigma \exp\Big(\frac{\beta}{\sqrt{N}} \sum_{1 \leq i < j \leq N} (\varepsilon_{ij} \sigma_i \sigma_j) K'_{ij}\Big), \tag{19}$$

where $\sum_\sigma$ denotes the sum over all 2^N choices of $\sigma_i = \pm 1$ for $1 \leq i \leq N$. Thus the normalization constant in Equation (1) for two replicas is $(Z_{N,\beta})^{-2}$ and the k-replica marginal distribution for $\{K'_{ij}\}$ is as follows, where $\mathrm{Prod}_N(\{\rho\})$ denotes $\mathrm{Prod}_N(\{\nu_{ij}\})$ with $\nu_{ij} \equiv \rho$, $\phi^k_{ij} = \varepsilon_{ij}(\sigma^1_i \sigma^1_j + \cdots + \sigma^k_i \sigma^k_j)$ and $C_{N,\beta}$ is a normalization constant depending on the ϕ^k_{ij}'s but not on the K'_{ij}'s.

$$d\tilde{\mu}^k_{N,\beta} = C_{N,\beta} (Z_{N,\beta}(\{K'_{ij}\}))^{-k} \exp\Big(\frac{\beta}{\sqrt{N}} \sum_{1 \leq i < j \leq N} \phi^k_{ij} K'_{ij}\Big)\, \mathrm{Prod}_N(\{\rho\}). \tag{20}$$

It is a standard fact about stochastic domination that if $\tilde{\mu}$ is a product probability measure on $\mathbf{R}^m$ and g is an increasing – i.e., coordinatewise nondecreasing – (respectively, decreasing) function on $\mathbf{R}^m$ (with $\int g d\tilde{\mu} < \infty$), then the probability measure $\tilde{\mu}_g$ defined as $d\tilde{\mu}_g(x_1, \ldots, x_m) = g(x_1, \ldots, x_m) d\tilde{\mu} / \int g d\tilde{\mu}$ satisfies $\tilde{\mu}_g \gg \tilde{\mu}$ (respectively, $\tilde{\mu}_g \ll \tilde{\mu}$). This follows from the fact that product measures satisfy the FKG inequalities – i.e., for f and g increasing $\int f g d\tilde{\mu} \geq \int f d\tilde{\mu} \int g d\tilde{\mu}$.

On the other hand, since each $\varepsilon_{ij}\sigma_i\sigma_j = \pm 1$, it follows from (19) that

$$Z^{\pm}_{N,\beta} \equiv Z_{N,\beta} e^{\pm(\beta/\sqrt{N}) \sum K'_{ij}} = \sum_{\sigma} e^{(\beta/\sqrt{N}) \sum \psi^{\pm}_{ij}(\sigma) K'_{ij}} \tag{21}$$

with each $\psi^{\pm}_{ij} = 0$ or ± 2 and hence each $\psi^{+}_{ij} \geq 0$ (respectively, each $\psi^{-}_{ij} \leq 0$). Thus, as a function of the K'_{ij}'s, $Z^{+}_{N,\beta}$ is increasing and $(Z^{+}_{N,\beta})^{-k}$ is decreasing while $Z^{-}_{N,\beta}$ is decreasing and $(Z^{-}_{N,\beta})^{-k}$ is increasing. Combining this with the previous discussion about stochastic domination and product measures, we see that

$$\begin{aligned} C^{-}_{N,\beta} e^{-(\beta/\sqrt{N}) \sum (-kK'_{ij} + \phi^k_{ij} K'_{ij})} \operatorname{Prod}_N(\{\rho\}) \ll \tilde{\mu}^k_{N,\beta} \ll \\ C^{+}_{N,\beta} e^{-(\beta/\sqrt{N}) \sum (+kK'_{ij} + \phi^k_{ij} K'_{ij})} \operatorname{Prod}_N(\{\rho\}) , \end{aligned} \tag{22}$$

which is just Equation (17) since the $C^{\pm}_{N,\beta}$ are normalization constants. □

The next two lemmas give stochastic domination inequalities for three different sets of occupation variables in terms of independent Bernoulli percolation variables. The first lemma covers the cases of CMR and TRFK variables involving two replicas and the second lemma deals with mixed CMR-FK variables involving three replicas. The parameters of the independent Bernoulli variables used for upper and lower bounds are denoted $p^{*,\natural}_{N,\beta}$, where $*$ denotes CMR or TRFK or 3 (for the mixed CMR-FK case) and $\natural$ denotes u (for upper bound) or ℓ (for lower bound) and are as follows.

$$p^{*,u}_{N,\beta} = \int_0^\infty g^*_{N,\beta}(x) d\rho^*_{N,\beta}(x) , \tag{23}$$

$$p^{*,\ell}_{N,\beta} = \int_0^\infty g^*_{N,\beta}(x) d\rho(x) , \tag{24}$$

where

$$g^{\text{CMR}}_{N,\beta}(x) = 1 - e^{-4(\beta/\sqrt{N})x} , \tag{25}$$

$$g^{\text{TRFK}}_{N,\beta}(x) = (1 - e^{-2(\beta/\sqrt{N})x})^2 , \tag{26}$$

$$g^3_{N,\beta}(x) = (1 - e^{-4(\beta/\sqrt{N})x})\,(1 - e^{-2(\beta/\sqrt{N})x}), \tag{27}$$

and

$$\rho^{\text{CMR}}_{N,\beta} = \rho^{\text{TRFK}}_{N,\beta} = \rho[4\beta/\sqrt{N}] , \tag{28}$$

$$\rho^3_{N,\beta} = \rho[6\beta/\sqrt{N}] . \tag{29}$$

Lemma 4.2. *For fixed $N, \beta, \sigma^1, \sigma^2$, we consider the conditional distributions $\hat{\mu}_{N,\beta}^{\mathrm{CMR}}$ of $\{\mathbf{b}_{i,j} \equiv \mathbf{b}_{i,j}^{12}\}_{1 \le i<j \le N}$ and $\hat{\mu}_{N,\beta}^{\mathrm{TRFK}}$ of $\{\mathbf{n}_{i,j} \equiv \mathbf{n}_{i,j}^1 \mathbf{n}_{i,j}^2\}_{1\le i<j\le N}$. Then*

$$\mathrm{Prod}_N(p_{i,j}^{*,\ell}\delta_1 + (1-p_{i,j}^{*,\ell})\delta_0) \ll \hat{\mu}_{N,\beta}^* \ll \mathrm{Prod}_N(p_{i,j}^{*,u}\delta_1 + (1-p_{i,j}^{*,u})\delta_0)\,, \tag{30}$$

where $p_{ij}^{*,\natural} = p_{N,\beta}^{*,\natural}$ if i,j are either both in Λ_a or both in Λ_d and $p_{ij}^{*,\natural} = 0$ otherwise. The asymptotic behavior as $N \to \infty$ of the parameters appearing in these inequalities is

$$p_{N,\beta}^{\mathrm{CMR},\natural} = \frac{2\beta}{\sqrt{N}}\Big(\int_{-\infty}^{+\infty} |x| d\rho(x)\Big)\Big(1 + O\Big(\frac{1}{\sqrt{N}}\Big)\Big)\,, \tag{31}$$

$$p_{N,\beta}^{\mathrm{TRFK},\natural} = \frac{2\beta^2}{N}\Big(1 + O\Big(\frac{1}{\sqrt{N}}\Big)\Big)\,. \tag{32}$$

Proof. We begin with some considerations for the general case of k replicas and arbitrary ε_{ij} before specializing to what is used in this lemma. The case $k=3$ will be used for the next lemma. Let $(\tilde{K}, \tilde{U})_N^k \equiv (\tilde{K}_{ij}, \tilde{U}_{ij}^1, \ldots, \tilde{U}_{ij}^k)_N$ denote random variables (with $1 \le i < j \le N$) whose joint distribution is the conditional distribution of $(K'_{ij}, U_{ij}^1, \ldots, U_{ij}^k)_N$ given $\sigma^1, \ldots, \sigma^k$. Note that the $\tilde{U}_{ij}^\ell$'s are independent of the $\tilde{K}_{ij}$'s and (like the U_{ij}^ℓ's) are independent mean one exponential random variables. Let $(K^{\mathrm{ind}}[\{\gamma_{ij}\}], U^{\mathrm{ind}})_N^k \equiv (K_{ij}^{\mathrm{ind}}[\gamma_{ij}], U_{ij}^{\mathrm{ind},1}, \ldots, U_{ij}^{\mathrm{ind},k})_N$ denote mutually independent random variables with $K_{ij}^{\mathrm{ind}}[\gamma_{ij}]$ distributed by $\rho[\gamma_{ij}]$ and each $U_{ij}^{\mathrm{ind},\ell}$ a mean one exponential random variable.

Lemma 4.1 says that

$$(K^{\mathrm{ind}}[\{\gamma_{ij}^{k,-}\}])_N^k \ll (\tilde{K})_N^k \ll (K^{\mathrm{ind}}[\{\gamma_{ij}^{k,+}\}])_N^k \tag{33}$$

and it immediately follows that

$$(K^{\mathrm{ind}}[\{\gamma_{ij}^{k,-}\}], -U^{\mathrm{ind}})_N^k \ll (\tilde{K}, -\tilde{U})_N^k \ll (K^{\mathrm{ind}}[\{\gamma_{ij}^{k,+}\}], -U^{\mathrm{ind}})_N^k\,. \tag{34}$$

We now take $k=2$, choose $\varepsilon_{ij} = \sigma_i^1\sigma_j^1$ and note that $\mathbf{n}_{ij} \equiv \mathbf{n}_{ij}^1\mathbf{n}_{ij}^2 = 0 = \mathbf{b}_{ij}$ *unless* i,j are either both in Λ_a or both in Λ_d, in which case we have $\sigma_i^1\sigma_j^1 = \sigma_i^2\sigma_j^2$, $\gamma_{ij}^{2,+} = 4\beta/\sqrt{N}$ and $\gamma_{ij}^{2,-} = 0$. Furthermore $\mathbf{n}_{ij}^\ell$ is the indicator of the event that $2\beta K'_{ij} + (-U_{ij}^\ell) \ge 0$ while $\mathbf{b}_{ij} \equiv \mathbf{b}_{ij}^{12}$ is the indicator of the event that $\mathbf{n}_{ij}^1 + \mathbf{n}_{ij}^2 \ge 1$ and $K'_{ij} \ge 0$, so that these occupation variables (for such i,j) are increasing functions of $(K', -U)_N^2$. Let us now define $\tilde{\mathbf{n}}_{ij}, \tilde{\mathbf{b}}_{ij}$ and $\mathbf{n}_{ij}^{\mathrm{ind},\pm}, \mathbf{b}_{ij}^{\mathrm{ind},\pm}$ as the same increasing functions, respectively, of $(\tilde{K}, -\tilde{U})_N^2$ and $(K^{\mathrm{ind}}[\{\gamma_{ij}^{k,\pm}\}], -U^{\mathrm{ind}})_N^2$ providing i,j are either both in Λ_a or both in Λ_d, and otherwise set these occupation variables to zero. Then, as a consequence of (34), we have

$$(\mathbf{n}^{\mathrm{ind},-})_N^2 \ll (\tilde{\mathbf{n}})_N^2 \ll (\mathbf{n}^{\mathrm{ind},+})_N^2\,, \tag{35}$$

and

$$(\mathbf{b}^{\mathrm{ind},-})_N^2 \ll (\tilde{\mathbf{b}})_N^2 \ll (\mathbf{b}^{\mathrm{ind},+})_N^2\,. \tag{36}$$

To complete the proof of Lemma 4.2, it remains to obtain the claimed formulas for the various Bernoulli occupation parameters. E.g., for the case $* =$ TRFK and $\natural = u$, we have for i, j either both in Λ_a or both in Λ_d,

$$\begin{aligned} p_{N,\beta}^{\mathrm{TRFK},u} &= \mathbf{P}(\mathbf{n}_{ij}^{\mathrm{ind},+} = 1) \\ &= \mathbf{P}(2\beta K_{ij}^{\mathrm{ind}} \gamma_{ij}^{2,+}/\sqrt{N} \geq U_{ij}^{\mathrm{ind},1}, U_{ij}^{\mathrm{ind},2}) \\ &= \int_0^\infty (1 - e^{-2(\beta/\sqrt{N})x})^2 d\rho[4\beta/\sqrt{N}](x), \end{aligned} \tag{37}$$

as given by Equations (23), (26) and (28). We leave the checking of the other three cases (for $k = 2$) and the straightforward derivation of the asymptotic behavior as $N \to \infty$ for all the parameters to the reader. □

Lemma 4.3. *For fixed $N, \beta, \sigma^1, \sigma^2, \sigma^3$, we consider the conditional distribution $\hat{\mu}_{N,\beta}^3$ of $\{\mathbf{b}_{i,j}\mathbf{n}_{i,j}^3\}_{1 \leq i < j \leq N}$. Then equation (30) with $* = 3$ remains valid, where $p_{ij}^{3,\natural} = p_{N,\beta}^{3,\natural}$ if i, j are either both in Λ_{aa} or both in Λ_{ad} or both in Λ_{da} or both in Λ_{dd} and $p_{ij}^{3,\natural} = 0$ otherwise. The asymptotic behavior as $N \to \infty$ of $p_{N,\beta}^{3,\natural}$ is*

$$p_{N,\beta}^{3,\natural} = \frac{4\beta^2}{N}\left(1 + O\left(\frac{1}{\sqrt{N}}\right)\right). \tag{38}$$

Proof. As in the proof of Lemma 4.2, we apply Lemma 4.1 in the guise of (34) with $\varepsilon_{ij} = \sigma_i^1 \sigma_j^1$ again, but this time with $k = 3$. Now $\mathbf{b}_{ij}\mathbf{n}_{ij}^3 = \mathbf{b}_{ij}^{12}\mathbf{n}_{ij}^3 = 0$ *unless* i, j are either both in Λ_{aa} or both in Λ_{ad} or both in Λ_{da} or both in Λ_{dd}, in which case we have $\sigma_i^1\sigma_j^1 = \sigma_i^2\sigma_j^2 = \sigma_i^3\sigma_j^3$, $\gamma_{ij}^{3,+} = 6\beta/\sqrt{N}$, $\gamma_{ij}^{3,-} = 0$, and $\mathbf{b}_{ij}^{12}\mathbf{n}_{ij}^3$ is an increasing function of $(K', -U)_N^3$. The remainder of the proof, which closely mimics that of Lemma 4.2, is straightforward. □

Proof of Theorem 1. We denote by $D_{\alpha,j}^{\mathrm{CMR}} = D_{\alpha,j}^{\mathrm{CMR}}(N, \beta)$ for $\alpha = a$ or d the density (fraction of N) of the jth largest CMR cluster in Λ_α and by $S_j^{\mathrm{RG}} = S_j^{\mathrm{RG}}(N, p_N)$ the number of sites in the largest cluster of the random graph $G(N, p_N)$. Recall that $D_\alpha = D_\alpha(N, \beta)$ denotes the density (fraction of N) of Λ_α and that $\max\{D_a, D_d\} = [1 + |Q|]/2$, $\min\{D_a, D_d\} = [1 - |Q|]/2$. Then Lemma 4.2 implies that, conditional on σ^1, σ^2,

$$N\, D_{\alpha,1}^{\mathrm{CMR}}(N, \beta) \gg S_1^{\mathrm{RG}}(N\, D_\alpha(N, \beta), (2\beta c_\rho/\sqrt{N}) + \mathrm{O}(1/N)) \tag{39}$$

with $c_\rho > 0$. By separating the cases of small and not so small $D_\alpha(N, \beta)$ and using the above stochastic domination combined with the facts that $m^{-1}S_1^{\mathrm{RG}}(m, p_m) \to \theta(1) = 1$ (in probability) when $m, mp_m \to \infty$, and that $D_{\alpha,1}^{\mathrm{CMR}} \leq D_\alpha$, it directly follows that $D_\alpha - D_{\alpha,1}^{\mathrm{CMR}} \to 0$ and hence also that $D_{\alpha,2}^{\mathrm{CMR}} \to 0$ (in probability). This then directly yields (3), (4) and (5). □

Proof of Theorem 2. The proof mimics that of Theorem 1 except that (39) is replaced by

$$S_1^{\mathrm{RG}}(N\,D_\alpha, \frac{2\beta^2}{N} + \mathrm{O}(N^{-3/2})) \gg N\,D_{\alpha,1}^{\mathrm{TRFK}} \gg S_1^{\mathrm{RG}}(N\,D_\alpha, \frac{2\beta^2}{N} + \mathrm{O}(N^{-3/2}))\,, \tag{40}$$

where the two terms correcting $2\beta^2/N$, while different, are both of order $\mathrm{O}(N^{-3/2})$. We then use the fact that $m^{-1}S_1^{\mathrm{RG}}(m,p_m) - \theta(mp_m) \to 0$ when $m \to \infty$ to conclude that $D_{\alpha,1}^{\mathrm{TRFK}}(N,\beta) - \theta(2\beta^2 D_\alpha(N,\beta))D_\alpha(N,\beta) \to 0$ which yields (7) and (8). To obtain (9), we need to show that $D_{\alpha,2}^{\mathrm{TRFK}} \to 0$. But this follows from what we have already showed, from the analogue of (40) with $D_{\alpha,1}^{\mathrm{TRFK}}$ replaced by $D_{\alpha,1}^{\mathrm{TRFK}} + D_{\alpha,2}^{\mathrm{TRFK}}$ and S_1^{RG} replaced by $S_1^{\mathrm{RG}} + S_2^{\mathrm{RG}}$, and by the fact that $m^{-1}S_2^{\mathrm{RG}}(m,p_m) \to 0$ in probability when $m \to \infty$. □

Proof of Theorem 4. Any limit in distribution of $(D_1^{\mathrm{CMR}}, D_2^{\mathrm{CMR}}, D_3^{\mathrm{CMR}})$ coincides, by Theorem 1, with some limit in distribution of $(\frac{1+|Q|}{2}, \frac{1-|Q|}{2}, 0)$. The claims of the theorem for $0 < \beta \leq 1$ then follow from Theorem 3 and for $1 < \beta < \infty$ from that theorem and Properties P1′ and P2′. □

Proof of Theorem 5. Any limit in distribution of $(D_1^{\mathrm{TRFK}}, D_2^{\mathrm{TRFK}}, D_3^{\mathrm{TRFK}})$ coincides, by Theorem 2, with some limit in distribution of $(\theta(\beta^2(1+|Q|))(\frac{1+|Q|}{2}),$ $\theta(\beta^2(1-|Q|))(\frac{1-|Q|}{2}), 0)$. The claims of the theorem for $0 < \beta \leq 1$ then follow from Theorem 3 and the fact that $\theta(c) = 0$ for $c \leq 1$ while the claims for $1 < \beta < \infty$ follow from Theorem 3 and Properties P1′ and P3′. Note that P3′ is relevant because $\theta(\beta^2(1-|Q|)) > 0$ if and only if $\beta^2(1-|Q|) > 1$ or equivalently $|Q| > 1 - (1/\beta^2)$. □

Proof of Theorem 6. This theorem is an immediate consequence of Theorem 1 (which can be applied to CMR percolation using any pair of replicas ℓ, m) since $D_1^{\ell m} - D_2^{\ell m} = |Q^{\ell m}|$. □

Proof of Theorem 7. The identity $Q^{12} = D_a - D_d$ which involves only two out of three replicas may be rewritten in terms of three-replica densities as

$$Q^{12} = D_{aa} + D_{ad} - D_{da} - D_{dd}\,. \tag{41}$$

The corresponding formulas for the other overlaps are

$$Q^{13} = D_{aa} - D_{ad} + D_{da} - D_{dd}\,, \tag{42}$$

$$Q^{23} = D_{aa} - D_{ad} - D_{da} + D_{dd}\,. \tag{43}$$

Combining these three equations with the identity $D_{aa} + D_{ad} + D_{da} + D_{dd} = 1$, one may solve for the $D_{\alpha\alpha'}$'s to obtain

$$D_{aa} = (Q^{12} + Q^{13} + Q^{23} + 1)/4\,, \tag{44}$$

$$D_{ad} = (Q^{12} - Q^{13} - Q^{23} + 1)/4\,, \tag{45}$$

$$D_{da} = (-Q^{12} + Q^{13} - Q^{23} + 1)/4\,, \tag{46}$$

$$D_{dd} = (-Q^{12} - Q^{13} + Q^{23} + 1)/4\,. \tag{47}$$

In the equilateral triangle case where $|Q^{12}| = |Q^{13}| = |Q^{23}|$ and also $Q^{12}Q^{13}Q^{23} > 0$ (from Property P5′), one sees that the corresponding values of the $D_{\alpha,\alpha'}$'s lie in $\mathbf{R}^{4,+}_{\text{ultra}}$ with $x_{(1)} > x_{(2)} = x_{(3)} = x_{(4)}$. In the isosceles triangle case (with $Q^{12}Q^{13}Q^{23} > 0$), one is again in $\mathbf{R}^{4,+}_{\text{ultra}}$, but this time with $x_{(1)} > x_{(2)} > x_{(3)} = x_{(4)}$. □

Proof of Theorem 8. Here we use Lemma 4.3 to study the densities (fractions of N) $D^3_{\alpha\alpha',j}$ (where α and α' are a or d) of the jth largest cluster in $\Lambda_{\alpha\alpha'}$ formed by the bonds $\{ij\}$ with $\mathbf{b}_{ij}\mathbf{n}^3_{ij} = 1$. Similarly to the proof of Theorem 5, we have

$$S_1^{\text{RG}}\left(N\,D_{\alpha\alpha'}, \frac{4\beta^2}{N} + \mathrm{O}\left(N^{-3/2}\right)\right) \gg N\,D^3_{\alpha\alpha',1} \gg S_1^{\text{RG}}(N\,D_{\alpha\alpha'}, \frac{4\beta^2}{N} + \mathrm{O}(N^{-3/2})) \tag{48}$$

and use this to conclude that $D^3_{\alpha\alpha',1} - \theta(4\beta^2 D_{\alpha\alpha'})\,D_{\alpha\alpha'} \to 0$ and $D^3_{\alpha\alpha',2} \to 0$. Noting that $D\,\theta(4\beta^2 D)$ is an increasing function of D, the rest of the proof follows as for Theorem 7. □

Acknowledgments

The research of CMN and DLS was supported in part by NSF grant DMS-06-04869. The authors thank Dmitry Panchenko, as well as Alexey Kuptsov, Michel Talagrand and Fabio Toninelli for useful discussions concerning the nonvanishing of the overlap for supercritical SK models. They also thank Pierluigi Contucci and Cristian Giardinà for useful discussions about Property P5 for three overlaps.

References

[1] C.M. Fortuin and P.W. Kasteleyn. On the random-cluster model. I. Introduction and relation to other models. *Physica*, 57:536–564, 1972.

[2] B. Bollobás, G. Grimmett, and S. Janson. The random-cluster model on the complete graph. *Prob. Theory Rel. Fields*, 104:283–317, 1996.

[3] J. Machta, C.M. Newman, and D.L. Stein. The percolation signature of the spin glass transition. *J. Stat. Phys*, 130:113–128, 2008.

[4] S. Edwards and P.W. Anderson. Theory of spin glasses. *J. Phys. F*, 5:965–974, 1975.

[5] D. Sherrington and S. Kirkpatrick. Solvable model of a spin glass. *Phys. Rev. Lett.*, 35:1792–1796, 1975.

[6] L. Chayes, J. Machta, and O. Redner. Graphical representations for Ising systems in external fields. *J. Stat. Phys.*, 93:17–32, 1998.

[7] O. Redner, J. Machta, and L.F. Chayes. Graphical representations and cluster algorithms for critical points with fields. *Phys. Rev. E*, 58:2749–2752, 1998.

[8] M. Mézard, G. Parisi, and M. Virasoro. *Spin Glass Theory and Beyond.* World Scientific, Singapore, 1987.

[9] M. Mézard, G. Parisi, N. Sourlas, G. Toulouse, and M. Virasoro. Nature of the spin-glass phase. *Phys. Rev. Lett.*, 52:1156–1159, 1984.

[10] P. Erdös and E. Rényi. On the evolution of random graphs. *Magyar Tud. Akad. Mat. Kutató Int. Közl*, 5:17–61, 1960.

[11] B. Bollobás. The evolution of random graphs. *Trans. Am. Math. Soc.*, 286:257–274, 1984.

[12] T. Łuczak. Component behavior near the critical point of the random graph process. *Rand. Struc. Alg.*, 1:287–310, 1990.

[13] S. Janson, D.E. Knuth, T. Łuczak, and B. Pittel. Component behavior near the critical point of random graph processes. *Rand. Struc. Alg.*, 4:233–358, 1993.

[14] D. Aldous. Brownian excursions, critical random graphs and the multiplicative coalescent. *Ann. Probab.*, 25:812–854, 1997.

[15] M. Talagrand. *Spin Glasses: A Challenge for Mathematicians.* Springer, Berlin, 2003.

[16] M. Aizenman, J.L. Lebowitz, and D. Ruelle. Some rigorous results on the Sherrington-Kirkpatrick spin glass model. *Commun. Math. Phys.*, 112:3–20, 1987.

[17] F. Guerra and L.T. Toninelli. The thermodynamic limit in mean field spin glass models. *Commun. Math. Phys.*, 230:71–79, 2002.

[18] F. Guerra. Broken replica symmetry bounds in the mean field spin glass. *Commun. Math. Phys.*, 233:1–12, 2003.

[19] P. Carmona and Y. Hu. Universality in Sherrington-Kirkpatrick's spin glass model. *Ann. I. H. Poincaré – PR*, 42:215–222, 2006.

[20] M. Talagrand. The Parisi formula. *Ann. Math.*, 163:221–263, 2006.

[21] M. Talagrand. Parisi measures. *J. Func. Anal.*, 231:269–286, 2006.

[22] D. Panchenko. On differentiabilty of the Parisi formula. ArXiv:0709.1514, 2007.

[23] F. Comets. A spherical bound for the Sherrington-Kirkpatrick model. *Astérisque*, 236:103–108, 1996.

[24] P. Contucci, C. Giardinà, C. Giberti, G. Parisi, and C. Vernia. Ultrametricity in the Edwards-Anderson model. *Phys. Rev. Lett.*, 99:057206, 2007.

J. Machta
Dept. of Physics
University of Massachusetts
Amherst, MA 01003, USA
e-mail: machta@physics.umass.edu

C.M. Newman
Courant Institute of Mathematical Sciences
New York University
New York, NY 10012, USA
e-mail: newman@courant.nyu.edu

D.L. Stein
Dept. of Physics and Courant Institute of Mathematical Sciences
New York University
New York, NY 10012, USA
e-mail: daniel.stein@nyu.edu

Progress in Probability, Vol. 60, 543–549

A Note on the Diffusivity of Finite-Range Asymmetric Exclusion Processes on $\mathbb{Z}$

Jeremy Quastel and Benedek Valkó

Abstract. The diffusivity $D(t)$ of finite-range asymmetric exclusion processes on $\mathbb{Z}$ with non-zero drift is expected to be of order $t^{1/3}$. Seppäläinen and Balázs recently proved this conjecture for the nearest neighbor case. We extend their results to general finite range exclusion by proving that the Laplace transform of the diffusivity is of the conjectured order. We also obtain a pointwise upper bound for $D(t)$ of the correct order.

Mathematics Subject Classification (2000). 60K35, 82C22.

Keywords. Asymmetric exclusion process, superdiffusivity.

1. Introduction

A finite-range exclusion process on the integer lattice $\mathbb{Z}$ is a system of continuous time, rate one random walks with finite-range jump law $p(\cdot)$, i.e., $p(z) \geq 0$, and $p(z) = 0$ for $|z| > R$ for some $R < \infty$, $\sum_z p(z) = 1$, interacting via *exclusion*: Attempted jumps to occupied sites are suppressed. We consider asymmetric exclusion process (AEP) with non-zero drift,

$$\sum_z zp(z) = b \neq 0. \tag{1.1}$$

The state space of the process is $\{0, 1\}^{\mathbb{Z}}$. Particle configurations are denoted by η, with $\eta_x \in \{0, 1\}$ indicating the absence, or presence, of a particle at $x \in \mathbb{Z}$. The infinitesimal generator of the process is given by

$$Lf(\eta) = \sum_{x,z\in\mathbb{Z}} p(z)\eta_x(1 - \eta_{x+z})(f(\eta^{x,x+z}) - f(\eta)) \tag{1.2}$$

where $\eta^{x,y}$ denotes the configuration obtained from η by interchanging the occupation variables at x and y.

The authors are supported by the Natural Sciences and Engineering Research Council of Canada. B. Valkó is partially supported by the Hungarian Scientific Research Fund grant K60708.

Bernoulli product measures π_ρ, $\rho \in [0,1]$, with $\pi_\rho(\eta_x = 1) = \rho$ form a one-parameter family of invariant measures for the process. The process starting from π_0 and π_1 are trivial and so we consider the stationary process obtained by starting with π_ρ for some $\rho \in (0,1)$. Although the fixed time marginal of this stationary process is easy to understand (since the η's are just independent Bernoulli(ρ) random variables), there are still lots of open questions about the full (space-time) distribution. Information about this process (and about the appropriate scaling limit) would be very valuable to understand such elusive objects as the Stochastic Burgers and the Kardar-Parisi-Zhang equations (see [QV] for a more detailed discussion).

We consider the two-point function,

$$S(x,t) = E[(\eta_x(t) - \rho)(\eta_0(0) - \rho)], \tag{1.3}$$

where the expectation is with respect to the stationary process obtained by starting from one of the invariant measures π_ρ. $S(x,t)$ satisfies the sum rules (see [PS])

$$\sum_x S(x,t) = \rho(1-\rho) = \chi, \qquad \frac{1}{\chi}\sum_x xS(x,t) = (1-2\rho)bt. \tag{1.4}$$

The diffusivity $D(t)$ is defined as

$$D(t) = (\chi t)^{-1} \sum_{x\in\mathbb{Z}} (x - (1-2\rho)bt)^2 S(x,t). \tag{1.5}$$

Using scaling arguments one conjectures [S],

$$S(x,t) \simeq t^{-2/3}\Phi(t^{-2/3}(x - (1-2\rho)bt)) \tag{1.6}$$

for some scaling function Φ, as $t \to \infty$. A reduced conjecture is that

$$D(t) \simeq Ct^{1/3}, \tag{1.7}$$

as $t \to \infty$. Note that this means that the process has a superdiffusive behavior, as the usual diffusive scaling would lead to $D(t) \to D$. It is known that the mean-zero jump law would lead to this case, see [V].

If $f(t) \simeq t^\rho$ as $t \to \infty$ then as $\lambda \to 0$,

$$\int_0^\infty e^{-\lambda t} t f(t) dt \simeq \lambda^{-(2+\rho)}. \tag{1.8}$$

If f satisfies (1.8) then we will say that $f(t) \simeq t^\rho$ *in the weak (Tauberian) sense.* Without some extra regularity for f (for example, lack of oscillations as $t \to \infty$), such a statement will not imply a strong version of $f(t) \simeq t^\rho$ as $t \to \infty$. However, it does capture the key scaling exponent. The weak (Tauberian) version of the conjecture (1.7) is $\int_0^\infty e^{-\lambda t} tD(t)dt \simeq \lambda^{-7/3}$.

The first non-trivial bound on $D(t)$ was given in [LQSY] using the so-called *resolvent approach*: the authors proved that $D(t) \geq Ct^{1/4}$ in a weak (Tauberian) sense. (They also proved the bound $D(t) \geq C(\log t)^{1/2}$ in $d = 2$, which was later improved to $D(t) \simeq C(\log t)^{2/3}$ in [Y].) This result shows that the stationary process is indeed superdiffusive, but does not provide the conjectured scaling exponent 1/3.

The identification of this exponent was given in the breakthrough paper of Ferrari and Spohn [FS]. They treated the case of the totally asymmetric simple exclusion process (TASEP) where the jump law is $p(1) = 1$, $p(z) = 0$, $z \neq 1$. The focus of [FS] is not the diffusivity, their main result is a scaling limit for the fluctuation at time t of a randomly growing discrete one dimensional interface $h_t(x)$ connected to the equilibrium process of TASEP. This random interface (the so-called *height function*) is basically the discrete integral of the function $\eta_x(t)$ in x. The scaling factor in their result is $t^{1/3}$ and the limiting distribution is connected to the Tracy-Widom distribution. The proof of Ferrari and Spohn is through a direct mapping between TASEP and a particular last passage percolation problem, using a combination of results from [BDJ], [J], [BR], [PS].

The diffusivity $D^{TASEP}(t)$ can be expressed using the variance of the height function (see [QV]):

$$D^{TASEP}(t) = (4\chi t)^{-1} \sum_{x\in\mathbb{Z}} Var(h_t(x)) - 4\chi|x - (1-2\rho)t|. \tag{1.9}$$

This identity, the results of [FS] and some additional tightness bounds would imply the existence of the limit $D^{TASEP}(t)t^{-1/3}$ and even the limiting constant can be computed (see [FS] and [QV] for details). Unfortunately, the needed estimates are still missing, but from [FS] one can at least obtain a lower bound of the right order:

$$D^{TASEP}(t) > Ct^{1/3} \tag{1.10}$$

with a positive constant C (see [QV] for the proof).

In [QV] the resolvent approach is used to prove the following comparison theorem.

Theorem 1 (QV, 2006). *Let $D_1(t), D_2(t)$ be the diffusivities of two finite range exclusion processes in $d = 1$ with non-zero drift. There exists $0 < \beta, C < \infty$ such that*

$$C^{-1}\int_0^\infty e^{-\beta\lambda t} tD_1(t)dt \;\; \leq \;\; \int_0^\infty e^{-\lambda t} tD_2(t)dt \leq C\int_0^\infty e^{-\beta^{-1}\lambda t} tD_1(t)dt \tag{1.11}$$

Combining Theorem 1 with with (1.10) it was shown in [QV] that

Theorem 2 (QV, 2006). *For any finite range exclusion process in $d = 1$ with non-zero drift, $D(t) \geq Ct^{1/3}$ in the weak (Tauberian) sense.*

[QV] also converts the Tauberian bound into pointwise bound in the nearest neighbor case to get $D(t) \geq Ct^{1/3}(\log t)^{-7/3}$.

Just a few months later Balázs and Seppäläinen [BS], building on ideas of Ferrari and Fontes [FF], Ferrari, Kipnis and Saada [FKS], and Cator and Groeneboom [CG], proved the following theorem:

Theorem 3 (Balázs–Seppäläinen, 2006). *For any nearest neighbor asymmetric exclusion process in $d = 1$ there exists a finite constant C such that for all $t \geq 1$,*

$$C^{-1}t^{1/3} \leq D(t) \leq Ct^{1/3}. \tag{1.12}$$

Their proof uses refined and ingenious couplings to give bounds on the tail-probabilities of the distribution of the second class particle.

The aim of the present short note is to show how one can extend the results of [QV] using Theorem 3. Once we have the correct upper and lower bounds for $D(t)$ from (1.12) in the nearest neighbor case, we can strengthen the statement of Theorem 2 using the comparison theorem:

Theorem 4. *For any finite range exclusion process in $d = 1$ with non-zero drift, $D(t) = \mathcal{O}(t^{1/3})$ in the weak (Tauberian) sense: there exists a constant $0 < C < \infty$ such that*

$$C^{-1}\lambda^{-7/3} \leq \int_0^\infty e^{-\lambda t} tD(t)dt \leq C\lambda^{-7/3}. \tag{1.13}$$

Getting strict estimates for a function using the asymptotic behavior of its Laplace transform usually requires some regularity and unfortunately very little is known qualitatively about $D(t)$. However, in our case (as noted in [QV]), one can get an upper bound for $D(t)$ using an inequality involving H_{-1} norms:

Theorem 5. *For any finite range exclusion process in $d = 1$ with non-zero drift, there exists $C > 0$ such that for all $t \geq 1$,*

$$D(t) \leq Ct^{1/3}. \tag{1.14}$$

Note that in all these statements the constant $0 < C < \infty$ depends on the jump law $p(\cdot)$ as well as the density $0 < \rho < 1$.

2. Proofs

Theorem 4 immediately follows from Theorem 1 using TASEP and a general finite range exclusion, together with Theorem 3. To prove Theorem 5 one needs the Green-Kubo formula which relates the diffusivity to the time integral of current-current correlation functions:

$$D(t) = \sum\nolimits_z z^2 p(z) + 2\chi t^{-1} \int_0^t \int_0^s \langle\langle w_0, e^{uL} w_0 \rangle\rangle du ds. \tag{2.1}$$

Here w_x is the normalized microscopic flux

$$w_x = \frac{1}{\rho(1-\rho)} \sum\nolimits_z p(z)(\eta_{x+z} - \rho)(\eta_x - \rho), \tag{2.2}$$

L is the generator of the exclusion process and the inner product $\langle\langle \cdot, \cdot \rangle\rangle$ is defined for mean zero local functions ϕ, ψ as

$$\langle\langle \phi, \psi \rangle\rangle = E\left[\phi \sum\nolimits_x \tau_x \psi\right], \tag{2.3}$$

with τ_x being the appropriate shift operator.

(2.1) is proved in [LOY] (in the special case $p(1) = 1$, but the proof for general AEP is the same.) A useful variant is obtained by taking the Laplace transform,

$$\int_0^\infty e^{-\lambda t} tD(t)dt = \lambda^{-2}\left(\sum_z z^2 p(z) + 2\chi \|w\|_{-1,\lambda}^2\right) \tag{2.4}$$

where the H_{-1} norm corresponding to L is defined on a core of local functions by

$$\|\phi\|_{-1,\lambda} = \langle\langle \phi, (\lambda - L)^{-1}\phi\rangle\rangle^{1/2}. \tag{2.5}$$

We also need the following inequality which has appeared in several similar versions in the literature (e.g., [LY],[KL]).

Lemma 1. *Let w be the current for a finite range exclusion process in $d = 1$ with non-zero drift. Then,*

$$t^{-1}\sum_x E[\int_0^t w_0(s)ds \int_0^t w_x(s)ds] \leq 12\|w_0\|_{-1,t^{-1}}^2. \tag{2.6}$$

Proof. We will use the notation

$$\|\phi\|^2 = \langle\langle \phi, \phi\rangle\rangle, \tag{2.7}$$

note that the left-hand side of (2.6) is equal to $t^{-1}\|\int_0^t w_0(s)ds\|^2$. Let $\lambda > 0$ and $u_\lambda = (\lambda - L)^{-1}w_0$. Using Ito's formula together with the identity $Lu_\lambda = \lambda u_\lambda - w_0$ we get

$$u_\lambda(t) = u_\lambda(0) - \int_0^t (\lambda u_\lambda - w_0)(s)ds + M_t \tag{2.8}$$

where M_t is a mean zero martingale with

$$\begin{aligned} \|M_t\|^2 &= \int_0^t \langle\langle u_\lambda(s), -Lu_\lambda(s)\rangle\rangle\, ds = t\langle\langle u_\lambda, -Lu_\lambda\rangle\rangle \\ &= t\langle\langle u_\lambda, (\lambda - L)u_\lambda\rangle\rangle - t\lambda\|u_\lambda\|^2. \end{aligned} \tag{2.9}$$

Rearranging (2.8), applying the Schwarz Lemma and using stationarity

$$\|\int_0^t w_0(s)ds\|^2 \leq 8\|u_\lambda\|^2 + 4\|M_t\|^2 + 4\lambda^2\|\int_0^t u_\lambda(s)ds\|^2 \tag{2.10}$$

To bound the last term of (2.10) we again use the Schwartz Lemma and stationarity to get

$$\|\int_0^t u_\lambda(s)ds\|^2 \leq t\int_0^t \|u_\lambda(s)\|^2 ds = t^2\|u_\lambda\|^2. \tag{2.11}$$

Putting our estimates together,

$$\|\int_0^t w_0(s)ds\|^2 \leq (8 - 4t\lambda + 4t^2\lambda^2)\|u_\lambda\|^2 + 4t\langle\langle u_\lambda, (\lambda - L)u_\lambda\rangle\rangle \tag{2.12}$$

Setting $\lambda = t^{-1}$ and dividing the previous inequality by t:

$$\begin{aligned} t^{-1}\left\| \int_0^t w_0(s)ds \right\|^2 &\le 8\lambda \|u_\lambda\|^2 + 4\langle\langle u_\lambda, (\lambda - L)u_\lambda \rangle\rangle \\ &\le 12\langle\langle u_\lambda, (\lambda - L)u_\lambda \rangle\rangle = 12\|w_0\|^2_{-1,\lambda^{-1}}, \end{aligned}$$

which proves the lemma. □

Proof of Theorem 5. Theorem 4 and identity (2.4) gives

$$\|w_0\|^2_{-1,\lambda} \le C\lambda^{-1/3} \tag{2.13}$$

if $\lambda > 0$ is sufficiently small. To bound $D(t)$ we use the Green-Kubo formula (2.1) noting that the second term on the right is equal to $\chi t^{-1}\| \int_0^t w_0(s)ds\|^2$. Using Lemma 1 and then (2.13) to estimate this term we get that

$$D(t) \le \sum_z z^2 p(z) + 12\chi \|w_0\|^2_{-1,t^{-1}} \le Ct^{1/3} \tag{2.14}$$

for large enough t. □

References

[BDJ] J. Baik, P. Deift, K. Johansson, *On the distribution of the length of the longest increasing subsequence of random permutations.* J. Amer. Math. Soc. 12 (1999), no. 4, 1119–1178.

[BR] J. Baik, E.M. Rains, *Limiting distributions for a polynuclear growth model with external sources*, J. Stat. Phys. **100** (2000), 523-542.

[BS] M. Balázs, T. Seppälläinen, *Order of current variance and diffusivity in the asymmetric simple exclusion process*, arxiv.org/math.PR/0608400

[CG] E. Cator, P. Groeneboom, *Second class particles and cube root asymptotics for Hammersley's process*, Ann. Probab. **34** (2006), no. 4, 1273–1295.

[FF] P.A. Ferrari, L.R. Fontes, *Current fluctuations for the asymmetric simple exclusion process*, Ann. Probab. **22** (1994), no. 2, 820–832.

[FKS] P.A. Ferrari, C. Kipnis, E. Saada, *Microscopic structure of travelling waves in the asymmetric simple exclusion process*, Ann. Probab. **19** (1991), no. 1, 226–244.

[FS] P.L. Ferrari and H. Spohn, *Scaling limit for the space-time covariance of the stationary totally asymmetric simple exclusion process*,

[J] K. Johansson, *Shape fluctuations and random matrices*, Comm. Math. Phys. **242** (2003), 277–329.

[KL] C. Kipnis and C. Landim, Scaling limits of interacting particle systems. Grundlehren der Mathematischen Wissenschaften 320, Springer-Verlag, Berlin, New York, 1999.

[LOY] C. Landim, S. Olla, H.T. Yau, *Some properties of the diffusion coefficient for asymmetric simple exclusion processes.* Ann. Probab. 24 (1996), no. 4, 1779–1808.

[LQSY] C. Landim, J. Quastel, M. Salmhofer and H.-T. Yau, *Superdiffusivity of asymmetric exclusion process in dimensions one and two*, Comm. Math. Phys. **244** (2004), no. 3, 455–481.

[LY] C. Landim and H.-T. Yau, *Fluctuation-dissipation equation of asymmetric simple exclusion processes*,Probab. Theory Related Fields **108** (1997), no. 3, 321–356.

[PS] M. Prähofer and H. Spohn, *Current fluctuations for the totally asymmetric simple exclusion process*, In and out of equilibrium (Mambucaba, 2000), 185–204, Progr. Probab. **51**, Birkhäuser Boston, Boston, MA, 2002.

[QV] J. Quastel, B. Valkó, $t^{1/3}$ *Superdiffusivity of Finite-Range Asymmetric Exclusion Processes on* $\mathbb{Z}$, Comm. Math. Phys. **273** (2007), no. 2, 379–394.

[S] H. Spohn, Large Scale Dynamics of Interacting Particles, Springer-Verlag, 1991.

[V] S.R.S. Varadhan, *Lectures on hydrodynamic scaling*, Hydrodynamic limits and related topics (Toronto, ON, 1998), 3–40, Fields Inst. Commun. **27**, Amer. Math. Soc., Providence, RI, 2000.

[Y] H.-T. Yau, $(\log t)^{2/3}$ *law of the two dimensional asymmetric simple exclusion process*, Ann. of Math. (2) **159** (2004), no. 1, 377–405.

Jeremy Quastel
Departments of Mathematics and Statistics,
University of Toronto
e-mail: `quastel@math.toronto.edu`

Benedek Valkó
Departments of Mathematics and Statistics,
University of Toronto
e-mail: `valko@math.toronto.edu`

Progress in Probability, Vol. 60, 551–557

On the Role of Spatial Aggregation in the Extinction of a Species

Rinaldo B. Schinazi

Abstract. We compare two spatial stochastic models. The first, introduced by Schinazi (2005), shows that spatial aggregation may cause the extinction of a species in catastrophic times. The second shows that, for a certain range of parameters, spatial aggregation may help the survival of a species in non catastrophic times.

Mathematics Subject Classification (2000). 60K35, 92B05.

Keywords. Mass extinction, Allee effect, spatial stochastic model, contact process.

1. Spatial aggregation may be bad in catastrophic times

There have been many documented mass extinctions of all sorts of animals in the last 60,000 years, see Martin and Klein (1984). In some cases the extinctions have occurred at a point in time suspiciously close to the arrival time of a new group of humans. One mass extinction theory is that human hunters waged a blitzkrieg against some species, quickly exterminating millions of animals. One such example is the extermination of Moa birds in New Zealand, see Diamond (2000) and Holdaway and Jacomb (2000). It seems that in a matter of a few decades, after the first settlements of Polynesians, all of the estimated 160,000 large flightless Moas disappeared. How could a few dozens of hunters provoke such a disaster? One hypothesis is that these animals were not afraid of humans and were therefore very easy to kill. Schinazi (2005) proposed a new hypothesis. In addition to their naive behavior these animals may have lived in very large flocks and once the flock was found by hunters it could be easily killed off. This could also explain how the Moas could be exterminated in one of the world's most rugged lands. Our hypothesis might be difficult to test on Moas. But there are documented examples of extinctions of animals living in huge flocks almost up to the end of

Partially supported by the National Science Foundation under grant 0701396.

their species, for instance passenger pigeons (see Austin (1983)) or the American bison.

Schinazi (2005) proposed a mathematical model that, at least in theory, shows that animals living in large flocks are more susceptible to mass extinctions than animals living in small flocks. More precisely, if the maximum flock size is above a certain threshold then the population is certain to get extinct while if the maximum flock size is below the threshold there is a strictly positive probability that the population will survive.

The model is spatial and stochastic on the lattice $\mathbb{Z}^d$, typically $d = 2$. Each site of the lattice may host a flock of up to N individuals. Each individual may give birth to a new individual at the same site at rate ϕ until the maximum of N individuals has been reached at the site. Once the flock reaches N individuals then, and only then, it starts giving birth on each of the $2d$ neighboring sites at rate λ. This rule is supposed to mimic the fact that individuals like to stay in a flock and will give birth outside the flock only when the flock attains the maximum number N that a site may support. Finally, disaster strikes at rate 1, that is, the whole flock disappears. This rule mimics an encounter with greedy hunters or a new disease. Both disasters seem to have stricken the American buffalo and the passenger pigeon.

We now write the above description mathematically. Each site x of $\mathbb{Z}^d$ may be in one of the states: 0, 1, 2,...,N and this state is the size of the flock at x. The model is a continuous time Markov process that we denote by η_t. Let $n_N(x, \eta_t)$ be the number of neighbors of site x, among its $2d$ nearest neighbors, that are in state N at time t.

Assume that the model is in configuration η, then the state at a given site x evolves as follows:

$$i \to i+1 \text{ at rate } i\phi + \lambda n_N(x,\eta) \text{ for } 0 \le i \le N-1$$
$$i \to 0 \text{ at rate } 1 \text{ for } 1 \le i \le N$$

We will have two models in this paper. We call the model above Model I. We now explain the transition rules in words. Assume that a site x at a given time has j neighbors in state N, $j = 0, 1, 2, 3$ or 4 in $d = 2$. Then, if site x is in state $i \le N-1$ there are two possibilities. Either, after an exponential random time T with rate $a = i\phi + j\lambda$ (i.e., $P(T > t) = \exp(-at)$), site x goes to state $i+1$ or after an exponential random time with rate 1 site x goes to state 0. The first of the two random times that occur determines the outcome. If site x is in state N then it may give birth to an individual on a neighboring site (provided the neighboring site is not full) with rate λ or it may go to state 0 with rate 1.

In the special case where there is a maximum of one individual per site (i.e., $N = 1$) this model is well known and is called the contact process (see Liggett (1999)). For the contact process, there exists a critical value λ_c (that depends on the dimension d of the lattice) such that the population dies out if and only if $\lambda \le \lambda_c$.

Theorem 1. *Assume that $\lambda > \lambda_c$ (the critical value of the contact process) and that $\phi > 0$. Then, there is a critical positive integer N_c, depending on λ and ϕ, such that for Model I on $\mathbb{Z}^d$, for any $d \geq 1$, the population dies out for $N > N_c$ and survives for $N < N_c$.*

Theorem 1 is a particular case of a result proved in Schinazi (2005), see Corollary 2 there.

2. Aggregation may be good in non catastrophic times

We now consider a model with the same rules for births (in particular a flock may give birth outside its site only when it has N individuals) but with a different rule for deaths: they now occur one by one. We call the following Model II.

$$i \to i+1 \text{ at rate } i\phi + \lambda n_N(x,\eta) \text{ for } 0 \leq i \leq N-1$$
$$i \to i-1 \text{ at rate } i \text{ for } 1 \leq i \leq N$$

This models the population in the absence of greedy hunters. As the reader will see below the role of N is strikingly different in Models I and II.

Theorem 2. *Consider Model II on Z^d for a dimension $d \geq 2$. Assume that $\lambda > 0$ and $\phi > 1$. There is a critical value $N_c(\lambda, \phi)$ depending on λ and ϕ such that if $N > N_c$ then, starting from any finite number of individuals, the population has a strictly positive probability of surviving.*

Theorem 2 tells us that survival is possible for any internal birth rate $\phi > 1$ and any external birth rate $\lambda > 0$ provided N is large enough. On the other hand, if $N = 1$ and $\lambda < \lambda_c$ then the species dies out. Hence, for some parameters spatial aggregation makes the species survive. This is consistent with the so-called Allee effect in Ecology. See, for instance, Stephens and Sutherland (1999).

We believe Theorem 2 holds in $d = 1$ as well but our (elementary) proof requires $d \geq 2$. This is so because we compare our stochastic model to a percolation model on $\mathbb{Z}^d$ for which there is percolation only for $d \geq 2$. In order to prove Theorem 2 in $d = 1$ we would need to use time as an extra dimension. This has been done many times since Bramson and Durrett (1988). For a model related to ours see the proof of Theorem 5 in Belhadji and Lanchier (2006) which holds for $d = 1$. Since we have in mind applications in $d = 2$ and the proof is simpler (and less known) in $d \geq 2$ we decided to skip the case $d = 1$.

As the next result shows, whether ϕ is larger or smaller than 1 plays a critical role in the behavior of this model.

Theorem 3. *Assume that $0 < \phi \leq 1$. If $N \geq 2d\lambda/\phi$ then, starting from any finite population, the population dies out for Model II on $\mathbb{Z}^d$.*

If the internal birth rate is less than 1 then excessive aggregation makes the species die out even in non catastrophic times.

Theorems 2 and 3 suggest the following conjecture. If $\phi > 1$ then the survival probability in Model II increases with N while if $\phi \leq 1$ then the survival probability decreases with N.

3. Proof of Theorem 2

There are two parts in our proof. In the first part we will show that if site x has N individuals and y is one of its $2d$ nearest neighbors then x will eventually give birth to an individual in y that will, by internal births alone, generate $N-1$ individuals and therefore make y reach state N. Moreover, we will show that the probability of the preceding event converges to 1 as N goes to infinity. In the second part we will compare the process to a percolation model.

First part. To prove the first part we need three steps.

Step 1. Let x be a site in the lattice $\mathbb{Z}^d$. Assume that x is in state N. Let R_N be the number of times site x returns to state N before dropping for the first time to state 0. We show that as N goes to infinity the probability that $\{R_N \geq N^2\}$ converges to 1. Moreover, we will show that this is true even if we ignore the external births (those with rate λ) and only consider internal births (those with rate ϕ).

Assume that site x has i individuals at some time, where $1 \leq i \leq N-1$. Then x goes to state $i+1$ at rate $i\phi$ or to $i-1$ at rate i. Hence, ignoring possible births from the outside, the number of individuals at x is a simple random walk where

$$i \to i+1 \text{ with probability } p = \frac{\phi}{\phi+1} \text{ for } 1 \leq i \leq N-1$$

$$i \to i-1 \text{ with probability } q = \frac{1}{\phi+1} \text{ for } 1 \leq i \leq N$$

The probability that, starting at $N-1$, this random walk returns to N before reaching 0 is given by the classical ruin problem formula:

$$a(N,\phi) = \frac{1-(q/p)^{N-1}}{1-(q/p)^N}.$$

See, for instance, (4.4) in I.4 in Schinazi (1999). By the Markov property we have

$$P(R_N \geq r) = a(N,\phi)^r.$$

Using that $a(N,\phi) > 1 - 2(q/p)^{N-1}$ we get

$$P(R_N \geq r) > \exp(-2r(q/p)^{N-1}).$$

Hence,

$$\lim_{N\to\infty} P(R_N \geq N^2) = 1,$$

where we use that since $\phi > 1$, $q < p$. Of course, the limit above holds for any power of N.

Step 2. We show that since site x is likely to return to state N at least N^2 times it will give birth on one of its $2d$ neighbors y at least $\sqrt{N}$ times. Let B_N be the number of births from site x to site y. We have

$$P(B_N \geq n) \geq P(B_N \geq n | R_N \geq N^2) P(R_N > N^2).$$

At each return to state N there are two possibilities either there is a birth from site x onto site y (at rate λ) or there is a death at site x at rate N. Thus, the probability of a birth at y before a death at x is

$$\frac{\lambda}{\lambda + N}.$$

Moreover, at each return what happens is independent of what happened at the preceding return. Conditioning on $\{R_N > N^2\}$ the number of births, B_N, from x onto y is therefore larger than a binomial random variable C_N with parameters N^2 and $\frac{\lambda}{\lambda+N}$. For every real $a > 0$ we have

$$\lim_{N \to \infty} \frac{\text{Var}(C_N)}{N^{1+2a}} = 0.$$

That is,

$$\frac{C_N - E(C_N)}{N^{1/2+a}}$$

converges to 0 in L^2 and hence in probability. In particular,

$$\lim_{N \to \infty} P(C_N > E(C_N) - N^{1/2+a}) = 1.$$

By picking a in $(0, 1/2)$ we get that

$$\lim_{N \to \infty} P(C_N > N^{1/2}) = 1.$$

Since,

$$\begin{aligned} P(B_N > N^{1/2}) &\geq P(B_N > N^{1/2} | R_N \geq N^2) P(R_N \geq N^2) \\ &\geq P(C_N > N^{1/2}) P(R_N > N^2) \end{aligned}$$

and that each probability on the r.h.s. converges to 1 as N goes to infinity, we have

$$\lim_{N \to \infty} P(B_N > N^{1/2}) = 1.$$

Step 3. We show that given that there are at least $N^{1/2}$ births at site y, at least one of these individuals generates, by internal births only, $N - 1$ individuals so that y eventually reaches state N. Every time there is a birth at y it starts a birth and death chain with transition rates:

$$\begin{aligned} &i \to i + 1 \text{ at rate } i\phi \text{ for } i \geq 1 \\ &i \to i - 1 \text{ at rate } i \text{ for } i \geq 1 \end{aligned}$$

Since $\phi > 1$, this birth and death chain is transient (see for instance Proposition I.4.1 in Schinazi (1999)) and therefore there is a positive probability, $q(\phi)$, that starting in state 1 it will never be in state 0 and will go on to infinity.

Thus, the probability that site y will reach N in Model II is at least as large as the probability that one of the birth and death chains is transient.

For x and y nearest neighbors, let E_{xy} be the event that, given that x starts in state N, it gives birth to at least one individual on y whose associated birth and death chain is transient. We have

$$P(E_{xy}) \geq P(E_{x,y}|B_N > N^{1/2})P(B_N > N^{1/2}) \geq (1-(1-q(\phi))^{N^{1/2}})P(B_N > N^{1/2}).$$

As N goes to infinity $P(E_{xy})$ approaches 1.

Second part. In this part we compare Model II to a percolation model. We follow closely Kuulasmaa (1982). Between any two nearest neighbors x and y in Z^d we draw a directed edge from x to y. We declare the directed edge open if the event E_{xy} happens. This defines a percolation model. Note that the probability of the directed edge xy be open is the same for all edges xy and can be made arbitrarily close to 1 by taking N large enough. By comparing this percolation model to a site percolation model it can be shown that there is a strictly positive probability of an infinite open path that starts at the origin of Z^d, provided the dimension d is at least 2. See the proof of Theorem 3.2 in Kuulasmaa (1982). It is important for this comparison to work that edges starting from different vertices are independently open. It is easy to check that if $x \neq z$ then the events E_{xy} and E_{zt} are independent for any y and t nearest neighbors of x and z, respectively.

An infinite open path starting from the origin is one way for the population, started with N individuals at the origin, to survive forever. This completes the proof of Theorem 2.

4. Proof of Theorem 3

Assume that at a certain time we have a total population of $n \geq 1$ individuals in Model II. The population may lose one individual or gain one. We start by examining the birth rate. Let k, $0 \leq k \leq n/N$, be the number of sites with N individuals each, this accounts for kN individuals. The other $n - kN$ individuals are in sites where there are $N - 1$ or less individuals per site.

If site x is in state N then it gives birth on nearest neighbor y with rate λ, provided y is not in state N. Since there are $2d$ nearest neighbors the birth rate from a site with N individuals is at most $2d\lambda$. The other $n - kN$ individuals all give birth with rate ϕ. Thus, the total birth rate for n individuals occupying k sites in Model II is at most

$$2d\lambda k + (n - kN)\phi = k(2d\lambda - N\phi) + n\phi \leq n\phi$$

since we assume that $2d\lambda \leq N\phi$.

The death rate for n individuals is n.

Consider now the birth and death chain with the following rates

$$n \to n+1 \text{ at rate } n\phi \text{ for } n \geq 1,$$
$$n \to n-1 \text{ at rate } n \text{ for } n \geq 1.$$

The total birth rate in Model II is less than this birth rate and the total death rate is the same. Hence, if this birth and death chain dies out with probability 1 so does the population in Model II. It is well known that this birth and death chain dies out if and only if $\phi \leq 1$ (see for instance Schinazi (1999)). This concludes the proof of Theorem 3.

References

[1] O.L. Austin Jr. (1983) *Birds of the world*, Golden Press, New York.

[2] L. Belhadji and N. Lanchier (2006). Individual versus cluster recoveries within a spatially structured population. Ann. Appl. Probab. 16, 403–422.

[3] M. Bramson and R. Durrett (1988). A simple proof of the stability criterion of Gray and Griffeath. Probability Theory and Related Fields, 80, 293–298.

[4] J. Diamond (2000). Blitzkrieg against the Moas. Science, 287, 2170–2172.

[5] R.N. Holdaway and C. Jacomb (2000). Rapid extinction of the Moas (Aves:Dinornithiformes):model, test and implications. Science, 287, 2250–2258.

[6] K. Kuulasmaa (1982). The spatial general epidemic and locally dependent random graphs. Journal of Applied Probability, 19, 745–758.

[7] T. Liggett (1999). *Stochastic interacting systems: contact, voter and exclusion processes.* Springer.

[8] P.S. Martin and R.G. Klein (1984). *Quarternary extinctions.* The University of Arizona Press, Tucson Arizona, USA.

[9] R.B. Schinazi (1999). *Classical and spatial stochastic processes.* Birkhäuser.

[10] R.B. Schinazi (2005) Mass extinctions: an alternative to the Allee effect. Annals of Applied Probability, 15, 984–991.

[11] P.A. Stephens and W.J. Sutherland (1999). Consequences of the Allee effect for behaviour, ecology and conservation. TREE, 14, 401–405.

Rinaldo B. Schinazi
University of Colorado at Colorado Springs
1420 Austin Bluffs Pkwy
Colorado Springs, CO 80933, USA
e-mail: rschinaz@uccs.edu

Progress in Probability, Vol. 60, 559–574

Systems of Random Equations. A Review of Some Recent Results

Mario Wschebor

Abstract. We review recent results on moments of the number of real roots of random systems of equations. The emphasis is on large polynomial systems, although the last section is on non-polynomial ones. We give no proofs, excepting for a short proof of the now classical Shub-Smale Theorem (1993) which illustrates, in this special case, about the methods based upon Rice formula and the theory of random fields.

Mathematics Subject Classification (2000). Primary 60G60, 14Q99; Secondary 30C15.

Keywords. Random polynomial systems, Rice formula.

1. Introduction

Let us consider m polynomials in m variables with real coefficients

$$X_i(t) = X_i(t_1, \ldots, t_m),\ i = 1, \ldots, m.$$

We use the notation

$$X_i(t) := \sum_{\|j\| \le d_i} a_j^{(i)}\, t^j, \tag{1.1}$$

where $j := (j_1, \ldots, j_m)$ is a multi-index of non-negative integers, $\|j\| := j_1 + \cdots + j_m$, $j! := j_1! \ldots j_m!$, $t^j := t_1^{j_1} \ldots t_m^{j_m}$, $a_j^{(i)} := a_{j_1,\ldots,j_m}^{(i)}$. The degree of the ith polynomial is d_i and we assume that $d_i \ge 1\,\forall i$.

Let $N^X(V)$ be the number of roots of the system of equations

$$X_i(t) = 0, \quad i = 1, \ldots, m. \tag{1.2}$$

lying in the Borel subset V of $\mathbb{R}^m$. We denote $N^X = N^X(\mathbb{R}^m)$.

Let us randomize the coefficients of the system. Generally speaking, little is known on the probability distribution of the random variables $N^X(V)$ or N^X, even for simple choices of the probability law on the coefficients. In the case of one equation in one variable, a certain number of results have been known since a long time, starting in the thirties with the work of Bloch and Polya [6] and Littlewood and Offord [12] and especially, of Marc Kac [10]. See for example the book by Bharucha-Reid and Sambandham [5].

In this paper we consider systems with $m > 1$, i.e., more than one equation, which appears to be quite different and much harder than one equation only. In fact, we will be especially interested in large systems, in the sense that $m \gg 1$. The first important result in this context is the Shub and Smale Theorem [15], in which the authors computed by means of a simple formula the expectation of N^X when the coefficients are Gaussian, centered independent random variables with certain specified variances (see Theorem 3.1 below and also the book by Blum, Cucker, Shub and Smale [7]). Extensions of their work, including new results for one polynomial in one variable, can be found in the review paper by Edelman and Kostlan [9]. See also Kostlan [11].

Our aim in this paper is to review some recent results which extend the above to other probability laws on the coefficients. Also, in the Kostlan-Shub-Smale model, we go beyond the computation of first moments and describe some results on the variance of the number of roots. The main tool is Rice formula to compute the factorial moments of the number of zeros of a random field (see section 2 below). For variances, we are only giving here some partial results on the asymptotic behavior of variances as $m \to +\infty$. A major problem is to show weak convergence of some renormalization of N^X, under the same asymptotic.

We are not giving proofs here, excepting a short proof of the Shub-Smale Theorem, aiming to show the kind of use one can make of Rice formula for this purpose, in a case in which the computations turn out to be quite simple.

In section 5 we consider "smooth analysis", which means that we start with a non-random system, perturb it with some noise, and the question is what can we say about the number of roots of the perturbed system, under some reasonable hypotheses on the relationship between "signal" and "noise".

Finally, in Section 6 we consider random systems having a probability law which is invariant under isometries and translations of the underlying Euclidean space, hence are non-polynomial. These systems are interesting by themselves and under some general conditions, one can use similar methods to compute the expected number of roots per unit volume, as well as to understand the behavior of the variance as the the number of unknowns m tends to infinity, which turns out to be strikingly opposite to the one in the Kostlan-Shub-Smale model for polynomial systems.

All the above concerns "square" systems. We have not included results on random systems having less equations than unknowns. If the system has n equations and m unknowns with $n < m$, generically the set of solutions will be $(m-n)$-dimensional, and the description of the geometry becomes more complicated (and more interesting) than for $m = n$. A recent contribution to the calculation of the expected value of certain parameters describing the geometry of the (random) set of solutions is in P. Bürgisser [8].

2. Rice formulas for the number of roots of random fields

In this section we review Rice formulas. The first theorem contains the formula for the expectation, the second one for higher order factorial moments. For proofs and details, see for example [2] (or [4], where a simpler self-contained proof is given).

Theorem 2.1 (Rice formula 1). *Let $Z : U \to \mathbb{R}^d$ be a random field, U an open subset of $\mathbb{R}^d$ and $u \in \mathbb{R}^d$ a fixed point in the codomain. Assume that:*

(i) *Z is Gaussian,*
(ii) *almost surely the function $t \rightsquigarrow Z(t)$ is of class $\mathcal{C}^1$,*
(iii) *for each $t \in U$, $Z(t)$ has a non-degenerate distribution (i.e., $\mathrm{Var}(Z(t))$ is positive definite),*
(iv) $\mathrm{P}\{\exists t \in U, Z(t) = u, \det(Z'(t)) = 0\} = 0$

Then, for every Borel set B contained in U, one has

$$\mathrm{E}\left(N_u^Z(B)\right) = \int_B \mathrm{E}\left(|\det(Z'(t))|/Z(t) = u\right)\, p_{Z(t)}(u)dt. \tag{2.1}$$

If B is compact, both sides in (2.1) are finite.

Theorem 2.2 (Rice formula 2). *Let k, $k \geq 2$ be an integer. Assume the same hypotheses as in Theorem 2.1 excepting (iii) which is replaced by:*

(iii′) *for $t_1, \ldots, t_k \in U$ pairwise different values of the parameter, the distribution of*

$$\left(Z(t_1), \ldots, Z(t_k)\right)$$

does not degenerate in $(\mathbb{R}^d)^k$. Then for every Borel set B contained in U, one has

$$\begin{aligned}\mathrm{E}&\left[\left(N_u^Z(B)\right)\left(N_u^Z(B)-1\right)\ldots\left(N_u^Z(B)-k+1\right)\right]\\ &= \int_{B^k} \mathrm{E}\Big(\prod_{j=1}^{k} |\det\left(Z'(t_j)\right)|/Z(t_1) = \cdots = Z(t_k) = u\Big)\\ &\qquad\qquad \cdot\, p_{Z(t_1),\ldots,Z(t_k)}(u,\ldots,u)dt_1\ldots dt_k,\end{aligned} \tag{2.2}$$

where both members may be infinite.

Remarks:

- In general, it may be non-trivial to verify conditions (iv) and (iii′) in the above theorems. However, in the case of the random systems that we will consider below, it is not hard to check them.
- Similar formulas hold true if instead of a Borel subset of an open set $U \subset \mathbb{R}^d$, the set B on which we count the number of roots of the random system $Z(t) = u$ is a Borel subset of a d-dimensional smooth manifold embedded in U, an open subset of $\mathbb{R}^{d'}$ with $d' > d$. In this case, one has to replace in formulas (2.1) and (2.2) the Lebesgue measure by the volume measure on the manifold and the free derivatives of the random field Z by the derivatives along the manifold. The proof of this extension is standard, since both formulas are local.

3. The Shub-Smale model

We say that (1.2) is a Shub-Smale system, if the coefficients

$$\{a_j^{(i)} : i = 1, \dots, m; \ \|j\| \le d_i\}$$

are centered independent Gaussian random variables, such that

$$\mathrm{Var}\big(a_j^{(i)}\big) = \binom{d_i}{j} = \frac{d_i!}{j!(d_i - \|j\|)!} \tag{3.1}$$

3.1. Expectation of N^X

Theorem 3.1 (Shub-Smale [15]**).** *Let the system* (1.2) *be a Shub-Smale system. Then,*

$$\mathrm{E}\big(N^X\big) \;=\; \sqrt{D} \tag{3.2}$$

where $D = d_1 \dots d_m$ is the Bézout number of the polynomial system.

Proof. For $i = 1, \dots, m$, let $\tilde{X}_i$ denote the homogeneous polynomial of degree d_i in $m+1$ variables associated to X_i, that is:

$$\tilde{X}_i(t_0, t_1, \dots, t_m) = \sum_{\sum_{h=0}^{m} j_h = d_i} a^{(i)}_{j_1,\dots,j_m} t_0^{j_0} t_1^{j_1} \dots t_m^{j_m}.$$

Y_i denotes the restriction of $\tilde{X}_i$ to the unit sphere S^m in $\mathbb{R}^{m+1}$. It is clear that

$$N^X \;=\; \frac{1}{2} N^Y(S^m). \tag{3.3}$$

A simple computation using (3.1) and the independence of the coefficients, shows that $\tilde{X}_1, \dots, \tilde{X}_m$ are independent Gaussian centered random fields, with covariances given by:

$$r^{\tilde{X}_i}(t,t') = \mathrm{E}\big(\tilde{X}_i(t)\tilde{X}_i(t')\big) = \langle t, t' \rangle^{d_i}, \ \ t, t' \in \mathbb{R}^{m+1}, \ \ i = 1, \dots, m. \tag{3.4}$$

Here $\langle ., . \rangle$ denotes the usual scalar product in $\mathbb{R}^{m+1}$.

For $\mathrm{E}(N^Y(S^m))$ we use Rice formula. On account of the second remark after the statement of this formula, we have:

$$\mathrm{E}\big(N^Y(S^m)\big) = \int_{S^m} \mathrm{E}\big(|\det(Y'(t))| / Y(t) = 0\big) \frac{1}{(2\pi)^{m/2}} \sigma_m(dt), \tag{3.5}$$

where $\sigma_m(dt)$ stands for the m-dimensional geometric measure on S^m. (3.5) follows easily from the fact that for each $t \in S^m$, the random variables $Y_1(t), \dots, Y_m(t)$ are i.i.d. standard normal.

Since $\mathrm{E}(Y_i^2(t)) = 1$ for all $t \in S^m$, on differentiating under the expectation sign, we see that for each $t \in S^m$, $Y(t)$ and $Y'(t)$ are independent, and the condition can be erased in the conditional expectation in the right-hand side of (3.5).

Since the law of $Y'(t)$ is invariant under the isometries of $\mathbb{R}^{m+1}$ it suffices to compute the integrand at one point of the sphere. Denote the canonical basis of $\mathbb{R}^{m+1}$ by $\{e_0, e_1, \dots, e_m\}$. Then:

$$\begin{aligned} \mathrm{E}\big(N^Y(S^m)\big) &= \sigma_m(S^m)\frac{1}{(2\pi)^{m/2}}\mathrm{E}\big(|\det(Y'(e_0))|\big) \\ &= \frac{2\pi^{(m+1)/2}}{\Gamma(\frac{m+1}{2})}\frac{1}{(2\pi)^{m/2}}\mathrm{E}\big(|\det(Y'(e_0))|\big). \end{aligned} \tag{3.6}$$

To compute the probability law of $Y'(e_0)$, let us write it as an $m \times m$ matrix with respect to the orthonormal basis $e_1, \dots, e_m$ of the tangent space to S^m at e_0. This matrix is

$$\Big(\Big(\frac{\partial \tilde{X}_i}{\partial t_j}(e_0)\Big)\Big)_{i,j=1,\dots,m}$$

and

$$\mathrm{E}\Big(\frac{\partial \tilde{X}_i}{\partial t_j}(e_0)\frac{\partial \tilde{X}_{i'}}{\partial t_{j'}}(e_0)\Big) = \delta_{ii'}\frac{\partial^2 r^{\tilde{X}_i}}{\partial t_j \partial t'_{j'}}\Big|_{t=t'=e_0} = d_i \delta_{ii'}\delta_{jj'}.$$

The last equality follows computing derivatives of the function $r^{\tilde{X}_i}$ given by (3.4). So,

$$\det(Y'(e_0)) = \sqrt{D}\det(G) \tag{3.7}$$

where G is an $m \times m$ matrix with i.i.d. Gaussian standard entries.

To finish, we only need to compute $\mathrm{E}(|\det(G)|)$. This is fairly standard. One way to do it, is to observe that $|\det(G)|$ is the volume (in $\mathbb{R}^m$) of the set

$$\{v \in \mathbb{R}^m : v = \sum_{k=1}^{m} \lambda_k\, g_k,\ 0 \le \lambda_k \le 1,\ k = 1, \dots, m\}$$

where $\{g_1, \dots, g_m\}$ are the columns of G. Then using the invariance of the standard normal law in $\mathbb{R}^m$ with respect to isometries, we get:

$$\mathrm{E}(|\det(G)|) = \prod_{k=1}^{m} \mathrm{E}(\|\eta_k\|)$$

where η_k is standard normal in $\mathbb{R}^k$. An elementary computation gives:

$$\mathrm{E}(\|\eta_k\|) = \sqrt{2}\frac{\Gamma((k+1)/2)}{\Gamma(k/2)},$$

which implies:

$$\mathrm{E}(|\det(G)|) = \frac{1}{\sqrt{2\pi}}\, 2^{(m+1)/2}\, \Gamma((m+1)/2).$$

Using (3.7), (3.6) and (3.3), we get the result. □

Remark 3.2. When the hypotheses of Theorem 3.1 are verified, and moreover all the degrees d_i $(i = 1, \dots, m)$ are equal, formula (3.2) was first proved by Kostlan. In what follows, we will call such a model the KSS (Kostlan-Shub-Smale) model.

3.2. Variance of the number of roots

For simplicity, we restrict this subsection to the KSS model. In this case, a few asymptotic results have been proved on variances, when the number m of unknowns tends to ∞. More precisely, consider the normalized random variable

$$n^X = \frac{N^X}{\sqrt{D}}.$$

It is an obvious consequence of Theorem 3.1 that $\mathrm{E}(n^X) = 1$. In [17] the following theorem was proved:

Theorem 3.3. *Assume that the random polynomial system* (1.1) *is a KSS system with common degree equal to d, $d \geq 2$, and the $d \leq d_0 < \infty$, where d_0 is some constant independent of m. Then,*

(a) $\limsup_{m\to+\infty} \mathrm{Var}(n^X) \leq 1$.

(b) *If $d \geq 3$, one has* $\lim_{m\to+\infty} \mathrm{Var}(n^X) = 0$.

Remarks on the statement of Theorem 3.3

- The results in Theorem 3.3 can be made more precise.

  One can prove that (b) holds true in all cases, i.e., for $d \geq 2$. In fact, it is possible to describe the equivalent of $\mathrm{Var}(n^X)$ as $m \to +\infty$, which depends on the common degree d. Denoting $\sigma^2_{m,d} = \mathrm{Var}(n^X)$, the results are the following:

  1 $\sigma^2_{m,2} \approx \frac{1}{2}\frac{\log\ m}{m}$.

  2 $\sigma^2_{m,3} \approx \frac{3}{2}\frac{\log\ m}{m^2}$.

  3 For $d \geq 4$ one has $\sigma^2_{m,d} \approx \frac{K_d}{m^{(d-2)\wedge 3}}$ where the constants K_d can be computed in explicit form and $K_d = K_5$ for $d \geq 5$.

  The proof of these results is based upon Rice formula for the second factorial moment of the number of roots, and requires somewhat long calculations ([18]).
- Notice that in the statement of Theorem 3.3 (and also, mutatis mutandis, in the improvements contained in the previous remark) the common degree d of the polynomials can vary with m provided it remains bounded.
- A corollary of the fact that $\mathrm{Var}(n^X)$ tends to zero as $m \to +\infty$ is that

$$n^X = \frac{N^X}{d^{m/2}}$$

  tends to 1 in probability. It is also possible to obtain the same type of result if we let d tend to infinity, slowly enough. For the time being, the validity of a central limit theorem as m tends to infinity remains an open question.

4. More general models

The probability law of the Shub-Smale model defined in Section 3 has the simplifying property of being invariant under isometries of the underlying Euclidean space

$\mathbb{R}^m$. In this section we present the extension of formula (3.2) to general systems which share the same invariance property. This paragraph follows [3], where one can find proofs.

More precisely, we require the polynomial random fields X_i $(i = 1, \dots, m)$ to be centered, Gaussian, independent and their covariances

$$r^{X_i}(s,t) = \mathrm{E}(X_i(s)X_i(t))$$

to be invariant under the orthogonal group of $\mathbb{R}^m$, i.e., $r^{X_i}(Us, Ut) = r^{X_i}(s,t)$ for any orthogonal linear transformation U and any pair $s, t \in \mathbb{R}^m$. This implies in particular that the coefficients $a_j^{(i)}$ remain independent for different i's but can now be correlated from one j to another for the same value of i. It is easy to check that this implies that for each $i = 1, \dots, m$, the covariance $r^{X_i}(s,t)$ is a function of the triple $(\langle s,t\rangle, \|s\|^2, \|t\|^2)$ ($\|.\|$ is Euclidean norm in $\mathbb{R}^m$). It can also be proved (Spivak [16]) that this function is in fact a polynomial with real coefficients, say $Q^{(i)}$

$$r^{X_i}(s,t) = Q^{(i)}(\langle s,t\rangle, \|s\|^2, \|t\|^2), \tag{4.1}$$

satisfying the symmetry condition

$$Q^{(i)}(u,v,w) = Q^{(i)}(u,w,v). \tag{4.2}$$

A natural question is which are the polynomials $Q^{(i)}$ such that the function in the right-hand side of (4.1) is a covariance, that is, non-negative definite. A simple way to construct a class of covariances of this type is to take

$$Q^{(i)}(u,v,w) = P(u, vw) \tag{4.3}$$

where P is a polynomial in two variables with non-negative coefficients. In fact, the functions $(s,t) \rightsquigarrow \langle s,t\rangle$ and $(s,t) \rightsquigarrow \|s\|^2\|t\|^2$ are covariances and the set of covariances is closed under linear combinations with non-negative coefficients as well as under multiplication, so that $P(\langle s,t\rangle, \|s\|^2\,\|t\|^2)$ is also the covariance of some random field.

The situation becomes simpler if one considers only functions of the scalar product, i.e.,

$$Q(u,v,w) = \sum_{k=0}^{d} c_k\; u^k.$$

The necessary and sufficient condition for $\sum_{k=0}^{d} c_k\;\langle s,t\rangle^k$ to be a covariance is that $c_k \geq 0\ \forall\ k = 0, 1, \dots, d$. In that case, it is the covariance of the random field $X(t) := \sum_{\|j\| \leq d} a_j\; t^j$ where the $a_j's$ are centered, Gaussian, independent random variables, $\mathrm{Var}(a_j) = c_{\|j\|}\frac{\|j\|!}{j!}$. The Shub-Smale model is the special case corresponding to the choice $c_k = \binom{d}{k}$.

The general description of the polynomial covariances which are invariant under isometries, is in Kostlan [11], part II.

We now state the extension of the Shub-Smale formula to the general case.

Theorem 4.1. *Assume that the X_i are independent centered Gaussian polynomial random fields with covariances $r^{X_i}(s,t) = Q^{(i)}(\langle s,t\rangle, \|s\|^2, \|t\|^2)$ $(i = 1,\dots,m)$.*

Let us denote by $Q_u^{(i)}, Q_w^{(i)}, Q_{uv}^{(i)}, \dots$ the partial derivatives of $Q^{(i)}$. We assume that for each $i = 1,\dots,m$, $Q^{(i)}(x,x,x)$ and $Q_u^{(i)}(x,x,x)$ do not vanish for $x \geq 0$. Set:

$$q_i(x) := \frac{Q_u^{(i)}}{Q^{(i)}}$$

$$r_i(x) := \frac{Q^{(i)}\big(Q_{uu}^{(i)} + 2Q_{uv}^{(i)} + 2Q_{uw}^{(i)} + 4Q_{vw}^{(i)}\big) - \big(Q_u^{(i)} + Q_v^{(i)} + Q_w^{(i)}\big)^2}{(Q^{(i)})^2}$$

where the functions in the right-hand sides are always computed at the triplet (x,x,x).

Put

$$h_i(x) := 1 + x\frac{r_i(x)}{q_i(x)}.$$

Then, for all Borel sets V we have

$$\mathrm{E}\big(N^X(V)\big) = \kappa_m \int_V \big(\prod_{i=1}^m q_i(\|t\|^2)\big)^{1/2} E_h(\|t\|^2)dt. \tag{4.4}$$

In this formula,

$$E_h(x) := \mathrm{E}\big((\sum_{i=1}^m h_i(x)\xi_i^2)^{1/2}\big)$$

where $\xi_1,\dots,\xi_m$ are i.i.d. standard normal in $\mathbb{R}$ and

$$\kappa_m = \frac{1}{\sqrt{2\pi}}\frac{\Gamma(m/2)}{\pi^{m/2}}.$$

4.1. Examples

1. Let $Q^{(i)}(u,v,w) = Q^{l_i}(u)$ for some polynomial Q. We get:

$$q_i(x) = l_i q(x) = l_i \frac{Q'(x)}{Q(x)}\ ,\ h_i(x) = h(x) = 1 - x\frac{Q'^2(x) - Q(x)Q''(x)}{Q(x)Q'(x)}.$$

Applying formula (4.4) with $V = \mathbb{R}^m$, and using polar coordinates:

$$\mathrm{E}(N^X) = \frac{2}{\sqrt{\pi}}\frac{\Gamma((m+1)/2)}{\Gamma(m/2)}\sqrt{l_1\dots l_m}\int_0^\infty \rho^{m-1}q(\rho^2)^{m/2}\sqrt{h(\rho^2)}d\rho. \tag{4.5}$$

2. If in example **1.** we put $Q(u) = 1+u$ we get the Shub-Smale model. Replacing in (4.5) an elementary computation reproduces (3.2).

3. A variant of Shub-Smale is taking $Q^{(i)}(u) = 1+u^d$ for all $i = 1,\dots,m$, which yields

$$q(x) = q_i(x) = \frac{du^{d-1}}{1+u^d}\ ;\ h(x) = h_i(x) = \frac{d}{1+u^d}$$

$$\mathrm{E}(N^X) = d^{(m-1)/2}$$

after an elementary computation. This differs by a factor $d^{-1/2}$ from the analogous Kostlan-Shub-Smale result. Even though the hypotheses of the previous theorem are not completely satisfied, one can easily adapt to it the present case.

4. Next, we show through a simple example a radically different behavior than the previous ones for large values of m. Following the same ideas, one can built more complicated ones of the same sort.
Consider a linear system with a quadratic perturbation

$$X_i(s) = \xi_i + \langle \eta_i, s\rangle + \zeta_i \|s\|^2,$$

where the ξ_i, ζ_i, η_i, $i = 1, \dots, m$ are independent and standard normal in $\mathbb{R}, \mathbb{R}$ and $\mathbb{R}^m$ respectively. This corresponds to the covariance $r^{X_i}(s,t) = 1 + \langle s,t\rangle + \|s\|^2\|t\|^2$. If there is no quadratic perturbation, it is obvious that the number of roots is almost surely equal to 1. For the perturbed system, applying formula (4.4), we obtain:

$$\mathrm{E}(N^X) = \frac{2}{\sqrt{\pi}} \frac{\Gamma((m+1)/2}{\Gamma(m/2)} \int_0^{+\infty} \frac{\rho^{m-1}(1+4\rho^2+\rho^4)^{\frac{1}{2}}}{(1+\rho^2+\rho^4)^{\frac{m}{2}+1}}\, d\rho.$$

An elementary computation shows that $\mathrm{E}(N^X) = o(1)$ as $m \to +\infty$ (geometrically fast). In other words, the probability that the perturbed system has no solution tends to 1 as $m \to +\infty$.

5. Consider again the case in which the polynomials $Q^{(i)}$ are all the same and the covariance depends only on the scalar product, i.e., $Q^{(i)}(u,v,w) = Q(u)$. We assume further that the roots of Q, that we denote $-\alpha_1, \dots, -\alpha_d$, are real ($0 < \alpha_1 \le \dots \le \alpha_d$). We get

$$q(x) = \sum_{h=1}^{d} \frac{1}{x+\alpha_h} \;;\; r(x) = \sum_{h=1}^{d} \frac{1}{(x+\alpha_h)^2} \;;\; h(x) = \frac{1}{q(x)} \sum_{h=1}^{d} \frac{\alpha_h}{(x+\alpha_h)^2}.$$

It is easy now to write an upper bound for the integrand in (4.4) and compute the remaining integral, thus obtaining the inequality

$$\mathrm{E}(N^X) \le \sqrt{\frac{\alpha_d}{\alpha_1}}\, d^{m/2},$$

which is sharp if $\alpha_1 = \dots = \alpha_d$.

5. Non-centered systems (smooth analysis)

The aim of this section is to remove the hypothesis that the coefficients have zero expectation. Let us start with a non-random system

$$P_i(t) = 0 \quad (i = 1, \dots, m), \tag{5.1}$$

and perturb it with a polynomial noise $\{X_i(t) : i = 1, \dots, m\}$, that is, we consider the new system:

$$P_i(t) + X_i(t) = 0 \quad (i = 1, \dots, m).$$

What can one say about the number of roots of the new system? Of course, to obtain results on $E(N^{P+X})$ we need a certain number of hypotheses both on the noise X and the polynomial "signal" P, especially the relation between the size of P and the probability distribution of X.

Some of these hypotheses are of technical nature, allowing to perform the computations. Beyond this, roughly speaking, Theorem 5.1 below says that if the relation signal over noise is neither too big nor too small, in a sense that we make precise later on, then there exist positive constants C, θ, $0 < \theta < 1$ such that

$$E(N^{P+X}) \leq C\,\theta^m E(N^X). \tag{5.2}$$

Inequality (5.2) becomes of interest if the starting non-random system (5.1) has a large number of roots, possibly infinite, and m is large. In this situation, the effect of adding polynomial noise is a reduction at a geometric rate of the expected number of roots, as compared to the centered case. In formula (5.2), $E(N^X)$ can be computed or estimated using the results in Sections 3 and 4 and bounds for the constants C, θ can be explicitly deduced from the hypotheses.

Before the statement we need to introduce some additional notations and hypotheses: H_1 and H_2 concern only the noise, H_3 and H_4 include relations between noise and signal.

The noise will correspond to polynomials $Q^{(i)}(u,v,w) = \sum_{k=0}^{d_i} c_k^{(i)}\, u^k$, $c_k^{(i)} \geq 0$, considered in Section 4, i.e., the covariances are only function of the scalar product. Also, each polynomial $Q^{(i)}$ has effective degree d_i, i.e.,

$$c_{d_i}^{(i)} > 0 \quad (i = 1, \dots, m).$$

and [illegible]es not vanish for $u \geq 0$, which amounts to saying that for each t the [illegible]tribution of $X_i(t)$ does not degenerate.

An elementary calculation then shows that for each polynomial $Q^{(i)}$, as $u \to +\infty$:

$$q_i(u) \sim \frac{d_i}{1+u} \tag{5.3a}$$

$$h_i(u) \sim \frac{c_{d_i-1}^{(i)}}{d_i c_{d_i}^{(i)}} \cdot \frac{1}{1+u}. \tag{5.3b}$$

Since we are interested in the large m asymptotics, the polynomials P, Q can vary with m and we will require somewhat more than relations (5.3a) and (5.3b), as specified in the following hypotheses:

H_1) h_i is independent of i ($i = 1, \dots, m$) (but may vary with m). We put $h = h_i$.
H_2) There exist positive constants D_i, E_i ($i = 1, \dots, m$) and $\underline{q}$ such that

$$0 \leq D_i - (1+u)q_i(u) \leq \frac{E_i}{1+u}, \quad \text{and} \quad (1+u)q_i(u) \geq \underline{q} \tag{5.4}$$

for all $u \geq 0$, and moreover

$$\max_{1 \leq i \leq m} D_i, \ \max_{1 \leq i \leq m} E_i$$

are bounded by constants $\overline{D}$, $\overline{E}$ respectively, which are independent of m. $\underline{q}$ is also independent of m.

Also, there exist positive constants $\underline{h}$, $\overline{h}$ such that

$$\underline{h} \leq (1+u)h(u) \leq \overline{h} \tag{5.5}$$

for $u \geq 0$.

Notice that the auxiliary functions q_i, r_i, h $(i = 1, \ldots, m)$ will also vary with m. To simplify somewhat the notations we are dropping the parameter m in P, Q, q_i, r_i, h. However, in H_2) the constants $\underline{h}$, $\overline{h}$ do not depend on m. One can check that these conditions imply that $h(u) \geq 0$ when $u \geq 0$.

Let us now describe the second set of hypotheses. Let P be a polynomial in m real variables with real coefficients having degree d and Q a polynomial in one variable with non-negative coefficients, having also degree d, $Q(u) = \sum_{k=0}^{d} c_k u^k$. We assume that Q does not vanish on $u \geq 0$ and $c_d > 0$. Define

$$H(P,Q) := \sup_{t \in \mathbb{R}^m} \left\{ (1 + \|t\|) \cdot \left\| \nabla \left(\frac{P}{\sqrt{Q(\|t\|^2)}} \right)(t) \right\| \right\}$$

$$K(P,Q) := \sup_{t \in \mathbb{R}^m \setminus \{0\}} \left\{ (1 + \|t\|^2) \cdot \left| \frac{\partial}{\partial \rho} \left(\frac{P}{\sqrt{Q(\|t\|^2)}} \right)(t) \right| \right\}$$

where $\frac{\partial}{\partial \rho}$ denotes the derivative in the direction defined by $\frac{t}{\|t\|}$, at each point $t \neq 0$.

For $r > 0$, put:

$$L(P,Q,r) := \inf_{\|t\| \geq r} \frac{P(t)^2}{Q(\|t\|^2)}.$$

One can check by means of elementary computations, that for each pair P, Q as above, one has

$$H(P,Q) < \infty, \;\; K(P,Q) < \infty.$$

With these notations, we introduce the following hypotheses on the systems P, Q, as m grows:

H_3)

$$A_m = \frac{1}{m} \cdot \sum_{i=1}^{m} \frac{H^2(P_i, Q^{(i)})}{i} = o(1) \quad \text{as } m \to +\infty \tag{5.6a}$$

$$B_m = \frac{1}{m} \cdot \sum_{i=1}^{m} \frac{K^2(P_i, Q^{(i)})}{i} = o(1) \quad \text{as } m \to +\infty. \tag{5.6b}$$

H_4) There exist positive constants r_0, l such that if $r \geq r_0$:

$$L(P_i, Q^{(i)}, r) \geq l \quad \text{for all } i = 1, \ldots, m.$$

Theorem 5.1. *Under the hypotheses H_1, H_2, H_3, H_4, one has*

$$E(N^{P+X}) \leq C\,\theta^m E(N^X) \tag{5.7}$$

where C, θ are positive constants, $0 < \theta < 1$.

For the proof, see [1].

5.1. Remarks on the statement of Theorem 5.1

1. It is obvious that our problem does not depend on the order in which the equations
$$P_i(t) + X_i(t) = 0 \quad (i = 1, \dots, m)$$
appear. However, conditions (5.6a) and (5.6b) in hypothesis H_3) do depend on the order. One can restate them saying that there exists an order $i = 1, \dots, m$ on the equations, such that (5.6a) and (5.6b) hold true.
2. Condition H_3) can be interpreted as a bound on the quotient signal over noise. In fact, it concerns the gradient of this quotient. In (5.6b) appears the radial derivative, which happens to decrease faster as $\|t\| \to \infty$ than the other components of the gradient.

   Clearly, if $H(P_i, Q^{(i)})$, $K(P_i, Q^{(i)})$ are bounded by fixed constants, (5.6a) and (5.6b) hold true. Also, some of them may grow as $m \to +\infty$ provided (5.6a) and (5.6b) remain satisfied.
3. Hypothesis H_4) goes in the opposite direction: for large values of $\|t\|$ we need a lower bound of the relation signal over noise.
4. A result of the type of Theorem 5.1 can not be obtained without putting some restrictions on the relation signal over noise. In fact consider the system
$$P_i(t) + \sigma X_i(t) = 0 \quad (i = 1, \dots, m) \tag{5.8}$$
where σ is a positive real parameter. As $\sigma \downarrow 0$ the expected value of the number of roots of (5.8) tends to the number of roots of $P_i(t) = 0$, $(i = 1, \dots, m)$, for which no a priori bound is available. In this case, the relation signal over noise tends to infinity. On the other hand, if we let $\sigma \to +\infty$, the relation signal over noise tends to zero and the expected number of roots will tend to $E(N^X)$.

5.2. Examples

5.2.1. Shub-Smale noise. Assume that the noise follows the Shub-Smale model. If the degrees d_i are uniformly bounded, one can easily check that H_1) and H_2) are satisfied.

For the signal, we give two simple examples. Let
$$P_i(t) = \|t\|^{d_i} - r^{d_i},$$
where d_i is even and $r > 0$ remains bounded as m varies. One has:
$$\frac{\partial}{\partial \rho}\left(\frac{P_i}{\sqrt{Q^{(i)}}}\right)(t) = \frac{d_i\,\|t\|^{d_i-1} + d_i\,r^{d_i}\,\|t\|}{(1+\|t\|^2)^{\frac{d_i}{2}+1}} \le \frac{d_i(1+r^{d_i})}{(1+\|t\|^2)^{3/2}}$$
$$\nabla\left(\frac{P_i}{\sqrt{Q^{(i)}}}\right)(t) = \frac{d_i\,\|t\|^{d_i-2} + d_i\,r^{d_i}}{(1+\|t\|^2)^{\frac{d_i}{2}+1}} \cdot t$$
which implies
$$\left\|\nabla\left(\frac{P_i}{\sqrt{Q^{(i)}}}\right)(t)\right\| \le \frac{d_i(1+r^{d_i})}{(1+\|t\|^2)^{3/2}}.$$

So, if the degrees $d_1, \dots, d_m$ are uniformly bounded, H_3) follows. H_4) also holds under the same hypothesis.

Notice that an interest in this choice of the P_i's lies in the fact that obviously the system $P_i(t) = 0$ $(i = 1, \dots, m)$ has infinite roots (all points in the sphere of radius r centered at the origin are solutions), but the expected number of roots of the perturbed system is geometrically smaller than the Shub-Smale expectation $\sqrt{D}$, when m is large.

Our second example of signal is as follows. Let T be a polynomial of degree d in one variable that has d distinct real roots. Define:

$$P_i(t_1, \dots, t_m) = T(t_i) \quad (i = 1, \dots, m).$$

One can easily check that the system verifies our hypotheses, so that there exist C, θ positive constants, $0 < \theta < 1$ such that

$$E(N^{P+X}) \leq C\,\theta^m d^{m/2}$$

where we have used the Kostlan-Shub-Smale formula. On the other hand, it is clear that $N^P = d^m$ so that the diminishing effect of the noise on the number of roots can be observed.

5.2.2. $Q^{(i)} = Q$, only real roots. Assume that all the $Q^{(i)}$'s are equal, $Q^{(i)} = Q$, and Q has only real roots. Since Q does not vanish on $u \geq 0$, all the roots should be strictly negative, say $-\alpha_1, \dots, -\alpha_d$ where $0 < \alpha_1 \leq \alpha_2 \leq \dots \leq \alpha_d$. With no loss of generality, we may assume that $\alpha_1 \geq 1$.

We will assume again that the degree d (which can vary with m) is bounded by a fixed constant $\overline{d}$ as well as the roots $\alpha_k \leq \overline{\alpha}$ $(k = 1, \dots, d)$ for some constant $\overline{\alpha}$. One verifies (5.4), choosing $D_i = d$, $E_i = d \cdot \max_{1 \leq k \leq d} (\alpha_k - 1)$. Similarly, a direct computation gives (5.5).

Again let us consider the particular example of signals:

$$P_i(t) = \|t\|^{d_i} - r^{d_i}$$

where d_i is even and r is positive and remains bounded as m varies.

$$\left| \frac{\partial}{\partial \rho} \left(\frac{P_i}{\sqrt{Q^{(i)}}} \right) \right| \leq d_i(\overline{\alpha} + r^{d_i}) \frac{1}{(1 + \|t\|^2)^{3/2}}$$

so that $K(P_i, Q^{(i)})$ is uniformly bounded. A similar computation shows that $H(P_i.Q^{(i)})$ is uniformly bounded. Finally, it is obvious that

$$L(P_i, Q^{(i)}, r) \geq \left(\frac{1}{1 + \overline{\alpha}} \right)^{\overline{d}}$$

for $i = 1, \dots, m$ and any $r \geq 1$. So the conclusion of Theorem 5.1 can be applied.

One can check that the second signal in the previous example also works with respect to this noise.

5.2.3. Some other examples. Assume that the covariance of the noise has the form of Example 4.1.1. Q is a polynomial in one variable having degree ν and positive coefficients, $Q(u) = \sum_{k=0}^{\nu} b_k u^k$. Q may depend on m, as well as the exponents $l_1, \dots, l_m$. Notice that $d_i = \nu \cdot l_i$ $(i = 1, \dots, m)$. One easily verifies that $H_1)$ is satisfied.

We assume that the coefficients $b_0, \dots, b_\nu$ of the polynomial Q verify the conditions

$$b_k \leq \frac{\nu - k + 1}{k} b_{k-1} \quad (k = 1, 2, \dots, \nu)$$

Moreover, $l_1, \dots, l_m, \nu$ are bounded by a constant independent of m and there exist positive constants $\underline{b}$, $\overline{b}$ such that

$$\underline{b} \leq b_0, b_1, \dots, b_\nu \leq \overline{b}.$$

Under these conditions, one can check that $H_2)$ holds true, with $D_i = d_i$ $(i = 1, \dots, m)$.

For the relation signal over noise, conditions are similar to the previous example.

Notice that already if $\nu = 2$, and we choose for Q the fixed polynomial:

$$Q(u) = 1 + 2a\,u + b\,u^2$$

with $0 < a \leq 1$, $\sqrt{b} > a \geq b > 0$, then the conditions in this example are satisfied, but the polynomial Q (hence Q^{d_i}) does not have real roots, so that it is not included in Example 5.2.2.

6. Systems having a law invariant under isometries and translations

In this section we assume that $X_i : \mathbb{R}^m \to \mathbb{R}$, $i = 1, \dots, m$ are independent centered Gaussian random fields with covariances having the form

$$r^{X_i}(s,t) = \gamma_i(\|t - s\|^2), \ (i = 1, \dots, m). \tag{6.1}$$

We will assume that γ_i is of class $\mathcal{C}^2$ and, with no loss of generality, that $\gamma_i(0) = 1$.

The computation of the expectation of the number of roots belonging to a Borel set V can be done using Rice formula (2.1), obtaining:

$$\mathrm{E}\big(N^X(V)\big) = (2\pi)^{-m/2}\mathrm{E}\big(|\det(X'(0))|\big)\lambda_m(V). \tag{6.2}$$

To obtain (6.2) we take into account that the law of the random field is invariant under translations and for each t, $X(t)$ and $X'(t)$ are independent. λ_m denotes Lebesgue measure in $\mathbb{R}^m$. Compute, for $i, \alpha, \beta = 1, \dots, m$

$$\mathrm{E}\left(\frac{\partial X_i}{\partial t_\alpha}(0)\frac{\partial X_i}{\partial t_\beta}(0)\right) = \left.\frac{\partial^2 r^{X_i}}{\partial s_\alpha \partial t_\beta}\right|_{t=s} = -2\gamma_i'(0)\delta_{\alpha\beta},$$

which implies, using a similar method to the one in the proof of Theorem 3.1:

$$\mathrm{E}\big(|\det(X'(0))|\big) = \frac{1}{\sqrt{\pi}} 2^m \Gamma((m+1)/2) \prod_{i=1}^{m} |\gamma_i'(0)|^{1/2}$$

and replacing in (6.2)

$$\mathrm{E}\big(N^X(V)\big) = \frac{1}{\sqrt{\pi}}\Big(\frac{2}{\pi}\Big)^{m/2}\Gamma((m+1)/2)\Big[\prod_{i=1}^{m}|\gamma_i'(0)|^{1/2}\Big]\lambda_m(V). \tag{6.3}$$

Next, let us consider the variance. One can prove that under certain additional technical conditions, the variance of the normalized number of roots:

$$n^X(V) = \frac{N^X(V)}{\mathrm{E}\big(N^X(V)\big)}$$

– which has obviously mean value equal to 1 – grows exponentially when the dimension m tends to infinity. This establishes a striking difference with respect to the results in section 3. In other words, one should expect to have large relative fluctuations of $n^X(V)$ around its mean for systems having large m.

Our additional requirements are the following:

1) All the γ_i coincide, $\gamma_i = \gamma$, $i = 1, \dots, m$,
2) the function γ is such that $(s,t) \rightsquigarrow \gamma(\|t-s\|^2)$ is a covariance for all dimensions m.

It is well known [14] that γ satisfies 2) and $\gamma(0) = 1$ if and only if there exists a probability measure G on $[0, +\infty)$ such that

$$\gamma(x) = \int_0^{+\infty} e^{-xw} G(dw) \quad \text{for all } x \geq 0. \tag{6.4}$$

Theorem 6.1. [3] *Let $r^{X_i}(s,t) = \gamma(\|t-s\|^2)$ for $i = 1, \dots, m$ where γ is of the form* (6.4). *We assume further that*

1. *G is not concentrated at a single point and*
$$\int_0^{+\infty} x^2 G(dx) < \infty.$$
2. *$\{V_m\}_{m=1,2\dots}$ is a sequence of Borel sets, $V_m \subset \mathbb{R}^m$, $\lambda_m(\partial V_m) = 0$ and there exist two positive constants δ, Δ such that for each m, V_m contains a ball with radius δ and is contained in a ball with radius Δ.*

Then,

$$\mathrm{Var}\big(n^X(V_m)\big) \to +\infty, \tag{6.5}$$

exponentially fast as $m \to +\infty$.

References

[1] Armentano, D. and Wschebor, M. (2006). Random systems of polynomial equations. The expected number of roots under smooth analysis. *submitted.*

[2] Azaïs J.-M. and Wschebor M. (2005a). On the Distribution of the Maximum of a Gaussian Field with d Parameters. *Annals Applied Probability*, 15 (1A), 254–278.

[3] Azaïs J-M. and Wschebor M. (2005b). On the roots of a random system of equations. The theorem of Shub and Smale and some extensions. *Found. Comp. Math.*, 125–144.

[4] Azaïs J-M. and Wschebor, M. (2006). A self contained proof of the Rice formula for random fields. Preprint available at http://www.lsp.ups-tlse.fr/Azais/publi/completeproof.pdf.

[5] Bharucha-Reid, A.T. and Sambandham, M (1986). *Random polynomials.* Probability and Mathematical Statistics. Academic Press Inc., Orlando, FL.

[6] Bloch, A. and Polya, G. (1932) On the number of real roots of a random algebraic equation. *Proc. Cambridge Phil. Soc.*, 33, 102–114.

[7] Blum, L., Cucker, F., Shub, M. and Smale, S. (1998) *Complexity and real computation.* Springer-Verlag, New York. With a foreword by Richard M. Karp.

[8] Bürgisser, P. (2006) Average volume, curvatures and Euler characteristic of random real algebraic varieties. *preprint*ArXiv: math.PR/0606755 v2.

[9] Edelman A. and Kostlan, E. (1995) How many zeros of a random polynomial are real? *Bull. Amer. Math. Soc.* (*N.S.*), 32 (1): 1–37.

[10] Kac, M. (1943) On the average number of real roots of a random algebraic equation. *Bull. Amer. Math. Soc.*, 49:314–320.

[11] Kostlan, E. (2002) On the expected number of real roots of a system of random polynomial equations. In *Foundations of computational mathematics* (*Hong Kong,* 2000), pages 149–188. World Sci. Publishing, River Edge, NJ.

[12] Littlewood J.E. and Offord A.C.(1938) On the number of real roots of a random algebraic equation. *J. London Math. Soc.*, 13, 288–295.

[13] Littlewood,J.E. and Offord, A.C. (1939) On the roots of certain algebraic equation. *Proc. London Math. Soc.*, 35, 133–148.

[14] Schoenberg, I.J. (1938) Metric spaces and completely monotone functions. *Ann. of Math.* (2), 39(4):811–841.

[15] Shub, M. and Smale, S. (1993) Complexity of Bézout's theorem. II. Volumes and probabilities. In *Computational algebraic geometry* (*Nice,* 1992), volume 109 of *Progr. Math.*, pages 267–285. Birkhäuser Boston, Boston, MA.

[16] Spivak, M. (1979) *A comprehensive introduction to differential geometry. Vol. V.* Publish or Perish Inc., Wilmington, Del., second edition.

[17] Wschebor, M. (2005) On the Kostlan-Shub-Smale model for random polynomial systems. Variance of the number of roots. *Journal of Complexity,* 21, 773–789.

[18] Wschebor, M. (2007) Large random polynomial systems, *preprint.*

Mario Wschebor
Centro de Matemática
Facultad de Ciencias
Universidad de la República
Calle Igua 4225
11400 Montevideo, Uruguay
e-mail: wschebor@cmat.edu.uy